Randy Q. Cron • Edward
Editors

Cytokine Storm Syndrome

Editors
Randy Q. Cron
UAB School of Medicine
University of Alabama at Birmingham
Birmingham, AL, USA

Edward M. Behrens
Perelman School of Medicine
University of Pennsylvania
Philadelphia, PA, USA

ISBN 978-3-030-22096-9 ISBN 978-3-030-22094-5 (eBook)
https://doi.org/10.1007/978-3-030-22094-5

This Springer imprint is published by the registered company Springer Nature Switzerland AG
The registered company address is: Gewerbestrasse 11, 6330 Cham, Switzerland

This textbook was necessitated by the broadening recognition of cytokine storm syndromes (CSS) and their import and impact on health worldwide. We dedicate this to the clinicians and scientists who explore CSS and to the patients and their families who allow us to learn how best to diagnose and treat CSS. We also would like to dedicate this work to our families who supported us during the process of preparing this textbook.

Foreword

The coeditors of the textbook Doctors Cron and Behrens have bravely embarked on the ambitious course of bringing together a group of international experts to dissect the current knowledge of cytokine storm syndrome (CSS) underlying both the primary (p) and secondary (s) haemophagocytic lymphohistiocytosis (HLH). The resulting state-of-the-art chapters cover the historic background, clinical and laboratory features, criteria for diagnosis, and current classification (Part I), pathophysiology (genetics, immunology, and murine models—Parts II, III, VII, respectively), different potential triggers (infections, rheumatic, various others—Parts IV, V, VI, respectively), and current treatment options (Part VIII).

Raising awareness of the CSS within the wide physician community is one of the textbook editors' major goals. The term 'cytokine storm' refers to an activation cascade of auto-amplifying cytokine production due to unregulated host immune response to different triggers, resulting in clinical presentation of a desperately ill patient with unremitting fever, hepatosplenomegaly, progressive liver failure with coagulopathy, cytopenias, hyperferritinemia, and often central nervous system (CNS) involvement. If this constellation of clinical features is not recognised and if adequate treatment is not promptly instituted, progression to multiple organ failure

(MOF) and eventual death is inevitable. Infections are the earliest recognised and by far the most common cause; other triggers include malignancy; rheumatic (autoimmune and autoinflammatory) disorders—where the process is termed macrophage activation syndrome (MAS)—as well as iatrogenic insults such as graft-versus-host disease (GvHD) in the course of haematopoietic stem cell transplantation (HSCT); and administration of different immunotherapeutic agents (e.g. monoclonal antibodies, chimeric antigen receptor (CAR) T cells, etc.).

Since the initial reports in the early 1990s of 'cytokine release syndrome' [1] and/or 'cytokine storm syndrome' [2] and the discovery of a number of underlying genetic mutations affecting primarily T and natural killer (NK) lymphocyte cytotoxicity in young children with rare hereditary diseases of immune dysregulation, the familial (primary) haemophagocytic lymphohistiocytosis (pHLH) [3], the understanding of the underlying pathogenesis continues [4]. It is well documented that other mechanisms besides failure of lymphocyte cytotoxicity, classical for pHLH-associated gene mutations, can lead to hyperinflammation, including primary macrophage activation (by accumulation of metabolites, constitutive activation, or dysregulation of inflammasomes), impaired control of common viruses (by impaired lymphocyte signalling, development, or function), defects in interferon signalling, or impaired autophagy [5]. Nevertheless, all patients lacking pHLH-associated gene mutations are still classified as sHLH, even those with proven genetic predisposition, such as an increasing number of other primary (genetic) immunodeficiency disorders (PID) (e.g. severe combined and combined immunodeficiency (SCID and CID, respectively), chronic granulomatous disease (CGD), autoimmune lymphoproliferative syndrome (ALPS), etc.) and certain hereditary metabolic disorders (e.g. lysosomal acid lipase deficiency, lysinuric protein intolerance). Moreover, the genetic predisposition to sHLH has been supported by an animal model implying that select combined variants in pHLH-associated genes (e.g. mutational burden) may be clinically relevant and by reports of variations of pHLH-associated gene mutations (e.g. biallelic hypomorphic mutations, mono-allelic or digenic mutations in genes affecting lymphocyte cytotoxicity) found in patients with underlying rheumatic disorders and malignancy, with or without MAS [6]. However, recent report of whole-exome sequencing (WES) study of a large patient group did not support a digenic model of susceptibility for HLH as the majority of such variants were present in general population. Nevertheless, the study pointed to the association between HLH and genetic variants in a group of dysregulated immune activation or proliferation (DIAP) genes, most of yet unknown significance but including significant associations for mono-allelic and biallelic variants in some of the inflammasome genes (e.g. *NLRC4* and *NLRP12*, and *NLRP4, NLRC3*, and *NLRP13*, respectively) [7].

The precise genetic diagnosis is primarily of major clinical importance as that information, alongside the progress in understanding the nature of the initiation and progression of the systemic (hyper)inflammatory process characterising CSS, enables clinical intervention relevant to and specific for individual patients, further paving the way towards precision medicine era [8]. This knowledge is already in practice, influencing important clinical decisions regarding appropriate treatments, such as targeting different and specific non- or malfunctioning pathways (e.g. failure of T- and NK-cell cytotoxicity, uncontrolled macrophage activation, overproduction of an array of different cytokines including interferon-gamma (IFN-γ), interleukin (IL)-1, IL-6, IL-18, tumour necrosis factor (TNF) [5]) and identifying the potential novel targets [9], as well as the role of allogeneic HSCT [10]. The multidisciplinary team (MDT) approach to the complex clinical management is paramount [11], both for recognising the underlying cause and for deciding about the most appropriate and specific treatment as demonstrated by our recent experience with a 2-year-old patient who initially presented to the haematology team with acute EBV-related liver and bone marrow failure and CNS involvement but failed to respond to HLH-2004 protocol. During the 2-month-long dramatic clinical course in paediatric intensive care unit, the diagnosis was confirmed as X-linked lymphoproliferative syndrome type 1 (XLP1), and in spite of escalating treatment regimen with an arsenal of classic (prolonged HLH-2004 protocol, anti-T-lymphocyte globulin (ATG)) and newly (at the time) emerging therapies (including rituximab, alemtuzumab, infliximab, and anakinra—the high-dose 'Cron regimen'), the immunology team only achieved transient and incomplete remission of the HLH process. However, and most importantly, this offered the window of opportunity for a curative allogeneic HSCT using a reduced-intensity conditioning regimen with alemtuzumab, treosulfan, and fludarabine [12].

As several overlapping themes are reviewed by the different authors, certain degree of repetition was inevitable. However, many topics, for example, the one on historic background, experienced and presented by the authors who 'lived through the journey' only enrich the reader's insight into the prolonged process, lasting over decades, to firstly recognise and subsequently define HLH as a new disease entity and to understand the underlying cytokine storm as 'a co-morbidity of another concurrent immunologic disease process' (Cron RQ and Behrens EM, Preface). No surprise that I reflected on my first experience with this devastating disease: back in the early 1980s with a trainee in Belgrade, former Yugoslavia, I looked after a dramatically sick young girl with very active systemic juvenile idiopathic arthritis (sJIA) who developed florid Epstein-Barr virus (EBV) infection—not that I knew at the time, of course, but only realised decades later that she was dying from macrophage activation syndrome!

Paradoxically, some of the challenges looming on the horizon may be directly induced by the unprecedented progress presented in this book. The precise genetic diagnosis [3, 13] may precipitate the 'information storm', and finding the right way out may not be easy. What is the actual clinical diagnosis of a complex phenotype including HLH features caused by multiple gene mutations defining different clinical entities [14]? What is the best management plan for asymptomatic children with underlying HLH confirmed by gene mutation (e.g. healthy siblings of a patient) [15]? The progress nevertheless continues, both in understanding the pathogenesis [3–5, 16] and in approaching the management [8, 11, 12] of CSS, irrespective of the numerous possible underlying causes.

Mario Abinun mario.abinun@ncl.ac.uk

Department of Paediatric Immunology, Great North Children's Hospital
Newcastle Upon Tyne Hospitals NHS Foundation Trust
Newcastle upon Tyne, UK

Primary Immunodeficiency Group, Institute of Cellular Medicine
Newcastle University
Newcastle upon Tyne, UK

References

1. Chatenoud, L., Ferran, C., Reuter, A., Legendre, C., Gevaert, Y., Kreis, H., et al. (1989). Systemic reaction to the anti-T-cell monoclonal antibody OKT3 in relation to serum levels of tumor necrosis factor and interferon-gamma [corrected]. *The New England Journal of Medicine, 320*(21), 1420–1421.
2. Ferrara, J. L., Abhyankar, S., & Gilliland, D. G. (1993). Cytokine storm of graft-versus-host disease: A critical effector role for interleukin-1. *Transplantation Proceedings, 25*(1 Pt 2), 1216–1217.
3. Picard, C., Gaspar, H. B., Al-Herz, W., Bousfiha, A., Casanova, J. L., Chatila, T., et al. (2018). International Union of Immunological Societies: 2017 primary immunodeficiency diseases committee report on inborn errors of immunity. *Journal of Clinical Immunology, 38*(1), 96–128.
4. Humblet-Baron, S., Franckaert, D., Dooley, J., Ailal, F., Bousfiha, A., Deswarte, C., et al. (2018). IFN-γ and CD25 drive distinct pathologic features during hemophagocytic lymphohistiocytosis. *Journal of Allergy and Clinical Immunology*. pii: S0091-6749(18)32773-8. https://doi.org/10.1016/j.jaci.2018.10.068. [Epub ahead of print].
5. Tesi, B., & Bryceson, Y. T. (2018). HLH: Genomics illuminates pathophysiological diversity. *Blood, 132*(1), 5–7.
6. Sepulveda, F. E., & de Saint Basile, G. (2017). Hemophagocytic syndrome: Primary forms and predisposing conditions. *Current Opinion in Immunology, 49*, 20–26.
7. Chinn, I. K., Eckstein, O. S., Peckham-Gregory, E. C., Goldberg, B. R., Forbes, L. R., Nicholas, S. K., et al. (2018). Genetic and mechanistic diversity in pediatric hemophagocytic lymphohistiocytosis. *Blood, 132*(1), 89–100.
8. Behrens, E. M., & Koretzky, G. A. (2017). Cytokine storm syndrome: Looking toward the precision medicine era. *Arthritis & Rheumatology (Hoboken, NJ), 69*(6), 1135–1143.
9. Zelic, M., Roderick, J. E., O'Donnell, J. A., Lehman, J., Lim, S. E., Janardhan, H. P., et al. (2018). RIP kinase 1-dependent endothelial necroptosis underlies systemic inflammatory response syndrome. *The Journal of Clinical Investigation, 128*(5), 2064–2075.
10. Allen, C. E., Marsh, R., Dawson, P., Bollard, C. M., Shenoy, S., Roehrs, P., et al. (2018). Reduced-intensity conditioning for hematopoietic cell transplant for HLH and primary immune deficiencies. *Blood, 132*(13), 1438–1451.
11. Halyabar, O., Chang, M. H., Schoettler, M. L., Schwartz, M. A., Baris, E. H., Benson, LA, et al. (2019). Calm in the midst of cytokine storm: A collaborative approach to the diagnosis

and treatment of hemophagocytic lymphohistiocytosis and macrophage activation syndrome. *Pediatric Rheumatology Online Journal*, *17*(1), 7.
12. Nikiforow, S. (2018). Finding "intermediate" ground in transplant and HLH. *Blood*, *132*(13), 1361–1333.
13. Heimall, J. R., Hagin, D., Hajjar, J., Henrickson, S. E., Hernandez-Trujillo, H. S., Itan, Y., et al. (2018). Use of genetic testing for primary immunodeficiency patients. *Journal of Clinical Immunology*, *38*(3), 320–329.
14. Boggio, E., Aricò, M., Melensi, M., Dianzani, I., Ramenghi, U., Dianzani, U., et al. (2013). Mutation of FAS, XIAP, and UNC13D genes in a patient with a complex lymphoproliferative phenotype. *Pediatrics*, *132*(4), e1052–e1058.
15. Lucchini, G., Marsh, R., Gilmour, K., Worth, A., Nademi, Z., Rao, A., et al. (2018). Treatment dilemmas in asymptomatic children with primary haemophagocytic lymphohistiocytosis. *Blood*, *132*(19), 2088–2096.
16. Crayne, C. B., Albeituni, S., Nichols, K. E., & Cron, R. Q. (2019). The immunology of macrophage activation syndrome. *Frontiers in Immunology*, *10*, 119.

Preface

'The beginning of knowledge is the discovery of something we do not understand'.

– Frank Herbert

Cytokine storm syndromes (CSS) are some of the scariest clinical scenarios for clinicians, patients, and their families. Many patients with CSS are in multi-organ system failure requiring intensive care, and mortality can be quite high. First and foremost, however, is the recognition of the CSS. The earlier the recognition of CSS and the earlier the treatment, the better the outcome. Unfortunately, CSS is frequently not recognized at all or until late in the process when therapy is less effective.

There are many aetiologies for poor recognition of CSS. Many clinicians, particularly of older generations, have never even heard of CSS. It was first described in the medical literature in 1952, with only 3 publications on the topic in the 1950s, 1 in the 1960s, and less than 20 in the 1970s. Thus, CSS is a relatively recently described phenomenon with over 6000 PubMed citations in a relatively short time frame (Fig. 1). Moreover, as it has been described by many different investigators in unique fields of study, CSS has many largely overlapping differently named entities. CSS, also referred to as cytokine release syndrome, is the systemic expression of a vast array of inflammatory mediators that impact the body as a whole. CSS is associated with hemophagocytic syndromes, including hemophagocytic lymphohistiocytosis (HLH). HLH is often divided into primary, or familial, and secondary, or acquired or reactive, forms of HLH. The genetic distinction between familial (fHLH) and secondary HLH (sHLH), however, is being blurred by the recognition in sHLH patients of contributory heterozygous mutations in genes also responsible for fHLH when they are present as homozygous or compound heterozygous mutations. Moreover, sHLH can result from infectious, oncologic, and rheumatic triggers, where it is termed macrophage activation syndrome (MAS). Thus, the current CSS nomenclature is partly responsible for its poor clinical recognition.

Another important clinical concept is that CSS is not a diagnosis of exclusion. Rather, it is often a co-morbidity of another concurrent immunologic disease process. Although a trigger of CSS is sometimes never identified, CSS is often the result of an underlying inflammatory disease with or without a recognized trigger

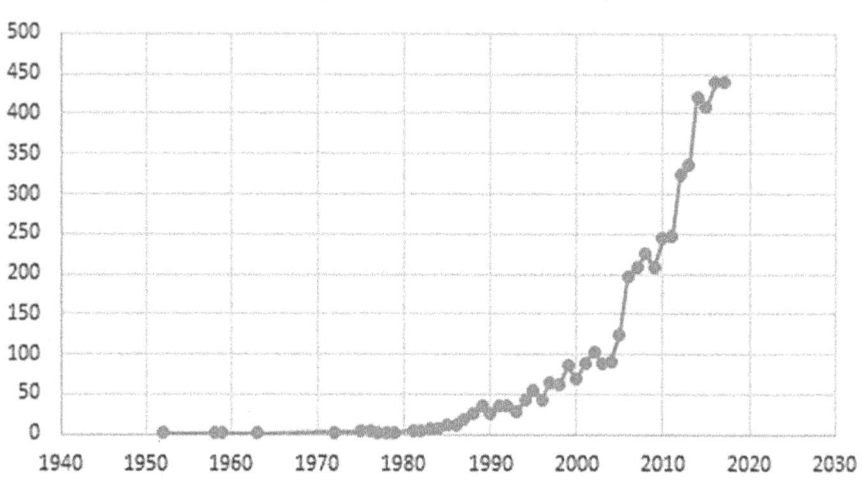

Fig. 1 Citations of CSS/MAS/HLH by year. PUBMED publication numbers (*Y* axis) reported by calendar year (*X* axis) by searching, "cytokine storm syndrome OR macrophage activation syndrome OR hemophagocytic lymphohistiocytosis OR haemophagocytic lymphohistiocytosis"

(often infectious). Thus, one can have bacterial sepsis and CSS on top of the infection. Similarly, a child with systemic juvenile idiopathic arthritis (sJIA) can have a severe flare of disease activity with or without associated CSS (MAS). This is also true of leukaemia or lymphoma, for which a subset of patients will have coincident CSS. In addition to diagnoses with a propensity to develop CSS (e.g. sJIA, T-cell leukaemia), various infections, particularly herpes virus family members such as EBV, are common triggers in these settings. Indeed, otherwise healthy individuals may also develop CSS after infection. Therefore, the astute clinician must be alert to the possibility of CSS in a variety of clinical scenarios, particularly in those strikingly ill febrile individuals.

Recently, attempts have been made to provide clinicians with diagnostic and classification criteria for CSS. HLH criteria are the most well-known, but the revitalized 2004 HLH criteria, while excellent for diagnosing fHLH, are often too restrictive and untimely for many of the CSS presenting as sHLH. In response, criteria have been proposed for recognizing CSS among some of the more commonly associated disorders, such as sJIA and systemic lupus erythematosus. While these criteria perform better than the 2004 HLH criteria in diagnosing CSS (MAS), they are specific to each disease for which they were generated and studied/validated. To overcome this lack of broad utility, the HScore was developed to recognize CSS associated with any form of sHLH. These criteria, while potentially broadly applicable, are rather cumbersome. Perhaps, simpler but less specific approaches, such as prolonged fever and hyperferritinemia, will serve as better screens with maximal sensitivity followed by additional testing which may then be employed to confirm the diagnosis following the initial suggestive findings.

CSS involve a broad array of clinicians, from emergency room doctors to intensivists, both paediatric and adult, to many in between. CSS, such as HLH, are often under the auspices of haemato-oncologists and bone marrow transplant physicians, particularly those cases associated with cancer, as well as those requiring bone marrow transplantation (familial HLH). Similarly, rheumatologists often care for those afflicted by secondary HLH CSS. Paediatric rheumatologists frequently treat MAS in children with sJIA, systemic lupus erythematosus (SLE), and Kawasaki disease, whereas adult rheumatologists recognize MAS in SLE, adult Still disease, and other forms of chronic arthritis. As infection, particularly herpes virus family members, often triggers CSS, infectious disease experts need to be aware of their role in identifying infectious agents and treating the underlying infection. Other subspecialists may also run across CSS in their patient populations, including neonatologists (fHLH), immunologists (immunodeficiencies), gastroenterologists (inflammatory bowel disease), cardiovascular surgeons (CSS associated with cardiac bypass circuit), geneticists (certain metabolic disorders), and others. Thus, educating physicians as a whole to the existence, diagnostic tools, and therapies available for CSS is critical.

Similarly, a broad array of research scientists, from immunologists to geneticists to cell biologists to pathologists to haematologists, all have and continue to contribute to a broader understanding of CSS pathogenesis. The works in murine models and human cells have often mirrored each other's findings. Much of the work has centred around genetic defects in the perforin-mediated cytolytic pathway employed by cytolytic CD8 T lymphocytes and natural killer cells in infants with fHLH (Fig. 2). The recognition of heterozygous defects in fHLH genes in many sHLH patients has expanded the potential roles of these mutations acting in dominant-negative or hypomorphic fashions to contribute to CSS. In addition, whole genome sequencing is also identifying mutations in non-exonic regulatory regions of fHLH genes in CSS patients. As the immune system has evolved not only to fight infection but also to regulate overexpansion of lymphocytes, defects in cytolysis of antigen-presenting cells (APC) can result in prolonged engagement between the cytolytic lymphocyte and the APC, resulting in a pro-inflammatory CSS. However, it would be a disservice and dangerous to assume that defective cytolysis is the only explanation for CSS physiology. Innovative, new work has demonstrated that beyond the perforin-mediated cytolytic pathway defects, there are other mechanisms which can result in similar cytokine storms (e.g. inflammasomopathies, autoinflammatory diseases), resulting in the end-common pathway of CSS. Many of these processes result in increased expression of common pro-inflammatory cytokines [e.g. interleukin-1 (IL-1), interferon-gamma (IFN-γ), IL-6, tumour necrosis factor (TNF), IL-18] that have been implicated in various aspects of the CSS and the resulting organ system failure that ensues.

As the cytokine storm itself appears to be responsible for much of the pathology in fHLH, sHLH (including MAS), and other forms of CSS, treatment aimed at dampening the pro-inflammatory cytokine storm is crucial for improving patient survival. Depending on the underlying illness, associated infections, genetic burden, and severity of the CSS, a variety of therapeutic approaches have evolved.

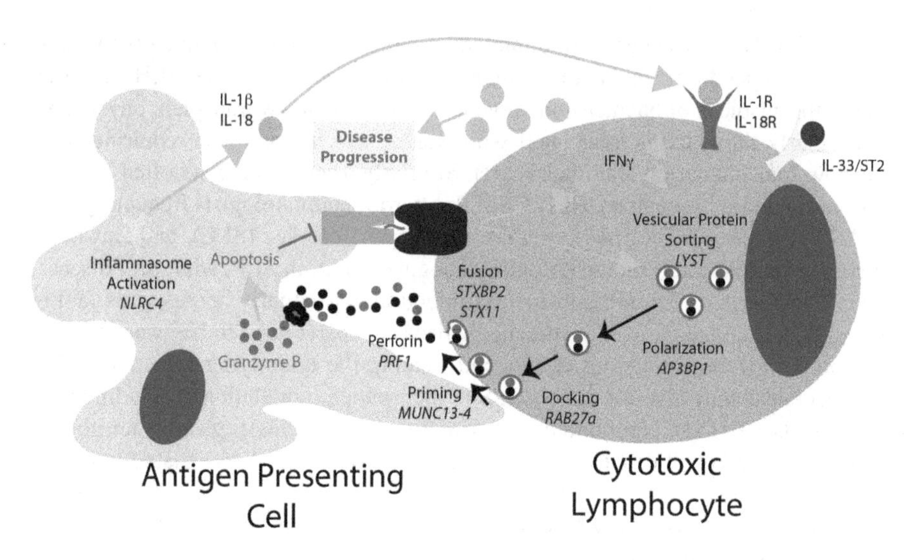

Fig. 2 Model of HLH pathogenesis. Depicted is a schematic summarizing some of the major mechanisms thought to lead to IFNγ production during HLH. Antigen presenting cells display peptide/MHC I complexes to the T-cell receptor of a CD8+ cytotoxic T lymphocyte. This results in a signal to produce and secrete IFNγ as well as a signal to activate the release of perforin and granzyme B. Release of perforin and granzyme B requires the coordination of a number of gene products to successfully accomplish sorting of these proteins to the correct vesicle, polarization of these vesicles to the immune synapse, docking of the vesicle with the plasma membrane, and priming and fusion of the vesicle. Loss of function of any of the genes required for this process (listed in italics underneath each function) results in the inability of perforin to reach the antigen presenting cell. Perforin normally would form a pore in the antigen presenting cell, allowing granzyme B to reach the cytoplasm and induce apoptosis, killing the cell and thereby eliminating further antigen presentation, stimulation of the CD8+ cytotoxic lymphocyte, and continued IFNγ production. In the absence of functional perforin capability, this does not happen and unchecked IFNγ production leads to disease manifestations. Increasingly, non-perforin related activities are being recognized as critical for this process as well, including the IL-1 family members IL-1β, IL-18, and IL-33. Genetic constitutive activation of the NLRC4 inflammasome leads to overproduction of IL-1β and IL-18, with IL-18 in particular leading to the secretion of IFNγ. IL-1β in other circumstances, such as Systemic Juvenile Idiopathic Arthritis, has been noted to be potentially part of disease progression. IL-33 has recently been established as being required in addition to T cell receptor signaling for the production of pathogenic IFNγ in HLH

Broadly, immunosuppressive approaches aimed at dampening the immune system and/or preparing for bone more transplant have been employed. These include chemotherapeutic agents, such as etoposide, lymphocyte targeting (e.g. cyclosporine), and high-dose corticosteroids. These therapies often deplete the immune system functionally and in numbers and, thus, increase the risk of secondary infections. Recently, the advent of biologic agents which target cytokines has led to alternative approaches to broadly immunosuppressive therapy. IL-1 blockade has proved useful in many sHLH patients, particularly those with sJIA, and IL-6 inhibition has proven valuable in cytokine release syndrome triggered by CAR-T cell therapy for refractory leukaemia. Clinical trials are currently exploring approaches to treat CSS by blocking IL-1, IL-18, or IFN-γ. As infections agents are frequent triggers of CSS, treatment of any underlying infection is also critical to help abate the CSS. This therapy may include antibiotics, antivirals, B-cell depletion of EBV-infected cells, and intravenous immunoglobulin. A host of other approaches, including plasmapheresis, T-cell depleting agents, and downstream inhibitors of cytokines (e.g. JAK inhibitors), are also being explored in treating CSS. The information in this textbook is designed to provide the reader with the fundamental basic and applied clinical science that represents the current state of the art in treating the various populations that develop CSS. We hope that with continued advancement, the not too distant future will entail not just these general principles but a truly personalized medicine approach to treat CSS. We look forward to the possibility of future editions of this work that will be able to outline all of the wonderful new advancements we believe are sure to come with the continued dedication of the physicians and scientists working on these issues.

Birmingham, AL, USA Randy Q. Cron
Philadelphia, PA, USA Edward M. Behrens

Contents

Part I
Background

History of Hemophagocytic Lymphohistiocytosis

Gritta E. Janka

First Descriptions

Hemophagocytic lymphohistiocytosis (HLH) is a severe cytokine storm syndrome (CSS) which was described in its primary, familial form (pHLH) in 1952 by Farquhar and Claireaux as "haemophagocytic reticulosis" [1]. Interestingly, hemophagocytosis, which gave HLH its name was not present during life in the two siblings but was found at autopsy only. Nevertheless, until not long ago hemophagocytosis was considered a sine qua non for the diagnosis of HLH.

Thirteen years earlier a clinical picture typical for HLH (fever, hepatosplenomegaly, cytopenias, and widespread histiocytic infiltration on autopsy) had been described in adults by Scott and Robb-Smith under the term "histiocytic medullary reticulosis" (HMR) [2]. Eight of these cases were later reviewed with more refined histological techniques. Five patients were found to have had malignant lymphomas; four were of T cell origin [3]. Lymphomas are now a well-known cause for secondary HLH (sHLH).

Another form of sHLH associated with viral infections (virus-associated hemophagocytic syndrome [VAHS]) was described in 1979 by Risdall and coworkers [4]. Interestingly, their cohort not only included adults under immunosuppression but also some children who may have had primary (familial) HLH (pHLH). Subsequent reports by other investigators linked sHLH to various infectious agents, including bacteria, protozoa, and fungi, though viruses, especially herpes viruses remain the most common trigger. The majority of Epstein–Barr virus-associated cases of HLH have been reported from Asia. The reason for this susceptibility remains mysterious to date.

G. E. Janka (✉)
Department of Pediatric Hematology and Oncology, University Medical Center Hamburg, Hamburg, Germany
e-mail: janka@uke.de

R. Q. Cron, E. M. Behrens (eds.), *Cytokine Storm Syndrome*,
https://doi.org/10.1007/978-3-030-22094-5_1

Risdall et al. emphasized that the histological picture in sHLH is distinct from malignant histiocytosis. Not all authors agreed, and the first case of sHLH due to Leishmania was published under the title "Systemic Leishmaniasis Mimicking Malignant Histiocytosis" [5]. Until the 1980s, malignant histiocytosis was considered a true histiocytic malignancy. Nearly all cases could later be reclassified as neoplasm of lymphoid origin, now called large-cell anaplastic lymphoma.

Since then, the list of conditions predisposing to the development of HLH has expanded, including metabolic diseases, other rare inborn immune defects, acquired immune deficiencies such as AIDS, and iatrogenic immunosuppression. The HLH-like picture in autoinflammatory/autoimmune diseases is usually called macrophage activation syndrome (MAS), a special subset of sHLH. pHLH is a rare disease with an estimated incidence of 1:50,000 live births [6]. Secondary forms are far more frequent. Between the first description of pHLH in 1952 and the turn of the century, PubMed lists only about 200 publications on HLH. Since then this syndrome has rapidly gained increased attention resulting in about 4600 additional publications of which more than 75% appeared in the last 10 years.

First Personal Experience

My first encounter with HLH was in the 1970s with four children of one family, three of whom died in early infancy [7]. The parents and an older sibling were healthy. The babies had unexplained high fever, hepatosplenomegaly, and cytopenias, the characteristic triad of HLH. The first child went from a cellular marrow to bone marrow aplasia after weeks of uncontrolled fever; lacking a better alternative at that time the final diagnosis was aplastic anemia. When the second child developed the same symptoms it became evident that this must be an inherited disease. Abundant histiocytes in the bone marrow and a liver biopsy prompted a search in the huge medical index called Index Medicus—a painstaking effort at a time when Internet research was far away. The search led to the publication by Farquhar and Claireaux. Since in their second patient a temporary remission was obtained with adrenocorticotropic hormone, our patient was treated with prednisolone. However, only some minor and transient improvement was achieved. When the third baby showed the same symptoms the family refused treatment but had the generosity to allow diagnostic procedures including a bone marrow aspirate and spinal tap, the latter showing a lymphocytic pleocytosis and increased protein level—my first encounter with central nervous system (CNS)-HLH. Unfortunately, no material was stored; thus, one can only speculate as to the type of genetic defect in this family.

Terminology

Various terms for pHLH were initially used such as familial reticulosis, lymphohistiocytic reticulosis, familial erythrophagocytic lymphohistiocytosis, familial histiocytosis, generalized lymphohistiocytic infiltration and others. After the first review on pHLH in 1983, which included 121 cases from the literature and 6 of my own cases [8], the term hemophagocytic lymphohistiocytosis was adopted by most authors. There has been some recent discussion whether to use a different name for HLH which takes into account that HLH is a severe hyperinflammatory syndrome and that lymphocytes (T cells) are not an absolute necessity for the syndrome to develop. However, as yet no consensus has been reached. sHLH is often called macrophage activation syndrome (MAS), especially in internal medicine. Pediatricians agree to reserve the term MAS for HLH in autoinflammatory/autoimmune diseases.

First HLH Symposium and Foundation of the FHL Study Group

Initially, there were only few pediatricians and pathologists who were interested in HLH. Through a colleague who had worked in Paris, I gained access to a doctoral thesis on pHLH ("histiocytose familiale") from Alain Fischer's group in which eight cases were described. Parts of it were later published [9]. The pathologist's description was presented by Christian Nezelof from Paris and Julio Goldberg, a visiting pathologist from Argentina. [10]. Another group led by Jan-Inge Henter in Stockholm had started to thoroughly investigate all Swedish cases. A third group, represented by Maurizio Aricó and Roberto Burgio, who also had a general interest in histiocytosis, finally organized a first international workshop on HLH in Pavia in 1988. One year later, the FLH (pHLH) Study Group was founded as a part of the international Histiocyte Society which had existed since 1985. FHL Study Group founding members were Maurizio Aricó, Göran Elinder, Blaise Favara, Jan-Inge Henter, Gritta Janka, Diane Komp, Christian Nezelof, and Jon Pritchard; at the HLH protocol meeting in 1994, several additional members joined.

In 1991, the FHL Study Group presented the first diagnostic guidelines for HLH [11]. The suggested diagnostic criteria consisted of five clinical and laboratory items which were easy to ascertain and which gained wide acceptance.

State of Knowledge Before Study HLH-94

When the first international HLH study started in 1994, the knowledge about HLH was still quite limited. It was evident that there were familial (genetic) and secondary forms; the latter due to infections or malignancies. However, the distinction between pHLH and infection-associated HLH was not possible when the family history was blank, since children with pHLH were found to have viral infections as well [8, 12]. Involvement of the central nervous system (CNS) was recognized as a serious complication; the common appearance on autopsy was that of "leptomeningitis," but parenchymal lesions were described as well [13]. With long-term survival, cognitive and psychosocial sequelae due to CNS involvement have now become a major concern in children with HLH [14]. CNS-HLH is still poorly understood, and treatment options are limited [15].

Although high levels of cytokines and soluble interleukin-2 receptor alpha chain [16–18] suggested uncontrolled activity of macrophages and lymphocytes, the etiology and pathogenesis of HLH remained elusive for a long time. Various, but inconsistent, immunological abnormalities were reported initially. The general assumption that HLH must be an immunodeficiency was finally confirmed by a profound deficit in natural killer (NK) cell activity [19]. Impaired NK cell activity had already previously been described in two immune deficiencies with partial albinism, namely, Griscelli syndrome (GS-2) and Chédiak–Higashi syndrome (CHS). GS-2 and CHS are frequently complicated by HLH and are now counted among the primary forms of HLH.

Study HLH-1994 could fall back only on limited experience regarding therapy of HLH. Various measures had been tried, such as splenectomy, exchange transfusions, corticosteroids, and cytotoxic drugs, but the prognosis of pHLH was dismal; only four children with prolonged survival had been reported in the first review [8]. A promising agent seemed to be the epipodophyllotoxin derivate, etoposide (VP-16), which produced longer remissions but could not prevent reactivations, including CNS relapses [20]. The efficacy of VP-16 combined with steroids and CNS directed therapy could be confirmed by other groups [7, 21]. Treatment with chemotherapy or immunotherapy, however, was only able to reverse disease activity for some period but was not curative in familial cases. Thus, it was a big step forward when the first patient receiving a hematopoietic stem cell transplant (HSCT) from his HLA-identical sibling remained free of disease without therapy [22].

There was another promising approach from a single institution with antithymocyte globulin (ATG), corticosteroids, and cyclosporin A [23] which, however, was not considered for the large international study due to less general experience with ATG and inferior availability.

HLH-94 Study

Between July 1994 and December 2003, study HLH-1994 recruited 249 patients fulfilling inclusion criteria from 25 countries. The protocol included an initial intensive therapy with dexamethasone and etoposide for 8 weeks. Dexamethasone was chosen due to its better penetration into the cerebrospinal fluid (CSF). Intrathecal methotrexate therapy was recommended for patients with progressive neurological symptoms and/or persisting CSF abnormalities. For patients with familial, persistent, or relapsing disease, continuation therapy, with cyclosporin A and periodic etoposide and dexamethasone pulses until HSCT, was recommended. Interim results were published in 2002 [24]. In the final report at 6.2-years median follow-up, probability of survival was 54% [25]. This was a very gratifying result, although at least about 20% of the patients must have had non familial disease, as evident from reactivation-free survival in 49 children without HSCT. Death was due to poor response to treatment within the first 8 weeks, reactivations before HSCT (median time to HSCT 6.1 months), and complications after HSCT, mostly due to the toxicity of myeloablative conditioning.

The second international study, HLH-2004, was based on the HLH-94 protocol with only few changes: Cyclosporin A was added up front, and intrathecal therapy was supplemented by corticosteroids. The revised diagnostic criteria also included impaired natural killer cell activity, a hallmark of the disease, hyperferritinemia, and increased levels of soluble interleukin-2 receptor alpha chain [26]. The results of study HLH-2004 have recently been published, reporting a 62% 5-year probability of survival [27].

Advances in Understanding HLH

Genetics

The description of the first genetic defect in pHLH [28] revolutionized our understanding of the pathogenesis of HLH. Previously, linkage analysis in a Pakistani family had revealed a putative disease gene on chromosome 9 [29]; thus, pHLH due to mutations in the *Perforin (PRF1)* gene was called FHL-2. Perforin, a component of cytolytic granules in cytotoxic cells is critical for the access of proteolytic enzymes to the target cell to initiate apoptotic death. Perforin is released into the immunological synapse between effector and target cell by cytolytic granules which traffic to the contact site, dock, and fuse with the plasma membrane. Notably, the cytotoxic effector response not only targets infected cells but also antigen-presenting cells (APCs). Elimination of APCs is an important negative feedback for the immune response. The identification of perforin deficiency as cause of pHLH showed the importance of this protein for controlling and terminating the immune response. The failure to contract the immune response in patients with pHLH explains the persistently high cytokine levels which are responsible for the symptomatology. Later, another critical role for perforin, which is immune surveillance, became evident from temperature-sensitive missense mutations where residual protein activity was present. These mutations were associated with late-onset pHLH and the occurrence of lymphomas or leukemia [30].

Within the next 10 years, three additional genetic defects for pHLH were described; all were found to also impair granule-dependent cytotoxic function. The genes are *UNC13D* (FHL-3) [31], *Syntaxin11* (FHL-4) [32], and *STXBP2* (FHL-5) [33, 34]. In addition, mutations in Griscelli syndrome *(RAB27A)* and Chédiak–Higashi syndrome *(LYST)* were found to also affect the cytolytic granule pathway of cytotoxic cells, and present as pHLH. Another rare disorder of lysosomal trafficking, Hermansky–Pudlak syndrome type 2 (HPS2), has only a low risk of developing HLH. Additionally, X-linked lymphoproliferative (XLP) disease types 1 and 2 were noted to have a high risk for developing HLH, particularly in the setting of EBV infection. Despite the fact that the XLP mutations do not affect the cytolytic pathway, they are now counted among the primary forms of HLH.

Disease onset and severity in primary forms of HLH are highly variable, depending on the gene involved and the type of mutation, determining complete or only partial loss of protein structure and function. When age at onset was taken as a surrogate marker for disease severity in patients with predicted severe protein impairment, FHL-2 patients had the earliest onset followed by GS-2, FHL-4, and CHS [35]. In FHL-5, patients with missense or splice-site mutations differ markedly in age of onset and presence of diarrhea, an atypical newly described symptom [36]. There are still a small number of patients with absent degranulation, indicative for biallelic mutations in the cytolytic pathway, but no mutation in the known genes.

Classical cases of pHLH are either autosomal recessive with biallelic mutations (FHL2–5, GS, CHS, and HPS) or hemizygous as in XLP-1 and XLP-2. Digenic inheritance with mutations within *PRF1* and a degranulation gene or within two genes of the degranulation pathway has also been described [37]. Reports on heterozygous mutations in patients with HLH have appeared with increasing frequency. They were found to have only reduced but not completely absent perforin expression or degranulation [38]. This suggests that these patients either have other unknown genetic factors outside the cytolytic pathway or environmental factors contributing to the development of HLH. It has to be emphasized that parents or heterozygous siblings of patients with pHLH typically do not show signs of HLH. Heterozygous carriers are accepted as donors for HSCT.

Mouse Models

Mouse models, created for pHLH and sHLH, have largely contributed to our understanding of how the genetic defects could be responsible for the aberrant immune response in HLH. Studies in perforin-deficient mice, infected with LCMV, have shown that activated CD8 T cells and interferon-γ (IFNγ) production play a central role in the pathogenesis of HLH. HLH could be prevented by neutralization of IFNγ [39]. This was confirmed also for Rab27a deficient mice [40]. In the *PRF1* deficiency mouse model, the therapeutic effect of etoposide involved deletion of activated T cells [41]. Polyclonal T cells are activated in both pHLH and sHLH [42], and INFγ is increased [16, 17]. However, the pathogenesis of HLH does not always involve T cells; patients with severe combined immunodeficiency and T cells <1000/μL can develop HLH, as well [43]. Activation of T cells as a prerequisite for HLH was also not required in a mouse model with a normal genetic background where repeated toll-like receptor 9 stimulation produced an HLH-like picture [44]. Innate immune stimulation may be an important pathway to develop HLH in patients without a cytotoxic defect.

A recent study in mouse and human cytotoxic cells demonstrated that failed target cell killing, leading to a prolonged synapse time, greatly amplifies cytokine secretion by NK cells and cytotoxic lymphocytes. Of note, blocking caspase in the target cell, important for the extrinsic pathway of apoptosis, phenocopied perforin deficiency regarding prolonged synapse time. This could be an explanation for sHLH in viral infections and malignancies since virus-infected or transformed cells often have defects in their apoptotic machinery [45].

The clinical and laboratory symptoms of HLH can all be explained by hypercytokinemia and organ infiltration by activated lymphocytes and histiocytes [46]. Macrophages secrete plasminogen, which is cleaved into plasmin, mediating fibrinolysis. A recent investigation in a murine model showed plasmin to be also an important regulator for the influx of inflammatory cells and the production of inflammatory cytokines leading to HLH [47].

Advances in HLH in Adults

Although the first cases of HLH had been described in adults [2], it took a long time until HLH received adequate interest in internal medicine. HLH was commonly regarded as are rare pediatric syndrome, and HLH patients outside of pediatric centers have been at high risk of not being diagnosed. Only recently, case reports and larger case series [48] of HLH in adults have been published with greater frequency. The recognition that mutations in HLH-relevant genes are also found in adults in a substantial number of patients [49] has further increased interest. Nevertheless, in most countries, a network of experts, as is present in pediatrics, is missing, although there is even a registry in one country (www.hlh-registry.org). Thus, HLH in adults is still very likely underdiagnosed.

The majority of HLH cases in adults are secondary to infections and malignancies; a smaller number is due to autoimmune diseases [48]. The true incidence of genetic cases is not known; mutations in these patients allow for residual protein expression and hence partially preserved cytotoxic function. There are no separate diagnostic criteria for HLH in adults. Usually the HLH-2004 criteria [26] or a recently developed score, the *HScore* [50], is used. Treatment of HLH in adults is challenging since older patients have a diminished hematopoietic reserve and may have comorbidities which limit treatment intensity. Lymphomas, often occult, have to be ruled out vigorously before treatment. Recommendations for the management of HLH in adults have been published [51].

Advances in Diagnostics

Diagnosis of HLH is based on a set of diagnostic and laboratory parameters; no single parameter, including hemophagocytosis, is sufficiently specific for HLH. pHLH and sHLH cannot be distinguished by these parameters, including NK cell activity. Differentiation between genetic and secondary forms, however, is crucial for early organization of a stem cell transplant. In 2012, a joint collaboration between five European countries showed that degranulation assays, a measure quantifying lytic granule exocytosis, were a reliable tool to identify patients with mutations in the cytolytic pathway [52]. In combination with intracellular measurement of perforin, SAP (XLP-1), and XIAP (XLP-2), also obtained by flow cytometry, this approach can give guidance whether a genetic analysis should be performed and a transplant should be prepared. A diagnostic algorithm how to proceed in patients with HLH has been proposed [53]. Recently, perforin and degranulation testing were shown to be superior to measurement of NK cell function for screening patients for genetic HLH [54]. Unfortunately, these assays have been established in larger HLH centers only and require rapid transport of material. Additionally, repeated testing may be necessary with equivocal results. In the future, with increasingly more rapid and less costly genetic testing, these functional assays may eventually be replaced by mutation analysis although functional testing will remain important as our knowledge of the entire genetic landscape of HLH remains incomplete. Moreover, a blurring of the distinction between pHLH and sHLH has been noted for patients with complete [55] or partial dominant-negative [56] heterozygous mutations in known HLH-associated genes.

Developments in Therapy

Stem Cell Transplantation

Like in other immunodeficiencies, HSCT is the only curative treatment for pHLH. In the absence of a suitable related or unrelated donor, haploidentical transplantation or cord blood HSCT constitute alternatives [57]. Post-transplant mortality after myeloablative conditioning (MAC) was high with a survival probability of only 49–64% in larger studies [58]. Liver (mainly veno-occlusive disease) and lung problems were leading causes of death. It was important to learn that sustained remissions could be achieved in all patients with a donor chimerism ≥20% [59]. In line with this, in the *PRF1* mouse model either a mixed hematopoietic or CD8 T cell chimerism of ~10–20% was sufficient for reestablishment of immune regulation [60]. The introduction of reduced-intensity conditioning was a big step forward; more than 80% of patients with pHLH can now be expected to survive HSCT [58, 61]. An increased incidence of mixed chimerism can be seen after RIC conditioning. In a large retrospective study from 23 bone marrow transplant centers, the protective effect of >20% donor cells against late reactivations could be confirmed. Interestingly, there were five patients living reactivation-free for 1.1–10 years (median 5.1) with ≤10% donor cells [62].

Whereas in XLP-2, HSCT should be reserved for patients with severe disease, there is a general indication for HSCT in patients with XLP-1, even before first exposure to EBV [57]. These patients not only have life-threatening EBV-associated HLH but also progressive hypogammaglobulinemia, and the risk of developing lymphomas.

Although HSCT should be reserved for patients with pHLH, there are some patients with EBV-associated HLH who may need a transplant. In a survey from Japan, fourteen pediatric patients with sHLH due to EBV were collected who failed HLH-2004 therapy and underwent HSCT [63]. Since such transplants were also successful with autologous stem cells, cells from an identical twin, and with graft failure, resetting of the adaptive immune response was suggested as mechanism for success, rather than replacement of a genetically defective immune system [63].

New Therapies

Etoposide-based treatment can currently be regarded as the standard of care for HLH. Two studies conducted in North America and Europe evaluated the combination of ATG, dexamethasone, and etoposide. The studies have been closed; results are not published yet. A French study is currently evaluating the role of the monoclonal anti-CD52-antibody alemtuzumab as first line treatment. Another promising approach is therapy with an anti-IFNγ antibody, which is currently tested in a phase II/III study. Recently, two groups used several mouse models to show that ruxolitinib, a Janus kinase inhibitor was not only successful in preventing [64] but also in treating manifest HLH [65]. As yet there are only two case reports in humans [66, 67]. A review on salvage therapy of HLH identified alemtuzumab as the only drug with data in a larger number of patients. A partial response was achieved in 14 of 22 patients [68]. Plasma exchange, a very old method, has received renewed interest [69]. It may still be of value, especially when therapy with etoposide is not possible due to renal failure. An interesting approach is cytokine absorption which has been successfully applied in some adult patients with HLH [70]. Specific targeting of pro-inflammatory cytokines, including IL-1, IL-6, and tumor necrosis factor is also being explored as therapy for CSS [71, 72].

Open Questions And Outlook

Our knowledge of HLH has increased rapidly within the last 20 years. However, our understanding is still incomplete, and many questions remain. Just to name a few: What other genes in the degranulation pathway are involved? What activates the T cells in children with pHLH where no readily identifiable infectious organism is found? What are the pathogenetic mechanisms of sHLH? Why do some genetic defects have a higher likelihood of CNS-HLH? Is there a place for drugs with good penetration into brain tissue such as thiotepa in refractory CNS disease? Why do some patients respond so poorly to frontline chemoimmunotherapy? Can efficacy and toxicity of frontline therapy be improved with the newer drugs? What compensatory mechanisms prevent recurring HLH in some patients with graft failure? Which patients in the ICU with severe hyperinflammation, fulfilling HLH criteria, could profit from immunosuppressive therapy?

HLH is no longer a disease of marginal existence but is increasingly being recognized as an important and dangerous syndrome. New drugs as supplement or substitution in initial therapy will hopefully improve response rates and long-term results. Gene therapy is presently being explored in animal models of HLH but will take time until introduction into the human setting.

References

1. Farquhar, J. W., & Claireaux, A. E. (1952). Familial haemophagocytic reticulosis. *Archives of Disease in Childhood, 27*, 519–525.
2. Scott, R. B., & Robb-Smith, A. H. T. (1939). Histiocytic medullary reticulosis. *Lancet, 2*, 194–198.
3. Falini, B., Pileri, S., De Solas, I., Martelli, M. F., Mason, D. Y., Delsol, G., et al. (1990). Peripheral T-cell lymphoma associated with hemophagocytic syndrome. *Blood, 75*, 434–444.
4. Risdall, R. J., McKenna, R. W., Nesbit, M. E., Krivit, W., Balfour Jr., H. H., Simmons, R. L., et al. (1979). Virus-associated hemophagocytic syndrome: A benign histiocytic proliferation distinct from malignant histiocytosis. *Cancer, 44*, 993–1002.
5. Matzner, Y., Behar, A., Beeri, E., Gunders, A. E., & Hershko, C. (1979). Systemic leishmaniasis mimicking malignant histiocytosis. *Cancer, 43*, 398–402.
6. Henter, J. I., Elinder, G., Söder, O., & Ost, A. (1991b). Incidence in Sweden and clinical features of hemophagocytic lymphohistiocytosis. *Acta Paediatrica Scandinavica, 80*, 428–435.
7. Janka, G. E. (1989). Familial hemophagocytic lymphohistiocyosis: Therapy in the German experience. *Pediatric Hematology and Oncology, 6*, 227–231.
8. Janka, G. E. (1983). Familial hemophagocytic lymphohistiocytosis. *European Journal of Pediatrics, 140*, 221–230.
9. Devictor, E., Fischer, A., Mamas, S., de Saint Basile, G., Durandy, A., Buriot, D., et al. (1982). Etude immunologique de la lymphohistiocytose familiale. *Archives Francaises de Pediatrie, 39*, 135–140.
10. Goldberg, J., & Nezelof, C. (1986). Lymphohistiocytosis: A multi-factorial syndrome of macrophagic activation clinico-pathological study of 38 cases. *Hematological Oncology, 4*, 275–289.

11. Henter, J. I., & Elinder, G. (1991). Diagnostic guidelines for hemophagoytic lymhohistiocytosis. The FHL study group of the histiocyte society. *Seminars in Oncology, 18*, 29–33.
12. Henter, J. I., Ehrnst, A., Anderson, J., & Elinder, G. (1993). Familial hemophagocytic lymphohistiocytosis and viral infections. *Acta Paediatrica, 82*, 369–372.
13. Akima, M., & Sumi, S. M. (1984). Neuropathology of familial erythrophagocytic lymphohistiocytosis. *Human Pathology, 15*, 161–168.
14. Jackson, J., Titman, P., Butler, S., Bond, K., Rao, A., Veys, P., et al. (2013). Cognitive and psychosocial function in post hematopoietic stem cell transplantation in children with hemophagocytic lymphohistiocytosis. *Journal of Allergy and Clinical Immunology, 132*, 889–895.
15. Horne, A. C., Wickström, R., Jordan, M. B., Yeh, E. A., Naqvi, A., Henter, J. I., et al. (2017). How to treat involvement of the central nervous system in hemophagocytic lymphohistiocytosis. *Current Treatment Options in Neurology, 19*, 3.
16. Henter, J. I., Elinder, G., Söder, O., Hansson, M., Andersson, B., & Andersson, U. (1991). Hypercytokinemia in familial hemophagocytic lymphohistiocytosis. *Blood, 78*, 2918–2922.
17. Imashuku, S., Ikushima, S., Esumi, N., Todo, S., & Saito, M. (1991). Serum levels of interferon-gamma, cytotoxic factor and soluble interleukin-2 receptor in childhood hemophagocytic syndromes. *Leukemia & Lymphoma, 3*, 287–292.
18. Komp, D. M., McNamara, J., & Buckley, P. (1989). Elevated soluble interleukin-2 receptor in childhood hemophagocyctic histiocytic syndromes. *Blood, 73*, 2128–2132.
19. Perez, N., Virelizier, J. L., Arenzana-Seisdedos, S., Fischer, A., & Griscelli, C. (1984). Impaired natural killer cell activity in lymphohistiocytosis syndrome. *The Journal of Pediatrics, 104*, 569–573.
20. Ambruso, D. R., Hays, T., Zwartjes, W. J., Tubergen, D. G., & Favara, B. E. (1980). Successful treatment of lymphohistiocytic reticulosis with phagocytosis with epipodophyllotoxin VP 16-213. *Cancer, 45*, 2516–2520.
21. Fischer, A., Virelizier, J. L., Arenzana-Seisdedos, F., Perez, N., Nezelof, C., & Griscelli, C. (1985). Treatment of four patients with erythrophagocytic lymphohistiocytosis by a combination of epipodophyllotoxin, steroids, intrathecal methotrexate, and cranial irradiation. *Pediatrics, 76*, 263–268.
22. Fischer, A., Cerf-Bensussan, N., Blanche, S., Le Deist, F., Bremard-Oury, C., Leverger, G., et al. (1986). Allogeneic bone marrow transplantation for erythrophagocytic lymphohistiocytosis. *The Journal of Pediatrics, 108*, 267–270.
23. Stéphan, J. L., Donadieu, J., Ledeist, F., Blanche, S., Griscelli, C., & Fischer, A. (1993). Treatment of familial hemophagocytic lymphohistiocytosis with antithymocyte globulins, steroids, and cyclosporin A. *Blood, 82*, 2319–2323.
24. Henter, I. J., Samuelsson-Horne, A. C., Aricò, M., Egeler, R. M., Elinder, G., Filopovich, A. H., et al. (2002). Treatment of hemophagocytic lymphohistiocytosis with HLH94 immunochemotherapy and bone marrow transplantation. *Blood, 100*, 2367–2373.
25. Trottestam, H., Horne, A. C., Aricò, M., Egeler, R. M., Filipovich, A. H., Gadner, H., et al. (2011). Chemoimmunotherapy for hemophagocytic lymphohistiocytosis: Long-term results of the HLH-94 treatment protocol. *Blood, 118*, 4577–4584.
26. Henter, J. I., Horne, A. C., Arićó, M., Egeler, R. M., Filopovich, A. H., Imashuku, S., et al. (2007). HLH-2004: Diagnostic and therapeutic guidelines for hemophagocytic lymphohistiocytosis. *Pediatric Blood & Cancer, 48*, 124–131.
27. Bergsten, E., Horne, A., Astigarraga, I., Egeler, R. M., Filipoivich, A. H., Ishii, E., et al. (2017). Confirmed efficacy of etoposide and dexamethasone in HLH treatment: Long-term results of the cooperative HLH-2004 study. *Blood, 130*, 2728–2738.
28. Stepp, S. E., Dufourcq-Lagelouse, R., Le Deist, F., Bhawan, S., Certain, S., Mathew, P. A., et al. (1999). Perforin gene defects in familial hemophagocytic lymphohistiocytosis. *Science, 286*, 1957–1959.
29. Ohadi, M., Lalloz, M. R., Sham, P., Zhao, J., Dearlove, A. M., Shiach, C., et al. (1999). Localization of a gene for familial hemophagocytic lymphohistiocytosis at chromosome 9q21.3-22 by homozygosity mapping. *American Journal of Human Genetics, 64*, 165–171.

30. Chia, J., Yeo, K. P., Whisstock, J. C., Dunstone, M. A., Trapani, J. A., & Voskoboinik, I. (2009). Temperature sensitivity of human perforin mutants unmasks subtotal loss of cytotoxicity, delayed FHL, and a predisposition to cancer. *Proceedings of the National Academy of Sciences of the United States of America, 106*, 9809–9814.
31. Feldmann, J., Callebaut, I., Raposo, G., Certain, S., Bacq, D., Dumont, C., et al. (2003). Munc13-4 is essential for cytolytic granules fusion and is mutated in a form of familial hemophagocytic lymphohistiocytosis (FHL3). *Cell, 115*, 461–473.
32. Zur Stadt, U., Schmidt, S., Kasper, B., Beutel, K., Diler, A. S., Henter, J. I., et al. (2005). Linkage of familial hemophagocytic lymphohistiocytosis (FHL) type-4 to chromosome 6q24 and identification of mutations in syntaxin 11. *Human Molecular Genetics, 14*, 827–834.
33. Côte, M., Ménager, M. M., Burgess, A., Mahlaoui, N., Picard, C., Schaffner, C., et al. (2009). Munc18-2 deficiency causes familial hemophagocytic lymphohistiocytosis type 5, and impairs cytotoxic granule exocytosis in patient NK cells. *The Journal of Clinical Investigation, 119*, 3765–3773.
34. Zur Stadt, U., Rohr, J., Seifert, W., Koch, F., Grieve, S., Pagel, J., et al. (2009). Familial hemophagocytic lymphohistiocytosis type 5 (FHL-5) is caused by mutations in Munc18-2 and impaired binding to syntaxin 11. *American Journal of Human Genetics, 85*, 482–492.
35. Jessen, B., Kögl, T., Sepulveda, F. E., de Saint Basile, G., Aichele, P., & Ehl, S. (2013). Graded defects in cytotoxicity determine severity of hemophagocytic lymphohistiocytosis in humans and mice. *Frontiers in Immunology, 4*, 448.
36. Pagel, J., Beutel, K., Lehmberg, K., Koch, F., Maul-Pavicic, A., Rohlfs, A. K., et al. (2012). Distinct mutations in STXBP2 are associated with variable clinical presentations in patients with familial hemophagocytic lymphohistiocytosis type 5 (FHL5). *Blood, 119*, 6016–6024.
37. Zhang, K., Chandrakasan, S., Chapman, H., Valencia, C. A., Husami, A., Kissell, D., et al. (2014). Synergistic defects of different molecules in the cytotoxic pathway lead to clinical familial hemophagocytic lymphohistiocytosis. *Blood, 124*, 1331–1334.
38. Cetica, V., Sieni, E., Pende, D., Danesino, C., De Fusco, C., Locatelli, F., et al. (2016). Genetic predisposition to hemophagocytic lymphohistiocytosis: Report on 500 patients from the Italian registry. *The Journal of Allergy and Clinical Immunology, 137*, 188–196.
39. Jordan, M. B., Hildeman, D., Kappler, J., & Marrack, P. (2004). An animal model of hemophagocytic lymphohistiocyosis (HLH): CD8+ T cells and interferon gamma are essential for the disorder. *Blood, 104*, 735–743.
40. Pachlopnik-Schmidt, J., Ho, C. H., Chrétien, F., Lefebvre, J. M., Pivert, G., Kosco-Vilbois, M., et al. (2009). Neutralization of IFNgamma defeats haemophagocytosis in LCMV-infected perforin- and Rab27a-deficient mice. *EMBO Molecular Medicine, 1*, 112–124.
41. Johnson, T. S., Terrell, C. E., Millen, S. H., Katz, J. D., Hildemann, D. A., & Jordan, M. B. (2014). Etoposide selectively ablates activated T cells to control the immunoregulatory disorder hemophagocytic lymphohistiocytosis. *Journal of Immunology, 192*, 84–91.
42. Amman, S., Lehmberg, K., zur Stadt, U., Janka, G., Rensing-Ehl, A., Klemann, C., et al. (2017). Primary and secondary hemophagocytic lymphohistiocytosis have different patterns of T-cell activation, differentiation and repertoire. *European Journal of Immunology, 47*, 364–373.
43. Bode, S. F. N., Ammann, S., Al-Herz, W., Bataneant, M., Dvorak, C. C., Gehring, S., et al. (2015). The syndrome of hemophagocytic lymophohistiocytosis in primary immunodeficiencies: Implications for differential diagnosis and pathogenesis. *Haematologica, 100*, 978–988.
44. Behrens, E. M., Canna, S. W., Slade, K., Rao, S., Kreiger, P. A., Paessler, M., et al. (2011). Repeated TLR9 stimulation results in macrophage activation syndrome-like disease in mice. *The Journal of Clinical Investigation, 121*, 2264–2277.
45. Jenkins, M. R., Rudd-Schmidt, J. A., Lopez, J. A., Ramsbottom, K. M., Mannering, S. I., Andrews, D. M., et al. (2015). Failed CTL/NK cell killing and cytokine hypersecretion are directly linked through prolonged synapse time. *The Journal of Experimental Medicine, 212*, 307–317.

46. Janka, G. E. (2012). Familial and acquired hemophagocytic lymphohistiocytosis. *Annual Review of Medicine, 63*, 233–246.
47. Shimazu, H., Munakata, S., Tashiro, Y., Salama, Y., Dhahri, D., Eiamboonsert, S., et al. (2017). Pharmacological targeting of plasmin prevents lethality in a murine model of macrophage activation syndrome. *Blood, 130*, 59–72.
48. Ramos-Casals, M., Brito-Zerón, P., López-Guillermo, A., Khamashta, M. A., & Bosch, X. (2014). Adult haemophagocytic syndrome. *Lancet, 383*, 1503–1516.
49. Zhang, K., Jordan, M. B., Marsh, R. A., Johnson, J. A., Kissell, D., Meller, J., et al. (2011). Hypomorphic mutations in PRF1, MUNC13-4, and STXBP2 are associated with adult-onset familial HLH. *Blood, 118*, 5794–5798.
50. Fardet, L., Galicier, L., Lambotte, O., Marzac, C., Aumont, C., Chahwan, D., et al. (2014). Development and validation of the HScore, a score for the diagnosis of reactive hemophagocytic syndrome. *Arthritis & Rhematology, 66*, 2613–2620.
51. La Rosée, P., Horne A. C., Hines M., von Bahr Greenwood T., Machowicz, R., Berliner, N., et al. (2019). Recommendations for the management of hemophagocytic lymphohistiocytosis in adults. *Blood, 133*, 2465–2477.
52. Bryceson, A. T., Pende, D., Maul-Pavicic, A., Gilmour, K. C., Ufheil, H., Vraetz, T., et al. (2012). A prospective evaluation of degranulation assays in the rapid diagnosis of familial hemophagocytic syndromes. *Blood, 119*, 2754–2763.
53. Lehmberg, K., & Ehl, S. (2013). Diagnostic evaluation of patients with suspected haemophagocytic lymphohistiocytosis. *British Journal of Haematology, 160*, 275–287.
54. Rubin, T. S., Zhang, K., Gifford, C., Lane, A., Choo, S., Bleesing, J. J., et al. (2017). Perforin and CD107a testing is superior to NK cell function testing for screening patients for genetic HLH. *Blood, 129*, 2993–2999.
55. Spessott, W. A., Sanmillan, M. L., McCormick, M. E., Patel, N., Villanueva, J., Zhang, K., et al. (2015). Hemophagocytic lymphohistiocytosis caused by dominant-negative mutations in STXBP2 that inhibit SNARE-mediated membrane fusion. *Blood, 125*, 1566–1577.
56. Zhang, M., Bracaglia, C., Prencipe, G., Bemrich-Stolz, C. J., Beukelman, T., Dimmitt, R. A., et al. (2016). A heterozygous RAB27A mutation associated with delayed cytolytic granule polarization and hemophagocytic lymphohistiocytosis. *Journal of Immunology, 196*, 2492–2503.
57. Janka, G. E., & Lehmberg, K. (2014). Hemophagocytic syndromes—an update. *Blood Reviews, 28*, 135–142.
58. Marsh, R. A., Jordan, M. B., & Filipovich, A. H. (2011). Reduced-intensity conditioning haematopoietic cell transplantation for haemophagocytic lymphohistiocytosis: An important step forward. *British Journal of Haematology, 154*, 556–563.
59. Ouachée-Chardin, M., Elie, C., de Saint Basile, G., Le Deist, F., Mahlaoui, N., Picard, C., et al. (2006). Hematopoietic stem cell transplantation in hemophagocytic lymphohistiocytosis: A single-center report of 48 patients. *Pediatrics, 117*, e743–e750.
60. Terrell, C. E., & Jordan, M. B. (2013). Mixed hematopoietic or T cell chimerism above a minimal threshold restores perforin-dependent immune regulation in perforin-deficient mice. *Blood, 122*, 2618–2621.
61. Lehmberg, K., Albert, M. H., Beier, R., Beutel, K., Gruhn, B., Kröger, N., et al. (2013). Treosulfan-based conditioning regimen for children and adolescents with hemophagocytic lymphohistiocytosis. *Haematologica, 99*, 180–184.
62. Hartz, B., Marsh, R., Rao, K., Henter, J. I., Jordan, M., Filipovich, L., et al. (2016). The minimum required level of donor chimerism in hereditary hemophagocytic lymphohistiocytosis. *Blood, 127*, 3281–3290.
63. Ohga, S., Kudo, K., Ishii, E., Honjo, S., Morimoto, A., Osugi, Y., et al. (2010). Hematopoietic stem cell transplantation for familial hemophagocytic lymphohistiocytosis and Epstein-Barr virus-associated hemophagocytic lymphohistiocytosis in Japan. *Pediatric Blood & Cancer, 54*, 299–306.

64. Das, R., Guan, P., Sprague, L., Verbist, K., Tedrick, P., An, Q. A., et al. (2016). Janus kinase inhibition lessens inflammation and ameliorates disease in murine models of hemophagocytic lymphohistiocytosis. *Blood, 127*, 1666–1675.
65. Maschalidi, S., Sepulveda, F. E., Garrigue, A., Fischer, A., & de Saint Basile, G. (2016). Therapeutic effect of JAK1/2 blockade on the manifestations of hemophagocytic lymphohistiocytosis in mice. 2016. *Blood, 128*, 60–71.
66. Broglie, L., Pommert, L., Rao, S., Thakar, M., Phelan, R., Margolis, D., et al. (2017). Ruxolitinib for treatment of refractory hemophagocytic lymphohistiocytosis. *Blood Advances, 1*, 1533–1536.
67. Sin, J. H., & Zangardi, M. L. (2017). Ruxolitinib for secondary hemophagocytic lymphohistiocytosis: First case report. *Hematology/Oncology and Stem Cell Therapy.*
68. Marsh, R. A., Jordan, M. B., Talano, J. A., Nichols, K. E., Kumar, A., Naqvi, S. R., et al. (2016). Salvage therapy for refractory hemophagocytic lymphohistiocytosis: A review of the published experience. *Pediatr. Blood Cancer, 64*, e26308.
69. Bosnak, M., Erdogan, S., Aktekin, E. H., & Bay, A. (2016). Therapeutic plasma exchange in primary hemophagocytic lymphohistiocytosis: Reports of two cases and a review of the literature. *Transfusion and Apheresis Science, 55*, 353–356.
70. Greil, C., Roether, F., La Rosée, P., Grimbacher, B., Duerschmied, D., & Warnatz, K. (2016). Rescue of cytokine storm due to HLH by hemoadsorption in a CTLA4-deficient patient. *Journal of Clinical Immunology, 37*, 273–276.
71. Maude, S. L., Barrett, D., Teachey, D. T., & Grupp, S. A. (2014). Managing cytokine release syndrome associated with novel T cell-engaging therapies. *Cancer Journal, 20*, 119–122.
72. Miettunen, P. M., Narendran, A., Jayanthan, A., Behrens, E. M., & Cron, R. Q. (2011). Successful treatment of severe paediatric rheumatic disease-associated macrophage activation syndrome with interleukin-1 inhibition following conventional immunosuppressive therapy: Case series with 12 patients. *Rheumatology (Oxford), 50*, 417–419.

The History of Macrophage Activation Syndrome in Autoimmune Diseases

Earl D. Silverman

The aim of this chapter is to present an overview of the history of macrophage activation syndrome (MAS) or secondary hemophagocytic lymphohistiocytosis (HLH).

In 1952, Farquhar et al. first described a familial disease that was clinically characterized by fever, hepatosplenomegaly, skin rash, lymphadenopathy, and central nervous system (CNS) involvement. Laboratory investigation showed a pancytopenia, a low ESR, abnormal liver function tests (LFTs), an abnormal prothrombin time (PT), and an abnormal cerebral spinal fluid examination (CSF) [1]. They termed this syndrome familial hemophagocytic reticulosis (FHR). The term was later changed to familial hemophagocytic lymphohistiocytosis (FHLH) and then just hemophagocytic lymphohistiocytosis (HLH). In 1997, the HLH Study Group defined primary HLH and secondary HLH. Both illnesses were characterized by activation of the mononuclear phagocytic system. Secondary HLH included virus-associated hemophagocytotic syndrome (VAHS) [now called infection-associated hemophagocytic syndrome (IAHS)], malignancy-associated hemophagocytic syndrome (MAHS), and HLH following prolonged intravenous nutrition, including administration of soluble lipids (fat overload syndrome). There was no mention of rheumatic diseases as a cause of secondary HLH [2]. HLH criteria were further updated in 2004, and these are the criteria that are currently used [3].

Secondary HLH

In 1939, Scott et al. described four patients with HLH and reviewed five patients found in the literature. These patients were described as having "atypical Hodgkin's disease." The patients had the following clinical features in common: fever, wasting,

E. D. Silverman (✉)
Hospital for Sick Children, University of Toronto, Toronto, ON, Canada
e-mail: earl.silverman@sickkids.ca

R. Q. Cron, E. M. Behrens (eds.), *Cytokine Storm Syndrome*,
https://doi.org/10.1007/978-3-030-22094-5_2

generalized lymphadenopathy, and hepatosplenomegaly with jaundice and purpura. Laboratory investigations showed anemia, thrombocytopenia and marked leukopenia. Post-mortem examination revealed hyperplasia of histiocytes throughout the lymphoreticular tissue, and, importantly, there was evidence of profound erythrophagocytosis by histiocytes. The cellular proliferation with active phagocytosis was most prominent in the medullary portion of node, spleen, and periportal areas of liver. The authors therefore termed this syndrome histiocytic medullary reticulosis (HMR) [4]. Even as late as 1973, a large review of histiocytic disorders still referred to HMR as a malignancy and felt it was a 'well differentiated form of reticulum cell sarcoma' [5]. It was not until 1979 that Risdall et al. introduced the term viral-associated hemophagocytic syndrome (VAHS) to differentiate HMR from malignant histiocytic disorders. It should be noted however, that 5 out of 19 patients described did not have a documented, or any evidence of, infection at the time of diagnosis of VAHS [6]. Despite this and other publications, even as late as in 1984 HMR was still occasionally called a malignant disorder, although the term "reactive" was beginning to gain usage [7]. It soon became recognized that his syndrome was associated with multiple hematologic malignancies as well as viruses (reviewed in [8]). However, the term macrophage activation syndrome (MAS) was not coined until 1993.

MAS in Rheumatic Diseases

Difficulty in Diagnosis of MAS in Rheumatic Diseases

The term MAS appears to be used only when secondary HLH is associated with an autoimmune disease and has led to confusion in terminology and diagnosis. Furthermore, it can be difficult to be sure of the diagnosis of secondary HLH in autoimmune diseases (MAS) as a result of the great overlap of the clinical and laboratory features between HLH and the underlying autoimmune disease. This is likely because activation of the immune system, including macrophages, is characteristic of autoimmune disease while the diagnostic criteria for HLH were based on the differentiation of this disease entity from other diseases of histiocytes rather than from patients with chronic activation of the immune system as seen in autoimmune diseases. Therefore, absolute levels of many of the diagnostic features may not be relevant in patients with autoimmune diseases. Furthermore, the criteria have not been validated in either pediatric or adult cases of MAS [9]. This was first recognized in patients with systemic juvenile idiopathic arthritis (sJIA) [10] but also applies to patients with other autoimmune diseases including systemic lupus erythematosus (SLE) which is frequently associated with levels of both soluble interleukin-2 receptor (sIL-2R) and ferritin (both part of the diagnostic criteria of HLH) within levels in the diagnostic criteria of HLH, while fibrinogen is often not below HLH diagnostic levels. The latter observation is likely the result of the elevation of baseline fibrinogen levels, as it is an acute phase reactant. Similarly, triglycerides maybe elevated in patients on corticosteroids. By contrast, elevated liver function tests (LFTs), generalized lymphadenopathy, hypoproteinemia, hyponatremia, and decreased HDL in MAS are frequently seen although only considered supportive of the diagnosis of HLH according to 2004 criteria [3].

MAS in Juvenile Idiopathic Arthritis (JIA)

The first use of the term "activated macrophages" in patients with what we now refer to as sJIA was in 1985 by Hadchouel et al. [11]. They described seven patients who developed sudden onset of fever with altered level of consciousness bleeding and hepatosplenomegaly. Laboratory investigations showed a fall in fibrinogen, hemoglobin, white blood cell count, platelets, and ESR with increased fibrin split products, and LFTs. In three patients, it was following a second injection of gold (vide infra), while in the other 4 there was either evidence of a recent infection, non-steroid anti-inflammatory drug (NSAID) use, or no known cause. Liver biopsy showed evidence of diffuse Kupffer cell hyperplasia. Many of the Kupffer cells contained ceroid or lipofuscin pigments but were without evidence of "excessive phagocytosis" and no erythrophagocytosis. Bone marrow examination in two patients showed large macrophages with phagocytosed material but no erythrophagocytosis. They concluded that the main histological feature was macrophage activation and hypothesized that these activated macrophages secreted enzymes that lead to the clinical picture. The first published paper which used the term macrophage activation syndrome in its title was in 1993 [12]. This paper described four patients with childhood-onset rheumatic diseases who clinically had fever, hepatosplenomegaly, pancytopenia, low ESR, abnormal LFTs, and hypofibrinogenemia. Bone marrow aspiration showed active hemophagocytosis. In two patients, there was evidence of high levels circulating cytokines. In 1994, it was then proposed that activated macrophages can lead to this syndrome, including the low fibrinogen and increased fibrin degradation products commonly seen in these patients [13].

Although these papers referred to the above defined MAS as a disease entity, previous publications described patients with sJIA who had this clinical syndrome. However, the authors used different names to describe the clinical and laboratory findings in these patients. It is likely that the first report was in 1971. This paper described seven patients with abnormal LFTs accompanied by a fall in ESR, platelets, hemoglobin, and/or white blood cell (WBC) count. Many of the patients had a new-onset macular rash which was different from the rash of sJIA and/or increased adenopathy, and/or hepatosplenomegaly. It is likely that at least some of these patients had MAS. In one patient each, the clinical syndrome was present a) shortly following diagnosis; b) following second injection of gold; and c) with concomitant EBV infection [14]. The next reports were not until 1983 when two papers appeared that described patients with sJIA who developed a picture of disseminated intravascular coagulation (DIC)/consumptive coagulopathy [15, 16].

Silverman et al. described seven patients with sJIA who had a fall in hemoglobin, platelet count, fibrinogen, and ESR with elevated LFTs, PT, and/or PTT, and fibrin split products. In two patients, this clinical picture occurred following the second injection of gold salts, while in the other five it was either at presentation or during a time of disease flare. In two patients, this syndrome recurred [16]. The other paper, by De Vere-Tyndall et al., described two patients with sJIA who developed a fall in hemoglobin, platelet count, ESR, with elevation of PT/PTT, fibrin degradation products, and evidence of active bleeding. In one patient, it followed a third injection of gold and in the other following an infection [15]. A paper from 1985 reported a patient with sJIA who developed what appears to be MAS during an episode of chickenpox. A bone marrow aspiration was reported as showing histiocytic medullary reticulosis (hemophagocytosis). The final diagnosis was VAHS [17]. In the same year, a patient with sJIA developed the clinical syndrome of MAS during a Coxsackie infection. Bone marrow aspiration showed hemophagocytosis, and the patient was reported to have histiocytic medullary reticulosis [8]. sJIA is now the most commonly cited cause of MAS and classification criteria have been proposed [18], although systemic lupus erythematosus (SLE) in MAS may account for more patients in total (see below).

Although MAS in patients with JIA is most commonly seen in patients with sJIA, it has also been described in other JIA subtypes [19].

MAS in Systemic Lupus Erythematosus (SLE)

Despite the recognition of MAS as early as 1985 in sJIA and in adult-onset Still disease (AOSD), it was until 1991 that the first recognized case of MAS in patients with SLE was described in the literature [20]. However, in 1979, Risdall et al. described a case of virus-associated MAS in a patient with SLE [6]. The reason for this late recognition in SLE is likely for two reasons: (1) As described above, there are many overlapping clinical and laboratory features of SLE and MAS; and (2) Unlike sJIA patients, SLE is known to have a significant mortality even at presentation. However, since then, there have many cases series and large cohorts of patients with both childhood-onset SLE (cSLE) and adult-onset SLE (aSLE) [21].

In adults, SLE is the most common rheumatic disease leading to MAS, and most reviews report that SLE is responsible for the majority of cases of MAS in adult rheumatic diseases (described below). The incidence of MAS in aSLE has been reported to be between 0.9 and 4.6% [22]. A report from France in 2017 reported 89 patients with aSLE and MAS [23]. A review of the literature of MAS in patients with rheumatic disease showed that the most common reported cause was sJIA with SLE second. However, similar to what is seen in aSLE, in our hospital, we see more cases of MAS secondary to SLE than sJIA (unpublished data), although others have reported that MAS occurs more frequently in sJIA than cSLE [24]. In childhood, MAS has been associated with Kikuchi's disease with and without associated SLE [25].

MAS in Kawasaki Disease (KD)

The first reported case in the literature of MAS secondary to KD appeared in a 1995 article that described a 32-month-old boy from Japan who developed HPS during KD that was resistant to treatment. It is interesting to note that this patient had elevated interferon-gamma (IFN-γ) levels at the time of MAS but not earlier during the course of his KD [26]. A systematic review of MAS in KD, as of September 2016, reviewed 67 cases of MAS in the literature, and two reported new cases. The first four case reports suggested that MAS occurred late following the diagnosis of KD (range 20–24 days after onset first symptom) and following failure of IVIG treatment (2–5 courses). The systemic review demonstrated that in KD, MAS either preceded (6%) or was simultaneous with KD presentation (21%). The initial observation that MAS in KD tended to be in older children was confirmed in a systemic review, as 34/69 KD patients were >5 years old [27]. Similar to what has been described in other causes of MAS, mild MAS may be seen in KD, and, in fact, Choi suggested that all patients with refractory KD should be considered to have 'occult' MAS [28]. A systematic review of MAS in rheumatic disease patients in 2012 reported that in 6% of patients MAS was associated with KD [29]. Conversely, MAS occured in 1.9–4.7% of KD patients [30, 31]. In 2016, it was suggested that HLH criteria may not be a good indicator of MAS in KD and suggested the possibility of testing the MAS using sJIA 2016 MAS criteria [32].

MAS in Adult Rheumatic Diseases

The first patient with adult rheumatoid arthritis (RA) with MAS was described in 1977 [33]. Interestingly, although the authors concluded it was HMR presenting as RA, the patient had a 4-year history of joint pains, but at the time of presentation the patient developed pancytopenia, a fall in ESR, hepatosplenomegaly, and generalized lymphadenopathy, and died. Postmortem examination of lymph nodes, liver, and spleen revealed histiocytic infiltrate with marked erythrophagocytosis. By 2007, 11 cases secondary to RA had been reported in the Japanese literature [34]. A review of 30 cases of MAS in adults showed that it occurred in order of frequency: SLE—18 cases, ASOD—3 cases, and 2 cases each in patients with RA, polymyositis/dermatomyositis (PM/DM), systemic sclerosis (SSc), and Sjögren syndrome (SS), and one case in a patient with vasculitis [35]. MAS has also been reported in adults with polyarteritis nodosa, mixed connective tissue disease, pulmonary sarcoidosis, Goodpasture disease, Behcet disease, granulomatosis with polyangiitis, and ankylosing spondylitis [29, 36–38]. It has been proposed that the EULAR/ACR/PRINTO classification criteria for MAS can be used in AOSD [39].

Therapy of MAS

As MAS was first recognized in rheumatic diseases in sJIA, the initial treatment was dictated by the treatment of sJIA and included steroid, high dose oral or pulse, and IVIG. If this failed, then treatment was either initiated with HLH 1994 (mostly etoposide) or cyclosporine alone [40, 41]. By the early 2000s, the drug of choice for patients who failed steroid and IVIG therapy was cyclosporine [42], although other authors reported success with anti-TNF agents [43]. Calcineurin inhibitors (cyclosporine or tacrolimus) remain alternatives in patients who fail or are dependent on high doses steroids and IVIG. A key to the use of a calcineurin inhibitor is likely its earlier introduction. In the mid-2000s, other therapies included the use of anti-thymocyte globulin [44, 45] or monoclonal anti-CD25 antibodies [46, 47], but these therapies are not routinely used. Currently, many investigators suggest that IL-1 inhibition, in particular, anakinra, with its short half-life, is the drug of choice when a calcineurin inhibitor does not control the MAS or is contra-indicated [48–51]. However, the doses of anakinra that are required to control MAS may exceed the standard daily dose, and more frequent dosing may be required [51, 52]. There is at least a theoretical word of caution for the use of IL-1 inhibition in SLE patients with MAS as blocking the IL-1-TNF (Th1 pathway) may lead to production of Th2 cytokines which have been implicated in the pathogenesis of SLE. This issue remains to be resolved.

In resistant cases, treatment with the HLH-2004 protocol should be considered. Hematologists who treat patients with HLH are frequently involved in the treatment of patients with MAS. They tend to advocate the use of dexamethasone rather than prednisone and the early use of the HLH-2004 protocol for steroid/IVIG failures. The rationale for the use of dexamethasone over prednisone is that dexamethasone penetrates the CSF better than prednisone and patients with HLH frequently have CNS involvement. Certainly, when CNS involvement is seen in patients with MAS then dexamethasone should be used but its routine use in MAS is not as clear. There have not been any studies that compare prednisone to dexamethasone in MAS. The use of the HLH-2004 protocol and, in particular, the use of etoposide is frequently advocated by hematologists and resisted by rheumatologist because of its significant immunosuppressive properties and the potential for the development of malignancy [3]. Patients treated with etoposide are at high risk for sepsis, including invasive infection with opportunistic organisms leading to death. However, severe unresponsive MAS also leads to death. The current protocol at our institution is the use of IVIG and high dose prednisone, including pulse methylprednisolone, with the early introduction of a calcineurin inhibitor. If response is unsatisfactory, then the rapid introduction of anakinra is advocated with escalating doses and decreasing time between doses to get a rapid response to avoid the requirement for life-support. The requirement for admission to an intensive care unit is associated with a poor prognosis, and all attempts should be made to prevent this, including the consideration of the use of the HLH-2004 protocol.

Newer, more specific therapies directed against other cytokines may available in the future. A mouse model of MAS showed there was a significant upregulation of the interferon-gamma (IFN-γ) pathway. Mice were treated with an anti-IFN-γ antibody showed a significant improvement in survival which was associated with a significantly decreased level of circulating chemokines, CXCL9 and CXCL10, and downstream pro-inflammatory cytokines [53]. In patients with active MAS and other secondary for types of HLH, levels of IFN-γ and of IFN-γ-induced chemokines were markedly elevated during times of active MAS. Furthermore, in patients with sJIA and active MAS, these levels were higher than in patients with active sJIA without MAS [54].

A second cytokine, interleukin-18 (IL-18), which is upstream of IFN-γ, may also be important in the pathogenesis of MAS, and blocking its action may be another potential therapeutic target. As early as 2001, elevated levels of IL-18 were found in patients with active AOSD [55]. A mutation in the NLRC4 gene leads to infantile enterocolitis and recurrent MAS, which is associated with increased production of IL-18, and treatment with recombinant IL-18 binding protein (IL-18BP) was associated with resolution of MAS [56, 57]. Furthermore, it was shown that very high serum IL-18 levels were associated with the development MAS in patients with sJIA [58]. In the CpG-treated IL-18BP(−/−) mouse model of MAS, levels of IFN-γ and IFN-γ -associated chemokines were significantly increased as compared to wild-type mice, and the use of IL-18BP decreased the severity of both MAS and the IFN-γ response [59].

Animal Models

Animal models can be useful to suggest important pathogenic mechanism of human disease and to test potential therapies. I will briefly review the history of animal of HLH and MAS.

In 2004, Jordan et al. developed a murine model of HLH by infecting a perforin-deficient (pfp−/−) mouse strain with lymphocytic choriomeningitic virus (LCMV). These mice had multiple clinical and histologic features of HLH. There was evidence of activated CD68+ macrophages and elevated levels of both IFN-γ and, to a lesser extent, IL-18. Depletion experiments showed that CD8+ T cells secreting IFN-γ were required for the development of the HLH-like syndrome and death [60]. This model was used to demonstrate that perforin is required for the functioning of a reciprocal interaction between CD8+ T cells and dendritic cells, whereby antigen-primed DCS activate T cells, while primed CD8+ T cells suppress DC function (T cell activation) [61]. An animal model of familial HLH type4 (FHL4) has also been developed in a Stx11-deficient mouse. This model was used to show the importance of T-cell exhaustion as an important factor for determination of disease severity in HLH [62]. The generation of doubly or triply heterozygous for mutations in HLH-associated genes (perforin, Rab27a, and syntaxin-11) demonstrated that the accumulation of multiple monoallelic mutations increased the risk of developing HLH [63].

The results of this study suggest that double heterozygotes for HLH genes may develop HLH.

The development of animal models for secondary HLH or MAS has been less successful. Models using either repeated stimulation of TLR9 or chronic typhoid fever have been less helpful in gaining insights in HLH/MAS [64, 65]. However, the mouse model using repeated stimulation of TLR9 was used to show the importance of tissue levels of IFN-γ, and that CXCL9 and CXCL10 blood levels were important, and that neutralization of tissue IFN-γ correlated with CXCL9 and CXCL10 blood level normalization [66]. A CMV model of secondary HLH developed only a mild phenotype, and depletion of CD8+ T cells could not inhibit or cure the HLH-like syndrome [67]. More promising was the development of EBV-associated hemophagocytic lymphohistiocytosis (EBV-HLH) in a human CD34+ cell-transplanted humanized mouse model [68].

In two different murine models of HLH, JAK1/2 inhibitor or anti-IFN-γ antibody treatment prevented full-blown HLH when given after the development of disease [69, 70]. In the CMV-prf−/− model, treatment with IL-18 binding protein (BP) decreased hemophagocytosis and reversed liver as well as spleen damage and cytokine production by CD8+ T and NK cells [71]. Although these results are encouraging, there have not been clinical trials of these treatments undertaken in MAS.

Genetics

The history of the genetics of HLH began as early as 1999 [72]. Subsequently, autosomal recessive defects in PRF1, UNC13D, STX11, and STXBP2 genes were shown to be responsible for types 2, 3, 4, and 5 types of primary or familial HLH (FHL). These four genes are important in perforin-mediated lymphocyte cytotoxicity. Similarly, autosomal recessive mutations in RAB27A and LYST genes were shown to lead to secondary HLH seen in Griscelli syndrome type 2 (GS2) and Chédiak–Higashi syndrome (CHS). Patients with X-linked lymphoproliferative disease and mutations in SH2D1A or XIAP will frequently present with secondary HLH that is often triggered by Epstein–Barr virus (EBV) infection (reviewed [73]). Monogenic mutations leading to macrophage activation and autoinflammatory disease have been associated with MAS. One example is a defect in the NLRC4 inflammasome that leads to an early onset systemic inflammatory disease with MAS [74]. MAS had been seen in multiple auto-inflammatory diseases, including cryopyrin-associated periodic syndrome, mevalonate kinase deficiency, familial Mediterranean fever, and tumor necrosis factor receptor-associated periodic syndrome [75].

Testing of FHL-associated genes in patients with sJIA has shown that there is an increased frequency of heterozygous, but not homozygous, mutations patients with a history of MAS as compared to those without a history of MAS [76–79]. These mutations were associated with decreased perforin gene expression and/or function [79]. Similarly a heterozygous mutation in the perforin gene was found in a patient with adult RA who died of MAS [73].

The recognition and understanding of MAS pathophysiology has come a long way in the last several decades. New diagnostic criteria and the availability of agents targeting inflammatory cytokines should help survival of patients with rheumatic diseases who go on to develop MAS. The future is looking more encouraging for these patients, in terms of preventing fatal outcomes.

References

1. Farquhar, J. W., & Claireaux, A. E. (1952). Familial haemophagocytic reticulosis. *Archives of Disease in Childhood, 27*, 519–525.
2. Henter, J. I., Arico, M., Egeler, R. M., et al. (1997). HLH-94: A treatment protocol for hemophagocytic lymphohistiocytosis. HLH study Group of the Histiocyte Society. *Medical and Pediatric Oncology, 28*, 342–347.
3. Henter, J. I., Horne, A., Arico, M., et al. (2007). HLH-2004: Diagnostic and therapeutic guidelines for hemophagocytic lymphohistiocytosis. *Pediatric Blood & Cancer, 48*, 124–131.
4. Scott, R., & Robb-Smith, A. (1939). Histiocytic medullary retiuculosis. *Lancet, 2*, 194–198.
5. Cline, M. J., & Golde, D. W. (1973). A review and reevaluation of the histiocytic disorders. *The American Journal of Medicine, 55*, 49–60.
6. Risdall, R. J., McKenna, R. W., Nesbit, M. E., et al. (1979). Virus-associated hemophagocytic syndrome A benign histiocytic proliferation distinct from malignant histiocytosis. *Cancer, 44*, 993–1002.
7. Stark, B., Hershko, C., Rosen, N., Cividalli, G., Karsai, H., & Soffer, D. (1984). Familial hemophagocytic lymphohistiocytosis (FHLH) in Israel. I. Description of 11 patients of Iranian-Iraqi origin and review of the literature. *Cancer, 54*, 2109–2121.
8. Heaton, D. C., & Moller, P. W. (1985). Still's disease associated with Coxsackie infection and haemophagocytic syndrome. *Annals of the Rheumatic Diseases, 44*, 341–344.
9. Otrock, Z. K., Daver, N., Kantarjian, H. M., & Eby, C. S. (2017). Diagnostic challenges of hemophagocytic lymphohistiocytosis. *Clinical Lymphoma, Myeloma & Leukemia, 17s*, S105–Ss10.
10. Parodi, A., Davi, S., Pringe, A. B., et al. (2009). Macrophage activation syndrome in juvenile systemic lupus erythematosus: A multinational multicenter study of thirty-eight patients. *Arthritis and Rheumatism, 60*, 3388–3399.
11. Hadchouel, M., Prieur, A. M., & Griscelli, C. (1985). Acute hemorrhagic, hepatic, and neurologic manifestations in juvenile rheumatoid arthritis: Possible relationship to drugs or infection. *The Journal of Pediatrics, 106*, 561–566.
12. Stephan, J. L., Zeller, J., Hubert, P., Herbelin, C., Dayer, J. M., & Prieur, A. M. (1993). Macrophage activation syndrome and rheumatic disease in childhood: A report of four new cases. *Clinical and Experimental Rheumatology, 11*, 451–456.
13. Prieur, A. M., & Stephan, J. L. (1994). Macrophage activation syndrome in rheumatic diseases in children. *Revue du Rhumatisme, 61*, 447–451.
14. Kornreich, H., Malouf, N. N., & Hanson, V. (1971). Acute hepatic dysfunction in juvenile rheumatoid arthritis. *The Journal of Pediatrics, 79*, 27–35.
15. De Vere-Tyndall, A., Macauley, D., & Ansell, B. M. (1983). Disseminated intravascular coagulation complicating systemic juvenile chronic arthritis ("Still's disease"). *Clinical Rheumatology, 2*, 415–418.
16. Silverman, E. D., Miller 3rd, J. J., Bernstein, B., & Shafai, T. (1983). Consumption coagulopathy associated with systemic juvenile rheumatoid arthritis. *The Journal of Pediatrics, 103*, 872–876.
17. Morris, J. A., Adamson, A. R., Holt, P. J., & Davson, J. (1985). Still's disease and the virus-associated haemophagocytic syndrome. *Annals of the Rheumatic Diseases, 44*, 349–353.

18. Ravelli, A., Minoia, F., Davi, S., et al. (2016). Classification criteria for macrophage activation syndrome complicating systemic juvenile idiopathic arthritis: A European League Against Rheumatism/American College of Rheumatology/Paediatric Rheumatology International Trials Organisation Collaborative Initiative. *Annals of the Rheumatic Diseases, 75*, 481–489.
19. Park, J. H., Seo, Y. M., Han, S. B., et al. (2016). Recurrent macrophage activation syndrome since toddler age in an adolescent boy with HLA B27 positive juvenile ankylosing spondylitis. *Korean Journal of Pediatrics, 59*, 421–424.
20. Wong, K. F., Hui, P. K., Chan, J. K., Chan, Y. W., & Ha, S. Y. (1991). The acute lupus hemophagocytic syndrome. *Annals of Internal Medicine, 114*, 387–390.
21. Borgia, R. E., Gerstein, M., Levy, D. M., Silverman, E. D., & Hiraki, L. T. (2018). Features, treatment, and outcomes of macrophage activation syndrome in childhood-onset systemic lupus erythematosus. *Arthritis & Rhematology, 70*, 616–624.
22. Granata, G., Didona, D., Stifano, G., Feola, A., & Granata, M. (2015). Macrophage activation syndrome as onset of systemic lupus erythematosus: A case report and a review of the literature. *Case Reports in Medicine, 2015*, 294041.
23. Gavand, P. E., Serio, I., Arnaud, L., et al. (2017). Clinical spectrum and therapeutic management of systemic lupus erythematosus-associated macrophage activation syndrome: A study of 103 episodes in 89 adult patients. *Autoimmunity Reviews, 16*, 743–749.
24. Aytac, S., Batu, E. D., Unal, S., et al. (2016). Macrophage activation syndrome in children with systemic juvenile idiopathic arthritis and systemic lupus erythematosus. *Rheumatology International, 36*, 1421–1429.
25. Khan, F. Y., Morad, N. A., & Fawzy, Z. (2007). Kikuchi's disease associated with hemophagocytosis. *Chang Gung Medical Journal, 30*, 370–373.
26. Ohga, S., Ooshima, A., Fukushige, J., & Ueda, K. (1995). Histiocytic haemophagocytosis in a patient with Kawasaki disease: Changes in the hypercytokinaemic state. *European Journal of Pediatrics, 154*, 539–541.
27. Garcia-Pavon, S., Yamazaki-Nakashimada, M. A., Baez, M., Borjas-Aguilar, K. L., & Murata, C. (2017). Kawasaki disease complicated with macrophage activation syndrome: A systematic review. *Journal of Pediatric Hematology/Oncology, 39*, 445–451.
28. Choi, U. Y., Han, S. B., Lee, S. Y., & Jeong, D. C. (2017). Should refractory Kawasaki disease be considered occult macrophage activation syndrome? *Seminars in Arthritis and Rheumatism, 46*, e17.
29. Atteritano, M., David, A., Bagnato, G., et al. (2012). Haemophagocytic syndrome in rheumatic patients. A systematic review. *European Review for Medical and Pharmacological Sciences, 16*, 1414–1424.
30. Kang, H. R., Kwon, Y. H., Yoo, E. S., et al. (2013). Clinical characteristics of hemophagocytic lymphohistiocytosis following Kawasaki disease: Differentiation from recurrent Kawasaki disease. *Blood Research, 48*, 254–257.
31. Latino, G. A., Manlhiot, C., Yeung, R. S., Chahal, N., & McCrindle, B. W. (2010). Macrophage activation syndrome in the acute phase of Kawasaki disease. *Journal of Pediatric Hematology/Oncology, 32*, 527–531.
32. Han, S. B., Lee, S. Y., Jeong, D. C., & Kang, J. H. (2016). Should 2016 criteria for macrophage activation syndrome be applied in children with Kawasaki disease, as well as with systemic-onset juvenile idiopathic arthritis? *Annals of the Rheumatic Diseases, 75*, e44.
33. Crow, J., & Gumpel, J. M. (1977). Histiocytic medullary reticulosis presenting as rheumatoid arthritis. *Proceedings of the Royal Society of Medicine, 70*, 632–634.
34. Katoh, N., Gono, T., Mitsuhashi, S., et al. (2007). Hemophagocytic syndrome associated with rheumatoid arthritis. *Internal Medicine, 46*, 1809–1813.
35. Fukaya, S., Yasuda, S., Hashimoto, T., et al. (2008). Clinical features of haemophagocytic syndrome in patients with systemic autoimmune diseases: Analysis of 30 cases. *Rheumatology, 47*, 1686–1691.
36. Dhote, R., Simon, J., Papo, T., et al. (2003). Reactive hemophagocytic syndrome in adult systemic disease: Report of twenty-six cases and literature review. *Arthritis and Rheumatism, 49*, 633–639.

37. Basnet, A., & Cholankeril, M. R. (2014). Hemophagocytic lymphohistiocytosis in a patient with Goodpasture's syndrome: A rare clinical association. *American Journal of Case Reports, 15*, 431–436.
38. Lou, Y. J., Jin, J., & Mai, W. Y. (2007). Ankylosing spondylitis presenting with macrophage activation syndrome. *Clinical Rheumatology, 26*, 1929–1930.
39. Ahn, S. S., Yoo, B. W., Jung, S. M., Lee, S. W., Park, Y. B., & Song, J. J. (2017). Application of the 2016 EULAR/ACR/PRINTO classification criteria for macrophage activation syndrome in patients with adult-onset still disease. *The Journal of Rheumatology, 44*, 996–1003.
40. Fishman, D., Rooney, M., & Woo, P. (1995). Successful management of reactive haemophagocytic syndrome in systemic-onset juvenile chronic arthritis. *British Journal of Rheumatology, 34*, 888.
41. Quesnel, B., Catteau, B., Aznar, V., Bauters, F., & Fenaux, P. (1997). Successful treatment of juvenile rheumatoid arthritis associated haemophagocytic syndrome by cyclosporin A with transient exacerbation by conventional-dose G-CSF. *British Journal of Haematology, 97*, 508–510.
42. Ravelli, A., Viola, S., De Benedetti, F., Magni-Manzoni, S., Tzialla, C., & Martini, A. (2001). Dramatic efficacy of cyclosporine A in macrophage activation syndrome. *Clinical and Experimental Rheumatology, 19*, 108.
43. Prahalad, S., Bove, K. E., Dickens, D., Lovell, D. J., & Grom, A. A. (2001). Etanercept in the treatment of macrophage activation syndrome. *The Journal of Rheumatology, 28*, 2120–2124.
44. Stabile, A., Bertoni, B., Ansuini, V., La Torraca, I., Salli, A., & Rigante, D. (2006). The clinical spectrum and treatment options of macrophage activation syndrome in the pediatric age. *European Review for Medical and Pharmacological Sciences, 10*, 53–59.
45. Ozturk, K., & Ekinci, Z. (2015). Successful treatment of macrophage activation syndrome due to systemic onset juvenile idiopathic arthritis with antithymocyte globulin. *Rheumatology International, 35*, 1779–1780.
46. Tomaske, M., Amon, O., Bosk, A., Handgretinger, R., Schneider, E. M., & Niethammer, D. (2002). Alpha-CD25 antibody treatment in a child with hemophagocytic lymphohistiocytosis. *Medical and Pediatric Oncology, 38*, 141–142.
47. Olin, R. L., Nichols, K. E., Naghashpour, M., et al. (2008). Successful use of the anti-CD25 antibody daclizumab in an adult patient with hemophagocytic lymphohistiocytosis. *American Journal of Hematology, 83*, 747–749.
48. Behrens, E. M., Kreiger, P. A., Cherian, S., & Cron, R. Q. (2006). Interleukin 1 receptor antagonist to treat cytophagic histiocytic panniculitis with secondary hemophagocytic lymphohistiocytosis. *The Journal of Rheumatology, 33*, 2081–2084.
49. Durand, M., Troyanov, Y., Laflamme, P., & Gregoire, G. (2010). Macrophage activation syndrome treated with anakinra. *The Journal of Rheumatology, 37*, 879–880.
50. Miettunen, P. M., Narendran, A., Jayanthan, A., Behrens, E. M., & Cron, R. Q. (2011). Successful treatment of severe paediatric rheumatic disease-associated macrophage activation syndrome with interleukin-1 inhibition following conventional immunosuppressive therapy: Case series with 12 patients. *Rheumatology, 50*, 417–419.
51. Shakoory, B., Carcillo, J. A., Chatham, W. W., et al. (2016). Interleukin-1 receptor blockade is associated with reduced mortality in sepsis patients with features of macrophage activation syndrome: Reanalysis of a prior phase III trial. *Critical Care Medicine, 44*, 275–281.
52. Kahn, P. J., & Cron, R. Q. (2013). Higher-dose Anakinra is effective in a case of medically refractory macrophage activation syndrome. *The Journal of Rheumatology, 40*, 743–744.
53. Prencipe, G., Caiello, I., Pascarella, A., et al. (2018). Neutralization of IFN-gamma reverts clinical and laboratory features in a mouse model of macrophage activation syndrome. *The Journal of Allergy and Clinical Immunology, 141*, 1439–1449.
54. Bracaglia, C., de Graaf, K., Pires Marafon, D., et al. (2017). Elevated circulating levels of interferon-gamma and interferon-gamma-induced chemokines characterise patients with macrophage activation syndrome complicating systemic juvenile idiopathic arthritis. *Annals of the Rheumatic Diseases, 76*, 166–172.

55. Kawashima, M., Yamamura, M., Taniai, M., et al. (2001). Levels of interleukin-18 and its binding inhibitors in the blood circulation of patients with adult-onset Still's disease. *Arthritis and Rheumatism, 44*, 550–560.
56. Canna, S. W., de Jesus, A. A., Gouni, S., et al. (2014). An activating NLRC4 inflammasome mutation causes autoinflammation with recurrent macrophage activation syndrome. *Nature Genetics, 46*, 1140–1146.
57. Canna, S. W., Girard, C., Malle, L., et al. (2017). Life-threatening NLRC4-associated hyperinflammation successfully treated with IL-18 inhibition. *The Journal of Allergy and Clinical Immunology, 139*, 1698–1701.
58. Shimizu, M., Nakagishi, Y., Inoue, N., et al. (2015). Interleukin-18 for predicting the development of macrophage activation syndrome in systemic juvenile idiopathic arthritis. *Clinical Immunology, 160*, 277–281.
59. Girard-Guyonvarc'h, C., Palomo, J., Martin, P., et al. (2018). Unopposed IL-18 signaling leads to severe TLR9-induced macrophage activation syndrome in mice. *Blood, 131*, 1430–1441.
60. Jordan, M. B., Hildeman, D., Kappler, J., & Marrack, P. (2004). An animal model of hemophagocytic lymphohistiocytosis (HLH): CD8+ T cells and interferon gamma are essential for the disorder. *Blood, 104*, 735–743.
61. Terrell, C. E., & Jordan, M. B. (2013). Perforin deficiency impairs a critical immunoregulatory loop involving murine CD8(+) T cells and dendritic cells. *Blood, 121*, 5184–5191.
62. Kogl, T., Muller, J., Jessen, B., et al. (2013). Hemophagocytic lymphohistiocytosis in syntaxin-11-deficient mice: T-cell exhaustion limits fatal disease. *Blood, 121*, 604–613.
63. Sepulveda, F. E., Garrigue, A., Maschalidi, S., et al. (2016). Polygenic mutations in the cytotoxicity pathway increase susceptibility to develop HLH immunopathology in mice. *Blood, 127*, 2113–2121.
64. Behrens, E. M., Canna, S. W., Slade, K., et al. (2011). Repeated TLR9 stimulation results in macrophage activation syndrome-like disease in mice. *The Journal of Clinical Investigation, 121*, 2264–2277.
65. Brown, D. E., McCoy, M. W., Pilonieta, M. C., Nix, R. N., & Detweiler, C. S. (2010). Chronic murine typhoid fever is a natural model of secondary hemophagocytic lymphohistiocytosis. *PLoS One, 5*, e9441.
66. Buatois, V., Chatel, L., Cons, L., et al. (2017). Use of a mouse model to identify a blood biomarker for IFNgamma activity in pediatric secondary hemophagocytic lymphohistiocytosis. *Translational Research, 180*, 37–52.e2.
67. Brisse, E., Imbrechts, M., Put, K., et al. (2016). Mouse cytomegalovirus infection in BALB/c mice resembles virus-associated secondary hemophagocytic lymphohistiocytosis and shows a pathogenesis distinct from primary hemophagocytic lymphohistiocytosis. *Journal of Immunology, 196*, 3124–3134.
68. Sato, K., Misawa, N., Nie, C., et al. (2011). A novel animal model of Epstein-Barr virus-associated hemophagocytic lymphohistiocytosis in humanized mice. *Blood, 117*, 5663–5673.
69. Maschalidi, S., Sepulveda, F. E., Garrigue, A., Fischer, A., & de Saint Basile, G. (2016). Therapeutic effect of JAK1/2 blockade on the manifestations of hemophagocytic lymphohistiocytosis in mice. *Blood, 128*, 60–71.
70. Das, R., Guan, P., Sprague, L., et al. (2016). Janus kinase inhibition lessens inflammation and ameliorates disease in murine models of hemophagocytic lymphohistiocytosis. *Blood, 127*, 1666–1675.
71. Chiossone, L., Audonnet, S., Chetaille, B., et al. (2012). Protection from inflammatory organ damage in a murine model of hemophagocytic lymphohistiocytosis using treatment with IL-18 binding protein. *Frontiers in Immunology, 3*, 239.
72. Stepp, S. E., Dufourcq-Lagelouse, R., Le Deist, F., et al. (1999). Perforin gene defects in familial hemophagocytic lymphohistiocytosis. *Science, 286*, 1957–1959.
73. Cetica, V., Sieni, E., Pende, D., et al. (2016). Genetic predisposition to hemophagocytic lymphohistiocytosis: Report on 500 patients from the Italian registry. *The Journal of Allergy and Clinical Immunology, 137*, 188–96.e4.

74. Mukda, E., Trachoo, O., Pasomsub, E., et al. (2017). Exome sequencing for simultaneous mutation screening in children with hemophagocytic lymphohistiocytosis. *International Journal of Hematology, 106*, 282–290.
75. Rigante, D., Emmi, G., Fastiggi, M., Silvestri, E., & Cantarini, L. (2015). Macrophage activation syndrome in the course of monogenic autoinflammatory disorders. *Clinical Rheumatology, 34*, 1333–1339.
76. Hazen, M. M., Woodward, A. L., Hofmann, I., et al. (2008). Mutations of the hemophagocytic lymphohistiocytosis-associated gene UNC13D in a patient with systemic juvenile idiopathic arthritis. *Arthritis and Rheumatism, 58*, 567–570.
77. Zhang, K., Biroschak, J., Glass, D. N., et al. (2008). Macrophage activation syndrome in patients with systemic juvenile idiopathic arthritis is associated with MUNC13-4 polymorphisms. *Arthritis and Rheumatism, 58*, 2892–2896.
78. Vastert, S. J., van Wijk, R., D'Urbano, L. E., et al. (2010). Mutations in the perforin gene can be linked to macrophage activation syndrome in patients with systemic onset juvenile idiopathic arthritis. *Rheumatology, 49*, 441–449.
79. Kaufman, K. M., Linghu, B., Szustakowski, J. D., et al. (2014). Whole-exome sequencing reveals overlap between macrophage activation syndrome in systemic juvenile idiopathic arthritis and familial hemophagocytic lymphohistiocytosis. *Arthritis & Rhematology, 66*, 3486–3495.

Clinical Features of Cytokine Storm Syndrome

Masaki Shimizu

Introduction

Cytokine storm syndrome (CSS) is a severe life-threating condition characterized by a clinical phenotype of overwhelming systemic inflammation, hyperferritinemia, hemodynamic instability, and multiple organ failure (MOF), and if it is untreated, it can potentially lead to death. The hallmark of CSS is an uncontrolled and dysfunctional immune response involving the continual activation and expansion of lymphocytes and macrophages, which secrete large amounts of cytokines, causing a cytokine storm. Many clinical features of CSS can be explained by the effects of proinflammatory cytokines, such as interferon (IFN)-γ, tumor necrosis factor (TNF), interleukin (IL)-1, IL-6, and IL-18 [1–7]. These cytokines are elevated in most patients with CSS, as well as in animal models of CSS [8, 9]. A constellation of symptoms, signs, and laboratory abnormalities occurs that depends on the severity of the syndrome, the underlying predisposing conditions, and the triggering agent.

Various infectious and noninfectious diseases have a causal relationship with CSS. Hemophagocytic lymphohistiocytosis (HLH) is a representative hyperinflammatory disease characterized by uncontrolled cytokine storm. HLH is classified as either primary (pHLH) or secondary HLH (sHLH) based on their etiology and pathogenesis. pHLH is an autosomal recessive, monogenic disorder caused by loss-of-function mutations in genes involved in the cytotoxic function of natural killer cells and $CD8^+$ T lymphocytes. sHLH is a similar clinical syndrome but lacks a known genetic basis. sHLH occurs in the context of an underlying immunological condition, including malignancy, infection, or autoimmune or autoinflammatory disease. sHLH associated with autoimmune or autoinflammatory disease is called macrophage activation syndrome (MAS).

M. Shimizu (✉)
Department of Pediatrics, Graduate School of Medical Sciences, Kanazawa University, Kanazawa, Japan
e-mail: shimizum@staff.kanazawa-u.ac.jp

R. Q. Cron, E. M. Behrens (eds.), *Cytokine Storm Syndrome*,
https://doi.org/10.1007/978-3-030-22094-5_3

CSS is a life-threatening disease; therefore, a timely and prompt diagnosis is essential to initiate life-saving treatment. Early recognition of the syndrome is essential for the selection of an appropriate therapeutic intervention in a timely fashion. Therefore, CSS should be considered in patients with unexplained and atypical symptoms suggesting MOF.

Clinical Manifestations in CSS

pHLH frequently occurs in patients aged <4 years, whereas sHLH may present in older patients. However, both pHLH and sHLH may present at any age. The common clinical features of CSS are sustained fever, splenomegaly, hepatomegaly with liver dysfunction, lymphadenopathy, coagulopathy, cytopenia, skin rash, and variable neurologic symptoms. The initial symptoms of CSS are nonspecific, with either acute or subacute (1–4 weeks) clinical presentation. CSS can affect all organ systems. Table 1 shows the clinical features indicating a diagnosis of CSS. Many of these features are seen in patients with severe systemic sepsis, pHLH, and sHLH as phenotypes of hyperinflammatory reactions. The clinical features of CSS may be difficult to distinguish from those of the underlying diseases and may vary from disease to disease (Table 2) [10–18]. Awareness of the disease associations and highly suspecting CSS in patients with some of the clinical features are critical to the diagnosis of CSS. Each feature alone is nonspecific; however, the combination of clinical signs and symptoms and laboratory abnormalities, their severity, and changes over time facilitates correct diagnosis.

Sustained fever is a cardinal feature of CSS, which is usually high-grade (>38.5 °C), prolonged, and unresponsive to anti-infective treatment. For example, in patients with systemic juvenile idiopathic arthritis (s-JIA)-associated MAS, the type of fever changes from daily spiking fever to continuous unremitting fever once MAS develops. Rapid weight loss may also occur.

As a manifestation of macrophage and T-lymphocyte proliferation and activation in the reticuloendothelial system, peripheral lymphadenopathy is seen in 17–51.4% of patients and is not always evident on physical exam [10–18]. Splenomegaly is seen in 27.2–98% of patients [10–18]. Reported hepatomegaly is also quite variable, ranging from 18.8% to 94% of patients [10–18]. Hepatosplenomegaly is usually progressive and is accompanied with severe liver dysfunction. Liver involvement may manifest as jaundice and portal hypertension. The elevation of aspartate aminotransferase (AST) and lactate dehydrogenase (LDH) levels, which is almost always observed in patients with CSS, does not only reflect hepatic injury but also represents systemic tissue damage induced by cytokines, such as TNF. AST levels are usually higher than alanine aminotransferase levels. Creatinine kinase levels are also often elevated in patients with CSS as a result of the tissue damage. In addition, hypoalbuminemia, hyponatremia, and high direct bilirubin level are often observed in association with tissue edema caused by capillary leakage.

Table 1 Clinical and laboratory features of cytokine storm syndrome

System	Clinical manifestations	Laboratory findings
General	Fever	Elevated C reactive protein
		Fall in erythrocyte sedimentation rate
		Elevated soluble interleukin 2 receptor
Hematological	Petechiae	Leukopenia
	Purpura	Anemia
	Ecchymoses	Thrombocytopenia
	Epistaxis	Hemophagocytosis in bone marrow aspiration
	Lymphadenopathy	Hyperferritinemia
Skin	Rash	
	Erythroderma	
	Edema	
Respiratory	Acute respiratory distress	
	Pulmonary infiltrates	
Cardiac		
Renal		Acute kidney injury
Gastrointestinal	Hematemesis	Transaminitis
	Rectal bleeding	Elevated bilirubin
	Hepatomegaly	Hypoalbuminemia
	Splenomegaly	Elevated ammonia
		Elevated triglycerides
Central nervous system	Altered mental state	Pleiocytosis in cerebrospinal fluid
	Seizures	
	Encephalopathy	
	Coma	

Table 2 Comparison of clinical features in each cytokine storm syndrome

Clinical features	FHL	s-JIA MAS	SLE MAS	EBVHLH
Fever	91%	96.1%	96.9–100%	100%
Hepatomegaly	94%	70.0%	18.80%	87.8%
Splenomegaly	98%	57.9%	27.2–56.3%	64.3%
Lymphadenopathy	17%	51.4%	31.30%	50.0%
CNS involvement	47–73%	35.0%	36.90%	18.4%
Hemophagocytosis in the bone marrow	85%	60.7%	60.7–100%	92.7%
References	[10–12]	[13]	[14, 15]	[16]

Cytopenia likely results from multiple causes. Some have implicated depression of hematopoiesis and others hemophagocytosis in the bone marrow as mechanisms, but none have truly been established as the cause of cytopenia. Most all patients with CSS have anemia, which is usually non-regenerative. Serum chemistry findings may suggest hemolysis, with hyperbilirubinemia and elevated LDH level. Thrombocytopenia is almost consistently present, occurs early in the course of CSS, and is usually profound. Leukopenia is less common, less severe, and occurs later in the course of the disease. In some reports, three of every four patients with CSS have pancytopenia and all have bicytopenia [8, 19].

Hemophagocytosis in bone marrow aspirate is a hallmark of HLH/MAS. Hemophagocytosis may not be observed in the initial stage of HLH/MAS and may evolve subsequently. However, hemophagocytes are not essential in the diagnosis of HLH/MAS and can be seen in patients with juvenile arthritis without overt MAS [20]. In addition, hemophagocytes are commonly found in a wide variety of CSS, including sepsis, and following bone marrow transplant [21, 22]. Thus, hemophagocytosis is not pathognomonic for CSS, being neither highly sensitive nor specific for the diagnosis.

Coagulopathy with elevated fibrin degradation products and d-dimer are often observed even in the absence of overt disseminated intravascular coagulation (DIC), indicating subclinical endothelial dysfunction. In patients with CSS, disruption of the homeostasis of endothelial function by angiopoietin-1 and -2 has been reported [23, 24]. High levels of plasminogen activator secreted by macrophages stimulate plasmin and cause hyperfibrinolysis [25]. Consequently, plasma fibrinogen levels decrease. Hypofibrinogenemia is present at diagnosis in approximately two-thirds of the patients [26]. Fibrinogen as an acute phase protein is usually increased in febrile illness. Therefore, a fibrinogen level below the normal range in a child with fever for several days may indicate CSS. A falling erythrocyte sedimentation rate (ESR), as a result of decreasing fibrinogen from consumption, in the setting of a rising C-reactive protein is common in CSS. Some have even suggested that the ratio of ferritin divided by the ESR can be used to identify MAS/CSS [27]. In addition, consumption of complement proteins can be seen in MAS/CSS [28]. DIC is a very important predictor of MOF development and when this occurs, it is associated with high mortality [29–32].

Many other organs may also be involved. Skin manifestations are seen in approximately 20–65% of the patients and are commonly in the form of transient maculopapular, nodular, or purpuric lesions [33]. Cases of erythroderma in patients with CSS have also been described [34]. Pulmonary involvement is frequent (in approximately 40% of the patients), and symptoms can include cough, dyspnea, and respiratory failure [35], especially in cases triggered by respiratory viruses. Pulmonary infiltrates are found in 20–30% of the patients [36]. Gastrointestinal symptoms are found in approximately 20% of the patients. Symptoms can include diarrhea, nausea, vomiting, and abdominal pain, with specific presentations including gastrointestinal hemorrhage and ulcerative bowel disease [35]. Acute pancreatitis can be also complicated because of severe hypertriglyceridemia. Cardiac involvement occurs less frequently. A case of an infant with HLH with severe transient left ventricular pseudohypertrophy, likely caused by interstitial edema followed by increased vascular permeability, has been reported [37]. Furthermore, a patient with Epstein–Barr virus (EBV)-associated HLH complicated with myocarditis and coronary artery aneurysm has also been reported [38]. Renal dysfunction including acute kidney injury (AKI) as a part of multiple organ dysfunction, nephrotic syndrome, and mild proteinuria has been reported in patients with HLH [39–42]. The most frequent renal manifestation is AKI. Aulagnon et al. reported that AKI occurred in 62% of patients with HLH [39]. Six-month survival of patients with AKI was lower (37%) compared with 56% in those without AKI. Most patients showed stage 2 or 3 AKI, and dialysis was required in 59% of them [39]. AKI was due to acute tubular necrosis (49%), hypoperfusion (46%), tumor lysis (29%), or glomerulopathy (17%). Nephrotic syndrome (NS) was also present. Approximately 32% of surviving patients had chronic kidney disease. Glomerulopathy and NS complicating HLH result from primary podocyte pathology. It has been reported that the underlying lesions proven by kidney biopsy were collapsing focal segmental glomerulosclerosis, minimal change disease, and thrombotic microangiopathy [40, 41]. Renal failure during CSS is usually a later finding, reflective of multiorgan failure.

Central nervous system (CNS) symptoms including coma, seizures, meningitis, encephalopathy, ataxia, hemiplegia, cranial nerve palsies, mental status changes, or simply irritability are commonly reported in patients with CSS [12, 35, 43]. These neurologic abnormalities are seen in up to 70% of patients at the time of diagnosis [12]. In severe cases, mechanical ventilation is required because of alterations in consciousness. Pleocytosis, increased protein, or both in the cerebrospinal fluid are observed in more than 50% of the patients [43]. On neurologic imaging studies, enhancing nodular parenchymal lesions, leptomeningeal enhancement, demyelinization, delayed myelination, parenchymal calcification, and atrophy have been reported and suggested to correlate with clinical symptoms [44]. Cases of patients with isolated CNS symptoms without accompanying systemic findings, which is known as cerebral HLH, have been reported [45, 46]. However, it is often challenging to recognize cytokine storm in these patients. In addition, therapy may contribute to posterior reversible encephalopathy syndrome in a subset of CSS patients [47].

The clinical presentation of CSS is different in neonates compared with those in older patients [48]. Fever is commonly absent; therefore, coagulopathy, hepatomegaly, and cytopenia should raise suspicion of CSS in neonates. A neonate with pHLH presenting as isolated fulminant liver failure has also been reported [43]. In this case, it may be difficult to distinguish CSS from neonatal hemochromatosis, although patients with neonatal hemochromatosis do not usually have fever, cytopenia, and hypertriglyceridemia.

Taken together, CSS constitutes a medical emergency at any age. Because of the nonspecific nature of its clinical presentation, this disease is often overlooked. The clinical findings in overt CSS are dramatic. Patients become acutely ill with persistent fever, mental status changes, splenomegaly, lymphadenopathy, hepatomegaly with liver dysfunction, easy bruising, mucosal bleeding, etc. These symptoms are closely related to precipitous fall in blood cell lines. Therefore, clinicians must highly suspect CSS in any patient with unexplained cytopenia and fever, so that appropriate testing can be rapidly conducted. It is important for physicians to recognize that CSS is not a diagnosis of exclusion and can occur in multiple settings (e.g., autoimmunity, cancer, infection).

Relationship Between Key Cytokines and Clinical Features of CSS

Many features of CSS could potentially be explained by the known effects of proinflammatory cytokines, including IFN-γ, TNF, IL-1β, IL-6, IL-10, and IL-18 [1–7]. The relationship between these key cytokines and characteristic clinical features of CSS are shown in Table 3. Fever and systemic illness is induced by IL-1β, IL-6, IFN-γ, and TNF-α. Cytopenia results from both hemophagocytosis in the bone marrow and depression of hematopoiesis by IFN-γ, IL-1β, and TNF [49]. TNF and IFN-γ production contributes to macrophage activation, resulting in hemophagocytosis. Coagulopathy is associated with fibrin deficiency due to liver dysfunction, and DIC develops as a result of IFN-γ and TNF overproduction. Liver dysfunction including cytolysis and cholestasis is frequently reported [6, 50]. IFN-γ can contribute to the development of cholestasis and also cause apoptosis and liver damage [51]. IFN-γ elevation also leads to hypoalbuminemia [52]. AKI may be related to

Table 3 Relationship between key cytokines and clinical features of cytokine storm syndrome

Key cytokines	Clinical manifestations
IFN-γ	Fever
	Depression of hematopoiesis
	Hemophagocytosis
	Macrophage activation
	Disseminated intravascular coagulation
	Hypoalbuminemia
TNF	Fever
	Cachexia
	Depression of hematopoiesis
	Hypertriglyceridemia
	Liver injury
	Disseminated intravascular coagulation
	Hypoalbuminemia
	Hyperferritinemia
	Neurological symptoms
IL-1β	Fever
	Acute phase proteins
	Depression of hematopoiesis
	Hyperferritinemia
IL-6	Fever
	Acute phase proteins
	Anemia
	Acute kidney Injury
	NK cell dysfunction
IL-18	Liver injury
	NK cell dysfunction

excess nephrotoxic IL-6 [53]. Hypertriglyceridemia has been reported to be the consequence of inhibition of lipoprotein lipase by TNF [54, 55]. Serum ferritin elevation may be caused by IL-1β and TNF elevation [56]. Thus, proinflammatory cytokines can disrupt homeostasis of many organ systems.

Immunologically, low or absent natural killer (NK) cell activity is a characteristic finding of HLH and is included as a criterion for its diagnosis [57]. Although the exact mechanism is still unknown, hypercytokinemia might be closely associated with NK cell dysfunction. IL-18 is a well-known stimulator of NK cell activity; however, a previous report has shown that high IL-18 plasma levels were significantly correlated with low NK cell activity in patients with MAS [58]. In a previous study, the response of NK cells to exogenous IL-18 was impaired in a patient, resulting from defective phosphorylation of the IL-18 receptor [59]. These findings indicate that continuous high IL-18 levels together with other cytokines, such as IL-6, may induce over-activation or exhaustion of NK cells. Recently, IL-6 elevation has been shown to lower NK cell function by lowering perforin and granzyme expression [60]. Reduced NK cell numbers and reduced cytotoxicity will lead to decreased killing of activated immune cells, thereby worsening the symptoms of HLH/MAS. Moreover, defective cytolytic activity results in prolonged engagement between the lytic lymphocyte and the target cell, and subsequent increased cross talk yielding elevated proinflammatory cytokines responsible for the CSS clinical feature [61].

The clinical presentations of CSS, although broad, can be similar to the background disease process; therefore, they may be difficult to distinguish from those of the underlying diseases. However, the cytokine release pattern may be different among patients with different backgrounds. Nevertheless, the clinical characteristics may bear a close resemblance [62]. For example, serum IL-18 levels in patients with s-JIA-associated MAS were significantly higher than those in patients with EBV-HLH or Kawasaki disease (KD). Serum IL-6 levels in patients with KD were higher than those in patients with EBV-HLH or s-JIA-associated MAS. Serum neopterin levels in patients with EBV-HLH were higher than those in patients with s-JIA-associated MAS or KD. Furthermore, a patient with systemic lupus erythematosus-associated MAS showed TNF dominant pattern [63]. These findings indicate that monitoring cytokine profile may be useful in differentiating backgrounds of patients with CSS. Future studies may also help identify broadly important elevated cytokines as biomarkers of disease in many associated forms of CSS.

Conclusion

The common clinical features of CSS are sustained fever, splenomegaly, hepatomegaly with liver dysfunction, lymphadenopathy, coagulopathy, cytopenia, skin rash, and variable neurologic symptoms. The clinical presentations of CSS can be similar to the underlying disease processes. Thus, CSS may be difficult to distinguish from the underlying diseases. Awareness of, and high suspicion for,

CSS in patients with some of the clinical features are critical to the diagnosis of CSS. Each feature alone is nonspecific; however, the combination of clinical signs and symptoms and laboratory abnormalities, their severity, and changes over time facilitates correct diagnosis. Ultimately, monitoring the cytokine profile may be useful in differentiating the backgrounds of patients with CSS, as well as tracking the disease course.

References

1. Sieni, E., Cetica, V., Mastrodicasa, E., Pende, D., Moretta, L., Griffiths, G., et al. (2012). Familial hemophagocytic lymphohistiocytosis: A model for understanding the human machinery of cellular cytotoxicity. *Cellular and Molecular Life Sciences, 69*, 29–40.
2. Henter, J. I., Elinder, G., Soder, O., Hansson, M., Andersson, B., & Andersson, U. (1991). Hyper-cytokinemia in familial hemophagocytic lymphohistiocytosis. *Blood, 78*, 2918–2922.
3. Créput, C., Galicier, L., Buyse, S., & Azoulay, E. (2008). Understanding organ dysfunction in hemophagocytic lymphohistiocytosis. *Intensive Care Medicine, 34*, 1177–1187.
4. Lachmann, H. J., Quartier, P., So, A., & Hawkins, P. N. (2011). The emerging role of interleukin- 1b in autoinflammatory diseases. *Arthritis and Rheumatism, 63*, 314–324.
5. Dinarello, C. A., Novick, D., Kim, S., & Kaplanski, G. (2013). Interleukin-18 and IL-18 binding protein. *Frontiers in Immunology, 4*, 289.
6. De Kerguenec, C., Hillaire, S., Molinié, V., Gardin, C., Degott, C., Erlinger, S., et al. (2001). Hepatic manifestations of hemophagocytic syndrome: A study of 30 cases. *The American Journal of Gastroenterology, 96*, 852–857.
7. Avau, A., Put, K., Wouters, C. H., & Matthys, P. (2014). Cytokine balance and cytokine-driven natural killer cell dysfunction in systemic juvenile idiopathic arthritis. *Cytokine & Growth Factor Reviews, 26*, 35–45.
8. Janka, G. E. (2007). Familial and acquired hemophagocytic lymphohistiocytosis. *European Journal of Pediatrics, 166*, 95–109.
9. Brisse, E., Wouters, C. H., & Matthys, P. (2015). Hemophagocytic lymphohistiocytosis (HLH): A heterogeneous spectrum of cytokine-driven immune disorders. *Cytokine & Growth Factor Reviews, 26*, 263–280.
10. Janka, G. E. (1983). Familial hemophagocytic lymphohistiocytosis. *European Journal of Pediatrics, 140*, 221–230.
11. Haddad, E., Sulis, M. L., Jabado, N., Blanche, S., Fischer, A., & Tardieu, M. (1997). Frequency and severity of central nervous system lesions in hemophagocytic lymphohistiocytosis. *Blood, 89*, 794–800.
12. Jovanovic, A., Kuzmanovic, M., Kravljanac, R., Micic, D., Jovic, M., Gazikalovic, S., et al. (2014). Central nervous system involvement in hemophagocytic lymphohistiocytosis: A single-center experience. *Pediatric Neurology, 50*, 233–237.
13. Minoia, F., Davì, S., Horne, A., Demirkaya, E., Bovis, F., Li, C., et al. (2014). Clinical features, treatment, and outcome of macrophage activation syndrome complicating systemic juvenile idiopathic arthritis: A multinational, multicenter study of 362 patients. *Arthritis & Rhematology, 66*, 3160–3169.
14. Aytaç, S., Batu, E. D., Ünal, Ş., Bilginer, Y., Çetin, M., Tuncer, M., et al. (2016). Macrophage activation syndrome in children with systemic juvenile idiopathic arthritis and systemic lupus erythematosus. *Rheumatology International, 36*, 1421–1429.
15. Gavand, P. E., Serio, I., Arnaud, L., Costedoat-Chalumeau, N., Carvelli, J., Dossier, A., et al. (2017). Clinical spectrum and therapeutic management of systemic lupus

erythematosus-associated macrophage activation syndrome: A study of 103 episodes in 89 adult patients. *Autoimmunity Reviews, 16*, 743–749.
16. Kogawa, K., Sato, H., Asano, T., Ohga, S., Kudo, K., Morimoto, A., et al. (2014). Prognostic factors of Epstein-Barr virus-associated hemophagocytic lymphohistiocytosis in children: Report of the Japan Histiocytosis Study Group. *Pediatric Blood & Cancer, 61*, 1257–1262.
17. Reiner, A. P., & Spivak, J. L. (1988). Hematophagic histiocytosis. A report of 23 new patients and a review of the literature. *Medicine (Baltimore), 67*, 369–388.
18. Shirono, K., & Tsuda, H. (1995). Virus-associated haemophagocytic syndrome in previously healthy adults. *European Journal of Haematology, 55*, 240–244.
19. Shabbir, M., Lucas, J., Lazarchick, J., & Shirai, K. (2011). Secondary hemophagocytic syndrome in adults: A case series of 18 patients in a single institution and a review of literature. *Hematological Oncology, 29*, 100–106.
20. Behrens, E. M., Beukelman, T., Paessler, M., & Cron, R. Q. (2007). Occult macrophage activation syndrome in patients with systemic juvenile idiopathic arthritis. *The Journal of Rheumatology, 34*, 1133–1138.
21. Kuwata, K., Yamada, S., Kinuwaki, E., Naito, M., & Mitsuya, H. (2006). Peripheral hemophagocytosis: An early indicator of advanced systemic inflammatory response syndrome/hemophagocytic syndrome. *Shock, 25*, 344–350.
22. Imahashi, N., Inamoto, Y., Ito, M., Koyama, D., Goto, T., Onodera, K., et al. (2012). Clinical significance of hemophagocytosis in BM clot sections during the peri-engraftment period following allogeneic hematopoietic SCT. *Bone Marrow Transplantation, 47*, 387–394.
23. Tasaki, Y., Shimizu, M., Inoue, N., Mizuta, M., Nakagishi, Y., Wada, T., et al. (2016). Disruption of vascular endothelial homeostasis in systemic juvenile idiopathic arthritis-associated macrophage activation syndrome: The dynamic roles of angiopoietin-1 and -2. *Cytokine, 80*, 1–6.
24. Fang, Y., Li, C., Shao, R., Yu, H., Zhang, Q., & Zhao, L. (2015). Prognostic significance of the angiopoietin-2/angiopoietin-1 and angiopoietin-1/Tie-2 ratios for early sepsis in an emergency department. *Critical Care, 19*, 367.
25. McClure, P. D., Strachan, P., & Saunders, E. F. (1974). Hypofibrinogenemia and thrombocytopenia in familial hemophagocytic reticulosis. *The Journal of Pediatrics, 85*, 67–70.
26. Janka, G. E. (2012). Familial and acquired hemophagocytic lymphohistiocytosis. *Annual Review of Medicine, 63*, 233–246.
27. Gorelik, M., Fall, N., Altaye, M., Barnes, M. G., Thompson, S. D., Grom, A. A., et al. (2013). Follistatin-like protein 1 and the ferritin/erythrocyte sedimentation rate ratio are potential biomarkers for dysregulated gene expression and macrophage activation syndrome in systemic juvenile idiopathic arthritis. *The Journal of Rheumatology, 40*, 1191–1199.
28. Gorelik, M., Torok, K. S., Kietz, D. A., & Hirsch, R. (2011). Hypocomplementemia associated with macrophage activation syndrome in systemic juvenile idiopathic arthritis and adult onset Still's disease: 3 cases. *The Journal of Rheumatology, 8*, 396–397.
29. Stéphan, F., Thiolière, B., Verdy, E., & Tulliez, M. (1997). Role of hemophagocytic histiocytosis in the etiology of thrombocytopenia in patients with sepsis syndrome or septic shock. *Clinical Infectious Diseases, 25*, 1159–1164.
30. Tong, H., Ren, Y., Liu, H., Xiao, F., Mai, W., Meng, H., et al. (2008). Clinical characteristics of T-cell lymphoma associated with hemophagocytic syndrome: Comparison of T-cell lymphoma with and without hemophagocytic syndrome. *Leukemia & Lymphoma, 49*, 81–87.
31. Tseng, Y. T., Sheng, W. H., Lin, B. H., Lin, C. W., Wang, J. T., Chen, Y. C., et al. (2011). Causes, clinical symptoms, and outcomes of infectious diseases associated with hemophagocytic lymphohistiocytosis in Taiwanese adults. *Journal of Microbiology, Immunology, and Infection, 44*, 191–119.
32. Han, A. R., Lee, H. R., Park, B. B., Hwang, I. G., Park, S., Lee, S. C., et al. (2007). Lymphoma-associated hemophagocytic syndrome: Clinical features and treatment outcome. *Annals of Hematology, 86*, 493–498.

33. Morrell, D. S., Pepping, M. A., Scott, J. P., Esterly, N. B., & Drolet, B. A. (2002). Cutaneous manifestations of hemophagocytic lymphohistiocytosis. *Archives of Dermatology, 138*, 1208–1212.
34. Lee, W. J., Lee, D. W., Kim, C. H., Won, C. H., Chang, S. E., Lee, M. W., et al. (2010). Dermatopathic lymphadenitis with generalized erythroderma in a patient with Epstein-Barr virus-associated hemophagocytic lymphohistiocytosis. *The American Journal of Dermatopathology, 32*, 357–361.
35. Karras, A., Thervet, E., Legendre, C., & the Groupe Cooperatif de transplantation d'Ile de France. (2004). Hemophagocytic syndrome in renal transplant recipients: Report of 17 cases and review of literature. *Transplantation, 77*, 238–243.
36. Ohta, H., Yumara-Yagi, K., Sakata, N., Inoue, M., & Kawa-Ha, K. (1994). Capillary leak syndrome in patients with hemophagocytic lymphohistiocytosis. *Acta Paediatrica, 83*, 1113–1114.
37. Kuzmanovic, M., Pasic, S., Prijic, S., Jovanovic, A., & Kosutic, J. (2012). Severe transient left ventricular pseudohypertrophy during treatment of hemophagocytic lymphohistiocytosis: A case report. *Journal of Pediatric Hematology/Oncology, 34*, 453–456.
38. Kawamura, Y., Miura, H., Matsumoto, Y., Uchida, H., Kudo, K., Hata, T., et al. (2016). A case of Epstein-Barr virus-associated hemophagocytic lymphohistiocytosis with severe cardiac complications. *BMC Pediatrics, 16*, 172.
39. Aulagnon, F., Lapidus, N., Canet, E., Galicier, L., Boutboul, D., Peraldi, M. N., et al. (2015). Acute kidney injury in adults with hemophagocytic lymphohistiocytosis. *American Journal of Kidney Diseases, 65*, 851–859.
40. Thaunat, O., Delahousse, M., Fakhouri, F., Martinez, F., Stephan, J. L., Noël, L. H., et al. (2006). Nephrotic syndrome associated with hemophagocytic syndrome. *Kidney International, 69*, 1892–1898.
41. Malaga-Dieguez, L., Ming, W., & Trachtman, H. (2015). Direct reversible kidney injury in familial hemophagocytic lymphohistiocytosis type 3. *Journal of the American Society of Nephrology, 26*, 1777–1780.
42. Landau, D., Gurevich, E., Kapelushnik, J., Tamary, H., Shelef, I., & Lazar, I. (2013). Association between childhood nephrotic syndromeand hemophagocytic lymphohistiocytosis. *Pediatric Nephrology, 28*, 2389–2392.
43. Fukaya, S., Yasuda, S., Hashimoto, T., Oku, K., Kataoka, H., Horita, T., et al. (2008). Clinical features of haemophagocytic syndrome in patients with systemic autoimmune diseases: Analysis of 30 cases. *Rheumatology (Oxford), 47*, 1686–1691.
44. Gurgey, A., Aytac, S., Balta, G., Oguz, K. K., & Gumruk, F. (2008). Central nervous system involvement in Turkish children with primary hemophagocytic lymphohistiocytosis. *Journal of Child Neurology, 23*, 1293–1299.
45. Shinoda, J., Murase, S., Takenaka, K., & Sakai, N. (2005). Isolated central nervous system hemophagocytic lymphohistiocytosis: Case report. *Neurosurgery, 56*, E187–E190.
46. Chong, K. W., Lee, J. H., Choong, C. T., Paeds, M. M., Chan, D. W., Fortier, M. V., et al. (2012). Hemophagocytic lymphohistiocytosis with isolated central nervous system reactivation and optic nerve involvement. *Journal of Child Neurology, 27*, 1336–1339.
47. Lee, G., Lee, S. E., Ryu, K. H., & Yoo, E. S. (2013). Posterior reversible encephalopathy syndrome in pediatric patients undergoing treatment for hemophagocytic lymphohistiocytosis: Clinical outcomes and putative risk factors. *Blood Research, 48*, 258–265.
48. Suzuki, N., Morimoto, A., Ohga, S., Kudo, K., Ishida, Y., Ishii, E., et al. (2009). Characteristics of hemophagocytic lymphohistiocytosis in neonates: A nationwide survey in Japan. *The Journal of Pediatrics, 155*, 235–238.
49. Hanada, T., Ono, I., Iinuma, S., & Nagai, Y. (1989). Pure red cell aplasia in association with virus associated haemophagocytic syndrome (VAHS). *British Journal of Haematology, 73*, 570–571.
50. Billiau, A. D., Roskams, T., Van DammeLombaerts, R., Matthys, P., & Wouters, C. (2005). Macrophage activation syndrome: Characteristic findings on liver biopsy illustrating the key

role of activated, IFN-gamma-producing lymphocytes and IL-6- and TNF alpha-producing macrophages. *Blood, 105*, 1648–1651.
51. Whiting, J. F., Green, R. M., Rosenbluth, A. B., & Gollan, J. L. (1995). Tumor necrosis factor alpha decreases hepatocyte bile salt uptake and mediates endotoxin-induced cholestasis. *Hepatology, 22*, 1273–1278.
52. Iso, O. N., Hashimoto, N., Tanaka, A., Sunaga, S., Oka, T., Kurokawa, K., et al. (1998). Cytokine-induced hypoalbuminemia in a patient with hemophagocytic syndrome: Direct in vitro evidence for the role of tumor necrosis factor-alpha. *Digestive Diseases and Sciences, 43*, 67–73.
53. Weber, J., Yang, J. C., Topalian, S. L., Parkinson, D. R., Schwartzentruber, D. S., Ettinghausen, S. E., et al. (1993). Phase I trial of subcutaneous interleukin-6 in patients with advanced malignancies. *Journal of Clinical Oncology, 11*, 499–506.
54. Dinarello, C. A., Gelfand, J. A., & Wolff, S. M. (1993). Anticytokine strategies in the treatment of the systemic inflammatory response syndrome. *Journal of the American Medical Association, 269*, 1829–1835.
55. Henter, J. I., Carlson, L. A., Soder, O., Nilsson-Ehle, P., & Elinder, G. (1991). Lipoprotein alterations and plasma lipoprotein lipase reduction in familial hemophagocytic lymphohistiocytosis. *Acta Paediatrica Scandinavica, 80*, 675–681.
56. Esumi, N., Ikushima, S., Hibi, S., Todo, S., & Imashuku, S. (1988). High serum ferritin level as a marker of malignant histiocytosis and virus-associated hemophagocytic syndrome. *Cancer, 61*, 2071–2076.
57. Henter, J. I., Horne, A., Aricó, M., Egeler, R. M., Filipovich, A. H., Imashuku, S., et al. (2007). HLH-2004: Diagnostic and therapeutic guidelines for hemophagocytic lymphohistiocytosis. *Pediatric Blood & Cancer, 48*, 124–131.
58. Shibatomi, K., Ida, H., Yamasaki, S., Nakashima, T., Origuchi, T., Kawakami, A., et al. (2001). A novel role for interleukin-18 in human natural killer cell death: High serum levels and low natural killer cell numbers in patients with systemic autoimmune diseases. *Arthritis and Rheumatism, 44*, 884–892.
59. de Jager, W., Vastert, S. J., Beekman, J. M., Wulffraat, N. M., Kuis, W., Coffer, P. J., et al. (2009). Defective phosphorylation of interleukin-18 receptor beta causes impaired natural killer cell function in systemic-onset juvenile idiopathic arthritis. *Arthritis and Rheumatism, 60*, 2782–2793.
60. Cifaldi, L., Prencipe, G., Caiello, I., Bracaglia, C., Locatelli, F., De Benedetti, F., et al. (2015). Inhibition of natural killer cell cytotoxicity by interleukin-6: Implications for the pathogenesis of macrophage activation syndrome. *Arthritis & Rhematology, 67*, 3037–3046.
61. Jenkins, M. R., Rudd-Schmidt, J. A., Lopez, J. A., Ramsbottom, K. M., Mannering, S. I., Andrews, D. M., et al. (2015). Failed CTL/NK cell killing and cytokine hypersecretion are directly linked through prolonged synapse time. *The Journal of Experimental Medicine, 212*, 307–317.
62. Shimizu, M., Yokoyama, T., Yamada, K., Kaneda, H., Wada, H., et al. (2010). Distinct cytokine profiles of systemic-onset juvenile idiopathic arthritis-associated macrophage activation syndrome with particular emphasis on the role of interleukin-18 in its pathogenesis. *Rheumatology (Oxford), 49*, 1645–1653.
63. Shimizu, M., Yokoyama, T., Tokuhisa, Y., Ishikawa, S., Sakakibara, Y., Ueno, K., et al. (2013). Distinct cytokine profile in juvenile systemic lupus erythematosus-associated macrophage activation syndrome. *Clinical Immunology, 146*, 73–76.

Laboratory Features and Pathology of the Cytokine Storm Syndromes

Flavia G. Rosado and Purva Gopal

The laboratory diagnosis of cytokine storm syndromes (CSS), that is, hemophagocytic lymphohistiocytosis (HLH) and macrophage activation syndrome (MAS), is often challenging. The laboratory features using routinely available tests lack specificity, while confirmatory testing is available in only few laboratories in the USA. The disease mechanisms are still largely unclear, particularly in adults. In this chapter, the pathogenesis of the CSS, its associated laboratory findings, and recommended diagnostic strategies are reviewed.

Pathogenesis

The pathogenesis of the cytokine storm syndromes (CSS) is better understood in the primary (genetic, familial) types of this disease, in which there are known associated genetic mutations. The mechanisms associated with other types of CSS are presumably similar, albeit less clearly defined [1]. It is based on an impaired function of cytotoxic T-lymphocytes and natural killer (NK) cells, resulting in an inability of the immune system to clear the original antigenic insult. The persistence of this insult leads to a continued and uncontrolled release of cytokines, which manifests as the sepsis-like clinical picture characteristic of this disease [2].

In unaffected individuals, certain types of antigenic insults (e.g., intracellular pathogens) activate the T-helper cell 1 (Th1)-type of inflammatory response, in

F. G. Rosado (✉)
Department of Pathology, University of Texas Southwestern Biocenter, Dallas, TX, USA
e-mail: Flavia.rosado@utsouthwestern.edu

P. Gopal
Department of Pathology, University of Texas Southwestern Medical Center, Dallas, TX, USA
e-mail: purva.gopal@utsouthwestern.edu

R. Q. Cron, E. M. Behrens (eds.), *Cytokine Storm Syndrome*,
https://doi.org/10.1007/978-3-030-22094-5_4

which cell-mediated immunity predominates. This type of inflammatory response is characterized by high levels of interferon γ (IFN-γ), tumor necrosis factor (TNF), interleukin-18, interleukin-1, and interleukin-6. These cytokines are strong activators of macrophages, involved in phagocytosis, and T-cells and NK-cells, effector cells containing cytotoxic granules. When T/NK-cells are activated through interactions with the target cells, their granules migrate and fuse with the cell membrane, and the cytotoxic molecules are released in the microenvironment [3].

Granzymes and perforin are the two main types of cytotoxic molecules. Granzymes are proteases that induce apoptosis by activating caspases that damage the membrane permeability of mitochondria leading to cell death. Granzymes also directly cause oxidative damage to intracellular pathogens [4]. Perforin is a protein that creates pores on the target cell cytoplasmic membrane leading to lethal osmotic instability and the delivery of granzyme B to trigger apoptosis of the target cell [5–8]. When this sequence of events is successful, macrophages phagocytize the cellular debris, the

Table 1 Pathogenesis of cytokine storm syndromes

Disease	Mutation	Potential mechanism	Phenotype
Familial HLH			
FHL1	Unknown	Potential locus on chromosome 9q23	
FHL2	PRF1	Deficiency of perforin	HLH in childhood
FHL3	UNC13-D	Impaired exocytosis	
FHL4	STX11	Impaired exocytosis	
FHL5	STXBP2	Impaired exocytosis	
Primary immunodeficiency HLH			
Chediak–Higashi	LYST	Dysregulation lysosome trafficking	Oculocutaneous albinism, immunodeficiency, HLH, neurologic symptoms
Griscelli Syndrome Type 2	RAB27A	Dysregulation lysosome trafficking	Skin/hair hypopigmentation, immunodeficiency, HLH
XLP1	SH2D1A (SAP)	Activation of T/NK cells	Immunodeficiency, HLH
XLP2 (Duncan Disease)	XIAP	Antiapoptotic protein	Susceptibility to EBV, EBV-HLH
Secondary HLH			
Infections		T-cell function suppression	
EBV, Herpes, CMV			
Malignancy		T/NK cell dysfunction?	
T/NK cell lymphoma			
Macrophage activation syndrome		Deficient secretion of perforin/inhibition of SAP gene	

XLP X-linked lymphoproliferative disorder, *HLH* hemophagocytic lymphohistiocytosis triggering agent is removed, and the stimulus to inflammation subsides [1]. In contrast, in individuals affected by the CSS, one of the steps above-described is defective, and clearing of the insulting trigger is not successful. The inflammatory stimulus persists, and the level of cytokines, particularly IFN-γ and TNF, rises uncontrolled [3]. The causes of T/NK cell dysfunction vary according to the type of CSS, as discussed below and summarized in Table 1.

Primary Hemophagocytic Lymphohistiocytosis

Primary HLH encompasses familial HLH and HLH related to primary immunodeficiencies. In the majority of the primary forms of HLH, T-cell/NK-cell function is impaired due to a mutation in genes involved in the release of cytotoxic granules. As a result, there is either decreased production of cytotoxic granules contents or defective exocytosis of these granules [3].

Familial HLH (FHL) is a rare inherited autosomal recessive disease. They are subtyped into FHL1–5, according to the gene mutated. FHL2 and FHL3 account for the majority of the familial cases worldwide, while FHL4 and 5 occur more frequently in some ethnic groups in Turkey and Central Europe [9].

FHL1 is caused by a mutation in a yet unknown gene located on chromosome 9q21 [10]. FHL2 is caused by a mutation in the perforin gene *PRF1* resulting in low levels of perforin. FHL3 is caused by a mutation in the gene *UNC13D* (also known as *MUNC13-4*) leading to an abnormal formation of microtubules and cytoskeleton proteins and ineffective granule exocytosis [11, 12]. In FHL4 and FHL5, mutations in genes *STX11* and *STXBP2* (also known as *UNC18B, or UNC18-2*), respectively, cause defective granule exocytosis through a mechanism similar to that of FHL3 [13–15]. Despite significant progress in identifying the genes that cause familial HLH, a number of patients presenting with primary HLH have no identifiable genetic alteration [16].

The primary immunodeficiencies more frequently associated with HLH are the rare Chediak–Higashi syndrome, Griscelli syndrome type 2, X-linked lymphoproliferative disorder type 1 and type 2 (known as Duncan disease). Other rare types of primary immunodeficiency disease that have been reported in association with HLH include Hermanksy–Pudlak syndrome type 2 and XMEN syndrome [17].

Chediak–Higashi syndrome is caused by a mutation in the *LYST* gene, which is involved in the regulation of lysosome trafficking, including cytotoxic granule trafficking [9]. Impaired granule exocytosis in these patients affects not only T/NK cell function but also exocytosis of melanosomes and vesicles in neurologic synapses. In consequence, patients with this syndrome characteristically show oculocutaneous albinism and signs of neurologic dysfunction, in addition to a strong predisposition to HLH. In Griscelli syndrome type 2, a mutation in gene *RAB27A* also results in defective lysosome trafficking. These patients also show cutaneous hypopigmentation and a characteristic silvery hair discoloration. In X-linked lymphoproliferative disorder (XLP) type 1, HLH occurs due to a mutation in *SH2D1A* gene (also known as *SAP* gene), which is needed for normal activation of T and NK cells. In XLP type 2, or Duncan disease, mutations in gene *XIAP* cause a defect in an antiapoptotic protein that appears to protect against Epstein Barr virus infections. Patients with Duncan disease are therefore prone to develop severe complications of Epstein Barr virus (EBV) infection, including development of EBV-related HLH [9].

Secondary Hemophagocytic Lymphohistiocytosis

In the secondary forms of HLH, the mechanisms of disease are not well understood. Numerous conditions have been reported in association with HLH, and include a variety of infections, malignancies, immunosuppressive therapy, autoimmune diseases and others [3]. The wide range of associations with secondary HLH, the lack of specificity of the HLH diagnostic criteria, and the known overlap between HLH and systemic inflammatory response syndrome (SIRS) have raised the possibility of an HLH overdiagnosis in the literature [18]. Nevertheless, it is also possible that the associations are truly random, as they simply represent a triggering mechanism in individuals with an underlying susceptibility to HLH. Alternatively, SIRS is also known to inhibit NK-cell function and could itself lead to HLH in susceptible individuals [19, 20]. In spite of this controversy, there are few secondary associations that appear to be more strongly associated with HLH than others. These include infections by EBV and other herpesviruses (e.g., herpes simplex and cytomegalovirus), certain types of T/NK-cell lymphomas, and some autoimmune disorders [21, 22].

EBV most frequently infects B-cells and CD4-positive T-cells, but occasionally CD8-positive cytotoxic T-cells are also infected [23, 24]. In infected T-cells, the viral-encoded latent membrane protein 1 (LMP-1) inhibits *SAP/SH2D1A* gene expression, with impaired T/NK-cell function and an increased risk of HLH [23, 25–28]. EBV also promotes survival of infected T-cells by blocking apoptotic signaling through activation of nuclear factor-κB signaling pathway and TNF /TNF receptor 1 [29].

Many types of solid and hematologic neoplasms have been reported in association with HLH, yet the disease mechanisms are largely unclear. The most frequent association is seen with malignant neoplasms of T-cells and NK-cells and non-Hodgkin B-cell lymphomas [30].

Autoimmune and autoinflammatory diseases have been associated with development of a an HLH-like syndrome, known as macrophage activation syndrome (MAS), particularly in children with systemic-onset juvenile idiopathic arthritis (Still disease), Kawasaki disease, systemic lupus erythematosus, and seronegative spondyloarthropathies [31]. The pathogenesis is linked to the decreased expression of perforin gene and *SAP* gene described in these patients [31–33]. In addition, heterozygous mutations in know HLH-associated genes (e.g., *PRF1, MUNC13-4*) have been associated with MAS in rheumatic diseases, and likely contribute to disease pathology as in primary HLH [34–36].

Laboratory Features

Patients with CSS show laboratory findings that are nonspecific and overlap with those of SIRS/sepsis [18]. The abnormalities included in the diagnostic criteria for HLH are cytopenias affecting least 2 lineages (anemia with hemoglobin <9 g/dL, or <10 g/dL for infants 1 month-old or less, thrombocytopenia of <100,000/μL or neutropenia <1000/μL), hypertriglyceridemia (>265 mg/dL), hypofibrinogenemia (<150 mg/dL), impaired NK cell activity, ferritin >500 ng/mL, elevated soluble CD25 (soluble IL-2 receptor alpha) by 2 standard deviations above age-adjusted reference values, and hemophagocytosis in bone marrow, liver, spleen, or other organs [3].

Ferritin is an acute phase reactant, and as such, is generally elevated in inflammatory responses of any type. However, in HLH, ferritin is elevated to levels higher than those expected in the majority of other inflammatory states. The majority of patients will show laboratory signs of liver dysfunction with mild to severe increase in liver enzymes (aspartate aminotransferase, alanine aminotransferase, gamma glutamyl transferase, and lactate dehydrogenase) and bilirubin. They also frequently show signs of consumptive coagulopathy with decreased fibrinogen and elevated D-dimers [37]. Hypertriglyceridemia may not be seen in the absence of liver disease [38].

In patients with neurologic symptoms, an analysis of the cerebrospinal fluid reveals a moderate increase in cell count and protein content in about 50% of pediatric cases [32]. Brain imaging studies may show vasogenic cerebral edema predominantly in the posterior cerebral hemispheres, hypodense areas suggestive of necrosis, and basal ganglia abnormalities [39, 40].

In the setting of macrophage activation syndrome (MAS), the laboratory values may differ from those seen in HLH. For example, as patients with sJIA typically have a baseline level of anemia and hyperferritinemia, a more severe change in hemoglobin and ferritin levels needs to be considered for the diagnosis of MAS [41, 42]. Moreover, the degree of cytopenia may be less pronounced, in part due to baseline leukocytosis [32, 42]. Similarly, baseline thrombocytosis may make it less likely to reach a threshold for thrombocytopenia.

Hemophagocytosis

The abnormal hemophagocytosis seen in HLH is characterized by an increase in number of phagocytizing macrophages containing intact cells, which can be red cells, white cells, platelets or hematopoietic precursors (Fig. 1). In addition to the bone marrow, hemophagocytosis may be identified in other hematolymphoid organs such as lymph nodes and spleen, as well as the liver, skin, lungs, meninges, and cerebrospinal fluid [43–46]. Nevertheless, hemophagocytosis is not pathognomonic for HLH or MAS, as it is only identified in approximately 60% of cases [47].

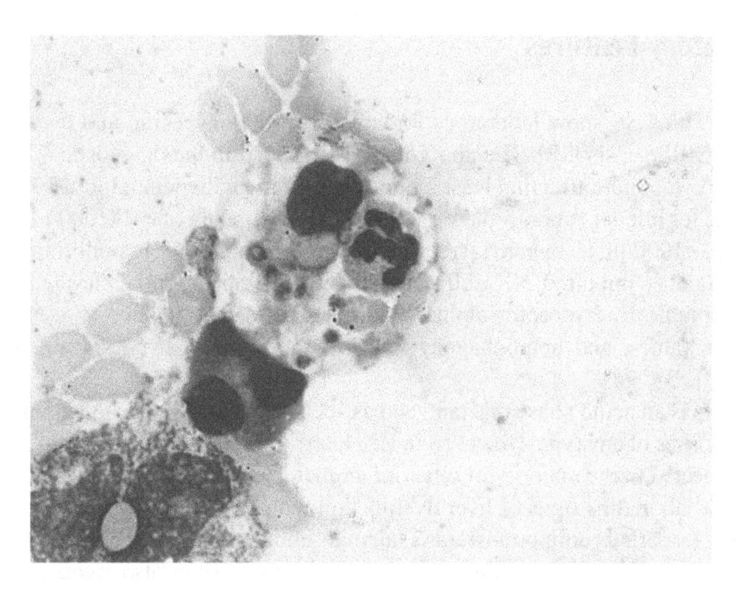

Fig. 1 Bone marrow aspirate smear shows a macrophage with phagocytized neutrophil, red cells, and platelets. Fungal yeast forms are also seen (Wright-Giemsa stain, ×1000)

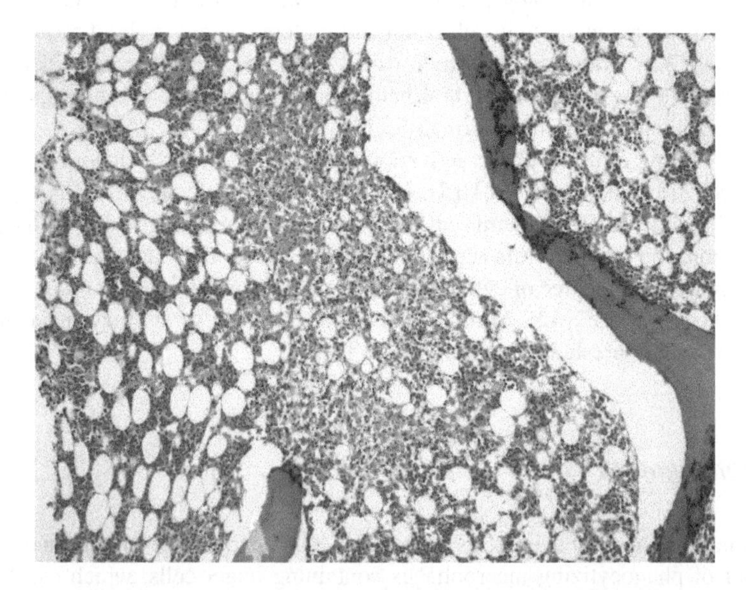

Fig. 2 Bone marrow core biopsy with a lymphohistiocytic interstitial infiltrate. Hemophagocytosis may be difficult to appreciate without immunohistochemistry (H&E, ×200)

In the bone marrow, macrophages are accompanied by small reactive lymphocytes, predominantly T-cells (Fig. 2). This process can be so pronounced that it may mask an underlying disease, such as a T-cell lymphoma [44]. In addition, bone

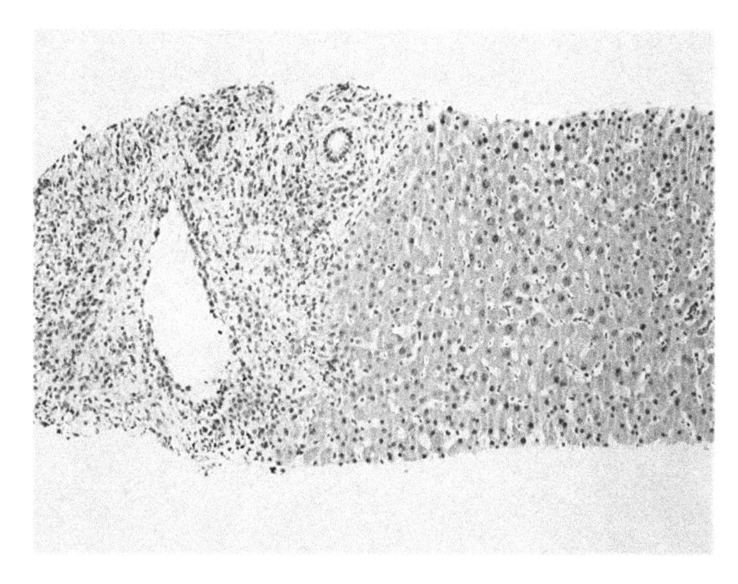

Fig. 3 Liver core biopsy shows portal lymphocytic infiltrate and endothelialitis of vein (H&E, ×100)

marrow biopsies may show granulocytic and erythroid hypoplasia, and variable degree of megakaryocytic hyperplasia [44]. The bone marrow may also show changes related to involvement by the triggering disease. For example, there may be an increase in immunoblast-type cells in a case of EBV infection, or granulomas may be present in fungal infections [44].

In the liver, there is typically Kupffer cell hyperplasia, and scattered macrophages infiltrating the sinusoids. The associated lymphocytic infiltrate tends to be predominantly portal and periportal (Fig. 3) and consists of predominantly small cytotoxic CD8-positive T-cells. Both Kupffer cells and intrasinusoidal macrophages may display hemophagocytosis (Fig. 4). Bile duct injury and endotheliitis may be seen in some cases [48–50]. Lymph node and splenic involvement is characterized by a similar CD8-positive lymphohistiocytic infiltrate that involves the nodal and splenic sinuses [50, 51]. In the skin, the infiltrate may be superficial and perivascular or may extend to the subcutaneous adipose tissue [52].

Pathophysiology

The majority of the SIRS-like symptoms seen in the cytokine storm syndromes are attributed to the high levels of cytokines seen in these patients [53]. The cytokines IFNγ and TNF are strong activators of macrophages, which then proliferate uncontrolled and infiltrate, along with activated lymphocytes, hematopoietic organs such as liver, spleen, lymph nodes and bone marrow. Hyperactivated macrophages

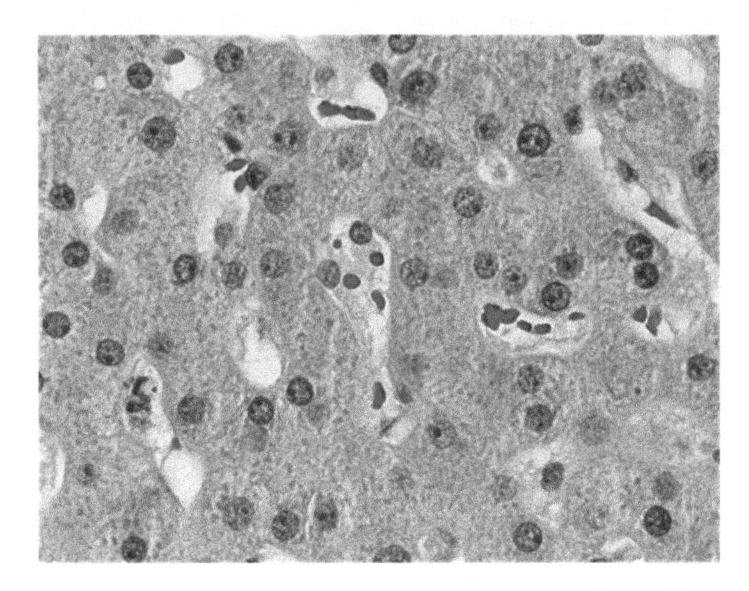

Fig. 4 Hemophagocytosis is noted in Kupffer cells lining the sinusoids (H&E, ×600)

phagocytize nearby blood cells independent from interactions with receptors on the surface of target cells [53, 54]. This type of abnormal phagocytosis does not induce apoptosis, and the phagocytized cells are thus seen intact within the macrophages (Fig. 1) [55].

High levels of IFNγ and TNF also cause direct apoptosis of hematopoietic precursors contributing to cytopenias [55]. The cytopenia is often associated with bone marrow hypoplasia, thus supporting intrinsic bone marrow injury as the likely underlying mechanism.

The lymphohistiocytic infiltration of the liver and the spleen manifests as hepatosplenomegaly and may contribute to the liver dysfunction, with increased serum levels of transaminases and bilirubin. Neurologic symptoms are attributed to both infiltration of the brain parenchyma and leptomeninges by macrophages and to the direct neurotoxicity of cytokines and free radicals [56].

Activated macrophages release plasminogen activators in the serum, resulting in hyperfibrinolysis and consumption of fibrinogen [57]. The resulting drop in fibrinogen can also lead to abnormally low erythrocyte sedimentation rate [58]. Hypercytokinemia inhibits lipoprotein lipase, then resulting in elevated triglycerides, which may also be a consequence of liver dysfunction [38]. Ferritin levels increase in response to an increased breakdown of heme by heme-oxygenase-1, an enzyme that is induced by tissue hypoxia and cytokines [59]. The elevated level of serum ferritin has also been attributed to increased secretion from activated macrophages, cells that normally store ferritin in tissues [60].

Laboratory Diagnosis

The diagnosis of hemophagocytic lymphohistiocytosis is based on the identification of either known homozygous (or compound heterozygous) HLH-associated genetic mutations, or five of eight HLH diagnostic criteria. The criteria are fever, splenomegaly, cytopenias, hypertriglyceridemia or hypofibrinogenemia, hyperferritinemia, tissue hemophagocytosis, low/absent NK cell function, and increased soluble CD25 [32]. These diagnostic criteria, validated for the diagnosis of primary HLH in children, have been criticized by their lack of sensitivity and specificity for the diagnosis of secondary HLH [61]. More recently, a scoring system was proposed in an attempt to facilitate the diagnosis of HLH in adults [62, 63]. Specific diagnostic criteria for the diagnosis of macrophage activation syndrome in the setting of sJIA and lupus have been proposed [64–67].

A step-wise approach for the laboratory diagnosis of HLH starts with an initial evaluation with noninvasive, inexpensive, and readily available tests. The goal of this initial assessment is to confirm the clinical concern for HLH [61]. Results indicating the possibility of HLH will support the request of additional testing that is invasive, not readily available, or expensive [61]. These include tissue biopsy, levels of sCD25, NK cell function testing, and flow cytometry. If there is concern for an underlying immunodeficiency, the evaluation may also include immunoglobulin levels and determination of lymphocyte subsets by flow cytometry. Genetic testing is recommended in selected adult patients and in most children [61]. Additional testing is also indicated to identify a potential triggering event [61]. In adults with suspected HLH, the identification of active EBV infection with EBV viral titers and the concurrent diagnosis of certain types of T-cell lymphomas, such as gamma delta T-cell lymphoma, may help to reinforce the clinical suspicion for HLH. Genetic testing for the XLP mutations is recommended in children with a documented EBV-associated HLH [11]. Testing strategies for the diagnosis of HLH are summarized in Table 2.

Table 2 Diagnostic testing for cytokine storm syndromes

Initial testing	Advantages	Limitations
CBC, triglycerides, fibrinogen	Available	Nonspecific
Liver function tests	Non-expensive	
Ferritin		
EBV levels[a]		
Search triggering event		
Bone marrow biopsy	Provides additional criteria for HLH dx	Nonspecific, invasive
sCD25		Send-out
NK-cell function assay		
Flow cytometry screening		
Genetic testing	Positive test confirmatory	
	Familial HLH	
	Primary HLH	

[a]Additional testing recommended to detect potential triggering events

Initial Testing

An initial assessment to identify the laboratory findings most commonly associated with the cytokine storm syndromes include a complete blood count, serum levels of ferritin, triglycerides and fibrinogen, and liver function tests (i.e., aspartate aminotransferase, alanine aminotransferase, gamma glutamyl transferase, lactate dehydrogenase, albumin, and bilirubin). These tests are readily available in the majority of health care facilities. Moreover, a trend of a rising CRP in the setting of a falling ESR over time is highly suggestive of MAS [68].

While a ferritin level greater than 500 ng/mL is required to be considered one of the diagnostic criteria for HLH, a level greater than 3000 ng/mL has been proposed as more specific for HLH [63]. Ferritin level of >10,000 ng/L is considered to be 90% sensitive and 96% specific for the diagnosis of HLH in children [69]. However, the significance of extremely high levels of ferritin for the diagnosis of HLH has been questioned in a study of adults [18]. In fact, a different study has shown that extremely high ferritin levels (greater than 50,000 ng/mL) in adults may actually impart reduced specificity for HLH in adults [70]. Results of one study suggests that the serum levels of hemeoxygenase-1 (HO-1) may help to interpret the significance of hyperferritinemia in this context, as HO-1 is increased in hyperferritinemia due to HLH but not in hyperferritinemia due to other causes [71].

Tissue Biopsy

A bone marrow biopsy may be indicated in patients with suspected HLH to identify abnormal hemophagocytosis. However, the sensitivity and specificity of bone marrow hemophagocytosis has been questioned in studies of adult patients [18, 72]. Hemophagocytosis may be absent in early cases, and may also be seen in patients with hemolysis, sepsis, and history of recurrent blood transfusions [73]. In the bone marrow, hemophagocytosis is best appreciated in bone marrow aspirates (Fig. 1). The core biopsy shows a lymphohistiocytic infiltrate with a variable number of small lymphocytes comprising predominantly CD8-positive T-cells [50]. In addition to the bone marrow, hemophagocytosis may also be identified in biopsies of the liver, lymph nodes, skin, and spleen. The utilization of immunohistochemical studies for histiocyte markers CD68 or CD163 facilitate the identification of hemophagocytosis in core biopsies [61]. In situ hybridization studies for EBV-encoded RNA protein (EBER) may be considered to determine an association with EBV [50].

Serum Soluble Receptor Testing (sCD25 and sCD163)

Activated T-lymphocytes express high levels of IL-2 receptor (CD25) on their cell surface and eventually release its soluble alpha-subunit in the serum. Increased sCD25 is useful as a diagnostic criterion for HLH as well as a marker of disease activity [74]. The ratio of level of sCD25 to ferritin has been also utilized to favor the diagnosis of a lymphoma-associated HLH [75].

The expression of hemoglobin-haptoglobin scavenger receptor CD163 also significantly increases in activated macrophages; therefore, the quantification of sCD163 in the serum has been proposed as a potential marker of HLH [76].

Flow Cytometry

Flow cytometric studies can be used to assess the level of surface or cytoplasmic expression of proteins involved in T/NK-cell function, such as perforin, granzyme B, XIAP, or SAP/SDH2D1A [77]. Some institutions have used this technique to screen for those cases in which subsequent confirmatory genetic testing is indicated [78]. This approach is limited in FHL2 because patients heterozygous for the PRF1 gene mutation may develop the disease, despite normal expression levels of perforin protein [79].

Peripheral blood flow cytometry to quantify the expression of surface CD107a (also known as lysosomal-associated membrane protein 1, LAMP-1) on mononuclear cells is another method to evaluate the effectiveness of NK/T-cell granule exocytosis. CD107a is a molecule that is highly expressed on the cell surface upon normal exocytosis [80]. A decreased CD107a expression following standardized stimulation indicates defective granule exocytosis [81].

51-Cr Release Assay

The 51-Cr release assay measures the quantity of radionuclide 51-Cr that is released in granules of previously labeled NK-cells. In patients with HLH, the release of the radioactive marker is decreased or absent, indicating an impaired granule exocytosis [82]. This method is regarded as the closest to a "gold standard" test to evaluate NK cell function; however, it is a labor-intensive test available in only few reference laboratories in the USA [1, 83]. More recently, flow cytometric (nonradioactive) approaches have been proposed assessing natural killer cell function [84].

Genetic Testing

Genetic testing is uniformly indicated in children with suspected HLH, since genetic forms of HLH are more likely to occur in this age group [32, 85]. This test may also be indicated for individuals with a strong family history of HLH [11]. The recommended workup includes testing for all known mutations associated with familial HLH (*PRF1, UNC13D or MUNC13-4, STX11,* and *STXBP2 or UNC18B*). In EBV-associated cases, or in males, testing for mutations associated with XLP syndromes, SH1D1A/SAP, and XIAP [11, 86]. Additional testing for and *LYST, RAB27A,* and

AP3B1 mutations may be recommended in HLH arising in the setting of suspected Chediak–Higashi, Griscelli syndrome type 2, and Hermansky–Pudlak syndrome, respectively. Using whole exome sequencing, heterozygous mutations have also been identified above the background rate in the population for many of the HLH-associated genes among patients with sHLH and MAS [35, 87]. Some of these heterozygous mutations have been shown to act in a dominant-negative fashion contributing to decreased NK cell function and sHLH pathophysiology [84, 88]. More recently, whole genome sequencing, or targeted sequencing, has identified mutations in non-coding regions of HLH-associated genes among patients with sHLH [89, 90]. There are likely other contributory genes to be identified which contribute to sHLH pathophysiology [35].

Cytokine Levels

A recent large study of HLH in adults has suggested that measuring levels of cytokines could be used to distinguish HLH from sepsis or acute EBV infection [91, 92]. In this study, the cutoff of IFNγ level of 75 pg/mL or greater and level of IL-10 of greater than 60 pg/mL yielded a sensitivity of 98.9% and specificity of 93% for the diagnosis of HLH. While measuring cytokine levels has not yet been incorporated in the diagnostic algorithms of HLH and is not routinely available in most laboratories, the data suggests a potential use for this parameter in the diagnosis of HLH, particularly in adults. Finally, the relative ratio of IL-18 to IL-6 may help to identify sJIA patients at risk MAS development [93].

References

1. Janka, G. (2009). Hemophagocytic lymphohistiocytosis: When the immune system runs amok. *Klinische Pädiatrie, 221*, 278–285.
2. Grom, A. A. (2004). Natural killer cell dysfunction: A common pathway in systemic-onset juvenile rheumatoid arthritis, macrophage activation syndrome, and hemophagocytic lymphohistiocytosis? *Arthritis and Rheumatism, 50*, 689–698.
3. Rosado, F. G. N., & Kim, A. S. (2013). Hemophagocytic lymphohistiocytosis: An update on diagnosis and pathogenesis. *American Journal of Clinical Pathology, 139*, 713–727.
4. Voskoboinik, I., Whisstock, J. C., & Trapani, J. A. (2015). Perforin and granzymes: Function, dysfunction and human pathology. *Nature Reviews Immunology, 15*, 388–400.
5. Goransdotter, E. K., Fadeel, B., Nilsson-Ardnor, S., Söderhäll, C., Samuelsson, A., Janka, G., et al. (2001). Spectrum of perforin gene mutations in familial hemophagocytic lymphohistiocytosis. *American Journal of Human Genetics, 68*, 590–597.
6. Stepp, S. E., Dufourcq-Lagelouse, R., Le Deist, F., Bhawan, S., Certain, S., Mathew, P. A., et al. (1999). Perforin gene defects in familial hemophagocytic lymphohistiocytosis. *Science, 286*, 1957–1959.
7. Ménasché, G., Feldmann, J., Fischer, A., & de Saint Basile, G. (2005). Primary hemophagocytic syndromes point to a direct link between lymphocyte cytotoxicity and homeostasis. *Immunological Reviews, 203*, 165–179.

8. Lichtenheld, M. G., Olsen, K. J., Lu, P., Lowrey, D. M., Hameed, A., Hengartner, H., et al. (1988). Structure and function of human perforin. *Nature, 335*, 448–451.
9. Zhang, K., Filipovich, A. H., Johnson, J., Marsh, R. A., & Villanueva, J. (2013). Hemophagocytic lymphohistiocytosis, familial. In M. P. Adam, H. H. Ardinger, R. A. Pagon, et al. (Eds.), *GeneReviews®* (pp. 1993–2017). Seattle, WA: University of Washington, Seattle.
10. Ohadi, M., Lalloz, M. R., Sham, P., Zhao, J., Dearlove, A. M., Shiach, C., et al. (1999). Localization of a gene for familial hemophagocytic lymphohistiocytosis at chromosome 9q21.3-22 by homozygosity mapping. *American Journal of Human Genetics, 64*, 165–171.
11. Johnson, T. S., Villanueva, J., Filipovich, A. H., Marsh, R. A., & Bleesing, J. J. (2011). Contemporary diagnostic methods for hemophagocytic lymphohistiocytic disorders. *Journal of Immunological Methods, 364*, 1–13.
12. Gholam, C., Grigoriadou, S., Gilmour, K. C., & Gaspar, H. B. (2011). Familial haemophagocytic lymphohistiocytosis: Advances in the genetic basis, diagnosis and management. *Clinical and Experimental Immunology, 163*, 271–283.
13. Feldmann, J., Callebaut, I., Raposo, G., Certain, S., Bacq, D., Dumont, C., et al. (2003). Munc13-4 is essential for cytolytic granules fusion and is mutated in a form of familial hemophagocytic lymphohistiocytosis (FHL3). *Cell, 115*, 461–473.
14. zur Stadt, U., Schmidt, S., Kasper, B., Beutel, K., Diler, A. S., Henter, J. I., et al. (2005). Linkage of familial hemophagocytic lymphohistiocytosis (FHL) type-4 to chromosome 6q24 and identification of mutations in syntaxin 11. *Human Molecular Genetics, 14*, 827–834.
15. zur Stadt, U., Rohr, J., Seifert, W., Koch, F., Grieve, S., Pagel, J., et al. (2009). Familial hemophagocytic lymphohistiocytosis type 5 (FHL-5) is caused by mutations in Munc18-2 and impaired binding to syntaxin 11. *American Journal of Human Genetics, 85*, 482–492.
16. Verbsky, J. W., & Grossman, W. J. (2006). Hemophagocytic lymphohistiocytosis: Diagnosis, pathophysiology, treatment, and future perspectives. *Annals of Medicine, 38*, 20–31.
17. Li, F. Y., Chaigne-Delalande, B., Su, H., Uzel, G., Matthews, H., & Lenardo, M. J. (2014 Apr). XMEN disease: A new primary immunodeficiency affecting Mg2+ regulation of immunity against Epstein-Barr virus. *Blood, 123*(14), 2148–2152.
18. Rosado, F. G., Rinker, E. B., Plummer, W. D., Dupont, W. D., Spradlin, N. M., Reichard, K. K., et al. (2016). The diagnosis of adult-onset haemophagocytic lymphohistiocytosis: Lessons learned from a review of 29 cases of bone marrow haemophagocytosis in two large academic institutions. *Journal of Clinical Pathology, 69*, 1–5.
19. Puente, J., Carvajal, T., Parra, S., Miranda, D., Sepulveda, C., Wolf, M. E., et al. (1993). In vitro studies of natural killer cell activity in septic shock patients: Response to a challenge with alpha-interferon and interleukin-2. *International Journal of Clinical Pharmacology, Therapy, and Toxicology, 31*, 271–275.
20. von Muller, L., Klemm, A., Durmus, N., Weiss, M., Suger-Wiedeck, H., Schneider, M., et al. (2007). Cellular immunity and active human cytomegalovirus infection in patients with septic shock. *The Journal of Infectious Diseases, 196*, 1288–1295.
21. Smith, M. C., Cohen, D. N., Greig, B., Yenamandra, A., Vnencak-Jones, C., Thompson, M. A., et al. (2014). The ambiguous boundary between EBV-related hemophagocytic lymphohistiocytosis and systemic EBV-driven T cell lymphoproliferative disorder. *International Journal of Clinical and Experimental Pathology, 7*, 5738–5749.
22. Henter, J. I., Elinder, G., & Ost, A. (1991). Diagnostic guidelines for hemophagocytic lymphohistiocytosis. The FHL Study Group of the Histiocyte Society. *Seminars in Oncology, 18*, 29–33.
23. Su, I. J., Chen, R. L., Lin, D. T., & Chen, C. (1994). Epstein-Barr virus (EBV) infects T lymphocytes in childhood EBV-associated hemophagocytic syndrome in Taiwan. *The American Journal of Pathology, 144*, 1219–1225.
24. Kasahara, Y., Yachie, A., Takei, K., Kanegane, C., Okada, K., Ohta, K., et al. (2001). Differential cellular targets of Epstein-Barr virus (EBV) infection between acute EBV-associated hemophagocytic lymphohistiocytosis and chronic active EBV infection. *Blood, 98*, 1882–1888.

25. Lay, J. D., Tsao, C. J., Chen, J. Y., Kadin, M. E., & Su, I. J. (1997). Upregulation of tumor necrosis factor-alpha gene by Epstein-Barr virus and activation of macrophages in Epstein-Barr virus-infected T cells in the pathogenesis of hemophagocytic syndrome. *The Journal of Clinical Investigation, 100*, 1969–1979.
26. Chuang, H. C., Lay, J. D., Hsieh, W. C., Wang, H. C., Chang, Y., Chuang, S. E., et al. (2005). Epstein-Barr virus LMP1 inhibits the expression of SAP gene and upregulates Th1 cytokines in the pathogenesis of hemophagocytic syndrome. *Blood, 106*, 3090–3096.
27. Kawaguchi, H., Miyashita, T., Herbst, H., Niedobitek, G., Asada, M., Tsuchida, M., et al. (1993). Epstein-Barr virus-infected T lymphocytes in Epstein-Barr virus-associated hemophagocytic syndrome. *The Journal of Clinical Investigation, 92*, 1444–1450.
28. Kasahara, Y., & Yachie, A. (2002). Cell type specific infection of Epstein-Barr virus (EBV) in EBV-associated hemophagocytic lymphohistiocytosis and chronic active EBV infection. *Critical Reviews in Oncology/Hematology, 44*, 283–294.
29. Chuang, H. C., Lay, J. D., Hsieh, W. C., & Su, I. J. (2007). Pathogenesis and mechanism of disease progression from hemophagocytic lymphohistiocytosis to Epstein-Barr virus-associated T-cell lymphoma: Nuclear factor-kappa B pathway as a potential therapeutic target. *Cancer Science, 98*, 1281–1287.
30. Wang, H., Xiong, L., Tang, W., Zhou, Y., & Li, F. (2017). A systematic review of malignancy-associated hemophagocytic lymphohistiocytosis that needs more attentions. *Oncotarget, 8*, 59977–59985.
31. Deane, S., Selmi, C., Teuber, S. S., & Gershwin, M. E. (2010). Macrophage activation syndrome in autoimmune disease. *International Archives of Allergy and Immunology, 153*, 109–120.
32. Janka, G. E. (2012). Familial and acquired hemophagocytic lymphohistiocytosis. *Annual Review of Medicine, 63*, 233–246.
33. Henter, J. I., Tondini, C., & Pritchard, J. (2004). Histiocyte disorders. *Critical Reviews in Oncology/Hematology, 50*, 157–174.
34. Zhang, M., Behrens, E. M., Atkinson, T. P., Shakoory, B., Grom, A. A., & Cron, R. Q. (2014). Genetic defects in cytolysis in macrophage activation syndrome. *Current Rheumatology Reports, 16*(9), 439.
35. Kaufman, K. M., Linghu, B., Szustakowski, J. D., Husami, A., Yang, F., Zhang, K., et al. (2014). Whole-exome sequencing reveals overlap between macrophage activation syndrome in systemic juvenile idiopathic arthritis and familial hemophagocytic lymphohistiocytosis. *Arthritis & Rhematology, 66*(12), 3486–3495.
36. Vastert, S. J., van Wijk, R., D'Urbano, L. E., de Vooght, K. M., de Jager, W., Ravelli, A., et al. (2010). Mutations in the perforin gene can be linked to macrophage activation syndrome in patients with systemic onset juvenile idiopathic arthritis. *Rheumatology (Oxford, England), 49*(3), 441–449.
37. Janka, G., & zur Stadt, U. (2005). Familial and acquired hemophagocytic lymphohistiocytosis. In ASH Education Book (Vol. 2005 no. 1, pp. 82–88).
38. Okamoto, M., Yamaguchi, H., Isobe, Y., Yokose, N., Mizuki, T., Tajika, K., et al. (2009). Analysis of triglyceride value in the diagnosis and treatment response of secondary hemophagocytic syndrome. *Internal Medicine, 48*(10), 775.
39. Henter, J. I., & Nennesmo, I. (1997). Neuropathologic findings and neurologic symptoms in twenty-three children with hemophagocytic lymphohistiocytosis. *The Journal of Pediatrics, 130*(3), 358.
40. Gratton, S. M., Powell, T. R., Theeler, B. J., Hawley, J. S., Amjad, F. S., & Tornatore, C. (2015). Neurological involvement and characterization in acquired hemophagocytic lymphohistiocytosis in adulthood. *Journal of the Neurological Sciences, 357*(1–2), 136–142.
41. Recalcati, S., Invernizzi, P., Arosio, P., & Cairo, G. (2008). New functions for an iron storage protein: The role of ferritin in immunity and autoimmunity. *Journal of Autoimmunity, 30*, 84–89.
42. Kelly, A., & Ramanan, A. V. (2007). Recognition and management of macrophage activation syndrome in juvenile arthritis. *Current Opinion in Rheumatology, 19*, 477–481.

43. Lai, S., Merritt, B. Y., Chen, L., Zhou, X., & Green, L. K. (2012). Hemophagocytic lymphohistiocytosis associated with influenza A (H1N1) infection in a patient with chronic lymphocytic leukemia: An autopsy case report and review of the literature. *Annals of Diagnostic Pathology, 16*, 477–484.
44. Foucar, K. (2010). Histiocytic disorders in bone marrow. In K. Foucar, K. Reichard, & D. Czuchlewski (Eds.), *Bone marrow pathology* (3rd ed.). Chicago, IL: American Society for Clinical Pathology.
45. Hsi, E. D. (2007). *Hematopathology: A volume in foundations in diagnostic pathology series*. London: Churchill Livingstone.
46. Aronson, I. K., & Worobec, S. M. (2010). Cytophagic histiocytic panniculitis and hemophagocytic lymphohistiocytosis: An overview. *Dermatologic Therapy, 23*, 389–402.
47. Minoia, F., Davì, S., Horne, A., Demirkaya, E., Bovis, F., Li, C., et al. (2014). Clinical features, treatment, and outcome of macrophage activation syndrome complicating systemic juvenile idiopathic arthritis: A multinational, multicenter study of 362 patients. *Arthritis & Rhematology, 66*(11), 3160–3169.
48. Chen, J. H., Fleming, M. D., Pinkus, G. S., Pinkus, J. L., Nichols, K. E., Mo, J. Q., et al. (2010). Pathology of the liver in familial hemophagocytic lymphohistiocytosis. *The American Journal of Surgical Pathology, 34*, 852–867.
49. Ost, A., Nilsson-Ardnor, S., & Henter, J. I. (1998). Autopsy findings in 27 children with haemophagocytic lymphohistiocytosis. *Histopathology, 32*(4), 310.
50. Rezk, S. A., Sullivan, J., & Woda, B. (2011). Non-neoplastic histiocytic proliferations of lymph nodes and bone marrow. In E. S. Jaffe, N. L. Harris, J. W. Vardiman, E. Campo, & D. A. Arber (Eds.), *Hematopathology*. Philadelphia, PA: Elsevier Saunders.
51. Hemophagocytic lymphohistiocytosis/hemophagocytic syndromes. In: RN Miranda, Joseph D. Khoury, L. Jeffrey Medeiros. Atlas of lymph node pathology, Springer Science+Business Media, New York, 2013
52. Millsop, J. W., Ho, B., Kiuru, M., Fung, M. A., & Sharon, V. R. (2016). Cutaneous hemophagocytosis: Bean bags from the bone. *JAMA Dermatology, 152*(8), 950–952.
53. Larroche, C., & Mouthon, L. (2004). Pathogenesis of hemophagocytic syndrome (HPS). *Autoimmunity Reviews, 3*, 69–75.
54. Schneider, E. M., Lorenz, I., Walther, P., & Janka-Schaub, G. E. (2003). Natural killer deficiency: A minor or major factor in the manifestation of hemophagocytic lymphohistiocytosis? *Journal of Pediatric Hematology/Oncology, 25*, 680–683.
55. Selleri, C., Sato, T., Anderson, S., Young, N. S., & Maciejewski, J. P. (1995). Interferon-gamma and tumor necrosis factor-alpha suppress both early and late stages of hematopoiesis and induce programmed cell death. *Journal of Cellular Physiology, 165*, 538–546.
56. Janka, G. E. (2007). Familial and acquired hemophagocytic lymphohistiocytosis. *European Journal of Pediatrics, 166*, 95–109.
57. Freeman, H. R., & Ramanan, A. V. (2011). Review of haemophagocytic lymphohistiocytosis. *Archives of Disease in Childhood, 96*, 688–693.
58. Gorelik, M., Fall, N., Altaye, M., Barnes, M. G., Thompson, S. D., Grom, A. A., et al. (2013). Follistatin-like protein 1 and the ferritin/erythrocyte sedimentation rate ratio are potential biomarkers for dysregulated gene expression and macrophage activation syndrome in systemic juvenile idiopathic arthritis. *The Journal of Rheumatology, 40*(7), 1191–1199.
59. Otterbein, L. E., Soares, M. P., Yamashita, K., & Bach, F. H. (2003). Heme oxygenase-1: Unleashing the protective properties of heme. *Trends in Immunology, 24*, 449–455.
60. Cohen, L. A., Gutierrez, L., Weiss, A., Leichtmann-Bardoogo, Y., Zhang, D., Crooks, D., et al. (2010). Serum ferritin is derived primarily from macrophages through a nonclassical secretory pathway. *Blood, 116*(9), 1574–1584.
61. Schram, A. M., & Berliner, N. (2015). How I treat hemophagocytic lymphohistiocytosis in the adult patient. *Blood, 125*, 2908–2914.
62. Fardet, L., Galicier, L., Lambotte, O., Marzac, C., Aumont, C., Chahwan, D., et al. (2014). Development and validation of the HScore, a score for the diagnosis of reactive hemophagocytic syndrome. *Arthritis and Rheumatism, 66*(9), 2613–2620.

63. Hejblum, G., Lambotte, O., Galicier, L., Coppo, P., Marzac, C., Aumontet, C., et al. (2014). A web-based delphi study for eliciting helpful criteria in the positive diagnosis of hemophagocytic syndrome in adult patients. *PLoS One, 9*(4), e94024.
64. Ravelli, A., Grom, A. A., Behrens, E. M., & Cron, R. Q. (2012). Macrophage activation syndrome as part of systemic juvenile idiopathic arthritis: Diagnosis, genetics, pathophysiology and treatment. *Genes and Immunity, 13*, 289–298.
65. Ravelli, A., Minoia, F., Davì, S., Horne, A., Bovis, F., Pistorio, A., et al. (2016). Classification criteria for macrophage activation syndrome complicating systemic juvenile idiopathic arthritis: A European league against rheumatism/american college of rheumatology/paediatric rheumatology international trials organisation collaborative initiative. *Arthritis & Rhematology, 68*(3), 566–576.
66. Ravelli, A., Minoia, F., Davì, S., Horne, A., Bovis, F., Pistorio, A., et al. (2016). Classification criteria for macrophage activation syndrome complicating systemic juvenile idiopathic arthritis: A European League against rheumatism/american college of rheumatology/paediatric rheumatology international trials organisation collaborative initiative. *Annals of the Rheumatic Diseases, 75*(3), 481–489.
67. Parodi, A., Davì, S., Pringe, A. B., Pistorio, A., Ruperto, N., Magni-Manzoni, S., et al. (2009). Macrophage activation syndrome in juvenile systemic lupus erythematosus: A multinational multicenter study of thirty-eight patients. *Arthritis and Rheumatism, 60*(11), 3388–3399.
68. Ravelli, A., Minoia, F., Davì, S., Horne, A. C., Bovis, F., Pistorio, A., et al. (2016). Expert consensus on dynamics of laboratory tests for diagnosis of macrophage activation syndrome complicating systemic juvenile idiopathic arthritis. *RMD Open, 2*(1), e000161.
69. Allen, C. E., Yu, X., Kozinetz, C. A., & KL, M. C. (2008). Highly elevated ferritin levels and the diagnosis of hemophagocytic lymphohistiocytosis. *Pediatric Blood & Cancer, 50*, 1227–1235.
70. Schram, A. M., Campigotto, F., Mullally, A., Fogerty, A., Massarotti, E., Neuberg, D., et al. (2015). Marked hyperferritinemia does not predict for HLH in the adult population. *Blood, 125*(10), 1548.
71. Kirino, Y., Takeno, M., Iwasaki, M., Ueda, A., Ohno, S., Shiraiet, A., et al. (2005). Increased serum HO-1 in hemophagocytic syndrome and adult-onset Still's disease: Use in the differential diagnosis of hyperferritinemia. *Arthritis Research & Therapy, 7*(3), R616–R624.
72. Ho, C., Yao, X., Tian, L., Li, F. Y., Podoltsev, N., & Xu, M. L. (2014). Marrow assessment for hemophagocytic lymphohistiocytosis demonstrates poor correlation with disease probability. *American Journal of Clinical Pathology, 141*(1), 62–71.
73. Goel, S., Polski, J. M., & Imran, H. (2012 Winter). Sensitivity and specificity of bone marrow hemophagocytosis in hemophagocytic lymphohistiocytosis. *Annals of Clinical and Laboratory Science, 42*(1), 21–25.
74. Bleesing, J., Prada, A., Siegel, D. M., Villanueva, J., Olson, J., Ilowite, N. T., et al. (2007). The diagnostic significance of soluble CD163 and soluble interleukin-2 receptor alpha-chain in macrophage activation syndrome and untreated new-onset systemic juvenile idiopathic arthritis. *Arthritis and Rheumatism, 56*, 965–971.
75. Tsuji, T., Hirano, T., Yamasaki, H., Tsuji, M., & Tsuda, H. (2014). A high sIL-2R/ferritin ratio is a useful marker for the diagnosis of lymphoma-associated hemophagocytic syndrome. *Annals of Hematology, 93*(5), 821–826.
76. Schaer, D. J., Schleiffenbaum, B., Kurrer, M., Imhof, A., Bächli, E., Fehr, J., et al. (2005). Soluble hemoglobin-haptoglobin scavenger receptor CD163 as a lineage-specific marker in the reactive hemophagocytic syndrome. *European Journal of Haematology, 74*(1), 6–10.
77. Tabata, Y., Villanueva, J., Lee, S. M., Zhang, K., Kanegane, H., Miyawaki, T., et al. (2005). Rapid detection of intracellular SH2D1A protein in cytotoxic lymphocytes from patients with X-linked lymphoproliferative disease and their family members. *Blood, 105*, 3066–3071.
78. Kogawa, K., Lee, S. M., Villanueva, J., Marmer, D., Sumegi, J., Filipovich, A. H., et al. (2002). Perforin expression in cytotoxic lymphocytes from patients with hemophagocytic lymphohistiocytosis and their family members. *Blood, 99*, 61–66.

79. Garcia-Astudillo, L. A., Fontalba, A., Mazorra, F., & Marin, M. J. (2009). Severe course of community-acquired pneumonia in an adult patient who is heterozygous for Q481P in the perforin gene: Are carriers of the mutation free of risk? *Journal of Investigational Allergology & Clinical Immunology, 19*, 311–316.
80. Bryceson, Y. T., Fauriat, C., Nunes, J. M., Wood, S. M., Björkström, N. K., Long, E. O., et al. (2010). Functional analysis of human NK cells by flow cytometry. *Methods in Molecular Biology, 612*, 335–352.
81. Marcenaro, S., Gallo, F., Martini, S., Santoro, A., Griffiths, G. M., Aricó, M., et al. (2006). Analysis of natural killer-cell function in familial hemophagocytic lymphohistiocytosis (FHL): Defective CD107a surface expression heralds Munc13-4 defect and discriminates between genetic subtypes of the disease. *Blood, 108*, 2316–2323.
82. Schneider, E. M., Lorenz, I., Muller-Rosenberger, M., Steinbach, G., Kron, M., Janka-Schaub, G. E., et al. (2002). Hemophagocytic lymphohistiocytosis is associated with deficiencies of cellular cytolysis but normal expression of transcripts relevant to killer-cell-induced apoptosis. *Blood, 100*, 2891–2898.
83. Somanchi, S. S., McCulley, K. J., Somanchi, A., Chan, L. L., & Lee, D. A. (2015). A novel method for assessment of natural killer cell cytotoxicity using image cytometry. *PLoS One, 10*(10), e0141074.
84. Zhang, M., Bracaglia, C., Prencipe, G., Bemrich-Stolz, C. J., Beukelman, T., Dimmitt, R. A., et al. (2016). A heterozygous RAB27A mutation associated with delayed cytolytic granule polarization and hemophagocytic lymphohistiocytosis. *Journal of Immunology, 196*(6), 2492–2503.
85. Janka, G. E., & Schneider, E. M. (2004). Modern management of children with haemophagocytic lymphohistiocytosis. *British Journal of Haematology, 124*, 4–14.
86. Cetica, V., Pende, D., Griffiths, G. M., & Aricò, M. (2010). Molecular basis of familial hemophagocytic lymphohistiocytosis. *Haematologica, 95*, 538–541.
87. Schulert, G. S., Zhang, M., Fall, N., Husami, A., Kissell, D., Hanosh, A., et al. (2016). Whole-exome sequencing reveals mutations in genes linked to hemophagocytic lymphohistiocytosis and macrophage activation syndrome in fatal cases of H1N1 influenza. *The Journal of Infectious Diseases, 213*(7), 1180–1188.
88. Spessott, W. A., Sanmillan, M. L., McCormick, M. E., Patel, N., Villanueva, J., Zhang, K., et al. (2015). Hemophagocytic lymphohistiocytosis caused by dominant-negative mutations in STXBP2 that inhibit SNARE-mediated membrane fusion. *Blood, 125*(10), 1566–1577.
89. Schulert, G. S., Zhang, M., Husami, A., Fall, N., Brunner, H., Zhang, K., et al. (2018). Novel UNC13D intronic variant disrupting a NFκB enhancer in a patient with recurrent macrophage activation syndrome and systemic juvenile idiopathic arthritis. *Arthritis & Rhematology*.
90. Cichocki, F., Schlums, H., Li, H., Stache, V., Holmes, T., Lenvik, T. R., et al. (2014). Transcriptional regulation of Munc13-4 expression in cytotoxic lymphocytes is disrupted by an intronic mutation associated with a primary immunodeficiency. *The Journal of Experimental Medicine, 211*(6), 1079–1091.
91. Tothova, Z., & Berliner, N. (2015). Hemophagocytic syndrome and critical illness: New insights into diagnosis and management. *Journal of Intensive Care Medicine, 30*(7), 401–412.
92. Xu, X. J., Tang, Y. M., Song, H., Yang, S. L., Xu, W. Q., Zhao, N., et al. (2012 Jun). Diagnostic accuracy of a specific cytokine pattern in hemophagocytic lymphohistiocytosis in children. *The Journal of Pediatrics, 160*(6), 984–990.
93. Shimizu, M., Nakagishi, Y., Inoue, N., Mizuta, M., Ko, G., Saikawa, Y., et al. (2015). Interleukin-18 for predicting the development of macrophage activation syndrome in systemic juvenile idiopathic arthritis. *Clinical Immunology, 160*(2), 277–281.

Criteria for Cytokine Storm Syndromes

Francesca Minoia, Sergio Davì, Alessandra Alongi, and Angelo Ravelli

Introduction

Cytokine storm syndromes (CSS) are a heterogeneous group of life-threatening hyperinflammatory disorders that are part of the spectrum of hemophagocytic lymphohistiocytosis (HLH). In the current classification of histiocytic disorders, HLH is subdivided into primary and secondary forms [1–3]. Primary HLH (also called familial HLH) refers to the cases associated with several inherited monogenic illnesses [4]. Secondary HLH (also known as acquired or reactive HLH) occurs as a complication numerous conditions, which include infections, malignancies, and rheumatic diseases. Conventionally, secondary HLH associated with rheumatic disorders is termed macrophage activation syndrome (MAS) [5–7]. This complication occurs most commonly in systemic juvenile idiopathic arthritis (JIA) and in its adult counterpart, adult-onset Still disease. However, it is reported with increasing frequency also in childhood-onset systemic lupus erythematosus (SLE) [8], Kawasaki disease [9, 10], and juvenile dermatomyositis. In spite of recent reports in periodic fever syndromes [11, 12], the occurrence of MAS in autoinflammatory diseases is rare.

F. Minoia
Fondazione IRCCS Ca' Granda Ospedale Maggiore Policlinico, Milan, Italy

S. Davì
IRCCS Istituto Giannina Gaslini, Genoa, Italy

A. Alongi
Università degli Studi di Genova, Genoa, Italy

A. Ravelli (✉)
IRCCS Istituto Giannina Gaslini, Genoa, Italy

Università degli Studi di Genova, Genoa, Italy
e-mail: angeloravelli@gaslini.org

R. Q. Cron, E. M. Behrens (eds.), *Cytokine Storm Syndrome*,
https://doi.org/10.1007/978-3-030-22094-5_5

Both primary and secondary HLH are potentially fatal and require immediate recognition to initiate prompt treatment and avoid deleterious outcome. However, because all forms of HLH bear close clinical similarities, their differentiation may be challenging. In addition, some clinical, laboratory, or histopathologic abnormalities of HLH may be seen in a number of non-hemophagocytic diseases. Differentiation among the various forms of HLH and between HLH and unrelated illnesses is essential to select the appropriate therapeutic interventions.

The difficulties in making the diagnosis, the recent therapeutic progress, and the advances in understanding pathophysiology of hemophagocytic syndromes highlight the need for diagnostic tools and well-established classification criteria. Well-validated and widely agreed upon diagnostic or classification criteria for CSS would be useful to enable timely diagnosis and correct classification of patients, to assist in the evaluation of different therapeutic approaches, and to facilitate communication among specialists in different fields who might be interested in hemophagocytic syndromes. Diagnostic or classification criteria would also be important for both educational and research purposes.

In recent years, there has been a great deal of effort to develop diagnostic guidelines or classification criteria for CSS. The purpose of this chapter is to provide a summary of the work accomplished thus far in this area of research.

The HLH-2004 Diagnostic Guidelines

In 1991, the Histiocyte Society published the first set of diagnostic guidelines for HLH, which were based on its most typical clinical, laboratory, and histopathologic findings [13]. Several years later, the recognition that these criteria did not allow recognition of cases with an atypical or insidious onset and that a number of patients could develop one of more of the diagnostic criteria over the course of the disease, together with the emerging knowledge on the clinical and laboratory features of the syndrome, prompted the Histiocyte Society to develop revised criteria, which were named HLH-2004 diagnostic guidelines [14].

All five items comprised in the 1991 guidelines (fever, splenomegaly, cytopenias affecting at least two of three lineages in the peripheral blood, hypertriglyceridemia and/or hypofibrinogenemia, and hemophagocytosis in the bone marrow, spleen, or lymph nodes) were retained in the new guidelines. In addition, the following three criteria were included: (1) low or absent NK-cell activity, (2) hyperferritinemia, and (3) high levels of soluble IL-2 receptor-alpha chain. The incorporation of these parameters was deemed necessary owing to the recent clinical and experimental evidence suggesting their diagnostic value in HLH [14]. It was established that the diagnosis of HLH would require the fulfillment of at least five of the eight criteria (Table 1). It was, however, acknowledged that patients with a molecular diagnosis consistent with HLH do not necessarily need to meet the diagnostic criteria [15, 16].

Table 1 HLH-2004 diagnostic guidelines

The diagnosis of HLH can be established if (A) or (B) below is fulfilled
(A) Molecular diagnosis consistent with HLH
(B) Diagnostic criteria for HLH fulfilled (5 of the 8 criteria below)
1. Fever
2. Splenomegaly
3. Cytopenia (affecting ≥2 lineages in peripheral blood):
• Hemoglobin <90 g/L (infants <4 weeks: hb < 100 g/L)
• Platelets <100 × 10^9/L
• Neutrophils<1.0 × 10^9/L
4. Hypertriglyceridemia and/or hypofibrinogenemia:
• Fasting triglycerides >3.0 mmol/L (i.e., >265 mg/dl)
• Fibrinogen <1.5 g/dL
5. Hemophagocytosis in bone marrow, spleen, or lymph nodes
6. Low or absent NK-cell activity
7. Ferritin ≥500 ng/mL
8. Soluble CD25 (i.e., IL-2 receptor-α) ≥ 2400 U/mL

Adapted from Henter JI et al. HLH-2004: diagnostic and therapeutic guidelines for hemophagocytic lymphohistiocytosis. Pediatr Blood Cancer 2007;48:124–31

The Preliminary Diagnostic Guidelines for MAS Complicating Systemic JIA

In 2005, preliminary diagnostic guidelines for MAS in systemic JIA were published [17]. The authors followed the classification criteria approach, which is based on the comparison of patients with the index disease with patients with a "confusable" disease. The former group included 74 patients with systemic JIA-associated MAS reported in the literature or seen by the authors, whereas the second group comprised 37 patients with systemic JIA seen by the authors who had 51 instances of "high disease activity" but not believed to have clinical MAS. The capacity of clinical, laboratory, and histopathologic variables in discriminating patients with MAS from patients with high disease activity was assessed by computing the sensitivity, specificity, area under the receiver operating characteristic curve, and diagnostic odds ratio. The combinations of variables that led to best separation between patients and control subjects were identified through "the number of criteria present" method.

Table 2 Preliminary Diagnostic Guidelines for MAS complicating sJIA[a]

Laboratory criteria:
1. Platelet count ≤262 × 10^9/L
2. Aspartate aminotransferase >59 U/L
3. White blood cell count ≤4 × 10^9 U/L
4. Fibrinogen ≤2.5 g/L
Clinical criteria:
1. CNS dysfunction (irritability, disorientation, lethargy, headache, seizures, coma)
2. Hemorrhages (purpura, easy bruising, mucosal bleeding)
3. Hepatomegaly (≥ 3 cm below the costal arch)
Histopathologic criterion:
Evidence of macrophage hemophagocytosis in the BM aspirate

Adapted from Ravelli A et al. Preliminary diagnostic guidelines for macrophage activation syndrome complicating systemic juvenile idiopathic arthritis. J Pediatr 2005;146:598–604
[a]Diagnostic rule: The diagnosis of MAS requires the presence of any two or more laboratory criteria or of any two or three or more clinical and/or laboratory criteria. A bone marrow aspirate for the demonstration of hemophagocytosis may be required only in doubtful cases

The final set of guidelines included four laboratory criteria and three clinical criteria (Table 2). Laboratory criteria included decreased platelet count (≤262 × 10^9/L), elevated levels of aspartate aminotransferase (>59 U/L), decreased white blood cell count (≤4.0 × 10^9/L), and hypofibrinogenemia (≤ 2.5 g/L). Clinical criteria were hepatomegaly, hemorrhagic manifestations, and central nervous system dysfunction. The diagnosis of MAS requires the presence of at least two laboratory criteria or the presence of at least one laboratory criterion and one clinical criterion. The demonstration of macrophage hemophagocytosis in the bone marrow aspirate is required only in doubtful cases or for the purpose of differential diagnosis.

HLH-2004 Guidelines Versus Preliminary MAS Guidelines in Systemic JIA-Associated MAS

Until recently, the optimal diagnostic criteria for MAS complicating systemic JIA were unclear. The recognition that the syndrome is clinically similar to primary HLH led some to recommend the use of the HLH-2004 diagnostic guidelines. An alternative approach was based on the application of the preliminary diagnostic guidelines for MAS complicating systemic JIA. However, although both guidelines are considered potentially suitable for detecting MAS in systemic JIA, it has been argued that each of them is limited by a number of potential shortcomings [18–20].

The main limitation of using the HLH criteria in diagnosing MAS is due to the fact that some criteria may not apply to patients with systemic JIA. Due to the prominent inflammatory nature of systemic JIA, the occurrence of a relative drop in white blood cell count, platelets, or fibrinogen rather than the absolute decrease required by the HLH criteria may be more useful to make an early diagnosis [21]. Patients with systemic JIA often have increased white blood cell and platelet counts as well as increased serum levels of fibrinogen as part of the underlying inflammatory process. Therefore, when these patients develop MAS, the degree of cytopenia and hypofibrinogenemia observed in HLH may only be reached in the late phase, when management may be more difficult. Furthermore, the minimum threshold level for hyperferritinemia required for the diagnosis of HLH (500 μg/L) is not suitable to detect MAS in children with sJIA. It is well known that many patients with active systemic JIA, in the absence of MAS, have ferritin levels above that threshold [22]. In the acute phase of MAS, ferritin levels may peak to more than 5000 μg/L. Thus, a 500 μg/L threshold may not discriminate MAS from a flare of systemic JIA. Other HLH criteria that are not readily applicable to MAS are the identification of low or absent natural killer cell activity or soluble IL-2 receptor-α chain above normal limits for age, as these tests are not routinely performed in most pediatric rheumatology centers.

Although the preliminary MAS guidelines have the merit of being data-driven, they have the disadvantage that the study that led to their development lacked several laboratory measurements in a number of patients and had insufficient data available for some important laboratory parameters of MAS, namely ferritin, triglycerides and lactic dehydrogenase.

The diagnostic performance of the two sets of guidelines was recently scrutinized using real patient data [23]. This study compared their potential to discriminate between MAS and two conditions potentially confusable with MAS, represented by active systemic JIA without evidence of MAS and febrile systemic infection requiring hospitalization. The preliminary MAS guidelines were found to perform better than the HLH-2004 guidelines in identifying MAS in the setting of sJIA, whereas the adapted HLH-2004 guidelines (with the exclusion of assessment of bone marrow hemophagocytosis, NK cell activity and sCD25 levels) and the preliminary MAS guidelines with the addition of a ferritin level > 500 ng/mL discriminated best between MAS and systemic infection. These findings led to the conclusion that the HLH-2004 guidelines are not appropriate for identification of MAS in children with sJIA.

The 2016 Classification Criteria for MAS Complicating Systemic JIA

An international collaborative effort aimed to develop a set of classification criteria for MAS in the context of systemic JIA has been accomplished recently [24, 25]. These criteria were based on a combination of expert consensus, available evidence from the medical literature and analysis of real patient data. The project was conducted through the following steps: (1) Delphi survey among international pediatric rheumatologists aimed to identify clinical, laboratory, and histopathologic features potentially suitable for inclusion in classification criteria; (2) large-scale data collection of patients with systemic JIA-associated MAS and two potentially confusable conditions; (3) web-based consensus surveys among experts; (4) selection of candidate criteria through statistical analyses; and (5) definition of final classification criteria in a consensus conference.

The purpose of the first step of the project was to identify candidate items using international consensus formation through the Delphi survey technique. This exercise involved 232 pediatric rheumatologists from 47 countries and led to the selection of the following nine features that were felt to be important as potential diagnostic criteria for MAS by the majority of the respondents: falling platelet count, hyperferritinemia, evidence of macrophage hemophagocytosis in the bone marrow, elevated serum liver enzymes, falling leukocyte count, persistent continuous fever greater than or equal to 38 °C, falling erythrocyte sedimentation rate (ESR), hypofibrinogenemia, and hypertriglyceridemia [26].

In the second phase, international pediatric rheumatologists and pediatric hematologists were invited to participate in a retrospective cohort study of patients with systemic JIA-associated MAS and with two conditions potentially confusable with MAS, represented by patients with active systemic JIA not complicated by MAS, and by children hospitalized for acute febrile infections [23, 27, 28]. The data of 1111 patients (362 with systemic JIA-associated MAS, 404 with active systemic JIA without MAS and 345 with systemic infection) were entered in the study website by 95 pediatric specialists practicing in 33 countries in five continents.

The final part of the project involved a panel of experts, composed of 20 pediatric rheumatologists and 8 pediatric hematologists with specific expertise in the care of children with MAS and related disorders. The participants were first asked to classify, by means of a series of web-based consensus procedures, a total of 428 patient profiles as having or not having MAS, based on the clinical and laboratory features recorded at disease onset. The profiles were selected randomly among the 1111 patients collected in the previous phase and comprised 161 patients with MAS, 140 patients with active sJIA without evidence of MAS and 127 patients with systemic infection. The experts were purposely kept unaware of the original diagnosis and of the overall course of the patient. The minimum level of agreement was set at 80%.

The best performing criteria were selected by means of univariate and multivariate statistical analyses based on their ability to classify individual patients as having or not having MAS, using the experts' diagnosis as the gold standard. Candidate classification criteria were partly derived from the literature and partly generated from the study data, through the combination of criteria approach or multivariable logistic regression. A total of 982 candidate classification criteria were tested. For each set of criteria, the sensitivity, specificity, area under the ROC curve (AUC-ROC), and k value for agreement between the classification yielded by the criteria and the diagnosis of the experts were calculated. It was established that to qualify

Table 3 2016 Classification Criteria for MAS complicating sJIA

A febrile patient with known or suspected sJIA is classified as having MAS if the following criteria are met:
Ferritin >648 ng/mL
and any two of the following:
Platelet count ≤181 × 10^9/μL
Aspartate aminotransferase >48 U/L
Triglycerides >156 mg/dL
Fibrinogen ≤360 mg/dL

Adapted from Ravelli A et al. 2016 Classification Criteria for Macrophage Activation Syndrome Complicating Systemic Juvenile Idiopathic Arthritis: A European League Against Rheumatism/American College of Rheumatology/Paediatric Rheumatology International Trials Organisation Collaborative Initiative. Ann Rheum Dis 2016;75:481–9 and Ravelli A et al. 2016 Classification Criteria for Macrophage Activation Syndrome Complicating Systemic Juvenile Idiopathic Arthritis: A European League Against Rheumatism/American College of Rheumatology/Paediatric Rheumatology International Trials Organisation Collaborative Initiative. Arthritis Rheumatol 2016;68:566–76

for inclusion in the expert voting procedures phase at the consensus conference, a set of classification criteria should demonstrate a kappa value ≥0.85, a sensitivity ≥0.80, a specificity ≥0.93, and an AUC-ROC ≥0.90. An exception was made for the historical literature criteria [14, 17], which were retained for further consideration even if they did not meet all statistical requirements. Forty-five criteria were selected for further evaluation in the consensus conference.

The International Consensus Conference on MAS Classification Criteria was held in Genoa, Italy, on 21–22 March 2014, and was attended by all 28 experts who participated in the web-consensus evaluations. Attendees were randomized into two equally sized nominal groups and, using the Nominal Group Technique, were asked to decide, independently of each other, which of the classification criteria that met the above statistical requirements were easiest to use and most credible (that is, had the best face/content validity). A series of repeated independent voting sessions was held until the top three classification criteria were selected by each voting group. Then, an 80% consensus was attained on the best (final) set of classification criteria in a session with the two tables of participants combined.

The 2016 Classification Criteria for MAS complicating systemic JIA are presented in Table 3.

Expert Consensus on Dynamics of Laboratory Tests for Diagnosis of MAS in Systemic JIA

All the criteria described above emphasize the achievement of a certain threshold in laboratory test value for the diagnosis of MAS. However, it has been argued that the relative change in laboratory values over time may be more relevant for making an early diagnosis than the decrease below, or increase above, a certain threshold, as stipulated by the criteria [18–20, 24, 25]. It is well known that patients with active systemic JIA often have elevated platelet counts as well as increased levels of ferritin or fibrinogen as part of the underlying inflammatory process [22, 29]. Thus, the occurrence of a relative decline (in the case of platelet count or fibrinogen) or elevation (in the case of ferritin) in these laboratory biomarkers, rather than the achievement of an absolute threshold required by the criteria, may be sufficient to herald the occurrence of MAS in the setting of sJIA [19, 27].

To address this issue, the 2016 MAS Classification Criteria project aimed to identify the laboratory tests whose change over time is most valuable for the timely diagnosis of MAS in the setting of systemic JIA. A multistep process, based on a combination of expert consensus and analysis of real patient data, was conducted. A panel of experts was first asked to evaluate 115 profiles of patients with MAS, which included the values of laboratory tests at the pre-MAS visit and at MAS onset, and the change in values between the two time points. The experts were asked to choose the five laboratory tests whose change was most important for the diagnosis of MAS and to rank the five selected tests in order of importance. The relevance of change in laboratory parameters was further discussed and ranked by the same experts at the consensus conference. Platelet count was the most frequently selected test, followed by ferritin, aspartate aminotransferase (AST), white blood cell count, neutrophil count, fibrinogen, and ESR. Ferritin was assigned most frequently the highest score. At the end of the process, platelet count, ferritin, and AST were the laboratory tests whose change over time was felt by the experts as most important [21].

Performance of the 2016 Classification Criteria in Instances of MAS Developing Under Biologic Therapy

Recently, several episodes of MAS have been observed in systemic JIA patients under treatment with the cytokine blockers canakinumab and tocilizumab in randomized controlled clinical trials and in postmarketing experience [30–32]. Because these agents inhibit the biologic effects of IL-1 and IL-6, respectively, which are among the proinflammatory cytokines involved in the physiopathology of MAS [7], it is conceivable that MAS episodes developing during treatment with these biologics may lack fever or some of the typical laboratory abnormalities of the syndrome. Clinical symptoms of patients with sJIA-associated MAS receiving tocilizumab were found to be milder than those of patients not receiving this medication [33].

Recently a systematic literature review identified 84 sJIA patients who developed MAS while treated with IL-1 and IL-6 blocking agents [34]. Clinical and laboratory features of 35 patients treated with canakinumab and 49 with tocilizumab has been compared to those of an historical retrospective cohort of sJIA associated MAS patients, that were not receiving biologic treatment at MAS onset [27].

The 2016 Classification Criteria were less likely to correctly classify MAS in patients treated with tocilizumab compared to the historical cohort and to patients treated with canakinumab (56.7% vs 78.5% and 84%, respectively; $p < 0.01$). Patients who developed MAS under canakinumab trended to have lower levels of serum ferritin at MAS onset than historical cohort, but not other differences were noted. Conversely, patients treated with tocilizumab at MAS onset were less likely febrile, had much lower ferritin levels, and trended towards lower platelet counts, lower fibrinogen and higher AST levels. These findings suggest that the observed alterations in MAS features may limit utility of the 2016 classification criteria for diagnosis of systemic JIA in patients treated with biologics. However, more data from the real world of clinical practice are needed to establish whether the criteria should be refined to increase their power to pick up the instances of MAS occurring during treatment with IL-1 and IL-6 inhibitors.

The MS Score: A Diagnostic Tool for Early Detection of MAS in Systemic JIA Based on an Improper Linear Model

An unconventional statistical approach was recently applied to create a diagnostic score for early diagnosis of MAS in patients with systemic JIA (Minoia F, unpublished). The clinical and laboratory features of 362 patients with systemic JIA-associated MAS and 404 patients with active systemic JIA without evidence of MAS collected in the aforementioned multinational collaborative project were employed in the analyses. Eighty-percent of the study population was used to develop the score and the remaining 20% constituted the validation sample. All features associated with the diagnosis of MAS in univariate analysis ($p < 0.05$) were further scrutinized in multivariable logistic regression procedures. By means of an improper linear model method, the variables that entered the best fit model of logistic regression were assigned a score based on their β-coefficient value (score 1 for β-coefficients ≤1, score 2 for β-coefficients >1, score 3 for β-coefficients ≥2). The total score was made up by summing the scores of each individual variable. The score cut-off that discriminated better MAS from active systemic JIA was computed by means of receiver operating characteristic (ROC) curve analysis. The sensitivity, specificity, area under the curve (AUC), and kappa value of the score were calculated for both developmental and validation samples.

The 12 variables that were most closely associated with a diagnosis of MAS in logistic regression analysis are presented in Table 4, together with their respective score. The final score ranged from 0 to 21. A cut-off value >7 performed best in discriminating MAS from active sJIA (sensitivity 84%, specificity 84%, AUC 0.91, kappa 0.69). The good performance of the score was confirmed in the validation sample (sensitivity 92%, specificity 88%, AUC 0.90, kappa 0.80).

Table 4 The MS score

	β-coefficient	Points
CNS involvement	2.9	3
Hemorrhagic manifestations	2.3	3
Platelet count ≤340 × 10^9/L	1.6	2
Triglycerides ≥145 mg/dL	1.5	2
Ferritin ≥1550 ng/mL	1.4	2
Lactic dehydrogenase ≥640 U/L	1.2	2
Albumin ≤3.3 g/dL	0.8	1
White blood cell count ≤14 × 10^9/L	0.8	1
Aspartate aminotransferase ≥48 U/L	0.7	1
Hepatomegaly	0.6	1
Absence of skin rash	−1.0	1
Absence of arthritis	−1.4	2

The new diagnostic score, provisionally named MAS/SJIA (MS) score, represents a powerful and feasible tool for the early detection of MAS in patients with active systemic JIA. Notably, the use of an improper linear model method may potentially allow the application of the score in different diseases.

Distinguishing MAS from Primary HLH: The MH Score

As mentioned above, because primary HLH and MAS bear close clinical similarities, their differentiation may be challenging. Distinction may be particularly difficult when MAS develops at onset of systemic JIA, and arthritis is not yet present or when primary HLH occurs at a later age [35]. Indeed, although pHLH typically develops in the first year of life, it is now understood that there are patients with a genetic basis for this illness who remain asymptomatic until the toddler or adolescent age [36]. Recognition of biallelic pathologic mutations in a disease-associated gene is the gold standard diagnostic test for primary HLH. However, these studies take weeks to complete and may not be available in resource-limited areas. In addition, although most cases of primary HLH can be molecularly verified, some cases still elude molecular diagnosis [37]. Finally, some genetic overlap between MAS and pHLH has been identified [38–42].

Timely differentiation between primary HLH and MAS is, nevertheless fundamental because primary HLH is often more severe than MAS and the management of the two conditions differs. Although both are treated with intravenous corticosteroids, the treatment protocol recommended for primary HLH (HLH-94) [43] includes dexamethasone and etoposide, with the optional addition of cyclosporine, whereas the first-line treatment for MAS in pediatric rheumatology settings almost always involves methylprednisolone and cyclosporine, often in combination with an IL-1 blocker [7]. Furthermore, primary HLH patients often require allogeneic hematopoietic stem cell transplantation [37], whereas MAS patients almost never need such treatment.

Until recently, the capacity of the HLH-2004 and MAS criteria to discriminate between the two conditions has never been investigated. An international collaborative initiative has led to devise a diagnostic score that may aid to distinguish between primary HLH and systemic JIA-associated MAS [44]. Clinical, laboratory, and histopathologic features of 362 patients with MAS and 258 patients with primary HLH were collected by both pediatric rheumatologists and pediatric hematologists. Eighty percent of the population was used to develop the score and the remaining 20% constituted a validation sample. Variables that entered the best fitted model of logistic regression were assigned a score, based on their statistical weight. The MAS/HLH (MH) score was composed of the individual scores of selected variables and comprised six items: age at onset, neutrophil count, fibrinogen, splenomegaly, platelet count, and hemoglobin (Table 5). The MH score ranged from 0 to 123 points, and its median value was 97 (1st–3rd quartile 75–123) and 12 (1st–3rd quartile 11–34) in primary HLH and MAS, respectively. The probability of a diagnosis of primary HLH ranged from <1% for a score < 11 to >99% for a score ≥ 123. By means of ROC curve analysis, a cut-off value ≥60 was found to perform best in discriminating primary HLH from MAS.

The MH score may assist physicians in differentiating timely of primary HLH and systemic JIA-associated MAS, facilitate the decision-making process regarding initial therapy and aid selection of patients who require further evaluation to diagnose suspected primary HLH. Its feasibility and easy applicability may foster its widespread use in different centers and in diverse countries.

Table 5 The MH score

	Points for scoring
Age at onset, years	0 (>1.6); 37 (≤1.6)
Neutrophil count, ×10^9/L	0 (>1.4); 37 (≤1.4)
Fibrinogen, mg/dL	0 (>131); 15 (≤131)
Splenomegaly	0 (no); 12 (yes)
Platelet count, ×10^9/L	0 (>78); 11 (≤78)
Hemoglobin, g/dL	0 (>8.3); 11 (≤8.3)

Adapted from Minoia F et al. Development and initial validation of the macrophage activation syndrome/ hemophagocytic lymphohistiocytosis score, a diagnostic tool that differentiates primary hemophagocytic lymphohistiocytosis from macrophage activation syndrome. J Pediatr 2017; 189:72–78

Diagnostic Criteria for MAS in Childhood-Onset SLE

As stated above, in recent years MAS has been reported with increasing frequency in childhood-onset SLE. Furthermore, it has been suggested that MAS in this disease may be underrecognized [45]. In 2009, Parodi et al. reported the demographic, clinical and histopathologic features of 38 patients with MAS as a part of childhood-onset SLE, collected in the context of a multinational survey or obtained from a systematic literature review [46]. Patients who had evidence of macrophage hemophagocytosis on bone marrow aspirate were considered to have definite MAS, and those who did not have such evidence were considered to have probable MAS. Patients with definite and probable MAS were comparable for all clinical and laboratory features of the syndrome, including the American College of Rheumatology lupus criteria at diagnosis, except for a greater frequency of lymphadenopathy, leukopenia, and thrombocytopenia in patients with definite MAS. Around two-thirds of the patients with MAS developed this complication within 1 month after diagnosis of childhood-onset SLE. The frequency of admission to the ICU was 43.7%, and the mortality rate was 11.4%. These figures indicate that MAS in childhood-onset SLE is a serious condition.

The clinical and laboratory features of the 38 childhood-onset SLE patients with MAS (20 definite and 18 probable) were compared with those of a control group composed of 416 patients with active childhood-onset SLE without MAS. The ability of each feature to discriminate MAS from active disease was evaluated by calculating sensitivity, specificity and AUC-ROC. Overall, clinical features revealed better specificity than sensitivity, except for fever, which was highly sensitive but poorly specificity. Among laboratory features, the best sensitivity and specificity were shown by hyperferritinemia, followed by increased levels of lactate dehydrogenase, hypertriglyceridemia, and hypofibrinogenemia. Both HLH-2004 diagnostic guidelines and preliminary diagnostic guidelines for MAS complicating systemic JIA were tested in the study sample, but were found to be inappropriate for detecting MAS in the context of childhood-onset SLE.

On the basis of the results of statistical analyses and using the "number of criteria present approach," a set of preliminary diagnostic guidelines for MAS in childhood-onset SLE was devised, which included five clinical criteria and six laboratory criteria. The best diagnostic performance was obtained by the simultaneous presence of any one or more clinical criteria and any two or more laboratory criteria, which yielded a sensitivity of 92.1% and a specificity of 90.9% (Table 6) Owing to the strong discriminatory ability shown by this definition, the demonstration of macrophage hemophagocytosis in the bone marrow aspirate was deemed necessary for diagnostic confirmation only in uncertain cases. The practical recommendation that came out from the study findings was that in the clinical setting, the occurrence of unexplained fever and cytopenia, when associated with hyperferritinemia, in a patient with childhood-onset SLE should raise the suspicion of MAS.

Table 6 Diagnostic criteria for MAS in childhood-onset SLE[a]

Clinical criteria:
1. Fever (38 °C)
2. Hepatomegaly (3 cm below the costal arch)
3. Splenomegaly (3 cm below the costal arch)
4. Hemorrhagic manifestations (purpura, easy bruising, or mucosal bleeding)
5. Central nervous system dysfunction (irritability, disorientation, lethargy, headache, seizures, or coma)
Laboratory criteria:
1. Cytopenia affecting two or more cell lineages (white blood cell count ≤4.0 × 10^9/L, hemoglobin ≤90 gm/L, or platelet count ≤150 × 10^9/L)
2. Increased aspartate aminotransferase (> 40 units/L)
3. Increased lactate dehydrogenase (> 567 units/L)
4. Hypofibrinogenemia (fibrinogen ≤1.5 g/L)
5. Hypertriglyceridemia (triglycerides >178 mg/dL)
6. Hyperferritinemia (ferritin >500 μg/L)
Histopathologic criterion:
Evidence of macrophage hemophagocytosis in the bone marrow aspirate

Adapted from Parodi A et al. Macrophage activation syndrome in juvenile systemic lupus erythematosus: a multinational multicenter study of thirty-eight patients. Arthritis Rheum 2009;60: 3388–99

[a]Diagnostic rule: the diagnosis of macrophage activation syndrome requires the simultaneous presence of at least one clinical criterion and at least two laboratory criteria. Bone marrow aspiration for evidence of macrophage hemophagocytosis may be required only in doubtful cases

Generic MAS Criteria

Recently, Fardet et al. developed and validated a weighted diagnostic score for the broader category of reactive hemophagocytic syndrome (HS), called the HScore [47] (Table 7). This score was created and tested using a multicenter retrospective cohort of 312 patients (many with oncologic conditions) who were classified by an expert panel to have reactive HS (n = 162), not to have reactive HS (n = 104), or in whom the diagnosis of reactive HS was undetermined (n = 46). For the construction of the score the population was divided in a developmental group and in a validation group. Ten explanatory variables, obtained from a previous Delphi survey involving 24 experts in HS from 13 countries, were analyzed in the study [48]. After the demonstration of their association with a positive diagnosis of HS in univariate analysis, the ten variables were included in a multivariate logistic regression model to assess their independent contribution to the outcome. Continuous variables were dichotomized and their threshold values were calculated by means of ROC curve analysis (for variables showing a linear relationship with the outcome), or by computing the lowest smoothing function and examining the resulting plot (for variables not showing a linear relationship with the outcome). The coefficients resulting from multiple

Table 7 HScore

Parameter	No. of points (criteria for scoring)
Known underlying immunosuppression[a]	0 (no) or 18 (yes)
Temperature (°C)	0 (<38.4), 33 (38.4–39.4), or 49 (>39.4)
Organomegaly	0 (no), 23 (hepatomegaly or splenomegaly), or 38 (hepatomegaly and splenomegaly)
No. of cytopenias[b]	0 (1 lineage), 24 (2 lineages), or 34 (3 lineages)
Ferritin (ng/mL)	0 (>2000), 35 (2000–6000), or 50 (>6000)
Triglyceride (mmoles/L)	0 (>1.5), 44 (1.5–4), or 64 (>4)
Fibrinogen (gm/L)	0 (>2.5) or 30 (≤2.5)
Serum glutamic oxaloacetic transaminase (IU/L)	0 (<30) or 19 (≥30)
Hemophagocytosis features on bone marrow aspirate	0 (no) or 35 (yes)

Adapted from Fardet L et al. Development and validation of the HScore, a score for the diagnosis of reactive hemophagocytic syndrome. Arthritis Rheumatol 2014;66:2613–20

[a]Human immunodeficiency virus positive or receiving long-term immunosuppressive therapy (i.e., glucocorticoids, cyclosporine, azathioprine)

[b]Defined as a hemoglobin level of ≤9.2 g/dL and/or a leukocyte count of ≤5000/mm^3 and/or a platelet count of ≤110,000/mm^3

logistic regression analysis were used to assign score points to each variable for construction of the HScore.

Nine variables, three clinical, five biologic, and one cytologic were included in the HScore. The maximum possible score assigned to each variable ranged from 18 for underlying immunosuppression to 64 for triglyceride level. In the validation set, the median HScore was 222 (interquartile range, 202–284) for patients with a positive diagnosis of reactive HS and 129 (interquartile range, 77–152) for patients with a negative diagnosis. The HScore revealed excellent diagnostic performance and discriminative ability in both developmental and validation data sets. The probability of having HS ranged from <1% with an HScore < 90 to >99% with an HScore ≥250.

General limitations of the HScore include the retrospective nature of the data collection, the heterogeneity of the underlying diseases and the small size (only 10% of the entire study population) of the validation data set. Because the patient sample included in the study of Fardet et al. was only composed of adults with HS, no information can be reliably drawn regarding the applicability of the HScore to pediatric patients with reactive HS, particularly those with systemic JIA-associated MAS. Moreover, it can be anticipated that the use in this condition of some individual criteria included in the HScore may be problematic. For instance, the item "known underlying immunosuppression" involves the use of some medications that are seldom prescribed in children with systemic JIA, such as cyclosporine A and azathioprine and does not include the newer cytokine antagonists that are a part of the treatment of this disease and have been associated with the occurrence of MAS [7, 30–34].

In addition, the threshold level for the platelet count (110,000/mm^3) is too low for a highly inflammatory condition like systemic JIA, which is typically characterized, in its acute phase, by marked thrombocytosis. As discussed above, due to the prominent inflammatory nature of systemic JIA, the occurrence of a relative drop in platelets, rather than an absolute decrease below a certain threshold, is more useful for an early diagnosis [21]. A further problem with the HScore is that the demonstration of hemophagocytosis in the bone marrow aspirate is rarely sought for in children with systemic JIA-associated MAS and does not constitute a mandatory requirement in either the HLH-2004 guidelines or the preliminary MAS guidelines.

Conclusions and Future Directions

Owing to the work performed in the past two decades, a variety of well-established diagnostic guidelines, classification criteria and diagnostic scores for CSS are now available. Altogether, these tools will increase the knowledge and awareness of these serious conditions among specialists in different fields, facilitate timely diagnosis in routine clinical settings, foster genetic, etiopathogenetic, and clinical research, and ensure inclusion of properly characterized patients in future therapeutic trials.

Although the most recent criteria have been developed using modern and rigorous methodologic and statistical procedures and appropriate and large patient samples, they have generally been based on retrospectively collected data. There is, thus, the need to scrutinize their validity in different patient populations evaluated prospectively. Furthermore, because some signs and symptoms of MAS in systemic JIA may be blurred under therapy with cytokine blockers, further studies of the diagnostic performance of existing criteria in patients treated with biologics are warranted. The validity of the 2016 Classification Criteria for MAS has been recently demonstrated in Japanese patients with systemic JIA [49].

Because MAS may occur in rheumatic diseases other than systemic JIA, particularly SLE and Kawasaki disease, there is the need to explore the validity of proposed criteria in these illnesses. As mentioned above, MAS is likely underrecognized in childhood-onset SLE [45]. Notably, it has been suggested that a fraction of children with Kawasaki disease who are refractory to intravenous immunoglobulin may have "occult" MAS and that the 2016 Classification Criteria for MAS in sJIA may be suitable to detect MAS in Kawasaki disease [50, 51]. Recently, the same criteria were found to be useful for the recognition of MAS and the identification of patients at risk for a poor outcome in patients with adult-onset Still disease, which is the adult equivalent of systemic JIA [52]. Finally, a facile calculation of the serum ferritin-to-ESR ratio may help to distinguish MAS from flares of disease in children with sJIA, and potentially other conditions with CSS [53].

References

1. Favara, B. E., Feller, A. C., Pauli, M., Jaffe, E. S., Weiss, L. M., Arico, M., et al. (1997). Contemporary classification of histiocytic disorders. The WHO Committee on Histiocytic/Reticulum Cell Proliferations. Reclassification Working Group of the Histiocyte Society. *Medical and Pediatric Oncology, 29*, 157–166.
2. Jordan, M. B., Allen, C. E., Weitzman, S., Filipovich, A. H., & McClain, K. L. (2011). How I treat hemophagocytic lymphohistiocytosis. *Blood, 118*, 4041–4052.
3. Janka, G. E. (2012). Familial and acquired hemophagocytic lymphohistiocytosis. *Annual Review of Medicine, 63*, 233–246.
4. Emile, J. F., Abla, O., Fraitag, S., Horne, A., Haroche, J., Donadieu, J., et al. (2016). Revised classification of histiocytoses and neoplasms of the macrophage-dendritic cell lineages. *Blood, 127*, 2672–2681.
5. Hadchouel, M., Prieur, A. M., & Griscelli, C. (1985). Acute hemorrhagic, hepatic, and neurologic manifestations in juvenile rheumatoid arthritis: Possible relationship to drugs or infection. *The Journal of Pediatrics, 106*, 561–566.
6. Sawhney, S., Woo, P., & Murray, K. (2001). Macrophage activation syndrome: A potentially fatal complication of rheumatic disorders. *Archives of Disease in Childhood, 85*, 421–426.
7. Ravelli, A., Grom, A. A., Behrens, E. M., & Cron, R. Q. (2012). Macrophage activation syndrome as part of systemic juvenile idiopathic arthritis: Diagnosis, genetics, pathophysiology and treatment. *Genes and Immunity, 13*, 289–298.
8. Parodi, A., Davi, S., Pringe, A. B., Pistorio, A., Ruperto, N., Magni-Manzoni, S., et al. (2009). Macrophage activation syndrome in juvenile systemic lupus erythematosus: A multinational multicenter study of thirty-eight patients. *Arthritis and Rheumatism, 60*(11), 3388–3399.
9. Avcin, T., Tse, S. M. L., Schneider, R., Ngan, B., & Silverman, E. D. (2006). Macrophage activation syndrome as the presenting manifestation of rheumatic diseases in childhood. *The Journal of Pediatrics, 148*, 683–686.
10. Simonini, G., Pagnini, I., Innocenti, L., Calabri, G. B., De Martino, M., & Cimaz, R. (2010). Macrophage activation syndrome/hemophagocytic lymphohistiocytosis and Kawasaki disease. *Pediatric Blood & Cancer, 55*(3), 592.
11. Rigante, D., Capoluongo, E., Bertoni, B., Ansuini, V., Chiaretti, A., Piastra, M., et al. (2007). First report of macrophage activation syndrome in hyperimmunoglobulinemia D with periodic fever syndrome. *Arthritis and Rheumatism, 56*, 658–661.
12. Rossi-Semerano, L., Hermeziu, B., Fabre, M., & Kone-Paut, I. (2011). Macrophage activation syndrome revealing familial Mediterranean fever. *Arthritis Care & Research, 63*, 780–783.
13. Henter, J. I., Elinder, G., Ost, A., & the FHL Study Group of the Histiocyte Society. (1991). Diagnostic guidelines for hemophagocytic lymphohistiocytosis. *Seminars in Oncology, 18*, 29–33.
14. Henter, J. I., Horne, A., Aricó, M., Egeler, R. M., Filipovich, A. H., Imashuku, S., et al. (2007). HLH-2004: Diagnostic and therapeutic guidelines for hemophagocytic lymphohistiocytosis. *Pediatric Blood & Cancer, 48*, 124–131.
15. Janka, G. E., & Schenider, E. M. (2004). Modern management of children with haemophagocytic lymphohistiocytosis. *British Journal of Haematology, 124*, 4–14.
16. Henter, J. I., Tondini, C., & Pritchard, J. (2004). Histiocytic syndromes. *Critical Reviews in Oncology/Hematology, 50*, 157–174.
17. Ravelli, A., Magni-Manzoni, S., Pistorio, A., Besana, C., Foti, T., Ruperto, N., et al. (2005). Preliminary diagnostic guidelines for macrophage activation syndrome complicating systemic juvenile idiopathic arthritis. *The Journal of Pediatrics, 146*, 598–604.
18. Ramanan, A. V., & Schneider, R. (2003). Macrophage activation syndrome—What's in a name! *The Journal of Rheumatology, 30*, 2513–2516.
19. Kelly, A., & Ramanan, A. V. (2007). Recognition and management of macrophage activation syndrome in juvenile arthritis. *Current Opinion in Rheumatology, 19*, 477–481.

20. Davi, S., Lattanzi, B., Demirkaya, E., Rosina, S., Bracciolini, G., Novelli, A., et al. (2012). Toward the development of new diagnostic criteria for macrophage activation syndrome in systemic juvenile idiopathic arthritis. *Annals of Paediatric Rheumatology, 1*, 1–7.
21. Ravelli, A., Minoia, F., Davì, S., Horne, A., Bovis, F., Pistorio, A., et al. (2016). Expert consensus on dynamics of laboratory tests for diagnosis of macrophage activation syndrome complicating systemic juvenile idiopathic arthritis. *RMD Open, 2*, e000161.
22. Pelkonen, P., Swanljung, K., & Siimes, M. A. (1986). Ferritinemia as an indicator of systemic disease activity in children with systemic juvenile rheumatoid arthritis. *Acta Paediatrica Scandinavica, 75*, 64–68.
23. Davi, S., Minoia, F., Pistorio, A., et al. (2014). Performance of current guidelines for diagnosis of macrophage activation syndrome complicating systemic juvenile idiopathic arthritis. *Arthritis & Rhematology, 66*, 2871–2880.
24. Ravelli, A., Minoia, F., Davì, S., Horne, A., Bovis, F., Pistorio, A., et al. (2016). Classification criteria for macrophage activation syndrome complicating systemic juvenile idiopathic arthritis: A European League Against Rheumatism/American College of Rheumatology/Paediatric Rheumatology International Trials Organisation Collaborative Initiative. *Annals of the Rheumatic Diseases, 75*, 481–489.
25. Ravelli, A., Minoia, F., Davì, S., Horne, A., Bovis, F., Pistorio, A., et al. (2016). Classification criteria for macrophage activation syndrome complicating systemic juvenile idiopathic arthritis: A European League Against Rheumatism/American College of Rheumatology/ Paediatric Rheumatology International Trials Organisation Collaborative Initiative. *Arthritis & Rheumatology, 68*, 566–576.
26. Davi, S., Consolaro, A., Guselnova, D., Pistorio, A., Ruperto, N., Martini, A., et al. (2011). An international consensus survey of diagnostic criteria for macrophage activation syndrome in systemic juvenile idiopathic arthritis. *The Journal of Rheumatology, 38*, 764–768.
27. Minoia, F., Davì, S., Horne, A., Demirkaya, E., Bovis, F., Li, C., et al. (2014). Clinical features, treatment, and outcome of macrophage activation syndrome complicating systemic juvenile idiopathic arthritis: A multinational, multicenter study of 362 patients. *Arthritis & Rhematology, 66*, 3160–3169.
28. Minoia, F., Davì, S., Horne, A., Bovis, F., Demirkaya, E., Akikusa, J., et al. (2015). Dissecting the heterogeneity of macrophage activation syndrome complicating systemic juvenile idiopathic arthritis. *The Journal of Rheumatology, 42*, 994–1001.
29. De Benedetti, F., Massa, M., Robbioni, P., Ravelli, A., Burgio, G. R., & Martini, A. (1991). Correlation of serum interleukin-6 levels with joint involvement and thrombocytosis in systemic juvenile rheumatoid arthritis. *Arthritis and Rheumatism, 34*, 1158–1163.
30. Ruperto, N., Brunner, H. I., Quartier, P., Constantin, T., Wulffraat, N., Horneff, G., et al. (2012). Two randomized trials of canakinumab in systemic juvenile idiopathic arthritis. *The New England Journal of Medicine, 367*, 2396 406.
31. De Benedetti, F., Brunner, H. I., Ruperto, N., Kenwright, A., Wright, S., Calvo, I., et al. (2012). Randomized trial of tocilizumab in systemic juvenile idiopathic arthritis. *The New England Journal of Medicine, 367*, 2385 95.
32. Nigrovic, P. A., Mannion, M., Prince, F. H., Zeft, A., Rabinovich, C. E., van Rossum, M. A., et al. (2011). Anakinra as first-line disease-modifying therapy in systemic juvenile idiopathic arthritis: Report of forty-six patients from an international multicenter series. *Arthritis and Rheumatism, 63*, 545 55.
33. Shimizu, M., Nakagishi, Y., Kasai, K., Yamasaki, Y., Miyoshi, M., Takei, S., et al. (2012). Tocilizumab masks the clinical symptoms of systemic juvenile idiopathic arthritis-associated macrophage activation syndrome: The diagnostic significance of interleukin-18 and interleukin-6. *Cytokine, 58*, 287–294.
34. Schulert, G. S., Minoia, F., Bohnsack, J., Cron, R. Q., Hashad, S., Kone-Paut, I., et al. (2018). Biologic therapy modifies clinical and laboratory features of macrophage activation syndrome associated with systemic juvenile idiopathic arthritis. *Arthritis Care & Research, 70*, 409–419.

35. Hayden, A., Park, S., Giustini, D., Lee, A. Y., & Chen, L. Y. (2016). Hemophagocytic syndromes (HPSs) including hemophagocytic lymphohistiocytosis (HLH) in adults: A systematic scoping review. *Blood Reviews, 30*, 411–420.
36. Wang, Y., Wang, Z., Zhang, J., Wei, Q., Tang, R., Qi, J., et al. (2014). Genetic features of late onset primary hemophagocytic lymphohistiocytosis in adolescence or adulthood. *PLoS One, 9*, e107386.
37. Degar, B. (2015). Familial hemophagocytic lymphohistiocytosis. *Hematology/Oncology Clinics of North America, 29*, 903–913.
38. Vastert, S. J., van Wijk, R., D'Urbano, L. E., de Jager, W., Ravelli, A., Magni-Manzoni, S., et al. (2010). Mutations in the perforin gene can be linked to macrophage activation syndrome in patients with systemic onset juvenile idiopathic arthritis. *Rheumatology (Oxford), 49*, 441–449.
39. Kaufman, K. M., Linghu, B., Szustakowski, J. D., Husami, A., Yang, F., Zhang, K., et al. (2014). Whole-exome sequencing reveals overlap between macrophage activation syndrome in systemic juvenile idiopathic arthritis and familial hemophagocytic lymphohistiocytosis. *Arthritis & Rhematology, 66*, 3486–3495.
40. Schulert, G. S., Zhang, M., Fall, N., Husami, A., Kissell, D., Hanosh, A., et al. (2016). Whole-exome sequencing reveals mutations in genes linked to hemophagocytic lymphohistiocytosis and macrophage activation syndrome in fatal cases of H1N1 influenza. *The Journal of Infectious Diseases, 213*, 1180–1188.
41. Zhang, M., Behrens, E. M., Atkinson, T. P., Shakoory, B., Grom, A. A., & Cron, R. Q. (2014). Genetic defects in cytolysis in macrophage activation syndrome. *Current Rheumatology Reports, 16*, 439.
42. Zhang, K., Jordan, M. B., Marsh, R. A., Johnson, J. A., Kissell, D., Meller, J., et al. (2011). Hypomorphic mutations in PRF1, MUNC13-4, and STXBP2 are associated with adult-onset familial HLH. *Blood, 118*, 5794–5798.
43. Henter, J. I., Aricò, M., Egeler, R. M., Elinder, G., Favara, B. E., Filipovich, A. H., et al. (1997). HLH-94: A treatment protocol for hemophagocytic lymphohistiocytosis. HLH study Group of the Histiocyte Society. *Medical and Pediatric Oncology, 28*, 342–347.
44. Minoia, F., Bovis, F., Davì, S., Insalaco, A., Lehmberg, K., Shenoi, S., et al. (2017). Development and initial validation of the macrophage activation syndrome/hemophagocytic lymphohistiocytosis score, a diagnostic tool that differentiates primary hemophagocytic lymphohistiocytosis from macrophage activation syndrome. *The Journal of Pediatrics, 189*, 72–78.
45. Pringe, A., Trail, L., Ruperto, N., Buoncompagni, A., Loy, A., Breda, L., et al. (2007). Macropahge activation syndrome in juvenile systemic lupus erythematosus: An underrecognized complication? *Lupus, 16*, 587–592.
46. Parodi, A., Davì, S., Pringe, A., Pistorio, A., Ruperto, N., Magni-Manzoni, S., et al. (2009). Macrophage activation syndrome in juvenile systemic lupus erythematosus: A multinational multicenter study of thirty-eight patients. *Arthritis and Rheumatism, 60*, 3388–3399.
47. Fardet, L., Galicier, L., Lambotte, O., Marzac, C., Aumont, C., Chahwan, D., et al. (2014). Development and validation of the HScore, a score for the diagnosis of reactive hemophagocytic syndrome. *Arthritis & Rhematology, 66*, 2613–2620.
48. Hejblum, G., Lambotte, O., Galicier, L., Coppo, P., Marzac, C., Aumont, C., et al. (2014). A web-based Delphi study for eliciting helpful criteria in the positive diagnosis of hemophagocytic syndrome in adult patients. *PLoS One, 7*, e94024.
49. Shimizu, M., Mizuta, M., Yasumi, T., Iwata, N., Okura, Y., Kinjo, N., et al. (2018). Validation of classification criteria of macrophage activation syndrome in Japanese patients with systemic juvenile idiopathic arthritis. *Arthritis Care & Research, 70*(9), 1412–1415.
50. Wang, W., Gong, F., Zhu, W., Fu, S., & Zhang, Q. (2015). Macrophage activation syndrome in Kawasaki disease: More common than we thought? *Seminars in Arthritis and Rheumatism, 44*, 405–410.
51. Han, S. B., Lee, S. Y., Jeong, D. C., & Kang, J. H. (2016). Should 2016 criteria for macrophage activation syndrome be applied in children with Kawasaki disease, as well as with systemic-onset juvenile idiopathic arthritis? *Annals of the Rheumatic Diseases, 75*, e44.

52. Ahn, S. S., Yoo, B. W., Jung, S. M., Lee, S. W., Park, Y. B., & Song, J. J. (2017). Application of the 2016 EULAR/ACR/PRINTO classification criteria for macrophage activation syndrome in patients with adult-onset Still disease. *The Journal of Rheumatology, 44*, 996–1003.
53. Gorelik, M., Fall, N., Altaye, M., Barnes, M. G., Thompson, S. D., Grom, A. A., et al. (2013). Follistatin-like protein 1 and the ferritin/erythrocyte sedimentation rate ratio are potential biomarkers for dysregulated gene expression and macrophage activation syndrome in systemic juvenile idiopathic arthritis. *The Journal of Rheumatology, 40*, 1191–1199.

Part II
Genetics of Cytokine Storm Syndromes

Genetics of Primary Hemophagocytic Lymphohistiocytosis

Spyridon Karageorgos and Hamid Bassiri

Introduction

Hemophagocytic lymphohistiocytosis (HLH) constitutes a rare, potentially life-threatening hyperinflammatory immune dysregulation syndrome that can present with a variety of clinical signs and symptoms including fever, hepatosplenomegaly, and abnormal laboratory and immunologic findings such as cytopenias, hyperferritinemia, hypofibrinogenemia, hypertriglyceridemia, elevated blood levels of soluble CD25 (IL-2 receptor α chain), or diminished natural killer (NK) cell cytotoxicity (reviewed in detail in chapter "Immunology of Cytokine Storm Syndromes: Natural Killer Cells" of this book). While HLH can be triggered by an inciting event (e.g., infections), certain single gene mutations have been associated with significantly elevated risk of development of HLH, or its recurrence in patients who have recovered from their disease episode. These monogenic predisposition syndromes are variably referred to as "familial" or "primary" HLH (henceforth referred to as "pHLH") and are the focus of this chapter. Conversely, secondary HLH (sHLH) often occurs in the absence of genetic lesions that are commonly associated with pHLH, and can be triggered by infections, malignancies or rheumatologic diseases; these triggers and the genetics associated with sHLH are discussed in more detail in other chapters in this book.

While the initial therapy of HLH is not dependent on the identification of an underlying genetic lesion, the subsequent confirmation of a mutation predisposing to pHLH is of clinical importance for several reasons. First, pHLH constitutes genetically heterogeneous disorders that are characterized by defective cytotoxic T lymphocyte (CTL) and natural killer (NK) cell cytolytic activity and include life-threatening immunodeficiencies, as well as predisposition to development of lym-

S. Karageorgos · H. Bassiri (✉)
Center for Childhood Cancer Research and Division of Infectious Diseases,
Children's Hospital of Philadelphia, Philadelphia, PA, USA
e-mail: BASSIRI@email.chop.edu

R. Q. Cron, E. M. Behrens (eds.), *Cytokine Storm Syndrome*,
https://doi.org/10.1007/978-3-030-22094-5_6

phoproliferative diseases (LPD). Therefore, it would be important for patients with pHLH to be started on infectious prophylaxis, and be periodically monitored for LPD. Second, the diagnosis of genetic lesions associated with pHLH bears important implications as to whether the patient should receive a hematopoietic stem cell transplant (HSCT) in order to avoid the future recurrence of HLH and/or LPD. Third, identification of genetic lesions in a patient may precipitate the genetic testing of blood relatives, in order to diagnose other affected family members.

The pathogenesis of HLH is described in greater detail in chapter "CD8[+] T Cell Biology in Cytokine Storm Syndromes" and will not be covered here. Yet, germane to this chapter, it is important to note that CTL and NK cells not only kill infected or transformed cells, they also cull antigen presenting cells (APCs), leading to the

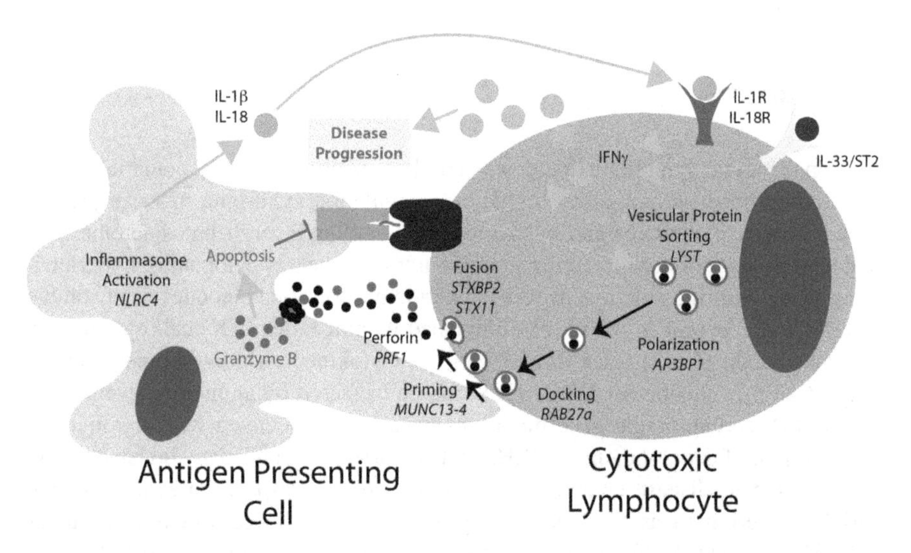

Fig. 1 Model of HLH pathogenesis. Depicted is a schematic summarizing some of the major mechanisms thought to lead to IFNγ production during HLH. Antigen presenting cells display peptide/MHC I complexes to the T-cell receptor of a CD8+ cytotoxic T lymphocyte. This results in a signal to produce and secrete IFNγ as well as a signal to activate the release of perforin and granzyme B. Release of perforin and granzyme B requires the coordination of a number of gene products to successfully accomplish sorting of these proteins to the correct vesicle, polarization of these vesicles to the immune synapse, docking of the vesicle with the plasma membrane, and priming and fusion of the vesicle. Loss of function of any of the genes required for this process (listed in italics underneath each function) results in the inability of perforin to reach the antigen presenting cell. Perforin normally would form a pore in the antigen presenting cell, allowing granzyme B to reach the cytoplasm and induce apoptosis, killing the cell and thereby eliminating further antigen presentation, stimulation of the CD8+ cytotoxic lymphocyte, and continued IFNγ production. In the absence of functional perforin capability, this does not happen and unchecked IFNγ production leads to disease manifestations. Increasingly, non-perforin related activities are being recognized as critical for this process as well, including the IL-1 family members IL-1β, IL-18, and IL-33. Genetic constitutive activation of the NLRC4 inflammasome leads to overproduction of IL-1β and IL-18, with IL-18 in particular leading to the secretion of IFNγ. IL-1β in other circumstances, such as Systemic Juvenile Idiopathic Arthritis, has been noted to be potentially part of disease progression. IL-33 has recently been established as being required in addition to T cell receptor signaling for the production of pathogenic IFNγ in HLH

removal of the depots of antigen, and the termination of ongoing immune responses (Fig. 1). CTL and NK cells perform these cytotoxic functions by either the exocytosis of granules containing perforin and granzymes, or via a non-secretory pathway that involves the upregulation of surface expression of death ligands such as Fas ligand (CD95L or CD178) or tumor necrosis factor (TNF)-receptor apoptosis-inducing ligand (TRAIL; CD278). Interestingly, many of the pHLH syndromes involve defects in the ability to induce cytotoxicity via the granule exocytosis pathway; these defects disrupt either the proper formation of full-length perforin or the trafficking, docking, or exocytosis of cytotoxic granules. This granule exocytosis pathway is not only used for the packaging and delivery of apoptosis-inducing proteins at the immune synapse, they are also used in melanocytes to deliver melanin pigment to keratinocytes; this shared biology therefore explains the overlap of certain pHLH syndromes with hypopigmentation and oculocutaneous albinism.

Finally, certain pHLH genetic mutations also appear to predispose patients to severe and even fulminant infections with Epstein–Barr virus (EBV), and/or EBV-mediated lymphoproliferative diseases (EBV LPD). In this chapter, we provide an overview of the pHLH syndromes and their relevant genetic mutations, including those associated with hypopigmentation and the syndromes associated with EBV LPD (summarized in Table 1).

Primary HLH Without Hypopigmentation

pHLH1

pHLH1-linked locus on chromosome 9q21.3 was identified in 1999 in four consanguineous Pakistani families by using homozygosity mapping [1]. The affected individuals all presented with clinical and laboratory criteria consistent with HLH. Although the authors of this study speculated on the involvement of two genes (*CKS2* and *GAS1*) from within the identified region, to date no causative genes have been definitively linked to this locus.

pHLH2: Perforin Deficiency

Perforin deficiency is caused by mutations in the perforin gene, which in humans is located on chromosome 10q21 (syntenic with mouse chromosome 10) [2]. Mutations in *PRF1* were the first to be associated with pHLH [3], and depending on ethnic origin, pHLH2 currently accounts for 20–30% of all pHLH cases worldwide [4, 5] and up to 50% of pHLH cases in African American patients [6, 7].

Even though the perforin gene sequence is homologous with that of the channel-forming, complement proteins (C6–C9), the perforin locus is not linked to that of

Table 1 Primary HLH syndromes

	Syndrome	Gene and chromosome	OMIM	Protein name (bold) and functions	Clinical features	Inheritance	Protein length and exons
pHLH without hypopigmentation	pHLH1	Unknown 9q21.3-22	267700	Unknown	HLH	AR	–
	pHLH2	*PRF1* 10q22.1	603553	**Perforin** Pore forming protein, apoptosis, cytotoxicity	Severe HLH, often with early onset	AR	555 aa 3 exons
	pHLH3	*UNC13D* 17q25.1	608898	**Munc13-4** Vesicle priming factor	Severe HLH, often with early onset and/or CNS involvement	AR	1090 aa 33 exons
	pHLH4	*STX11* 6q24.2	603552	**Syntaxin 11** Vesicle fusion with the cell membrane	HLH, potentially with later onset & lower severity than pHLH2/3; congenital cardiac defects in some	AR	287 aa 9 exons
	pHLH5	*STXBP2* 19p13.2	613101	**Munc 18-2/ Syntaxin-binding protein** Vesicle fusion with cell membrane	Diarrhea &colitis, hypogammaglobulinemia, sensorineural hearing loss, bleeding diatheses, neurologic abnormalities	AR	604 aa 21 exons
pHLH with partial albinism	Griscelli syndrome	*RAB27A* 15q2.3	607624	**Rab27a** GTPase, promotes vesicle docking to the cell membrane	Recurrent HLH with waxing/waning phases and neurologic sequelae, hair with large irregular melanin granules	AR	221 aa 12 exons
	Chediak-Higashi	*LYST* 1q42.3	214500	**Lyst** Lysosomal trafficking, protein sorting	Giant lysosomes in leukocytes, pyogenic infections, gingival/periodontal disease, bleeding, neurologic manifestations	AR	3801 aa 61 exons
	Hermansky-Pudlak 2	*AP3B1* 5q14.1	608233	**AP3βchain** Trafficking from Golgi to granules	Facial dysmorphisms, platelet defects & bleeding, neurologic manifestations, interstitial lung disease, colitis, neutropenia	AR	1094 aa 29 exons

pHLH with EBV-LPD	ITK deficiency	*ITK* 5q33.3	613011	**Itk** IL-2 inducible T cell kinase, T cell signaling	Progressive CD4 T cell loss and hypogammaglobulinemia, Hodgkin's lymphoma	AR	620 aa 18 exons
	CD27 deficiency	*CD27* 12p13.31	615122	**CD27** TNFR member, lymphocyte co-stimulation	Combined immunodeficiency with susceptibility to viral infections	AR	260 aa 6 exons
	MAGT1 deficiency	*MAGT1* Xq21.1	300853	**Magnesium transporter 1** T cell activation via TCR	Combined immunodeficiency with loss of T cells, but elevations in CD19+ B cells	AR	367 aa 10 exons
	XLP 1	*SH2D1A* Xq25	308240	**SAP** Activation of lymphocytes	Hypogammaglobulinemia, fulminant EBV infectious mononucleosis	X-linked	128 aa 4 exons
pHLH without EBV-LPD	XLP 2	*XIAP* Xq25	300635	**XIAP** Inhibition of apoptosis	Recurrent HLH, colitis	X-linked	497 aa 9 exons
	LPI	*SLC7A7* 14q11.2	222700	**SLC7A7** Amino acid transport	Gastrointestinal symptoms, Intellectual disability, hyperammonemia	AR	511 aa 12 exons

Abbreviations: *pHLH* primary hemophagocytic lymphohistiocytosis, *LPD* lymphoproliferative disease, *AR* autosomal recessive, *aa* amino acids. All protein lengths and exon counts were obtained from gene searches on useast.ensembl.org and ncbi.nlm.nih.gov/gene respectively the genes involved in the terminal complement pathway [2]. The perforin gene consists of 3 exons; exons 2 and 3 are translated and encode a 555-amino acid polypeptide, whereas the exon 1 is untranslated [8]. The perforin protein consists of three domains: the membrane attack complex/cholesterol-dependent cytolysin (MACPF/CDC) domain, the epidermal growth factor (EGF-like) domain and the carboxyl-terminal (C2) domain [9]. Full length perforin forms a soluble, pore-forming protein that is normally stored within the cytotoxic granules of a variety of T and NK cell populations, and gets secreted in a targeted manner at the immune synapse between these cytolytic effector lymphocytes and cells undergoing immune surveillance [9]. Mutations in any of the three domains of perforin result in either decreased protein expression or the complete absence of intact protein in the granules that are necessary for T and NK cell cytolytic function.

More than 90 different mutations in the *PRF1* gene have been described thus far in pHLH2 patients, including missense, nonsense, and microdeletion mutations along the length of the chromosome. Although not absolute, the age of onset for patients with nonsense mutations tends to be earlier than that of patients with missense mutations [10–12]. Missense mutations fall into three subclasses based on expression of human perforin in rat basophilic leukemia (RBL-1) cells, as first presented by Risma et al. in 2006 [13]. Class 1 mutations result in partial maturation of perforin and variable cytotoxic function, while class 2 mutations abrogate the maturation of perforin and class 3 mutations lead to protein misfolding and increased degradation [13]. Clinically, patients with class 1 mutations present with later disease onset compared with the other two classes. Patients class 2 and 3 mutations present with severe HLH within the first 6 months of life, whereas patients with decreased perforin function (atypical form of pHLH2) present with other clinical manifestations (aplastic anemia, recurrent fever, sterile encephalitis) and later onset of disease [14], some as late as the seventh decade of life.

Interestingly, some *PRF1* mutations follow a geographical distribution, as recently shown by Willenbring et al. [15]. Moreover, some mutations recur in certain populations, suggesting a possible founder effect. For instance, the L364 frameshift (1090delCT) mutations is present in Japanese families, the Trp374 stop (G1122A) is found frequently in Turkish origin families and is associated with early-onset disease [12], and in South Asian population, missense mutations resulting in Val145Ala, Ala211Val, Ala437Val, and Arg232His substitutions have been frequently described, whereas in Latino populations, the substitutions Ile266Val and Gly149Ser are the most frequent. The L17 frameshift (50delIT) alteration along with missense mutations Arg4His and Val135Met constitute more than 50% of pHLH2 mutations in patients of African descent; these patients also appear to present with very early disease onset [7]. Finally, Ala91Val (A91V) substitution mutation, resulted by the common polymorphism C272T, results in a 50% reduction in perforin activity and is present in 8–9% of the Caucasian population [16]. The majority of homozygous $PRF1^{A91V/A91V}$ patients present with severe immunodeficiency, whereas A91V heterozygosity ($PRF1^{A91V/+}$) has been linked with immune disorders and hematologic malignancies (including leukemia/lymphoma), and compound heterozygosity of A91V and a mutant allele with "null" activity results in pHLH [16].

pHLH3: Munc13-4 Deficiency

pHLH3 was first described by Feldmann et al. in 2003 [17] in ten patients from seven different families. pHLH3 was shown to be caused by homozygous or compound heterozygous mutations in the *UNC13D* gene (located on chromosome 17q25 in humans), which encodes for Munc13-4 protein [17]. All patients presented with fever, hepatosplenomegaly, pancytopenia, coagulation abnormalities and signs of hemophagocytosis in the bone marrow, associated with increased numbers of activated T cells [17]. *UNC13D* gene includes 33 exons and encodes a 123 kDa protein that contains two calcium-binding (C2) domains divided by two Munc13 homology domains (MHD1, MHD2) [18]. Mutations in *UNC13D* include deletion, insertion, missense, nonsense, or splice-site alterations, as well as those affecting mRNA splicing; these mutations typically lead to significant changes in either the protein structure or its expression. Consistent with this, in a recent study Schulert et al. described a novel *UNC13D* gene intronic variant (c.117 + 143A>G), which was found to reduce Munc13-4 transcript levels and to also alter the binding of NFκB to a transcriptional enhancer element; they further demonstrated that a partial reduction in Munc13-4 was associated with defective NK cell degranulation [19]. Overall, mutations in the *UNC13D* gene are responsible for about 30% of pHLH cases [5]. While a correlation has been suggested to exist between the 847A>G (I283V) mutation and African American patients, no other significant correlations between ethnic groups and specific mutations have been definitively demonstrated.

Munc 13-4 is essential for the release of cytolytic granules from the cytoplasmic membrane at the immune synapse. In fact, T lymphocytes that lack Munc13-4 still have cytotoxic granules docked at the cytoplasmic membrane, but these granules fail to be released, supporting that notion that defects in Munc13-4 is necessary for granule exocytosis [17]. Moreover, lack of Munc13-4 does not result in defects in the secretion of IFN-γ and in conjugation and polarization of lymphocytes with target cells. Furthermore, perforin expression is either normal or increased in patients with pHLH3 [17]. Of note, Munc13-4 differs from the other Munc13 proteins in that it is not expressed in the brain, although it is highly expressed in T and B lymphocytes, monocytes, and in non-hematopoietic tissues such as the lungs and placenta. Despite its lack of expression in the brain, CNS involvement is observed more frequently with pHLH3 than that in pHLH2 [20, 21].

HLH in patients with pHLH3 typically presents in the first year of life. Phenotypically, patients with pHLH3 are similar to patients with perforin deficiency (pHLH2). Of interest, prenatal onset in utero has also been reported [22]. Moreover, several atypical manifestations including increased susceptibility to infections, hypogammaglobulinemia, granulomatous disease of the lung and liver have been described, especially in patients with splice site or missense mutations [23, 24].

These clinical manifestations resemble those seen in common variable immunodeficiency (CVID). Patients with recurrent episodes of HLH may develop a CVID-like manifestations, triggered by the impact of chronic activation of macrophages and T cells on B cell differentiation in lymphoid organs. Interestingly, Munc13-4 deficiency may also affect mast cell degranulation and platelet function, but no clinical correlations have been reported [25].

pHLH4: Syntaxin 11 Deficiency

Mutations in the Syntaxin 11 gene (*STX11*; located on chromosome 6q24 in humans), are associated with the development of pHLH4. This syndrome was first described using genome-wide homozygosity in a large consanguineous Kurdish origin family in 2005 [26]. The *STX11* gene consists of nine exons and spans a 37 kb genomic interval. Only exon 2 is translated and gives rise to a protein that consists of 287 amino acid residues [26].

Syntaxin 11 is a member of the target membrane-associated soluble N-ethylmaleimide sensitive factor-attachment protein receptor (tSNARE) family [26]. These proteins play a vital role in granule membrane fusion, thus regulating the release of cytotoxic granules. STX11 is expressed in monocytes, NK, and $CD8^+$ T cells and participates in vesicle priming and membrane fusion [27]. Mutations identified in *STX11* gene are mostly null mutations and result in defective degranulation and abrogation of NK cell cytotoxicity. Importantly, interleukin 2 (IL-2) stimulation has been shown to partially restore NK cell degranulation and cytotoxicity. Finally, T cell cytotoxicity in patients with pHLH4 may also be spared [28]. The precise step of the cytotoxic pathway that is regulated by STX11 is yet to be further characterized. The IL-2-mediated amelioration of STX11 function and the lineage-specific differential effects of STX11 loss may collectively explain the less severe clinical course observed in pHLH4 patients.

Currently, pHLH4 represents approximately 20% of all pHLH cases. The homozygous mutation Val1124fsX60 was found in all patients of the Kurdish family and was associated with an early termination codon and complete loss of STX11 protein. Furthermore, a 5 bp deletion, a large 19.2 kb deletion spanning the entire coding region of STX11 exon 2, and a nonsense mutation that lead to a premature stop codon were also detected in five families with pHLH4 [26, 29, 30]. While many pHLH4 mutations have been described in families with Turkish and Kurdish origin, Caucasian patients and those of Hispanic origin have also been shown to harbor pHLH4 mutations [31].

In patients with pHLH4, age of onset of HLH is usually later than in pHLH2 and 3, typically after the second year of life. In a study published in 2006, one patient with pHLH4 was noted to have developed lymphoma [29]. The clinical course of pHLH4 also appears to be less severe than that of patients with perforin deficiency. Of interest, in a study performed in three patients of Caucasian and Hispanic origin, two novel mutations of STX11 were detected [31]. The first two patients, born to consanguineous Hispanic parents, had the homozygous nonsense 73G>T (E25X) mutation that resulted in lack of STX11 protein. In both patients poor NK cell degranulation and cytotoxicity, accompanied by early onset of disease were reported. The third patient from a Caucasian family, possessed two biallelic heterozygous missense mutations, the first to be described in STX11 gene, the 106G>C (E36Q) that is located in the N-terminal region of STX11 and the 616G>A (E206K) that results in a non-conservative replacement of glutamic acid located proximal to the SNARE core motif [31]; interestingly, this mutated STX11 protein still allowed for normal NK cell degranulation and in vitro cytotoxicity, which could explain the patient's later disease onset and more favorable response to therapy. Finally, complex congenital heart defects were observed in some patients [31], which may be associated with the expression of STX11 in the heart tissue and needs to be further studied.

pHLH5: Munc18-2 Deficiency

Homozygous or compound heterozygous mutations of the syntaxin-binding-protein 2 (*STXBP2*) gene are associated with the development of pHLH5, first described by two independent groups in 2009 [32, 33]. The *STXBP2* gene is located on chromosome 19p13, which encodes for the mammalian uncoordinated protein 18-2 (Munc18-2) [32]. Munc18-2 protein is a member of the Munc-18/Sec1 family of fusion accessory proteins, which along with SNARE proteins play a crucial role in membrane fusion. Similar to syntaxin 11, Munc18-2 is variably expressed [32, 33].

Zur Stadt et al. studied 15 patients with HLH (14 Turkish origin, 1 Saudi Arabia origin) from consanguineous family backgrounds that were known to not harbor mutations in PRF1, UNC13D or STX11. Using autozygosity mapping they detected 9 homozygous and heterozygous mutations in 12 patients, five missense mutations affecting highly conserved residues, one in-frame and two frameshift deletions, as well as a mutation at the splice acceptor site of exon 15 of the *STXBP2* gene [32]. They also observed allelic heterogeneity at this locus, as the same mutation was found in Turkish and Saudi Arabian families even after a founder effect was excluded. In another study by Cote et al. complete absence of Munc18-2 due to a P477L mutation was associated with early onset and more severe disease course [33]. On the contrary, a hypomorphic mutation was related to delayed disease onset [33].

Mutations in the *STXBP2* gene affect protein stability and expression [26, 33]. Interestingly, lymphoblasts that are Munc18-2 deficient are found to have low STX11 levels, suggesting that STX11 is not only the main partner of Munc18-2 in lymphocytes but that the former is required for stable expression of the latter. Cytotoxicity assays have shown that Munc18-2 deficient NK cells and CTL have decreased killing due to impaired cytotoxic granule exocytosis, as evidenced by decreased or absent surface CD107a expression [33]. Of note, although the precise mechanism of action of Munc18-2 is still being elucidated, defects in cytotoxicity and degranulation are partially reversible via IL-2 stimulation.

The majority of patients with *STXBP2* gene mutations develop HLH in first 6 months of life. On the other hand, certain patients with features of CVID present with milder clinical manifestations and later onset of HLH; this appears to be associated with c.1247-1G > C splice site mutations in exon 15 [26, 33]. Interestingly, in one study, chronic diarrhea that necessitated parenteral nutrition was present in 14 out of 37 patients with early-onset pHLH5. Diarrhea preceded the pHLH5 diagnosis and persisted even after HSCT in six out of eight patients [34]. Moreover, sensorineural hearing loss was also observed in six patients (between 4 and 17 years of age), whereas other neurological manifestations, bleeding disorders, and hypogammaglobulinemia were also reported [35]. Finally, Hodgkin disease was diagnosed at age 9 years in a patient with biallelic *STXBP2* mutations [35].

Primary HLH with Hypopigmentation

Griscelli Syndrome 2

Biallelic mutations in the *RAB27a* gene, located on the 15q21 chromosome, are responsible for the development of Griscelli syndrome 2 (GS2). The *RAB27a* gene gives rise to a 221-amino acid polypeptide with 25 kDa molecular mass [36, 37]. The *RAB27a* gene encodes the Rab27a protein, a small GTP-binding GTPase protein, that is ubiquitously expressed [36]. Thus far, 45 mutations including nonsense mutations, deletions, splice-site alterations, single-nucleotide insertions and few missense mutations have been identified in the *RAB27a* gene in approximately 100 patients [36]. Most mutations typically result in loss-of-function of Rab27a protein via early protein truncation of the carboxyl-terminal motif that participates in protein geranyl-geranylation. Notably, in some patients with GS2 no disease-causing mutations could be identified.

Griscelli syndrome 2 was initially described in 1978 in patients who presented with partial albinism and immunodeficiency—manifestations resembling Chediak–Higashi syndrome—but were distinguished by a different pattern of hypopigmentation and absence of giant lysosomal granules in leukocytes [38]. Patients with GS2 present with cutaneous pigmentary dilution, silvery-gray hair and accumulation of melanosomes in melanocytes [39–41]. Defects in the protein complex formed by Rab27a (along with melanophilin and myosin-Va) have been shown to be responsible for capturing mature melanosomes and transferring these to keratinocytes. This defect explains the accumulation of melanosomes in melanocytes. In addition to defects in Rab27a, pigmentary dilution with primary neurological manifestations are also found in GS1, which is caused by mutations in the myosin-Va (*MYO5A*) gene (chromosome 15q21.2 in humans); finally, GS3 is caused by mutations in the melanophilin (*MLPH*) gene (chromosome 2q37.3 in humans).

Importantly, patients with GS2 may develop recurrent HLH with waxing and waning phases. Infections are typically involved as a trigger for the development of HLH in these patients by eliciting continuous activation and proliferation of T cells and macrophages. Of note, activated lymphocytes and macrophages can infiltrate the brain of patients with GS2 and give rise to secondary neurological manifestations such as seizures, meningitis or coma [36, 41]. Conversely, the neurological manifestations seen in patients with GS1 appear to be primary in origin, as opposed to the secondary neurological manifestations that present with CNS involvement of activated lymphocytes in GS2 [36]. Lastly, mutations in the *RAB27a* gene predispose to the development of HLH by their effects on granule exocytosis. Variable defects in cellular immunity have also been reported. GS1 and GS3 present with similar defects in their melanosome, but do not have defective cytotoxicity; hence, they do not predispose to HLH [39].

Rab27a-deficient NK cells and CTLs demonstrate impaired exocytosis of cytotoxic granules. Remarkably, interactions between RAB27a and Munc13-4 are critical for the exocytosis of lytic granule by coordinating docking and priming of lytic granules [42, 43]. Indeed, RAB27a deficient cytotoxic granules cannot reach the immune synapse to dock to the plasma membrane. Lastly, polarization of cytotoxic granules can be delayed by a heterozygous RAB27a mutation [44].

Chediak–Higashi Syndrome

Mutations in lysosomal trafficking regulator (*LYST*) gene, also known as *CHS1* gene, are responsible for the development of Chediak-Higashi syndrome (CHS). The *LYST* gene, located on chromosome 1q43 in humans, consists of 61 exons and gives rise to a ubiquitously expressed cytosolic protein (425 kDa) consisting of 3801 amino acids [45, 46].

The CHS1/LYST protein constitutes a member of the vesicle trafficking regulatory proteins named "beige and Chediak–Higashi" (BEACH) proteins [47]. This gene was first described in mice that had an altered beige coat color [48]. BEACH proteins are reported to play important roles in membrane dynamics and receptor signaling [49, 50]. The significant length of the LYST gene and the exonic complexity makes the *LYST* gene prone to mutations. Of note, the majority of *LYST* genetic abnormalities identified thus far are nonsense or frameshift mutations that result in the truncation of the protein and the impairment of protein function [51]. Genotype–phenotype correlations for CHS were first described in 2002 by Karim et al. [51]. It appears that patients with homozygous mutations that lead to truncated protein present with early onset of severe clinical manifestations of CHS and HLH with an "accelerated phase," which is typically fatal without bone marrow transplantation. This accelerated phase is characterized by fever, increasing hepatosplenomegaly and lymphadenopathy, bleeding, and worsening pancytopenia; this phase can be triggered by viruses including EBV infections, which can predispose to development of all of the features of pHLH. On the other hand, phenotypic heterogeneity has been documented even in patients with homozygous mutations within the same family [51]. Missense mutations have been shown to give rise to milder clinical aspects of CHS, typically in adults who may or may not present with HLH.

Clinically, patients with CHS typically present with hypopigmentation of hair and skin, and a predisposal to recurrent pyogenic infections (especially of the skin, respiratory tract and mucous membranes), bleeding diathesis, and progressive impairment of neurological function. Other manifestations include oral ulcers, gingivitis and periodontal disease [52], and enterocolitis [53]. The time of diagnosis is typically within the first ten years of life, and is usually precipitated by an uncontrolled EBV infection, although some cases are diagnosed in adulthood. Importantly, patients with CHS develop neurological manifestations either early in the disease course or after childhood. Notably, in CHS patients the neurological manifestations are primary and are not believed to be caused by the infiltration of the central nervous system with macrophages and activated lymphocytes, as has been described for GS2 syndrome. Hence, it is clinically important to distinguish between patients with CHS and GS2.

Hermansky–Pudlak Syndrome 2

Homozygous or compound heterozygous mutations in the gene that encodes the beta-3A subunit of the AP3 complex (AP3B1*)* gene, located on chromosome 5q14.1 in humans, result in Hermansky–Pudlak syndrome 2 (HPS2), an autosomal recessive lysosomal trafficking disorder [54]. HPS is an autosomal recessive disorder in which abnormal biogenesis of lysosome-related organelles lysosomal storage of ceroid lipofuscin is believed to result in a number of multisystemic clinical manifestations. There are ten human genetic disorders in the HPS group. While HPS is generally rare, type 1 HPS appears to have an elevated incidence in those of Puerto Rican descent [55], with most of the carriers and the affected patients carrying the same 16-bp duplication in exon 15 of *HPS1* (chromosome 10q23) [56].

The HPS types share common clinical features such as tyrosinase positive hypopigmentation and oculocutaneous albinism, as well as platelet defects due to abnormalities in dense (delta) granules [57, 58]. An additional clinical manifestation in some patients is chronic neutropenia; while this is reversible in some patients by administration of granulocyte colony stimulating factor (G-CSF) [59], it can predispose to recurrent infections, and bleeding. Patients with HPS2 may also have facial dysmorphisms, developmental delays, nystagmus, and neurological manifestations including hearing loss. Finally, some patients with HPS2 also develop pulmonary fibrosis and interstitial lung disease, as well as granulomatous colitis [60, 61].

The AP3 complexes are either ubiquitous or neuron-specific cytoplasmic complexes that consists of multiple subunits. In HPS, the beta-3A protein, which is a component of the AP3B1 complex, is defective. This complex shuttles cargo proteins from the trans-Golgi and tubular-endosomal compartment to endosome-lysosome related organelles [62, 63]. Thus, AP3 plays a vital role in protein sorting to lysosomes. Defects in the AP3 complex result in disruption of the complex and rapid degradation of the subunits.

While HPS can be caused by several gene defects, only HPS2 is associated with pHLH. HPS2 was first described in 1999 by Dell'Angelica et al. in two male siblings that were found to have heterozygous mutations in the gene that encodes the β3A subunit of the AP3 adaptor complex [54]. Later, heterozygous nonsense mutations in the AP3B1 gene were detected in a patient with severe clinical presentation consistent with HPS2 [54]. Finally, a homozygous deletion was reported in two patients from a consanguineous family of Turkish origin twelve years after the original report of the patients using genetic linkage analysis and targeted gene sequencing [57, 64].

AP3 deficiency is associated with abrogation of cytotoxicity in CTL and NK cells due to impaired biogenesis and degranulation of lytic granules that contain perforin. Clark et al. found that AP3 deficient CTLs derived from patients with HPS2 showed reduced cytotoxicity when exposed to target cells, and demonstrated that this defect was due to a secretory defect [65]. Despite this defect in cytotoxicity and degranulation, few patients with HPS2 patients appear to develop HLH, suggesting that risk of developing HLH is lower in HPS2, as compared to other pHLH syndromes such as GS2 or CHS. Interestingly, in one of the patients with HPS2 that developed HLH, a heterozygous *RAB27a* mutation was also detected, begging the question of whether some HPS2 patients may harbor additional mutations that may further predispose to the development of HLH [66].

Primary HLH Associated with EBV LPD

Primary Epstein–Barr virus (EBV) infection can induce the development of EBV-HLH. Even though the exact mechanism of EBV-HLH has not been elucidated, it has been reported that EBV-infected B cells have the ability to induce the robust proliferation of $CD8^+$ cytotoxic T lymphocytes. This T cell activation is believed to lead to the exuberant activation of macrophages and hypercytokinemia ("cytokine storm") [67, 68]. In certain instances, EBV can has also been demonstrated to infect NK and T cells; this appears to occur more commonly in Asian and South American populations, and leads to chronic viremia, infiltration of organs with virally infected lymphocytes, lymphoproliferative disorders (LPDs), and/or EBV T/NK cell lymphomas [69].

ITK Deficiency

First described in two sisters of consanguineous Turkish descent in 2009 [70], ITK (interleukin-2-inducible tyrosine kinase) deficiency results from biallelic mutations in the *ITK* gene (on chromosome 5q in humans). Since 2009, a number of patients from a variety of other ethnic backgrounds have been described [70–76]. Most of these patients presented clinically with massive EBV LPD, manifesting as fever, lymphadenopathy, hepatosplenomegaly, EBV viremia, and in some, pulmonary nodules [75]. In other patients, additional viral infections were also noted (e.g., cytomegalovirus or varicella), in addition to nephritis, thyroiditis, and predisposition to *Pneumocystis jirovecii* infections and pHLH.

The *ITK* gene comprises 18 exons that encode for a 620-amino acid 71 kDa protein tyrosine kinase of the TEC/BTK family [71]. These kinases are important for development and signaling in specific lymphoid lineages. Structurally ITK resembles Bruton's tyrosine kinase (BTK), consisting from the N-terminus of a pleckstrin homology domain, a Tec homology domain, an Src homology 3 (SH3) and an Src homology 2 (SH2) domain, and a C-terminus catalytic kinase domain. Mutations in the pleckstrin homology, the SH2, and the catalytic domains have all been described [71].

While most of these mutations do not appear to decrease ITK protein levels, affected T cells have been shown to display decreased Ca^{2+} mobilization [71]. This is consistent with the way in which ITK is believed to function in T cells. Upon T cell receptor ligation, ITK is phosphorylated and activated by another Src family kinase, LCK, allowing ITK to phosphorylate phospholipase C gamma 1 (PLCγ1), which leads to cleavage of phosphatidyl inositol substrates and results in Ca^{2+} mobilization and further phosphorylation events via ERK and MAP kinase pathways, and the translocation of cytosolic nuclear factor of activated T cells (NFAT) to the nucleus to allow for new gene transcription, cytokine production, and T cell differentiation and clonal expansion [77, 78]. In the absence of ITK, patients develop certain common immunologic features, including progressive loss of $CD4^+$ T cells and onset of hypogammaglobulinemia. Of note, in a study from ITK-deficient mice it was shown that in the absence of ITK, the $CD8^+$ T-cell expansion and maturation to CTLs is impaired and results in decreased $CD8^+$ T-cell cytotoxic responses [79]. Also, ITK deficient patient have deficient response in EBV infections. Although the exact mechanism is not yet delineated, it could be explained by the defective maturation and expansion of $CD8^+$ T cells against EBV.

CD27 Deficiency

CD27 deficiency is caused by biallelic mutations in the *CD27* gene, which in humans is located on chromosome 12 (12p13.31). The *CD27* gene encodes the costimulatory protein tumor necrosis factor superfamily receptor (TNFSFR) 7, which is highly expressed on T cells and memory B cells [69]; in fact, lack of CD27-expressing B cells is often seen in B cell differentiation abnormalities, including CVID.

CD27 binds to its ligand, CD70 (TNFSF7), to enhance T-cell survival and expansion. CD27-CD70 interactions are also important for lymphocyte effector functions, including those necessary for control of EBV by $CD8^+$ T cells. As such, CD27 deficiency is the latest of 11 genetic disorders described thus far that are associated with immunodeficiency and EBV LPD. While CD27 deficiency and four other gene mutations discussed here (XLP1 and 2, ITK deficiency, and MAGT1 deficiency) share susceptibility to pHLH and EBV LPD, five other genetic syndromes (including mutations in *WAS* [Wiskott–Aldrich syndrome] [80], *CORO1A* [Coronin actin binding protein 1A] [81], *MST1/STK4* [Mammalian sterile 20-like kinase 1/Serine-threonine protein kinase 4] [82], *DCLRE1C* [DNA cross-link repair 1C] [83], and *IL10* [interleukin 10] [84]) include EBV LPD in the absence of HLH.

The first 10 patients with CD27 deficiency were identified from 4 kindreds; these patients all harbored homozygous mutations (c.G24A/p.W8X and c.G158A/p.C53Y [85, 86], but other novel mutations have since been described, including one patient with deficiency caused by compound heterozygous mutations [87].

Immunologic analyses in these latter patients revealed a decreased percentage of central memory cytotoxic T cells and T_H17-like $CD3^+$ cells, but normal frequencies of T_{REG}, invariant natural killer T (iNKT) cells and recent thymic emigrants; some individuals had expanded $CD8^+$ T cells, leading to inverted CD4:CD8 T cell ratios. Additionally, the majority also displayed impaired NK cell function. To date almost all CD27 deficient patients that have been described presented with EBV LPD (B-cell lymphoma, Hodgkin's lymphoma) and EBV-triggered HLH [87]. Finally, uveitis and oral ulcers have also been described [87, 88].

While predisposition to EBV-mediated diseases are prominent in CD27 deficiency [85, 89], humans and mice in whom CD27-CD70 interactions were disrupted also displayed impaired immunity against other pathogens including influenza virus, lymphocytic choriomeningitis virus (LCMV), vesicular stomatitis virus (VSV), and even *Listeria monocytogenes* [90–92].

MAGT1 Deficiency

First described in 2011, MAGT1 (magnesium transporter 1) deficiency results in XMEN syndrome (X-linked immunodeficiency with magnesium defect, EBV infection, and neoplasia), due to hemizygous mutations in *MAGT1* gene [93]. In humans, the MAGT1 gene sits on the X chromosome (Xq21.1) and encompasses 10 exons, allowing for 3 protein coding isoforms, consisting of 367, 355, and 134 amino acids [93, 94], with the full-length protein comprising of a predicted signaling sequence in the amino terminal segment and a consensus phosphorylation site in the carboxy terminus. Thus far, there are 11 male patients described in the literature with MAGT1 deficiency [93, 95]. The mutations described appear to be randomly distributed in the various exons, although mutations in the intronic sequence and the 3′ UTR have also been described [96]. Typically, patients present with splenomegaly, $CD4^+$ T cell lymphopenia (with reversal of CD4:CD8 T cell ratio), chronic EBV infection with elevated numbers of EBV-infected B cells, and eventually the development of EBV-associated B-cell lymphomas. The immunodeficiency in these patients usually manifests with frequent sinopulmonary infections, epiglottitis, and diarrhea. Other less common manifestations include presentation with HHV-8 Kaposi's sarcoma [97]. Copy number variations in MAGT1 have also been associated with skin disorders [98] and intellectual disabilities [99], although the latter association has been questioned by a more recent analysis by the NHLBI of a large number of X chromosomes from "normal" subjects [100].

The *MAGT1* gene encodes a ubiquitously expressed transmembrane transporter which participates in the maintenance of free basal intracellular Mg^{2+} pools. MAGT1 is evolutionarily conserved and does not have structural similarity to other Mg^{2+} transport proteins [96]. The tumor suppressor TUSC3 is a human gene paralog to *MAGT1* [101], although TUSC3 has more limited tissue distribution and appears to be more permissible in its substrate specificity [102].

The MAGT1 protein localizes to the plasma membrane and mediates voltage-dependent Mg^{2+} transport in a selective manner [102]. Specifically, after T cell receptor engagement, MAGT1 mediates a transient influx of Mg^{2+} that is required for the activation of phospholipase C gamma 1 (PLCγ1), which subsequently drives a rise in Ca^{2+} and downstream signaling [93]. As such, abolishment of MAGT-1 function has been shown to impair TCR-mediated signaling in T cells [93]. In addition, MAGT1 appears to also be required for the expression and function of NKG2D (natural killer receptor group 2 member D) and DAP10 (DNAX activation protein 10), which are expressed by NK, γδ T, and $CD8^+$, and certain subsets of $CD4^+$ T cells [103]. Importantly, NKG2D recognizes self ligands that are induced by cellular stress (e.g., senescence, infection, or malignancy) and are important for clearance of such cells. Consistent with this function, NKG2D ligands are upregulated on EBV-infected B cells, and MAGT1-deficient $CD8^+$ T cells display defective cytotoxic activity against EBV-transformed B cells [104]. Notably, NKG2D expression, cell cytotoxicity, and immunity to EBV in MAGT1-deficient patients are restored by supplementation of magnesium in vivo and in vitro [103]. Curiously, while MAGT1-deficient B cells also demonstrated defective Mg^{2+} homeostasis, when stimulated via the immunoglobulin receptor, these cells display a heightened influx of Ca^{2+} and increased phosphorylation of B cell receptor-associated signaling proteins, resulting in elevated frequencies of $CD19^+$ and marginal zone B cells and decreased proportions of plasma cells [105], implying that MAGT1 deficiency may contribute to the array of disorders seen in patients with XMEN syndrome.

X-linked Lymphoproliferative Disease (XLP)

XLP encompasses a rare inherited immunodeficiency with an approximate incidence of one in a million males. XLP, was initially identified as Duncan disease by Purtilo et al. in 1970s [106]. Patients were described as boys who presented with cytopenias, hypogammaglobulinemia, and fulminant EBV mononucleosis or EBV-LPD. Since this initial description, however, two different genetic mutations (XLP1 and XLP2) have been described that associate with the clinical features typical of XLP syndrome.

XLP1: SAP Deficiency

The gene responsible for the development of XLP1 was initially described using positional and functional cloning methods by three different research teams in 1998 [107–109]. The SH2 domain protein 1A (*SH2D1A*) gene is located on chromosome Xq25 in humans, consists of 4 exons, and encodes for a 128-amino acid protein, consisting of a 5-amino acid N-terminal sequence, an SH2 domain and a 25-amino acid C-terminal tail. The protein encoded by *SH2D1A* is called the signaling-lymphocytic-activation molecule (SLAM)-associated protein (SAP) [110]. Defects in SAP have been described to arise from a variety of mutations including insertions, deletions, single nucleotide substitutions and splice-site abnormalities. XLP1 is inherited as an X-linked recessive syndrome, with de novo mutations being rare. Recently, the Arg55stop mutation was identified in 4 out of 21 families of Japanese origin with XLP1, suggesting that it may be a mutational hotspot. Yet there is no strong racial or ethnic predispositions for XLP1 [111, 112] and genotype–phenotype correlations appear to be poor, with considerable phenotypic variability being observed in family members harboring the same genetic mutation [113]. While XLP1 is a disease of males, and female carriers are typically asymptomatic, some females can be partially or fully affected due to skewed or unbalanced X-inactivation or lyonization [114]. Finally, certain patients display spontaneous somatic reversion [115]—further complicating phenotype–genotype correlations.

The three most common clinical manifestations of XLP1 are fulminant infectious mononucleosis (FIM) due to EBV (over 50% of patients), dysgammaglobulinemia (in 20–30% of patients), and development of lymphoproliferative disease (LPD; in up to 30% of patients) [116]. While the EBV-induced FIM appears to be the most common manifestation, the dysgammaglobulinemia and LPD can occur in the absence of prior exposure to EBV [117, 118]. FIM is associated with HLH physiology and can be severe, although HLH can occur in response to other triggers. While the LPD is most commonly either due to diffuse large B cell lymphoma (DLBCL) or Burkitt's lymphoma, T cell lymphoma was described in one patient [117]. Other less common clinical features include lymphocytic vasculitis, lymphomatoid granulomatosis, and aplastic anemia [119]. Much of what we now know about XLP1 arises from a worldwide registry that was started in 1980 [120].

SAP is produced exclusively in various T, NK, and invariant natural killer T (iNKT) cells and acts as a key regulator protein of normal immune function [109, 121]. SAP interacts with the cytoplasmic tail of most (but not all) members of the SLAM family (SLAMF) receptors including SLAM itself (CD150; SLAMF1), Ly9 (CD229; SLAMF3), 2B4 (CD244; SLAMF4), CD84 (SLAMF5), NTB-A (CD352; SLAMF6), and CRACC (CD319; SLAMF7). These SLAMF members bind each other homotypically (except for 2B4, which binds CD48) [121, 122]. These interactions are postulated to improve cell–cell adhesion and costimulate certain lymphocyte functions. Specifically, 2B4 has been demonstrated to increase NK cell cytotoxicity [123, 124].

SAP may function in three non-mutually exclusive manners in lymphocytes. First, via its SH2 domain, SAP binds to immunoreceptor tyrosine-based switch motifs (ITSM) in the cytoplasmic domains of SLAMF receptors. In the case of its association with SLAM (but not other SLAMF receptors), it appears that the ITSM tyrosines do not necessarily need to be phosphorylated, although phosphorylation of these residues further stabilizes SAP:ITSM binding. In this manner, SAP acts as an adapter molecule to allow for recruitment of the Src family protein, tyrosine kinase Fyn, thereby promoting downstream phosphorylation events [125, 126]. Second, SAP competes with another SH2-containing protein, SHP2 (a phosphatase), to restrict the access for the latter protein to the ITSMs [109]. Loss of SAP expression and/or function results in reduced T, NK, and iNKT cell cytotoxicity [127, 128], poor germinal center formation due to impaired follicular T helper (T_{FH}) cell function and altered T_H2 cytokine production [129, 130]. Moreover, patients with SAP deficiency have impaired development of memory B cells [131]. Finally, SAP enhances apoptosis in T and B cells, especially upon restimulation [132–134]. Thus, the lack of SAP may lead to decreased apoptosis of CD8+ T cells responding to EBV [135]. This reduction in apoptosis, combined with decreased cytolytic activity, likely contributes to an over-exuberant proliferation of T cells that are unable to clear the virus, culminating in FIM and HLH.

Primary HLH Without EBV-LPD

XLP2: XIAP Deficiency

In 2006, a second gene was identified in close proximity to the *SH2D1A* gene on Xq25 in humans [136]; this gene consists of 9 exons and spans 42 kb. Importantly, hemizygous mutations in this gene, called the X-linked inhibitor of apoptosis (*XIAP*), were present in patients with clinical manifestations of XLP who lacked mutations in the *SH2D1A* gene [136]. Thus far, more than 100 patients with mutations in *XIAP* have been described [137]. Mutations in *XIAP* include missense and nonsense changes, as well as deletions and insertions, and these mutations have been identified in all exons; these mutations can result in loss or decreased expression of the XIAP protein. Although the patients with *XIAP* mutations are described as having XLP2, this may be a misclassification, as thus far no patients with XLP2 have been reported to develop lymphomas. These patients do, however, develop recurrent episodes of HLH [138, 139]. As such, this syndrome may be better classified as "X-linked pHLH" rather than "XLP."

In contrast to SAP, XIAP is ubiquitously expressed, rather than being restricted to certain lineages of the hematopoietic system [138]; this likely contributes to the difference in clinical manifestations between XLP1 and XLP2. Similarly to XLP1, patients with XLP2 commonly present with HLH in the context of EBV infection. In addition to pHLH, XLP1 and XLP2 patients can both develop hypogammaglobulinemia (more common in XLP1 than XLP2). On the other hand, XLP2 predisposes to other clinical manifestations that are not common in XLP1; these include chronic colitis (with features of Crohn's disease), and other autoinflammatory symptoms including uveitis, erythema nodosum, and nephritis [140, 141]. Lastly, patients with XLP2 rarely develop neurologic manifestations [139].

XIAP encodes for a 497-amino acid antiapoptotic molecule [142–144], known as baculoviral inhibitor of apoptosis (IAP)-repeat-containing 4 (BIRC4) protein. In addition to containing baculoviral IAP-repeat (BIR) domains, this protein also contains an UBA domain (which allows this protein to bind ubiquitin) and a carboxy terminal RING (Really Interesting New Gene) finger domain with E3 ubiquitin ligase activity [145]. The main function of the XIAP protein is the inhibition of apoptosis via its direct interaction with caspases. Specifically, the BIR domain interacts with and inhibit caspases 3, 7, and 9 [144, 146]. Moreover, BIRC4 is involved in a number of innate and adaptive immune cell and non-immune cell signaling pathways, including that of transforming growth factor-beta receptor (TGFβR), Notch, MAP kinase, c-Jun N-terminal kinase (JNK), and NFκB signaling pathways. However, it is still not fully understood how mutations in *XIAP*, and the disinhibition of apoptosis, give rise to the clinical manifestations of XLP2 [116].

In contrast to XLP1, NK and CTL cell cytotoxic activity is preserved in patients with XLP2. Yet, following T-cell receptor (TCR) activation, T cells from XLP2 patients show heightened activation-induced cell death (AICD) [136]; similarly, there is increased apoptosis to ligation of death receptors such as FAS (CD95) and TRAIL receptor [136, 147]. These findings may explain the less severe HLH manifestations in patients with XLP2, while increased apoptosis in B cells may contribute to the development of hypogammaglobulinemia.

Lysinuric Protein Intolerance (LPI)

Initially described in 1965 as an inborn error of metabolism [148], lysinuric protein intolerance (LPI)I was later shown by Lauteala et al. [149] to be due to biallelic mutations in the *SLC7A7* gene using genome-wide linkage analysis in patients of

Finnish origin. One year later, this group also found similar mutations in chromosome 14 in non-Finnish patients with LPI (the majority of whom were of Italian descent) [150]. Since this time, more than 200 patients have been reported with LPI, with one-third being of Finnish origin [151]. Consistent with this ethnic predisposition, a homozygous founder mutation was detected in Finnish patients by Torrents et al. [152]; this was later shown to be due to a splice acceptor mutation that led to a frameshift and premature termination of translation [153]. In addition to this Finnish mutation, a founder mutation (p.R410∗ mutation) was also described in patients of Japanese origin [154–156]. In a recent study in Japanese patients with LPI, pR410∗ was the most frequent mutation identified [156]. Despite these ethnic associations, no genotype–phenotype correlations have been established thus far in LPI patients [156].

The *SLC7A7* gene encodes for the y(+)L amino acid transporter-1 (y(+) LAT-1) protein [152, 153]. The y(+) LAT-1 protein normally heterodimerizes with the 4F2 heavy chain (4F2hc), which is encoded by *SLC3A2*, in order to form a cationic amino acid exchanger (CAA) [157]. CAA are localized mainly at the basolateral membranes of the tubular kidney and small bowel cells. Mutations in *SLC7A7* gene lead to defects in CAA structures, which result in leakage of arginine, lysine, and ornithine [158].

Clinically, patients with LPI present with variable clinical characteristic; these include gastrointestinal symptoms which can be severe enough to result in failure to thrive [159], urea cycle dysfunction and hyperammonemia stemming from the low levels of arginine and ornithine in hepatocytes [158, 159], and neurological impairment. Correction of hyperammonemia leads to improvement of patient outcomes [160]. On the other hand, protein intolerance, splenomegaly, hepatomegaly, lung disease, kidney failure, and immunological disorders (autoimmunity and HLH) may be attributed to the expression of y(+) LAT-1 in cells of the lung and spleen, as well as hematopoietic cells such as monocytes and macrophages [160].

References

1. Ohadi, M., Lalloz, M. R., Sham, P., Zhao, J., Dearlove, A. M., Shiach, C., et al. (1999). Localization of a gene for familial hemophagocytic lymphohistiocytosis at chromosome 9q21.3-22 by homozygosity mapping. *American Journal of Human Genetics, 64*, 165–171.
2. Fink, T. M., Zimmer, M., Weitz, S., Tschopp, J., Jenne, D. E., & Lichter, P. (1992). Human perforin (PRF1) maps to 10q22, a region that is syntenic with mouse chromosome 10. *Genomics, 13*, 1300–1302.
3. Stepp, S. E., Dufourcq-Lagelouse, R., Le Deist, F., Bhawan, S., Certain, S., Mathew, P. A., et al. (1999). Perforin gene defects in familial hemophagocytic lymphohistiocytosis. *Science, 286*, 1957–1959.
4. Voskoboinik, I., Thia, M. C., De Bono, A., Browne, K., Cretney, E., Jackson, J. T., et al. (2004). The functional basis for hemophagocytic lymphohistiocytosis in a patient with co-inherited missense mutations in the perforin (PFN1) gene. *The Journal of Experimental Medicine, 200*, 811–816.

5. Ishii, E., Ueda, I., Shirakawa, R., Yamamoto, K., Horiuchi, H., Ohga, S., et al. (2005). Genetic subtypes of familial hemophagocytic lymphohistiocytosis: Correlations with clinical features and cytotoxic T lymphocyte/natural killer cell functions. *Blood, 105*, 3442–3448.
6. Molleran Lee, S., Villanueva, J., Sumegi, J., Zhang, K., Kogawa, K., Davis, J., et al. (2004). Characterisation of diverse PRF1 mutations leading to decreased natural killer cell activity in North American families with haemophagocytic lymphohistiocytosis. *Journal of Medical Genetics, 41*, 137–144.
7. Lee, S. M., Sumegi, J., Villanueva, J., Tabata, Y., Zhang, K., Chakraborty, R., et al. (2006). Patients of African ancestry with hemophagocytic lymphohistiocytosis share a common haplotype of PRF1 with a 50delT mutation. *The Journal of Pediatrics, 149*, 134–137.
8. Lichtenheld, M. G., Olsen, K. J., Lu, P., Lowrey, D. M., Hameed, A., Hengartner, H., et al. (1988). Structure and function of human perforin. *Nature, 335*, 448–451.
9. Law, R. H., Lukoyanova, N., Voskoboinik, I., Caradoc-Davies, T. T., Baran, K., Dunstone, M. A., et al. (2010). The structural basis for membrane binding and pore formation by lymphocyte perforin. *Nature, 468*, 447–451.
10. Clementi, R., Emmi, L., Maccario, R., Liotta, F., Moretta, L., Danesino, C., et al. (2002). Adult onset and atypical presentation of hemophagocytic lymphohistiocytosis in siblings carrying PRF1 mutations. *Blood, 100*, 2266–2267.
11. Feldmann, J., Le Deist, F., Ouachee-Chardin, M., Certain, S., Alexander, S., Quartier, P., et al. (2002). Functional consequences of perforin gene mutations in 22 patients with familial haemophagocytic lymphohistiocytosis. *British Journal of Haematology, 117*, 965–972.
12. Zur Stadt, U., Beutel, K., Kolberg, S., Schneppenheim, R., Kabisch, H., Janka, G., et al. (2006). Mutation spectrum in children with primary hemophagocytic lymphohistiocytosis: Molecular and functional analyses of PRF1, UNC13D, STX11, and RAB27A. *Human Mutation, 27*, 62–68.
13. Risma, K. A., Frayer, R. W., Filipovich, A. H., & Sumegi, J. (2006). Aberrant maturation of mutant perforin underlies the clinical diversity of hemophagocytic lymphohistiocytosis. *The Journal of Clinical Investigation, 116*, 182–192.
14. Solomou, E. E., Gibellini, F., Stewart, B., Malide, D., Berg, M., Visconte, V., et al. (2007). Perforin gene mutations in patients with acquired aplastic anemia. *Blood, 109*, 5234–5237.
15. Willenbring, R. C., Ikeda, Y., Pease, L. R., & Johnson, A. J. (2018). Human perforin gene variation is geographically distributed. *Molecular Genetics & Genomic Medicine, 6*, 44–55.
16. Voskoboinik, I., Sutton, V. R., Ciccone, A., House, C. M., Chia, J., Darcy, P. K., et al. (2007). Perforin activity and immune homeostasis: The common A91V polymorphism in perforin results in both presynaptic and postsynaptic defects in function. *Blood, 110*, 1184–1190.
17. Feldmann, J., Callebaut, I., Raposo, G., Certain, S., Bacq, D., Dumont, C., et al. (2003). Munc13-4 is essential for cytolytic granules fusion and is mutated in a form of familial hemophagocytic lymphohistiocytosis (FHL3). *Cell, 115*, 461–473.
18. Koch, H., Hofmann, K., & Brose, N. (2000). Definition of Munc13-homology-domains and characterization of a novel ubiquitously expressed Munc13 isoform. *The Biochemical Journal, 349*, 247–253.
19. Schulert, G. S., Zhang, M., Husami, A., Fall, N., Brunner, H., Zhang, K., et al. (2018). Brief report: Novel UNC13D intronic variant disrupting an NF-kappaB enhancer in a patient with recurrent macrophage activation syndrome and systemic juvenile idiopathic arthritis. *Arthritis & Rhematology, 70*, 963–970.
20. Rudd, E., Bryceson, Y. T., Zheng, C., Edner, J., Wood, S. M., Ramme, K., et al. (2008). Spectrum, and clinical and functional implications of UNC13D mutations in familial haemophagocytic lymphohistiocytosis. *Journal of Medical Genetics, 45*, 134–141.
21. Santoro, A., Cannella, S., Bossi, G., Gallo, F., Trizzino, A., Pende, D., et al. (2006). Novel Munc13-4 mutations in children and young adult patients with haemophagocytic lymphohistiocytosis. *Journal of Medical Genetics, 43*, 953–960.
22. Bechara, E., Dijoud, F., de Saint Basile, G., Bertrand, Y., & Pondarre, C. (2011). Hemophagocytic lymphohistiocytosis with Munc13-4 mutation: A cause of recurrent fatal hydrops fetalis. *Pediatrics, 128*, e251–e254.

23. Sieni, E., Cetica, V., Piccin, A., Gherlinzoni, F., Sasso, F. C., Rabusin, M., et al. (2012). Familial hemophagocytic lymphohistiocytosis may present during adulthood: Clinical and genetic features of a small series. *PLoS One, 7*, e44649.
24. Sieni, E., Cetica, V., Santoro, A., Beutel, K., Mastrodicasa, E., Meeths, M., et al. (2011). Genotype-phenotype study of familial haemophagocytic lymphohistiocytosis type 3. *Journal of Medical Genetics, 48*, 343–352.
25. Neeft, M., Wieffer, M., de Jong, A. S., Negroiu, G., Metz, C. H., van Loon, A., et al. (2005). Munc13-4 is an effector of rab27a and controls secretion of lysosomes in hematopoietic cells. *Molecular Biology of the Cell, 16*, 731–741.
26. zur Stadt, U., Schmidt, S., Kasper, B., Beutel, K., Diler, A. S., Henter, J. I., et al. (2005). Linkage of familial hemophagocytic lymphohistiocytosis (FHL) type-4 to chromosome 6q24 and identification of mutations in syntaxin 11. *Human Molecular Genetics, 14*, 827–834.
27. Bryceson, Y. T., Rudd, E., Zheng, C., Edner, J., Ma, D., Wood, S. M., et al. (2007). Defective cytotoxic lymphocyte degranulation in syntaxin-11 deficient familial hemophagocytic lymphohistiocytosis 4 (FHL4) patients. *Blood, 110*, 1906–1915.
28. Macartney, C. A., Weitzman, S., Wood, S. M., Bansal, D., Steele, M., Meeths, M., et al. (2011). Unusual functional manifestations of a novel STX11 frameshift mutation in two infants with familial hemophagocytic lymphohistiocytosis type 4 (FHL4). *Pediatric Blood & Cancer, 56*, 654–657.
29. Rudd, E., Goransdotter Ericson, K., Zheng, C., Uysal, Z., Ozkan, A., Gurgey, A., et al. (2006). Spectrum and clinical implications of syntaxin 11 gene mutations in familial haemophagocytic lymphohistiocytosis: Association with disease-free remissions and haematopoietic malignancies. *Journal of Medical Genetics, 43*, e14.
30. Muller, M. L., Chiang, S. C., Meeths, M., Tesi, B., Entesarian, M., Nilsson, D., et al. (2014). An N-terminal missense mutation in STX11 causative of FHL4 abrogates syntaxin-11 binding to Munc18-2. *Frontiers in Immunology, 4*, 515.
31. Marsh, R. A., Satake, N., Biroschak, J., Jacobs, T., Johnson, J., Jordan, M. B., et al. (2010). STX11 mutations and clinical phenotypes of familial hemophagocytic lymphohistiocytosis in North America. *Pediatric Blood & Cancer, 55*, 134–140.
32. zur Stadt, U., Rohr, J., Seifert, W., Koch, F., Grieve, S., Pagel, J., et al. (2009). Familial hemophagocytic lymphohistiocytosis type 5 (FHL-5) is caused by mutations in Munc18-2 and impaired binding to syntaxin 11. *American Journal of Human Genetics, 85*, 482–492.
33. Cote, M., Menager, M. M., Burgess, A., Mahlaoui, N., Picard, C., Schaffner, C., et al. (2009). Munc18-2 deficiency causes familial hemophagocytic lymphohistiocytosis type 5 and impairs cytotoxic granule exocytosis in patient NK cells. *The Journal of Clinical Investigation, 119*, 3765–3773.
34. Pagel, J., Beutel, K., Lehmberg, K., Koch, F., Maul-Pavicic, A., Rohlfs, A. K., et al. (2012). Distinct mutations in STXBP2 are associated with variable clinical presentations in patients with familial hemophagocytic lymphohistiocytosis type 5 (FHL5). *Blood, 119*, 6016–6024.
35. Meeths, M., Entesarian, M., Al-Herz, W., Chiang, S. C., Wood, S. M., Al-Ateeqi, W., et al. (2010). Spectrum of clinical presentations in familial hemophagocytic lymphohistiocytosis type 5 patients with mutations in STXBP2. *Blood, 116*, 2635–2643.
36. Menasche, G., Pastural, E., Feldmann, J., Certain, S., Ersoy, F., Dupuis, S., et al. (2000). Mutations in RAB27A cause Griscelli syndrome associated with haemophagocytic syndrome. *Nature Genetics, 25*, 173–176.
37. Chen, D., Guo, J., Miki, T., Tachibana, M., & Gahl, W. A. (1997). Molecular cloning and characterization of rab27a and rab27b, novel human rab proteins shared by melanocytes and platelets. *Biochemical and Molecular Medicine, 60*, 27–37.
38. Griscelli, C., Durandy, A., Guy-Grand, D., Daguillard, F., Herzog, C., & Prunieras, M. (1978). A syndrome associating partial albinism and immunodeficiency. *The American Journal of Medicine, 65*, 691–702.
39. Menasche, G., Ho, C. H., Sanal, O., Feldmann, J., Tezcan, I., Ersoy, F., et al. (2003). Griscelli syndrome restricted to hypopigmentation results from a melanophilin defect (GS3) or a MYO5A F-exon deletion (GS1). *The Journal of Clinical Investigation, 112*, 450–456.

40. Pastural, E., Ersoy, F., Yalman, N., Wulffraat, N., Grillo, E., Ozkinay, F., et al. (2000). Two genes are responsible for Griscelli syndrome at the same 15q21 locus. *Genomics, 63*, 299–306.
41. Pastural, E., Barrat, F. J., Dufourcq-Lagelouse, R., Certain, S., Sanal, O., Jabado, N., et al. (1997). Griscelli disease maps to chromosome 15q21 and is associated with mutations in the myosin-Va gene. *Nature Genetics, 16*, 289–292.
42. Elstak, E. D., Neeft, M., Nehme, N. T., Voortman, J., Cheung, M., Goodarzifard, M., et al. (2011). The munc13-4-rab27 complex is specifically required for tethering secretory lysosomes at the plasma membrane. *Blood, 118*, 1570–1578.
43. Menager, M. M., Menasche, G., Romao, M., Knapnougel, P., Ho, C. H., Garfa, M., et al. (2007). Secretory cytotoxic granule maturation and exocytosis require the effector protein hMunc13-4. *Nature Immunology, 8*, 257–267.
44. Zhang, M., Bracaglia, C., Prencipe, G., Bemrich-Stolz, C. J., Beukelman, T., Dimmitt, R. A., et al. (2016). A heterozygous RAB27A mutation associated with delayed cytolytic granule polarization and hemophagocytic lymphohistiocytosis. *Journal of Immunology, 196*, 2492–2503.
45. Nagle, D. L., Karim, M. A., Woolf, E. A., Holmgren, L., Bork, P., Misumi, D. J., et al. (1996). Identification and mutation analysis of the complete gene for Chediak-Higashi syndrome. *Nature Genetics, 14*, 307–311.
46. Barbosa, M. D., Nguyen, Q. A., Tchernev, V. T., Ashley, J. A., Detter, J. C., Blaydes, S. M., et al. (1996). Identification of the homologous beige and Chediak-Higashi syndrome genes. *Nature, 382*, 262–265.
47. Martens, S., & McMahon, H. T. (2008). Mechanisms of membrane fusion: Disparate players and common principles. *Nature Reviews. Molecular Cell Biology, 9*, 543–556.
48. Spritz, R. A. (1998). Genetic defects in Chediak-Higashi syndrome and the beige mouse. *Journal of Clinical Immunology, 18*, 97–105.
49. Gil-Krzewska, A., Wood, S. M., Murakami, Y., Nguyen, V., Chiang, S. C. C., Cullinane, A. R., et al. (2016). Chediak-Higashi syndrome: Lysosomal trafficking regulator domains regulate exocytosis of lytic granules but not cytokine secretion by natural killer cells. *The Journal of Allergy and Clinical Immunology, 137*, 1165–1177.
50. Gebauer, D., Li, J., Jogl, G., Shen, Y., Myszka, D. G., & Tong, L. (2004). Crystal structure of the PH-BEACH domains of human LRBA/BGL. *Biochemistry, 43*, 14873–14880.
51. Karim, M. A., Suzuki, K., Fukai, K., Oh, J., Nagle, D. L., Moore, K. J., et al. (2002). Apparent genotype-phenotype correlation in childhood, adolescent, and adult Chediak-Higashi syndrome. *American Journal of Medical Genetics, 108*, 16–22.
52. Bailleul-Forestier, I., Monod-Broca, J., Benkerrou, M., Mora, F., & Picard, B. (2008). Generalized periodontitis associated with Chediak-Higashi syndrome. *Journal of Periodontology, 79*, 1263–1270.
53. Ishii, E., Matui, T., Iida, M., Inamitu, T., & Ueda, K. (1987). Chediak-Higashi syndrome with intestinal complication. Report of a case. *Journal of Clinical Gastroenterology, 9*, 556–558.
54. Dell'Angelica, E. C., Shotelersuk, V., Aguilar, R. C., Gahl, W. A., & Bonifacino, J. S. (1999). Altered trafficking of lysosomal proteins in Hermansky-Pudlak syndrome due to mutations in the beta 3A subunit of the AP-3 adaptor. *Molecular Cell, 3*, 11–21.
55. Santiago Borrero, P. J., Rodriguez-Perez, Y., Renta, J. Y., Izquierdo, N. J., Del Fierro, L., Munoz, D., et al. (2006). Genetic testing for oculocutaneous albinism type 1 and 2 and Hermansky-Pudlak syndrome type 1 and 3 mutations in Puerto Rico. *The Journal of Investigative Dermatology, 126*, 85–90.
56. Gahl, W. A., Brantly, M., Kaiser-Kupfer, M. I., Iwata, F., Hazelwood, S., Shotelersuk, V., et al. (1998). Genetic defects and clinical characteristics of patients with a form of oculocutaneous albinism (Hermansky-Pudlak syndrome). *The New England Journal of Medicine, 338*, 1258–1264.

57. Jung, J., Bohn, G., Allroth, A., Boztug, K., Brandes, G., Sandrock, I., et al. (2006). Identification of a homozygous deletion in the AP3B1 gene causing Hermansky-Pudlak syndrome, type 2. *Blood, 108*, 362–369.
58. Hermansky, F., & Pudlak, P. (1959). Albinism associated with hemorrhagic diathesis and unusual pigmented reticular cells in the bone marrow: Report of two cases with histochemical studies. *Blood, 14*, 162–169.
59. Huizing, M., & Gahl, W. A. (2002). Disorders of vesicles of lysosomal lineage: The Hermansky-Pudlak syndromes. *Current Molecular Medicine, 2*, 451–467.
60. Shotelersuk, V., Dell'Angelica, E. C., Hartnell, L., Bonifacino, J. S., & Gahl, W. A. (2000). A new variant of Hermansky-Pudlak syndrome due to mutations in a gene responsible for vesicle formation. *The American Journal of Medicine, 108*, 423–427.
61. Seward Jr., S. L., & Gahl, W. A. (2013). Hermansky-Pudlak syndrome: Health care throughout life. *Pediatrics, 132*, 153–160.
62. Dell'Angelica, E. C., Ohno, H., Ooi, C. E., Rabinovich, E., Roche, K. W., & Bonifacino, J. S. (1997). AP-3: An adaptor-like protein complex with ubiquitous expression. *The EMBO Journal, 16*, 917–928.
63. Dell'Angelica, E. C., Ooi, C. E., & Bonifacino, J. S. (1997). Beta3A-adaptin, a subunit of the adaptor-like complex AP-3. *The Journal of Biological Chemistry, 272*, 15078–15084.
64. Kotzot, D., Richter, K., & Gierth-Fiebig, K. (1994). Oculocutaneous albinism, immunodeficiency, hematological disorders, and minor anomalies: A new autosomal recessive syndrome? *American Journal of Medical Genetics, 50*, 224–227.
65. Clark, R. H., Stinchcombe, J. C., Day, A., Blott, E., Booth, S., Bossi, G., et al. (2003). Adaptor protein 3-dependent microtubule-mediated movement of lytic granules to the immunological synapse. *Nature Immunology, 4*, 1111–1120.
66. Enders, A., Zieger, B., Schwarz, K., Yoshimi, A., Speckmann, C., Knoepfle, E. M., et al. (2006). Lethal hemophagocytic lymphohistiocytosis in Hermansky-Pudlak syndrome type II. *Blood, 108*, 81–87.
67. Ishii, E. (2016). Hemophagocytic lymphohistiocytosis in children: Pathogenesis and treatment. *Frontiers in Pediatrics, 4*, 47.
68. Moss, D. J., Burrows, S. R., Khanna, R., Misko, I. S., & Sculley, T. B. (1992). Immune surveillance against Epstein-Barr virus. *Seminars in Immunology, 4*, 97–104.
69. Latour, S., & Winter, S. (2018). Inherited immunodeficiencies with high predisposition to Epstein-Barr virus-driven lymphoproliferative diseases. *Frontiers in Immunology, 9*, 1103.
70. Huck, K., Feyen, O., Niehues, T., Ruschendorf, F., Hubner, N., Laws, H. J., et al. (2009). Girls homozygous for an IL-2-inducible T cell kinase mutation that leads to protein deficiency develop fatal EBV-associated lymphoproliferation. *The Journal of Clinical Investigation, 119*, 1350–1358.
71. Linka, R. M., Risse, S. L., Bienemann, K., Werner, M., Linka, Y., Krux, F., et al. (2012). Loss-of-function mutations within the IL-2 inducible kinase ITK in patients with EBV-associated lymphoproliferative diseases. *Leukemia, 26*, 963–971.
72. Mansouri, D., Mahdaviani, S. A., Khalilzadeh, S., Mohajerani, S. A., Hasanzad, M., Sadr, S., et al. (2012). IL-2-inducible T-cell kinase deficiency with pulmonary manifestations due to disseminated Epstein-Barr virus infection. *International Archives of Allergy and Immunology, 158*, 418–422.
73. Serwas, N. K., Cagdas, D., Ban, S. A., Bienemann, K., Salzer, E., Tezcan, I., et al. (2014). Identification of ITK deficiency as a novel genetic cause of idiopathic CD4+ T-cell lymphopenia. *Blood, 124*, 655–657.
74. Cipe, F. E., Aydogmus, C., Serwas, N. K., Tugcu, D., Demirkaya, M., Bicici, F. A., et al. (2015). ITK deficiency: How can EBV be treated before lymphoma? *Pediatric Blood & Cancer, 62*, 2247–2248.
75. Cagdas, D., Erman, B., Hanoglu, D., Tavil, B., Kuskonmaz, B., Aydin, B., et al. (2017). Course of IL-2-inducible T-cell kinase deficiency in a family: Lymphomatoid granulomato-

sis, lymphoma and allogeneic bone marrow transplantation in one sibling; and death in the other. *Bone Marrow Transplantation, 52*, 126–129.
76. Shadur, B., Abuzaitoun, O., NaserEddin, A., Even-Or, E., Zaidman, I., & Stepensky, P. (2019). Management of XLP-1 and ITK deficiency: The challenges posed by PID with an unpredictable spectrum of disease manifestations. *Clinical Immunology, 198*, 39–45.
77. Andreotti, A. H., Schwartzberg, P. L., Joseph, R. E., & Berg, L. J. (2010). T-cell signaling regulated by the Tec family kinase, Itk. *Cold Spring Harbor Perspectives in Biology, 2*, a002287.
78. Liu, K. Q., Bunnell, S. C., Gurniak, C. B., & Berg, L. J. (1998). T cell receptor-initiated calcium release is uncoupled from capacitative calcium entry in Itk-deficient T cells. *The Journal of Experimental Medicine, 187*, 1721–1727.
79. Kapnick, S. M., Stinchcombe, J. C., Griffiths, G. M., & Schwartzberg, P. L. (2017). Inducible T cell kinase regulates the acquisition of cytolytic capacity and degranulation in CD8(+) CTLs. *Journal of Immunology, 198*, 2699–2711.
80. Ochs, H. D., & Thrasher, A. J. (2006). The Wiskott-Aldrich syndrome. *The Journal of Allergy and Clinical Immunology, 117*, 725–738. quiz 739.
81. Moshous, D., Martin, E., Carpentier, W., Lim, A., Callebaut, I., Canioni, D., et al. (2013). Whole-exome sequencing identifies Coronin-1A deficiency in 3 siblings with immunodeficiency and EBV-associated B-cell lymphoproliferation. *The Journal of Allergy and Clinical Immunology, 131*, 1594–1603.
82. Nehme, N. T., Schmid, J. P., Debeurme, F., Andre-Schmutz, I., Lim, A., Nitschke, P., et al. (2012). MST1 mutations in autosomal recessive primary immunodeficiency characterized by defective naive T-cell survival. *Blood, 119*, 3458–3468.
83. Moshous, D., Pannetier, C., Chasseval Rd, R., Deist Fl, F., Cavazzana-Calvo, M., Romana, S., et al. (2003). Partial T and B lymphocyte immunodeficiency and predisposition to lymphoma in patients with hypomorphic mutations in Artemis. *The Journal of Clinical Investigation, 111*, 381–387.
84. Helminen, M. E., Kilpinen, S., Virta, M., & Hurme, M. (2001). Susceptibility to primary Epstein-Barr virus infection is associated with interleukin-10 gene promoter polymorphism. *The Journal of Infectious Diseases, 184*, 777–780.
85. Salzer, E., Daschkey, S., Choo, S., Gombert, M., Santos-Valente, E., Ginzel, S., et al. (2013). Combined immunodeficiency with life-threatening EBV-associated lymphoproliferative disorder in patients lacking functional CD27. *Haematologica, 98*, 473–478.
86. Seidel, M. G. (2012). CD27: A new player in the field of common variable immunodeficiency and EBV-associated lymphoproliferative disorder? *The Journal of Allergy and Clinical Immunology, 129*, 1175. author reply 1175-1176.
87. Alkhairy, O. K., Perez-Becker, R., Driessen, G. J., Abolhassani, H., van Montfrans, J., Borte, S., et al. (2015). Novel mutations in TNFRSF7/CD27: Clinical, immunologic, and genetic characterization of human CD27 deficiency. *The Journal of Allergy and Clinical Immunology, 136*, 703–712.e710.
88. Usui, M., & Sakai, J. (1990). Three cases of EB virus-associated uveitis. *International Ophthalmology, 14*, 371–376.
89. van Montfrans, J. M., Hoepelman, A. I., Otto, S., van Gijn, M., van de Corput, L., de Weger, R. A., et al. (2012). CD27 deficiency is associated with combined immunodeficiency and persistent symptomatic EBV viremia. *The Journal of Allergy and Clinical Immunology, 129*, 787–793.e786.
90. Hendriks, J., Gravestein, L. A., Tesselaar, K., van Lier, R. A., Schumacher, T. N., & Borst, J. (2000). CD27 is required for generation and long-term maintenance of T cell immunity. *Nature Immunology, 1*, 433–440.
91. Peperzak, V., Veraar, E. A., Keller, A. M., Xiao, Y., & Borst, J. (2010). The Pim kinase pathway contributes to survival signaling in primed CD8+ T cells upon CD27 costimulation. *Journal of Immunology, 185*, 6670–6678.

92. Schildknecht, A., Miescher, I., Yagita, H., & van den Broek, M. (2007). Priming of CD8+ T cell responses by pathogens typically depends on CD70-mediated interactions with dendritic cells. *European Journal of Immunology, 37*, 716–728.
93. Li, F. Y., Chaigne-Delalande, B., Kanellopoulou, C., Davis, J. C., Matthews, H. F., Douek, D. C., et al. (2011). Second messenger role for Mg2+ revealed by human T-cell immunodeficiency. *Nature, 475*, 471–476.
94. Ravell, J., Chaigne-Delalande, B., & Lenardo, M. (2014). X-linked immunodeficiency with magnesium defect, Epstein-Barr virus infection, and neoplasia disease: A combined immune deficiency with magnesium defect. *Current Opinion in Pediatrics, 26*, 713–719.
95. Dhalla, F., Murray, S., Sadler, R., Chaigne-Delalande, B., Sadaoka, T., Soilleux, E., et al. (2015). Identification of a novel mutation in MAGT1 and progressive multifocal leucoencephalopathy in a 58-year-old man with XMEN disease. *Journal of Clinical Immunology, 35*, 112–118.
96. Trapani, V., Shomer, N., & Rajcan-Separovic, E. (2015). The role of MAGT1 in genetic syndromes. *Magnesium Research, 28*, 46–55.
97. Brigida, I., Chiriaco, M., Di Cesare, S., Cittaro, D., Di Matteo, G., Giannelli, S., et al. (2017). Large deletion of MAGT1 gene in a patient with classic kaposi sarcoma, CD4 lymphopenia, and EBV infection. *Journal of Clinical Immunology, 37*, 32–35.
98. Qiao, Y., Mondal, K., Trapani, V., Wen, J., Carpenter, G., Wildin, R., et al. (2014). Variant ATRX syndrome with dysfunction of ATRX and MAGT1 genes. *Human Mutation, 35*, 58–62.
99. Molinari, F., Foulquier, F., Tarpey, P. S., Morelle, W., Boissel, S., Teague, J., et al. (2008). Oligosaccharyltransferase-subunit mutations in nonsyndromic mental retardation. *American Journal of Human Genetics, 82*, 1150–1157.
100. Piton, A., Redin, C., & Mandel, J. L. (2013). XLID-causing mutations and associated genes challenged in light of data from large-scale human exome sequencing. *American Journal of Human Genetics, 93*, 368–383.
101. Bashyam, M. D., Bair, R., Kim, Y. H., Wang, P., Hernandez-Boussard, T., Karikari, C. A., et al. (2005). Array-based comparative genomic hybridization identifies localized DNA amplifications and homozygous deletions in pancreatic cancer. *Neoplasia, 7*, 556–562.
102. Quamme, G. A. (2010). Molecular identification of ancient and modern mammalian magnesium transporters. *American Journal of Physiology. Cell Physiology, 298*, C407–C429.
103. Chaigne-Delalande, B., Li, F. Y., O'Connor, G. M., Lukacs, M. J., Jiang, P., Zheng, L., et al. (2013). Mg2+ regulates cytotoxic functions of NK and CD8 T cells in chronic EBV infection through NKG2D. *Science, 341*, 186–191.
104. Pappworth, I. Y., Wang, E. C., & Rowe, M. (2007). The switch from latent to productive infection in epstein-barr virus-infected B cells is associated with sensitization to NK cell killing. *Journal of Virology, 81*, 474–482.
105. Gotru, S. K., Gil-Pulido, J., Beyersdorf, N., Diefenbach, A., Becker, I. C., Vogtle, T., et al. (2018). Cutting edge: Imbalanced cation homeostasis in MAGT1-deficient B cells dysregulates B cell development and signaling in mice. *Journal of Immunology, 200*, 2529–2534.
106. Purtilo, D. T., DeFlorio Jr., D., Hutt, L. M., Bhawan, J., Yang, J. P., Otto, R., et al. (1977). Variable phenotypic expression of an X-linked recessive lymphoproliferative syndrome. *The New England Journal of Medicine, 297*, 1077–1080.
107. Nichols, K. E., Harkin, D. P., Levitz, S., Krainer, M., Kolquist, K. A., Genovese, C., et al. (1998). Inactivating mutations in an SH2 domain-encoding gene in X-linked lymphoproliferative syndrome. *Proceedings of the National Academy of Sciences of the United States of America, 95*, 13765–13770.
108. Coffey, A. J., Brooksbank, R. A., Brandau, O., Oohashi, T., Howell, G. R., Bye, J. M., et al. (1998). Host response to EBV infection in X-linked lymphoproliferative disease results from mutations in an SH2-domain encoding gene. *Nature Genetics, 20*, 129–135.
109. Sayos, J., Wu, C., Morra, M., Wang, N., Zhang, X., Allen, D., et al. (1998). The X-linked lymphoproliferative-disease gene product SAP regulates signals induced through the co-receptor SLAM. *Nature, 395*, 462–469.

110. Tangye, S. G. (2014). XLP: Clinical features and molecular etiology due to mutations in SH2D1A encoding SAP. *Journal of Clinical Immunology, 34*, 772–779.
111. Sumegi, J., Huang, D., Lanyi, A., Davis, J. D., Seemayer, T. A., Maeda, A., et al. (2000). Correlation of mutations of the SH2D1A gene and epstein-barr virus infection with clinical phenotype and outcome in X-linked lymphoproliferative disease. *Blood, 96*, 3118–3125.
112. Nichols, K. E., Ma, C. S., Cannons, J. L., Schwartzberg, P. L., & Tangye, S. G. (2005). Molecular and cellular pathogenesis of X-linked lymphoproliferative disease. *Immunological Reviews, 203*, 180–199.
113. Sumegi, J., Seemayer, T. A., Huang, D., Davis, J. R., Morra, M., Gross, T. G., et al. (2002). A spectrum of mutations in SH2D1A that causes X-linked lymphoproliferative disease and other Epstein-Barr virus-associated illnesses. *Leukemia & Lymphoma, 43*, 1189–1201.
114. Woon, S. T., Ameratunga, R., Croxson, M., Taylor, G., Neas, K., Edkins, E., et al. (2008). Follicular lymphoma in a X-linked lymphoproliferative syndrome carrier female. *Scandinavian Journal of Immunology, 68*, 153–158.
115. Wada, T., & Candotti, F. (2008). Somatic mosaicism in primary immune deficiencies. *Current Opinion in Allergy and Clinical Immunology, 8*, 510–514.
116. Yang, X., Miyawaki, T., & Kanegane, H. (2012). SAP and XIAP deficiency in hemophagocytic lymphohistiocytosis. *Pediatrics International, 54*, 447–454.
117. Booth, C., Gilmour, K. C., Veys, P., Gennery, A. R., Slatter, M. A., Chapel, H., et al. (2011). X-linked lymphoproliferative disease due to SAP/SH2D1A deficiency: A multicenter study on the manifestations, management and outcome of the disease. *Blood, 117*, 53–62.
118. Brandau, O., Schuster, V., Weiss, M., Hellebrand, H., Fink, F. M., Kreczy, A., et al. (1999). Epstein-Barr virus-negative boys with non-Hodgkin lymphoma are mutated in the SH2D1A gene, as are patients with X-linked lymphoproliferative disease (XLP). *Human Molecular Genetics, 8*, 2407–2413.
119. Gaspar, H. B., Sharifi, R., Gilmour, K. C., & Thrasher, A. J. (2002). X-linked lymphoproliferative disease: Clinical, diagnostic and molecular perspective. *British Journal of Haematology, 119*, 585–595.
120. Seemayer, T. A., Gross, T. G., Egeler, R. M., Pirruccello, S. J., Davis, J. R., Kelly, C. M., et al. (1995). X-linked lymphoproliferative disease: Twenty-five years after the discovery. *Pediatric Research, 38*, 471–478.
121. Ma, C. S., Nichols, K. E., & Tangye, S. G. (2007). Regulation of cellular and humoral immune responses by the SLAM and SAP families of molecules. *Annual Review of Immunology, 25*, 337–379.
122. Chan, B., Lanyi, A., Song, H. K., Griesbach, J., Simarro-Grande, M., Poy, F., et al. (2003). SAP couples Fyn to SLAM immune receptors. *Nature Cell Biology, 5*, 155–160.
123. Garni-Wagner, B. A., Purohit, A., Mathew, P. A., Bennett, M., & Kumar, V. (1993). A novel function-associated molecule related to non-MHC-restricted cytotoxicity mediated by activated natural killer cells and T cells. *Journal of Immunology, 151*, 60–70.
124. Mathew, P. A., Garni-Wagner, B. A., Land, K., Takashima, A., Stoneman, E., Bennett, M., et al. (1993). Cloning and characterization of the 2B4 gene encoding a molecule associated with non-MHC-restricted killing mediated by activated natural killer cells and T cells. *Journal of Immunology, 151*, 5328–5337.
125. Bassiri, H., Janice Yeo, W. C., Rothman, J., Koretzky, G. A., & Nichols, K. E. (2008). X-linked lymphoproliferative disease (XLP): A model of impaired anti-viral, anti-tumor and humoral immune responses. *Immunologic Research, 42*, 145–159.
126. Cannons, J. L., Tangye, S. G., & Schwartzberg, P. L. (2011). SLAM family receptors and SAP adaptors in immunity. *Annual Review of Immunology, 29*, 665–705.
127. Dupre, L., Andolfi, G., Tangye, S. G., Clementi, R., Locatelli, F., Arico, M., et al. (2005). SAP controls the cytolytic activity of CD8+ T cells against EBV-infected cells. *Blood, 105*, 4383–4389.
128. Das, R., Bassiri, H., Guan, P., Wiener, S., Banerjee, P. P., Zhong, M. C., et al. (2013). The adaptor molecule SAP plays essential roles during invariant NKT cell cytotoxicity and lytic synapse formation. *Blood, 121*, 3386–3395.

129. Engel, P., Eck, M. J., & Terhorst, C. (2003). The SAP and SLAM families in immune responses and X-linked lymphoproliferative disease. *Nature Reviews. Immunology, 3*, 813–821.
130. Grierson, H. L., Skare, J., Hawk, J., Pauza, M., & Purtilo, D. T. (1991). Immunoglobulin class and subclass deficiencies prior to Epstein-Barr virus infection in males with X-linked lymphoproliferative disease. *American Journal of Medical Genetics, 40*, 294–297.
131. Tangye, S. G., Ma, C. S., Brink, R., & Deenick, E. K. (2013). The good, the bad and the ugly—TFH cells in human health and disease. *Nature Reviews. Immunology, 13*, 412–426.
132. Nagy, N., Matskova, L., Kis, L. L., Hellman, U., Klein, G., & Klein, E. (2009). The proapoptotic function of SAP provides a clue to the clinical picture of X-linked lymphoproliferative disease. *Proceedings of the National Academy of Sciences of the United States of America, 106*, 11966–11971.
133. Snow, A. L., Marsh, R. A., Krummey, S. M., Roehrs, P., Young, L. R., Zhang, K., et al. (2009). Restimulation-induced apoptosis of T cells is impaired in patients with X-linked lymphoproliferative disease caused by SAP deficiency. *The Journal of Clinical Investigation, 119*, 2976–2989.
134. Katz, G., Krummey, S. M., Larsen, S. E., Stinson, J. R., & Snow, A. L. (2014). SAP facilitates recruitment and activation of LCK at NTB-A receptors during restimulation-induced cell death. *Journal of Immunology, 192*, 4202–4209.
135. Chen, G., Tai, A. K., Lin, M., Chang, F., Terhorst, C., & Huber, B. T. (2007). Increased proliferation of CD8+ T cells in SAP-deficient mice is associated with impaired activation-induced cell death. *European Journal of Immunology, 37*, 663–674.
136. Rigaud, S., Fondaneche, M. C., Lambert, N., Pasquier, B., Mateo, V., Soulas, P., et al. (2006). XIAP deficiency in humans causes an X-linked lymphoproliferative syndrome. *Nature, 444*, 110–114.
137. Aguilar, C., & Latour, S. (2015). X-linked inhibitor of apoptosis protein deficiency: More than an X-linked lymphoproliferative syndrome. *Journal of Clinical Immunology, 35*, 331–338.
138. Filipovich, A. H., Zhang, K., Snow, A. L., & Marsh, R. A. (2010). X-linked lymphoproliferative syndromes: Brothers or distant cousins? *Blood, 116*, 3398–3408.
139. Pachlopnik Schmid, J., Canioni, D., Moshous, D., Touzot, F., Mahlaoui, N., Hauck, F., et al. (2011). Clinical similarities and differences of patients with X-linked lymphoproliferative syndrome type 1 (XLP-1/SAP deficiency) versus type 2 (XLP-2/XIAP deficiency). *Blood, 117*, 1522–1529.
140. Speckmann, C., Lehmberg, K., Albert, M. H., Damgaard, R. B., Fritsch, M., Gyrd-Hansen, M., et al. (2013). X-linked inhibitor of apoptosis (XIAP) deficiency: The spectrum of presenting manifestations beyond hemophagocytic lymphohistiocytosis. *Clinical Immunology, 149*, 133–141.
141. Aguilar, C., Lenoir, C., Lambert, N., Begue, B., Brousse, N., Canioni, D., et al. (2014). Characterization of Crohn disease in X-linked inhibitor of apoptosis-deficient male patients and female symptomatic carriers. *The Journal of Allergy and Clinical Immunology, 134*, 1131–1141.e1139.
142. Uren, A. G., Pakusch, M., Hawkins, C. J., Puls, K. L., & Vaux, D. L. (1996). Cloning and expression of apoptosis inhibitory protein homologs that function to inhibit apoptosis and/or bind tumor necrosis factor receptor-associated factors. *Proceedings of the National Academy of Sciences of the United States of America, 93*, 4974–4978.
143. Liston, P., Roy, N., Tamai, K., Lefebvre, C., Baird, S., Cherton-Horvat, G., et al. (1996). Suppression of apoptosis in mammalian cells by NAIP and a related family of IAP genes. *Nature, 379*, 349–353.
144. Duckett, C. S., Nava, V. E., Gedrich, R. W., Clem, R. J., Van Dongen, J. L., Gilfillan, M. C., et al. (1996). A conserved family of cellular genes related to the baculovirus iap gene and encoding apoptosis inhibitors. *The EMBO Journal, 15*, 2685–2694.
145. Bassiri, H., & Nichols, K. E. (2012). X-linked lymphoproliferative disease (XLP). *Atlas of Genetics and Cytogenetics in Oncology and Haematology, 16*, 685–688.
146. Holcik, M., Gibson, H., & Korneluk, R. G. (2001). XIAP: Apoptotic brake and promising therapeutic target. *Apoptosis, 6*, 253–261.

147. Schimmer, A. D., Dalili, S., Batey, R. A., & Riedl, S. J. (2006). Targeting XIAP for the treatment of malignancy. *Cell Death and Differentiation, 13*, 179–188.
148. Perheentupa, J., & Visakorpi, J. K. (1965). Protein intolerance with deficient transport of basic aminoacids. Another inborn error of metabolism. *Lancet, 2*, 813–816.
149. Lauteala, T., Sistonen, P., Savontaus, M. L., Mykkanen, J., Simell, J., Lukkarinen, M., et al. (1997). Lysinuric protein intolerance (LPI) gene maps to the long arm of chromosome 14. *American Journal of Human Genetics, 60*, 1479–1486.
150. Lauteala, T., Mykkanen, J., Sperandeo, M. P., Gasparini, P., Savontaus, M. L., Simell, O., et al. (1998). Genetic homogeneity of lysinuric protein intolerance. *European Journal of Human Genetics, 6*, 612–615.
151. Sperandeo, M. P., Andria, G., & Sebastio, G. (2008). Lysinuric protein intolerance: Update and extended mutation analysis of the SLC7A7 gene. *Human Mutation, 29*, 14–21.
152. Torrents, D., Mykkanen, J., Pineda, M., Feliubadalo, L., Estevez, R., de Cid, R., et al. (1999). Identification of SLC7A7, encoding y+LAT-1, as the lysinuric protein intolerance gene. *Nature Genetics, 21*, 293–296.
153. Borsani, G., Bassi, M. T., Sperandeo, M. P., De Grandi, A., Buoninconti, A., Riboni, M., et al. (1999). SLC7A7, encoding a putative permease-related protein, is mutated in patients with lysinuric protein intolerance. *Nature Genetics, 21*, 297–301.
154. Noguchi, A., Shoji, Y., Koizumi, A., Takahashi, T., Matsumori, M., Kayo, T., et al. (2000). SLC7A7 genomic structure and novel variants in three Japanese lysinuric protein intolerance families. *Human Mutation, 15*, 367–372.
155. Koizumi, A., Shoji, Y., Nozaki, J., Noguchi, A., E, X., Dakeishi, M., et al. (2000). A cluster of lysinuric protein intolerance (LPI) patients in a northern part of Iwate, Japan due to a founder effect. The Mass Screening Group. *Human Mutation, 16*, 270–271
156. Noguchi, A., Nakamura, K., Murayama, K., Yamamoto, S., Komatsu, H., Kizu, R., et al. (2016). Clinical and genetic features of lysinuric protein intolerance in Japan. *Pediatrics International, 58*, 979–983.
157. Fotiadis, D., Kanai, Y., & Palacin, M. (2013). The SLC3 and SLC7 families of amino acid transporters. *Molecular Aspects of Medicine, 34*, 139–158.
158. Ogier de Baulny, H., Schiff, M., & Dionisi-Vici, C. (2012). Lysinuric protein intolerance (LPI): A multi organ disease by far more complex than a classic urea cycle disorder. *Molecular Genetics and Metabolism, 106*, 12–17.
159. Sebastio, G., Sperandeo, M. P., & Andria, G. (2011). Lysinuric protein intolerance: Reviewing concepts on a multisystem disease. *American Journal of Medical Genetics. Part C, Seminars in Medical Genetics, 157C*, 54–62.
160. Mauhin, W., Habarou, F., Gobin, S., Servais, A., Brassier, A., Grisel, C., et al. (2017). Update on lysinuric protein intolerance, a multi-faceted disease retrospective cohort analysis from birth to adulthood. *Orphanet Journal of Rare Diseases, 12*, 3.

Genetics of Acquired Cytokine Storm Syndromes

Grant S. Schulert and Kejian Zhang

Introduction

Secondary hemophagocytic lymphohistiocytosis (sHLH) is typically defined as HLH occurring in the setting of triggers leading to strong immunological activation, without known genetic predilection [1, 2]. This is in contrast to primary or familial HLH (pHLH), which is caused by defined genetic mutations affecting lymphocyte cytotoxic functions (see chapter "Genetics of Primary Hemophagocytic Lymphohistiocytosis"). Secondary HLH can occur in the setting of numerous severe infections, rheumatic disorders, and various malignancies, most notably lymphoma. However, these conditions are relatively common compared to the incidence of HLH, and many patients with even severe manifestations of these do not develop sHLH. Indeed, this suggests there may exist underlying genetic factors which may synergize with these disease and/or environmental triggers, leading to sHLH. Here, we review reported genetic contributions to the various forms of sHLH in adults and children (Table 1). Finally, we use these findings to discuss how the consequences of this emerging understanding of HLH genetics may support reexamining the primary vs. secondary dichotomy.

G. S. Schulert
Division of Rheumatology, University of Cincinnati, College of Medicine, Cincinnati, OH, USA

K. Zhang (✉)
Division of Human Genetics, Children's Hospital Medical Center, University of Cincinnati, College of Medicine, Cincinnati, OH, USA
e-mail: kejian.zhang@cchmc.org

R. Q. Cron, E. M. Behrens (eds.), *Cytokine Storm Syndrome*,
https://doi.org/10.1007/978-3-030-22094-5_7

Table 1 Genes associated with sHLH and/or MAS

Gene function/pathway	Gene	Function	Gene function alteration	sHLH triggers	References
Granule-mediated cytolytic pathway	*PRF1*	Pore formation	Decreased/absent	Inf, AIF, AI, Mal	[22, 30, 57, 96]
	UNC13D	Granule priming	Decreased/absent	AIF, AI, Mal	[43, 65, 66, 97, 98]
	STX11	Granule fusion	Decreased/absent	AIF, Mal	[43, 56, 69, 98]
	STXBP2	Granule fusion	Decreased/absent	AIF, Mal	[43, 56, 98]
	Rab27a	Granule docking	Decreased/absent	AIF	[34]
	LYST	Granule trafficking	Decreased/absent	Inf, AIF	[22, 98]
Microtubule organization	*CCDC141*	Migration	Unknown	AIF	[98]
	MICAL2	Actin depolymerization	Unknown	AIF	[98]
	ARHGAP21	Regulates actin dynamics	Unknown	AIF	[98]
	XIRP2	Actin cytoskeleton stabilization	Unknown	Inf, AIF	[22, 98]
Cytokine production/signaling pathway	*TGFB*	Immunoregulation	Unknown	Inf	[10]
	IFNGR1	IFNγ receptor	Decreased/absent	Inf	[99]
	IFNGR2	IFNγ receptor	Decreased/absent	Inf	[99]
	IL-10	Immunoregulation	Unknown	Inf	[100]
	MEFV	Pyrin inflammasome, cytokine production	Unknown	Inf, AI, AIF	[22, 33, 34]
	NLRC4	NOD-like receptor, cytokine production	Activated	AIF	[48, 49]
	IRF5	Interferon regulatory factor transcription factor	Unknown	Inf, AI	[9]
	CADPS2	Regulation of exocytosis	Unknown	AIF	[98]
	FKBPL	Immunoregulation and cell cycle control	Unknown	AIF	[98]
	GDI1	Vesicular trafficking between organelles	Unknown	AIF	[98]
	FAM160A2	FTS/Hook/FHIP complex	Unknown	AIF	[98]

NK cell receptors	*KIR2DS5*	Immunoglobulin-like receptor	Unknown	Inf	[101]
	KIR3DS1	Immunoglobulin-like receptor	Unknown	Inf	[101]
Cell signaling	*ALK*	Receptor tyrosine kinase	Activating	Mal	[73, 74]
	SH2D1A	Signaling in T and NK cells	Decreased/absent	Inf	[11]
	XIAP	Apoptotic suppressor protein	Decreased/absent	Inf	[11]
	CD27	TNF-receptor superfamily	Decreased/absent	Inf	[13]
	CD70	CD27 ligand	Decreased/absent	Inf	[14]
	MAGT1	Magnesium transporter, N-glycosylation	Decreased/absent	Inf	[15]
	ITK	Intracellular tyrosine kinase in T cells	Decreased/absent	Inf, Mal	[12]
Gene expression/transcriptional regulation	*GATA2*	Zinc-finger transcription factor	Decreased/absent	Inf, Mal	[76]
	EZH2	Maintaining transcriptional repression	Decreased/absent	Mal	[77]
	MYST3-CREBBP fusion	Histone acetyltransferases	Activating	Mal	[72, 75]

Inf infection, *AI* autoimmune, *AIF* autoinflammatory, *Mal* malignancy

Genetics and Genomics in Infection-Associated CSS

SHLH can occur in the course of a severe and uncontrolled infection, including those caused by viral, bacterial, fungal, protozoal (leishmaniasis and malaria), rickettsial, visceral leishmaniasis, or mycobacterial pathogens [3]. Infection can serve as triggers for the full clinical spectrum of CSS including pHLH, but in cases where there is no evidence of familial recurrence, they are referred to as "sporadic" or sHLH. In a proportion of these cases, one may find cytolytic defects similar (although typically less profound) to those seen in patients with pHLH, but by definition no biallelic mutations are detected in known HLH associated genes that are typically inherited in an autosomal recessive fashion.

Infection, most often by virus, is the most common identified cause for sHLH, and is a statistically significant predictor of poor prognosis [4]. Of these, herpes viruses, and particularly Epstein–Barr virus (EBV), are the most frequently identified. Cattaneo et al. examined 35 adult patients diagnosed with sHLH based on HLH-2004 criteria and found infection by EBV in 28.6% of cases. With more than 90% of the world's population infected with EBV, primary EBV infection is asymptomatic in most people. Rarely, however, EBV causes life threatening infection in the form of EBV-associated HLH and chronic active EBV (CAEBV) infection. It typically affects older patients with no central nervous system (CNS) involvement, in contrast to pHLH. However, Magaki et al. [5] reported a case of fatal EBV-associated HLH with severe involvement of the CNS showing florid hemophagocytosis in the choroid plexus, with extensive neuron loss and gliosis in the cerebrum, cerebellum, and brainstem. Latent asymptomatic EBV infection is established for life in most immunocompetent individuals in B cells and nasopharyngeal epithelial cells. In contrast, in EBV-associated sHLH, the predominant EBV-infected cells are CD8-positive T lymphocytes, whereas in CAEBV, EBV infects mainly CD4 or CD8 positive T lymphocytes and NK cells [6]. Acute cytomegalovirus (CMV) associated sHLH in the immunocompetent host has been rarely reported, and most of these only partially met the HLH-2004 criteria [7]. Patients with infection-associated HLH other than EBV-HLH often enter remission when they are treated with CS, IVIG, and/or CSA, in addition to specific treatment for the infectious disease [8]. Genetic causes are largely not investigated and largely unknown until recently. While the pathogenesis of EBV-HLH remains unclear, sequence variations in cytokine production, cell proliferation and programed cell death (apoptosis) signaling related genes (Table 1) have been associated with susceptibility to this condition [9, 10]. Other inherited immune disorders are known to be associated with EBV-HLH, such as X-linked lymphoproliferative disease type1 (XLP1-*SH2D1A*) and type 2 (XIAP-*BIRC4*) [11], IL-2-inducible T cell kinase deficiency (ITK) [12], CD27 deficiency [13], CD70 deficiency [14], and magnesium transporter gene (*XMEN-MAGT1*) [15]. They should be ruled out clinically, as they may warrant more aggressive therapy, such as bone marrow transplantation, for a better clinical outcome.

Of particular note is influenza A virus, which until recently was only rarely reported in sHLH, and mostly in coinfection with other viruses such as EBV [16]. However, the recent global outbreak H1N1 influenza affecting children and young adults included many patients with clinical features resembling sHLH and MAS, and with a significantly high fatality rate. Several reports of fatal infection demonstrated liver dysfunction, cytopenias, coagulopathy, hyperferritinemia, and hemophagocytosis, which were successfully treated with HLH specific therapy [17, 18]. The potential contribution of H1N1 influenza to the development of sHLH and associated genetic and genomic defects has not been studied extensively. None of the reported patients had prior histories suggestive of immunodeficiency, but all required intensive care with mechanical ventilation and circulatory support [19–21]. Interestingly, a comprehensive genetic study using whole exome sequencing (WES) in fatal cases of H1N1 influenza identified mutations in genes have been associated with pHLH and MAS [22]. This analysis found sequence variants in the genes that encode perforin (*PRF1*-A91V), the granule trafficking protein *LYST*, and other genes that associate with MAS. These genes include Xin actin-binding protein 2 (*XIRP2*), leucine-rich repeats and guanylate kinase-domain containing protein (*LRGUK*), nipped-B homolog (*NIPBL*) and *FAM220A*. These *de novo* changes were predicted to alter the protein function and disrupt lymphocyte cytolytic function and contribute to the development of MAS/HLH like clinical syndrome in these H1N1 infected patients. In addition, two in cis MEFV variants, the causative gene of familial Mediterranean fever, also were identified in one H1N1 patient along with the A91V-*PRF1* allele (discussed further below). With supportive functional studies of NK cell cytotoxicity, this suggests that pHLH and MAS associated gene variants indeed contributed to CSS in fatal H1N1 influenza, although further functional studies are needed to confirm a causal link. Together, this evidence suggests that sHLH/MAS may be a common complication in fatal H1N1 influenza infection in patients, in particular those with certain genetic backgrounds.

There are sporadic cases reported of sHLH associate with bacterial, fungal, and other types of infectious triggers [23–25], but the genetic cause of these cases were not known and/or not well studied.

Genetics and Genomics in Rheumatic Disease Associated CSS, Often Referred to as Macrophage Activation Syndrome

HLH occurring in the setting of rheumatic diseases has historically been categorized as macrophage activation syndrome (MAS, see chapter "The History of Macrophage Activation Syndrome in Autoimmune Diseases"). However, MAS is now broadly considered to be a form of secondary HLH. sHLH/MAS occurs most frequently in children with systemic juvenile idiopathic arthritis, the genetics of which are discussed separately (see chapter "Genetics of Macrophage Activation Syndrome in Systemic Juvenile Idiopathic Arthritis"). However, this can occur albeit less commonly in children and adults with a variety of other rheumatic disorders, including both autoimmune and autoinflammatory syndromes. In general, sHLH/MAS most often complicates rheumatic diseases in periods of high or persistent disease activity, and is associated with substantial morbidity and mortality.

Other than systemic JIA, sHLH/MAS is most frequently reported in patients with systemic lupus erythematosus (SLE) [26]. While the vast majority of SLE patients do not appear to develop sHLH/MAS, it represents a major cause of death in this disease, associated with a substantial increase in in-hospital mortality for febrile SLE patients [27]. The largest series of sHLH/MAS episodes in patients with SLE found that nearly half of episodes occurred at time of SLE diagnosis, and that 15% of patients with sHLH/MAS had relapse or recurrent episodes [28]. However, little is known regarding potential genetic contributors to sHLH/MAS in patients with SLE. There is scant data regarding the occurrence of variants in pHLH-associated genes in patients with SLE and sHLH/MAS, as with few exceptions [29], sequencing is either not performed or not reported. There is a single report of a patient with atypical biallelic *PRF1* missense mutations who developed early-onset lupus as well as HLH [30]. Several large genome-wide association studies have identified *IRF5* haplotypes as significant risk factors for development of lupus [31, 32]. Interestingly, Yanagimachi and colleagues identified a *IRF5* haplotype that was significantly associated with development of sHLH, although none of the patients in that study were identified as having lupus [9]. Another recent report described a patient with SLE-associated MAS and a heterozygous P369S-R408Q variant in *MEFV*, which encodes the pyrin inflammasome and is the causative gene of familial Mediterranean fever (FMF) [33]. Although these MEFV variants are relatively common (minor allele frequency 0.5–2% depending on the population), they have also been identified in multiple other patients with sHLH/MAS [22, 34]. This is particularly intriguing in light of recent findings regarding the role of inflammasome, macrophage-intrinsic genetic defects such as *NLRC4* in triggering HLH/MAS (see below and chapter "The Intersections of Autoinflammation and Cytokine Storm").

Kawasaki disease is a self-limited vasculitis of unknown etiology, and one of the most common vasculitides during childhood. This includes a myocarditis in many patients, with late development of coronary artery aneurysms in up to 25% of untreated patients [35]. There are numerous reported cases of sHLH/MAS complicating the acute phase of Kawasaki disease, with a recent cross-sectional study suggesting it occurs in at least 1% of cases [36]. Most often, Kawasaki disease-associated sHLH/MAS occurs in patients with prolonged and treatment refractory courses, with a high incidence of coronary abnormalities and death [36, 37]. However, little is known regarding genetic risks for sHLH/MAS in patients with Kawasaki disease. Genetic analysis of pHLH genes is reported in a small number of cases and all have been negative [36–39]. There are no additional putative genetic variants that have been reported in patients with sHLH/MAS complicating Kawasaki disease.

There are scattered reports of sHLH/MAS occurring in patients with various other rheumatic diseases, including nonsystemic subtypes of JIA, rheumatoid arthritis, spondyloarthritis, dermatomyositis, mixed connective tissue disease, Sjogren's syndrome, and antiphospholipid antibody syndrome [26, 40, 41]. sHLH/MAS appears to be a very rare complication of these disorders, and to most often occur during periods of persistently high disease activity. Two recent reports have highlighted patients with spondyloarthropathy and uveitis complicated by sHLH/MAS. Genetic testing found these patients carried heterozygous variants in the pHLH genes *PRF1* [42], *UNC13D*, and *RAB27A* [43]. Potential genetic contributions to sHLH/MAS in other rheumatic diseases is largely unexplored. Several case reports have described negative genetic testing of pHLH genes in patients with sHLH/MAS and dermatomyositis [44] and Sjogren's [45] syndrome.

In addition to the autoimmune disorders discussed above, sHLH/MAS also occurs in patients with autoinflammatory disorders such as the monogenic periodic fever syndromes [46]. Autoinflammatory syndromes are characterized by seemingly unprovoked episodes of inflammation in the absence of high-titer autoantibodies or autoreactive lymphocytes, and generally believed to involve defects in innate immunity [47]. The links between sHLH/MAS and autoinflammation were highlighted particularly by the recent discovery of gain-of-function variants in the inflammasome component NLRC4 in patients with recurrent episodes of CSS resembling pHLH [48, 49]. The phenotype of these patients suggests a linkage between macrophage-intrinsic defects and development of CSS. Indeed, there is some evidence that sHLH/MAS may occur relatively frequently in other autoinflammatory disorders. In one large cohort of patients with mevalonate kinase deficiency, 6% had a history of sHLH/MAS [50]. In support of this, an infant with severe mevalonate kinase deficiency and recurrent sHLH/MAS was found to have a rare heterozygous *PRF1* variant [51]. Whether similar variants in pHLH genes occur in patients with other autoinflammatory syndromes is unknown.

Genetics and Genomics in Malignancy Associated CSS

Malignancy is the single most common trigger for HLH in adults, being present in nearly 50% of all adult cases [52]. When considering all pediatric HLH, including hereditary forms, malignancy-associated sHLH is less common, but still may be a trigger in approximately 10% of cases [53, 54]. sHLH is most commonly reported in association with non-Hodgkin's, B-cell or T-cell lymphoma, but have also been reported in Hodgkin's lymphoma, acute leukemia, and a variety of solid tumors [52]. While overall sHLH is a rare complication of hematologic malignancies, it can occur in up to 20% of some lymphoma subtypes [52]. The occurrence of sHLH in Hodgkin's lymphoma is particularly notable given the association between this malignancy and EBV, which as discussed above and elsewhere (see chapter "Cytokine Storm Syndromes Associated with Epstein–Barr Virus") is a key infectious trigger of HLH more broadly [55]. While a majority of malignancy-associated sHLH cases were diagnosed either at onset or immediately prior to identification of cancer, a substantial number of cases occurred during treatment, particularly with antileukemic therapies [56]. Malignancy and sHLH have multiple shared pathologic mechanisms, including dysfunction of immune surveillance and immunomodulation. As such, identifying shared genetic risks is an area of urgent clinical need.

Perforin is a key cytolytic protein that is essential for cytotoxic CD8 cell (CTL) and natural killer (NK) cell function, including killing of transformed cells. As discussed in Arico chapter, perforin deficiency is also the most common identified cause of pHLH. The role of perforin at the intersection of malignancy and HLH is controversial, with multiple studies with contradictory findings. In 2005, Clementi et al. first reported four lymphoma patients with biallelic perforin mutations, all of which had evidence of sHLH [57]. Concurrently, a large Italian study found that the perforin A91V polymorphism, which leads to a more mild impairment in NK cell function [22, 58, 59], was overrepresented in patients with acute lymphoblastic leukemia (ALL) [60]. However, a subsequent large study from the Children's Oncology Group could not confirm this association with ALL [61]. Other small cohort studies have found perforin A91V in association with anaplastic large cell lymphoma [62], B-cell lymphoma, melanoma [63], and *BCR-ABL* positive ALL [61], although in none of these reports was the coexistence of sHLH discussed. There are also several reports of known pathogenic perforin variants found in patients with lymphoma and signs of sHLH [30, 57, 64]. Together, the totality of evidence supports a role for perforin as a risk factor both for certain malignancies and malignancy-associated sHLH.

In support of this pathologic linkage, there are several reports of patients with malignancy-associated sHLH and variants in other cytolytic pathway genes linked to pHLH. Chang and colleagues described a 3-year-old who developed sHLH with heterozygous variant in *UNC13D*, which causes pHLH3 [65]. Four months after recovery, this child developed acute monoblastic leukemia. Other patients with *UNC13D* variants in association with malignancy and sHLH have also been reported [43, 65, 66]. Mutations in syntaxin 11 cause the hemophagocytic syndrome pHLH4 [67]. Recently, syntaxin 11 was proposed to function as a tumor suppressor gene in T-cell lymphomas, with several reports of patients with genomic deletions and genetic variants [68]. In support of this, there are reports of syntaxin 11 variants in patients with both leukemia and T-cell lymphoma who also developed malignancy-associated sHLH [56, 69]. Finally, one patient has been described with biallelic variants in *STXBP2*, which causes pHLH5, who developed sHLH in association with nodular sclerosing Hodgkin's lymphoma [56]. Most intriguingly, Lofstedt and colleagues recently found that first-degree relatives of Swedish children with pHLH had a significantly increased risk of cancer compared to matched controls [70].

Several other genetic factors have been reported in patients with malignancy-associated sHLH. Chromosomal abnormalities including pericentric inversion 12 and deletions of the long arm of chromosome 6 have been reported in patients with leukemia who presented with sHLH [66, 71]. More specifically, several fusion genes have been described in infants and children with hematologic malignancies and sHLH, including *MYST3-CREBBP* fusion at t(8;16)(p11;p13) and *ALK* fusions [72–75]. In particular, the *MYST3-CREBBP* fusion, which has histone acetylase activities from both genes and activates numerous cell cycle control genes, is associated with hemophagocytosis in a majority of cases at diagnosis [72, 75]. Two immunodeficiency syndromes, with loss of the IL-2-inducible T-cell kinase ITK and the transcription factor GATA2, cause susceptibility to EBV including EBV-associated cancers and sHLH [12, 76]. Finally, mutations in *EZH2*, which cause Weaver syndrome and are associated with hematologic malignancies, have also been reported in leukemia-associated sHLH [77]. Taken together, multiple genetic mutations and chromosomal alterations that affect cell cycle and leukocyte proliferation may also contribute to development of sHLH.

Genetics and Genomics of Other Causes of Acquired CSS

Several other disorders associated with sHLH have been reported. Selective immunoglobulin M deficiency (sIgMD) is a rare form of dysgammaglobulinemia characterized by an isolated low level of serum IgM. Agarwal et al. reported an adult case of primary sIgMD with absent B cells and sHLH who presented with recurrent infections, fever, splenomegaly, hemophagocytosis in bone marrow, and pancytopenia [78]. However, genetic studies were not carried out in this case for either pHLH or B cells/antibody deficiency. There is also a report of a patient who developed fatal sHLH in the setting of X-linked chronic granulomatous disease (CGD), which is caused by defects in the phagocyte NADPH oxidase [79]. In this case, the patient was found to have a novel heterozygous *PRF1* variant, although no functional studies were performed to determine the consequence of this change.

Transplant-related sHLH have been reported in the setting of HSCT, umbilical cord blood [80], and organ transplants such as kidney [81] and liver [82] transplants. In these cases, CSS is induced by multiple factors such as tissue damage due to conditioning, cytokine production from hematopoietic cells that proliferate at engraftment, immunological interactions between host antigen presenting cells (APC) and donor lymphocytes, and reaction of latent virus. Recently, many reports have suggested reduced intensity of HLH-specific treatment for the patients with less immunosuppressive therapy may lead to better outcome [83]. Genetic causes of these transplantation related HLH are not known.

Some congenital metabolic diseases have been reported to have a clinical complication similar to clinical sHLH, including LCHAD [84], lysinuric protein intolerance [85], multiple sulfatase deficiency [86], galactosemia [87], Gaucher disease [88], Pearson syndrome [89], and galactosialidosis [90]. These cases typically carried mutations in genes associated with the underlying metabolic conditions. Genetic studies of HLH associated genes have not identified any defects and/or susceptibility factor led to the development of clinical HLH. There are also patients with gastroparesis and associated sHLH found to have pathogenic variants in *STXBP2* [43].

Finally, another therapy that has been recently associated with sHLH is use of novel immunotherapeutics. Several reports described a "cytokine release syndrome" after using chimeric antigen receptor-modified (CAR) T-cells, bispecific T-cell engagers and cytotoxic T calls in cancer patients [91, 92]. These conditions presented a CSS-like pattern of elevated ferritin, sIL2Ra, IFNγ, IL-6, IL-8, and IL-10, but whether there is a specific genetic susceptibility to this "cytokine release syndrome" remains an open question. These conditions can be managed by cytokine-directed therapy.

Summary

HLH has historically been described in binary terms as “primary” or “familial,” indicating Mendelian inheritance of genetic mutations resulting in cytotoxic lymphocyte dysfunction, or “secondary” indicating an acquired reactive disorder without strong genetic component. This remains a critical distinction clinically, as it largely directs patient management. In many cases of infection associated sHLH, the cytokine storm resolves with appropriate specific antimicrobial therapy and no need for other immunosuppressive therapy. Even for patients with sHLH or MAS who do require aggressive therapy directed at the cytokine storm, such as the etoposide/dexamethasone protocol, once in remission, they typically remain disease free. Therefore, it is important to distinguish pHLH from sHLH, so that HSCT can be undertaken quickly for patients with pHLH and, just as importantly, patients with sHLH are spared unnecessarily aggressive therapy.

However, as discussed here and proposed by others [93, 94], increasing evidence has revealed HLH as a more complex phenomenon, resulting from specific immune activation in patients with a susceptible genetic background (Fig. 1). The clinical

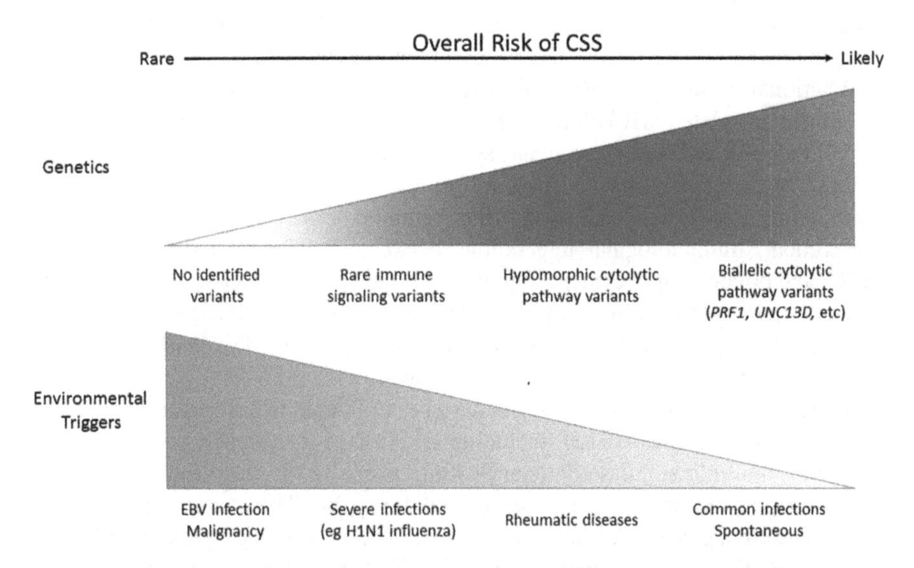

Fig. 1 Interplay of genetic background and environmental triggers leading to CSS. It is increasingly clear that both genetics and specific environmental factors or diseases together contribute to risk of CSS such as HLH. In the setting of biallelic, loss of function mutations in the perforin cytolytic pathway, common infections or even no identified trigger are sufficient to induce fulminant HLH. In contrast, hypomorphic variants in these genes, as well as numerous other genes involved in immune signaling or cytokine production, are increasingly identified in patients with sHLH/MAS associated with rheumatic disease or severe infections. In other cases, including those with significant immune dysregulation such as malignant transformation or EBV infection, no predisposing genetic variants can be identified. The collective burden of these genetic variants may predict the level of environmental trigger required to trigger CSS

development of HLH can be considered as being pushed over the edge of a cliff into a potentially catastrophic drop by an environmental trigger [95]. If the push (systemic immune activation by an infection, autoimmune/autoinflammatory disorder, or malignant transformation) is powerful enough, anybody can be made to fall. However, if the edge is lowered (through certain genetic variants with mild defects in lymphocyte cytolytic function), a much weaker push might be sufficient to produce the same result. This is most strikingly apparent in children with pHLH and biallelic mutations in genes such as *PRF1*, where life-threatening cytokine storm can be caused by trivial or even imperceptible triggers.

At the same time, most patients with sHLH have not undergone comprehensive genetic testing, even of known HLH-associated genes such as *PRF1*. Even for those who have had some genetic testing, one wonders whether patients with sporadic HLH in whom no mutation in pHLH-related genes are found might have mutations in other genes, such as those that control the production or secretion of IFNγ, or even in regulatory or noncoding genomic regions of known pHLH-related genes. The expanded use of clinical genetic testing, and particularly the availability of whole-exome sequencing and now whole genome sequencing, may provide compelling answers to this crucial question.

References

1. Henter, J. I., Horne, A., Arico, M., Egeler, R. M., Filipovich, A. H., Imashuku, S., et al. (2007). HLH-2004: Diagnostic and therapeutic guidelines for hemophagocytic lymphohistiocytosis. *Pediatric Blood and Cancer, 48*, 124–131.
2. Schram, A. M., & Berliner, N. (2015). How I treat hemophagocytic lymphohistiocytosis in the adult patient. *Blood, 125*(19), 2908–2914.
3. Trottestam, H., Horne, A., Aricò, M., Egeler, R. M., Filipovich, A. H., Gadner, H., et al. (2011). Chemoimmunotherapy for hemophagocytic lymphohistiocytosis: Long-term results of the HLH-94 treatment protocol. *Blood, 118*, 4577–4584.
4. Cattaneo, C., Oberti, M., Skert, C., Passi, A., Farina, M., Re, A., et al. (2016). Adult onset hemophagocytic lymphohistiocytosis prognosis is affected by underlying disease and coexisting viral infection: Analysis of a single institution series of 35 patients. *Hematological Oncology*.
5. Magaki, S., Ostrzega, N., Ho, E., Yim, C., Wu, P., & Vinters, H. V. (2017). Hemophagocytic lymphohistiocytosis associated with Epstein-Barr virus in the central nervous system. *Human Pathology, 59*, 108–112.
6. Kasahara, Y., Yachie, A., Takei, K., Kanegane, C., Okada, K., Ohta, K., et al. (2001). Differential cellular targets of Epstein-Barr virus (EBV) infection between acute EBV-associated hemophagocytic lymphohistiocytosis and chronic active EBV infection. *Blood, 98*, 1882–1888.
7. Bonnecaze, A. K., Willeford, W. G., Lichstein, P., & Ohar, J. (2017). Acute cytomegalovirus (CMV) infection associated with hemophagocytic lymphohistiocytosis (HLH) in an immunocompetent host meeting all eight HLH 2004 diagnostic criteria. *Cureus, 9*, e1070.
8. Morimoto, A., Nakazawa, Y., & Ishii, E. (2016). Hemophagocytic lymphohistiocytosis: Pathogenesis, diagnosis, and management. *Pediatrics International, 58*, 817–825.
9. Yanagimachi, M., Goto, H., Miyamae, T., Kadota, K., Imagawa, T., Mori, M., et al. (2011). Association of IRF5 polymorphisms with susceptibility to hemophagocytic lymphohistiocytosis in children. *Journal of Clinical Immunology, 31*, 946–951.

10. Hatta, K., Morimoto, A., Ishii, E., Kimura, H., Ueda, I., Hibi, S., et al. (2007). Association of transforming growth factor-beta1 gene polymorphism in the development of Epstein-Barr virus-related hematologic diseases. *Haematologica, 92*, 1470–1474.
11. Filipovich, A. H. (2011). The expanding spectrum of hemophagocytic lymphohistiocytosis. *Current Opinion in Allergy and Clinical Immunology, 11*, 512–516.
12. Stepensky, P., Weintraub, M., Yanir, A., Revel-Vilk, S., Krux, F., Huck, K., et al. (2011). IL-2-inducible T-cell kinase deficiency: Clinical presentation and therapeutic approach. *Haematologica, 96*, 472–476.
13. Salzer, E., Daschkey, S., Choo, S., Gombert, M., Santos-Valente, E., Ginzel, S., et al. (2013). Combined immunodeficiency with life-threatening EBV-associated lymphoproliferative disorder in patients lacking functional CD27. *Haematologica, 98*(3), 473–478.
14. Izawa, K., Martin, E., Soudais, C., Bruneau, J., Boutboul, D., Rodriguez, R., et al. (2017). Inherited CD70 deficiency in humans reveals a critical role for the CD70-CD27 pathway in immunity to Epstein-Barr virus infection. *The Journal of Experimental Medicine, 214*(1), 73–89.
15. Li, F. Y., Chaigne-Delalande, B., Su, H., Uzel, G., Matthews, H., & Lenardo, M. J. (2014). XMEN disease: A new primary immunodeficiency affecting Mg2+ regulation of immunity against Epstein-Barr virus. *Blood, 123*(14), 2148–2152.
16. Bohne, S., Kentouche, K., Petersen, I., Fritzenwanger, M., Pletz, M. W., Lehmberg, K., et al. (2013). Fulminant Epstein-Barr virus-associated hemophagocytic lymphohistiocytosis. *The Laryngoscope, 123*, 362–365.
17. Henter, J.-I., Palmkvist-Kaijser, K., Holzgraefe, B., Bryceson, Y. T., & Palmér, K. (2010). Cytotoxic therapy for severe swine flu A/H1N1. *Lancet, 376*, 2116.
18. Unal, S., Gökçe, M., Aytaç-Elmas, S., Karabulut, E., Altan, I., Ozkaya-Parlakay, A., et al. (2010). Hematological consequences of pandemic influenza H1N1 infection: A single center experience. *The Turkish Journal of Pediatrics, 52*, 570–575.
19. Tang, J. W., Shetty, N., & Lam, T. T.-Y. (2010). Features of the new pandemic influenza A/H1N1/2009 virus: Virology, epidemiology, clinical and public health aspects. *Current Opinion in Pulmonary Medicine, 16*, 235–241.
20. Randolph, A. G., Vaughn, F., Sullivan, R., Rubinson, L., Thompson, B. T., Yoon, G., et al. (2011). Critically ill children during the 2009-2010 influenza pandemic in the United States. *Pediatrics, 128*, e1450–e1458.
21. Yöntem, Y., Ilker, D., Yeşim, O., Ayşen, T., Gülcihan, O., Özgür, C., et al. (2013). Analysis of fatal cases of pandemic influenza A (H1N1) virus infections in pediatric patients with leukemia. *Pediatric Hematology and Oncology, 30*, 437–444.
22. Schulert, G. S., Zhang, M., Fall, N., Husami, A., Kissell, D., Hanosh, A., et al. (2016). Whole-exome sequencing reveals mutations in genes linked to hemophagocytic lymphohistiocytosis and macrophage activation syndrome in fatal cases of H1N1 influenza. *Journal of Infectious Diseases, 213*, 1180–1188.
23. Kumar, V. S., & Sharma, S. (2016). A case of HLH secondary to visceral leishmaniasis. *The Journal of the Association of Physicians of India, 64*(1), 108–109.
24. Foley, J. M., Borders, H., & Kurt, B. A. (2016). A diagnostic dilemma: Similarity of neuroradiological findings in central nervous system hemophagocytic lymphohistiocytosis and aspergillosis. *Pediatric Blood and Cancer, 63*(7), 1296–1299.
25. Zhang, Y., Liang, G., Qin, H., Li, Y., & Zeng, X. (2017). Tuberculosis-associated hemophagocytic lymphohistiocytosis with initial presentation of fever of unknown origin in a general hospital: An analysis of 8 clinical cases. *Medicine (Baltimore), 96*(16), e6575.
26. Atteritano, M., David, A., Bagnato, G., Beninati, C., Frisina, A., Iaria, C., et al. (2012). Haemophagocytic syndrome in rheumatic patients. A systematic review. *European Review for Medical and Pharmacological Sciences, 16*(10), 1414–1424.
27. Ahn, S. S., Yoo, B.-W., Jung, S. M., Lee, S.-W., Park, Y.-B., Song, J. J. (2017). In-hospital mortality in febrile lupus patients based on 2016 EULAR/ACR/PRINTO classification criteria for macrophage activation syndrome. *Seminars in Arthritis and Rheumatism.*

28. Gavand, P.-E., Serio, I., Arnaud, L., Costedoat-Chalumeau, N., Carvelli, J., Dossier, A., et al. (2017). Clinical spectrum and therapeutic management of systemic lupus erythematosus-associated macrophage activation syndrome: A study of 103 episodes in 89 adult patients. *Autoimmunity Reviews, 16*(7), 743–749.
29. Lee, W.-I., Chen, S.-H., Hung, I.-J., Yang, C.-P., Jaing, T.-H., Chen, C.-J., et al. (2009). Clinical aspects, immunologic assessment, and genetic analysis in Taiwanese children with hemophagocytic lymphohistiocytosis. *The Pediatric Infectious Disease Journal, 28*, 30–34.
30. Tesi, B., Chiang, S. C. C., El-Ghoneimy, D., Hussein, A. A., Langenskiöld, C., Wali, R., et al. (2015). Spectrum of atypical clinical presentations in patients with biallelic PRF1 missense mutations. *Pediatric Blood and Cancer, 62*, 2094–2100.
31. Graham, R. R., Kozyrev, S. V., Baechler, E. C., Reddy, M. V. P. L., Plenge, R. M., Bauer, J. W., et al. (2006). A common haplotype of interferon regulatory factor 5 (IRF5) regulates splicing and expression and is associated with increased risk of systemic lupus erythematosus. *Nature Genetics, 38*, 550–555.
32. Kottyan, L. C., Zoller, E. E., Bene, J., Lu, X., Kelly, J. A., Rupert, A. M., et al. (2015). The IRF5–TNPO3 association with systemic lupus erythematosus has two components that other autoimmune disorders variably share. *Human Molecular Genetics, 24*, 582–596.
33. Shimizu, M., Yokoyama, T., Tokuhisa, Y., Ishikawa, S., Sakakibara, Y., Ueno, K., et al. (2013). Distinct cytokine profile in juvenile systemic lupus erythematosus-associated macrophage activation syndrome. *Clinical Immunology, 146*(2), 73–76.
34. Zhang, M., Bracaglia, C., Prencipe, G., Bemrich-Stolz, C. J., Beukelman, T., Dimmitt, R. A., et al. (2016). A heterozygous RAB27A mutation associated with delayed cytolytic granule polarization and hemophagocytic lymphohistiocytosis. *Journal of Immunology, 196*, 2492–2503.
35. Son, M. B. F., & Newburger, J. W. (2013). Kawasaki Disease. *Pediatrics in Review, 34*, 151–162.
36. Wang, W., Gong, F., Zhu, W., Fu, S., & Zhang, Q. (2015). Macrophage activation syndrome in Kawasaki disease: More common than we thought? *Seminars in Arthritis and Rheumatism, 44*, 405–410.
37. Kang, H.-R., Kwon, Y.-H., Yoo, E.-S., Ryu, K.-H., Kim, J. Y., Kim, H.-S., et al. (2013). Clinical characteristics of hemophagocytic lymphohistiocytosis following Kawasaki disease: Differentiation from recurrent Kawasaki disease. *Blood Research, 48*, 254–257.
38. Titze, U., Janka, G., Schneider, E. M., Prall, F., Haffner, D., & Classen, C. F. (2009). Hemophagocytic lymphohistiocytosis and Kawasaki disease: Combined manifestation and differential diagnosis. *Pediatric Blood and Cancer, 53*, 493–495.
39. Cummings, C., McCarthy, P., van Hoff, J., & Porter, G. (2008). Kawasaki disease associated with reactive hemophagocytic lymphohistiocytosis. *The Pediatric Infectious Disease Journal, 27*, 1116–1118.
40. Lin, C. I., Yu, H. H., Lee, J. H., Wang, L. C., Lin, Y. T., Yang, Y. H., et al. (2012). Clinical analysis of macrophage activation syndrome in pediatric patients with autoimmune diseases. *Clinical Rheumatology, 31*, 1223–1230.
41. Lou, Y.-J., Jin, J., & Mai, W.-Y. (2007). Ankylosing spondylitis presenting with macrophage activation syndrome. *Clinical Rheumatology, 26*, 1929–1930.
42. Filocamo, G., Petaccia, A., Torcoletti, M., Sieni, E., Ravelli, A., & Corona, F. (2016). Recurrent macrophage activation syndrome in spondyloarthritis and monoallelic missense mutations in PRF1: A description of one paediatric case. *Clinical and Experimental Rheumatology, 34*, 719.
43. Zhang, M., Behrens, E. M., Atkinson, T. P., Shakoory, B., Grom, A. A., & Cron, R. Q. (2014). Genetic defects in cytolysis in macrophage activation syndrome. *Current Rheumatology Reports, 16*, 439.
44. Thomas, A., Appiah, J., Langsam, J., Parker, S., & Christian, C. (2013). Hemophagocytic lymphohistiocytosis associated with dermatomyositis: A case report. *Connecticut Medicine, 77*, 481–485.

45. García-Montoya, L., Sáenz-Tenorio, C. N., Janta, I., Menárguez, J., López-Longo, F. J., Monteagudo, I., et al. (2017). Hemophagocytic lymphohistiocytosis in a patient with Sjögren's syndrome: Case report and review. *Rheumatology International, 37*, 663–669.
46. Rigante, D., Emmi, G., Fastiggi, M., Silvestri, E., & Cantarini, L. (2015). Macrophage activation syndrome in the course of monogenic autoinflammatory disorders. *Clinical Rheumatology, 34*(8), 1333–1339.
47. Masters, S. L., Simon, A., Aksentijevich, I., & Kastner, D. L. (2009). Horror autoinflammaticus: The molecular pathophysiology of autoinflammatory disease (*). *Annual Review of Immunology, 27*, 621–668.
48. Canna, S. W., de Jesus, A. A., Gouni, S., Brooks, S. R., Marrero, B., Liu, Y., et al. (2014). An activating NLRC4 inflammasome mutation causes autoinflammation with recurrent macrophage activation syndrome. *Nature Genetics, 46*, 1140–1146.
49. Liang, J., Alfano, D. N., Squires, J. E., Riley, M. M., Parks, W. T., Kofler, J., et al. (2017). Thrombotic vasculopathy, and congenital anemia and ascites. *Pediatric and Developmental Pathology, 20*(6), 498–505.
50. Bader-Meunier, B., Florkin, B., Sibilia, J., Acquaviva, C., Hachulla, E., Grateau, G., et al. (2011). Mevalonate kinase deficiency: A survey of 50 patients. *Pediatrics, 128*, e152–e159.
51. Schulert, G. S., Bove, K., McMasters, R., Campbell, K., Leslie, N., & Grom, A. A. (2014). Mevalonate kinase deficiency associated with recurrent liver dysfunction, macrophage activation syndrome and perforin gene polymorphism. *Arthritis Care & Research, 67*, 1173–1179.
52. Ramos-Casals, M., Brito-Zerón, P., López-Guillermo, A., Khamashta, M. A., & Bosch, X. (2014). Adult haemophagocytic syndrome. *Lancet, 383*, 1503–1516.
53. Veerakul, G., Sanpakit, K., Tanphaichitr, V. S., Mahasandana, C., & Jirarattanasopa, N. (2002). Secondary hemophagocytic lymphohistiocytosis in children: An analysis of etiology and outcome. *Journal of the Medical Association of Thailand, 85*(Suppl 2), S530–S541.
54. Lehmberg, K., Sprekels, B., Nichols, K. E., Woessmann, W., Muller, I., Suttorp, M., et al. (2015). Malignancy-associated haemophagocytic lymphohistiocytosis in children and adolescents. *British Journal of Haematology, 170*(4), 539–549.
55. Ménard, F., Besson, C., Rincé, P., Lambotte, O., Lazure, T., Canioni, D., et al. (2008). Hodgkin lymphoma-associated hemophagocytic syndrome: A disorder strongly correlated with Epstein-Barr virus. *Clinical Infectious Diseases, 47*, 531–534.
56. Lehmberg, K., Sprekels, B., Nichols, K. E., Woessmann, W., Müller, I., Suttorp, M., et al. (2015). Malignancy-associated haemophagocytic lymphohistiocytosis in children and adolescents. *British Journal of Haematology, 170*, 539–549.
57. Clementi, R., Locatelli, F., Dupré, L., Garaventa, A., Emmi, L., Bregni, M., et al. (2005). A proportion of patients with lymphoma may harbor mutations of the perforin gene. *Blood, 105*, 4424–4428.
58. Trambas, C., Gallo, F., Pende, D., Marcenaro, S., Moretta, L., De Fusco, C., et al. (2005). A single amino acid change, A91V, leads to conformational changes that can impair processing to the active form of perforin. *Blood, 106*, 932–937.
59. Chia, J., Yeo, K. P., Whisstock, J. C., Dunstone, M. A., Trapani, J. A., & Voskoboinik, I. (2009). Temperature sensitivity of human perforin mutants unmasks subtotal loss of cytotoxicity, delayed FHL, and a predisposition to cancer. *Proceedings of the National Academy of Sciences of the United States of America, 106*, 9809–9814.
60. Santoro, A., Cannella, S., Trizzino, A., Lo Nigro, L., Corsello, G., & Aricò, M. (2005). A single amino acid change A91V in perforin: A novel, frequent predisposing factor to childhood acute lymphoblastic leukemia? *Haematologica, 90*, 697–698.
61. Mehta, P. A., Davies, S. M., Kumar, A., Devidas, M., Lee, S., Zamzow, T., et al. (2006). Perforin polymorphism A91V and susceptibility to B-precursor childhood acute lymphoblastic leukemia: A report from the Children's Oncology Group. *Leukemia, 20*, 1539–1541.
62. Cannella, S., Santoro, A., Bruno, G., Pillon, M., Mussolin, L., Mangili, G., et al. (2007). Germline mutations of the perforin gene are a frequent occurrence in childhood anaplastic large cell lymphoma. *Cancer, 109*, 2566–2571.

63. Trapani, J. A., Thia, K. Y. T., Andrews, M., Davis, I. D., Gedye, C., Parente, P., et al. (2013). Human perforin mutations and susceptibility to multiple primary cancers. *Oncoimmunology, 2*, e24185.
64. El Abed, R., Bourdon, V., Voskoboinik, I., Omri, H., Youssef, Y. B., Laatiri, M. A., et al. (2011). Molecular study of the perforin gene in familial hematological malignancies. *Hereditary Cancer in Clinical Practice, 9*, 9.
65. Ciambotti, B., Mussolin, L., d'Amore, E. S. G., Pillon, M., Sieni, E., Coniglio, M. L., et al. (2014). Monoallelic mutations of the perforin gene may represent a predisposing factor to childhood anaplastic large cell lymphoma. *Journal of Pediatric Hematology/Oncology, 36*, e359–e365.
66. Moritake, H., Kamimura, S., Nunoi, H., Nakayama, H., Suminoe, A., Inada, H., et al. (2014). Clinical characteristics and genetic analysis of childhood acute lymphoblastic leukemia with hemophagocytic lymphohistiocytosis: A Japanese retrospective study by the Kyushu–Yamaguchi Children's Cancer Study Group. *International Journal of Hematology, 100*, 70–78.
67. zur Stadt, U., Schmidt, S., Diler, A. S., Henter, J. I., Kabisch, H., Schneppenheim, R., et al. (2005). Linkage of familial hemophagocytic lymphohistiocytosis (FHL) type-4 to chromosome 6q24 and identification of mutations in syntaxin 11. *Human Molecular Genetics, 14*, 827–834.
68. Yoshida, N., Tsuzuki, S., Karube, K., Takahara, T., Suguro, M., Miyoshi, H., et al. (2015). *STX11* functions as a novel tumor suppressor gene in peripheral T-cell lymphomas. *Cancer Science, 106*, 1455–1462.
69. Rudd, E., Göransdotter Ericson, K., Zheng, C., Uysal, Z., Ozkan, A., Gürgey, A., et al. (2005). Spectrum and clinical implications of syntaxin 11 gene mutations in familial haemophagocytic lymphohistiocytosis: Association with disease-free remissions and haematopoietic malignancies. *Journal of Medical Genetics, 43*, e14–e14.
70. Lofstedt, A., Chiang, S. C., Onelov, E., Bryceson, Y. T., Meeths, M., & Henter, J. I. (2015). Cancer risk in relatives of patients with a primary disorder of lymphocyte cytotoxicity: A retrospective cohort study. *The Lancet. Haematology., 2*(12), e536–e542.
71. Goi, K., Sugita, K., Nakamura, M., Miyamoto, N., Karakida, N., Iijima, K., et al. (1999). Development of acute lymphoblastic leukemia with translocation (4;11) in a young girl with familial pericentric inversion 12. *Cancer Genetics and Cytogenetics, 110*, 124–127.
72. Coenen, E. A., Zwaan, C. M., Reinhardt, D., Harrison, C. J., Haas, O. A., de Haas, V., et al. (2013). Pediatric acute myeloid leukemia with t(8;16)(p11;p13), a distinct clinical and biological entity: A collaborative study by the International-Berlin-Frankfurt-Munster AML-study group. *Blood, 122*, 2704–2713.
73. Tokuda, K., Eguchi-Ishimae, M., Yagi, C., Kawabe, M., Moritani, K., Niiya, T., et al. (2014). CLTC-ALK fusion as a primary event in congenital blastic plasmacytoid dendritic cell neoplasm. *Genes, Chromosomes and Cancer, 53*, 78–89.
74. Shah, M., Karnik, L., Nadal-Melsió, E., Reid, A., Ahmad, R., & Bain, B. J. (2015). ALK-positive anaplastic large cell lymphoma presenting with hemophagocytic lymphohistiocytosis. *American Journal of Hematology, 90*, 746.
75. Andrade, F. G., Noronha, E. P., Baseggio, R. M., Fonseca, T. C. C., Freire, B. M. R., Quezado Magalhaes, I. M., et al. (2016). Identification of the MYST3-CREBBP fusion gene in infants with acute myeloid leukemia and hemophagocytosis. *Revista Brasileira de Hematologia e Hemoterapia, 38*, 291–297.
76. Cohen, J. I., Dropulic, L., Hsu, A. P., Zerbe, C. S., Krogmann, T., Dowdell, K., et al. (2016). Association of GATA2 deficiency With severe primary Epstein-Barr Virus (EBV) infection and EBV-associated cancers. *Clinical Infectious Diseases, 63*, 41–47.
77. Usemann, J., Ernst, T., Schäfer, V., Lehmberg, K., & Seeger, K. (2016). EZH2 mutation in an adolescent with Weaver syndrome developing acute myeloid leukemia and secondary hemophagocytic lymphohistiocytosis. *American Journal of Medical Genetics Part A, 170A*, 1274–1277.

78. Agarwal, A., Sharma, S., & Airun, M. (2016). Symptomatic primary selective IgM immunodeficiency—B lymphoid cell defect in adult man with secondary HLH syndrome. *The Journal of the Association of Physicians of India, 64*(7), 91–93.
79. van Montfrans, J. M., Rudd, E., van de Corput, L., Henter, J.-I., Nikkels, P., Wulffraat, N., et al. (2009). Fatal hemophagocytic lymphohistiocytosis in X-linked chronic granulomatous disease associated with a perforin gene variant. *Pediatric Blood and Cancer, 52*, 527–529.
80. Long, B., Cheng, L., Lai, S.-P., Zhang, J.-W., Sun, Y.-L., Lai, W.-X., et al. (2017). Tuberculosis-associated hemophagocytic lymphohistiocytosis in an umbilical cord blood transplant recipient. *Clinica Chimica Acta, 468*, 111–113.
81. Filippone, E. J., & Farber, J. L. (2016). Hemophagocytic lymphohistiocytosis: An update for nephrologists. *International Urology and Nephrology, 48*, 1291–1304.
82. Amir, A. Z., Ling, S. C., Naqvi, A., Weitzman, S., Fecteau, A., Grant, D., et al. (2016). Liver transplantation for children with acute liver failure associated with secondary hemophagocytic lymphohistiocytosis. *Liver Transplantation, 22*, 1245–1253.
83. Haytoglu, Z., Yazici, N., & Erbay, A. (2017). Secondary hemophagocytic lymphohistiocytosis: Do we really need chemotherapeutics for all patients? *Journal of Pediatric Hematology/Oncology, 39*, e106–e109.
84. Erdol, S., Ture, M., Baytan, B., Yakut, T., & Saglam, H. (2016). An unusual case of LCHAD deficiency presenting with a clinical picture of hemophagocytic lymphohistiocytosis: Secondary HLH or coincidence? *Journal of Pediatric Hematology/Oncology, 38*, 661–662.
85. Duval, M., Fenneteau, O., Doireau, V., Faye, A., Emilie, D., Yotnda, P., et al. (1999). Intermittent hemophagocytic lymphohistiocytosis is a regular feature of lysinuric protein intolerance. *The Journal of Pediatrics, 134*(2), 236–239.
86. Pinto, M. V., Esteves, I., Bryceson, Y., & Ferrao, A. (2013). Hemophagocytic syndrome with atypical presentation in an adolescent. *BMJ Case Reports, 2013*.
87. Marcoux, M. O., Laporte-Turpin, E., Alberge, C., Fournie-Gardini, E., Castex, M. P., Rolland, M., et al. (2005). Congenital galactosaemia: An unusual presentation. *Archives de Pédiatrie, 12*, 160–162.
88. Ben Turkia, H., Tebib, N., Azzouz, H., Abdelmoula, M. S., Ben Chehida, A., Caillaud, C., et al. (2009). Phenotypic continuum of type 2 Gaucher's disease: An intermediate phenotype between perinatal-lethal and classic type 2 Gaucher's disease. *Journal of Perinatology, 29*, 170–172.
89. Topaloğlu, R., Lebre, A. S., Demirkaya, E., Kuşkonmaz, B., Coşkun, T., Orhan, D., et al. (2008). Two new cases with Pearson syndrome and review of Hacettepe experience. *The Turkish Journal of Pediatrics, 50*, 572–576.
90. Olcay, L., Gümrük, F., Boduroğlu, K., Coşkun, T., & Tunçbilek, E. (1998). Anaemia and thrombocytopenia due to haemophagocytosis in a 7-month-old boy with galactosialidosis. *Journal of Inherited Metabolic Disease, 21*, 679–680.
91. Papadopoulou, A., Krance, R. A., Allen, C. E., Lee, D., Rooney, C. M., Brenner, M. K., et al. (2014). Systemic inflammatory response syndrome after administration of unmodified T lymphocytes. *Molecular Therapy, 22*, 1134–1138.
92. Teachey, D. T., Rheingold, S. R., Maude, S. L., Zugmaier, G., Barrett, D. M., Seif, A. E., et al. (2013). Cytokine release syndrome after blinatumomab treatment related to abnormal macrophage activation and ameliorated with cytokine-directed therapy. *Blood, 121*, 5154–5157.
93. Canna, S. W., & Behrens, E. M. (2012). Making sense of the cytokine storm: A conceptual framework for understanding, diagnosing, and treating hemophagocytic syndromes. *Pediatric Clinics of North America, 59*, 329–344.
94. Brisse, E., Wouters, C. H., & Matthys, P. (2016). Advances in the pathogenesis of primary and secondary haemophagocytic lymphohistiocytosis: Differences and similarities. *British Journal of Haematology, 174*(2), 203–217.
95. Cetica, V., Sieni, E., Pende, D., Danesino, C., De Fusco, C., Locatelli, F., et al. (2016). Genetic predisposition to hemophagocytic lymphohistiocytosis: Report on 500 patients from the Italian registry. *The Journal of Allergy and Clinical Immunology, 137*(1), 188–196.e184.

96. Vastert, S. J., van Wijk, R., D'Urbano, L. E., de Vooght, K. M., de Jager, W., Ravelli, A., et al. (2010). Mutations in the perforin gene can be linked to macrophage activation syndrome in patients with systemic onset juvenile idiopathic arthritis. *Rheumatology (Oxford), 49*, 441–449.
97. Zhang, K., Biroschak, J., Glass, D. N., Thompson, S. D., Finkel, T., Passo, M. H., et al. (2008). Macrophage activation syndrome in patients with systemic juvenile idiopathic arthritis is associated with MUNC13-4 polymorphisms. *Arthritis and Rheumatism, 58*, 2892–2896.
98. Kaufman, K. M., Linghu, B., Szustakowski, J. D., Husami, A., Yang, F., Zhang, K., et al. (2014). Whole exome sequencing reveals overlap between macrophage activation syndrome in systemic juvenile idiopathic arthritis and familial hemophagocytic lymphohistiocytosis. *Arthritis and Rheumatology*.
99. Tesi, B., Sieni, E., Neves, C., Romano, F., Cetica, V., Cordeiro, A. I., et al. (2015). Hemophagocytic lymphohistiocytosis in 2 patients with underlying IFN-gamma receptor deficiency. *The Journal of Allergy and Clinical Immunology, 135*(6), 1638–1641.
100. Wang, Y., Ai, J., Xie, Z., Qin, Q., Wu, L., Liu, Y., et al. (2016). IL-10-592 A/C polymorphisms is associated with EBV-HLH in Chinese children. *Hematology, 21*(2), 95–98.
101. Qiang, Q., Zhengde, X., Chunyan, L., Zhizhuo, H., Junmei, X., Junhong, A., et al. (2012). Killer cell immunoglobulin-like receptor gene polymorphisms predispose susceptibility to Epstein-Barr virus associated hemophagocytic lymphohistiocytosis in Chinese children. *Microbiology and Immunology, 56*(6), 378–384.

Genetics of Macrophage Activation Syndrome in Systemic Juvenile Idiopathic Arthritis

Alexei A. Grom

Macrophage activation syndrome (MAS) is a life-threatening episode of hyperinflammation driven by excessive activation and expansion of T cells (mainly CD8) and hemophagocytic macrophages producing proinflammatory cytokines [1, 2]. MAS has been reported in association with almost every rheumatic disease, but it is by far the most common in systemic juvenile idiopathic arthritis (sJIA). Clinically, MAS is similar to familial or primary hemophagocytic lymphohistiocytosis (pHLH), a group of rare autosomal recessive disorders [3, 4] linked to various genetic defects all affecting the perforin-mediated cytolytic pathway employed by NK cells and cytotoxic CD8 T lymphocytes [5–9]. Decreased cytolytic activity in pHLH patients leads to prolonged survival of target cells associated with increased production of proinflammatory cytokines that overstimulate macrophages [10–14]. The resulting cytokine storm is believed to be responsible for the frequently fatal multiorgan system failure see in MAS.

Initially HLH was described as a group of rare autosomal recessive immunodeficiency disorders of early childhood [4]. More recently, HLH has been recognized as both a familial disorder (pHLH) and as a sporadic one that is usually referred as secondary HLH (sHLH) [3, 4]. sHLH may occur at any age and is typically associated with malignancies, infections, or rheumatologic disorders [1–4]. When sHLH occurs in association with a rheumatic disease, it is usually referred to as macrophage activation syndrome (MAS) [1, 2].

Primary HLH is not a single disease but rather a constellation of rare autosomal recessive immunodeficiency disorders linked to various genetic defects all affecting the perforin-mediated cytolytic pathway [3, 4]. In about 30% of pHLH patients the cytolytic dysfunction is due to loss of function mutations in the gene encoding perforin (*PRF1*), a protein which cytolytic cells utilize to induce apoptosis of target

A. A. Grom (✉)
Department of Pediatrics, Cincinnati Children's Hospital Medical Center, Cincinnati, OH, USA
e-mail: Alexi.Grom@cchmc.org

R. Q. Cron, E. M. Behrens (eds.), *Cytokine Storm Syndrome*,
https://doi.org/10.1007/978-3-030-22094-5_8

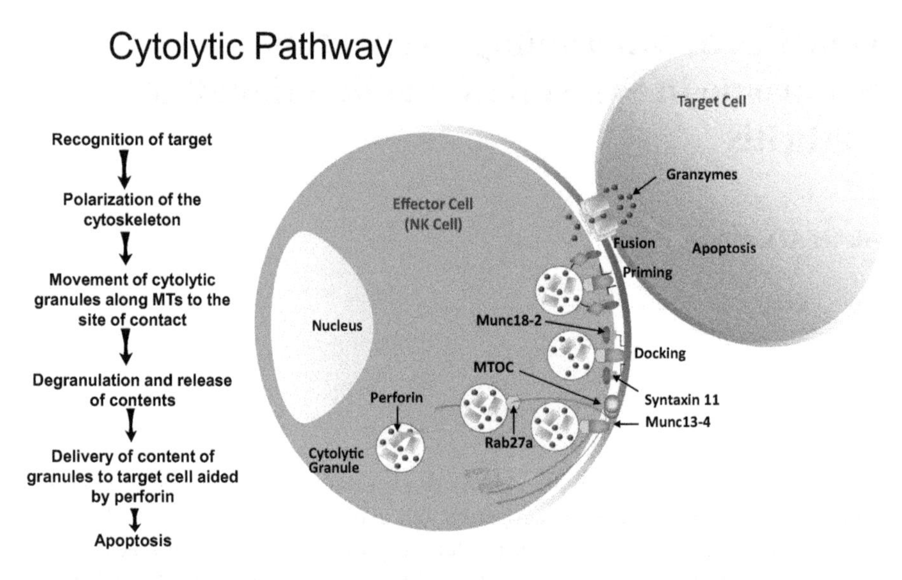

Fig. 1 Cytolytic cells cause destruction of target cells by delivering granules that contain perforin and granzymes (Modified from Jordan, et al. Blood 2011 [3])

cells [5]. When released at the surface interface with the target cell, also called the "immune synapse," peforin self-polymerizes creating pores in the plasma membrane (Fig. 1). Granzymes then pass through the perforin pores into the target cell to trigger apoptosis. The genes implicated in three other types of pHLH (*MUNC13-4, STX11, and MUNC18-2*) encode proteins involved in the intracellular transport of perforin- and granzyme-containing granules to the immune synapse [6–9]. The cytolytic cells in these patients with pHLH produce sufficient amounts of perforin, but the poor ability to release perforin into the immune synapse with a target cell leads to profoundly decreased cytolytic activity. Although mutations in *PRF1, MUNC13-4*, *STX11*, and *MUNC18-2* explain the disease in the majority of patients with pHLH, about 30–40% of clearly familial cases are still awaiting molecular definition [3] suggesting that there are likely other genes involved in defective cytolytic granule transport. Depressed cytolytic function due to abnormal movements of intracellular granules also contributes to the development of pHLH in Griscelli syndrome type 2 (GS2) and Chediak–Higashi syndrome, caused by mutations in *RAB27A* and *LYST*, respectively [15, 16].

In normal physiologic conditions, cytolytic cells, such as cytotoxic CD8 T lymphocytes or natural killer (NK) cells, induce apoptosis of cells infected with viruses or cells undergoing malignant transformation. It has been shown that even moderate defects in the cytolytic pathway may lead to prolonged survival of target cells, ultimately leading to overproduction of proinflammatory cytokines [10]. In many pHLH patients however, extensive workup often fails to identify infectious triggers, suggesting that the cytolytic function may be important in controlling T cell homoeostasis even in the absence of apparent infectious stimuli [11]. Thus cytolytic cells may also be directly involved in the termination of immune responses by inducing apoptosis of overly activated immune cells [12–14]. Based on these observations, it has been postulated that in HLH, failure to induce apoptosis of target cells by cytotoxic T cells and/or NK cells may delay the contraction stage of the immune response leading to persistent expansion of activated T cells and macrophages and escalating production of proinflammatory cytokines creating a cytokine storm [3]. Similar mechanisms may be responsible for the development of HLH in the primary immunodeficiencies, X-linked Lymphoproliferative Syndromes type 1 and 2, caused by mutations in *SH2D1A* and *XIAP* (or *BIRC4*), respectively [17, 18]. HLH in these patients is usually triggered by EBV infection causing rapid expansion of activated lymphocytes. Genetic defects in XLP type 1 interrupt activation-induced apoptosis of immune cells, leading to their prolonged survival and increased production of cytokines. Although lymphoproliferation plays a role in the development of HLH in XLP type 2 as well, these patients may also have additional intrinsic abnormalities in the myeloid cells. More specifically, XIAP has been shown to modulate NOD2 signaling, and thus it may affect inflammasome activity [19, 20].

Like pHLH, rheumatologic patients with MAS also have profoundly depressed cytolytic function, although it tends to improve with better control of the activity of the underlying rheumatic disease [21, 22]. The development of cytolytic dysfunction in sHLH appears complex and is influenced by the inflammatory activity of the underlying disease and by a genetic component [1]. To explore the genetic component in sJIAs-associated MAS patients, Kaufman et al., used *whole exome sequencing* (WES) [23]. This methodology allows detecting rare variants both in the genes localized to a specific locus and in the genes from multiple loci involved in the same pathway. First, this methodology was used to identify novel and previously reported rare protein-altering SNPs/indels in the known pHLH-associated genes. Overall, rare protein-altering variants in pHLH-associated genes were present in 36% of sJIA/MAS patients [23], similar to the percentage of sJIA patients at risk for MAS [24, 25]. Remarkably, these patients were more likely to have recurrent episodes of MAS [23]. In the same study, subsequent genome-wide recessive homozygosity and compound heterozygosity analysis identified additional potentially pathogenic variants in other genes encoding proteins important for intracellular vesicle transport that need to be explored further. Targeted sequencing of pHLH-associated genes in patients with MAS presenting as a complication of various rheumatic diseases performed by other groups [26, 27] revealed an even higher proportion of patients with heterozygous variants (above 40%).

Marked enrichment for rare variants in the genes involved in the intracellular transport of perforin-containing granules as well as in the perforin gene itself was seen in sHLH cohorts, but not in sJIA patients with no MAS history, or in healthy individuals [23, 26, 27], suggesting causal roles in disease pathogenesis. Indeed, many of these variants have been clearly associated with pHLH, and therefore are likely pathogenic. The functional and clinical relevance of many other variants, however, is not clear. In the pHLH literature, it has been suggested that the degree of cytolytic defect resulting from specific pLH gene mutations correlates with severity of disease in murine models of HLH and in patients with pHLH [28]. It has been also documented that heterozygous mutations in the pHLH gene, *MUNC18-2*, can function as dominant-negative proteins resulting in lethal sHLH [29].

Along these lines, Zhang et al. have recently characterized the impact on NK cell lytic activity of a novel heterozygous *RAB27A* p.A87P missense mutation identified in two teenagers with sHLH/MAS responsive to immunosuppression [30]. Introduction of the *RAB27A* p.A87P mutation into a human NK cell line by lentiviral transduction resulted in decreased NK cell degranulation and cytolytic activity against K562 target cells. Confocal microscopy revealed delayed cytolytic granule trafficking/polarization to the immunologic synapse in *RAB27A* p.A87P expressing NK cells (Fig. 2a). This delay in cytolytic granule transport and defective cytolysis from the *RAB27A* A87P mutation prolonged the interaction of the lytic lymphocyte and its antigen presenting cell (APC) target and was associated with increased production of IFN-γ (Fig. 2b) [30]. This was highly consistent with the recent report demonstrating a fivefold increase in the duration of the contact between the lytic lymphocytes and the antigen-presenting cells resulting in a proinflammatory cytokine storm and subsequent clinical sHLH [10].

Interestingly, the same *RAB27A* p.A87P mutation was present in each of the patients' fathers who were both asymptomatic. NK function assessment in one of the fathers revealed decreased NK cell lytic activity and degranulation. Furthermore, he also was noted to have a markedly elevated baseline serum ferritin value (a marker of MAS) [30], suggesting that he may be at risk for MAS after encountering certain infections or a rheumatic inflammatory disease. Thus, either a modifying gene(s) and/or an inflammatory trigger (e.g., certain infections) are required for MAS/sHLH to present clinically in those with heterozygous dominant-negative

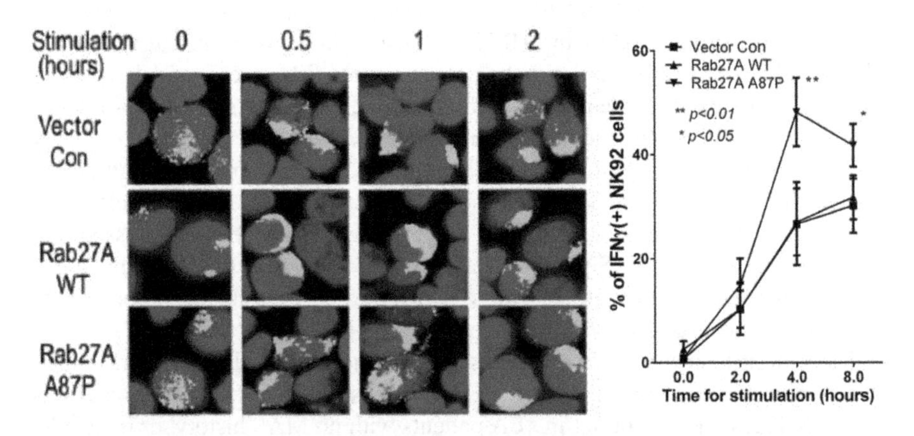

Fig. 2 (**a**) *RAB27A* p.A87P mutation delays granzyme B polarization to the immunologic synapse. Vector control (top row), *RAB27A* WT (middle), and *RAB27A* p.A87P mutant (bottom) transduced NK-92 cells were stimulated with K562 cells for the indicated time periods prior to intracellular analysis of granzyme B polarization assessed by confocal microscopy. One representative experiment is shown, where green represents granzyme B staining and blue denotes cell nuclei (DAPI stain). (**b**) *Increased IFN-γ expression* by mutant *RAB27A* p.A87P compared with WT Rab27a-expressing NK-92 cells following incubation with K562 target cells

pHLH gene mutations. Alternatively, perhaps the vesicular transport defects that affects perforin function only mildly may also affect other cellular functions yet to be explored (such as cytokine secretion, for instance), and these are the true cause of both MAS and sJIA pathophysiology.

Schulert et al. used a custom enrichment assays targeting the extended haplotype on chromosome 17 region of approximately 1 Mbps that has been associated with MAS in patients with sJIA [31, 32]. While there were no pathogenic variants in the protein-coding region, examination of these highly conserved regions known to bind transcription factors including some introns revealed several interesting variants [33]. One example is the variant c.118-308C>T in Intron 1 of the *MUNC13-4* gene. This variant is identical to what has been recently described by another group which shown that c.118-308C>T mutation abrogates transcription of an alternative form of MUNC13-4 specifically in cytotoxic lymphocytes [34]. Residual transcripts of the conventional MUNC13-4 isoform appear insufficient for supporting any cytotoxic lymphocyte degranulation. This variant has now been linked to the development of pHLH [34]. MUNC13-4 expression in PBMCs of sJIA patients with this variant was decreased as determined by RT-PCR compared to healthy controls [33]. This intronic variant partially disrupted NFκB engagement relative to the WT sequence. Moreover, reporter genes containing the variant sequence demonstrated diminished transcriptional enhancer activity consistent with decreased MUNC13-4 message as seen in the patient. These observations suggest that relevant pathogenic variants in sJIA patients with MAS can be present in both coding and non-coding regions.

Genetic variations predisposing to MAS may not be limited to only the cytolytic pathway. Thus, in patients with gain-of-function mutations in the *NLRC4* gene, MAS-like clinical presentation seems to be induced by a macrophage-intrinsic defect in the absence of primary cytotoxic abnormalities [35, 36] suggesting that in the future the search for pathogenic variants should be extended beyond the cytolytic pathway and include macrophage activation pathways as well. Another informative example is XLP type 2 where the occurrence of HLH could be linked to abnormal lymphoproliferation as well as intrinsic abnormalities in the myeloid cells related to the ability of XIAP to regulate NOD2 signaling and inflammasome activation [19, 20]. Of note, these genetic diseases are characterized by strikingly high circulating levels of IL-18, a feature that they both share with sJIA, raising the possibility of both pathophysiologic and genetic overlaps.

Additionally, a recent study from Japan identified IRF5 (interferon regulatory factor 5) gene polymorphisms as risk factors for MAS development in patients with sJIA [37]. Given the presumed role that IFN-γ has in the pathophysiology of MAS [38], this observation is very intriguing and needs to be confirmed in other ethnic groups.

In summary, MAS is a life-threatening condition that affects 10–40% of children with sJIA [1–4]. Unfortunately, at present, MAS is frequently diagnosed at the later stages when it may be fatal. However, if sHLH/MAS is diagnosed in a timely fashion, immunosuppressive therapy, including corticosteroids, cyclosporine, and anti-proinflammatory cytokine therapy, can markedly improve outcomes [1–4]. We are just beginning to recognize the breadth of genetic risk factors which predispose to MAS development in sJIA and other rheumatic diseases. A better knowledge of variants in cytolytic pathway genes and the degree to which they confer susceptibility to MAS development will be crucial to improved and earlier recognition of MAS other populations. The experience with MAS in patients with gain-of-function mutations in the *NLRC4* gene as well as patients with XLP2 also suggests that the search for pathogenic variants should be extended beyond the cytolytic pathway and should include macrophage activation pathways. This type of personalized medicine approach (knowing one's MAS genetic risk) will also be helpful in predicting outcomes, potential for MAS recurrence, and those most likely to benefit from bone marrow transplantation therapy.

References

1. Grom, A. A., Horne, A. C., & De Benedetti, F. (2016). Macrophage activation syndrome in the era of biologic therapy: Clues to pathogenesis and impact on diagnostic approaches. *Nature Reviews. Rheumatology, 12*, 259–268.
2. Mouy, R., Stephan, J. L., Pillet, P., et al. (1996). Efficacy of cyclosporine a in the treatment of macrophage activation syndrome in juvenile arthritis: Report of five cases. *The Journal of Pediatrics, 129*, 750–754.
3. Jordan, M. B., Allen, C. E., Weitzman, S., et al. (2011). How I treat hemophagocytic lymphohistiocytosis. *Blood, 118*, 4041–4052.
4. Henter, J. I., Horne, A., Arico, M., Egeler, R. M., Filipovich, A. H., et al. (2007). HLH-2004: Diagnostic and therapeutic guidelines for hemophagocytic lymphohistiocytosis. *Pediatric Blood & Cancer, 48*, 124–131.
5. Stepp, S. E., Dufourcq-Lagelouse, R., Le Deist, F., et al. (1999). Perforin gene defects in familial hemophagocytic lymphohistiocytosis. *Science, 286*, 1957–1959.
6. Feldmann, J., Callebaut, I., Raposo, G., et al. (2003). Munc13-4 is essential for cytolytic granules fusion and is mutated in a form of familial hemophagocytic lymphohistiocytosis (FHL3). *Cell, 115*, 461–473.
7. zur Stadt, U., Schmidt, S., Diler, A. S., et al. (2005). Linkage of familial hemophagocytic lymphohistiocytosis (FHL) type-4 to chromosome 6q24 and identification of mutations in syntaxin 11. *Human Molecular Genetics, 14*, 827–834.
8. zur Stadt, U., Rohr, J., Seifert, W., et al. (2009). Familial hemophagocytic lymphohistiocytosis type 5 (FHL-5) is caused by mutations in Munc18-2 and impaired binding to Syntaxin 11. *American Journal of Human Genetics, 85*, 482–492.
9. Zhang, K. J., Jordan, M. B., Marsh, R., et al. (2011). Hypomorphic mutations in *PRF1*, *MUNC13-4*, and *STXBP2* are associated with adult-onset familial HLH. *Blood, 118*, 5794–5798.
10. Jenkins, M. R., Rudd-Schmidt, J. A., Lopez, J. A., Ramsbottom, K. M., Mannering, S. I., Andrews, D. M., et al. (2015). Failed CTL/NK cell killing and cytokine hypersecretion are directly linked through prolonged synapse time. *The Journal of Experimental Medicine, 212*, 307–317.
11. Heeg, M., Ammann, S., Klemann, C., Panning, M., Falcone, V., Hengel, H., et al. (2018). Is an infectious trigger always required for primary hemophagocytic lymphohistiocytosis? Lessons from in utero and neonatal disease. *Pediatric Blood & Cancer, 65*, e27344. https://doi.org/10.1002/pbc.27344
12. Kagi, D., Odermatt, B., & Mak, T. W. (1999). Homeostatic regulation of CD8 T cells by perforin. *European Journal of Immunology, 29*, 3262–3272.

13. Menasche, G., Feldmann, J., Fischer, A., & de Saint Basile, G. (2005). Primary hemophagocytic syndromes point to a direct link between lymphocyte cytotoxicity and homeostasis. *Immunological Reviews, 203*, 165–179.
14. Lykens, J. E., Terrell, C. E., Zoller, E. E., Risma, K., & Jordan, M. B. (2011). Perforin is a critical physiologic regulator of T-cell activation. *Blood, 118*, 618–626.
15. Barbosa, M. D., Nguyen, Q. A., Tchernev, V. T., Ashley, J. A., Detter, J. C., Blaydes, S. M., et al. (1996). Identification of the homologous beige and Chediak-Higashi syndrome genes. *Nature, 382*, 262–265.
16. Menasche, G., Pastural, E., Feldmann, J., Certain, S., Ersoy, F., Dupuis, S., et al. (2000). Mutations in RAB27A cause Griscelli syndrome associated with haemophagocytic syndrome. *Nature Genetics, 25*, 173–176.
17. Coffey, A. J., Brooksbank, R. A., Brandau, O., Oohashi, T., Howell, G. R., Bye, J. M., et al. (1998). Host response to EBV infection in X-linked lymphoproliferative disease results from mutations in an SH2-domain encoding gene. *Nature Genetics, 20*, 129–135.
18. Marsh, R. A., Madden, L., Kitchen, B. J., Mody, R., McClimon, B., Jordan, M. B., et al. (2010). XIAP deficiency: A unique primary immunodeficiency best classified as X-linked familial hemophagocytic lymphohistiocytosis and not as X-linked lymphoproliferative disease. *Blood, 116*, 1079–1082.
19. Wada, T., Kanegane, H., Ohta, K., Katoh, F., Imamura, T., Nakazawa, Y., et al. (2014). Sustained elevation of serum interleukin-18 and its association with hemophagocytic lymphohistiocytosis in XIAP deficiency. *Cytokine, 65*(1), 74–78. https://doi.org/10.1016/j.cyto.2013.09.007
20. Chirieleison, S. M., Marsh, R. A., Kumar, P., Rathkey, J. K., Dubyak, G. R., & Abbott, D. W. (2017). Nucleotide-binding oligomerization domain (NOD) signaling defects and cell death susceptibility cannot be uncoupled in X-linked inhibitor of apoptosis (XIAP)-driven inflammatory disease. *The Journal of Biological Chemistry, 292*(23), 9666–9679. https://doi.org/10.1074/jbc.M117.781500
21. Grom, A. A., Villanueva, J., Lee, S., Goldmuntz, E. A., Passo, M. H., & Filipovich, A. (2003). Natural killer cell dysfunction in patients with systemic-onset juvenile rheumatoid arthritis and macrophage activation syndrome. *The Journal of Pediatrics, 142*, 292–296.
22. Vastert, S. J., van Wijk, R., D'Urbano, L. E., de Vooght, K. M., de Jager, W., Ravelli, A., et al. (2010). Mutations in the perforin gene can be linked to macrophage activation syndrome in patients with systemic onset juvenile idiopathic arthritis. *Rheumatology (Oxford), 49*, 441–449.
23. Kaufman, K. M., Linghu, B., Szustakowski, J. D., Husami, A., Yang, F., Zhang, K., et al. (2014). Whole exome sequencing reveals overlap between macrophage activation syndrome in systemic juvenile idiopathic arthritis and familial hemophagocytic lymphohistiocytosis. *Arthritis and Rheumatism, 66*, 3486–3495.
24. Behrens, E. M., Beukelman, T., Paessler, M., & Cron, R. Q. (2007). Occult macrophage activation syndrome in patients with systemic juvenile idiopathic arthritis. *The Journal of Rheumatology, 34*, 1133–1138.
25. Bleesing, J., Prada, A., Siegel, D. M., Villanueva, J., Olson, J., Ilowite, N. T., et al. (2007). The diagnostic significance of soluble CD163 and soluble interleukin-2 receptor alpha-chain in macrophage activation syndrome and untreated new-onset systemic juvenile idiopathic arthritis. *Arthritis and Rheumatism, 56*, 965–971.
26. Bracaglia, C., Sieni, E., Da Ros, M., De Fusco, C., Micalizzi, C., Cetica, V., et al. (2014). Mutations of familial hemophagocytic lymphohistiocytosis related genes and abnormalities of cytotoxicity function tests in patients with macrophage activation syndrome (MAS) occurring in systemic juvenile idiopathic arthritis. *Pediatric Rheumatology, 12*(Suppl 1), P53.
27. Zhang, M., Behrens, E. M., Atkinson, T. P., Shakoory, B., Grom, A. A., & Cron, R. Q. (2014). Genetic defects in cytolysis in macrophage activation syndrome. *Current Rheumatology Reports, 16*, 439–446.

28. Jessen, B., Kogl, T., Sepulveda, F. E., de Saint Basile, G., Aichele, P., & Ehl, S. (2013). Graded defects in cytotoxicity determine severity of hemophagocytic lymphohistiocytosis in humans and mice. *Frontiers in Immunology, 4*, 448.
29. Spessott, W. A., Sanmillan, M. L., McCormick, M. E., Patel, N., Villanueva, J., Zhang, K., et al. (2015). Hemophagocytic lymphohistiocytosis caused by dominant-negative mutations in STXBP2 that inhibit SNARE-mediated membrane fusion. *Blood, 125*, 1566–1577.
30. Zhang, M., Bracaglia, C., Prencipe, G., Bemrich-Stolz, C. J., Beukelman, T., Dimmitt, R. A., et al. (2016). A heterozygous RAB27A mutation associated with delayed cytolytic granule polarization and hemophagocytic lymphohistiocytosis. *Journal of Immunology, 196*, 2492–2503.
31. Zhang, K., Biroscak, J., Glass, D. N., Thompson, S., Finkel, T., Murray, P., et al. (2008). Macrophage activation syndrome in systemic juvenile idiopathic arthritis is associated with MUNC13D gene polymorphisms. *Arthritis and Rheumatism, 58*, 2892–2896.
32. Bruno, G., Cannella, S., Trizzino, A., Mosa, C., Faruggia, P., Guggino, G., et al. (2012). PRF1 A91V mutation associated with MUNC13-4 polymorphisms predispose to haemophagocytic lymphohistiocytosis. *Journal of Clinical Immunology, 32*(Suppl 1), 631.
33. Schulert, G., Zhang, M., Husami, A., Fall, N., Brunner, H., Zhang, K., et al. (2018). Novel NC13D intronic variant disrupting a NFkB enhancer in a patient with recurrent macrophage activation syndrome and systemic JIA. *Arthritis Rheumatol 70*(6), 963–970.
34. Cichocki, F., Schlums, H., Li, H., et al. (2014). Transcriptional regulation of MUNC13-4 expression in cytotoxic lymphocytes is disrupted by an intronic mutation associated with primary immunodeficiency. *The Journal of Experimental Medicine, 211*, 1079–1091.
35. Canna, S. W., de Jesus, A. A., Gouni, S., Brooks, S. R., Marrero, B., Liu, Y., et al. (2014). An activating NLRC4 inflammasome mutation causes autoinflammation with recurrent macrophage activation syndrome. *Nature Genetics, 46*, 1140–1146.
36. Romberg, N., Al Moussawi, K., Nelson-Williams, C., Stiegler, A. L., Loring, E., Choi, M., et al. (2014). Mutation of NLRC4 causes a syndrome of enterocolitis and autoinflammation. *Nature Genetics, 46*, 1135–1139.
37. Yanagimachi, M., Naruto, T., Miyamae, T., Hara, T., Kikuchi, M., Hara, R., et al. (2011). Association of IRF5 polymorphisms with susceptibility to macrophage activation syndrome in patients with juvenile idiopathic arthritis. *The Journal of Rheumatology, 38*, 769–774.
38. Bracaglia, C., Kathy de Graaf, K., Marafon, D. P., D'Ario, G., Guilhot, F., Ferlin, W., et al. (2017). Elevated circulating levels of interferon-γ and interferon- induced chemokines characterize patients with macrophage activation syndrome complicating systemic JIA. *Annals of the Rheumatic Diseases, 76*, 166–172.

Part III
Immunology of Cytokine Storm Syndromes

CD8^{+} T Cell Biology in Cytokine Storm Syndromes

Takuya Sekine, Donatella Galgano, Giovanna P. Casoni, Marie Meeths, and Yenan T. Bryceson

Introduction

Cytokine storm syndromes (CSS) represent a group of potentially fatal hyperinflammatory syndromes. Hemophagocytic lymphohistiocytosis (HLH) is characterized by unremitting fever, hepatosplenomegaly, hyperferritinemia, cytopenia, and sometimes hemophagocytosis. Familial forms of the HLH are typically early-onset and often triggered by infections (e.g., herpes viruses) that lead to acute, fulminant, and unremitting inflammation [1, 2]. High levels of pro-inflammatory cytokines, including interferon (IFN)-γ, tumor necrosis factor (TNF), interleukin (IL)-6, IL-12, and IL-18, as well as the anti-inflammatory cytokine IL-10, have been reported in HLH [3–6]. In one study high levels of IFN-γ, interferon-inducible protein-10 (IP-10/CXCL10, a chemokine induced by IFN-γ), and IL-10 were

T. Sekine · D. Galgano · G. P. Casoni
Department of Medicine, Center for Hematology and Regenerative Medicine,
Karolinska Institutet, Stockholm, Sweden

M. Meeths
Childhood Cancer Research Unit, Department of Women's and Children's Health,
Karolinska Institutet, Karolinska University Hospital Solna, Stockholm, Sweden

Clinical Genetics Unit, Department of Molecular Medicine and Surgery, and Center
for Molecular Medicine, Karolinska Institutet, Karolinska University Hospital,
Solna, Stockholm, Sweden

Y. T. Bryceson (✉)
Department of Medicine, Center for Hematology and Regenerative Medicine,
Karolinska Institutet, Stockholm, Sweden

Department of Clinical Sciences, Broegelmann Research Laboratory, University of Bergen,
Bergen, Norway

Department of Medicine, Center for Hematology and Regenerative Medicine,
Karolinska Institutet, Huddinge, Sweden
e-mail: Yenan.Bryceson@ki.se

R. Q. Cron, E. M. Behrens (eds.), *Cytokine Storm Syndrome*,
https://doi.org/10.1007/978-3-030-22094-5_9

associated with early mortality [7]. Histologically, HLH is characterized by activated macrophages and expansion of CD8+ T cells that infiltrate tissues [8, 9]. According to current clinical criteria, at least five of eight defined criteria need to be fulfilled for the diagnosis of HLH [10]. Clinically, macrophage activation syndrome (MAS) has similar diagnostic criteria as, and can be difficult to distinguish from, HLH [11]. A clinical scoring system can assist discrimination of HLH from MAS [12]. Providing a biological marker potentially differentiating HLH from MAS, chronic elevation of IL-18, unbound from the IL-18 binding protein, has recently been associated with an increased risk of MAS [13]. Another form of CSS has been reported following T cell-directed therapies (e.g., upon administration of agonistic anti-CD28 antibodies or treatment of B-precursor acute lymphoblastic leukemia with CD19/CD3-bispecific T cell receptor engaging antibodies) [14, 15]. The pathophysiology of CSS after immunotherapy is still poorly understood, but is linked to massive T cell-mediated release of cytokines following strong therapeutic stimuli. In turn, these cytokines may trigger activation of innate immune cells, exacerbating inflammation.

Here, we focus on the role of CD8+ T cells in the pathophysiology of CSS, with a particular emphasis on congenital forms of HLH noted for CD8+ T cell dysfunction.

Genetic Hemophagocytic Lymphohistiocytosis Predispositions Implicating CD8+ T Cells

Through seminal studies of patients with familial forms of HLH by Kumar and colleagues, biallelic loss-of-function mutations in the gene encoding perforin, *PRF1*, were identified as a cause of primary HLH [16]. This study thereby provided a definitive link between perforin-mediated cytotoxicity and disease, demonstrating that lymphocyte cytotoxicity required not only for clearing abnormal cells but also for controlling the magnitude of immune responses. Perforin is specifically expressed by cytotoxic lymphocytes and is stored together with apoptosis-inducing granzyme proteins in cytotoxic granules, a form of secretory lysosome. Of note, Fas ligand is another apoptosis-inducing transmembrane protein that localizes to cytotoxic granules in unstimulated cells [17]. Defects in Fas ligand-mediated killing of lymphocytes, typically caused by somatic, dominant-negative mutations in the Fas receptor are also associated with lymphoproliferative disease [18], termed autoimmune lymphoproliferative syndrome (ALPS). However, in contrast to HLH, ALPS is typically characterized by chronic, non-infectious lymphoadenopathy or splenomegaly, in addition to an elevated frequency of CD3+TCRαβ+CD4−CD8− double-negative T cells in peripheral blood [19].

In subsequent studies of other families with familial HLH, biallelic loss-of-function mutations in *UNC13D*, *STX11*, and *STXBP2* were also identified as causative of disease [20–23]. These genes encode the cytosolic proteins Munc13-4, syntaxin-11 (Stx11), and Munc18-2, respectively, which are widely expressed in the immune system as well as other hematopoietic cell types and certain tissues. Studies of lymphocytes from patients with biallelic nonsense mutations in these genes have established that that their protein products are required for exocytosis of perforin-containing cytotoxic granules [20, 22–24], providing an explanation for why such mutations may give rise to syndromes that clinically phenocopy perforin deficiency. In addition, Griscelli syndrome type 2 (GS2) and Chediak–Higashi syndrome (CHS), caused by mutations in *RAB27A* and *LYST*, respectively, display partial albinism and are also associated with development of HLH [25, 26]. Both GS2 and CHS patients display impaired lymphocyte cytotoxicity due to defective cytotoxic granule exocytosis [25, 27]. Furthermore, patients with X-linked lymphoproliferative syndrome (XLP) caused by mutations in *SH2D1A* and *XIAP* [28–31], respectively, may also present with HLH although an overt defect in lymphocyte cytotoxicity *per se* is not usually observed [32].

Altogether, genetic studies have provided strong links between mutations in genes required for lymphocyte cytotoxicity and hyperinflammatory syndromes such as HLH, representing an example of how rare human diseases can provide molecular insights to crucial physiological processes. Retrospective analyses have demonstrated that patients with biallelic nonsense mutations in *PRF1* or *UNC13D* invariably present with HLH within their two first years of life [33–35]. Nonsense mutations in *STXBP2*, *RAB27A* and in particular *STX11* may present later in childhood. The incidence of such autosomal recessively inherited HLH in infancy and childhood has been estimated to 1/50,000 live-births [36], which is for example comparable to that of severe combined immunodeficiency [37]. Warranting functional studies of patients [32], disease-causing mutations are not necessarily detected by sequencing the coding regions of genes. Noncoding mutations can explain a high proportion of early-onset HLH cases [38–41]. These studies have revealed a lymphocyte-specific intronic enhancer and alternative transcriptional start site of *UNC13D* controlled by ETS family transcription factor binding [42]. In the same intron, another mutation affecting NFκB transcription factor binding has been associated with impaired *UNC13D* transcription in a patient diagnosed with recurrent MAS [43]. Moreover, deletions of the *RAB27A* promoter have been reported in GS2 [44, 45], in certain cases deleting a lymphocyte-specific promoter that thereby impaired lymphocyte cytotoxicity without affecting the pigmentation of melanocytes. Results from multiple lines of investigation suggest that only mutations that severely reduce the function of specific proteins required for lymphocyte cytotoxicity are associated with familial HLH, whereas hypomorphic variants can be associated with later onset of disease and reduced penetrance [46, 47]. In accordance with these observations, mouse studies of perforin-mediated immune regulation in the context of mixed chimerism indicate that 10–20% of perforin-expressing lymphocytes is sufficient to avoid development of hyperinflammatory diseases upon viral challenge in an animal model of HLH [48]. Similarly, clinical data suggest more than 20% donor cells suffice to prevent HLH reactivation in transplanted patients [49]. Nonetheless, biallelic hypomorphic variants in HLH-associated genes have also been associated with presentation of HLH, more commonly in adults [33, 50, 51]. The degree to which monoallelic mutations in HLH-associated genes may cause disease is not clear [50, 52]. Mouse models suggest that heterozygous mutations in different HLH-associated may add up in terms of HLH susceptibility [53], although analyses of HLH patient exomes do not suggest that polygenic effects are a major cause of HLH [54]. Adding to the complexity of HLH genetics, two recent reports have suggested that certain mutations in HLH-associated genes may act in a dominant negative manner. Two rare heterozygous *STXBP2* missense variants have been described in children with HLH [55], albeit with incomplete penetrance. In vitro membrane fusion assays revealed that these variants arrest the late steps of soluble N-ethylmaleimide-sensitive factor attachment protein receptor (SNARE)-complex assembly, explaining reduced cytotoxic granule exocytosis and potentially HLH susceptibility. Furthermore, two unrelated adolescents with HLH have been described to carry a specific heterozygous *RAB27A* missense variant, again with incomplete penetrance [56]. This Rab27a variant displayed impaired interactions with Munc13-4 and overexpression of the variant reduced lymphocyte degranulation.

Together these genetic findings suggest a strong association between lymphocyte cytotoxicity, in part mediated by CD8$^+$ T cells, and HLH. Genetic associations highlight a continuum of mutations that also includes heterozygous, dominant negative variants. It can be assumed that only a fraction of individuals carrying hypomorphic variants in HLH-associated genes actually present with HLH. Given the prevalence of loss-of-function variants in HLH-associated genes, it is quite possible that many individuals present with other diseases as a consequence of impaired lymphocyte cytotoxicity (e.g., cancer) [46, 57–59].

Key Roles of CD8$^+$ T Cells in the Pathophysiology of Familial Forms of Hemophagocytic Lymphohistiocytosis

In human peripheral blood, natural killer (NK) cells and differentiated CD8$^+$ T cells represent the major perforin-expressing cell subsets [60, 61]. In agreement with the finding that autosomal recessive loss-of-function mutations in perforin cause HLH, a characteristic laboratory finding of HLH is defective NK cell-mediated cytotoxicity [62, 63]. Animal models have offered more detailed understanding of the cellular processes underlying HLH. Although both NK cells and T cells can play important roles in protection against infected or transformed cells, perforin-deficient mice do not develop HLH-like symptoms unless they are infected [64–66]. Moreover, a murine cytomegalovirus (MCMV) susceptibility screen of N-ethyl-N-nitrosourea-mutagenized mice identified an *Unc13d* splice mutation, with animals susceptible to HLH-like disease upon lymphocytic choriomeningitis (LCMV) infection, but not upon infection with MCMV or the intracellular parasite *Listeria monocytogenes* [67]. Notably, a seminal study by Jordan and colleagues infecting perforin-deficient mice on the C57BL/6 background with LCMV revealed that disease could be ameliorated by depletion of CD8+ cells or neutralization of IFN-γ, but not other cytokines including TNF, IL-12, and IL-18, suggesting central roles for CD8+ T cells and IFN-γ in HLH pathogenesis [68]. In this mouse model, administration of anti-IFN-γ could ameliorate disease, indicating therapeutic potential [69]. Providing insights to the cellular basis for pathophysiology, the perforin-deficient mouse model displays greater CD4+ and CD8+ T cell expansions upon LCMV infection [65]. In this context, CD8+ T cells mediate a negative feedback loop involving perforin-dependent elimination of antigen-presenting dendritic cells (DCs) [70]. Genetic interference of *Itgb2*, encoding the β2-integrin important for lymphocyte adhesion, or *Tnf* upon LCMV infection of Munc13-4-deficient mice demonstrated that neither cell contact nor TNF was essential for development of disease [71]. However, studies of perforin-deficient mice have indicated that defective lymphocyte cytotoxicity mechanistically results in prolonged interactions between cytotoxic effector cell and their target cell, greatly amplifying the quanta of IFN-γ secreted by CD8+ T cells and NK cells [72]. Disruption of *Myd88*, encoding a cytosolic adaptor protein for inflammatory signalling pathways, in Munc13-4-deficient mice infected with LCMV blocked development of HLH-like disease, revealing a key role for MyD88 in promoting myeloid and lymphoid proliferation [71]. IL-33 is a cytokine belonging to the IL-1β and IL-18 family that signals via a MyD88-dependent pathway. Interestingly, blockade of IL-33 receptor, ST2, improves survival of LCMV-infected perforin-deficient mice, reducing serum levels of IFN-γ [73]. This finding suggests that danger signals such as IL-33 released from tissues can act as amplifiers of immune dysregulation in viral-triggered forms of familial HLH.

Together, these studies of the LCMV-infection based mouse model of HLH point to key roles of CD8+ T cells and IFN-γ in driving the pathogenesis of HLH. In this context, it should be noted that mutations in genes besides those implicated in lymphocyte cytotoxicity have also been linked to HLH susceptibility [54, 74, 75]. Notably, T cell-deficient severe combined immunodeficiency patients, harboring mutations in genes required for T cell development, can develop hyperinflammatory disease that fulfills HLH criteria [74]. Supporting the notion of CD8+ T cell-independent hyperinflammatory disease, persistent exposure of mice to Toll-like receptor (TLR)9 ligand CpG resulted in an HLH- or MAS-like disease which could develop independently of T cells [76]. Such a TLR9-triggered cytokine storm was also partially mediated by IFN-γ and dependent on the presence of conventional DCs, while depletion of NK cells ameliorated disease and blockade of IL-10 receptors exacerbated disease. Moreover, MCMV infection of the BALB/c strain of mice induces an HLH-like disease that can also develop upon depletion of CD8+ T cells specifically or in animals lacking lymphocytes altogether [77, 78]. Notably, activating mutations in *NLRC4* can cause severe MAS-like autoinflammation in humans and mice, mimicking HLH [79, 80]. Moreover, monoallelic variants in *NLRP12* and biallelic variants in *NLRP4*, *NLRC3*, and *NLRP13*, encoding different inflammasomes, have been linked to HLH susceptibility [54]. In settings of inflammasome activation, IL-18 and other pro-inflammatory innate cytokines could in combination induce IFN-γ production by T cells and NK cells [81, 82].

Elegant studies have demonstrated that IFN-γ plays a critical role in the pathogenesis of HLH, driving macrophage activation, eliciting hemophagocytosis and thereby leading to cytopenias affecting multiple cell lineages [83]. Potentially being one of the major sources of IFN-γ, $CD8^+$ T cells may represent cellular culprits of IFN-γ-dependent pathogenicity. However, in a model of hemophagocytosis driven by TLR9 stimulation, immunopathology was comparable between wild-type and *Ifng*-deficient mice, demonstrating that fulminant MAS can arise independently of IFN-γ [84]. Importantly, this study suggested that dysregulation of erythropoiesis rather than hemophagocytosis may cause anemia in hyperinflammatory settings [84]. In line with these observations, two patients with IFN-γ receptor deficiency have been reported to fulfill HLH criteria in the context of severe viral and mycobacterial infections, with a notable lack of hemophagocytosis [85]. Interestingly, antibody-mediated neutralization of TNF in MCMV-infected *Prf1* knock-out mice, a model of HLH that also involves the antiviral activity of NK cells, suggested a central role for TNF in pathogenesis [66]. Interestingly, in pre-immunized mice, pathology induced by LCMV infection is less dependent on IFN-γ and more mediated by TNF than in non-immunized, pathogen-naïve mice [86]. Thus, other cell types and cytokines distinct from IFN-γ (e.g., TNF) may also contribute to disease.

Besides $CD8^+$ T cells, NK cells represent a major subset of perforin-expressing lymphocytes. Interestingly, in the mouse LCMV infection model, NK cells can eliminate activated $CD4^+$ T cells, thereby tuning $CD8^+$ T cell responses [87]. In this model, mice with $CD8^+$ T cell-specific perforin deficiency (i.e., with intact NK cell cytotoxicity) displayed reduced hyperinflammatory manifestations and improved survival without reduction in viral loads relative to complete perforin-deficient mice (i.e., also lacking NK cell cytotoxicity) [88]. In contrast, mice treated with anti-NK1.1 antibody to deplete NK cells displayed exacerbated inflammation [88]. These experiments reveal a crucial contribution of NK cells to immunoregulation and maintenance of immune homeostasis upon viral challenge via control of T cell activation, rather than mediating HLH pathogenesis. It is also possible that NK cells contribute to pruning of activated DCs [89].

Remarkably, *Stx11* knock-out mice develop all clinical signs of HLH-like disease upon LCMV infection, but unlike *Prf1* knock-out mice do not progress to fatal disease [90]. Survival of Stx11-deficient mice was determined by exhaustion of antigen-specific $CD8^+$ T cells, characterized by expression of inhibitory receptors, and sequential loss of effector functions leading to T cell deletion. Notably, in this model, blockade of inhibitory receptors on T cells in Stx11-deficient mice resulted in fatal HLH, potentially identifying T cell exhaustion as an important factor determining HLH disease severity. In familial HLH patients, polyclonal $CD8^+$ T cells typically display high expression of the activation marker HLA-DR and check-point inhibitory receptor PD-1 [91]. By comparison, expression of activation markers was similar in secondary HLH patients associated with viral infection but less in other secondary HLH patients. Notably, possibly reflecting different pathways of pathogenesis, a high proportion of polyclonal $CD4^+CD127^-$ T cells expressing HLA-DR, CD57, and perforin was a signature of infants with familial HLH, distinguishing them from patients with virus-associated secondary HLH. Highlighting an interplay between CD4^{+} and CD8^{+} T cells in development of HLH, the LCMV infection perforin-deficient mouse model of HLH has demonstrated that excessive activation of CD8^{+} T cells consumes IL-2, resulting in a collapse in regulatory CD4^{+} T cell numbers [92]. Diminished CD4^{+} T cell numbers were also observed in patients with HLH flares [92]. With most insights to HLH pathophysiology stemming from mouse models of disease, it is important to note that standard laboratory mice differ in several ways from human as well as murine counterparts exposed to a natural environment that provides a greater diversity of immune challenges. Laboratory mice typically display a T cell population with features of differentiation reminiscent of newborn humans, whereas mice exposed to a more natural environment have more differentiated T cells [93]. A recent study demonstrated that prior antigen exposure of *Prf1* knock-out mice exacerbated the course of HLH-like disease upon LCMV infection [86]. Thus, some mouse models may display reduced severity of disease and, with relevance to patients, a more differentiated T cell compartment may worsen the outcome of HLH.

In summary, relevant models of HLH pathogenesis in the setting of defective lymphocyte cytotoxicity indicate that CD8^{+} T cells can represent drivers of disease, releasing cytokines that promote pathology and killing target cells that further exacerbate the release of pro-inflammatory cytokines from innate immune cells (Fig. 1). However, other forms of disease, such as MAS, may develop independently of CD8^{+} T cells.

Molecular Mechanisms of CD8^{+} T Cell Killing of Target Cells

More than 30 years ago, prior to genetic associations with HLH, perforin was recognized as a pore-forming molecule implicated in killing of target cells [94]. In light of its central role in lymphocyte cytotoxicity, numerous studies have been focused on understanding how perforin itself mediates and how other HLH-associated cytosolic proteins facilitates target cell killing [95, 96]. Here, we focus on recent insights on the molecular regulation of cytotoxic lymphocyte exocytosis and perforin-mediated target cell killing.

Perforin-mediated lysis of target cells is believed to occur either through direct osmotic lysis or by facilitating entry of granzymes that subsequently induce target cell death [97]. Live-cell imaging studies have demonstrated that, upon encounter of susceptible target cells, cytotoxic lymphocytes can release cytotoxic granules and form pores in the target cell membrane within seconds [98]. Such pores on the target cell membrane are rapidly repaired, with signs of target cell apoptosis nonetheless appearing within minutes of perforin-mediated permeabilization [98]. Although the process is not completely understood, cytotoxic lymphocyte detachment from dying target cells are caspase-dependent [72]. The virtue of individual cytotoxic lymphocytes to unidirectionally kill target cells without themselves undergoing apoptosis, yet retain sensitivity to attack mediated by other cytotoxic lymphocytes, has been

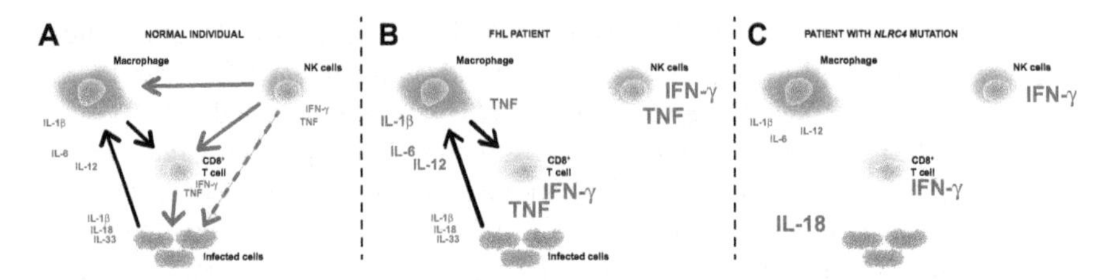

Fig. 1 Involvement of CD8+ T cells in different genetic forms of cytokine storm syndromes. (**a**) In a normal individual, cellular infection triggers release of pro-inflammatory cytokines such as IL-1, IL-18, and IL-33 as well as chemokines from cells in tissue. In turn, antigen presenting cells such as macrophages and dendritic cells are recruited and activated by the cytokine environment and by direct recognition of pathogen-derived molecules. Antigen presenting cells can produce innate cytokines, including IL-1β, IL-6, and IL-12, as well as chemokines, alerting and recruiting other immune cells, including CD8+ T cell and NK cells. T cells can be primed through interactions with antigen-presenting cells and in turn kill infected cells. A major role for NK cells is in maintaining immune homeostasis through killing of activated immune cells. In certain settings, NK cells may also kill infected cells. Activation of CD8+ T cells and NK cells via direct target cell interactions as well as cytokine stimulation can induce production and release of cytokines such as IFN-γ and TNF. (**b**) In a patient with mutations in genes required for lymphocyte cytotoxicity and associated with familial forms of HLH, cellular infection triggers an excessive inflammatory response. Defective CD8+ T cells are unable to kill infected cells, leading to high levels of pro-inflammatory cytokines. High levels of pro-inflammatory cytokines, as well as prolonged interactions with infected cells by cytotoxic lymphocytes unable to kill, resulting in massive expansions of CD8+ T cells and over-production of cytokines such as IFN-γ and TNF. These cytokines in turn drive pathology, ultimately leading to multiorgan failure and death. Defective NK cell killing of activated immune cells, including CD8+ T cells, can contribute to the exacerbated immune responses and pathology. (**c**) Activating mutations in *NLRC4*, an inflammasome component, result in high constitutive release of IL-18 from tissue cells, which may occur without any infection. IL-18 secretion leads to immune activation, with CD8+ T cell and NK cells capable of producing large amounts of IFN-γ in response to IL-18 in combination with certain other cytokines. IFN-γ. Black arrows indicate activation of immune cells and red arrows indicate interactions that result in target cell killing. Dashed arrows represent possible cytotoxic activity. The font size of cytokines indicates their magnitude in different settings

enigmatic. A possible explanation for this feature may be concomitant surface expression of CD107a (also known as lysosome-associated membrane protein [LAMP]-1) upon cytotoxic granule release. CD107a is a highly glycosylated membrane integral protein localized to cytotoxic granules and is transiently expressed on the surface of cytotoxic lymphocytes following cytotoxic granule exocytosis. Besides representing a useful marker for quantifying the frequency and intensity of cytotoxic granule exocytosis [32], CD107a surface expression may also facilitate granule biogenesis and protect cytotoxic lymphocytes from self-destruction [99, 100]. Most mutations in *PRF1* associated with HLH can be explained by an inability of cytotoxic lymphocytes to express perforin capable of forming pores in target cell membranes [101]. Of note, granzyme-deficiencies have so far not been associated with HLH, suggesting functional redundancy among different members of the granzyme family of proteins for induction of target cell death [97]. Granzyme B, an effector of target cell apoptosis and major constituent of cytotoxic granules, is activated through proteolytic cleavage by cathepsin C [102]. Papillon–Lefevre syndrome is caused by mutations in *CTSC*, encoding the cathepsin C peptidase. Although NK cells from Papillon–Lefevre syndrome patients display impaired cytotoxicity, they have not been reported to develop HLH. Rather, the syndrome is characterized by early onset of peridontitis, which is in some cases thought to be viral in origin [103]. The lack of HLH in Papillon–Lefevre syndrome patients might be explained by observations that stimulation of NK cells with IL-2 restored the ability to process granzyme B [102]. Consistent with this, another study that did not find any cytotoxic defect in cultured NK cells from Papillon–Lefevre patients [104].

Apart from perforin and other cytotoxic granule constituents, several studies have also attempted to elucidate how deficiency in cytosolic proteins LYST, Rab27a, Munc13-4, Stx11 and Munc18-2 mechanistically cause disease. LYST is a large, membrane integral protein. LYST-deficiency in CD8+ T cells results in enlarged cytotoxic granules incapable of fusing with the plasma membrane [105]. In LYST-deficient cells, the enlarged organelle represents a hybrid compartment, suggesting a role for LYST in fission and biogenesis of cytotoxic granules, thus affecting effector protein compartmentalization and thereby cytotoxic granule fusion with the plasma membrane [106, 107]. LYST is one of several members of a BEACH-domain containing protein family. Notably, whereas LYST-deficient CD8+ T cells and NK cells display similar defects in cytotoxic granule exocytosis, they differ in that CD8+ T cells typically have several enlarged lysosomes while NK cells typically only contain a single enlarged lysosome [108]. Subtle differences in the function of LYST-deficient CD8+ T cells versus NK cells may be of clinical interest. In line with a prominent role for CD8+ T cells in HLH pathogenesis, studies of CHS patients have also suggested that *LYST* mutations preferentially affecting NK cell cytotoxicity, rather than both CTL and NK cell cytotoxicity, may be associated with milder disease and reduced predisposition to HLH, although this study could not ascribe clear genotype–phenotype correlations [109]. Of note, adaptor protein-3 deficiency also leads to enlarged cytotoxic granules and impaired exocytosis; however, patients with *AP3BP1* mutations typically do not develop HLH [110, 111]. Rab27a supports anterograde transport of cytotoxic granules on actin filaments to the plasma membrane through binding of Slp3 and kinesin-1 [112]. Munc13-4 is an effector of GTP-bound Rab27a [113], sensing intracellular Ca^{2+} concentrations that are elevated upon recognition of sensitive target cells via two C2 protein domains [114, 115]. Interestingly, neither Munc13-4 nor Rab27a is constitutively associated with cytotoxic granules. Rather, highlighting a complex series of fusion events preceding cytotoxic granule exocytosis, Munc13-4 and Rab27a are recruited to cytotoxic granules downstream of signals from distinct plasma membrane receptors [116, 117].

Stx11 and Munc18-2, in addition to SNAP-23 represent a plasma membrane complex facilitating exocytosis of cytotoxic granules. Unlike most other syntaxin family members, Stx11 lacks a C-terminal transmembrane domain but rather is anchored to membrane through S-acetylation of C-terminal cysteine residues [118]. Stx11 resides on recycling endosomes that are recruited to the immune synapse prior to cytotoxic granule exocytosis [119], depositing Stx11 to the plasma membrane [120]. Stx11 interacts with Munc18-2, with Munc18-2 deficiency resulting in loss of Stx11 expression [22, 23]. Recent structural and mutagenesis studies of Munc18-2, as well as mutagenesis studies of Stx11, reveal that Stx11 and Munc18-2 interactions depend on both the N-terminal peptide (corresponding to the first few amino acids) and Habc domain of Stx11 [121, 122], similar to the binary interaction described between Stx1 and Munc18-1 in neurons [123, 124]. Not only Munc18-2 acts as a chaperone of Stx11, but it also directly promotes SNARE complex assembly, facilitating membrane fusion [125]. Moreover, via its SNARE domain Stx11 interacts with SNAP-23 [118, 126], a plasma membrane anchored protein with two SNARE domains. Interestingly, unlike Munc13-4 and Munc18-2-deficiency, cytokine stimulation of Stx11 deficient cytotoxic lymphocytes can result in a partial gain of exocytic capacity, explaining the later onset and somewhat milder disease progression of patients carrying biallelic *STX11* mutations [24, 32, 33]. Besides Stx11, Munc18-2 can also bind Stx3, albeit with a binding affinity 20-fold lower than that for Stx11 [122]. Thus, it is possible that Munc18-2 binding to Stx3 can contribute to cytotoxic granule exocytosis in activated cells, providing some level of redundancy for cytotoxic granule exocytosis [122]. Furthermore, Stx11 and its interaction partner SNAP-23 may well form the t-SNARE on the plasma membrane for cytotoxic granule fusion. The identity of the v-SNARE for lytic granule exocytosis is not clear. Studies of murine CTL indicate that that both *Vamp2* and *Vamp8*-deficient mice display impaired cytotoxic granule exocytosis [119, 127, 128]. In mice, VAMP2 localizes to cytotoxic granules [129], but it is not expressed in human cytotoxic lymphocytes [120]. In human $CD8^+$ T cells, VAMP8 localizes to $Rab11a^+$ recycling endosomes and mediates T cell receptor-triggered fusion at the plasma membrane, preceding and facilitating that of cytotoxic granules [120]. Knockdown of Stx4, a plasma-membrane localized t-SNARE, in $CD8^+$ T cells attenuated exocytosis of $Rab11a^+$ recycling endosomes and surface expression of CD107a, resulting in diminished cytotoxic activity [130]. Thus, Stx4, likely together with SNAP-23, serves as a cognate plasma membrane t-SNARE for recycling endosome exocytosis. Thus far, mutations in *STX4*, *VAMP2* or *VAMP8* have not been associated with development of HLH. *Vamp8* knockout mice are viable put display severe pancreatic defects and die early [131], whereas *Vamp2* knockout mice die immediately after birth [132].

In summary, studies of proteins implicated in cytotoxic granule exocytosis have unraveled the intricate regulation of cytotoxic lymphocyte-mediated target cell killing (Fig. 2). Nonetheless, further studies are warranted to gain deeper understanding of the molecular mechanism underlying lymphocyte cytotoxicity. For example, it remains unclear how signals from transmembrane receptors precisely spatially regulate exocytosis at the immune synapse and how endocytosis and recycling of membrane-anchored proteins is regulated for serial exocytosis and target cell killing [133].

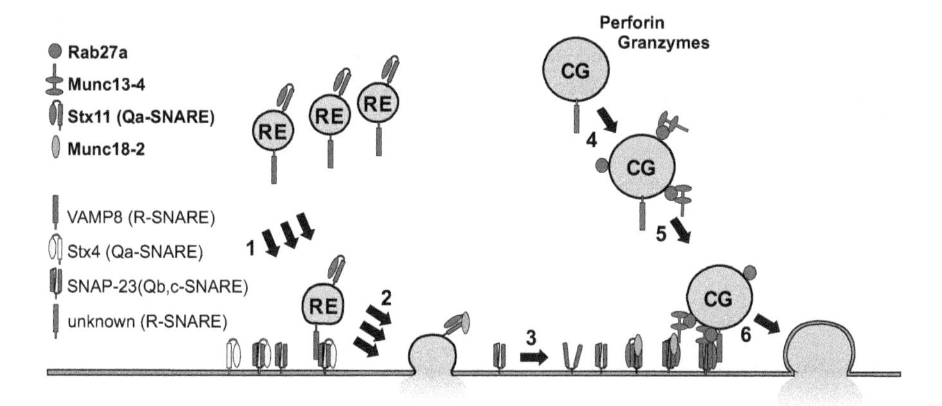

Fig. 2 Mechanisms of cytotoxic granule release in cytotoxic CD8^{+} T cells. LYST has been implicated in biogenesis of cytotoxic granules (CGs), regulating fission and fusion (not depicted). After target cell recognition, numerous small recycling endosomes (RE) carrying VAMP8 and Stx11 (1) traffic to the CD8^{+} T cell immune synapse and (2) undergo VAMP8-dependent fusion with the plasma membrane, through interaction with plasma membrane Stx4 and SNAP-23 SNARE complex partners. Such exocytosis of recycling endosomes deposits high amounts of Stx11 molecules in the plasma membrane, which may (3) lead to the formation of Stx11 and SNAP-23 SNARE complexes. Munc18-2 interacts with Stx11, promoting stability and regulating Stx11 conformation. Initial activity at the immune synapse thus forms "active zones" where docking of cytotoxic granules containing perforin and granzymes may occur. Engagement of and signaling from the T cell receptor induces (4) recruitment of Rab27a and Munc13-4 to cytotoxic granules, facilitating clustering of granules to the microtubule organizing center and polarization of this structure toward the plasma membrane. At the plasma membrane, larger CGs can (5) dock through interactions of R-SNARE proteins with plasma membrane Stx11–SNAP-23 complexes, facilitating (6) Munc13-4-mediated CG exocytosis that releases cargo necessary for target cell killing. Proteins in which mutations have been associated with familial forms of HLH are highlighted in red or orange. The figure is adapted from reference [120]

Pathophysiology of Cytokine Storm Syndromes in T Cell-Targeted Immunotherapy

There is currently a strong interest in harnessing T cell function, in particular that of cytotoxic CD8^{+} T cells, for cellular immunotherapy of cancer. Different approaches have been taken. One promising avenue of investigation that already has reached the clinic is chimeric antigen receptor (CAR)-modified T cells [134]. Potentially fatal CSS represent a common complication of both epidermal growth factor receptor 2 and B cell receptor targeted T cells [135–137]. Usually occurring within days of CAR T cell infusion, the severity of CSS correlate with tumor load. In the clinic, these severe hyperinflammatory reactions have been treated with TNF blockade (etanercept) and anti-IL-6 receptor antibody (tocilizumab) [134, 138]. A mouse model of potentially lethal CSS developing 2–3 days after CAR T cell infusion has recently provided pathophysiological insights with potentially therapeutic implications. In this model, a range of cytokines including IL-6 and IFN-γ were elevated [139]. The CSS in this model was ameliorated by IL-6 receptor blockade. The severity in this model was not mediated by CAR T cell-derived cytokines, but by IL-6, IL-1 and nitric oxide (NO) produced by recipient macrophages. Remarkably, IL-1Ra blockade protected from severe CSS without compromising antitumor efficacy. These results demonstrate how innate immune cells can greatly exacerbate pathology instigated by strong T cell responses [139]. Importantly, together with studies of innate cytokines in HLH [73], they provide a rationale for targeting the innate immune system for treatment of diverse CSS.

Summary and Therapeutic Perspectives

A large number of studies ranging from animal models to clinical observations in CSS patient samples implicate CD8+ T cells as key culprits in propagating hyperinflammatory disease, both in settings of familial HLH as well as in T cell-mediated immunotherapy. Their role appears more peripheral in more autoinflammatory CSS forms such as MAS. Recent reports have highlighted the release of cytokines from innate cells as an important component of the inflammatory cascade both downstream of inflammasome activation and of T cell-related tissue destruction in distinct forms of CSS. Such novel mechanistic insights are providing new, promising therapeutic targets.

An etoposide and dexamethasone-based regiment has proven efficacy in treating HLH [140], effectively ablating activated T cells [141], and thereby bridging familial HLH patients to curative hematopoietic stem cell transplantation. However, more effective therapies based on immunological knowledge are called for [142]. IL-1β antagonists have successfully been used to treat MAS patients [143]. Molecular insights to MAS caused by activating *NLRC4* mutation have uncovered IL-18 inhibition as a promising disease target [144]. Animal models of familial HLH have provided a rational for neutralization of CD8+ T cell-derived IFN-γ [68, 69], although this may not have any effect in some forms of HLH and may be less effective in older individuals that have a more mature, experienced immune system [85, 86]. Potentially attenuating pro-inflammatory cytokine signalling and production in both CD8+ T cells and innate immune cells, the JAK1/2 inhibitor ruxolitinib has demonstrated remarkably efficacy in models of both familial HLH and MAS [145, 146]. These findings hold promise for broadly targeting hyperactivated cells in a variety of CSS, thereby reducing patient morbidity and mortality.

Acknowledgments The authors would like to thank members of the Bryceson laboratory as well as colleagues in the HLH research field for insightful discussions.

References

1. Henter, J. I., Ehrnst, A., Andersson, J., & Elinder, G. (1993). Familial hemophagocytic lymphohistiocytosis and viral infections. *Acta Paediatrica, 82*, 369–372.
2. Grossman, W. J., Radhi, M., Schauer, D., Gerday, E., Grose, C., & Goldman, F. D. (2005). Development of hemophagocytic lymphohistiocytosis in triplets infected with HHV-8. *Blood, 106*, 1203–1206.
3. Henter, J. I., Elinder, G., Soder, O., Hansson, M., Andersson, B., & Andersson, U. (1991). Hypercytokinemia in familial hemophagocytic lymphohistiocytosis. *Blood, 78*, 2918–2922.
4. Takada, H., Ohga, S., Mizuno, Y., Suminoe, A., Matsuzaki, A., Ihara, K., et al. (1999). Oversecretion of IL-18 in haemophagocytic lymphohistiocytosis: A novel marker of disease activity. *British Journal of Haematology, 106*, 182–189.
5. Akashi, K., Hayashi, S., Gondo, H., Mizuno, S., Harada, M., Tamura, K., et al. (1994). Involvement of interferon-gamma and macrophage colony-stimulating factor in pathogenesis of haemophagocytic lymphohistiocytosis in adults. *British Journal of Haematology, 87*, 243–250.
6. Osugi, Y., Hara, J., Tagawa, S., Takai, K., Hosoi, G., Matsuda, Y., et al. (1997). Cytokine production regulating Th1 and Th2 cytokines in hemophagocytic lymphohistiocytosis. *Blood, 89*, 4100–4103.
7. My, L. T., Lien le, B., Hsieh, W. C., Imamura, T., Anh, T. N., Anh, P. N., et al. (2010). Comprehensive analyses and characterization of haemophagocytic lymphohistiocytosis in Vietnamese children. *British Journal of Haematology, 148*, 301–310.
8. Vaiselbuh, S. R., Bryceson, Y. T., Allen, C. E., Whitlock, J. A., & Abla, O. (2014). Updates on histiocytic disorders. *Pediatric Blood & Cancer, 61*, 1329–1335.
9. Janka, G. E. (2012). Familial and acquired hemophagocytic lymphohistiocytosis. *Annual Review of Medicine, 63*, 233–246.
10. Henter, J. I., Horne, A., Arico, M., Egeler, R. M., Filipovich, A. H., Imashuku, S., et al. (2007). HLH-2004: Diagnostic and therapeutic guidelines for hemophagocytic lymphohistiocytosis. *Pediatric Blood & Cancer, 48*, 124–131.
11. Davi, S., Consolaro, A., Guseinova, D., Pistorio, A., Ruperto, N., Martini, A., et al. (2011). An international consensus survey of diagnostic criteria for macrophage activation syndrome in systemic juvenile idiopathic arthritis. *The Journal of Rheumatology, 38*, 764–768.
12. Minoia, F., Bovis, F., Davi, S., Insalaco, A., Lehmberg, K., Shenoi, S., et al. (2017). Development and initial validation of the macrophage activation syndrome/primary hemophagocytic lymphohistiocytosis score, a diagnostic tool that differentiates primary hemophagocytic lymphohistiocytosis from macrophage activation syndrome. *The Journal of Pediatrics, 189*, 72–78.e73.
13. Weiss, E. S., Girard-Guyonvarc'h, C., Holzinger, D., de Jesus, A. A., Tariq, Z., Picarsic, J., et al. (2018). Interleukin-18 diagnostically distinguishes and pathogenically promotes human and murine macrophage activation syndrome. *Blood, 131*, 1442–1455.
14. Teachey, D. T., Rheingold, S. R., Maude, S. L., Zugmaier, G., Barrett, D. M., Seif, A. E., et al. (2013). Cytokine release syndrome after blinatumomab treatment related to abnormal macrophage activation and ameliorated with cytokine-directed therapy. *Blood, 121*, 5154–5157.
15. Suntharalingam, G., Perry, M. R., Ward, S., Brett, S. J., Castello-Cortes, A., Brunner, M. D., et al. (2006). Cytokine storm in a phase 1 trial of the anti-CD28 monoclonal antibody TGN1412. *The New England Journal of Medicine, 355*, 1018–1028.
16. Stepp, S. E., Dufourcq-Lagelouse, R., Le Deist, F., Bhawan, S., Certain, S., Mathew, P. A., et al. (1999). Perforin gene defects in familial hemophagocytic lymphohistiocytosis. *Science, 286*, 1957–1959.
17. Bossi, G., & Griffiths, G. M. (1999). Degranulation plays an essential part in regulating cell surface expression of Fas ligand in T cells and natural killer cells. *Nature Medicine, 5*, 90–96.

18. Fisher, G. H., Rosenberg, F. J., Straus, S. E., Dale, J. K., Middleton, L. A., Lin, A. Y., et al. (1995). Dominant interfering Fas gene mutations impair apoptosis in a human autoimmune lymphoproliferative syndrome. *Cell, 81*, 935–946.
19. Oliveira, J. B., Bleesing, J. J., Dianzani, U., Fleisher, T. A., Jaffe, E. S., Lenardo, M. J., et al. (2010). Revised diagnostic criteria and classification for the autoimmune lymphoproliferative syndrome (ALPS): Report from the 2009 NIH International Workshop. *Blood, 116*, e35–e40.
20. Feldmann, J., Callebaut, I., Raposo, G., Certain, S., Bacq, D., Dumont, C., et al. (2003). Munc13-4 is essential for cytolytic granules fusion and is mutated in a form of familial hemophagocytic lymphohistiocytosis (FHL3). *Cell, 115*, 461–473.
21. zur Stadt, U., Schmidt, S., Kasper, B., Beutel, K., Diler, A. S., Henter, J. I., et al. (2005). Linkage of familial hemophagocytic lymphohistiocytosis (FHL) type-4 to chromosome 6q24 and identification of mutations in syntaxin 11. *Human Molecular Genetics, 14*, 827–834.
22. zur Stadt, U., Rohr, J., Seifert, W., Koch, F., Grieve, S., Pagel, J., et al. (2009). Familial hemophagocytic lymphohistiocytosis type 5 (FHL-5) is caused by mutations in Munc18-2 and impaired binding to syntaxin 11. *American Journal of Human Genetics, 85*, 482–492.
23. Cote, M., Menager, M. M., Burgess, A., Mahlaoui, N., Picard, C., Schaffner, C., et al. (2009). Munc18-2 deficiency causes familial hemophagocytic lymphohistiocytosis type 5 and impairs cytotoxic granule exocytosis in patient NK cells. *The Journal of Clinical Investigation, 119*, 3765–3773.
24. Bryceson, Y. T., Rudd, E., Zheng, C., Edner, J., Ma, D., Wood, S. M., et al. (2007). Defective cytotoxic lymphocyte degranulation in syntaxin-11 deficient familial hemophagocytic lymphohistiocytosis 4 (FHL4) patients. *Blood, 110*, 1906–1915.
25. Menasche, G., Pastural, E., Feldmann, J., Certain, S., Ersoy, F., Dupuis, S., et al. (2000). Mutations in RAB27A cause Griscelli syndrome associated with haemophagocytic syndrome. *Nature Genetics, 25*, 173–176.
26. Barbosa, M. D., Nguyen, Q. A., Tchernev, V. T., Ashley, J. A., Detter, J. C., Blaydes, S. M., et al. (1996). Identification of the homologous beige and Chediak-Higashi syndrome genes. *Nature, 382*, 262–265.
27. Roder, J. C., Haliotis, T., Klein, M., Korec, S., Jett, J. R., Ortaldo, J., et al. (1980). A new immunodeficiency disorder in humans involving NK cells. *Nature, 284*, 553–555.
28. Coffey, A. J., Brooksbank, R. A., Brandau, O., Oohashi, T., Howell, G. R., Bye, J. M., et al. (1998). Host response to EBV infection in X-linked lymphoproliferative disease results from mutations in an SH2-domain encoding gene. *Nature Genetics, 20*, 129–135.
29. Rigaud, S., Fondaneche, M. C., Lambert, N., Pasquier, B., Mateo, V., Soulas, P., et al. (2006). XIAP deficiency in humans causes an X-linked lymphoproliferative syndrome. *Nature, 444*, 110–114.
30. Nichols, K. E., Harkin, D. P., Levitz, S., Krainer, M., Kolquist, K. A., Genovese, C., et al. (1998). Inactivating mutations in an SH2 domain-encoding gene in X-linked lymphoproliferative syndrome. *Proceedings of the National Academy of Sciences of the United States of America, 95*, 13765–13770.
31. Sayos, J., Wu, C., Morra, M., Wang, N., Zhang, X., Allen, D., et al. (1998). The X-linked lymphoproliferative-disease gene product SAP regulates signals induced through the co-receptor SLAM. *Nature, 395*, 462–469.
32. Bryceson, Y. T., Pende, D., Maul-Pavicic, A., Gilmour, K. C., Ufheil, H., Vraetz, T., et al. (2012). A prospective evaluation of degranulation assays in the rapid diagnosis of familial hemophagocytic syndromes. *Blood, 119*, 2754–2763.
33. Sepulveda, F. E., Debeurme, F., Menasche, G., Kurowska, M., Cote, M., Pachlopnik Schmid, J., et al. (2013). Distinct severity of HLH in both human and murine mutants with complete loss of cytotoxic effector PRF1, RAB27A, and STX11. *Blood, 121*, 595–603.
34. Horne, A., Ramme, K. G., Rudd, E., Zheng, C., Wali, Y., al-Lamki, Z., et al. (2008). Characterization of PRF1, STX11 and UNC13D genotype-phenotype correlations in familial hemophagocytic lymphohistiocytosis. *British Journal of Haematology, 143*, 75–83.

35. Sieni, E., Cetica, V., Santoro, A., Beutel, K., Mastrodicasa, E., Meeths, M., et al. (2011). Genotype-phenotype study of familial haemophagocytic lymphohistiocytosis type 3. *Journal of Medical Genetics, 48*, 343–352.
36. Meeths, M., Horne, A., Sabel, M., Bryceson, Y. T., & Henter, J. I. (2015). Incidence and clinical presentation of primary hemophagocytic lymphohistiocytosis in Sweden. *Pediatric Blood & Cancer, 62*, 346–352.
37. Verbsky, J. W., Baker, M. W., Grossman, W. J., Hintermeyer, M., Dasu, T., Bonacci, B., et al. (2012). Newborn screening for severe combined immunodeficiency; the Wisconsin experience (2008-2011). *Journal of Clinical Immunology, 32*, 82–88.
38. Meeths, M., Chiang, S. C., Wood, S. M., Entesarian, M., Schlums, H., Bang, B., et al. (2011). Familial hemophagocytic lymphohistiocytosis type 3 (FHL3) caused by deep intronic mutation and inversion in UNC13D. *Blood, 188*, 5783–5793.
39. Seo, J. Y., Song, J. S., Lee, K. O., Won, H. H., Kim, J. W., Kim, S. H., et al. (2012). Founder effects in two predominant intronic mutations of UNC13D, c.118-308C>T and c.754-1G>C underlie the unusual predominance of type 3 familial hemophagocytic lymphohistiocytosis (FHL3) in Korea. *Annals of Hematology, 92*, 357–364.
40. Entesarian, M., Chiang, S. C., Schlums, H., Meeths, M., Chan, M. Y., Mya, S. N., et al. (2013). Novel deep intronic and missense UNC13D mutations in familial haemophagocytic lymphohistiocytosis type 3. *British Journal of Haematology, 162*, 415–418.
41. Qian, Y., Johnson, J. A., Connor, J. A., Valencia, C. A., Barasa, N., Schubert, J., et al. (2014). The 253-kb inversion and deep intronic mutations in UNC13D are present in North American patients with familial hemophagocytic lymphohistiocytosis 3. *Pediatric Blood & Cancer, 61*, 1034–1040.
42. Cichocki, F., Schlums, H., Li, H., Stache, V., Holmes, T., Lenvik, T. R., et al. (2014). Transcriptional regulation of Munc13-4 expression in cytotoxic lymphocytes is disrupted by an intronic mutation associated with a primary immunodeficiency. *The Journal of Experimental Medicine, 211*, 1079–1091.
43. Schulert, G. S., Zhang, M., Husami, A., Fall, N., Brunner, H., Zhang, K., et al. (2018). Brief report: Novel UNC13D intronic variant disrupting an NF-kappaB enhancer in a patient with recurrent macrophage activation syndrome and systemic juvenile idiopathic arthritis. *Arthritis & Rhematology, 70*, 963–970.
44. Tesi, B., Rascon, J., Chiang, S. C. C., Burnyte, B., Lofstedt, A., Fasth, A., et al. (2018). A RAB27A 5′ untranslated region structural variant associated with late-onset hemophagocytic lymphohistiocytosis and normal pigmentation. *The Journal of Allergy and Clinical Immunology, 142*, 317–321 e318.
45. Grandin, V., Sepulveda, F. E., Lambert, N., Al Zahrani, M., Al Idrissi, E., Al-Mousa, H., et al. (2017). A RAB27A duplication in several cases of Griscelli syndrome type 2: An explanation for cases lacking a genetic diagnosis. *Human Mutation, 38*, 1355–1359.
46. Chia, J., Yeo, K. P., Whisstock, J. C., Dunstone, M. A., Trapani, J. A., & Voskoboinik, I. (2009). Temperature sensitivity of human perforin mutants unmasks subtotal loss of cytotoxicity, delayed FHL, and a predisposition to cancer. *Proceedings of the National Academy of Sciences of the United States of America, 106*, 9809–9814.
47. Rudd, E., Bryceson, Y. T., Zheng, C., Edner, J., Wood, S. M., Ramme, K., et al. (2008). Spectrum, and clinical and functional implications of UNC13D mutations in familial hemophagocytic lymphohistiocytosis. *Journal of Medical Genetics, 45*, 134–141.
48. Terrell, C. E., & Jordan, M. B. (2013). Mixed hematopoietic or T-cell chimerism above a minimal threshold restores perforin-dependent immune regulation in perforin-deficient mice. *Blood, 122*, 2618–2621.
49. Hartz, B., Marsh, R., Rao, K., Henter, J. I., Jordan, M., Filipovich, L., et al. (2016). The minimum required level of donor chimerism in hereditary hemophagocytic lymphohistiocytosis. *Blood, 127*, 3281–3290.

50. Zhang, K., Jordan, M. B., Marsh, R. A., Johnson, J. A., Kissell, D., Meller, J., et al. (2011). Hypomorphic mutations in PRF1, MUNC13-4, and STXBP2 are associated with adult-onset familial hemophagocytic lymphohistiocytosis. *Blood.*
51. Meeths, M., Entesarian, M., Al-Herz, W., Chiang, S. C., Wood, S. M., Al-Ateeqi, W., et al. (2010). Spectrum of clinical presentations in familial hemophagocytic lymphohistiocytosis (FHL) type 5 patients with mutations in STXBP2. *Blood, 116*, 2635–2643.
52. Tesi, B., Lagerstedt-Robinson, K., Chiang, S. C., Bdira, E. B., Abboud, M., Belen, B., et al. (2015). Targeted high-throughput sequencing for genetic diagnostics of hemophagocytic lymphohistiocytosis. *Genome Medicine, 7*, 130.
53. Sepulveda, F. E., Garrigue, A., Maschalidi, S., Garfa-Traore, M., Menasche, G., Fischer, A., et al. (2016). Polygenic mutations in the cytotoxicity pathway increase susceptibility to develop HLH immunopathology in mice. *Blood, 127*, 2113–2121.
54. Chinn, I. K., Eckstein, O. S., Peckham-Gregory, E. C., Goldberg, B. R., Forbes, L. R., Nicholas, S. K., et al. (2018). Genetic and mechanistic diversity in pediatric hemophagocytic lymphohistiocytosis. *Blood, 132*, 89–100.
55. Spessott, W. A., Sanmillan, M. L., McCormick, M. E., Patel, N., Villanueva, J., Zhang, K., et al. (2015). Hemophagocytic lymphohistiocytosis caused by dominant-negative mutations in STXBP2 that inhibit SNARE-mediated membrane fusion. *Blood, 125*, 1566–1577.
56. Zhang, M., Bracaglia, C., Prencipe, G., Bemrich-Stolz, C. J., Beukelman, T., Dimmitt, R. A., et al. (2016). A heterozygous RAB27A mutation associated with delayed cytolytic granule polarization and hemophagocytic lymphohistiocytosis. *Journal of Immunology, 196*, 2492–2503.
57. Tesi, B., Chiang, S. C., El-Ghoneimy, D., Hussein, A. A., Langenskiold, C., Wali, R., et al. (2015). Spectrum of atypical clinical presentations in patients with biallelic PRF1 missense mutations. *Pediatric Blood & Cancer.*
58. Lofstedt, A., Chiang, S. C., Onelov, E., Bryceson, Y. T., Meeths, M., & Henter, J. I. (2015). Cancer risk in relatives of patients with a primary disorder of lymphocyte cytotoxicity: A retrospective cohort study. *The Lancet. Haematology, 2*, e536–e542.
59. Chaudhry, M. S., Gilmour, K. C., House, I. G., Layton, M., Panoskaltsis, N., Sohal, M., et al. (2016). Missense mutations in the perforin (PRF1) gene as a cause of hereditary cancer predisposition. *Oncoimmunology, 5*, e1179415.
60. Chiang, S. C., Theorell, J., Entesarian, M., Meeths, M., Mastafa, M., Al-Herz, W., et al. (2013). Comparison of primary human cytotoxic T-cell and natural killer cell responses reveal similar molecular requirements for lytic granule exocytosis but differences in cytokine production. *Blood, 121*, 1345–1356.
61. Chattopadhyay, P. K., Betts, M. R., Price, D. A., Gostick, E., Horton, H., Roederer, M., et al. (2009). The cytolytic enzymes granyzme A, granzyme B, and perforin: Expression patterns, cell distribution, and their relationship to cell maturity and bright CD57 expression. *Journal of Leukocyte Biology, 85*, 88–97.
62. Perez, N., Virelizier, J. L., Arenzana-Seisdedos, F., Fischer, A., & Griscelli, C. (1984). Impaired natural killer activity in lymphohistiocytosis syndrome. *The Journal of Pediatrics, 104*, 569–573.
63. Schneider, E. M., Lorenz, I., Muller-Rosenberger, M., Steinbach, G., Kron, M., & Janka-Schaub, G. E. (2002). Hemophagocytic lymphohistiocytosis is associated with deficiencies of cellular cytolysis but normal expression of transcripts relevant to killer-cell-induced apoptosis. *Blood, 100*, 2891–2898.
64. Binder, D., van den Broek, M. F., Kagi, D., Bluethmann, H., Fehr, J., Hengartner, H., et al. (1998). Aplastic anemia rescued by exhaustion of cytokine-secreting CD8+ T cells in persistent infection with lymphocytic choriomeningitis virus. *The Journal of Experimental Medicine, 187*, 1903–1920.
65. Matloubian, M., Suresh, M., Glass, A., Galvan, M., Chow, K., Whitmire, J. K., et al. (1999). A role for perforin in downregulating T-cell responses during chronic viral infection. *Journal of Virology, 73*, 2527–2536.

66. van Dommelen, S. L., Sumaria, N., Schreiber, R. D., Scalzo, A. A., Smyth, M. J., & Degli-Esposti, M. A. (2006). Perforin and granzymes have distinct roles in defensive immunity and immunopathology. *Immunity, 25*, 835–848.
67. Crozat, K., Hoebe, K., Ugolini, S., Hong, N. A., Janssen, E., Rutschmann, S., et al. (2007). Jinx, an MCMV susceptibility phenotype caused by disruption of Unc13d: A mouse model of type 3 familial hemophagocytic lymphohistiocytosis. *The Journal of Experimental Medicine, 204*, 853–863.
68. Jordan, M. B., Hildeman, D., Kappler, J., & Marrack, P. (2004). An animal model of hemophagocytic lymphohistiocytosis (HLH): CD8+ T cells and interferon gamma are essential for the disorder. *Blood, 104*, 735–743.
69. Pachlopnik Schmid, J., Ho, C. H., Chretien, F., Lefebvre, J. M., Pivert, G., Kosco-Vilbois, M., et al. (2009). Neutralization of IFNgamma defeats haemophagocytosis in LCMV-infected perforin- and Rab27a-deficient mice. *EMBO Molecular Medicine, 1*, 112–124.
70. Terrell, C. E., & Jordan, M. B. (2013). Perforin deficiency impairs a critical immunoregulatory loop involving murine CD8+ T cells and dendritic cells. *Blood, 121*, 5184–5191.
71. Krebs, P., Crozat, K., Popkin, D., Oldstone, M. B., & Beutler, B. (2011). Disruption of MyD88 signaling suppresses hemophagocytic lymphohistiocytosis in mice. *Blood, 117*, 6582–6588.
72. Jenkins, M. R., Rudd-Schmidt, J. A., Lopez, J. A., Ramsbottom, K. M., Mannering, S. I., Andrews, D. M., et al. (2015). Failed CTL/NK cell killing and cytokine hypersecretion are directly linked through prolonged synapse time. *The Journal of Experimental Medicine, 212*, 307–317.
73. Rood, J. E., Rao, S., Paessler, M., Kreiger, P. A., Chu, N., Stelekati, E., et al. (2016). ST2 contributes to T-cell hyperactivation and fatal hemophagocytic lymphohistiocytosis in mice. *Blood, 127*, 426–435.
74. Bode, S. F., Ammann, S., Al-Herz, W., Bataneant, M., Dvorak, C. C., Gehring, S., et al. (2015). The syndrome of hemophagocytic lymphohistiocytosis in primary immunodeficiencies: Implications for differential diagnosis and pathogenesis. *Haematologica, 100*, 978–988.
75. Tesi, B., & Bryceson, Y. T. (2018). HLH: Genomics illuminates pathophysiological diversity. *Blood, 132*, 5–7.
76. Behrens, E. M., Canna, S. W., Slade, K., Rao, S., Kreiger, P. A., Paessler, M., et al. (2011). Repeated TLR9 stimulation results in macrophage activation syndrome-like disease in mice. *The Journal of Clinical Investigation, 121*, 2264–2277.
77. Brisse, E., Imbrechts, M., Put, K., Avau, A., Mitera, T., Berghmans, N., et al. (2016). Mouse cytomegalovirus infection in BALB/c mice resembles virus-associated secondary hemophagocytic lymphohistiocytosis and shows a pathogenesis distinct from primary hemophagocytic lymphohistiocytosis. *Journal of Immunology, 196*, 3124–3134.
78. Brisse, E., Imbrechts, M., Mitera, T., Vandenhaute, J., Berghmans, N., Boon, L., et al. (2018). Lymphocyte-independent pathways underlie the pathogenesis of murine cytomegalovirus-associated secondary haemophagocytic lymphohistiocytosis. *Clinical and Experimental Immunology, 192*, 104–119.
79. Canna, S. W., de Jesus, A. A., Gouni, S., Brooks, S. R., Marrero, B., Liu, Y., et al. (2014). An activating NLRC4 inflammasome mutation causes autoinflammation with recurrent macrophage activation syndrome. *Nature Genetics, 46*, 1140–1146.
80. Kitamura, A., Sasaki, Y., Abe, T., Kano, H., & Yasutomo, K. (2014). An inherited mutation in NLRC4 causes autoinflammation in human and mice. *The Journal of Experimental Medicine, 211*, 2385–2396.
81. Yoshimoto, T., Takeda, K., Tanaka, T., Ohkusu, K., Kashiwamura, S., Okamura, H., et al. (1998). IL-12 up-regulates IL-18 receptor expression on T cells, Th1 cells, and B cells: Synergism with IL-18 for IFN-gamma production. *Journal of Immunology, 161*, 3400–3407.
82. Fehniger, T. A., Shah, M. H., Turner, M. J., VanDeusen, J. B., Whitman, S. P., Cooper, M. A., et al. (1999). Differential cytokine and chemokine gene expression by human NK cells following activation with IL-18 or IL-15 in combination with IL-12: Implications for the innate immune response. *Journal of Immunology, 162*, 4511–4520.

83. Zoller, E. E., Lykens, J. E., Terrell, C. E., Aliberti, J., Filipovich, A. H., Henson, P. M., et al. (2011). Hemophagocytosis causes a consumptive anemia of inflammation. *The Journal of Experimental Medicine, 208*, 1203–1214.
84. Canna, S. W., Wrobel, J., Chu, N., Kreiger, P. A., Paessler, M., & Behrens, E. M. (2013). Interferon-gamma mediates anemia but is dispensable for fulminant toll-like receptor 9-induced macrophage activation syndrome and hemophagocytosis in mice. *Arthritis and Rheumatism, 65*, 1764–1775.
85. Tesi, B., Sieni, E., Neves, C., Romano, F., Cetica, V., Cordeiro, A. I., et al. (2015). Hemophagocytic lymphohistiocytosis in 2 patients with underlying IFN-gamma receptor deficiency. *The Journal of Allergy and Clinical Immunology*.
86. Taylor, M. D., Burn, T. N., Wherry, E. J., & Behrens, E. M. (2018). CD8 T cell memory increases immunopathology in the perforin-deficient model of hemophagocytic lymphohistiocytosis secondary to TNF-alpha. *ImmunoHorizons, 2*, 67–73.
87. Waggoner, S. N., Cornberg, M., Selin, L. K., & Welsh, R. M. (2012). Natural killer cells act as rheostats modulating antiviral T cells. *Nature, 481*, 394–398.
88. Sepulveda, F. E., Maschalidi, S., Vosshenrich, C. A., Garrigue, A., Kurowska, M., Menasche, G., et al. (2015). A novel immunoregulatory role for NK-cell cytotoxicity in protection from HLH-like immunopathology in mice. *Blood, 125*, 1427–1434.
89. Ferlazzo, G., & Munz, C. (2009). Dendritic cell interactions with NK cells from different tissues. *Journal of Clinical Immunology, 29*, 265–273.
90. Kogl, T., Muller, J., Jessen, B., Schmitt-Graeff, A., Janka, G., Ehl, S., et al. (2013). Hemophagocytic lymphohistiocytosis in syntaxin-11-deficient mice: T-cell exhaustion limits fatal disease. *Blood, 121*, 604–613.
91. Ammann, S., Lehmberg, K., Zur Stadt, U., Janka, G., Rensing-Ehl, A., Klemann, C., et al. (2017). Primary and secondary hemophagocytic lymphohistiocytosis have different patterns of T-cell activation, differentiation and repertoire. *European Journal of Immunology, 47*, 364–373.
92. Humblet-Baron, S., Franckaert, D., Dooley, J., Bornschein, S., Cauwe, B., Schonefeldt, S., et al. (2016). IL-2 consumption by highly activated CD8 T cells induces regulatory T-cell dysfunction in patients with hemophagocytic lymphohistiocytosis. *The Journal of Allergy and Clinical Immunology, 138*, 200–209 e208.
93. Beura, L. K., Hamilton, S. E., Bi, K., Schenkel, J. M., Odumade, O. A., Casey, K. A., et al. (2016). Normalizing the environment recapitulates adult human immune traits in laboratory mice. *Nature, 532*, 512–516.
94. Podack, E. R., & Konigsberg, P. J. (1984). Cytolytic T cell granules. Isolation, structural, biochemical, and functional characterization. *The Journal of Experimental Medicine, 160*, 695–710.
95. de Saint Basile, G., Menasche, G., & Fischer, A. (2010). Molecular mechanisms of biogenesis and exocytosis of cytotoxic granules. *Nature Reviews. Immunology, 10*, 568–579.
96. de la Roche, M., Asano, Y., & Griffiths, G. M. (2016). Origins of the cytolytic synapse. *Nature Reviews. Immunology, 16*, 421–432.
97. Trapani, J. A., & Smyth, M. J. (2002). Functional significance of the perforin/granzyme cell death pathway. *Nature Reviews. Immunology, 2*, 735–747.
98. Lopez, J. A., Susanto, O., Jenkins, M. R., Lukoyanova, N., Sutton, V. R., Law, R. H., et al. (2013). Perforin forms transient pores on the target cell plasma membrane to facilitate rapid access of granzymes during killer cell attack. *Blood, 121*, 2659–2668.
99. Cohnen, A., Chiang, S. C., Stojanovic, A., Schmidt, H., Claus, M., Saftig, P., et al. (2013). Surface CD107a/LAMP-1 protects natural killer cells from degranulation-associated damage. *Blood, 122*, 1411–1418.
100. Krzewski, K., Gil-Krzewska, A., Nguyen, V., Peruzzi, G., & Coligan, J. E. (2013). LAMP1/CD107a is required for efficient perforin delivery to lytic granules and NK-cell cytotoxicity. *Blood, 121*, 4672–4683.

101. Abdalgani, M., Filipovich, A. H., Choo, S., Zhang, K., Gifford, C., Villanueva, J., et al. (2015). Accuracy of flow cytometric perforin screening for detecting patients with FHL due to PRF1 mutations. *Blood, 126*, 1858–1860.
102. Meade, J. L., de Wynter, E. A., Brett, P., Sharif, S. M., Woods, C. G., Markham, A. F., et al. (2006). A family with Papillon-Lefevre syndrome reveals a requirement for cathepsin C in granzyme B activation and NK cell cytolytic activity. *Blood, 107*, 3665–3668.
103. Orange, J. S. (2006). Human natural killer cell deficiencies. *Current Opinion in Allergy and Clinical Immunology, 6*, 399–409.
104. Pham, C. T., Ivanovich, J. L., Raptis, S. Z., Zehnbauer, B., & Ley, T. J. (2004). Papillon-Lefevre syndrome: Correlating the molecular, cellular, and clinical consequences of cathepsin C/dipeptidyl peptidase I deficiency in humans. *Journal of Immunology, 173*, 7277–7281.
105. Baetz, K., Isaaz, S., & Griffiths, G. M. (1995). Loss of cytotoxic T lymphocyte function in Chediak-Higashi syndrome arises from a secretory defect that prevents lytic granule exocytosis. *Journal of Immunology, 154*, 6122–6131.
106. Sepulveda, F. E., Burgess, A., Heiligenstein, X., Goudin, N., Menager, M. M., Romao, M., et al. (2015). LYST controls the biogenesis of the endosomal compartment required for secretory lysosome function. *Traffic, 16*, 191–203.
107. Gil-Krzewska, A., Wood, S. M., Murakami, Y., Nguyen, V., Chiang, S. C., Cullinane, A. R., et al. (2016). Chediak-Higashi syndrome: Lysosomal trafficking regulator domains regulate exocytosis of lytic granules but not cytokine secretion by natural killer cells. *The Journal of Allergy and Clinical Immunology, 137*, 1165–1177.
108. Chiang, S. C. C., Wood, S. M., Tesi, B., Akar, H. H., Al-Herz, W., Ammann, S., et al. (2017). Differences in granule morphology yet equally impaired exocytosis among cytotoxic T cells and NK cells from Chediak-Higashi syndrome patients. *Frontiers in Immunology, 8*, 426.
109. Jessen, B., Maul-Pavicic, A., Ufheil, H., Vraetz, T., Enders, A., Lehmberg, K., et al. (2011). Subtle differences in CTL cytotoxicity determine susceptibility to hemophagocytic lymphohistiocytosis in mice and humans with Chediak-Higashi syndrome. *Blood.*
110. Jessen, B., Kogl, T., Sepulveda, F. E., de Saint Basile, G., Aichele, P., & Ehl, S. (2013). Graded defects in cytotoxicity determine severity of hemophagocytic lymphohistiocytosis in humans and mice. *Frontiers in Immunology, 4*, 448.
111. Clark, R. H., Stinchcombe, J. C., Day, A., Blott, E., Booth, S., Bossi, G., et al. (2003). Adaptor protein 3-dependent microtubule-mediated movement of lytic granules to the immunological synapse. *Nature Immunology, 4*, 1111–1120.
112. Kurowska, M., Goudin, N., Nehme, N. T., Court, M., Garin, J., Fischer, A., et al. (2012). Terminal transport of lytic granules to the immune synapse is mediated by the kinesin-1/Slp3/Rab27a complex. *Blood, 119*, 3879–3889.
113. Shirakawa, R., Higashi, T., Tabuchi, A., Yoshioka, A., Nishioka, H., Fukuda, M., et al. (2004). Munc13-4 is a GTP-Rab27-binding protein regulating dense core granule secretion in platelets. *The Journal of Biological Chemistry, 279*, 10730–10737.
114. Chicka, M. C., Ren, Q., Richards, D., Hellman, L. M., Zhang, J., Fried, M. G., et al. (2016). Role of Munc13-4 as a Ca2+–dependent tether during platelet secretion. *The Biochemical Journal, 473*, 627–639.
115. Bin, N. R., Ma, K., Tien, C. W., Wang, S., Zhu, D., Park, S., et al. (2018). C2 domains of Munc13-4 are crucial for Ca(2+)-dependent degranulation and cytotoxicity in NK cells. *Journal of Immunology, 201*, 700–713.
116. Menager, M. M., Menasche, G., Romao, M., Knapnougel, P., Ho, C. H., Garfa, M., et al. (2007). Secretory cytotoxic granule maturation and exocytosis require the effector protein hMunc13-4. *Nature Immunology, 8*, 257–267.
117. Wood, S. M., Meeths, M., Chiang, S. C., Bechensteen, A. G., Boelens, J. J., Heilmann, C., et al. (2009). Different NK cell-activating receptors preferentially recruit Rab27a or Munc13-4 to perforin-containing granules for cytotoxicity. *Blood, 114*, 4117–4127.
118. Hellewell, A. L., Foresti, O., Gover, N., Porter, M. Y., & Hewitt, E. W. (2014). Analysis of familial hemophagocytic lymphohistiocytosis type 4 (FHL-4) mutant proteins reveals that

S-acylation is required for the function of syntaxin 11 in natural killer cells. *PLoS One, 9*, e98900.
119. Halimani, M., Pattu, V., Marshall, M. R., Chang, H. F., Matti, U., Jung, M., et al. (2013). Syntaxin11 serves as a t-SNARE for the fusion of lytic granules in human cytotoxic T lymphocytes. *European Journal of Immunology*.
120. Marshall, M. R., Pattu, V., Halimani, M., Maier-Peuschel, M., Muller, M. L., Becherer, U., et al. (2015). VAMP8-dependent fusion of recycling endosomes with the plasma membrane facilitates T lymphocyte cytotoxicity. *The Journal of Cell Biology, 210*, 135–151.
121. Muller, M. L., Chiang, S. C., Meeths, M., Tesi, B., Entesarian, M., Nilsson, D., et al. (2014). An N-terminal missense mutation in STX11 causative of FHL4 abrogates syntaxin-11 bnding to Munc18-2. *Frontiers in Immunology, 4*, 515.
122. Hackmann, Y., Graham, S. C., Ehl, S., Honing, S., Lehmberg, K., Arico, M., et al. (2013). Syntaxin binding mechanism and disease-causing mutations in Munc18-2. *Proceedings of the National Academy of Sciences of the United States of America, 110*, E4482–E4491.
123. Rickman, C., Medine, C. N., Bergmann, A., & Duncan, R. R. (2007). Functionally and spatially distinct modes of munc18-syntaxin 1 interaction. *The Journal of Biological Chemistry, 282*, 12097–12103.
124. Dulubova, I., Khvotchev, M., Liu, S., Huryeva, I., Sudhof, T. C., & Rizo, J. (2007). Munc18-1 binds directly to the neuronal SNARE complex. *Proceedings of the National Academy of Sciences of the United States of America, 104*, 2697–2702.
125. Spessott, W. A., Sanmillan, M. L., McCormick, M. E., Kulkarni, V. V., & Giraudo, C. G. (2017). SM protein Munc18-2 facilitates transition of Syntaxin 11-mediated lipid mixing to complete fusion for T-lymphocyte cytotoxicity. *Proceedings of the National Academy of Sciences of the United States of America, 114*, E2176–E2185.
126. Valdez, A. C., Cabaniols, J. P., Brown, M. J., & Roche, P. A. (1999). Syntaxin 11 is associated with SNAP-23 on late endosomes and the trans-Golgi network. *Journal of Cell Science, 112*, 845–854.
127. Loo, L. S., Hwang, L. A., Ong, Y. M., Tay, H. S., Wang, C. C., & Hong, W. (2009). A role for endobrevin/VAMP8 in CTL lytic granule exocytosis. *European Journal of Immunology, 39*, 3520–3528.
128. Dressel, R., Elsner, L., Novota, P., Kanwar, N., & Fischer von Mollard, G. (2010). The exocytosis of lytic granules is impaired in Vti1b- or Vamp8-deficient CTL leading to a reduced cytotoxic activity following antigen-specific activation. *Journal of Immunology, 185*, 1005–1014.
129. Matti, U., Pattu, V., Halimani, M., Schirra, C., Krause, E., Liu, Y., et al. (2013). Synaptobrevin2 is the v-SNARE required for cytotoxic T-lymphocyte lytic granule fusion. *Nature Communications, 4*, 1439.
130. Spessott, W. A., Sanmillan, M. L., Kulkarni, V. V., McCormick, M. E., & Giraudo, C. G. (2017). Syntaxin 4 mediates endosome recycling for lytic granule exocytosis in cytotoxic T-lymphocytes. *Traffic, 18*, 442–452.
131. Wang, C. C., Ng, C. P., Lu, L., Atlashkin, V., Zhang, W., Seet, L. F., et al. (2004). A role of VAMP8/endobrevin in regulated exocytosis of pancreatic acinar cells. *Developmental Cell, 7*, 359–371.
132. Schoch, S., Deak, F., Konigstorfer, A., Mozhayeva, M., Sara, Y., Sudhof, T. C., et al. (2001). SNARE function analyzed in synaptobrevin/VAMP knockout mice. *Science, 294*, 1117–1122.
133. Chang, H. F., Mannebach, S., Beck, A., Ravichandran, K., Krause, E., Frohnweiler, K., et al. (2018). Cytotoxic granule endocytosis depends on the Flower protein. *The Journal of Cell Biology, 217*, 667–683.
134. Grupp, S. A., Kalos, M., Barrett, D., Aplenc, R., Porter, D. L., Rheingold, S. R., et al. (2013). Chimeric antigen receptor-modified T cells for acute lymphoid leukemia. *The New England Journal of Medicine, 368*, 1509–1518.
135. Grupp, S. A., Prak, E. L., Boyer, J., McDonald, K. R., Shusterman, S., Thompson, E., et al. (2012). Adoptive transfer of autologous T cells improves T-cell repertoire diversity

and long-term B-cell function in pediatric patients with neuroblastoma. *Clinical Cancer Research, 18*, 6732–6741.
136. Morgan, R. A., Yang, J. C., Kitano, M., Dudley, M. E., Laurencot, C. M., & Rosenberg, S. A. (2010). Case report of a serious adverse event following the administration of T cells transduced with a chimeric antigen receptor recognizing ERBB2. *Molecular Therapy, 18*, 843–851.
137. Brentjens, R., Yeh, R., Bernal, Y., Riviere, I., & Sadelain, M. (2010). Treatment of chronic lymphocytic leukemia with genetically targeted autologous T cells: Case report of an unforeseen adverse event in a phase I clinical trial. *Molecular Therapy, 18*, 666–668.
138. Davila, M. L., Riviere, I., Wang, X., Bartido, S., Park, J., Curran, K., et al. (2014). Efficacy and toxicity management of 19-28z CAR T cell therapy in B cell acute lymphoblastic leukemia. *Science Translational Medicine, 6*, 224ra225.
139. Giavridis, T., van der Stegen, S. J. C., Eyquem, J., Hamieh, M., Piersigilli, A., & Sadelain, M. (2018). CAR T cell-induced cytokine release syndrome is mediated by macrophages and abated by IL-1 blockade. *Nature Medicine, 24*, 731–738.
140. Bergsten, E., Horne, A., Arico, M., Astigarraga, I., Egeler, R. M., Filipovich, A. H., et al. (2017). Confirmed efficacy of etoposide and dexamethasone in HLH treatment: Long-term results of the cooperative HLH-2004 study. *Blood, 130*, 2728–2738.
141. Johnson, T. S., Terrell, C. E., Millen, S. H., Katz, J. D., Hildeman, D. A., & Jordan, M. B. (2014). Etoposide selectively ablates activated T cells to control the immunoregulatory disorder hemophagocytic lymphohistiocytosis. *Journal of Immunology, 192*, 84–91.
142. Ehl, S. (2017). Etoposide for HLH: The limits of efficacy. *Blood, 130*, 2692–2693.
143. Miettunen, P. M., Narendran, A., Jayanthan, A., Behrens, E. M., & Cron, R. Q. (2011). Successful treatment of severe paediatric rheumatic disease-associated macrophage activation syndrome with interleukin-1 inhibition following conventional immunosuppressive therapy: Case series with 12 patients. *Rheumatology, 50*, 417–419.
144. Canna, S. W., Girard, C., Malle, L., de Jesus, A., Romberg, N., Kelsen, J., et al. (2017). Life-threatening NLRC4-associated hyperinflammation successfully treated with IL-18 inhibition. *The Journal of Allergy and Clinical Immunology, 139*, 1698–1701.
145. Das, R., Guan, P., Sprague, L., Verbist, K., Tedrick, P., An, Q. A., et al. (2016). Janus kinase inhibition lessens inflammation and ameliorates disease in murine models of hemophagocytic lymphohistiocytosis. *Blood, 127*, 1666–1675.
146. Maschalidi, S., Sepulveda, F. E., Garrigue, A., Fischer, A., & de Saint Basile, G. (2016). Therapeutic effect of JAK1/2 blockade on the manifestations of hemophagocytic lymphohistiocytosis in mice. *Blood, 128*, 60–71.

Immunology of Cytokine Storm Syndromes: Natural Killer Cells

Anthony R. French and Megan A. Cooper

Introduction

Natural killer (NK) cells are innate immune lymphocytes first recognized more than 40 years ago for their ability to kill tumor cells without prior sensitization [1]. Since that time, it has been recognized that NK cell activation is tightly regulated by cytokines and through a repertoire of germ line-encoded receptors that allow NK cells to distinguish self- and nonself cells. NK cells also play a critical role in the early innate immune response by production of cytokines, in particular serving as an early source of interferon-gamma (IFN-γ), a cytokine important for the elimination of intracellular organisms and activation of an adaptive immune response. While classified as "innate" lymphocytes, NK cells have memory-like properties with altered responses based on prior experiences. Thus, our understanding of the biology of NK cells has changed over the last four decades, as has our appreciation of their significance in human disease.

Here we focus on the fundamental biology of human NK cells, and how alterations in NK cell function relate to cytokine storm syndrome (CSS) in primary and secondary HLH (pHLH and sHLH). NK cell cytotoxic function is frequently low or absent in both pHLH and sHLH, as well as relatives of patients with HLH [2]. In pHLH, impaired NK cell cytotoxicity is reflective of underlying genetic defects affecting mechanisms of lymphocyte killing. In sHLH, including macrophage activation syndrome (MAS), it is not always certain whether defects in NK cell cytotoxicity are due solely to the immune status and dysregulated cytokine levels of patients with CSS, or related to genetic risk factor(s).

A. R. French · M. A. Cooper (✉)
Division of Rheumatology/Immunology, Department of Pediatrics, Washington University School of Medicine, St. Louis, MO, USA
e-mail: French_a@wustl.edu; cooper_m@wustl.edu

R. Q. Cron, E. M. Behrens (eds.), *Cytokine Storm Syndrome*,
https://doi.org/10.1007/978-3-030-22094-5_10

NK Cell Functional Responses

NK cells have two primary roles in the immune response, production of cytokines and killing of cancer and virally infected cells (Fig. 1). There are many parallels between mouse and human NK cell differentiation, maintenance, and functional responses, and much of our knowledge of NK cell biology comes from seminal studies in the mouse system. However, there are some distinct differences in phenotypic markers and receptors expressed by mouse and human cells, and here we focus primarily on human NK cells and their role in CSS, with some discussion of key biologic findings from the murine system.

Production of Cytokines by NK Cells

In response to cytokine and target cell stimulation, NK cells rapidly produce multiple cytokines within hours, the most well characterized being IFN-γ [1]. Other cytokines released by human NK cells include proinflammatory and regulatory cytokines and chemokines such as tumor necrosis factor alpha (TNF-α), interleukin (IL)-10, granulocyte-macrophage colony-stimulating factor (GM-CSF), chemokine C-C motif ligand 3 (CCL3), CCL4, and CCL5 [3–6]. Following cytokine stimulation by antigen presenting cells (APCs) including macrophages and dendritic cells, NK cells provide an early innate immune source of IFN-γ and may also kill highly activated APCs (Fig. 1) [7–12]. Studies in the mouse demonstrate that NK cell-derived IFN-γ is important for the host response to tumors and viruses [13, 14].

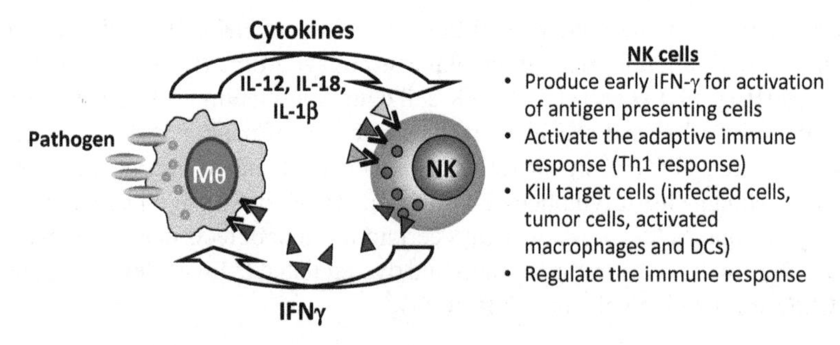

Fig. 1 NK cell and macrophage cytokine cross talk during a normal immune response. NK cells constitutively express multiple cytokine receptors and in response to infected and/or activated macrophages NK cells produce abundant IFN-γ. Early NK cell production of IFN-γ is important for activation of macrophages and elimination of intracellular organisms. Activated NK cells are capable of killing target cells, and NK cell IFN-γ influences the differentiation of a Th1 adaptive immune response. In addition, there is evidence that NK cells recognize and kill highly activated macrophages and dendritic cells through recognition of stress ligands upregulated on these cells, thereby helping to limit the immune response

IFN-γ is significantly elevated in the serum of patients with pHLH and sHLH (MAS) associated with systemic JIA (sJIA) and has been proposed to play a pathogenic role, leading to ongoing clinical trials of anti-IFN-γ monoclonal antibody therapy for children with these disorders [[15, 16] and ClinicalTrials.gov #NCT01818492 and #NCT03311854]. Data from multiple animal models of primary and secondary HLH supports pathogenicity of IFN-γ, with neutralization or genetic deletion of IFN-γ leading to amelioration of disease in several different models [17–21]. The source of IFN-γ and relative contribution of NK cell versus T cell-derived IFN-γ in HLH remains unclear. Uncontrolled cytokine cross talk between NK cells and macrophages, as in Fig. 1, has the potential to result in excessive and pathogenic cytokines, including IFN-γ. While evidence points toward the importance of IFN-γ in mediating CSS, this cytokine is unlikely to be the sole driver of disease, as evidenced by two patients with different genetic deficiencies in components of the IFN-γ receptor who presented with CSS in the context of systemic mycobacterial and viral (EBV and CMV) infections [22]. It is worth noting that while these patients fulfilled criteria for HLH, hemophagocytosis was not identified and may be dependent on IFN-γ receptor signaling in macrophages.

Interestingly, NK cells constitutively express transcript for IFN-γ and require activation by cytokines or through their receptors for translation and secretion of protein. The tight control of IFN-γ protein production by NK cells (and T lymphocytes) is regulated at multiple levels, including epigenetic modifications of the *IFNG* locus, upregulation of transcription after stimulation, posttranscriptional regulation of mRNA, and metabolism-dependent production and release of IFN-γ protein [23, 24]. The importance of regulation of constitutive *Ifng* transcript is highlighted by a murine model in which the 3′ untranslated region (UTR) of *Ifng* is disrupted, leading to high systemic levels of protein and associated autoimmunity [25]. It is unknown whether alterations in the *IFNG* gene or the regulatory mechanisms controlling constitutive IFN-γ transcript in NK cells production (i.e., posttranscriptional and metabolic) are associated with aberrant NK cell-derived IFN-γ in HLH.

NK Cell Receptors and Cytotoxicity

NK cells utilize a fixed assortment of germ line-encoded activating and inhibitory receptors (NKRs) to interact with their environment [26]. While the receptors in mouse and human are structurally dissimilar, they serve the same purpose, an interesting example of divergent evolution within the immune system [1]. Inhibitory and activating killer immunoglobulin-like receptors (KIR) are unique to humans and recognize classical MHC class I molecules. Conserved C-type lectin receptors (NKG2A and NKG2C) partner with CD94 to recognize nonclassical HLA molecules, such as HLA-E, and signal for activation (CD94/NKG2C) or inhibition (CD94/NKG2A) of NK cell cytotoxicity. Other major activating receptors include

CD16, an Fc receptor which directs NK-mediated antibody-dependent cellular cytotoxicity (ADCC), and NKG2D receptor which recognizes stress-induced self-ligands, both of which are conserved between mice and humans. The natural cytotoxicity receptor family (NCRs) in humans is comprised of NKp46, NKp44, and NKp30 which predominantly signal for NK cell activation in response to a diverse array of ligands, including some expressed by tumors [27]. Inhibitory KIR and NKG2A signal through immunoreceptor tyrosine-based inhibitor motifs (ITIMs), whereas activating receptors partner with adaptor molecules such as DAP12, TCR-ζ, or FcεRI-γ which express immunoreceptor tyrosine-based activating motifs (ITAMs), or DAP10 which signals through a YINM motif [26].

NK cell cytotoxicity requires the integration and overall balance of signals from inhibitory and activating receptors, as well as second signals from other adhesion molecules and receptors, resulting in a commitment to NK cell cytotoxicity. After identification of target cells, the process of NK cell killing is highly coordinated and requires the formation of an immunologic synapse between the NK cell and target cell, proper polarization of NK cell lytic granules containing perforin and granzymes to this synapse, and degranulation across the synapse toward the target cell [28]. Almost all genetic causes of pHLH cause impaired NK cell killing (Table 1 and Section "NK Cell Defects in Primary HLH (pHLH)"), either by deficiency of perforin or in the ability to form an effective synapse with target cells and directionally target lytic granules. It is unknown whether there are any associations with NKR genotype and HLH, as seen in other diseases including susceptibility to cancer, viral infection (Hepatitis C and HIV), and pregnancy complications [29–31].

Table 1 NK cell function in primary HLH (pHLH)

Disease	*GENE*/protein	NK killing	NK degranulation	Function
FHLH1	Unknown	–	–	–
FHLH2	*PRF1*/perforin	Absent	Normal	Pore-forming
FHLH3	*UNC13D*/Munc13-4	Decreased	Absent	Vesicle priming
FHLH4	*STX11*/syntaxin-11	Decreased	Absent	Vesicle fusion
FHLH5	*STXBP2*/syntaxin binding protein 2	Decreased	Absent	Vesicle fusion
XLP-1	*SH2D1A*/SH2 domain-containing protein 1A	Decreased	Normal	Signaling
XLP-2	*BIRC4*/XIAP	Normal	Normal	Signaling
Griscelli-2	*RAB27A*/Rab-27A	Decreased	Absent	Vesicle fusion
Chediak–Higashi	*LYST*/lysosomal-trafficking regulator	Absent	Absent	Vesicle trafficking
Hermansky–Pudlak type 2	*AP3B1*/adapter related protein complex 3 beta 1 subunit	Absent	Absent	Vesicle trafficking
NLRC4 inflammasomopathy	*NLRC4*/NLRC4	Variable	Unknown	Inflammasome

FHLH familial HLH
References: [96–108]

NK cell destruction of healthy-self cells should be prevented by recognition of self-MHC Class I molecules by inhibitory receptors. However, expression of inhibitory NKRs is stochastic and not linked genetically to the MHC locus, and in both humans and mice NK cells lacking inhibitory receptors specific for self-MHC class I are found. Such NK cells are actually hypofunctional, and do not require inhibition of their killing activity. The explanation for this phenomena comes from studies demonstrating a requirement for NK cell "education" via their inhibitory receptors, such that only those NK cells expressing inhibitory receptors recognizing self-MHC class I molecules acquire functional competence [32, 33]. Interestingly, NK cell education is not fixed, and can be altered by the environment, including cytokines and presence or absence of MHC Class I molecules. For example, during viral infection in mice, and following cytokine activation of human NK cells, "unlicensed" NK cells acquire the capacity for functional competence [34, 35]. Thus, it is possible with the high levels of proinflammatory cytokines in HLH normally tolerant NK cells, lacking appropriate self-MHC class I inhibitory receptors, might acquire functional competence and exhibit increased cytokine production and killing in response to interactions with self-cells. There have not yet been any studies of human NK cell education and loss of tolerance in the context of HLH, and this will be an interesting area to investigate.

NK Cell Subsets

Human NK cells are traditionally phenotypically identified as peripheral blood lymphocytes expressing CD56 and lacking other lineage markers including the T cell receptor CD3. With technological advances in protein and mRNA profiling, and investigation of NK cells from other tissues, we now recognize an array of human NK cell subsets with diverse phenotypes [4]. Indeed, NK cells are now classified as one member of a broader group of innate lymphoid cells (ILCs), the other members of which have distinct transcription factors and cytokine profiles, but lack inhibitory KIR and cytotoxic capacity.

CD56^{bright} and CD56^{dim} Peripheral Blood NK Cells

The first subset of human NK cells to be identified were CD56^{bright} and CD56^{dim} NK cells in the 1980s, based on cell-surface density of CD56 by flow cytometry. While the role of CD56 in NK cell biology is still largely unknown, approximately 5–15% of peripheral blood NK cells express high levels of this cell-surface receptor. CD56^{bright} NK cells mostly lack the activating Fc receptor CD16, have low expression of cytotoxic granules, and express high amounts of inhibitory NKG2A with low expression of KIRs and activating receptors. CD56^{dim} NK cells are more abundant in the peripheral blood and have higher expression of CD16, KIRs, and cytotoxic granules. These subsets have unique functional properties, with CD56^{bright} NK cells producing abundant IFN-γ and other cytokines in response to cytokine stimulation with generally poor cytotoxic capacity, whereas CD56^{dim} NK cells have enhanced cytotoxicity and cytokine responses when triggered through their activating NKRs [6, 7, 36]. While these subsets appear to fulfill distinct roles in the immune response, experimental evidence suggests that there is a developmental relationship with CD56^{bright} NK cells being precursors to more mature CD56^{dim} NK cells [37–39]. Relevant to HLH, in vitro stimulation of CD56^{bright} NK cells with the cytokines interleukin (IL)-2, 12, or 15 causes them to acquire phenotypic and functional properties of CD56^{dim} cells [37]. Decreased percentages of peripheral blood CD56^{bright} NK cells have been observed in patients with HLH and sJIA [40]. The functional significance of this finding is uncertain, as is whether this alteration is a consequence of excess cytokines or an inherent abnormality in NK cell differentiation.

Tissue Resident NK Cells

NK cells represent approximately 5–15% of peripheral blood lymphocytes, which are only a fraction of our total body NK cells. Tissue-resident NK cells (trNK), such as those found in lymph nodes, bone marrow, liver, spleen, gastrointestinal tract, and uterus exhibit distinct developmental origins, phenotypes, and functions [4, 41]. Phenotypically, trNK cells express variable amounts of CD56. For example, while in the peripheral blood CD56^{bright} cells are the minority, but they may actually represent a majority of NK cells in our body and are the predominant NK cell type in lymph nodes, the gastrointestinal tract, liver, uterus, visceral adipose tissue, and inflamed tissues. trNK cells appear to have unique roles in different tissues, and have phenotypic and functional differences when compared to peripheral blood CD56 subsets, that is, ability to produce cytokines and kill target cells [4]. The role and function of trNKs in HLH is unexplored. Given the diversity in the functional and phenotypic properties of trNK cells and peripheral blood NK cells, it seems relevant to examine the function of NK cells, particularly in affected tissues, such as the bone marrow and liver, to examine mechanisms by which trNK cells might modulate disease.

Adaptive NK Cells

NK cells with adaptive properties were first recognized in the murine system with experiments demonstrating that NK cells can mediate hapten-specific contact hypersensitivity [42]. There are three major differentiation pathways of NK cell memory responses identified to date, including antigen-specific liver NK cell responses in the mouse, and cytomegalovirus (CMV)-adapted NK cell memory and cytokine-induced NK cell memory-like responses in mice and humans [42–44]. In all of these models, following an initial activation event, NK cells display long-lived enhanced effector functions not dependent on continual stimulation.

A history of infection with CMV, a common trigger of CSS, is associated with an increased percentage of peripheral blood NK cells expressing high levels of the activating CD94/NKG2C [45, 46]. NKG2C^{high} NK cells with a mature phenotype (CD56^{dim}CD57^{+}NKG2A^{-}) and loss of intracellular FcεRγ expand during acute CMV and reactivation of latent CMV, and have been referred to as "adaptive" or "memory-like" [47–51]. Acute infection with other viruses has also been associated with an increased percentage of NKG2C^{high} NK cells, but only in those individuals with a history of CMV infection [52]. CMV-adapted NK cells have enhanced proliferation and production of IFN-γ in response to antibody stimulation through CD16 [46, 50]. Epigenetic remodeling has been demonstrated in CMV-adapted NK cells, suggesting a molecular basis for their differentiation and function [50, 51, 53].

Cytokine induction of memory-like NK cells describes the phenomena by which a single exposure to cytokines, in particular IL-12 and IL-18, leads to the differentiation of NK cells in mouse and human with long-term enhanced capacity for IFN-γ production and antitumor properties [43, 54–56]. Human memory-like NK cells upregulate CD25 (IL-2Rα) expression (resulting in expression of the heterotrimeric high affinity IL-2Rαβγ) and respond with enhanced proliferation, IFN-γ, and cytotoxicity to low-doses of IL-2 [57]. Cytokine-induced memory-like NK cells have antitumor responses in a mouse model, and were recently reported as adoptive immunotherapy in patients with acute myeloid leukemia [55, 58]. It remains to be identified how cytokine-induced memory-like NK cells develop during a physiologic response in vivo, for example, in the context of infection or CSS.

NK cell memory in the context of HLH has not been investigated, and it is unknown whether there are changes in NKG2C^{high} CMV-adaptive NK cells in this disorder. It is interesting to speculate that the cytokine environment of HLH might induce the differentiation of cytokine-induced memory-like NK cells, which may or may not be beneficial to the host given their high capacity for IFN-γ production, a pathogenic cytokine in HLH.

NK Cells in Human Disease

While the focus here is on the role of NK cells in CSS, some of the most well-defined roles for NK cells in human health is their protective role against infections and cancer. Perhaps the best examples of the importance of NK cells in human health are genetically defined primary immunodeficiencies in which patients lack NK cells or have significantly impaired NK cell function, either in isolation or as part of a larger syndrome. Patients with NK cell deficiencies (NKDs) uniformly have difficulties with herpesviruses, in particular CMV, stressing the importance of NK cells in our immune response to this class of viruses [59]. The nonredundant role of NK cells in host defense against viruses has been definitively confirmed in mice where depletion of NK cells renders mice very susceptible to viral infections, including MCMV [60]. There is also evidence supporting NK cell KIR/HLA haplotypes associated with risk or severity of infection with human immunodeficiency virus (HIV) and hepatitis C, providing further support for an antiviral role of human NK cells [31, 61].

While numerous studies in the mouse have demonstrated NK cell antitumor responses, the important role of human NK cells for tumor immunosurveillance is illustrated by an increased risk of cancer in patients with low NK cell function [29, 62]. NK cells are targets for cancer immunotherapy, including mismatch of NK inhibitory receptor and MHC ligand interactions in the context of hematopoietic cell transplantation (HCT) for acute myeloid leukemia (AML), NK cell adoptive immunotherapy (autologous and allogenic), and administration of antibodies, cytokines, or drugs aimed at enhancing NK cell function [63–65].

There is also correlative evidence for NK cells in other states of human health, including autoimmunity and pregnancy health [66, 67]. Human studies in rheumatoid arthritis (RA), multiple sclerosis (MS), and systemic lupus erythematosus (SLE) provide provocative but incomplete evidence for contributions of NK cells in the onset or progression of autoimmunity (reviewed in [67] and [68]).

NK Cell Defects in Primary HLH (pHLH)

Intrinsic disorders of NK cell cytotoxicity seen in pHLH may contribute to the pathogenesis of HLH through the inability to kill virally infected cells and/or hyperstimulated immune cells. Defects in cell-mediated cytotoxicity found in pHLH and sHLH result in impaired NK cell killing of target cells, a test used clinically to aid in the diagnosis of patients with HLH.

The genetic mutations that underlie most forms of pHLH lead to impaired NK and CD8 T lymphocyte cytotoxicity (NK defects summarized in Table 1) due to disrupted perforin expression or defective cytotoxic granule movement, docking, or fusion with the plasma membrane. The mechanisms and genetics of pHLH are described in detail elsewhere in this text, and the relevant NK cell abnormalities are summarized here and in Table 1. Perforin (*PRF1*, FHL2) is hydrophobic protein contained in the cytotoxic granules of NK and CD8 T cells. It polymerizes into a pore-forming structure in the plasma membrane or endosomal membrane of target cells and facilitates the entry of cytotoxic effector molecules. MUNC13-4 (*UNC13D*, FHL3) is involved in the vesicle-priming that follows granule docking but precedes fusion of the granule membrane with the plasma membrane. Syntaxin 11 (*STX11*, FHL4) is a SNARE protein (N-ethylmaleimide-sensitive factor attachment protein receptor) and with its binding partner syntaxin binding protein 2 (*STXBP2*, FHL5) facilitates fusion of the granule with the plasma membrane. In addition, there are several additional proteins (including *SH2D1A*, *RAB27A*, *LYST*, and *AP3B1*; Table 1) associated with immunodeficiencies that result in defective granule biogenesis or movement. Interestingly, heterozygous mutations in several of these HLH-associated genes have been shown to be sufficient to disrupt cytotoxic granule movement/degranulation and cause HLH [69, 70].

Biallelic (or hemizygous) deleterious mutations in the genes encoding these 8 proteins result in defective cell-mediated cytotoxicity. A gradation in severity of mutations in these proteins (assessed by age of HLH onset) was observed with perforin (early onset) > Rab-27a > syntaxin-11 (later onset) [71, 72]. pHLH typically presents in the first several years life; however adult-onset HLH has been associated with hypomorphic mutations (missense mutations or splicing variations) with later onset and a more indolent course [73].

Beyond measurement of NK cell cytotoxic capacity during acute HLH, there is little known about the functional role of NK cells in human pHLH. However, findings in STXBP2-deficient patients directly implicate defective NK cell cytolysis in HLH pathogenesis [74]. There is substantially higher expression of STXBP2 in human NK cells than CD8 T cells, and defects in degranulation have been observed in NK cells but not in CD8 T cells from STXBP2-deficient patients [74]. An immunoregulatory role for NK cells in HLH is also supported by a murine model of LMCV-induced HLH where cytotoxic NK cells were sufficient to protect mice from HLH immunopathology (discussed in more detail below in Section "Dysregulation of NK Cells During sHLH"; [75]).

NK Cell Defects in Secondary HLH

Secondary HLH (sHLH) is typically associated with infection, underlying rheumatologic disease, and/or malignancy. While defective NK cell killing is a clinical diagnostic criterion commonly used for defining sHLH in patients, it remains unclear whether NK cell dysfunction is critical to the pathogenesis of sHLH versus the result of the inflammatory environment in CSS in these patients. However, heterozygous mutations in HLH-associated genes have been reported to contribute to sHLH.

NK Cell Function in Systemic JIA

The rheumatic disease most commonly associated with sHLH is systemic juvenile idiopathic arthritis (sJIA). Low numbers of peripheral blood NK cells and decreased cytotoxicity have been identified in patients with sJIA [40, 76–78]. Furthermore, a subset of sJIA patients had nearly a complete absence of circulating CD56^{bright} NK cells [40, 79], a NK cell subset that accumulates in the synovial fluid of inflamed joints in adult rheumatoid arthritis (RA) patients [80, 81]. The impaired NK cell functional responses in sJIA patients have been linked to defective IL-18Rβ phosphorylation [79], low levels of perforin [40, 77, 78], and heterozygous pHLH mutations in components of the cytolytic pathway [82–84].

Heterozygous genetic variants in pHLH-associated genes in sJIA patients is intriguing and suggests that subtler defects in cytotoxicity may contribute to susceptibility to sJIA. Several sJIA patients with and without a history of sHLH have been identified with MUNC13-4 mutations including an 8 year old female with sJIA without sHLH who had compound heterozygous mutations [83] and 2 sJIA patients with biallelic MUNC13-4 mutations in a study of 18 sJIA patients who had previously had sHLH [84]. A larger European study of 56 sJIA patients (of which 15 had a history of sHLH) performed targeted sequencing of *PRF1* and found heterozygous mutations in perforin in 11 of the patients (20%) [82]. More recently a whole-exome sequencing study of 14 sJIA patients with a history of sHLH and their parents found that 5 of the patients (36%) had heterozygous mutations in known pHLH genes (of which one had heterozygous mutations in 2 different pHLH genes) [85]. In addition, this study found 3 novel recessive pairs of genes for which the parents were heterozygous and 20 heterozygous rare variants that occurred in at least 2 patients. Most of these variants were involved in intracellular vesicle transport, actin stabilization, or microtubule organization [85]. Two studies reported a high percentage of sJIA and other patients with heterozygous mutations identified by targeted sequencing of *PRF1*, *UNC13d*, *STX11*, *STXBP2*, and *RAB27a*. The first found that 13 out of 17 pediatric patients with a history of sHLH (76%) had a heterozygous mutation in at least one HLH-associated gene [86]. The second study found monoallelic mutations in 11 of 31 sJIA patients with a history of sHLH (36%) with 3 of the patients having heterozygous mutations in 2 genes [87]. Collectively, these studies suggest that sJIA patients are enriched for heterozygous mutations in pHLH-associated genes and that these mutations may potentially be contributing to presentation of sJIA as well as to the high incidence of sHLH seen in sJIA patients.

Dysregulation of NK Cells During sHLH

There are several lines of evidence suggesting that human NK cell dysfunction might contribute to the pathogenesis of sHLH. First, NK cells can down-modulate the inflammatory response by killing highly activated autologous macrophages, dendritic cells, and T cells [9–12]. For example, coculture of human NK cells with LPS-activated, but not naïve, macrophages leads to NK cell killing of these highly activated macrophages [9]. NK cell killing of activated macrophages was dependent on the activating NKG2D receptor, and macrophages upregulated NKG2D ligands. In murine models of lymphocytic choriomeningitis (LCMV) infection, NK cells indirectly limit cytotoxic CD8 T cell responses, including IFN-γ production and proliferation, by perforin-dependent killing of activated CD4 T cells [11, 12]. Together, these findings suggest that NK cells have the capacity to limit immunopathology associated with dysregulated immune cell activation, and the hypothesis that a failure of NK cells to limit immunopathology might be associated with sHLH. Second, high levels of IL-6, as seen in sJIA and CSS, inhibit NK cell function. Evidence for this comes from a study evaluating NK cell function in mice overexpressing IL-6 and in human NK cells cultured with IL-6. In both cases, NK cells exhibited decreased cytotoxic proteins with normal degranulation but impaired killing of target cells [88]. This suggests that increased levels of IL-6 in CSS might impair NK cell function, resulting in diminished capacity of NK cells to limit immunopathology. Finally, many cases of infection-associated MAS/sHLH are associated with herpesviral infections [89]. As discussed earlier, NK cells are particularly important for control of herpesviral infections, especially CMV and EBV. In a mouse model of murine CMV infection, depletion of NK cells led not only to increased susceptibility to infection, but also caused exaggerated T cell responses and IFN-γ production [90].

These studies all suggest that intact NK cell function may limit immunopathology in HLH, although more definitive studies in patients are lacking. Murine models of HLH have demonstrated that NK cells are not sufficient to cause disease. In a murine model of TLR9-triggered MAS, NK cells were dispensable for disease, although mice lacking T, B, and NK cells had less disease [20]. A perforin-deficient murine model of viral-driven (LCMV) HLH demonstrated that CD8 T cells, but not NK cells, are required for development of disease [17]. However, while NK cells are not sufficient to drive disease, they may still play an important regulatory role. A bone marrow chimera model, in which either cytotoxic T cells or NK cells lacked perforin, demonstrated that functionally competent wild-type NK cells can control LCMV-induced HLH immunopathology observed with perforin-deficient CD8 T cells [75]. In this model, cytotoxic NK cells controlled macrophage and cytotoxic T cell activation, but not viral burden, suggesting a protective regulatory role for NK cell cytotoxic function in this context.

NK Cell Function as Diagnostic Criteria for HLH

Low or absent NK cytotoxicity is one of eight diagnostic criteria in the 2004 International HLH Study Diagnostic Guidelines [91] that function as the standard in diagnosing both pHLH and sHLH. However, NK cytotoxicity testing is a functional assay that is not readily available at most centers. Ravelli and colleagues [92] recently proposed an abbreviated criteria to identify sHLH (MAS) in sJIA patients that did not include NK cytotoxicity. However, the utility of evaluating NK cytotoxicity may take on greater significance in the diagnosis of sHLH as more sJIA patients are treated with tocilizumab (anti-IL-6 receptor monoclonal antibody) which suppresses fever and ferritin levels [93], two of the primary criteria in the recently proposed sHLH classification scheme. However, NK killing assays do not discriminate between pHLH and sHLH, and several groups have advocated for the use of CD107a degranulation assays as an adjunct or alternative indirect assessment of NK cell cytotoxic capacity [94, 95].

Bryceson and colleagues studied 494 patients referred to three centers for evaluation of suspected HLH and demonstrated that CD107a degranulation studies had a high sensitivity and specificity for the 209 patients with a biallelic genetic diagnosis of HLH [94]. For example, 97% percent of the patients with FHLH3-5 (Table 1) had abnormal NK cell degranulation. However, 23% of the 266 patients with infection-associated or incomplete HLH had borderline degranulation while 15% had low degranulation, including two patients with heterozygous *UNC13D* mutations. Interestingly, the specificity was higher in those patients presenting with HLH before the age of 2 (97% compared to 81%). Thus, this study demonstrated that degranulation assays have the potential to rapidly and reliably identify patients with biallelic pHLH mutations, although their utility in sHLH is not as clear.

A subsequent study by Rubin and colleagues directly compared sensitivity and specificity of NK cytotoxicity, CD107a degranulation, and perforin expression in 1614 patients referred to a single center for targeted HLH genetic testing of seven genes (FHLH2-5, RAB27A, Lyst, and AP3B1) who also had at least 1 of the following tests performed: perforin protein expression, NK cell cytotoxicity, and/or CD107a degranulation [95]. They demonstrated that the combination of perforin and CD107a degranulation was more sensitive and no less specific than NK cytotoxicity testing in screening for pHLH in patients with biallelic mutations in one of the 7 tested HLH genes. Interestingly, 57% of the 635 patients with suspected HLH without biallelic mutations in this panel of 7 HLH genes had low NK cell cytotoxicity. Specifically, 20 of 38 (53%) patients with monoallelic likely disease-causing mutations, 48 of 75 (64%) of patients with variants of unknown clinical significance (VUCS), and 296 of 522 (57%) of patients without mutations in these 7 HLH genes had low cytotoxicity. In contrast, degranulation studies were low in almost all patients (30 of 32, 94%) with biallelic mutations in one of the tested HLH genes, and normal in the majority of patients (197 of 326, 60%) without identified biallelic defects. Abnormal degranulation was observed in patients with monoallelic mutations (3 of 11, 27%), VUCS (8 of 16, 50%), and no potentially deleterious variants (118 of 299, 39.5%) in the 7 tested HLH genes.

Thus, the approach of combining perforin and degranulation studies may have utility in rapidly identifying patients with biallelic mutations in pHLH genes and facilitating rapid treatment decisions. However, it may not be as useful in evaluating sHLH, even in patients with monoallelic mutations in pHLH genes thought to be associated with disease. In addition, this testing may not identify sHLH patients with genetic defects in currently unknown genes that compromise NK cytotoxicity but not degranulation, as demonstrated by higher percentage of abnormal cytotoxicity compared to abnormal degranulation in HLH patients without identified pHLH mutations.

Conclusion

The past four decades have seen tremendous advances in our understanding of NK cell biology and the cellular and molecular mechanisms responsible for their development and function. Genetic deficiencies in cytotoxic function are associated with pHLH, and clinical assays of NK cell killing are helpful for the diagnosis of these disorders. While defects in cytotoxic T cells have been the focus of much investigation of the pathogenesis of HLH, there is now increasing appreciation that dysregulation of NK cell cytotoxicity and failure to limit the immune response may contribute to disease. Further investigation of the functional, phenotypic, and molecular changes in peripheral blood and tissue resident NK cells has the potential to yield important insight into the pathogenesis and treatment of HLH.

Acknowledgments Work in French's laboratory is supported by NIH RO1AI078994 and a grant from the Strategic Pharma-Academic Research Consortium. Work in Cooper's laboratory is supported by NIH RO1AI127752.

References

1. Yokoyama, W. M. (2013). Natural killer cells. In W. Paul (Ed.), *Fundamental immunology* (pp. 395–430). Philadelphia: Lippincott, Williams &Wilkins.
2. Sullivan, K. E., Delaat, C. A., Douglas, S. D., & Filipovich, A. H. (1998). Defective natural killer cell function in patients with hemophagocytic lymphohistiocytosis and in first degree relatives. *Pediatric Research, 44*, 465–468.
3. Caligiuri, M. A. (2008). Human natural killer cells. *Blood, 112*, 461–469.
4. Freud, A. G., Mundy-Bosse, B. L., Yu, J., & Caligiuri, M. A. (2017). The broad spectrum of human natural killer cell diversity. *Immunity, 47*, 820–833.
5. Fehniger, T. A., Shah, M. H., Turner, M. J., VanDeusen, J. B., Whitman, S. P., Cooper, M. A., et al. (1999). Differential cytokine and chemokine gene expression by human NK cells following activation with IL-18 or IL-15 in combination with IL-12: Implications for the innate immune response. *Journal of Immunology, 162*, 4511–4520.
6. Fauriat, C., Long, E. O., Ljunggren, H. G., & Bryceson, Y. T. (2010). Regulation of human NK-cell cytokine and chemokine production by target cell recognition. *Blood, 115*, 2167–2176.
7. Cooper, M. A., Fehniger, T. A., Turner, S. C., Chen, K. S., Ghaheri, B. A., Ghayur, T., et al. (2001). Human natural killer cells: A unique innate immunoregulatory role for the CD56(bright) subset. *Blood, 97*, 3146–3151.
8. Lucas, M., Schachterle, W., Oberle, K., Aichele, P., & Diefenbach, A. (2007). Dendritic cells prime natural killer cells by trans-presenting interleukin 15. *Immunity, 26*, 503–517.
9. Nedvetzki, S., Sowinski, S., Eagle, R. A., Harris, J., Vely, F., Pende, D., et al. (2007). Reciprocal regulation of human natural killer cells and macrophages associated with distinct immune synapses. *Blood, 109*, 3776–3785.
10. Moretta, A. (2002). Natural killer cells and dendritic cells: Rendezvous in abused tissues. *Nature Reviews. Immunology, 2*, 957–964.
11. Waggoner, S. N., Cornberg, M., Selin, L. K., & Welsh, R. M. (2011). Natural killer cells act as rheostats modulating antiviral T cells. *Nature, 481*, 394–398.
12. Lang, P. A., Lang, K. S., Xu, H. C., Grusdat, M., Parish, I. A., Recher, M., et al. (2012). Natural killer cell activation enhances immune pathology and promotes chronic infection by limiting CD8+ T-cell immunity. *Proceedings of the National Academy of Sciences of the United States of America, 109*, 1210–1215.
13. O'Sullivan, T., Saddawi-Konefka, R., Vermi, W., Koebel, C. M., Arthur, C., White, J. M., et al. (2012). Cancer immunoediting by the innate immune system in the absence of adaptive immunity. *The Journal of Experimental Medicine, 209*, 1869–1882.
14. Loh, J., Chu, D. T., O'Guin, A. K., Yokoyama, W. M., & Virgin, H. W. (2005). Natural killer cells utilize both perforin and gamma interferon to regulate murine cytomegalovirus infection in the spleen and liver. *Journal of Virology, 79*, 661–667.

15. Henter, J. I., Elinder, G., Soder, O., Hansson, M., Andersson, B., & Andersson, U. (1991). Hypercytokinemia in familial hemophagocytic lymphohistiocytosis. *Blood, 78*, 2918–2922.
16. Bracaglia, C., de Graaf, K., Pires Marafon, D., Guilhot, F., Ferlin, W., Prencipe, G., et al. (2017). Elevated circulating levels of interferon-gamma and interferon-gamma-induced chemokines characterise patients with macrophage activation syndrome complicating systemic juvenile idiopathic arthritis. *Annals of the Rheumatic Diseases, 76*, 166–172.
17. Jordan, M. B., Hildeman, D., Kappler, J., & Marrack, P. (2004). An animal model of hemophagocytic lymphohistiocytosis (HLH): CD8+ T cells and interferon gamma are essential for the disorder. *Blood, 104*, 735–743.
18. Pachlopnik Schmid, J., Ho, C. H., Chretien, F., Lefebvre, J. M., Pivert, G., Kosco-Vilbois, M., et al. (2009). Neutralization of IFNgamma defeats haemophagocytosis in LCMV-infected perforin- and Rab27a-deficient mice. *EMBO Molecular Medicine, 1*, 112–124.
19. Prencipe, G., Caiello, I., Pascarella, A., Grom, A. A., Bracaglia, C., Chatel, L., et al. (2017). Neutralization of IFN-gamma reverts clinical and laboratory features in a mouse model of macrophage activation syndrome. *The Journal of Allergy and Clinical Immunology*.
20. Behrens, E. M., Canna, S. W., Slade, K., Rao, S., Kreiger, P. A., Paessler, M., et al. (2011). Repeated TLR9 stimulation results in macrophage activation syndrome-like disease in mice. *The Journal of Clinical Investigation, 121*, 2264–2277.
21. Prencipe, G., Caiello, I., Pascarella, A., Grom, A. A., Bracaglia, C., Chatel, L., et al. (2018). Neutralization of IFN-gamma reverts clinical and laboratory features in a mouse model of macrophage activation syndrome. *The Journal of Allergy and Clinical Immunology, 141*, 1439–1449.
22. Tesi, B., Sieni, E., Neves, C., Romano, F., Cetica, V., Cordeiro, A. I., et al. (2015). Hemophagocytic lymphohistiocytosis in 2 patients with underlying IFN-gamma receptor deficiency. *The Journal of Allergy and Clinical Immunology, 135*, 1638–1641.
23. Schoenborn, J. R., Dorschner, M. O., Sekimata, M., Santer, D. M., Shnyreva, M., Fitzpatrick, D. R., et al. (2007). Comprehensive epigenetic profiling identifies multiple distal regulatory elements directing transcription of the gene encoding interferon-gamma. *Nature Immunology, 8*, 732–742.
24. Mah, A. Y., & Cooper, M. A. (2016). Metabolic regulation of natural killer cell IFN-g production. *Critical Reviews in Immunology, 36*, 131–147.
25. Hodge, D. L., Berthet, C., Coppola, V., Kastenmuller, W., Buschman, M. D., Schaughency, P. M., et al. (2014). IFN-gamma AU-rich element removal promotes chronic IFN-gamma expression and autoimmunity in mice. *Journal of Autoimmunity, 53*, 33–45.
26. Lanier, L. L. (2008). Up on the tightrope: Natural killer cell activation and inhibition. *Nature Immunology, 9*, 495–502.
27. Pazina, T., Shemesh, A., Brusilovsky, M., Porgador, A., & Campbell, K. S. (2017). Regulation of the functions of natural cytotoxicity receptors by interactions with diverse ligands and alterations in splice variant expression. *Frontiers in Immunology, 8*, 369.
28. Mace, E. M., Dongre, P., Hsu, H. T., Sinha, P., James, A. M., Mann, S. S., et al. (2014). Cell biological steps and checkpoints in accessing NK cell cytotoxicity. *Immunology and Cell Biology, 92*, 245–255.
29. Malmberg, K. J., Carlsten, M., Bjorklund, A., Sohlberg, E., Bryceson, Y. T., & Ljunggren, H. G. (2017). Natural killer cell-mediated immunosurveillance of human cancer. *Seminars in Immunology, 31*, 20–29.
30. Colucci, F. (2017). The role of KIR and HLA interactions in pregnancy complications. *Immunogenetics, 69*, 557–565.
31. Khakoo, S. I., Thio, C. L., Martin, M. P., Brooks, C. R., Gao, X., Astemborski, J., et al. (2004). HLA and NK cell inhibitory receptor genes in resolving hepatitis C virus infection. *Science, 305*, 872–874.
32. Elliott, J. M., & Yokoyama, W. M. (2011). Unifying concepts of MHC-dependent natural killer cell education. *Trends in Immunology, 32*, 364–372.
33. Cooper, M. A. (2016). Teach your NK cells well. *Immunity, 45*, 229–231.

34. Orr, M. T., Murphy, W. J., & Lanier, L. L. (2010). 'Unlicensed' natural killer cells dominate the response to cytomegalovirus infection. *Nature Immunology, 11*, 321–327.
35. Wagner, J. A., Berrien-Elliott, M. M., Rosario, M., Leong, J. W., Jewell, B. A., Schappe, T., et al. (2017). Cytokine-induced memory-like differentiation enhances unlicensed natural killer cell antileukemia and FcgammaRIIIa-triggered responses. *Biology of Blood and Marrow Transplantation, 23*, 398–404.
36. Cooper, M. A., Fehniger, T. A., & Caligiuri, M. A. (2001). The biology of human natural killer-cell subsets. *Trends in Immunology, 22*, 633–640.
37. Romagnani, C., Juelke, K., Falco, M., Morandi, B., D'Agostino, A., Costa, R., et al. (2007). CD56brightCD16- killer Ig-like receptor- NK cells display longer telomeres and acquire features of CD56dim NK cells upon activation. *Journal of Immunology, 178*, 4947–4955.
38. Chan, A., Hong, D. L., Atzberger, A., Kollnberger, S., Filer, A. D., Buckley, C. D., et al. (2007). CD56bright human NK cells differentiate into CD56dim cells: Role of contact with peripheral fibroblasts. *Journal of Immunology, 179*, 89–94.
39. Freud, A. G., & Caligiuri, M. A. (2006). Human natural killer cell development. *Immunological Reviews, 214*, 56–72.
40. Villanueva, J., Lee, S., Giannini, E. H., Graham, T. B., Passo, M. H., Filipovich, A., et al. (2005). Natural killer cell dysfunction is a distinguishing feature of systemic onset juvenile rheumatoid arthritis and macrophage activation syndrome. *Arthritis Research & Therapy, 7*, R30–R37.
41. Melsen, J. E., Lugthart, G., Lankester, A. C., & Schilham, M. W. (2016). Human circulating and tissue-resident CD56(bright) natural killer cell populations. *Frontiers in Immunology, 7*, 262.
42. O'Leary, J. G., Goodarzi, M., Drayton, D. L., & von Andrian, U. H. (2006). T cell- and B cell-independent adaptive immunity mediated by natural killer cells. *Nature Immunology, 7*, 507–516.
43. Cooper, M. A., Elliott, J. M., Keyel, P. A., Yang, L., Carrero, J. A., & Yokoyama, W. M. (2009). Cytokine-induced memory-like natural killer cells. *Proceedings of the National Academy of Sciences of the United States of America, 106*, 1915–1919.
44. Sun, J. C., Beilke, J. N., & Lanier, L. L. (2009). Adaptive immune features of natural killer cells. *Nature, 457*, 557–561.
45. Guma, M., Angulo, A., Vilches, C., Gomez-Lozano, N., Malats, N., & Lopez-Botet, M. (2004). Imprint of human cytomegalovirus infection on the NK cell receptor repertoire. *Blood, 104*, 3664–3671.
46. Rolle, A., & Brodin, P. (2016). Immune adaptation to environmental influence: The case of NK cells and HCMV. *Trends in Immunology, 37*, 233–243.
47. Kuijpers, T. W., Baars, P. A., Dantin, C., van den Burg, M., van Lier, R. A., & Roosnek, E. (2008). Human NK cells can control CMV infection in the absence of T cells. *Blood, 112*, 914–915.
48. Lopez-Verges, S., Milush, J. M., Schwartz, B. S., Pando, M. J., Jarjoura, J., York, V. A., et al. (2011). Expansion of a unique CD57(+)NKG2Chi natural killer cell subset during acute human cytomegalovirus infection. *Proceedings of the National Academy of Sciences of the United States of America, 108*, 14725–14732.
49. Foley, B., Cooley, S., Verneris, M. R., Pitt, M., Curtsinger, J., Luo, X., et al. (2012). Cytomegalovirus reactivation after allogeneic transplantation promotes a lasting increase in educated NKG2C+ natural killer cells with potent function. *Blood, 119*, 2665–2674.
50. Lee, J., Zhang, T., Hwang, I., Kim, A., Nitschke, L., Kim, M., et al. (2015). Epigenetic modification and antibody-dependent expansion of memory-like NK cells in human cytomegalovirus-infected individuals. *Immunity, 42*, 431–442.
51. Schlums, H., Cichocki, F., Tesi, B., Theorell, J., Beziat, V., Holmes, T. D., et al. (2015). Cytomegalovirus infection drives adaptive epigenetic diversification of NK cells with altered signaling and effector function. *Immunity, 42*, 443–456.

52. Malmberg, K. J., Beziat, V., & Ljunggren, H. G. (2012). Spotlight on NKG2C and the human NK-cell response to CMV infection. *European Journal of Immunology, 42*, 3141–3145.
53. Luetke-Eversloh, M., Hammer, Q., Durek, P., Nordstrom, K., Gasparoni, G., Pink, M., et al. (2014). Human cytomegalovirus drives epigenetic imprinting of the IFNG locus in NKG2Chi natural killer cells. *PLoS Pathogens, 10*, e1004441.
54. Keppel, M. P., Yang, L., & Cooper, M. A. (2013). Murine NK cell intrinsic cytokine-induced memory-like responses are maintained following homeostatic proliferation. *Journal of Immunology, 190*, 4754–4762.
55. Romee, R., Rosario, M., Berrien-Elliott, M. M., Wagner, J. A., Jewell, B. A., Schappe, T., et al. (2016). Cytokine-induced memory-like natural killer cells exhibit enhanced responses against myeloid leukemia. *Science Translational Medicine, 8*(357), 357ra123.
56. Romee, R., Schneider, S. E., Leong, J. W., Chase, J. M., Keppel, C. R., Sullivan, R. P., et al. (2012). Cytokine activation induces human memory-like NK cells. *Blood, 120*, 4751–4760.
57. Leong, J. W., Chase, J. M., Romee, R., Schneider, S. E., Sullivan, R. P., Cooper, M. A., et al. (2014). Preactivation with IL-12, IL-15, and IL-18 induces CD25 and a functional high-affinity IL-2 receptor on human cytokine-induced memory-like natural killer cells. *Biology of Blood and Marrow Transplantation, 20*, 463–473.
58. Ni, J., Miller, M., Stojanovic, A., Garbi, N., & Cerwenka, A. (2012). Sustained effector function of IL-12/15/18-preactivated NK cells against established tumors. *The Journal of Experimental Medicine, 209*, 2351–2365.
59. Mace, E. M., & Orange, J. S. (2016). Genetic causes of human NK cell deficiency and their effect on NK cell subsets. *Frontiers in Immunology, 7*, 545.
60. Brown, M. G., Dokun, A. O., Heusel, J. W., Smith, H. R., Beckman, D. L., Blattenberger, E. A., et al. (2001). Vital involvement of a natural killer cell activation receptor in resistance to viral infection. *Science, 292*, 934–937.
61. Martin, M. P., & Carrington, M. (2013). Immunogenetics of HIV disease. *Immunological Reviews, 254*, 245–264.
62. Imai, K., Matsuyama, S., Miyake, S., Suga, K., & Nakachi, K. (2000). Natural cytotoxic activity of peripheral-blood lymphocytes and cancer incidence: An 11-year follow-up study of a general population. *Lancet, 356*, 1795–1799.
63. Knorr, D. A., Bachanova, V., Verneris, M. R., & Miller, J. S. (2014). Clinical utility of natural killer cells in cancer therapy and transplantation. *Seminars in Immunology, 26*, 161–172.
64. Berrien-Elliott, M. M., Romee, R., & Fehniger, T. A. (2015). Improving natural killer cell cancer immunotherapy. *Current Opinion in Organ Transplantation, 20*, 671–680.
65. Ruggeri, L., Capanni, M., Urbani, E., Perruccio, K., Shlomchik, W. D., Tosti, A., et al. (2002). Effectiveness of donor natural killer cell alloreactivity in mismatched hematopoietic transplants. *Science, 295*, 2097–2100.
66. Hiby, S. E., Walker, J. J., O'Shaughnessy, K. M., Redman, C. W., Carrington, M., Trowsdale, J., et al. (2004). Combinations of maternal KIR and fetal HLA-C genes influence the risk of preeclampsia and reproductive success. *The Journal of Experimental Medicine, 200*, 957–965.
67. Kulkarni, S., Martin, M. P., & Carrington, M. (2008). The Yin and Yang of HLA and KIR in human disease. *Seminars in Immunology, 20*, 343–352.
68. Fogel, L. A., Yokoyama, W. M., & French, A. R. (2013). Natural killer cells in human autoimmune disorders. *Arthritis Research & Therapy, 15*, 216.
69. Spessott, W. A., Sanmillan, M. L., McCormick, M. E., Patel, N., Villanueva, J., Zhang, K., et al. (2015). Hemophagocytic lymphohistiocytosis caused by dominant-negative mutations in STXBP2 that inhibit SNARE-mediated membrane fusion. *Blood, 125*, 1566–1577.
70. Zhang, M., Bracaglia, C., Prencipe, G., Bemrich-Stolz, C. J., Beukelman, T., Dimmitt, R. A., et al. (2016). A heterozygous RAB27A mutation associated with delayed cytolytic granule polarization and hemophagocytic lymphohistiocytosis. *Journal of Immunology, 196*, 2492–2503.

71. Sepulveda, F. E., Debeurme, F., Menasche, G., Kurowska, M., Cote, M., Pachlopnik Schmid, J., et al. (2013). Distinct severity of HLH in both human and murine mutants with complete loss of cytotoxic effector PRF1, RAB27A, and STX11. *Blood, 121*, 595–603.
72. Jessen, B., Kogl, T., Sepulveda, F. E., de Saint, B. G., Aichele, P., & Ehl, S. (2013). Graded defects in cytotoxicity determine severity of hemophagocytic lymphohistiocytosis in humans and mice. *Frontiers in Immunology, 4*, 448.
73. Zhang, K., Jordan, M. B., Marsh, R. A., Johnson, J. A., Kissell, D., Meller, J., et al. (2011). Hypomorphic mutations in PRF1, MUNC13-4, and STXBP2 are associated with adult-onset familial HLH. *Blood, 118*, 5794–5798.
74. Cote, M., Menager, M. M., Burgess, A., Mahlaoui, N., Picard, C., Schaffner, C., et al. (2009). Munc18-2 deficiency causes familial hemophagocytic lymphohistiocytosis type 5 and impairs cytotoxic granule exocytosis in patient NK cells. *The Journal of Clinical Investigation, 119*, 3765–3773.
75. Sepulveda, F. E., Maschalidi, S., Vosshenrich, C. A., Garrigue, A., Kurowska, M., Menasche, G., et al. (2015). A novel immunoregulatory role for NK-cell cytotoxicity in protection from HLH-like immunopathology in mice. *Blood, 125*, 1427–1434.
76. Wouters, C. H., Ceuppens, J. L., & Stevens, E. A. (2002). Different circulating lymphocyte profiles in patients with different subtypes of juvenile idiopathic arthritis. *Clinical and Experimental Rheumatology, 20*, 239–248.
77. Wulffraat, N. M., Rijkers, G. T., Elst, E., Brooimans, R., & Kuis, W. (2003). Reduced perforin expression in systemic juvenile idiopathic arthritis is restored by autologous stem-cell transplantation. *Rheumatology (Oxford), 42*, 375–379.
78. Grom, A. A., Villanueva, J., Lee, S., Goldmuntz, E. A., Passo, M. H., & Filipovich, A. (2003). Natural killer cell dysfunction in patients with systemic-onset juvenile rheumatoid arthritis and macrophage activation syndrome. *The Journal of Pediatrics, 142*, 292–296.
79. de Jager, W., Vastert, S. J., Beekman, J. M., Wulffraat, N. M., Kuis, W., Coffer, P. J., et al. (2009). Defective phosphorylation of interleukin-18 receptor beta causes impaired natural killer cell function in systemic-onset juvenile idiopathic arthritis. *Arthritis and Rheumatism, 60*, 2782–2793.
80. Dalbeth, N., & Callan, M. F. (2002). A subset of natural killer cells is greatly expanded within inflamed joints. *Arthritis and Rheumatism, 46*, 1763–1772.
81. Pridgeon, C., Lennon, G. P., Pazmany, L., Thompson, R. N., Christmas, S. E., & Moots, R. J. (2003). Natural killer cells in the synovial fluid of rheumatoid arthritis patients exhibit a CD56bright,CD94bright,CD158negative phenotype. *Rheumatology (Oxford), 42*, 870–878.
82. Vastert, S. J., van Wijk, R., D'Urbano, L. E., de Vooght, K. M., de Jager, W., Ravelli, A., et al. (2010). Mutations in the perforin gene can be linked to macrophage activation syndrome in patients with systemic onset juvenile idiopathic arthritis. *Rheumatology (Oxford), 49*, 441–449.
83. Hazen, M. M., Woodward, A. L., Hofmann, I., Degar, B. A., Grom, A., Filipovich, A. H., et al. (2008). Mutations of the hemophagocytic lymphohistiocytosis-associated gene UNC13D in a patient with systemic juvenile idiopathic arthritis. *Arthritis and Rheumatism, 58*, 567–570.
84. Zhang, K., Biroschak, J., Glass, D. N., Thompson, S. D., Finkel, T., Passo, M. H., et al. (2008). Macrophage activation syndrome in patients with systemic juvenile idiopathic arthritis is associated with MUNC13-4 polymorphisms. *Arthritis and Rheumatism, 58*, 2892–2896.
85. Kaufman, K. M., Linghu, B., Szustakowski, J. D., Husami, A., Yang, F., Zhang, K., et al. (2014). Whole-exome sequencing reveals overlap between macrophage activation syndrome in systemic juvenile idiopathic arthritis and familial hemophagocytic lymphohistiocytosis. *Arthritis & Rhematology, 66*, 3486–3495.
86. Zhang, M., Behrens, E. M., Atkinson, T. P., Shakoory, B., Grom, A. A., & Cron, R. Q. (2014). Genetic defects in cytolysis in macrophage activation syndrome. *Current Rheumatology Reports, 16*, 439.

87. Bracaglia, C., Prencipe, G., & De Benedetti, F. (2017). Macrophage activation syndrome: Different mechanisms leading to a one clinical syndrome. *Pediatric Rheumatology Online Journal, 15*, 5.
88. Cifaldi, L., Prencipe, G., Caiello, I., Bracaglia, C., Locatelli, F., De Benedetti, F., et al. (2015). Inhibition of natural killer cell cytotoxicity by interleukin-6: Implications for the pathogenesis of macrophage activation syndrome. *Arthritis & Rhematology, 67*, 3037–3046.
89. Brisse, E., Wouters, C. H., Andrei, G., & Matthys, P. (2017). How viruses contribute to the pathogenesis of hemophagocytic lymphohistiocytosis. *Frontiers in Immunology, 8*, 1102.
90. Su, H. C., Nguyen, K. B., Salazar-Mather, T. P., Ruzek, M. C., Dalod, M. Y., & Biron, C. A. (2001). NK cell functions restrain T cell responses during viral infections. *European Journal of Immunology, 31*, 3048–3055.
91. Henter, J. I., Horne, A., Arico, M., Egeler, R. M., Filipovich, A. H., Imashuku, S., et al. (2007). HLH-2004: Diagnostic and therapeutic guidelines for hemophagocytic lymphohistiocytosis. *Pediatric Blood & Cancer, 48*, 124–131.
92. Ravelli, A., Minoia, F., Davi, S., Horne, A., Bovis, F., Pistorio, A., et al. (2016). 2016 classification criteria for macrophage activation syndrome complicating systemic juvenile idiopathic arthritis: A European League Against Rheumatism/American College of Rheumatology/ Paediatric Rheumatology International Trials Organisation Collaborative Initiative. *Annals of the Rheumatic Diseases, 75*, 481–489.
93. Schulert, G. S., Minoia, F., Bohnsack, J., Cron, R. Q., Hashad, S., Kone-Paut, I., et al. (2017). Biologic therapy modifies clinical and laboratory features of macrophage activation syndrome associated with systemic juvenile idiopathic arthritis. *Arthritis Care & Research.*
94. Bryceson, Y. T., Pende, D., Maul-Pavicic, A., Gilmour, K. C., Ufheil, H., Vraetz, T., et al. (2012). A prospective evaluation of degranulation assays in the rapid diagnosis of familial hemophagocytic syndromes. *Blood, 119*, 2754–2763.
95. Rubin, T. S., Zhang, K., Gifford, C., Lane, A., Choo, S., Bleesing, J. J., et al. (2017). Perforin and CD107a testing is superior to NK cell function testing for screening patients for genetic HLH. *Blood, 129*, 2993–2999.
96. Weitzman, S. (2011). Approach to hemophagocytic syndromes. *Hematology. American Society of Hematology. Education Program, 2011*, 178–183.
97. Rigaud, S., Fondaneche, M. C., Lambert, N., Pasquier, B., Mateo, V., Soulas, P., et al. (2006). XIAP deficiency in humans causes an X-linked lymphoproliferative syndrome. *Nature, 444*, 110–114.
98. Shabrish, S., Gupta, M., & Madkaikar, M. (2016). A modified NK cell degranulation assay applicable for routine evaluation of NK cell function. *Journal of Immunology Research, 2016*, 3769590.
99. Cetica, V., Santoro, A., Gilmour, K. C., Sieni, E., Beutel, K., Pende, D., et al. (2010). STXBP2 mutations in children with familial haemophagocytic lymphohistiocytosis type 5. *Journal of Medical Genetics, 47*, 595–600.
100. Fontana, S., Parolini, S., Vermi, W., Booth, S., Gallo, F., Donini, M., et al. (2006). Innate immunity defects in Hermansky-Pudlak type 2 syndrome. *Blood, 107*, 4857–4864.
101. Marcenaro, S., Gallo, F., Martini, S., Santoro, A., Griffiths, G. M., Arico, M., et al. (2006). Analysis of natural killer-cell function in familial hemophagocytic lymphohistiocytosis (FHL): Defective CD107a surface expression heralds Munc13-4 defect and discriminates between genetic subtypes of the disease. *Blood, 108*, 2316–2323.
102. Argov, S., Johnson, D. R., Collins, M., Koren, H. S., Lipscomb, H., & Purtilo, D. T. (1986). Defective natural killing activity but retention of lymphocyte-mediated antibody-dependent cellular cytotoxicity in patients with the X-linked lymphoproliferative syndrome. *Cellular Immunology, 100*, 1–9.
103. Marsh, R. A., Madden, L., Kitchen, B. J., Mody, R., McClimon, B., Jordan, M. B., et al. (2010). XIAP deficiency: A unique primary immunodeficiency best classified as X-linked familial hemophagocytic lymphohistiocytosis and not as X-linked lymphoproliferative disease. *Blood, 116*, 1079–1082.

104. Jessen, B., Bode, S. F., Ammann, S., Chakravorty, S., Davies, G., Diestelhorst, J., et al. (2013). The risk of hemophagocytic lymphohistiocytosis in Hermansky-Pudlak syndrome type 2. *Blood, 121*, 2943–2951.
105. Romberg, N., Al Moussawi, K., Nelson-Williams, C., Stiegler, A. L., Loring, E., Choi, M., et al. (2014). Mutation of NLRC4 causes a syndrome of enterocolitis and autoinflammation. *Nature Genetics, 46*, 1135–1139.
106. Canna, S. W., de Jesus, A. A., Gouni, S., Brooks, S. R., Marrero, B., Liu, Y., et al. (2014). An activating NLRC4 inflammasome mutation causes autoinflammation with recurrent macrophage activation syndrome. *Nature Genetics, 46*, 1140–1146.
107. Zhang K, Wakefield E, Marsh R. 2016. Lymphoproliferative disease, X-linked. In Adam MP, Ardinger HH, Pagon RA, Wallace SE, Bean LJH, Stephens K, Amemiya A GeneReviews®. Seattle, WA: University of Washington, Seattle
108. Meazza, R., Tuberosa, C., Cetica, V., Falco, M., Parolini, S., Grieve, S., et al. (2014). Diagnosing XLP1 in patients with hemophagocytic lymphohistiocytosis. *The Journal of Allergy and Clinical Immunology, 134*(1381–7), e7.

Myeloid Cells in the Immunopathogenesis of Cytokine Storm Syndromes

Lehn K. Weaver

Introduction

The innate immune system is composed of a diverse collection of cell types that protect the host from infectious challenges and tissue damage, and ultimately preserve host fitness by returning the system to homeostasis. Neutrophils, monocytes, macrophages, and dendritic cells patrol the blood and/or populate broad host tissues making them positioned to respond to infectious challenges and alterations in homeostasis. Each of these innate immune cell types expresses an ensemble of receptors that are stimulated by endogenous and environmental triggers released by tissue damage or infection. Upon activation, innate immune cells assume a wide variety of distinct functional states that converts them from dormant sentinels into effector cells that generate inflammation or exert immunoregulatory functions. How these myeloid cells and their effector responses lead to dysregulated and hyperinflammatory immune responses is of importance to understanding the diverse mechanisms that contribute to cytokine storm pathogenesis.

A diversity of inciting triggers activates a systemic inflammatory response leading to widespread immunopathology in CSSs. The characteristic immune-mediated collateral damage that occurs in CSSs results in widespread endothelial cell activation and damage, tissue ischemia, cell injury, consumptive coagulopathy, and multisystem organ failure. Distinct myeloid-specific functions initiate and propagate this maladaptive systemic host response in CSSs. Activated myeloid cells produce pro- and anti-inflammatory cytokines, exert anti-pathogen responses, and activate surrounding immune and non-immune cells to amplify ongoing inflammatory responses in CSSs. As potent antigen-presenting cells, myeloid cells initiate antigen-

L. K. Weaver (✉)
Children's Hospital of Philadelphia, Division of Rheumatology, Philadelphia, PA, USA
e-mail: WEAVERL1@email.chop.edu

R. Q. Cron, E. M. Behrens (eds.), *Cytokine Storm Syndrome*,
https://doi.org/10.1007/978-3-030-22094-5_11

specific adaptive immune responses leading to further amplification of pathogenic immune responses in certain CSSs. In contrast, recent evidence suggests that myeloid cells have immunoregulatory and tissue-healing roles in CSSs. These diverse effector responses of myeloid cells position them to be key contributors to all phases of cytokine storm physiology from initiation to resolution of the systemic inflammatory response.

Despite the growing circumstantial evidence that myeloid cells contribute to CSS pathogenesis, there is no direct evidence of their role in CSS pathogenesis in human systems, and only limited evidence of their role in murine models. Therefore, this chapter contextualizes how multiple effector functions of myeloid cells have the potential to contribute to host inflammatory responses in CSSs. Although inherently speculative in nature, this chapter outlines how these diverse functions of myeloid cells make them rational targets for the design of novel therapeutics for the treatment of CSSs.

Pattern Recognition Receptors Initiate Cytokine Storm Physiology

Myeloid cells respond to inflammatory triggers through their expression of a restricted number of germ-line encoded pattern recognition receptors (PRRs) that sense a diverse range of endogenous and exogenous activating signals. PRR expression allows myeloid cells to be "first responders" to the diverse inflammatory triggers known to initiate CSSs following trauma, burns, surgery, infection, or the development of a rheumatic disease. Chronic or exaggerated activation of myeloid cells through their PRRs initiates a dysregulated and hyperinflammatory response that is detrimental to the host. Myeloid cells are at the forefront of systemic inflammatory responses because of their widespread tissue distribution and their ability to recognize and respond to the diverse triggers of CSSs.

Toll-like receptors (TLRs) are the best-characterized PRRs and have been shown to play important roles in initiating immune responses to diverse inflammatory triggers [1]. As membrane-bound receptors, TLRs are expressed at the cell surface or in the endolysosomal compartment of immune cells where they are activated by endogenous or exogenous "danger" signals [1]. Clinical data supports the role of TLRs in driving cytokine storm immunopathology in macrophage activation syndrome (MAS), a type of CSS seen in patients with underlying rheumatologic diseases. Patients with systemic juvenile idiopathic arthritis (SJIA), the rheumatic disease with the greatest predisposition to the development of MAS, have an interleukin (IL)-1 and TLR gene expression signature in their peripheral blood mononuclear cells [2]. Furthermore, SJIA patients with polymorphisms in IRF5, a signaling molecule downstream of TLR activation, have a fourfold higher risk of MAS [3]. The IRF5 haplotype associated with this higher risk of MAS in SJIA patients creates a donor splice site in an alternate exon 1 of IRF5 allowing expression of higher levels of IRF5 and the production of several unique IRF5 isoforms

[4]. These clinical data implicate signaling pathways downstream of TLRs as important contributors to MAS pathogenesis.

The NOD-like receptor (NLR) family of PRRs is also implicated in the development of CSSs. NLRs are cytosolic proteins that can initiate immune responses to cellular stress or invading microorganisms [5]. Stimulation of NLRs leads to inflammasome activation and the generation of the pro-inflammatory cytokines IL-1 and IL-18 in addition to pyroptotic cell death, a form of cell death contributing to pro-inflammatory immune responses [5]. Mutations in the NLR family member NLRC4 leads to the spontaneous development of an autoinflammatory syndrome in humans predisposing to MAS immunopathology, termed NLRC4-MAS [6, 7]. Patients with mutations in NLRC4 spontaneously develop early-onset enterocolitis, erythroderma, elevated acute phase reactants, and features of MAS including cytopenias, splenomegaly, hemophagocytosis, and hyperferritinemia [6, 7]. These disease features of NLRC4-MAS are thought to occur because of the spontaneous cleavage of caspase-1, heightened production of IL-1β and IL-18, and increased cell death in macrophages harboring NLRC4-activating mutations [6, 7]. This leads to an excessive accumulation of pathogenic IL-1β and IL-18, as combined blockade of both IL1β and IL-18 was sufficient to ameliorate severe disease in a critically ill infant with NLRC4-MAS [8]. IL-18 may contribute to cytokine storm pathogenesis in NLRC4-MAS by inducing the production of interferon (IFN)γ, as high levels of serum IL-18 correlated with an IFNγ-associated gene expression network in whole blood transcriptional analysis from a patient with NLRC4-MAS [8], and use of an IFNγ-neutralizing antibody was effective in the treatment of another patient with NLRC4-MAS [9]. The link between IL-18 and IFNγ production is of particular interest, as other cytokine storm syndromes are also characterized by high levels of IFNγ [10, 11], which will be discussed in detail later in this chapter. These clinical data link spontaneous activation of NLRC4 to overproduction of IL1β, IL-18, and IFNγ, which in turn drive cytokine storm immunopathology.

Epidemiologic observations suggest that additional factors influence the development of MAS. For example, patients with SJIA develop common manifestations of autoinflammation including daily fevers, rash, and arthritis. However, only a subset of SJIA patients develop MAS with an incidence of 10–30% [12, 13], which suggests the inflammatory milieu created by SJIA is not sufficient to drive MAS immunopathology in all patients. Similarly, identical NLRC4-activating mutations cause a spectrum of disease ranging from life-threatening to self-limited inflammatory disease in infancy [7]. As early-onset enterocolitis is a hallmark of NLRC4-MAS, it is tempting to speculate that the gut microbiota of the host plays a role in driving cytokine storm immunopathology in these patients. Future efforts will be needed to determine if this is the case and whether environmental factors play a role in other CSSs.

These clinical data suggest that PRRs contribute to the immunopathogenesis of MAS. As myeloid cells are the main cells that express PRRs, their effector functions downstream of PRR activation are prime suspects in driving systemic inflammation and immunopathology in CSSs. The following section describes how myeloid cell responses to PRR activation contribute to the cytokine milieu that drives cytokine storm immunopathology.

Myeloid Cells and the Pathogenic Cytokine Milieu in CSSs

As immune sentinels, activated myeloid cells mediate pathogen uptake and killing, secrete effector molecules that have local and systemic effects, and activate adaptive immune responses through their antigen presentation capabilities. These functions make myeloid cells important initiators and amplifiers of immune responses that are detrimental to host survival in CSSs. This section highlights how myeloid cell effector functions direct the production of pathogenic cytokines contributing to cytokine storm immunopathology in MAS and primary hemophagocytic lymphohistiocytosis (HLH).

Macrophage Activation Syndrome

Soluble mediators produced during immune responses are capable of producing diverse and widespread changes to the homeostatic landscape of the host. In patients with SJIA who are predisposed to the development of MAS, aberrant production of pro-inflammatory cytokines can be detected through serum cytokine analysis. Shimizu et al. detected two dominant patterns of elevated pro-inflammatory cytokines in patients with SJIA based on elevated levels of IL-6 and IL-18 [14], both of which are known to be produced by myeloid cells. Intriguingly, patients with high levels of IL-6 are more prone to the development of arthritis, while patients with high levels of IL-18 are more susceptible to MAS [14]. IL-18 is also a marker of disease severity in MAS, as it correlates with hyperferritinemia, elevated transaminases, and is associated with clinical features of MAS flare [15]. Additional markers of SJIA-MAS include other myeloid cell-derived products including neopterin and soluble TNFRI/II [15], while MAS associated with systemic lupus erythematosus is marked by elevations in TNF [16], another cytokine derived from activated myeloid cells. In contrast, IFNγ is a lymphocyte-derived cytokine and has been linked to the pathogenesis of MAS, as SJIA patients with MAS have an IFNγ signature that correlates with MAS disease activity and severity [10]. Pathogenic IFNγ may also be indirectly dependent on the functions of myeloid cells as they are the main producers of IL-12 and IL-18, which stimulate lymphocyte-derived IFNγ [17]. Therefore, myeloid cells not only directly contribute to the production of pathogenic cytokines in MAS, but also control the production of bystander lymphocytes to produce pathogenic IFNγ. These data suggest that heightened myeloid cell activation is central to MAS immunopathology, as cytokines produced directly or indirectly downstream of myeloid cell activation correlate with MAS disease severity.

Myeloid cells are directly implicated in mediating cytokine storm immunopathology in murine models of MAS. In the first described model of MAS, repeated activation of TLR9 drives IFNγ-dependent cytopenias, hepatosplenomegaly, hepatitis, dysregulated thermoregulation, hyperferritinemia, hypercytokinemia, and hemophagocytosis [18]. This model is particularly relevant to MAS in SJIA patients given the clinical evidence linking IFNγ to MAS disease activity and severity [10]. Although it remains unclear how IFNγ mediates end-organ damage in mice or humans with MAS, the pathways leading to the induction of IFNγ are beginning to be elucidated. In the TLR9-mediated model of MAS, elevated levels of IFNγ are dependent on IL-12 production [19], which links myeloid cells to disease pathogenesis, as inflammatory monocytes are the main producers of IL-12 in this model [20]. The heightened levels of IL-12 in TLR9-mediated MAS correlate with the accumulation of inflammatory monocytes and enhanced extramedullary myelopoiesis [20]. Therefore, chronic TLR9 activation is thought to generate a feed-forward inflammatory loop, whereby the induction of inflammatory myelopoiesis accelerates the production of new inflammatory monocytes that are necessary to amplify and perpetuate pathogenic cytokine production. As the JAK inhibitor ruxolitinib is able to inhibit immunopathology in this model of TLR9-mediated MAS [21], it is intriguing to speculate that JAK-activating, myeloid-specific growth factors are critical for this feed-forward inflammatory loop in MAS. These data also suggest that JAK inhibition may be effective in the prevention and/or treatment of MAS. Definitive insight into this possibility may come from a clinical trial testing the efficacy of the JAK inhibitor tofacitinib in SJIA patients (clinicaltrials.gov; NCT03000439) (Table 1).

A second murine model of MAS also implicates a TLR-driven pathway in the induction of cytokine storm and adds additional evidence for myeloid cell involvement in the pathogenesis of MAS. Strippoli et al. demonstrate that a single dose of lipopolysaccharide or Poly I:C, TLR4 and TLR3 agonists, respectively, leads to decreased survival in transgenic mice that overexpress the cytokine IL-6 [22]. Macrophages from IL-6-transgenic mice and human monocytes treated with exogenous IL-6 have heightened capacity to produce pro-inflammatory cytokines [22]. This suggests that high levels of IL-6, which accumulate in SJIA patients [16], may be linked to enhanced myeloid cell-derived cytokine production and MAS immunopathology. However, SJIA patients treated with tocilizumab developed MAS despite marked improvements in their SJIA-related symptoms [23]. These data indicate that IL-6 blockade is not sufficient to prevent the development of MAS in all SJIA patients, at least at the doses of tocilizumab employed (Table 2).

A third model of MAS demonstrates that a single dose of complete Freund's adjuvant (CFA) is sufficient to induce MAS in mice [24]. CFA is a solution of mycobacterial antigen emulsified in oil that creates a chronic depot of multiple TLR and PRR agonists when injected subcutaneously into mice leading to chronic activation of PRRs on myeloid cells. Intriguingly, Avau et al. demonstrate that CFA-treated IFNγ$^{-/-}$ mice develop more severe MAS-like disease and arthritis than their wild-type counterparts [24]. This IFNγ-independent model of MAS is dependent on innate and adaptive lymphoid cell-derived IL-17, a cytokine not previously implicated in the pathogenesis of cytokine storm [24]. These data suggest that chronic

Table 1 Murine models of HLH/MAS implicating myeloid cells in disease pathogenesis

Mouse	Cytokine storm trigger	Pathogenic cytokines	Immunomodulatory cytokines	Role(s) of myeloid cells in disease	References
MAS					
Wild-type C57BL/6 mice	CpG1826 (TLR9 agonist)	IL-12, IFNγ	IL-10	TLR9-induced production of pathogenic IL-12; hemophagocyte production of immunomodulatory IL-10	[18–20, 36]
IL-6 transgenic mice	LPS (TLR4 agonist) or Poly I:C (TLR3 agonist)	IL-6, IFNγ	Unknown	Mediate responses to LPS through expression of PRRs	[22, 42]
Wild-type and IFNγ-deficient Balb/c mice	CFA (TLR2, TLR4, and TLR9 agonists)	IL-17, IL-12/ IL-23	Unknown	Mediate responses to CFA through expression of PRRs	[24]
HLH					
Perforin-deficient mice	LCMV	IFNγ, IL-33	IL-10	Enhanced antigen presentation and T cell activation; hemophagocyte production of immunomodulatory IL-10	[28, 31, 36, 43]
Perforin-deficient, DC-specific Fas-deficient mice	Spontaneous	IFNγ	Unknown	Induction of CD8 T cell-dependent IFNγ production	[32]

TLR activation drives IL-17-mediated MAS in the absence of IFNγ. These findings may be relevant to MAS patients that are refractory to IFNγ neutralizing antibody therapy, which is being tested for efficacy in the treatment of CSSs (clinicaltrials.gov; NCT01818492). These findings may also be relevant in rare cases of HLH that develop in patients lacking the IFNγ receptor [25], which indicates IFNγ-dependent signaling is not required for the development of cytokine storm in humans. Therefore, both MAS patients refractory to IFNγ neutralization and IFNγR-deficient patients that develop HLH may benefit from IL-17 blocking therapies. IFNγ-independent cytokine storm is also described in mice treated with repeated TLR9 activation in combination with IL-10 receptor blockade [18]. These murine models of disease demonstrate that TLR-dependent pathways generate pathogenic cytokine storm-mediated immunopathology in the absence of IFNγ.

Table 2 Putative myeloid effector functions that influence cytokine storm pathogenesis

	Evidence in HLH/MAS	References
Drivers of HLH/MAS		
Myeloid cell functions		
Activation of pattern recognition receptors results in pro-inflammatory cytokine production to drive MAS-like physiology in mice	MAS	[20, 22, 24]
Release of IL-12 and IL-18 augments IFNγ production by lymphocytes	MAS	[8, 17, 20]
Prolonged immunologic synapse time augments CD8 T cell effector functions	HLH	[28–30]
Modulators of HLH/MAS		
Myeloid cell functions		
Production of immunomodulatory IL-10	MAS and HLH	[18, 36]
Patrolling monocytes remove cell debris after TLR-induced damage to endothelial cells	No	[39]
Patrolling monocytes spontaneously produce IL-1RA	No	[40]

Collectively, these data link chronic TLR signals to the hyper-responsiveness of myeloid cells as a pathogenic mechanism driving MAS in mice and humans. Myeloid cells are capable of responding to the initial TLR signals to amplify the production of pathogenic cytokines to mediate MAS immunopathology. The next section will highlight how myeloid cells function as potent antigen-presenting cells to generate a feed-forward inflammatory response in primary hemophagocytic lymphohistiocytosis (HLH).

Antigen Presentation by Myeloid Cells Contributes to Immune Dysregulation in Primary HLH

Myeloid cells initiate adaptive immune responses through their ability to present antigens when they are encountered in the context of a "danger" signal. PRR activation of myeloid cells upregulates the ability of these cells to engulf, process, and present endogenous and exogenous antigens on major histocompatibility complex (MHC) I and II, respectively. These peptide–MHC complexes are transported to the cell surface where they are recognized by their cognate T cell receptor (TCR). Myeloid cells direct the outcome of MHC:TCR engagement through the expression of co-stimulatory molecules needed for robust T cell activation and through the production of cytokines that influence T cell effector functions. These antigen-presenting capabilities of myeloid cells are critical for the activation of T cells, and failure to downregulate the antigen presentation capabilities of myeloid cells contributes to heightened immune responses and immunopathology in primary HLH.

Primary HLH is triggered in patients with genetic defects in perforin or other proteins critical for exocytosis of cytotoxic granules, which result in defective CD8

T cell- and NK cell-directed cytotoxicity [26]. Patients with primary HLH are healthy until they are infected by a virus, which triggers an ineffective yet hyperinflammatory systemic immune response [27]. Upon infection, myeloid cells are activated to present viral peptides complexed with MHC I to activate CD8 T cells to eliminate virally infected cells. However, because primary HLH patients' CD8 T cells have defective cytotoxicity, they fail to eliminate the virus resulting in chronic viremia [27]. CD8 T cells with defective cytotoxicity also fail to eliminate virally infected antigen-presenting cells [28], a negative regulatory mechanism that inhibits continuous and exaggerated antigen presentation during healthy immune responses. Failure of target cell death leads to immunologic synapse engagement lasting fivefold longer than normal, which is directly linked to hypersecretion of pathogenic pro-inflammatory cytokines by T cells and myeloid cells [29]. These findings are likely clinically relevant, as a heterozygous mutation in the primary HLH gene *RAB27A* results in delayed cytotoxic granule polarization and was sufficient to produce susceptibility to HLH in two patients [30]. Therefore, in primary HLH, CD8 T cells are unable to kill infected antigen-presenting cells [28], which results in more robust antigen-presenting cell–cytotoxic T cell interactions and heightened production of pathogenic pro-inflammatory cytokines that drive disease. This feed-forward inflammatory loop is critical for HLH pathogenesis, as depletion of CD8 T cells is sufficient to reduce overproduction of IFNγ and ameliorate clinical manifestations of HLH in mice [31]. These data highlight that myeloid cell antigen presentation is central to the initiation and perpetuation of hyperinflammatory CD8 T cell responses characteristic of primary HLH, and IFNγ is a prime therapeutic target in this CSS that is currently being tested in the clinic (clinicaltrials.gov; NCT01818492).

Termination of antigen presentation by cytotoxic lymphocyte-mediated killing of myeloid cells is also critical to prevent HLH in mice. In mice harboring defects in both perforin-mediated cytotoxicity and Fas-dependent cell death of CD11c-expressing myeloid cells, features of HLH develop in uninfected mice including cytopenias, elevated IFNγ levels, hyperactivation of T cells, hepatitis, splenomegaly, and early mortality [32]. This spontaneous model of HLH further highlights the importance of lymphocyte-mediated killing of myeloid cells to terminate continuous antigen presentation, which is an immunomodulatory mechanism preventing the aberrant hyperactivation of T cells capable of triggering cytokine storm.

Immunoregulatory Functions of Myeloid Cells in Cytokine Storm Syndromes

Although myeloid cells have multiple functions that initiate and perpetuate cytokine storm immunopathology, their cellular plasticity makes them prime agents in mediating immunoregulatory and tissue-healing functions in the resolution phase of CSSs. These myeloid cell functions are likely important clinically, as more patients are surviving the first few days of the characteristic hyperinflammatory response of

CSSs owing to improved diagnosis, therapeutic interventions and supportive care. These individuals often have persistent hyperinflammatory responses, develop a state of immunoparalysis predisposing to secondary infectious challenges, and develop multisystem organ failure in the setting of a systemic catabolic state. The etiology of these secondary effects remains largely unknown, yet they remain clinically important features of CSSs, as they contribute significantly to morbidity and mortality. Future attempts to capitalize on the plasticity of myeloid cell functionality may offer novel therapeutic approaches to speed the resolution of inflammation and aid in the return to homeostasis in patients with CSSs.

Evidence for an immunomodulatory role of myeloid cells in CSSs includes the recognition that hemophagocytosing histiocytes may not play a pathologic role in MAS and HLH, but rather have anti-inflammatory functions. Hemophagocytes are myeloid cells that have phagocytosed hematopoietic cells and are commonly found in the bone marrow and tissues of patients with CSSs. Schaer et al. were the first to identify that hemophagocytosing CD163^{+} myeloid cells in patients with CSSs express high levels of the anti-inflammatory molecule heme oxygenase (HO)-1 [33]. HO-1 catalyzes the degradation of heme into biliverdin, ferrous iron, and carbon monoxide, which have known anti-inflammatory, antiapoptotic, and antioxidative effects [34]. As hemophagocytes often engulf whole red blood cells, they contain sufficient sources of heme to be degraded into its anti-inflammatory constituents in the presence of HO-1. Furthermore, Canna et al. used laser capture microscopy to identify hemophagocytes for transcriptomic analysis, and demonstrated significant gene set enrichment of genes expressed by M2 macrophages [35]. M2 macrophages are known to have functions that decrease inflammation and support tissue repair indicating that hemophagocytic myeloid cells may have beneficial roles in resolving inflammation during CSSs. Ohyagi et al. bring functional relevance to these observations by demonstrating that hemophagocytes are an important source of IL-10, an anti-inflammatory cytokine, downstream of TLR and PRR activation in mice [36]. Furthermore, blockade of hemophagocytosis or defective IL-10 production by hemophagocytes increased virally induced mortality in a murine model of a CSS [36]. The production of IL-10 by hemophagocytes may not be surprising, as HO-1-mediated degradation of heme is known to induce IL-10 production from myeloid cells [37]. These data support a role for IL-10 production and hemophagocytosis as immunomodulatory functions of myeloid cells in preventing collateral immune-mediated damage in CSSs. Future efforts to identify the IL-10 producers and responders mediating the immunomodulatory effect of IL-10 in CSSs may reveal opportunities to mitigate the overwhelming hyperinflammatory response that leads to life-threatening disease in HLH and MAS.

In addition to their immunomodulatory functions, myeloid cells may be directly involved in reparative processes that restore homeostasis following immune-mediated damage in CSSs. Patrolling monocytes, also known as nonclassical monocytes, are well-positioned to function in this role, as they are closely associated with vascular endothelial cells, cells that are commonly damaged in CSSs. Patrolling monocytes maintain close contacts with vascular endothelial cells where they detect damaged endothelial cells [38]. Upon encounter of TLR agonists, patrolling monocytes increase their interactions with small vessel endothelial cells in inflamed tissues and mediate the in situ removal of cellular debris following TLR-mediated endothelial cell death [39]. Human patrolling monocytes also spontaneously produce the anti-inflammatory molecule IL-1RA and induce lower levels of pro-inflammatory cytokine upon activation by TLR agonists compared to inflammatory monocytes [40], which indicate that patrolling monocytes have immunomodulatory rather than pro-inflammatory functions. The differentiation of inflammatory monocytes into reparative patrolling monocytes has also been reported in a model of sterile liver injury, whereby they prevent excess tissue damage through IL-4- and IL-10-dependent mechanisms [41]. These data suggest that patrolling monocytes are poised to respond to damaged tissues and aid in the removal of cellular debris to promote wound healing. These myeloid cell functions are likely critical for restoring homeostasis in patients suffering from systemic tissue injury in the resolution phase of CSSs.

Conclusions

Myeloid cells have diverse and malleable functions that empower them to influence all phases of cytokine storm physiology from initiation to resolution making them prime therapeutic targets in CSSs. Future efforts will be needed to identify the critical myeloid specific effector functions downstream of "danger" signal recognition that contribute to widespread tissue damage in CSSs. Evidence linking myeloid-derived inflammatory cytokines to specific cytokine storms may offer novel cytokine-directed treatments for CSSs. Identification of the molecular interactions between myeloid-derived antigen-presenting cells and pathogenic T cells could yield effective approaches to break the perpetual cycle of antigen presentation that drives pathogenic immune responses in primary HLH. Finally, the capacity of myeloid cells to gain immunomodulatory and tissue-healing properties during systemic inflammatory responses could be manipulated to accelerate the resolution of inflammation for therapeutic benefit in CSSs. Such myeloid cell-directed interventions would be welcome to combat the devastating hyperinflammatory responses that result in severe morbidity and mortality across the spectrum of CSSs.

References

1. Bryant, C. E., Gay, N. J., Heymans, S., Sacre, S., Schaefer, L., & Midwood, K. S. (2015). Advances in Toll-like receptor biology: Modes of activation by diverse stimuli. *Critical Reviews in Biochemistry and Molecular Biology, 50*, 359–379.
2. Fall, N., Barnes, M., Thornton, S., Luyrink, L., Olson, J., Ilowite, N. T., et al. (2007). Gene expression profiling of peripheral blood from patients with untreated new-onset systemic juvenile idiopathic arthritis reveals molecular heterogeneity that may predict macrophage activation syndrome. *Arthritis and Rheumatism, 56*, 3793–3804.

3. Yanagimachi, M., Naruto, T., Miyamae, T., Hara, T., Kikuchi, M., Hara, R., et al. (2011). Association of IRF5 polymorphisms with susceptibility to macrophage activation syndrome in patients with juvenile idiopathic arthritis. *The Journal of Rheumatology, 38*, 769–774.
4. Graham, R. R., Kozyrev, S. V., Baechler, E. C., Reddy, M. V., Plenge, R. M., Bauer, J. W., et al. (2006). A common haplotype of interferon regulatory factor 5 (IRF5) regulates splicing and expression and is associated with increased risk of systemic lupus erythematosus. *Nature Genetics, 38*, 550–555.
5. Kufer, T. A., Nigro, G., & Sansonetti, P. J. (2016). Multifaceted functions of NOD-like receptor proteins in myeloid cells at the intersection of innate and adaptive immunity. *Microbiology Spectrum, 4*.
6. Canna, S. W., de Jesus, A. A., Gouni, S., Brooks, S. R., Marrero, B., Liu, Y., et al. (2014). An activating NLRC4 inflammasome mutation causes autoinflammation with recurrent macrophage activation syndrome. *Nature Genetics, 46*, 1140–1146.
7. Romberg, N., Al Moussawi, K., Nelson-Williams, C., Stiegler, A. L., Loring, E., Choi, M., et al. (2014). Mutation of NLRC4 causes a syndrome of enterocolitis and autoinflammation. *Nature Genetics, 46*, 1135–1139.
8. Canna, S. W., Girard, C., Malle, L., de Jesus, A., Romberg, N., Kelsen, J., et al. (2017). Life-threatening NLRC4-associated hyperinflammation successfully treated with Interleukin-18 inhibition. *The Journal of Allergy and Clinical Immunology, 139*, 1698–1701.
9. Bracaglia, C., Prencipe, G., Gatto, A., Pardeo, M., Lapeyre, G., Raganelli, L., et al. (2015). Anti interferon-gamma (IFNg) Monoclonal antibody treatment in a child with NLRC4-related disease and severe Hemophagocytic Lymphohistiocytosis (HLH) [abstract]. *Arthritis and Rheumatology, 67* (suppl 10).
10. Bracaglia, C., de Graaf, K., Pires Marafon, D., Guilhot, F., Ferlin, W., Prencipe, G., et al. (2017). Elevated circulating levels of interferon-gamma and interferon-gamma-induced chemokines characterise patients with macrophage activation syndrome complicating systemic juvenile idiopathic arthritis. *Annals of the Rheumatic Diseases, 76*, 166–172.
11. Takada, H., Takahata, Y., Nomura, A., Ohga, S., Mizuno, Y., & Hara, T. (2003). Increased serum levels of interferon-gamma-inducible protein 10 and monokine induced by gamma interferon in patients with haemophagocytic lymphohistiocytosis. *Clinical and Experimental Immunology, 133*, 448–453.
12. Moradinejad, M. H., & Ziaee, V. (2011). The incidence of macrophage activation syndrome in children with rheumatic disorders. *Minerva Pediatrica, 63*, 459–466.
13. Behrens, E. M., Beukelman, T., Paessler, M., & Cron, R. Q. (2007). Occult macrophage activation syndrome in patients with systemic juvenile idiopathic arthritis. *The Journal of Rheumatology, 34*, 1133–1138.
14. Shimizu, M., Nakagishi, Y., & Yachie, A. (2013). Distinct subsets of patients with systemic juvenile idiopathic arthritis based on their cytokine profiles. *Cytokine, 61*, 345–348.
15. Shimizu, M., Yokoyama, T., Yamada, K., Kaneda, H., Wada, H., Wada, T., et al. (2010). Distinct cytokine profiles of systemic-onset juvenile idiopathic arthritis-associated macrophage activation syndrome with particular emphasis on the role of interleukin-18 in its pathogenesis. *Rheumatology (Oxford), 49*, 1645–1653.
16. Shimizu, M., Yokoyama, T., Tokuhisa, Y., Ishikawa, S., Sakakibara, Y., Ueno, K., et al. (2013). Distinct cytokine profile in juvenile systemic lupus erythematosus-associated macrophage activation syndrome. *Clinical Immunology, 146*, 73–76.
17. Tominaga, K., Yosimoto, T., Torigoe, K., Kurimoto, M., Matsui, K., Hada, T., et al. (2000). IL-12 synergizes with IL-18 or IL-1beta for IFN-gamma production from human T cells. *International Immunology, 12*(2), 151–160.
18. Behrens, E. M., Canna, S. W., Slade, K., Rao, S., Kreiger, P. A., Paessler, M., et al. (2011). Repeated TLR9 stimulation results in macrophage activation syndrome-like disease in mice. *The Journal of Clinical Investigation, 121*, 2264–2277.
19. Canna, S. W., Wrobel, J., Chu, N., Kreiger, P. A., Paessler, M., & Behrens, E. M. (2013). Interferon-gamma mediates anemia but is dispensable for fulminant toll-like receptor

9-induced macrophage activation syndrome and hemophagocytosis in mice. *Arthritis and Rheumatism, 65*, 1764–1775.
20. Weaver, L. K., Chu, N., & Behrens, E. M. (2016). TLR9-mediated inflammation drives a Ccr2-independent peripheral monocytosis through enhanced extramedullary monocytopoiesis. *Proceedings of the National Academy of Sciences of the United States of America, 113*, 10944–10949.
21. Das, R., Guan, P., Sprague, L., Verbist, K., Tedrick, P., An, Q. A., et al. (2016). Janus kinase inhibition lessens inflammation and ameliorates disease in murine models of hemophagocytic lymphohistiocytosis. *Blood, 127*, 1666–1675.
22. Strippoli, R., Carvello, F., Scianaro, R., De Pasquale, L., Vivarelli, M., Petrini, S., et al. (2012). Amplification of the response to Toll-like receptor ligands by prolonged exposure to interleukin-6 in mice: Implication for the pathogenesis of macrophage activation syndrome. *Arthritis and Rheumatism, 64*, 1680–1688.
23. De Benedetti, F., Brunner, H. I., Ruperto, N., Kenwright, A., Wright, S., Calvo, I., et al. (2012). Randomized trial of tocilizumab in systemic juvenile idiopathic arthritis. *New England Journal of Medicine, 367*, 2385–2395.
24. Avau, A., Mitera, T., Put, S., Put, K., Brisse, E., Filtjens, J., et al. (2014). Systemic juvenile idiopathic arthritis-like syndrome in mice following stimulation of the immune system with Freund's complete adjuvant: Regulation by interferon-gamma. *Arthritis & Rhematology, 66*, 1340–1351.
25. Tesi, B., Sieni, E., Neves, C., Romano, F., Cetica, V., Cordeiro, A. I., et al. (2015). Hemophagocytic lymphohistiocytosis in 2 patients with underlying IFN-γ receptor deficiency. *The Journal of Allergy and Clinical Immunology, 135*(6), 1638–1641.
26. Morimoto, A., Nakazawa, Y., & Ishii, E. (2016). Hemophagocytic lymphohistiocytosis: Pathogenesis, diagnosis, and management. *Pediatrics International, 58*, 817–825.
27. Jordan, M. B., Allen, C. E., Weitzman, S., Filipovich, A. H., & McClain, K. L. (2011). How I treat hemophagocytic lymphohistiocytosis. *Blood, 118*, 4041–4052.
28. Terrell, C. E., & Jordan, M. B. (2013). Perforin deficiency impairs a critical immunoregulatory loop involving murine CD8(+) T cells and dendritic cells. *Blood, 121*, 5184–5191.
29. Jenkins, M. R., Rudd-Schmidt, J. A., Lopez, J. A., Ramsbottom, K. M., Mannering, S. I., Andrews, D. M., et al. (2015). Failed CTL/NK cell killing and cytokine hypersecretion are directly linked throuhg prolonged synapse time. *The Journal of Experimental Medicine, 212*(3), 307–317.
30. Zhang, M., Bracaglia, C., Prencipe, G., Bemrich-Stolz, C. J., Beukelman, T., Dimmitt, R. A., et al. (2016). A heterozygous RAB27A mutation associated with delayed cytolytic granule polarization and hemophagocytic lymphohistiocytosis. *Journal of Immunology, 196*(6), 2492–2503.
31. Jordan, M. B., Hildeman, D., Kappler, J., & Marrack, P. (2004). An animal model of hemophagocytic lymphohistiocytosis (HLH): CD8+ T cells and interferon gamma are essential for the disorder. *Blood, 104*, 735–743.
32. Chen, M., Felix, K., & Wang, J. (2012). Critical role for perforin and Fas-dependent killing of dendritic cells in the control of inflammation. *Blood, 119*, 127–136.
33. Schaer, D. J., Schaer, C. A., Schoedon, G., Imhof, A., & Kurrer, M. O. (2006). Hemophagocytic macrophages constitute a major compartment of heme oxygenase expression in sepsis. *European Journal of Haematology, 77*, 432–436.
34. Ryter, S. W., & Choi, A. M. (2016). Targeting heme oxygenase-1 and carbon monoxide for therapeutic modulation of inflammation. *Translational Research, 167*, 7–34.
35. Canna, S. W., Costa-Reis, P., Bernal, W. E., Chu, N., Sullivan, K. E., Paessler, M. E., et al. (2014). Brief report: Alternative activation of laser-captured murine hemophagocytes. *Arthritis & Rhematology, 66*, 1666–1671.
36. Ohyagi, H., Onai, N., Sato, T., Yotsumoto, S., Liu, J., Akiba, H., et al. (2013). Monocyte-derived dendritic cells perform hemophagocytosis to fine-tune excessive immune responses. *Immunity, 39*, 584–598.

37. Hull, T. D., Agarwal, A., & George, J. F. (2014). The mononuclear phagocyte system in homeostasis and disease: A role for heme oxygenase-1. *Antioxidants & Redox Signaling, 20*, 1770–1788.
38. Auffray, C., Fogg, D., Garfa, M., Elain, G., Join-Lambert, O., Kayal, S., et al. (2007). Monitoring of blood vessels and tissues by a population of monocytes with patrolling behavior. *Science, 317*, 666–670.
39. Carlin, L. M., Stamatiades, E. G., Auffray, C., Hanna, R. N., Glover, L., Vizcay-Barrena, G., et al. (2013). Nr4a1-dependent Ly6C(low) monocytes monitor endothelial cells and orchestrate their disposal. *Cell, 153*, 362–375.
40. Cros, J., Cagnard, N., Woollard, K., Patey, N., Zhang, S. Y., Senechal, B., et al. (2010). Human CD14dim monocytes patrol and sense nucleic acids and viruses via TLR7 and TLR8 receptors. *Immunity, 33*, 375–386.
41. Dal-Secco, D., Wang, J., Zeng, Z., Kolaczkowska, E., Wong, C. H., Petri, B., et al. (2015). A dynamic spectrum of monocytes arising from the in situ reprogramming of CCR2+ monocytes at a site of sterile injury. *The Journal of Experimental Medicine, 212*, 447–456.
42. Prencipe, G., Caiello, I., Pascarella, A., Grom, A. A., Bracaglia, C., Chatel, L., et al. (2018). Neutralization of IFN-γ reverts clinical and laboratory features in a mouse model of macrophage activation syndrome. *The Journal of Allergy and Clinical Immunology, 141*(4), 1439–1449.
43. Rood, J. E., Rao, S., Paessler, M., Kreiger, P. A., Chu, N., Stelekati, E., et al. (2016). *Blood, 127*(4), 426–435.

Cytokines in Cytokine Storm Syndrome

Edward M. Behrens

Introduction

As the name directly implies, cytokine storm syndrome (CSS) in general is characterized by the elaboration of large amounts of multiple cytokines that are in large part responsible for the organ toxicity of the disease. The identity of the cytokines produced often varies depending on the underlying cause or trigger of CSS. Additionally, the response to these cytokines is likely modulated by host factors including genetics, resulting in different interpretations of different cytokine levels depending on context. The advent of clinical "cytokine panel" testing allows the clinician to survey the landscape of cytokine production with the hope that this information may provide cues toward both diagnosis and therapy. However, the utility of serum cytokine testing is limited by a number of factors: cytokine burden is often in the tissue and therefore not detected by sampling the blood, cytokine levels may change rapidly in a nonlinear fashion throughout disease course rendering a single time-point difficult to interpret, and therapy may change cytokine levels in unpredictable and unknown ways. Nonetheless, work in both animal models and human studies have determined some stereotypical cytokine patterns associated with disease state and pathogenesis, and indeed in some circumstances, cytokine targeted therapeutic approaches. Table 1 summarizes the key cytokines that are discussed in this chapter. Figure 1 presents this information graphically.

E. M. Behrens (✉)
The Children's Hospital of Philadelphia, Philadelphia, PA, USA
e-mail: BEHRENS@email.chop.edu

R. Q. Cron, E. M. Behrens (eds.), *Cytokine Storm Syndrome*,
https://doi.org/10.1007/978-3-030-22094-5_12

Table 1 Major cytokines associated with CSS

Cytokine	Type of evidence	Benefits of targeting	References
IFNγ	Murine model of pHLH, MAS Human studies of pHLH, MAS	Potent efficacy in multiple murine models	[1, 3–8]
IL-1	Case-series level reports in humans	Typically excellent safety profile	[15–17]
IL-18	Murine model of MAS Case report on human NLRC4-MAS	May target a more proximal mediator of disease than IFNγ	[24–28]
IL-33	Murine model of pHLH	May target a more proximal mediator of disease than IFNγ	[31]
IL-6	Murine of MAS Human studies of CAR T-cell CRS	May target a different, specific population of CSS patients that have a different pathophysiology from pHLH or MAS	[34, 37, 38]
TNF	Conflicting case reports in humans	May be a proximate cause in a select population of CSS	[41–43]
IL-10	Negative regulator of disease in murine MAS model Found to be elevated in human pHLH	May be able to enhance negative regulation across multiple different scenarios	[5, 44]
IL-4	Murine model of hemophagocytic syndrome	Unknown	[48]

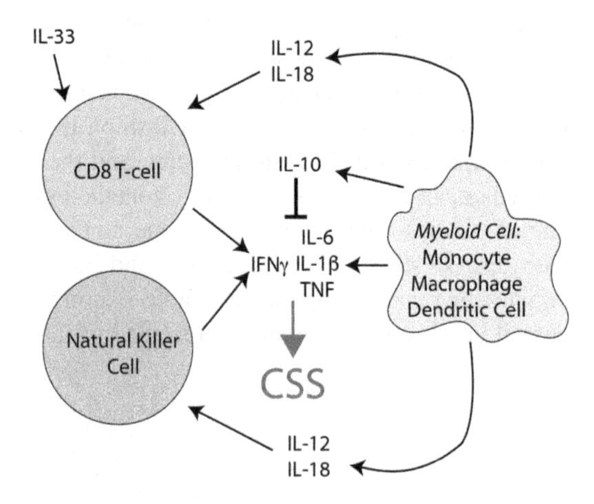

Fig. 1 Schematic of cytokine networks in CSS

Interferon Gamma

Perhaps the best studied mediator of CSS, Interferon gamma (IFNγ) has been identified as a common mediator of CSS in multiple scenarios. In seminal work by Jordan et al. using the perforin deficient mouse to model primary hemophagocytic lymphohistiocytosis (pHLH), it was shown that neutralization of IFNγ ameliorates disease [1]. Subsequent work has shown that this IFNγ is produced by overstimulated CD8 T-cells that are unable to clear stimulating antigen [2]. This has led to the advent of clinical trials of IFNγ neutralization in HLH that are currently ongoing (ClinicalTrials.gov Identifier: NCT03312751 & NCT03311854). IFNγ leads to the expression of hundreds of so-called interferon stimulated genes (ISG), providing the opportunity to direct a complex effector response. It is not yet clear which cell types, and which IFNγ stimulated gene products lead to organ toxicity. However, one particular ISG, CXCL9 has proven to be an effective biomarker of IFNγ activity in HLH [3]. Furthermore, patients with lymphoma associated HLH had much higher CXCL9 levels compared to septic patients, and levels decreased concordant with therapeutic response [4].

IFNγ has also been implicated in forms of secondary HLH (sHLH) in both human and murine models of disease. Using a murine model of Macrophage Activation Syndrome (MAS) predicated on repeated toll-like receptor 9 stimulation (TLR9-MAS), neutralization of IFNγ has been shown to ameliorate disease, much as it has in the perforin deficient model [5, 6]. Subsequent work has shown elevated of levels CXCL9 in this model acting as a biomarker similar in HLH [7]. This has been translated to human disease, where IFNγ and CXCL9 if found in patients with MAS that is not present in patients with active Systemic Juvenile Idiopathic Arthritis (SJIA) [8], distinguishing these two often confusable conditions. In the setting of both the murine MAS models, as well as in human disease, the source of IFNγ does not appear to be the same as in pHLH. Cytotoxic killing defects are not universally present in this setting like in pHLH, thus CD8 T-cell antigen stimulated IFNγ release does not appear to always be the source of IFNγ in sHLH models. Natural killer (NK) cells are often invoked in SJIA/MAS pathogenesis [9], and these cells may be another potent producer of IFNγ, although evidence for their involvement in SJIA is mixed [10]. In the TLR9-MAS model, IL-12 is essential for IFNγ production while this is dispensable in the perforin deficient pHLH model. IL-18 (discussed in greater detail below) may be another important IFNg-inducing factor in autoinflammatory MAS.

While increased levels of IFNγ are present in serum of both pHLH and MAS patients, sampling of serum may significantly underestimate the amount of IFNγ actually present. Treatment of pHLH patients with an IFNγ neutralizing antibody show increases of 100- to 100-fold more total IFNγ in the serum after antibody administration, presumably as the antibody complexes with IFNγ in the tissues and is cleared in the blood [11]. This highlights the fact that while blood is generally the only clinically accessible tissue to sample for cytokine levels, much of the important cytokine response may be happening in the tissue.

Notably, as central as IFNγ seems to be in multiple forms of CSS, there are also both murine and human examples of CSS occurring in absence of IFNγ. In the TLR9-MAS murine model, blockade of IL-10 signaling allows all symptoms of MAS to proceed, with the exception anemia, despite neutralization of IFNγ [12]. Multiple cases of humans with IFNγ receptor deficiencies presenting with HLH syndromes have been reported [13, 14]. It therefore seems that while IFNγ is certainly a critical cytokine for many CSS scenarios, it is not universally necessary to develop many of the clinical features of CSS.

Interleukin-1 Family Cytokines

The interleukin 1 (IL-1) family cytokines include 11 related gene products, of which IL-1β, IL-18, and IL-33, have known relevance to CSS and will be discussed herein. IL-1 family members signal through receptors that utilize the MyD88 adaptor, and downstream NFκB and ERK signals to direct inflammatory responses. IL-1β and IL-18 are activated from their pre-protein forms by the inflammasomes, a class of multimolecular complexes that activate caspase-1 to cleave the inactive pro-forms of IL-1β and IL-18 to their active forms. Inflammasomes act as sensors for cytosolic "danger" signals, moieties made by pathogens that enter the cytoplasm of mammalian cells. A number of genetic lesions of inflammasome function have been described that lead to constitutive activity, and therefore overproduction of IL-1β and/or IL-18 that in some cases result in a CSS.

IL-1β

IL-1β has been implicated in MAS due to a number of reports of the successful use of the drug anakinra, recombinant IL-1 receptor antagonist, to treat MAS and related CSS syndromes [15–17]. The strong association of MAS with SJIA [18] and the well-described success of IL-1β blockade in the treatment of active SJIA [19–21] provide an additional link to this cytokine and MAS. Which effector functions of IL-1β that might drive the CSS in these scenarios has not been yet elucidated. In contrast to these reports of the successful use of anakinra to treat MAS, in a large trial of the IL-1β neutralizing antibody canakinumab in the treatment of SJIA, treatment with drug failed to show a significant protection from the development of MAS [22]. This may be due to this study not having power to detect as small effect size, as there was a trend toward protection, or may be due to differences in IL-1β neutralization compared to the receptor blockade provided by anakinra, which would block both IL-1β and IL-1α. It has also been suggested that the doses of canakinumab used were too low to prevent MAS, since much higher doses of anakinra seemed to help MAS like presentations of sepsis in a retrospective reanalysis of anakinra in sepsis trial [23]. Alternatively, there may be differences in the role of IL-1β in initiating disease versus propagating disease, explaining why blockade might not prevent MAS, and yet still show efficacy in treating it once it has occurred. In the context of these data, it is important to note that genetic activating lesions of the NLRP3 inflammasome result in the production of large amounts of IL-1β release that is effectively treated by anakinra, and yet these patients do not seem to be at greater risk for MAS, suggesting that IL-1β alone may be insufficient for MAS development. However, a clinical trial of IL-1 blockade for treatment of MAS is currently underway (ClinicalTrials.gov Identifier: NCT02780583).

IL-18

The other major cytokine of inflammasome origin is IL-18. While like many other cytokines it has many diverse effector functions, from the view point of CSS pathophysiology, its most notable function is the induction of IFNγ from both CD8 T-cells and NK cells. High serum levels of IL-18 are associated with increased risk of MAS in SJIA [24]. In the TLR9-MAS animal, deletion of the antagonizing IL-18 binding protein exacerbated disease, suggesting that IL-18 does play a direct role in disease activity [25]. Recently, activating mutations of the NLRC4 inflammasome have been associated with an MAS-like syndrome [26, 27]. These patients have extremely high levels of serum IL-18 not seen in the NLRP3 activating mutations. Accordingly, a patient with NLRC4 mutation treated with a recombinant IL-18 binding protein that antagonizes IL-18 activity was reported to have a dramatic immediate and prolonged resolution of her MAS symptoms [28]. IL-18 blockade was associated with a decrease in the genetic signature of IFNγ response, consistent with a model of IL-18 induced IFNγ production driving MAS. Loss of function mutations in XIAP also cause an HLH syndrome accompanied by elevated IL-18 [29]. XIAP is a negative regulator of inflammasome function, and represses the production of inflammasome activated cytokines. A trial of IL-18 blockade with IL-18 binding protein in XIAP and NLRC4-MAS is currently underway (ClinicalTrials.gov Identifier: NCT03113760).

IL-18 has been studied in murine models of pHLH. Unlike the profound effect reported in NLRC-MAS, in these murine models IL-18 blockade did not improve survival [30]. The mice did have some improvement in liver inflammation, but IL-18 was clearly not central to disease pathogenesis. This likely speaks to differences in underlying pathophysiology between MAS and pHLH, and the different contributors to the pathogenic IFNγ in each case and/or inherent differences in among mice and between humans.

IL-33

IL-33 biology is different from IL-1β and IL-18 in that it is not inflammasome produced. Rather, it is present as a nuclear protein in endothelial and epithelial cells that is released upon inflammatory cell death. While it has traditionally been thought to be a cytokine that drives the atopic "Th2" cytokine response (IL-4/13), it also has a role similar to IL-18 in inducing IFNγ from CD8 T-cells. In the perforin deficient murine model of pHLH, IL-33 was critically required for pathogenic IFNγ production and blockade the IL-33 receptor ameliorated disease to an extent similar to IFNγ blockade [31]. There is not yet human data that assesses a role of IL-33; however, it remains possible that IL-33 plays a role in pHLH similar to what IL-18 plays in MAS. Thus, there may be a role for IL-33 blockade in different human CSS subsets.

Interleukin-6

IL-6 is perhaps one of the most pleotropic cytokines, produced by and acting on immune and nonimmune cells, with effects across multiple organ systems. Blockade of IL-6 receptor is effective for the treatment of SJIA; however, reports suggest that is may not be completely efficacious in SJIA-MAS, masking inflammatory symptoms rather than actually resolving them [32]. Unlike IL-18, IL-6 elevation does not correlate with MAS risk in SJIA [33]. There are no formal trials of IL-6 blockade in either human or murine systems of pHLH or MAS. However, an IL-6 transgenic mouse that constitutively produces the cytokine develops an MAS like syndrome after treatment with the TLR4 ligand LPS, much like the repeated TLR9 model [34]. Additional studies have shown that IL-6 acts on NK cells to suppress the cytolytic activity [8, 35]. Thus, it has been postulated that IL-6 may be an environmental exacerbating factor that could mimic or even enhance the genetic factors that impair cytotoxicity and lead to pHLH. It has been speculated that it is perhaps a combination of hypomorphic genetic lesions that impair cytotoxicity at a subclinical level but are then further impaired by IL-6 high environments, leading to secondary HLH or MAS [8, 35, 36].

Patients that experience cytokine release syndrome, a CSS that occurs in patients with B-cell leukemia treated with chimeric antigen receptor T-cell therapy or with the CD19/CD3 bispecific antibody blinatumomab develop extremely high IL-6 levels [37, 38]. Accordingly, the cytokine release syndrome appears to be very sensitive to IL-6 receptor blockade with tocilizumab. Thus, it would appear that in this context, IL-6 is indeed a central cytokine in the pathogenesis of the iatrogenic CSS of cytokine release syndrome.

Tumor Necrosis Factor

Tumor necrosis factor alpha (TNF) has long been associated with the inflammatory response of sepsis, although its blockade did not improve outcomes in clinical trials [39, 40]. Case reports of TNF blockade as both the cause of, and successful treatment for SJIA-MAS are present in the literature [41–43]. In murine models of perforin deficient pHLH, blockade of TNF had no effect on outcome [1]. There is little high quality evidence for or against TNF blockade in human disease; however, given that there are already established targets in other pathways, TNF blockade is not likely to emerge as candidate for a population level trial in CSS in the near future. Perhaps as we learn more about the specific individual lesions that lead to CSS, TNF blockade may reemerge as an opportunity for a personalized medicine approach.

Interleukin-10

IL-10 is an archetypical example of an anti-inflammatory cytokine acting on both myeloid and lymphoid cells to regulate and restrict their effector functions. Notably, along with IFNγ, it is highly upregulated in pHLH, presumably as a futile attempt at compensatory control of the inflammation. High levels of concomitant serum IL-10 and IFNγ distinguished pHLH patients from septic patents, who instead tended instead to have higher IL-6 serum levels [44]. In the murine model of perforin deficient pHLH, blockade of IL-10 had no effect on disease progression; however, in the repeated TLR9 model of MAS, IL-10 blockade significantly worsened morbidity and mortality. This may reflect the relative ability of IL-10 to control CSS mediated inflammation during different triggers of the cytokine storm state. The relevant sources of IL-10 and responding cells remain poorly understood, although some reports in both human and murine settings suggest that hemophagocytes themselves may be a source of IL-10 [45, 46]. While enhancing IL-10 levels or function may be a possible therapeutic target in the future, there are no currently available means to address this pathway clinically.

Hematopoietic Cytokines

Granulocyte monocyte colony stimulating factor (GM-CSF), IL-3, and stem cell factor (SCF) are all cytokines the promote hematopoiesis of myeloid cell lineages. In order to study xenografted human bone marrow in mice, an attempt to increase myeloid engraftment using triple transgenic mice expressing all three of these cytokines were made. Remarkably, these developed an MAS-like phenotype after engraftment with human umbilical cord blood that did not require lymphocytes, but was responsive to IL-6 receptor blockade [47]. Although this murine model is not directly related to any human clinical scenario of CSS, the implication that overexpression of myelopoietic cytokines opens up the possibility that these pathways may play some role in certain forms of CSS particularly in cases that are IL-6 blockade responsive.

Interleukin-4

IL-4 is traditionally considered to be a central mediator of "Th2" inflammation, the immune response associated with atopic and antiparasitic responses. However, mice that were exposed to chronic high levels of IL-4 using an osmotic pump developed many of the features of pHLH, primarily cytopenia and hemophagocytosis [48]. These effects did not depend on lymphocytes, nor on IFNγ. Much like the GM-CSF/IL-3/SCF model, there is not a direct clinical correlate for this system; however, this raises the possibility that in cases of severe Th2 responses, some features of CSS may become present. For instance, drug reaction with eosinophilia and systemic symptoms (DRESS) syndrome, may be a clinical example of an over exaggerated Th2 response that results in a CSS-like picture.

Conclusion

Given that the CSS spectrum consists of a heterogeneous group of genetic and environmental triggers, it should be expected that there will be a diverse array of cytokines that drive the initiation and propagation of disease. Further basic investigation will no doubt clarify which mechanisms lead to which cytokine profiles, and will inform therapeutic strategies. Improved technology to sample the cytokine milieu in the serum as well as target tissues may be needed to make use of this knowledge. Alternatively, biomarkers of cytokine response, such as CXCL9 for IFNγ, may prove useful diagnostically as well as for measurement of treatment response. Nonetheless, it remains remarkable that in many scenarios, blockade of a single, key cytokine is often able to produce dramatic clinical benefit, highlighting the importance of cytokine biology and the need for continued therapeutic development for these patients.

References

1. Jordan, M. B., Hildeman, D., Kappler, J., & Marrack, P. (2004). An animal model of hemophagocytic lymphohistiocytosis (HLH): CD8+ T cells and interferon gamma are essential for the disorder. *Blood, 104*, 735–743.
2. Lykens, J. E., Terrell, C. E., Zoller, E. E., Risma, K., & Jordan, M. B. (2011). Perforin is a critical physiologic regulator of T-cell activation. *Blood, 118*, 618–626.
3. Takada, H., Takahata, Y., Nomura, A., Ohga, S., Mizuno, Y., & Hara, T. (2003). Increased serum levels of interferon-gamma-inducible protein 10 and monokine induced by gamma interferon in patients with haemophagocytic lymphohistiocytosis. *Clinical and Experimental Immunology, 133*, 448–453.
4. Maruoka, H., Inoue, D., Takiuchi, Y., Nagano, S., Arima, H., Tabata, S., et al. (2014). IP-10/CXCL10 and MIG/CXCL9 as novel markers for the diagnosis of lymphoma-associated hemophagocytic syndrome. *Annals of Hematology, 93*, 393–401.

5. Behrens, E. M., Canna, S. W., Slade, K., Rao, S., Kreiger, P. A., Paessler, M., et al. (2011). Repeated TLR9 stimulation results in macrophage activation syndrome-like disease in mice. *The Journal of Clinical Investigation, 121*, 2264–2277.
6. Prencipe, G., Caiello, I., Pascarella, A., Grom, A. A., Bracaglia, C., Chatel, L., et al. (2018). Neutralization of IFN-gamma reverts clinical and laboratory features in a mouse model of macrophage activation syndrome. *The Journal of Allergy and Clinical Immunology, 141*, 1439–1449.
7. Buatois, V., Chatel, L., Cons, L., Lory, S., Richard, F., Guilhot, F., et al. (2017). Use of a mouse model to identify a blood biomarker for IFNgamma activity in pediatric secondary hemophagocytic lymphohistiocytosis. *Translational Research, 180*, 37–52 e32.
8. Zhang, M., Bracaglia, C., Prencipe, G., Bemrich-Stolz, C. J., Beukelman, T., Dimmitt, R. A., et al. (2016). A heterozygous RAB27A mutation associated with delayed cytolytic granule polarization and hemophagocytic lymphohistiocytosis. *Journal of Immunology, 196*, 2492–2503.
9. Villanueva, J., Lee, S., Giannini, E. H., Graham, T. B., Passo, M. H., Filipovich, A., et al. (2005). Natural killer cell dysfunction is a distinguishing feature of systemic onset juvenile rheumatoid arthritis and macrophage activation syndrome. *Arthritis Research & Therapy, 7*, R30–R37.
10. Put, K., Vandenhaute, J., Avau, A., Van Nieuwenhuijze, A., Brisse, E., Dierckx, T., et al. (2016). Inflammatory gene expression profile and defective IFN-gamma and granzyme K in natural killer cells of systemic juvenile idiopathic arthritis patients. *Arthritis & Rhematology, 69*(1), 213–224.
11. Jordan, M., Locatelli, F., Allen, C., De Benedetti, F., Grom, A. A., Ballabio, M., et al. (2015). A novel targeted approach to the treatment of hemophagocytic lymphohistiocytosis (HLH) with an anti-interferon gamma (IFNγ) monoclonal antibody (mAb), NI-0501: First results from a pilot phase 2 study in children with primary HLH. *Blood, 126*, LBA–LB3.
12. Canna, S. W., Wrobel, J., Chu, N., Kreiger, P. A., Paessler, M., & Behrens, E. M. (2013). Interferon-gamma mediates anemia but is dispensable for fulminant toll-like receptor 9-induced macrophage activation syndrome and hemophagocytosis in mice. *Arthritis and Rheumatism, 65*, 1764–1775.
13. Staines-Boone, A. T., Deswarte, C., Venegas Montoya, E., Sanchez-Sanchez, L. M., Garcia Campos, J. A., Muniz-Ronquillo, T., et al. (2017). Multifocal recurrent osteomyelitis and hemophagocytic lymphohistiocytosis in a boy with partial dominant IFN-gammaR1 deficiency: Case report and review of the literature. *Frontiers in Pediatrics, 5*, 75.
14. Tesi, B., Sieni, E., Neves, C., Romano, F., Cetica, V., Cordeiro, A. I., et al. (2015). Hemophagocytic lymphohistiocytosis in 2 patients with underlying IFN-gamma receptor deficiency. *The Journal of Allergy and Clinical Immunology, 135*, 1638–1641.
15. Miettunen, P. M., Narendran, A., Jayanthan, A., Behrens, E. M., & Cron, R. Q. (2011). Successful treatment of severe paediatric rheumatic disease-associated macrophage activation syndrome with interleukin-1 inhibition following conventional immunosuppressive therapy: Case series with 12 patients. *Rheumatology (Oxford), 50*, 417–419.
16. Behrens, E. M., Kreiger, P. A., Cherian, S., & Cron, R. Q. (2006). Interleukin 1 receptor antagonist to treat cytophagic histiocytic panniculitis with secondary hemophagocytic lymphohistiocytosis. *The Journal of Rheumatology, 33*, 2081–2084.
17. Kelly, A., & Ramanan, A. V. (2008). A case of macrophage activation syndrome successfully treated with anakinra. *Nature Clinical Practice. Rheumatology, 4*, 615–620.
18. Behrens, E. M., Beukelman, T., Paessler, M., & Cron, R. Q. (2007). Occult macrophage activation syndrome in patients with systemic juvenile idiopathic arthritis. *The Journal of Rheumatology, 34*, 1133–1138.
19. Ruperto, N., Brunner, H. I., Quartier, P., Constantin, T., Wulffraat, N., Horneff, G., et al. (2012). Two randomized trials of canakinumab in systemic juvenile idiopathic arthritis. *The New England Journal of Medicine, 367*, 2396–2406.

20. Quartier, P., Allantaz, F., Cimaz, R., Pillet, P., Messiaen, C., Bardin, C., et al. (2011). A multicentre, randomised, double-blind, placebo-controlled trial with the interleukin-1 receptor antagonist anakinra in patients with systemic-onset juvenile idiopathic arthritis (ANAJIS trial). *Annals of the Rheumatic Diseases, 70*, 747–754.
21. Nigrovic, P. A., Mannion, M., Prince, F. H., Zeft, A., Rabinovich, C. E., van Rossum, M. A., et al. (2011). Anakinra as first-line disease-modifying therapy in systemic juvenile idiopathic arthritis: Report of forty-six patients from an international multicenter series. *Arthritis and Rheumatism, 63*, 545–555.
22. Grom, A. A., Ilowite, N. T., Pascual, V., Brunner, H. I., Martini, A., Lovell, D., et al. (2016). Rate and clinical presentation of macrophage activation syndrome in patients with systemic juvenile idiopathic arthritis treated with canakinumab. *Arthritis & Rhematology, 68*, 218–228.
23. Shakoory, B., Carcillo, J. A., Chatham, W. W., Amdur, R. L., Zhao, H., Dinarello, C. A., et al. (2016). Interleukin-1 receptor blockade is associated with reduced mortality in sepsis patients with features of macrophage activation syndrome: Reanalysis of a prior phase III trial. *Critical Care Medicine, 44*, 275–281.
24. Shimizu, M., Nakagishi, Y., Inoue, N., Mizuta, M., Ko, G., Saikawa, Y., et al. (2015). Interleukin-18 for predicting the development of macrophage activation syndrome in systemic juvenile idiopathic arthritis. *Clinical Immunology, 160*, 277–281.
25. Girard-Guyonvarc'h, C., Palomo, J., Martin, P., Rodriguez, E., Troccaz, S., Palmer, G., et al. (2018). Unopposed IL-18 signaling leads to severe TLR9-induced macrophage activation syndrome in mice. *Blood, 131*, 1430–1441.
26. Canna, S. W., de Jesus, A. A., Gouni, S., Brooks, S. R., Marrero, B., Liu, Y., et al. (2014). An activating NLRC4 inflammasome mutation causes autoinflammation with recurrent macrophage activation syndrome. *Nature Genetics, 46*, 1140–1146.
27. Romberg, N., Al Moussawi, K., Nelson-Williams, C., Stiegler, A. L., Loring, E., Choi, M., et al. (2014). Mutation of NLRC4 causes a syndrome of enterocolitis and autoinflammation. *Nature Genetics, 46*, 1135–1139.
28. Canna, S. W., Girard, C., Malle, L., de Jesus, A., Romberg, N., Kelsen, J., et al. (2017). Life-threatening NLRC4-associated hyperinflammation successfully treated with IL-18 inhibition. *The Journal of Allergy and Clinical Immunology, 139*(5), 1698–1701.
29. Wada, T., Kanegane, H., Ohta, K., Katoh, F., Imamura, T., Nakazawa, Y., et al. (2014). Sustained elevation of serum interleukin-18 and its association with hemophagocytic lymphohistiocytosis in XIAP deficiency. *Cytokine, 65*, 74–78.
30. Chiossone, L., Audonnet, S., Chetaille, B., Chasson, L., Farnarier, C., Berda-Haddad, Y., et al. (2012). Protection from inflammatory organ damage in a murine model of hemophagocytic lymphohistiocytosis using treatment with IL-18 binding protein. *Frontiers in Immunology, 3*, 239.
31. Rood, J. E., Rao, S., Paessler, M., Kreiger, P. A., Chu, N., Stelekati, E., et al. (2016). ST2 contributes to T-cell hyperactivation and fatal hemophagocytic lymphohistiocytosis in mice. *Blood, 127*, 426–435.
32. Shimizu, M., Nakagishi, Y., Kasai, K., Yamasaki, Y., Miyoshi, M., Takei, S., et al. (2012). Tocilizumab masks the clinical symptoms of systemic juvenile idiopathic arthritis-associated macrophage activation syndrome: The diagnostic significance of interleukin-18 and interleukin-6. *Cytokine, 58*, 287–294.
33. Shimizu, M., Yokoyama, T., Yamada, K., Kaneda, H., Wada, H., Wada, T., et al. (2010). Distinct cytokine profiles of systemic-onset juvenile idiopathic arthritis-associated macrophage activation syndrome with particular emphasis on the role of interleukin-18 in its pathogenesis. *Rheumatology (Oxford), 49*, 1645–1653.
34. Strippoli, R., Carvello, F., Scianaro, R., De Pasquale, L., Vivarelli, M., Petrini, S., et al. (2012). Amplification of the response to Toll-like receptor ligands by prolonged exposure to interleukin-6 in mice: Implication for the pathogenesis of macrophage activation syndrome. *Arthritis and Rheumatism, 64*, 1680–1688.

35. Cifaldi, L., Prencipe, G., Caiello, I., Bracaglia, C., Locatelli, F., De Benedetti, F., et al. (2015). Inhibition of natural killer cell cytotoxicity by interleukin-6: Implications for the pathogenesis of macrophage activation syndrome. *Arthritis & Rhematology, 67*, 3037–3046.
36. Strippoli, R., Caiello, I., & De Benedetti, F. (2013). Reaching the threshold: A multilayer pathogenesis of macrophage activation syndrome. *The Journal of Rheumatology, 40*, 761–767.
37. Grupp, S. A., Kalos, M., Barrett, D., Aplenc, R., Porter, D. L., Rheingold, S. R., et al. (2013). Chimeric antigen receptor-modified T cells for acute lymphoid leukemia. *The New England Journal of Medicine, 368*, 1509–1518.
38. Teachey, D. T., Rheingold, S. R., Maude, S. L., Zugmaier, G., Barrett, D. M., Seif, A. E., et al. (2013). Cytokine release syndrome after blinatumomab treatment related to abnormal macrophage activation and ameliorated with cytokine-directed therapy. *Blood, 121*, 5154–5157.
39. Abraham, E., Wunderink, R., Silverman, H., Perl, T. M., Nasraway, S., Levy, H., et al. (1995). Efficacy and safety of monoclonal antibody to human tumor necrosis factor alpha in patients with sepsis syndrome. A randomized, controlled, double-blind, multicenter clinical trial. TNF-alpha MAb Sepsis Study Group. *JAMA, 273*, 934–941.
40. Clark, M. A., Plank, L. D., Connolly, A. B., Streat, S. J., Hill, A. A., Gupta, R., et al. (1998). Effect of a chimeric antibody to tumor necrosis factor-alpha on cytokine and physiologic responses in patients with severe sepsis—a randomized, clinical trial. *Critical Care Medicine, 26*, 1650–1659.
41. Maeshima, K., Ishii, K., Iwakura, M., Akamine, M., Hamasaki, H., Abe, I., et al. (2012). Adult-onset Still's disease with macrophage activation syndrome successfully treated with a combination of methotrexate and etanercept. *Modern Rheumatology, 22*, 137–141.
42. Makay, B., Yilmaz, S., Turkyilmaz, Z., Unal, N., Oren, H., & Unsal, E. (2008). Etanercept for therapy-resistant macrophage activation syndrome. *Pediatric Blood & Cancer, 50*, 419–421.
43. Stern, A., Riley, R., & Buckley, L. (2001). Worsening of macrophage activation syndrome in a patient with adult onset Still's disease after initiation of etanercept therapy. *Journal of Clinical Rheumatology, 7*, 252–256.
44. Xu, X. J., Tang, Y. M., Song, H., Yang, S. L., Xu, W. Q., Zhao, N., et al. (2012). Diagnostic accuracy of a specific cytokine pattern in hemophagocytic lymphohistiocytosis in children. *The Journal of Pediatrics, 160*, 984–990.e981.
45. Ohyagi, H., Onai, N., Sato, T., Yotsumoto, S., Liu, J., Akiba, H., et al. (2013). Monocyte-derived dendritic cells perform hemophagocytosis to fine-tune excessive immune responses. *Immunity, 39*, 584–598.
46. Schaer, D. J., Schaer, C. A., Schoedon, G., Imhof, A., & Kurrer, M. O. (2006). Hemophagocytic macrophages constitute a major compartment of heme oxygenase expression in sepsis. *European Journal of Haematology, 77*, 432–436.
47. Wunderlich, M., Stockman, C., Devarajan, M., Ravishankar, N., Sexton, C., Kumar, A. R., et al. (2016). A xenograft model of macrophage activation syndrome amenable to anti-CD33 and anti-IL-6R treatment. *JCI Insight, 1*, e88181.
48. Milner, J. D., Orekov, T., Ward, J. M., Cheng, L., Torres-Velez, F., Junttila, I., et al. (2010). Sustained IL-4 exposure leads to a novel pathway for hemophagocytosis, inflammation, and tissue macrophage accumulation. *Blood, 116*, 2476–2483.

Primary Immunodeficiencies and Cytokine Storm Syndromes

David A. Hill and Neil Romberg

Abbreviations

ADA	Adenosine deaminase
ALPS	Autoimmune lymphoproliferative syndrome
APC	Antigen-presenting cell
APDS	Activated phosphoinositide 3-kinase δ syndrome
AT	Ataxia–telangiectasia
CGD	Chronic granulomatous disease
CRAC	Calcium release activated channel
CSS	Cytokine storm syndrome
CTL	Cytotoxic T lymphocyte
CTLA-4	Cytotoxic T lymphocytic antigen-4
CVID	Common variable immunodeficiency
DC	Dyskeratosis congenita
DNT	Double negative T cells
EBV	Epstein–Barr virus
FOXP3	Forkhead box protein 3
HLH	Hemophagocytic lymphohistiocytosis
HSCT	Hematopoietic stem cell transplant
IL	Interleukin
IPEX	Immune dysregulation, polyendocrinopathy, enteropathy, X-linked syndrome
ITK	Interleukin 2-inducible T cell kinase

D. A. Hill · N. Romberg (✉)
Division of Allergy and Immunology, Department of Pediatrics,
Children's Hospital of Philadelphia, Philadelphia, PA, USA

Department of Pediatrics, The Perelman School of Medicine at the University
of Pennsylvania, Philadelphia, PA, USA
e-mail: hilld3@email.chop.edu; rombergn@email.chop.edu

R. Q. Cron, E. M. Behrens (eds.), *Cytokine Storm Syndrome*,
https://doi.org/10.1007/978-3-030-22094-5_13

LCMV	Lymphocytic choriomeningitis virus
MSMD	Mendelian susceptibility to mycobacterial diseases
mTOR	Mammalian target of rapamycin
NEMO	Nuclear factor-kappa B essential modulator
NFκβ	Nuclear factor-κβ
NK	Natural killer
PID	Primary immunodeficiency disease
SCID	Severe combined immunodeficiency
TACI	Transmembrane activator and CAML interactor
T_H	T helper cell
TLR	Toll-like receptor
Treg	Regulatory T cell
Tregs	T regulatory cells
WAS	Wiscott–Aldrich syndrome
XLA	X-linked agammaglobulinemia
XLP	X-linked lymphoproliferative
XMEN	X-linked immunodeficiency with magnesium defect, Epstein–Barr virus infection, and neoplasia

Introduction

Primary immunodeficiency diseases (PID) are a diverse and growing category of more than 300 chronic disorders that share a susceptibility to infections. Traditionally, the immune systems of PID patients were considered essentially inert, but more recently the field has identified new PIDs (or recognized new phenotypes of existing PIDs) that demonstrate prominent inflammatory features. Whether characterized by excessive lymphoproliferation, defective granule-dependent cytotoxicity, or an overwhelming infection representing a unique antigenic challenge, PIDs can result in the development of cytokine storm syndrome (CSS).

This chapter reviews CSS pathophysiology as it relates to PIDs (Table 1). For each PID, we provide an overview of the immunologic defect and how it promotes or discourages CSS phenomena. We highlight PID-associated molecular defects in pathways that are postulated to be critical to CSS physiology (i.e., interferon gamma, interleukin (IL)-6, IL-12, and TNF-alpha), and review strategies for treating CSS in PID patients with molecularly directed therapies.

Defects in Granule-Dependent Cytotoxic Killing

Cytotoxic killing is a key host mechanism for eliminating infected or malignant cells. Impaired cytotoxicity is a general feature of PIDs predisposed to CSS events, but there are several genetic deficiencies of specific components of the granule-dependent cytotoxic machinery that acutely demonstrate this concept. As the associated diseases, including *familial hemophagocytic lymphohistiocytosis types 1–5*,

Table 1 Association of PIDs with cytokine storm syndromes

Disease	Genes involved	References
Associated with cytokine storms		
Autoimmune lymphoproliferative syndrome	*FAS, FASL*	[28, 29]
Autosomal dominant hyper-IgE syndrome	*STAT3*	[113]
CD27 deficiency	*CD27*	[2–5]
Ataxia–telangiectasia	*ATM*	[49]
Chronic granulomatous disease	*CYBA*, *CYBB*, *NCF1*, *NCF2*	[49, 142–144, 146]
Chronic mucocutaneous candidiasis	*STAT1* GOF	[112]
CTLA-4 haploinsufficiency	*CTLA4*	[11, 12]
Dyskeratosis congenita	*TERT, TERC, DKC1, TINF2*	[49]
GATA2 deficiency	*GATA2*	[152, 153]
Immune dysregulation, polyendocrinopathy, enteropathy, X-linked	*FOXP3*, *CD25*, *IL10*, *IL10R*	[121]
ITK deficiency	*ITK*	[55, 59]
Mendelian susceptibility to mycobacterial disease	*IL12R, IL12, IFNg, STAT1*	[104–106]
NEMO deficiency	*IKBKG*	[74, 75]
Severe combined immunodeficiency (SCID)	*IL2RG*, *IL7R*, others	[47–51]
Wiskott–Aldrich syndrome	*WASP*	[49, 70]
X-linked agammaglobulinemia	*BTK*	[132]
XMEN and other channel defects	*MAGT1, ORAI1*	[89–91]
Unclear association with cytokine storms		
Autosomal dominant hyper-IgE syndrome	*STAT3*	[113]
Chronic mucocutaneous candidiasis	*STAT1* GOF	[112]
Immune dysregulation, polyendocrinopathy, enteropathy, X-linked	*FOXP3*, *CD25*, *IL10*, *IL10R*	[121]
22q11.2 deletion syndrome (DiGeorge, velocardiofacial syndrome)	*TBX1*	[63–65]
Not associated with cytokine storms		
Activated phosphoinositide 3-kinase δ syndrome	*PIK3CD, PIK3R1*	
ADA SCID	*ADA*	
CD40/CD40L deficiency	*CD40, CD40L*	
CD70 deficiency	*CD70*	
STIM1 deficiency	*STIM1*	
TACI deficiency	*TACI*	
MyD88 and IRAK4 deficiency	*MyD88, IRAK4*	

Chediak–Higashi syndrome, *Griscelli syndrome type 2*, *Hermansky–Pudlak syndrome type 2*, and *X-linked lymphoproliferative syndrome type 2* are covered in detail by other chapter authors, we will not review them further.

Lymphoproliferative Disorders

PID patients unable to control excessive effector lymphocyte expansion or malignant/infected target cell proliferation are susceptible to CSS events.

X-linked Lymphoproliferative Syndrome Types 1

X-linked lymphoproliferative syndrome type 1 (XLP1) is caused by Src homology 2 domain protein 1A (SH2D1A) deficiency and is characterized by a predilection for Epstein–Barr virus (EBV)-lymphoproliferative disease. Sixty percent of XLP1 patients experience recurrent hemophagocytic lymphohistiocytosis (HLH) episodes, often related to EBV infection. XLP1 patients also demonstrate subtle hypogammaglobulinemia, impaired cytotoxicity, and are predisposed to lymphoma. XLP1 is covered in detail in other book chapters. We mention it (briefly) here because it is the prototypical lymphoproliferative disease and because it may be confused with other PIDs, specifically common variable immune deficiency (CVID).

CD27 and CD70 Deficiencies

CD27 deficiency and CD70 deficiency are related autosomal recessive combined immune deficiencies. Both CD27 and CD70 are TNF-receptor/TNF superfamily members and are an exclusive receptor–ligand pair (Fig. 1). In healthy controls, CD70 expression on malignant and EBV-infected B cells activates $CD8^+$ cytotoxic T

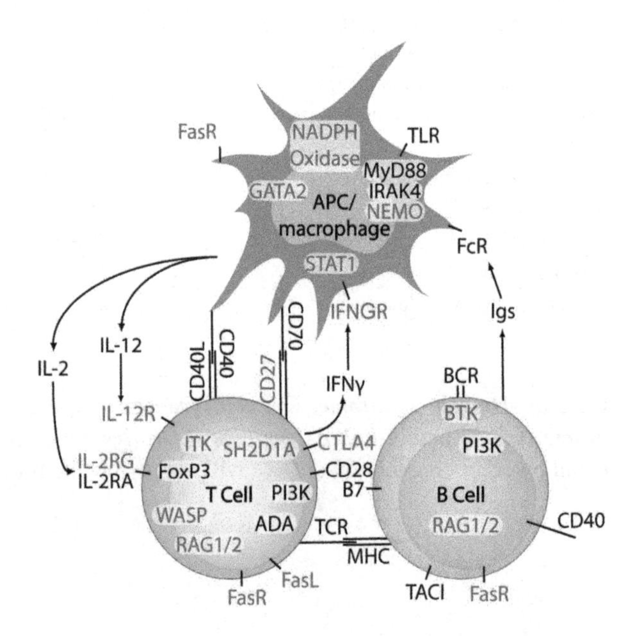

Fig. 1 Simplified schematic: mediators of cytokine storm-associated primary immunodeficiency diseases. Antigen-presenting cell (APC), T cell, and B cell are depicted. Major secreted, membrane-associated, and intracellular proteins are labeled. CSS-related immune molecules discussed in the chapter are denoted in red. Molecules absent in familial hemophagocytic lymphohistiocytosis patients with defective granule-dependent cytotoxic killing are not depicted

lymphocytes (CTLs) through interaction with CD27 [1]. CD27 deficient patients display a combined immunodeficiency characterized by hypogammaglobulinemia and EBV infection-related complications including HLH and malignant lymphoma [2–5]. Interestingly, despite increased EBV susceptibility, CSS events have not been reported in CD70 deficient patients. The reason for this apparent discrepancy is unknown.

CTLA-4 Haploinsufficiency

Cytotoxic T lymphocytic antigen-4 (CTLA-4) haploinsufficiency is an autosomal dominantly inherited PID. Incomplete penetrance is common. CTLA-4 is an important negative regulator of T-cell responses [6]. It is constitutively expressed on T regulatory cells (Tregs), and upregulated in conventional helper T cells after their activation. CTLA-4 acts in trans through competition with the T-cell co-activation molecule CD28 for B7 ligands, and in cis by directly antagonizing T-cell receptor signaling [7, 8]. CTLA-4-deficient mice experience fatal infiltrative lymphoproliferation [9]. CTLA-4 haploinsufficiency in humans manifests similarly to the murine model of deficiency, but is also frequently associated with hypogammaglobulinemia [10]. CTLA-4 haploinsufficient patients responding to significant antigenic challenges have experienced CSS events. For example, HLH episodes have been reported in CTLA-4 haploinsufficient patients with EBV-induced Hodgkin lymphoma [11] and Aspergillus fungemia [12]. The later patient was successfully treated with abatacept, a fusion protein of the IgG1 Fc-region and the extracellular CLTA-4 domain. Chloroquine/hydroxychloroquine, which augments CTLA-4 expression by inhibiting its lysosomal degradation, may also hold promise for restoring immune equilibrium in CTLA-4 haploinsufficient patients [13].

Autoimmune Lymphoproliferative Syndrome

Autoimmune lymphoproliferative syndrome (ALPS) type 1A is caused by deleterious germ line or somatic mutations in the gene encoding FasR. FasR is a type I-membrane protein expressed by activated or infected lymphocytes including EBV infected B cells [14]. Its ligand (FasL) is predominantly expressed by CTLs and natural killer (NK) cells. FasR–FasL interactions initiate caspase-8-mediated apoptosis [15]. Absent FasR/FasL interactions, ALPS patients are unable to maintain lymphocyte homeostasis through apoptosis, and often display abnormal, expanded populations of $CD4^{-}CD8^{-}$double negative T cell receptor alpha beta T cells (DNT). ALPS manifests clinically with impressive lymphadenopathy, splenomegaly, and autoimmune cytopenias [16–18]. Urticarial rashes, pulmonary fibrosis, and systemic lupus erythematosus have also been reported [19, 20]. Although ALPS patients are characteristically hypergammaglobulinemic, they are at increased risk of EBV-related infections, and lymphomas [21]. ALPS has also been reported in patients with mutations in genes encoding Caspase 8, Caspase 10, NRAS, and KRAS [22–26]. A genetic defect cannot be identified in 20–30% of ALPS patients [27].

CSS episodes in undiagnosed ALPS patients have been mistaken for familial HLH [28, 29]. In confounding cases, laboratory testing including cytotoxicity assessments, DNT enumeration, and ultimately gene sequencing are critical tools for clinicians wishing to discriminate between these two conditions.

Other Lymphoproliferative Syndromes

It is notable that CSS events have not been reported in patients carrying heterozygous mutations in the gene encoding transmembrane activator and CAML interactor (TACI). Such mutations are exceedingly common and are highly associated with a lymphoproliferative form of CVID [30–34]. This may be because TACI is expressed on human B cells, not T cells, and therefore leave cytotoxicity unaffected. Interestingly, TACI is also expressed in murine macrophages where it downregulates genes classically characteristic of a M2 activation state including IL-4Rα, CD206, CCL22, IL-10, Arg1, IL1RN, and FIZZ1 [35]. Polarization of macrophages away from the M1 activation state, which is associated with HLH [36], may also explain why TACI deficient patients have not experienced CSS events despite an affinity for lymphoproliferation.

Similarly, activated phosphoinositide 3-kinase δ syndrome (APDS), caused by gain-of-function mutations in genes encoding PIK3CD or PIK3R1, is also characterized by humoral immunodeficiency, EBV or CMV-induced lymphoproliferation, and a risk of B cell lymphoma, but not CSS episodes [37–40]. *PIK3CD or PIK3R1* mutations cause chronic mammalian target of rapamycin (mTOR) hyperactivation and resultant T-cell exhaustion. It is possible that exhausted *PIK3CD or PIK3R1* mutated T cells are less capable of mounting a systemic CSS immune response [41].

T-Cell Deficiency Disorders

Severe Combined Immunodeficiency

Severe combined immunodeficiency (SCID) encompasses a group of primary disorders characterized by severe T-cell lymphopenia and poor T cell proliferative responses. Several monogenic forms of SCID have been identified, and can be transmitted by X-linked or autosomal recessive inheritance. SCID may also occur in the context of thymic aplasia. The most common SCID causing mutations occur in the genes encoding the IL-2 receptor gamma chain (IL-2RG), adenosine deaminase (ADA), IL-7 receptor (IL7RA), and several gene products involved in DNA repair (RAG, ARTEMIS, etc.) [42]. Hypomorphic mutations in SCID genes, especially those involved in DNA repair, can result in the oligoclonal expansion of autoreactive T cells. This phenomenon, called Ommen syndrome, presents with characteristic erythroderma and diffuse alopecia from infiltrating T cells [43–45]. T cell mediated inflammation in Ommen syndrome also effects other organs including the lung, liver, and intestines.

Favorable SCID outcomes correlate with early age at hematopoietic stem cell transplant (HSCT), and are inversely related to the number of pre-transplant infections [46]. Prior to the introduction of universal newborn screening assays for T cell lymphopenia, SCID patients typically presented late in infancy with severe opportunistic infections and related pulmonary disease, diarrhea, and failure-to-thrive. At the time of publication, most US states screen all newborns for congenital lymphopenia allowing rapid identification, timely infectious prophylaxis initiation, and early transplantation.

HLH episodes have been reported in IL-2RG, RAG-1, IL7RA, and CD3ε-deficient patients, primarily in infectious contexts. Linked pathogens include common viruses (CMV, EBV, Rhinovirus, Adenovirus), enteric bacteria (*Enterobacter*, gram-negative sepsis), and opportunistic microbes (*P. jirovecii*, *P. aeruginosa*, *Candida*) [47–51]. In many cases, patient bone marrow aspirates characteristically display extensive hemophagocytosis, but absent T lymphocytes. This finding clearly suggests that T cell contributions to CSS episodes are nonessential in this setting. It is interesting to the authors' that CSS episodes have not been reported in adenosine deaminase (ADA) deficient SCID, whereas elevated serum ADA concentrations are reported in immune competent HLH patients [52]. Adenosine possesses several dose-dependent, anti-inflammatory properties and it is possible that accumulation of extracellular adenosine simultaneously promotes SCID and prevents CSS immunopathology [53].

Interleukin 2-Inducible T Cell Kinase Deficiency

Interleukin 2-inducible T cell Kinase (ITK) deficiency is an autosomal recessive combined immunodeficiency. ITK is a member of the TEC family, and functions as a non-receptor protein-tyrosine kinase in the development and signaling of lymphoid cells [54]. ITK-deficient patients exhibit hypogammaglobulinemia and progressive loss of $CD4^+$ T cells [55] which predisposes them to bacterial and viral infections. Like XLP1, ITK-deficient patients are especially susceptible to EBV-induced B cell proliferation and Hodgkin lymphoma [56–58]. EBV related HLH has been reported in multiple ITK-deficient patients [55, 59].

Chromosome 22q11.2 Deletion Syndrome

22q11.2 deletion syndrome (also known as DiGeorge syndrome, as well as velocardiofacial syndrome) is a multisystem developmental disorder. Although most 22q11.2 deletions occur de novo, they can also be inherited in an autosomal dominant manner. Many features of the syndrome are caused by haploinsufficiency of TBX1 which is essential to forming the third and fourth pharyngeal arches during embryogenesis. The result is midline developmental defects including cleft palate, conotruncal cardiac anomalies, thymic aplasia, and hypoparathyroidism [60, 61]. Most chromosome 22q11.2 deletion syndrome patients have sufficient thymic tissue to generate an adequate number of T cells for immune protection. Patients who do not have sufficient thymic tissue require thymic, not hematopoietic, transplantation [62].

There are five reports of 22q11.2 deletion syndrome patients with CSS-like episodes. One with thymic aplasia and reduced cytotoxicity assessments presented with HLH [63], although others with thymic tissue and/or normal T/B-cell function and NK cell activity presented similarly [64]. More recently, a 25-year-old man with 22q11.2 deletion syndrome developed EBV-associated T-cell lymphoma and hemophagocytosis [65]. 22q11.2 deletion and HLH have been proposed by some to be non-randomly associated, but given the large number of 22q11.2 deletion syndrome patients (incidence = 1:3000 births) and the limited number of HLH cases reported in the literature, it is likely that these conditions are independent events that casually co-occur.

Wiscott–Aldrich Syndrome

Wiscott–Aldrich Syndrome (WAS) is an X-linked PID that is characterized by thrombocytopenia, eczema, and a variable degree of combined immunodeficiency [66, 67]. WAS results from a deficiency in the WAS-protein (WASp), which is expressed in hematopoietic cells and is an important regulator of actin polymerization [68]. Because it is expressed in all hematopoietic cells, the immunologic consequences of WASp deficiency can be broad and include impaired migration, phagocytosis, and immune synapse formation in T cells, B cells, and innate immune cells. Autoimmunity is also common, and presents in early childhood with antibody-mediated cytopenias and vasculitis [69]. B-cell lymphoma (often EBV-positive) and leukemias are common in adolescent WAS patients, and EBV and CMV-associated HLH episodes have been reported [49, 70]. WASp is essential for NK cells to form a cytotoxic synapse with a target cell [71], possibly accounting for the difficulty WAS patient have clearing viral infections and controlling immune responses.

NEMO Deficiency

NF-kappa-B essential modulator (NEMO) deficiency is an X-linked combined immune deficiency. NEMO, encoded by the X-linked gene *IKBKG*, is an essential component of the nuclear factor-κβ (NFκβ) activation complex. As NFκB is biologically ubiquitous, NEMO deficiency manifests as a multi-system disease with features including ectodermal dysplasia, osteopetrosis, lymphedema, and immune deficiency [72]. In the context of immunity, NEMO is positioned downstream of many cell surface receptors including the B/T-cell receptors, toll-like receptors

(TLRs), CD40, and cytokine receptors that utilize JAK-STAT signaling, including those for IL-1 and TNF. Although first described as form of hyper-IgM syndrome, NEMO deficient patients display wide-ranging alterations in innate and adaptive immunity, including impaired NK cell cytotoxicity [73]. Like other diseases associated with impaired cytotoxic killing, NEMO deficiency appears to confer a risk of HLH [74, 75].

CD40/CD40L Deficiency

CD40 deficiency is an autosomal recessive combined immune deficiency, and CD40L deficiency is an X-linked combined immune deficiency; they are otherwise indistinguishable. CD40, a TNF receptor superfamily member, is expressed on B cells and other professional APCs. Its ligand (CD40L, also known as CD154) is transiently expressed on the surface of activated $CD4^+$ T cells [76]. CD40/CD40L interactions mediate key aspects of humoral immunity like class-switch recombination and somatic hypermutation [77–79]. CD40/CD40L interactions are also essential for T helper cell (T_H) type 1 priming of cellular immunity. Hence, CD40/CD40L-deficient patients are highly susceptible to both sinopulmonary and opportunistic infections. CD40/CD40L-deficient patients also develop inflammatory diseases including cytopenias, arthritis, inflammatory bowel disease, and development of organ-specific autoantibodies, highlighting the importance of this signaling pathway for normal immune tolerance [80]. Despite the tendency to develop autoimmune disease, CSS episodes are yet to be reported in CD40/CD40L-deficient patients. This may suggest that CD40/CD40L co-activation is an important component of CSS physiology. Alternatively, independent of CD40, CD40L deficient B cells have demonstrated resistance to EBV lymphoblastic transformation, and may also be resistant to malignant transformation [81].

Ataxia–Telangiectasia

Ataxia–telangiectasia (AT) is an autosomal recessive DNA repair disorder. AT is caused by mutations in the tumor suppressor gene *ATM* which encodes an eponymous protein that stalls cell cycle progression upon detection of double-stranded DNA breaks [82, 83]. Absent ATM, VDJ recombination is highly inefficient, leading to a combined immunodeficiency [84]. ATM-deficient cells also accumulate somatic mutations, increasing the risk of malignant transformation. As a result, lymphomas and acute leukemias occur commonly in AT patients. HLH has also been reported in at least one AT patient infected with EBV and parvovirus infections [49].

Dyskeratosis Congenita

Dyskeratosis congenita (DC) describes a group of primary disorders caused by mutations in genes that maintain telomeres [85, 86]. DC patients present with dysplastic nails, lacy reticular pigmentation of the upper chest and/or neck, and oral leukoplakia. Accelerated telomere shortening limits lymphocyte replicative lifespan causing progressive bone marrow failure and combined immunodeficiency [87]. Squamous cell carcinomas commonly occur in DC patients. HLH has also been reported in at least one DC patient infected with EBV [49].

Calcium and Magnesium Channel Defects

Monogenic ion channel defects can present with cellular immunodeficiency and CSS episodes. One such disorder is caused by biallelic mutations in the gene encoding ORAI1, the pore subunit of the Calcium release activated channel (CRAC) [88]. ORAI1-deficient T cells, display altered calcium flux and attenuated cytotoxicity. CMV-infection associated HLH has been reported in an ORAI1-deficient patient [89]. Similarly, magnesium transporter (MAGT1) deficiency is an X-linked immunodeficiency with magnesium defect, Epstein–Barr virus infection, and neoplasia (XMEN). Absent magnesium flux through MAGT1, XMEN T cells cannot eliminate chronic EBV infections or EBV-associated tumors. A fatal episode of HLH was reported in an XMEN patient with chronic EBV infection. He succumbed shortly after HSCT [90, 91]. Finally, STIM1 deficiency is an autosomal recessive combined immunodeficiency caused by defective cellular store-operated calcium entry [92]. STIM1 deficient T cells proliferate poorly and produce few cytokines yet display normal cytotoxicity. Unlike ORAI1 or MAGT1-deficient patients, CSS episodes have not been reported in STIM1-deficient patients [93].

Defects in Specific T-Helper Cell Subsets

There are at least four commonly recognized subsets of $CD4^+$ T cells, each with a distinct cytokine expression profile and function.

T_H1 cells assist with clearance of intracellular infections. They are characterized by expression of the transcription factor T-bet, are induced by IL-12, and produce IL-2, IFNγ, and TNFα [94]. IFNγ induces high and sustained IL-12 production by macrophages and dendritic cells, establishing a positive feedback loop known as the IL-12/IFNγ axis. STAT1 serves the IFNγ receptor.

T_H2 cells mediate humoral immune responses. They are characterized by expression of the transcription factor GATA3, are induced by IL-4, and produce IL-4, IL-5 and IL-13 [95]. STAT6 serves the IL-4 receptor.

T_H17 cells control fungal and bacterial infections at mucosal sites. They are characterized by expression of the transcription factor RORγt, are induced by TGFβ, IL-6, IL-21, IL-1β, and IL-23, and produce IL-17, IL-21, and IL-22 [94]. Upon detection of fungal pathogens, mucosal epithelial cells secrete IL-6 and IL-23 that bind to receptors on naïve T cells and induce differentiation into T_H17 cells. IL-17 licenses mucosal epithelial cells to secrete bactericidal particles, among other effector functions.

Treg cells regulate/suppress other T-helper cell and CTL responses. They are characterized by expression of the transcription factor FOXP3, are induced by TGFβ and IL-2, and produce IL-10. IL-2 is critical to Treg function. CD25, the high affinity IL-2 receptor, is highly expressed on Tregs allowing these cells to act as IL-2 sinks [96].

Several PID phenotypes are categorized according to similar T helper cell-subset defects.

Mendelian Susceptibility to Mycobacterial Disease

There are several monogenic PIDs that manifest with a narrow infectious susceptibility to nonpathogenic mycobacteria, including the Bacillus Calmette–Guerin vaccine strain. These disorders are collectively named Mendelian susceptibility to mycobacterial diseases (MSMD). MSMD genetic defects are primarily deleterious autosomal recessive mutations in genes encoding components of the IL-12/IFNγ axis. For instance, patients that are IL-12 receptor deficient, IL-12 deficient, IFNγ receptor deficient, and STAT1 deficient have been identified [97, 98].

Given the importance of the IL-12/IFNγ axis to mouse models of inflammatory disease, and the relative efficacy of anti-IFNγ antibodies in treating human inflammatory diseases [99–103], one might expect patients with MSMD to be resistant to CSS. Yet, CSS has been reported in IL-12 receptor, IFNγ receptor, and STAT1 deficient patients [104–106]. Although this may appear counter-intuitive, it is consistent with the mouse model of TLR9 agonist-induced CSS. In this model, both wild-type and IFNγ-deficient mice develop CSS in the absence of IL-10 signaling [107], indicating that IFNγ is not required for some aspects of CSS physiology.

Chronic Mucocutaneous Candidiasis

There are several monogenic PIDs that manifest with chronic mucocutaneous candidiasis (CMC) of the nails, skin, oral cavity, and genital mucosae. About half of CMC patients harbor heterozygous *STAT1* gain-of-function (GOF) mutations [108, 109]. It is thought that increased STAT1 activity limits the signals from STAT3-dependent receptors like the IL-6R and IL-23R which are critical to Th17 cell development and function [108, 110, 111]. Despite the role of STAT1 in the IL-12/IFNγ axis, STAT1 gain-of-function patients appear to develop CSS episodes less frequently than MSMD patients; a 2016 review of 274 immune-deficient STAT1 gain-of-function patients uncovered only one with a HLH/CSS history [112].

CMC is also a common finding in autosomal dominant hyper-IgE syndrome, a disease caused by heterozygous loss-of-function *STAT3* mutations. As STAT3 serves the IL-6 receptor, and IL-6 is considered a key CSS mediator, it is interesting that at least one patient carrying a loss of function *STAT3* mutation experienced a CSS episode secondary to Histoplasmosis [113]. Similarly, CSS episodes have not been yet described in the growing number of patients carrying heterozygous *STAT3* gain-of-function mutations [114, 115]. Both observations suggest IL-6 may be expendable in CSS physiology.

T Regulatory Cells

There are several monogenic PIDs that manifest with immune dysregulation due to Treg deficiency. The prototypical Treg disorder is immune dysregulation, polyendocrinopathy, enteropathy, X-linked (IPEX) syndrome. IPEX syndrome is caused by deleterious mutations in the gene encoding FOXP3—the master Treg transcriptional regulator [116, 117]. FOXP3 expression is driven by IL-2 through the high-affinity heterodimeric IL-2 receptor comprised of IL-2RA and IL-2RG. The receptor is serviced by STAT5b. IL-2RA and STAT5b deficient patients also present with an IPEX-like disease [118, 119].

Many IPEX and IPEX-like patients develop symptoms of autoimmune disease in utero that become apparent shortly after birth. The most common IPEX disease features include severe diarrhea and failure to thrive, diabetes mellitus, eczema, hemolytic anemia, autoimmune thrombocytopenia, and hypothyroidism [118, 120]. Considering the dramatic autoimmune phenotype, the immunodeficiency associated with these conditions is comparably mild, but often exaggerated by immune suppressive therapy. HSCT offers a potential cure for IPEX patients yet, unfortunately, in the IPEX context, the procedure has been associated with significant morbidity and mortality [121]. Without HCST, patients succumb during childhood to infectious disease.

There is evidence from animal models suggesting that functional Tregs may prevent CSS episodes by acting as IL-2 sinks. When lymphocytic choriomeningitis virus (LCMV) infected perforin-deficient mice develop HLH, Treg frequencies decreases leaving excess free IL-2 to activate CTLs. Similarly, transient Treg deficiencies have also been noted in HLH patients early in CSS episodes. Despite these interesting observations, there is only one report of a CSS episode in an IPEX patient, and it occurred 29 months after a fully engrafted HSCT [121]. Given the paucity of cases, the importance of Tregs in preventing CSS episodes remains unclear at this time.

B Cell Defects

Undifferentiated CVID

Common Variable Immunodeficiency is the most common, clinically significant immunodeficiency. CVID onset typically occurs after four years of age and can develop in adult patients. Characterized by a deficiency in antibody production, CVID is a genetically heterogeneous group of disorders that, unlike the majority of PIDs, is likely largely polygenic in nature. Monogenic causes currently account for approximately 20% of patients. In addition to infection susceptibility, CVID is complicated by autoimmune and lymphoproliferative complications [34, 122].

CVID-associated CSS episodes have primary been reported in CTLA-4 deficient patients (discussed above) or in undifferentiated CVID patients who lack a specific genetic diagnosis. Notably, several CSS cases have been reported in undifferentiated male CVID patients affected by EBV infections or lymphoma [123, 124]. Given these reoccurring features and the absence of molecular diagnoses, alternative X-linked diagnoses such as combined immune deficiencies or lymphoproliferative disorders, may better explain the CSS episodes in these patients.

X-Linked Agammaglobulinemia

X-linked agammaglobulinemia (XLA) patients are absent B cells and hypogammaglobulinemic. XLA is caused by deleterious mutations in the gene encoding Bruton's tyrosine kinase (BTK) [125–128]. BTK is critical for B cell receptor (BCR) signaling, and essential to B cell development [129]. XLA patients are predisposed to sinopulmonary infections and enteroviral meningoencephalitis [130]. BTK has been shown to interact with MyD88 and TRIF to promote TLR signaling, providing an additional mechanism by which BTK may provide protection against disseminated enteroviral infections [131].

CSS episodes do not occur commonly in XLA, but have been reported in two brothers who developed HLH [132]. One brother was treated with dexamethasone, cyclosporine, and etoposide. He succumbed to his disease. Adenovirus was identified in the bronchoalveolar lavage fluid of the second brother. He was successfully treated with cidofovir and IVIG. These two cases emphasize the importance of identifying the trigger of CSS episodes, and tailoring therapy accordingly.

Granulocyte and Other Innate Defects

Chronic Granulomatous Disease

Chronic granulomatous disease (CGD) encompass X-linked and autosomal recessive inherited deficiencies of NADPH oxidase components [133, 134]. NADPH oxidase mediates the oxidative burst, a process used by phagocytes to destroy engulfed microbes [135]. CGD patients are particularly susceptible to bacterial and fungal infections of the lung, skin, and lymph nodes. Catalase, which counters oxidative burst stress, is expressed by many organisms infecting CGD patients including *S. aureus*, *Aspergillus* species, *B. cepacia*, *S. marcescens*, and *Nocardia* spp. [136, 137]. CGD is traditionally managed with antimicrobial prophylaxis alone [138, 139], although HSCT is now becoming a preferred therapeutic modality [140].

CGD patients can develop inflammatory and autoimmune complications including inflammatory bowel disease, lupus, and arthritis [141]. They also can experience CSS episodes. Many HLH cases have been reported in CGD patients, typically in the context of infections [49, 142–144]. Without the ability to kill engulfed microbes, CGD phagocytes are continuously activated resulting in a state of systemic hyper-inflammation associated with elevated IL-1, IL-6, and TNFα concentrations [145]. Given this suspected pathophysiology, steroids and high-dose IVIG have been added to targeted antibiotic therapies to treat CGD-associated CSS episodes with good effect [143, 146]. Treatment with the IL-1 receptor antagonist, anakinra, may also be a beneficial adjuvant therapy [145].

GATA Binding Protein 2 Deficiency

Heterozygous germ line mutations in the gene encoding GATA2 have been associated with a broad phenotype, including myelodysplasia, congenital deafness, and lymphedema (classically termed Emberger syndrome), and convey an increased risk of developing myelodysplastic syndrome/acute myeloid leukemia [147]. GATA2 is a zinc-finger transcription factor that controls early erythroid development, thrombopoiesis, and myeloid/monocytic/dendritic cell maturation. GATA2-haploinsufficient patients may be specifically monocytopenic, and present with MSMD [148]. They may also be congenitally neutropenic [149]; demonstrate aplastic anemia [150]; or display isolated B-cell immunodeficiency [151]. Finally, GATA2-deficient patients can lack dendritic cells, monocytes, B cells, and NK cells, and present with recurrent EBV infection, EBV-associated lymphomas, and HLH [152, 153].

MyD88 and IRAK4 Deficiency

MyD88 deficiency and IRAK4 deficiency are autosomal recessive innate immunodeficiencies. Like NEMO, IRAK4 and MyD88 mediate TLR and interleukin-1 receptor-mediated immunity, but they have a comparatively more limited role in adaptive immunity [154–157]. Accordingly, MyD88 deficient and IRAK4 deficient children struggle with invasive and noninvasive bacterial infections until they can develop adaptive immunological memory. Subsequently, infections typically cease.

Whereas Unc13d deficient mice, which exhibit impaired NK and CD8(+) T cells cytotoxicity, develop CSS with LCMV infection, Unc13d/MyD88 double knockout mice do not [158]. Consistent with this animal model, CSS episodes have not been described in MyD88 or IRAK4 deficient patients. Together, these observations suggest that chronic innate immune system activation by recognition of persistent pathogen-associated molecular patterns may be key drivers of CSS physiology.

Conclusion

PIDs featuring impaired cytotoxic function and/or a lymphoproliferative proclivity are often complicated by CSS episodes. Additionally, tumors and pathogens that thrive in the altered environment provided by PID patients may present insurmountable antigenic challenges that provoke systemic inflammatory events. Compared with the immune competent host, establishing the underlying immunologic defect in PID patients is critical to understanding the cause and therefore the best treatment of CSS episodes.

References

1. Jacobs, J., Deschoolmeester, V., Zwaenepoel, K., Rolfo, C., Silence, K., Rottey, S., et al. (2015). CD70: An emerging target in cancer immunotherapy. *Pharmacology & Therapeutics, 155*, 1–10.
2. Abolhassani, H., Edwards, E. S., Ikinciogullari, A., Jing, H., Borte, S., Buggert, M., et al. (2017). Combined immunodeficiency and Epstein-Barr virus-induced B cell malignancy in humans with inherited CD70 deficiency. *The Journal of Experimental Medicine, 214*, 91–106.
3. Izawa, K., Martin, E., Soudais, C., Bruneau, J., Boutboul, D., Rodriguez, R., et al. (2017). Inherited CD70 deficiency in humans reveals a critical role for the CD70-CD27 pathway in immunity to Epstein-Barr virus infection. *The Journal of Experimental Medicine, 214*, 73–89.
4. van Montfrans, J. M., Hoepelman, A. I., Otto, S., van Gijn, M., van de Corput, L., de Weger, R. A., et al. (2012). CD27 deficiency is associated with combined immunodeficiency and persistent symptomatic EBV viremia. *The Journal of Allergy and Clinical Immunology, 129*, 787–793.e6.
5. Salzer, E., Daschkey, S., Choo, S., Gombert, M., Santos-Valente, E., Ginzel, S., et al. (2013). Combined immunodeficiency with life-threatening EBV-associated lymphoproliferative disorder in patients lacking functional CD27. *Haematologica, 98*, 473–478.

6. Brunet, J. F., Denizot, F., Luciani, M. F., Roux-Dosseto, M., Suzan, M., Mattei, M. G., et al. (1987). A new member of the immunoglobulin superfamily—CTLA-4. *Nature, 328*, 267–270.
7. Walunas, T. L., Bakker, C. Y., & Bluestone, J. A. (1996). CTLA-4 ligation blocks CD28-dependent T cell activation. *The Journal of Experimental Medicine, 183*, 2541–2550.
8. Krummel, M. F., & Allison, J. P. (1995). CD28 and CTLA-4 have opposing effects on the response of T cells to stimulation. *The Journal of Experimental Medicine, 182*, 459–465.
9. Waterhouse, P., Penninger, J. M., Timms, E., Wakeham, A., Shahinian, A., Lee, K. P., et al. (1995). Lymphoproliferative disorders with early lethality in mice deficient in Ctla-4. *Science, 270*, 985–988.
10. Schubert, D., Bode, C., Kenefeck, R., Hou, T. Z., Wing, J. B., Kennedy, A., et al. (2014). Autosomal dominant immune dysregulation syndrome in humans with CTLA4 mutations. *Nature Medicine, 20*, 1410–1416.
11. Greil, C., Roether, F., La Rosee, P., Grimbacher, B., Duerschmied, D., & Warnatz, K. (2017). Rescue of cytokine storm due to HLH by hemoadsorption in a CTLA4-deficient patient. *Journal of Clinical Immunology, 37*, 273–276.
12. Lee, S., Moon, J. S., Lee, C. R., Kim, H. E., Baek, S. M., Hwang, S., et al. (2016). Abatacept alleviates severe autoimmune symptoms in a patient carrying a de novo variant in CTLA-4. *The Journal of Allergy and Clinical Immunology, 137*, 327–330.
13. Lo, B., Zhang, K., Lu, W., Zheng, L., Zhang, Q., Kanellopoulou, C., et al. (2015). AUTOIMMUNE DISEASE. Patients with LRBA deficiency show CTLA4 loss and immune dysregulation responsive to abatacept therapy. *Science, 349*, 436–440.
14. Le Clorennec, C., Ouk, T. S., Youlyouz-Marfak, I., Panteix, S., Martin, C. C., Rastelli, J., et al. (2008). Molecular basis of cytotoxicity of Epstein-Barr virus (EBV) latent membrane protein 1 (LMP1) in EBV latency III B cells: LMP1 induces type II ligand-independent autoactivation of CD95/Fas with caspase 8-mediated apoptosis. *Journal of Virology, 82*, 6721–6733.
15. Nagata, S. (1999). Fas ligand-induced apoptosis. *Annual Review of Genetics, 33*, 29–55.
16. Fisher, G. H., Rosenberg, F. J., Straus, S. E., Dale, J. K., Middleton, L. A., Lin, A. Y., et al. (1995). Dominant interfering Fas gene mutations impair apoptosis in a human autoimmune lymphoproliferative syndrome. *Cell, 81*, 935–946.
17. Magerus-Chatinet, A., Stolzenberg, M. C., Lanzarotti, N., Neven, B., Daussy, C., Picard, C., et al. (2013). Autoimmune lymphoproliferative syndrome caused by a homozygous null FAS ligand (FASLG) mutation. *The Journal of Allergy and Clinical Immunology, 131*, 486–490.
18. Rieux-Laucat, F., Le Deist, F., Hivroz, C., Roberts, I. A., Debatin, K. M., Fischer, A., et al. (1995). Mutations in Fas associated with human lymphoproliferative syndrome and autoimmunity. *Science, 268*, 1347–1349.
19. Hansford, J. R., Pal, M., Poplawski, N., Haan, E., Boog, B., Ferrante, A., et al. (2013). In utero and early postnatal presentation of autoimmune lymphoproliferative syndrome in a family with a novel FAS mutation. *Haematologica, 98*, e38–e39.
20. Deutsch, M., Tsopanou, E., & Dourakis, S. P. (2004). The autoimmune lymphoproliferative syndrome (Canale-Smith) in adulthood. *Clinical Rheumatology, 23*, 43–44.
21. Straus, S. E., Jaffe, E. S., Puck, J. M., Dale, J. K., Elkon, K. B., Rosen-Wolff, A., et al. (2001). The development of lymphomas in families with autoimmune lymphoproliferative syndrome with germline Fas mutations and defective lymphocyte apoptosis. *Blood, 98*, 194–200.
22. Chun, H. J., Zheng, L., Ahmad, M., Wang, J., Speirs, C. K., Siegel, R. M., et al. (2002). Pleiotropic defects in lymphocyte activation caused by caspase-8 mutations lead to human immunodeficiency. *Nature, 419*, 395–399.
23. Wang, J., Zheng, L., Lobito, A., Chan, F. K., Dale, J., Sneller, M., et al. (1999). Inherited human Caspase 10 mutations underlie defective lymphocyte and dendritic cell apoptosis in autoimmune lymphoproliferative syndrome type II. *Cell, 98*, 47–58.
24. Oliveira, J. B., Bidere, N., Niemela, J. E., Zheng, L., Sakai, K., Nix, C. P., et al. (2007). NRAS mutation causes a human autoimmune lymphoproliferative syndrome. *Proceedings of the National Academy of Sciences of the United States of America, 104*, 8953–8958.

25. Takagi, M., Shinoda, K., Piao, J., Mitsuiki, N., Takagi, M., Matsuda, K., et al. (2011). Autoimmune lymphoproliferative syndrome-like disease with somatic KRAS mutation. *Blood, 117*, 2887–2890.
26. Niemela, J. E., Lu, L., Fleisher, T. A., Davis, J., Caminha, I., Natter, M., et al. (2011). Somatic KRAS mutations associated with a human nonmalignant syndrome of autoimmunity and abnormal leukocyte homeostasis. *Blood, 117*, 2883–2886.
27. Li, P., Huang, P., Yang, Y., Hao, M., Peng, H., & Li, F. (2016). Updated understanding of Autoimmune Lymphoproliferative Syndrome (ALPS). *Clinical Reviews in Allergy and Immunology, 50*, 55–63.
28. Kuijpers, T. W., Baars, P. A., aan de Kerk, D. J., Jansen, M. H., Dors, N., van Lier, R. A., et al. (2011). Common variable immunodeficiency and hemophagocytic features associated with a FAS gene mutation. *The Journal of Allergy and Clinical Immunology, 127*, 1411–4.e2.
29. Rudman Spergel, A., Walkovich, K., Price, S., Niemela, J. E., Wright, D., Fleisher, T. A., et al. (2013). Autoimmune lymphoproliferative syndrome misdiagnosed as hemophagocytic lymphohistiocytosis. *Pediatrics, 132*, e1440–e1444.
30. Salzer, U., Chapel, H. M., Webster, A. D., Pan-Hammarstrom, Q., Schmitt-Graeff, A., Schlesier, M., et al. (2005). Mutations in TNFRSF13B encoding TACI are associated with common variable immunodeficiency in humans. *Nature Genetics, 37*, 820–828.
31. Castigli, E., Wilson, S. A., Garibyan, L., Rachid, R., Bonilla, F., Schneider, L., et al. (2005). TACI is mutant in common variable immunodeficiency and IgA deficiency. *Nature Genetics, 37*, 829–834.
32. Seshasayee, D., Valdez, P., Yan, M., Dixit, V. M., Tumas, D., & Grewal, I. S. (2003). Loss of TACI causes fatal lymphoproliferation and autoimmunity, establishing TACI as an inhibitory BLyS receptor. *Immunity, 18*, 279–288.
33. Yan, M., Wang, H., Chan, B., Roose-Girma, M., Erickson, S., Baker, T., et al. (2001). Activation and accumulation of B cells in TACI-deficient mice. *Nature Immunology, 2*, 638–643.
34. Romberg, N., Chamberlain, N., Saadoun, D., Gentile, M., Kinnunen, T., Ng, Y. S., et al. (2013). CVID-associated TACI mutations affect autoreactive B cell selection and activation. *The Journal of Clinical Investigation, 123*, 4283–4293.
35. Allman, W. R., Dey, R., Liu, L., Siddiqui, S., Coleman, A. S., Bhattacharya, P., et al. (2015). TACI deficiency leads to alternatively activated macrophage phenotype and susceptibility to Leishmania infection. *Proceedings of the National Academy of Sciences of the United States of America, 112*, E4094–E4103.
36. Zoller, E. E., Lykens, J. E., Terrell, C. E., Aliberti, J., Filipovich, A. H., Henson, P. M., et al. (2011). Hemophagocytosis causes a consumptive anemia of inflammation. *The Journal of Experimental Medicine, 208*, 1203–1214.
37. Lucas, C. L., Kuehn, H. S., Zhao, F., Niemela, J. E., Deenick, E. K., Palendira, U., et al. (2014). Dominant-activating germline mutations in the gene encoding the PI(3)K catalytic subunit p110delta result in T cell senescence and human immunodeficiency. *Nature Immunology, 15*, 88–97.
38. Angulo, I., Vadas, O., Garcon, F., Banham-Hall, E., Plagnol, V., Leahy, T. R., et al. (2013). Phosphoinositide 3-kinase delta gene mutation predisposes to respiratory infection and airway damage. *Science, 342*, 866–871.
39. Deau, M. C., Heurtier, L., Frange, P., Suarez, F., Bole-Feysot, C., Nitschke, P., et al. (2014). A human immunodeficiency caused by mutations in the PIK3R1 gene. *The Journal of Clinical Investigation, 124*, 3923–3928.
40. Lucas, C. L., Zhang, Y., Venida, A., Wang, Y., Hughes, J., McElwee, J., et al. (2014). Heterozygous splice mutation in PIK3R1 causes human immunodeficiency with lymphoproliferation due to dominant activation of PI3K. *The Journal of Experimental Medicine, 211*, 2537–2547.
41. Elkaim, E., Neven, B., Bruneau, J., Mitsui-Sekinaka, K., Stanislas, A., Heurtier, L., et al. (2016). Clinical and immunologic phenotype associated with activated phosphoinositide 3-kinase delta syndrome 2: A cohort study. *The Journal of Allergy and Clinical Immunology, 138*, 210–218.e9.
42. Shearer, W. T., Dunn, E., Notarangelo, L. D., Dvorak, C. C., Puck, J. M., Logan, B. R., et al. (2014). Establishing diagnostic criteria for severe combined immunodeficiency disease

(SCID), leaky SCID, and Omenn syndrome: The Primary Immune Deficiency Treatment Consortium experience. *The Journal of Allergy and Clinical Immunology, 133*, 1092–1098.
43. Al-Herz, W., & Nanda, A. (2011). Skin manifestations in primary immunodeficient children. *Pediatric Dermatology, 28*, 494–501.
44. Villa, A., Santagata, S., Bozzi, F., Imberti, L., & Notarangelo, L. D. (1999). Omenn syndrome: A disorder of Rag1 and Rag2 genes. *Journal of Clinical Immunology, 19*, 87–97.
45. Couedel, C., Roman, C., Jones, A., Vezzoni, P., Villa, A., & Cortes, P. (2010). Analysis of mutations from SCID and Omenn syndrome patients reveals the central role of the Rag2 PHD domain in regulating V(D)J recombination. *The Journal of Clinical Investigation, 120*, 1337–1344.
46. Pai, S. Y., Logan, B. R., Griffith, L. M., Buckley, R. H., Parrott, R. E., Dvorak, C. C., et al. (2014). Transplantation outcomes for severe combined immunodeficiency, 2000-2009. *The New England Journal of Medicine, 371*, 434–446.
47. Patiroglu, T., Haluk Akar, H., van den Burg, M., Unal, E., Akyildiz, B. N., Tekerek, N. U., et al. (2014). X-linked severe combined immunodeficiency due to a novel mutation complicated with hemophagocytic lymphohistiocytosis and presented with invagination: A case report. *European Journal of Microbiology and Immunology, 4*, 174–176.
48. Schmid, I., Reiter, K., Schuster, F., Wintergerst, U., Meilbeck, R., Nicolai, T., et al. (2002). Allogeneic bone marrow transplantation for active Epstein-Barr virus-related lymphoproliferative disease and hemophagocytic lymphohistiocytosis in an infant with severe combined immunodeficiency syndrome. *Bone Marrow Transplantation, 29*, 519–521.
49. Bode, S. F., Ammann, S., Al-Herz, W., Bataneant, M., Dvorak, C. C., Gehring, S., et al. (2015). The syndrome of hemophagocytic lymphohistiocytosis in primary immunodeficiencies: Implications for differential diagnosis and pathogenesis. *Haematologica, 100*, 978–988.
50. Dvorak, C. C., Sandford, A., Fong, A., Cowan, M. J., George, T. I., & Lewis, D. B. (2008). Maternal T-cell engraftment associated with severe hemophagocytosis of the bone marrow in untreated X-linked severe combined immunodeficiency. *Journal of Pediatric Hematology/Oncology, 30*, 396–400.
51. Grunebaum, E., Zhang, J., Dadi, H., & Roifman, C. M. (2000). Haemophagocytic lymphohistiocytosis in X-linked severe combined immunodeficiency. *British Journal of Haematology, 108*, 834–837.
52. Chen, W., Zhang, S., Zhang, W., Yang, X., Xu, J., Qiu, H., et al. (2015). Elevated serum adenosine deaminase levels in secondary hemophagocytic lymphohistiocytosis. *International Journal of Laboratory Hematology, 37*, 544–550.
53. Kuno, M., Seki, N., Tsujimoto, S., Nakanishi, I., Kinoshita, T., Nakamura, K., et al. (2006). Anti-inflammatory activity of non-nucleoside adenosine deaminase inhibitor FR234938. *European Journal of Pharmacology, 534*, 241–249.
54. Readinger, J. A., Mueller, K. L., Venegas, A. M., Horai, R., & Schwartzberg, P. L. (2009). Tec kinases regulate T-lymphocyte development and function: New insights into the roles of Itk and Rlk/Txk. *Immunological Reviews, 228*, 93–114.
55. Ghosh, S., Bienemann, K., Boztug, K., & Borkhardt, A. (2014). Interleukin-2-inducible T-cell kinase (ITK) deficiency—clinical and molecular aspects. *Journal of Clinical Immunology, 34*, 892–899.
56. Huck, K., Feyen, O., Niehues, T., Ruschendorf, F., Hubner, N., Laws, H. J., et al. (2009). Girls homozygous for an IL-2-inducible T cell kinase mutation that leads to protein deficiency develop fatal EBV-associated lymphoproliferation. *The Journal of Clinical Investigation, 119*, 1350–1358.
57. Linka, R. M., Risse, S. L., Bienemann, K., Werner, M., Linka, Y., Krux, F., et al. (2012). Loss-of-function mutations within the IL-2 inducible kinase ITK in patients with EBV-associated lymphoproliferative diseases. *Leukemia, 26*, 963–971.
58. Mansouri, D., Mahdaviani, S. A., Khalilzadeh, S., Mohajerani, S. A., Hasanzad, M., Sadr, S., et al. (2012). IL-2-inducible T-cell kinase deficiency with pulmonary manifestations due to disseminated Epstein-Barr virus infection. *International Archives of Allergy and Immunology, 158*, 418–422.

59. Stepensky, P., Weintraub, M., Yanir, A., Revel-Vilk, S., Krux, F., Huck, K., et al. (2011). IL-2-inducible T-cell kinase deficiency: Clinical presentation and therapeutic approach. *Haematologica, 96*, 472–476.
60. Shprintzen, R. J., Goldberg, R. B., Young, D., & Wolford, L. (1981). The velo-cardio-facial syndrome: A clinical and genetic analysis. *Pediatrics, 67*, 167–172.
61. Ryan, A. K., Goodship, J. A., Wilson, D. I., Philip, N., Levy, A., Seidel, H., et al. (1997). Spectrum of clinical features associated with interstitial chromosome 22q11 deletions: A European collaborative study. *Journal of Medical Genetics, 34*, 798–804.
62. Markert, M. L., Devlin, B. H., & McCarthy, E. A. (2010). Thymus transplantation. *Clinical Immunology, 135*, 236–246.
63. Cesaro, S., Messina, C., Sainati, L., Danesino, C., & Arico, M. (2003). Del 22Q11.2 and hemophagocytic lymphohistiocytosis: A non-random association. *American Journal of Medical Genetics. Part A, 116A*, 208–209.
64. Arico, M., Bettinelli, A., Maccario, R., Clementi, R., Bossi, G., & Danesino, C. (1999). Hemophagocytic lymphohistiocytosis in a patient with deletion of 22q11.2. *American Journal of Medical Genetics, 87*, 329–330.
65. Itoh, S., Ohno, T., Kakizaki, S., & Ichinohasama, R. (2011). Epstein-Barr virus-positive T-cell lymphoma cells having chromosome 22q11.2 deletion: An autopsy report of DiGeorge syndrome. *Human Pathology, 42*, 2037–2041.
66. Aldrich, R. A., Steinberg, A. G., & Campbell, D. C. (1954). Pedigree demonstrating a sex-linked recessive condition characterized by draining ears, eczematoid dermatitis and bloody diarrhea. *Pediatrics, 13*, 133–139.
67. Wiskott, A. (1936). Familia¨rer, angeborener morbus Werlhofii. *Monatschrift Kinderheil, 68*, 21.
68. Thrasher, A. J., & Burns, S. O. (2010). WASP: A key immunological multitasker. *Nature Reviews. Immunology, 10*, 182–192.
69. Dupuis-Girod, S., Medioni, J., Haddad, E., Quartier, P., Cavazzana-Calvo, M., Le Deist, F., et al. (2003). Autoimmunity in Wiskott-Aldrich syndrome: Risk factors, clinical features, and outcome in a single-center cohort of 55 patients. *Pediatrics, 111*, e622–e627.
70. Pasic, S., Micic, D., & Kuzmanovic, M. (2003). Epstein-Barr virus-associated haemophagocytic lymphohistiocytosis in Wiskott-Aldrich syndrome. *Acta Paediatrica, 92*, 859–861.
71. Orange, J. S., Ramesh, N., Remold-O'Donnell, E., Sasahara, Y., Koopman, L., Byrne, M., et al. (2002). Wiskott-Aldrich syndrome protein is required for NK cell cytotoxicity and colocalizes with actin to NK cell-activating immunologic synapses. *Proceedings of the National Academy of Sciences of the United States of America, 99*, 11351–11356.
72. Picard, C., Casanova, J. L., & Puel, A. (2011). Infectious diseases in patients with IRAK-4, MyD88, NEMO, or IkappaBalpha deficiency. *Clinical Microbiology Reviews, 24*, 490–497.
73. Orange, J. S., Brodeur, S. R., Jain, A., Bonilla, F. A., Schneider, L. C., Kretschmer, R., et al. (2002). Deficient natural killer cell cytotoxicity in patients with IKK-gamma/NEMO mutations. *The Journal of Clinical Investigation, 109*, 1501–1509.
74. Ricci, S., Romano, F., Nieddu, F., Picard, C., & Azzari, C. (2017). OL-EDA-ID syndrome: A novel hypomorphic NEMO mutation associated with a severe clinical presentation and transient HLH. *Journal of Clinical Immunology, 37*, 7–11.
75. Pachlopnik Schmid, J. M., Junge, S. A., Hossle, J. P., Schneider, E. M., Roosnek, E., Seger, R. A., et al. (2006). Transient hemophagocytosis with deficient cellular cytotoxicity, monoclonal immunoglobulin M gammopathy, increased T-cell numbers, and hypomorphic NEMO mutation. *Pediatrics, 117*, e1049–e1056.
76. Foy, T. M., Aruffo, A., Bajorath, J., Buhlmann, J. E., & Noelle, R. J. (1996). Immune regulation by CD40 and its ligand GP39. *Annual Review of Immunology, 14*, 591–617.
77. Allen, R. C., Armitage, R. J., Conley, M. E., Rosenblatt, H., Jenkins, N. A., Copeland, N. G., et al. (1993). CD40 ligand gene defects responsible for X-linked hyper-IgM syndrome. *Science, 259*, 990–993.
78. Aruffo, A., Farrington, M., Hollenbaugh, D., Li, X., Milatovich, A., Nonoyama, S., et al. (1993). The CD40 ligand, gp39, is defective in activated T cells from patients with X-linked hyper-IgM syndrome. *Cell, 72*, 291–300.

79. Ferrari, S., Giliani, S., Insalaco, A., Al-Ghonaium, A., Soresina, A. R., Loubser, M., et al. (2001). Mutations of CD40 gene cause an autosomal recessive form of immunodeficiency with hyper IgM. *Proceedings of the National Academy of Sciences of the United States of America, 98*, 12614–12619.
80. Jesus, A. A., Duarte, A. J., & Oliveira, J. B. (2008). Autoimmunity in hyper-IgM syndrome. *Journal of Clinical Immunology, 28*(Suppl 1), S62–S66.
81. Imadome, K., Shirakata, M., Shimizu, N., Nonoyama, S., & Yamanashi, Y. (2003). CD40 ligand is a critical effector of Epstein-Barr virus in host cell survival and transformation. *Proceedings of the National Academy of Sciences of the United States of America, 100*, 7836–7840.
82. Savitsky, K., Bar-Shira, A., Gilad, S., Rotman, G., Ziv, Y., Vanagaite, L., et al. (1995). A single ataxia telangiectasia gene with a product similar to PI-3 kinase. *Science, 268*, 1749–1753.
83. Shiloh, Y. (2003). ATM and related protein kinases: Safeguarding genome integrity. *Nature Reviews. Cancer, 3*, 155–168.
84. Bredemeyer, A. L., Huang, C. Y., Walker, L. M., Bassing, C. H., & Sleckman, B. P. (2008). Aberrant V(D)J recombination in ataxia telangiectasia mutated-deficient lymphocytes is dependent on nonhomologous DNA end joining. *Journal of Immunology, 181*, 2620–2625.
85. Alter, B. P., Rosenberg, P. S., Giri, N., Baerlocher, G. M., Lansdorp, P. M., & Savage, S. A. (2012). Telomere length is associated with disease severity and declines with age in dyskeratosis congenita. *Haematologica, 97*, 353–359.
86. Savage, S. A. (1993). Dyskeratosis congenita. In R. A. Pagon, M. P. Adam, H. H. Ardinger, S. E. Wallace, A. Amemiya, L. J. H. Bean, T. D. Bird, N. Ledbetter, H. C. Mefford, R. J. H. Smith, & K. Stephens (Eds.), *GeneReviews(R)*. Seattle: University of Washington. GeneReviews is a registered trademark of the University of Washington, Seattle. All rights reserved, Seattle (WA).
87. Fernandez Garcia, M. S., & Teruya-Feldstein, J. (2014). The diagnosis and treatment of dyskeratosis congenita: A review. *Journal of Blood Medicine, 5*, 157–167.
88. Feske, S., Gwack, Y., Prakriya, M., Srikanth, S., Puppel, S. H., Tanasa, B., et al. (2006). A mutation in Orai1 causes immune deficiency by abrogating CRAC channel function. *Nature, 441*, 179–185.
89. Klemann, C., Ammann, S., Heizmann, M., Fuchs, S., Bode, S. F., Heeg, M., et al. (2017). Hemophagocytic lymphohistiocytosis as presenting manifestation of profound combined immunodeficiency due to an ORAI1 mutation. *The Journal of Allergy and Clinical Immunology*.
90. Li, F. Y., Lenardo, M. J., & Chaigne-Delalande, B. (2011). Loss of MAGT1 abrogates the Mg2+ flux required for T cell signaling and leads to a novel human primary immunodeficiency. *Magnesium Research, 24*, S109–S114.
91. Mukda, E., Trachoo, O., Pasomsub, E., Tiyasirichokchai, R., Iemwimangsa, N., Sosothikul, D., et al. (2017). Exome sequencing for simultaneous mutation screening in children with hemophagocytic lymphohistiocytosis. *International Journal of Hematology, 106*, 282–290.
92. Picard, C., McCarl, C. A., Papolos, A., Khalil, S., Luthy, K., Hivroz, C., et al. (2009). STIM1 mutation associated with a syndrome of immunodeficiency and autoimmunity. *The New England Journal of Medicine, 360*, 1971–1980.
93. Fuchs, S., Rensing-Ehl, A., Speckmann, C., Bengsch, B., Schmitt-Graeff, A., Bondzio, I., et al. (2012). Antiviral and regulatory T cell immunity in a patient with stromal interaction molecule 1 deficiency. *Journal of Immunology, 188*, 1523–1533.
94. Zhu, J., Yamane, H., & Paul, W. E. (2010). Differentiation of effector CD4 T cell populations (*). *Annual Review of Immunology, 28*, 445–489.
95. Nakayama, T., Hirahara, K., Onodera, A., Endo, Y., Hosokawa, H., Shinoda, K., et al. (2017). Th2 cells in health and disease. *Annual Review of Immunology, 35*, 53–84.
96. Josefowicz, S. Z., Lu, L. F., & Rudensky, A. Y. (2012). Regulatory T cells: Mechanisms of differentiation and function. *Annual Review of Immunology, 30*, 531–564.
97. Newport, M. J., Huxley, C. M., Huston, S., Hawrylowicz, C. M., Oostra, B. A., Williamson, R., et al. (1996). A mutation in the interferon-gamma-receptor gene and susceptibility to mycobacterial infection. *The New England Journal of Medicine, 335*, 1941–1949.

98. Jouanguy, E., Altare, F., Lamhamedi, S., Revy, P., Emile, J. F., Newport, M., et al. (1996). Interferon-gamma-receptor deficiency in an infant with fatal bacille Calmette-Guerin infection. *The New England Journal of Medicine, 335*, 1956–1961.
99. Reinhardt, R. L., Liang, H. E., Bao, K., Price, A. E., Mohrs, M., Kelly, B. L., et al. (2015). A novel model for IFN-gamma-mediated autoinflammatory syndromes. *Journal of Immunology, 194*, 2358–2368.
100. Jordan, M. B., Hildeman, D., Kappler, J., & Marrack, P. (2004). An animal model of hemophagocytic lymphohistiocytosis (HLH): CD8+ T cells and interferon gamma are essential for the disorder. *Blood, 104*, 735–743.
101. Baker, K. F., & Isaacs, J. D. (2017). Novel therapies for immune-mediated inflammatory diseases: What can we learn from their use in rheumatoid arthritis, spondyloarthritis, systemic lupus erythematosus, psoriasis, Crohn's disease and ulcerative colitis? *Annals of the Rheumatic Diseases*.
102. Cui, D., Huang, G., Yang, D., Huang, B., & An, B. (2013). Efficacy and safety of interferon-gamma-targeted therapy in Crohn's disease: A systematic review and meta-analysis of randomized controlled trials. *Clinics and Research in Hepatology and Gastroenterology, 37*, 507–513.
103. Bracaglia, C., Gatto, A., Pardeo, M., Lapeyre, G., Ferlin, W., Nelson, R., et al. (2015). Anti interferon-gamma (IFNγ) monoclonal antibody treatment in a patient carrying an *NLRC4* mutation and severe hemophagocytic lymphohistiocytosis. *Pediatric Rheumatology, 13*, O68.
104. Staines-Boone, A. T., Deswarte, C., Venegas Montoya, E., Sanchez-Sanchez, L. M., Garcia Campos, J. A., Muniz-Ronquillo, T., et al. (2017). Multifocal recurrent osteomyelitis and hemophagocytic lymphohistiocytosis in a boy with partial dominant IFN-gammaR1 deficiency: Case report and review of the literature. *Frontiers in Pediatrics, 5*, 75.
105. Tesi, B., Sieni, E., Neves, C., Romano, F., Cetica, V., Cordeiro, A. I., et al. (2015). Hemophagocytic lymphohistiocytosis in 2 patients with underlying IFN-gamma receptor deficiency. *The Journal of Allergy and Clinical Immunology, 135*, 1638–1641.
106. Muriel-Vizcaino, R., Yamazaki-Nakashimada, M., Lopez-Herrera, G., Santos-Argumedo, L., & Ramirez-Alejo, N. (2016). Hemophagocytic lymphohistiocytosis as a complication in patients with MSMD. *Journal of Clinical Immunology, 36*, 420–422.
107. Canna, S. W., Wrobel, J., Chu, N., Kreiger, P. A., Paessler, M., & Behrens, E. M. (2013). Interferon-gamma mediates anemia but is dispensable for fulminant toll-like receptor 9-induced macrophage activation syndrome and hemophagocytosis in mice. *Arthritis and Rheumatism, 65*, 1764–1775.
108. Liu, L., Okada, S., Kong, X. F., Kreins, A. Y., Cypowyj, S., Abhyankar, A., et al. (2011). Gain-of-function human STAT1 mutations impair IL-17 immunity and underlie chronic mucocutaneous candidiasis. *The Journal of Experimental Medicine, 208*, 1635–1648.
109. van de Veerdonk, F. L., Plantinga, T. S., Hoischen, A., Smeekens, S. P., Joosten, L. A., Gilissen, C., et al. (2011). STAT1 mutations in autosomal dominant chronic mucocutaneous candidiasis. *The New England Journal of Medicine, 365*, 54–61.
110. Uzel, G., Sampaio, E. P., Lawrence, M. G., Hsu, A. P., Hackett, M., Dorsey, M. J., et al. (2013). Dominant gain-of-function STAT1 mutations in FOXP3 wild-type immune dysregulation-polyendocrinopathy-enteropathy-X-linked-like syndrome. *The Journal of Allergy and Clinical Immunology, 131*, 1611–1623.
111. Takezaki, S., Yamada, M., Kato, M., Park, M. J., Maruyama, K., Yamazaki, Y., et al. (2012). Chronic mucocutaneous candidiasis caused by a gain-of-function mutation in the STAT1 DNA-binding domain. *Journal of Immunology, 189*, 1521–1526.
112. Toubiana, J., Okada, S., Hiller, J., Oleastro, M., Gomez, M. L., Becerra, J. C. A., et al. (2016). Heterozygous STAT1 gain-of-function mutations underlie an unexpectedly broad clinical phenotype. *Blood, 127*, 3154–3164.
113. Odio, C. D., Milligan, K. L., McGowan, K., Rudman Spergel, A. K., Bishop, R., Boris, L., et al. (2015). Endemic mycoses in patients with STAT3-mutated hyper-IgE (Job) syndrome. *The Journal of Allergy and Clinical Immunology, 136*, 1411–3.e1-2.

114. Milner, J. D., Vogel, T. P., Forbes, L., Ma, C. A., Stray-Pedersen, A., Niemela, J. E., et al. (2015). Early-onset lymphoproliferation and autoimmunity caused by germline STAT3 gain-of-function mutations. *Blood, 125*, 591–599.
115. Flanagan, S. E., Haapaniemi, E., Russell, M. A., Caswell, R., Allen, H. L., De Franco, E., et al. (2014). Activating germline mutations in STAT3 cause early-onset multi-organ autoimmune disease. *Nature Genetics, 46*, 812–814.
116. Brunkow, M. E., Jeffery, E. W., Hjerrild, K. A., Paeper, B., Clark, L. B., Yasayko, S. A., et al. (2001). Disruption of a new forkhead/winged-helix protein, scurfin, results in the fatal lymphoproliferative disorder of the scurfy mouse. *Nature Genetics, 27*, 68–73.
117. Chatila, T. A., Blaeser, F., Ho, N., Lederman, H. M., Voulgaropoulos, C., Helms, C., et al. (2000). JM2, encoding a fork head-related protein, is mutated in X-linked autoimmunity-allergic disregulation syndrome. *The Journal of Clinical Investigation, 106*, R75–R81.
118. Goudy, K., Aydin, D., Barzaghi, F., Gambineri, E., Vignoli, M., Ciullini Mannurita, S., et al. (2013). Human IL2RA null mutation mediates immunodeficiency with lymphoproliferation and autoimmunity. *Clinical Immunology, 146*, 248–261.
119. Caudy, A. A., Reddy, S. T., Chatila, T., Atkinson, J. P., & Verbsky, J. W. (2007). CD25 deficiency causes an immune dysregulation, polyendocrinopathy, enteropathy, X-linked-like syndrome, and defective IL-10 expression from CD4 lymphocytes. *The Journal of Allergy and Clinical Immunology, 119*, 482–487.
120. Bennett, C. L., & Ochs, H. D. (2001). IPEX is a unique X-linked syndrome characterized by immune dysfunction, polyendocrinopathy, enteropathy, and a variety of autoimmune phenomena. *Current Opinion in Pediatrics, 13*, 533–538.
121. Baud, O., Goulet, O., Canioni, D., Le Deist, F., Radford, I., Rieu, D., et al. (2001). Treatment of the immune dysregulation, polyendocrinopathy, enteropathy, X-linked syndrome (IPEX) by allogeneic bone marrow transplantation. *The New England Journal of Medicine, 344*, 1758–1762.
122. Maglione, P. J. (2016). Autoimmune and lymphoproliferative complications of common variable immunodeficiency. *Current Allergy and Asthma Reports, 16*, 19. https://doi.org/10.1007/s11882-016-0597-6
123. Bajaj, P., Clement, J., Bayerl, M. G., Kalra, N., Craig, T. J., & Ishmael, F. T. (2014). High-grade fever and pancytopenia in an adult patient with common variable immune deficiency. *Allergy and Asthma Proceedings, 35*, 78–82.
124. Malkan, U. Y., Gunes, G., Aslan, T., Etgul, S., Aydin, S., & Buyukasik, Y. (2015). Common variable immune deficiency associated Hodgkin's lymphoma complicated with EBV-linked hemophagocytic lymphohistiocytosis: A case report. *International Journal of Clinical and Experimental Medicine, 8*, 14203–14206.
125. Rawlings, D. J., Saffran, D. C., Tsukada, S., Largaespada, D. A., Grimaldi, J. C., Cohen, L., et al. (1993). Mutation of unique region of Bruton's tyrosine kinase in immunodeficient XID mice. *Science, 261*, 358–361.
126. Thomas, J. D., Sideras, P., Smith, C. I., Vorechovsky, I., Chapman, V., & Paul, W. E. (1993). Colocalization of X-linked agammaglobulinemia and X-linked immunodeficiency genes. *Science, 261*, 355–358.
127. Tsukada, S., Saffran, D. C., Rawlings, D. J., Parolini, O., Allen, R. C., Klisak, I., et al. (1993). Deficient expression of a B cell cytoplasmic tyrosine kinase in human X-linked agammaglobulinemia. *Cell, 72*, 279–290.
128. Vetrie, D., Vorechovsky, I., Sideras, P., Holland, J., Davies, A., Flinter, F., et al. (1993). The gene involved in X-linked agammaglobulinaemia is a member of the src family of protein-tyrosine kinases. *Nature, 361*, 226–233.
129. Corneth, O. B., Klein Wolterink, R. G., & Hendriks, R. W. (2016). BTK Signaling in B cell differentiation and autoimmunity. *Current Topics in Microbiology and Immunology, 393*, 67–105.
130. Bearden, D., Collett, M., Quan, P. L., Costa-Carvalho, B. T., & Sullivan, K. E. (2016). Enteroviruses in X-linked agammaglobulinemia: Update on epidemiology and therapy. *The Journal of Allergy and Clinical Immunology. In Practice, 4*, 1059–1065.

131. Liu, X., Zhan, Z., Li, D., Xu, L., Ma, F., Zhang, P., et al. (2011). Intracellular MHC class II molecules promote TLR-triggered innate immune responses by maintaining activation of the kinase Btk. *Nature Immunology, 12*, 416–424.
132. Schultz, K. A., Neglia, J. P., Smith, A. R., Ochs, H. D., Torgerson, T. R., & Kumar, A. (2008). Familial hemophagocytic lymphohistiocytosis in two brothers with X-linked agammaglobulinemia. *Pediatric Blood & Cancer, 51*, 293–295.
133. Berendes, H., Bridges, R. A., & Good, R. A. (1957). A fatal granulomatosus of childhood: The clinical study of a new syndrome. *Minnesota Medicine, 40*, 309–312.
134. Janeway, C. A., Craig, J., Davidson, M., Downey, W., Gitlin, D., & Sullivan, J. C. (1954). Hypergammaglobulinemia associated with severe, recurrent and chronic non-specific infection. *American Journal of Diseases of Children, 88*, 388–392.
135. Kuhns, D. B., Alvord, W. G., Heller, T., Feld, J. J., Pike, K. M., Marciano, B. E., et al. (2010). Residual NADPH oxidase and survival in chronic granulomatous disease. *The New England Journal of Medicine, 363*, 2600–2610.
136. van den Berg, J. M., van Koppen, E., Ahlin, A., Belohradsky, B. H., Bernatowska, E., Corbeel, L., et al. (2009). Chronic granulomatous disease: The European experience. *PLoS One, 4*, e5234.
137. Segal, B. H., Leto, T. L., Gallin, J. I., Malech, H. L., & Holland, S. M. (2000). Genetic, biochemical, and clinical features of chronic granulomatous disease. *Medicine (Baltimore), 79*, 170–200.
138. Margolis, D. M., Melnick, D. A., Alling, D. W., & Gallin, J. I. (1990). Trimethoprim-sulfamethoxazole prophylaxis in the management of chronic granulomatous disease. *The Journal of Infectious Diseases, 162*, 723–726.
139. Gallin, J. I., Alling, D. W., Malech, H. L., Wesley, R., Koziol, D., Marciano, B., et al. (2003). Itraconazole to prevent fungal infections in chronic granulomatous disease. *The New England Journal of Medicine, 348*, 2416–2422.
140. Cole, T., Pearce, M. S., Cant, A. J., Cale, C. M., Goldblatt, D., & Gennery, A. R. (2013). Clinical outcome in children with chronic granulomatous disease managed conservatively or with hematopoietic stem cell transplantation. *The Journal of Allergy and Clinical Immunology, 132*, 1150–1155.
141. Magnani, A., Brosselin, P., Beaute, J., de Vergnes, N., Mouy, R., Debre, M., et al. (2014). Inflammatory manifestations in a single-center cohort of patients with chronic granulomatous disease. *The Journal of Allergy and Clinical Immunology, 134*, 655–662.e8.
142. Valentine, G., Thomas, T. A., Nguyen, T., & Lai, Y. C. (2014). Chronic granulomatous disease presenting as hemophagocytic lymphohistiocytosis: A case report. *Pediatrics, 134*, e1727–e1730.
143. Parekh, C., Hofstra, T., Church, J. A., & Coates, T. D. (2011). Hemophagocytic lymphohistiocytosis in children with chronic granulomatous disease. *Pediatric Blood & Cancer, 56*, 460–462.
144. Maignan, M., Verdant, C., Bouvet, G. F., Van Spall, M., & Berthiaume, Y. (2013). Undiagnosed chronic granulomatous disease, Burkholderia cepacia complex pneumonia, and acquired hemophagocytic lymphohistiocytosis: A deadly association. *Case Reports in Pulmonology, 2013*, 874197.
145. Schappi, M. G., Jaquet, V., Belli, D. C., & Krause, K. H. (2008). Hyperinflammation in chronic granulomatous disease and anti-inflammatory role of the phagocyte NADPH oxidase. *Seminars in Immunopathology, 30*, 255–271.
146. Alvarez-Cardona, A., Rodriguez-Lozano, A. L., Blancas-Galicia, L., Rivas-Larrauri, F. E., & Yamazaki-Nakashimada, M. A. (2012). Intravenous immunoglobulin treatment for macrophage activation syndrome complicating chronic granulomatous disease. *Journal of Clinical Immunology, 32*, 207–211.
147. Wlodarski, M. W., Hirabayashi, S., Pastor, V., Stary, J., Hasle, H., Masetti, R., et al. (2016). Prevalence, clinical characteristics, and prognosis of GATA2-related myelodysplastic syndromes in children and adolescents. *Blood, 127*, 1387–1397. quiz 1518.
148. Hsu, A. P., Sampaio, E. P., Khan, J., Calvo, K. R., Lemieux, J. E., Patel, S. Y., et al. (2011). Mutations in GATA2 are associated with the autosomal dominant and sporadic monocytopenia and mycobacterial infection (MonoMAC) syndrome. *Blood, 118*, 2653–2655.

149. Pasquet, M., Bellanne-Chantelot, C., Tavitian, S., Prade, N., Beaupain, B., Larochelle, O., et al. (2013). High frequency of GATA2 mutations in patients with mild chronic neutropenia evolving to MonoMac syndrome, myelodysplasia, and acute myeloid leukemia. *Blood, 121*, 822–829.
150. Ganapathi, K. A., Townsley, D. M., Hsu, A. P., Arthur, D. C., Zerbe, C. S., Cuellar-Rodriguez, J., et al. (2015). GATA2 deficiency-associated bone marrow disorder differs from idiopathic aplastic anemia. *Blood, 125*, 56–70.
151. Novakova, M., Zaliova, M., Sukova, M., Wlodarski, M., Janda, A., Fronkova, E., et al. (2016). Loss of B cells and their precursors is the most constant feature of GATA-2 deficiency in childhood myelodysplastic syndrome. *Haematologica, 101*, 707–716.
152. Cohen, J. I., Dropulic, L., Hsu, A. P., Zerbe, C. S., Krogmann, T., Dowdell, K., et al. (2016). Association of GATA2 deficiency with severe primary Epstein-Barr Virus (EBV) infection and EBV-associated cancers. *Clinical Infectious Diseases, 63*, 41–47.
153. Dickinson, R. E., Griffin, H., Bigley, V., Reynard, L. N., Hussain, R., Haniffa, M., et al. (2011). Exome sequencing identifies GATA-2 mutation as the cause of dendritic cell, monocyte, B and NK lymphoid deficiency. *Blood, 118*, 2656–2658.
154. Janssen, R., van Wengen, A., Hoeve, M. A., ten Dam, M., van der Burg, M., van Dongen, J., et al. (2004). The same IkappaBalpha mutation in two related individuals leads to completely different clinical syndromes. *The Journal of Experimental Medicine, 200*, 559–568.
155. McDonald, D. R., Mooster, J. L., Reddy, M., Bawle, E., Secord, E., & Geha, R. S. (2007). Heterozygous N-terminal deletion of IkappaBalpha results in functional nuclear factor kappaB haploinsufficiency, ectodermal dysplasia, and immune deficiency. *The Journal of Allergy and Clinical Immunology, 120*, 900–907.
156. Picard, C., Puel, A., Bonnet, M., Ku, C. L., Bustamante, J., Yang, K., et al. (2003). Pyogenic bacterial infections in humans with IRAK-4 deficiency. *Science, 299*, 2076–2079.
157. von Bernuth, H., Picard, C., Jin, Z., Pankla, R., Xiao, H., Ku, C. L., et al. (2008). Pyogenic bacterial infections in humans with MyD88 deficiency. *Science, 321*, 691–696.
158. Krebs, P., Crozat, K., Popkin, D., Oldstone, M. B., & Beutler, B. (2011). Disruption of MyD88 signaling suppresses hemophagocytic lymphohistiocytosis in mice. *Blood, 117*, 6582–6588.

Part IV
Infectious Triggers of Cytokine Storm Syndromes

Infectious Triggers of Cytokine Storm Syndromes: Herpes Virus Family (Non-EBV)

Daniel Dulek and Isaac Thomsen

The ability of infectious agents to trigger cytokine storm syndromes (CSS) was first reported in 1979, with the recognition of a series of cases of hemophagocytic lymphohistiocytosis (HLH) with apparent viral triggers [1]. Notably, in this this series, 14 of the 15 cases with viral triggers were associated with members of the *Herpesviridae* family, also known as the herpesviruses. In the nearly 40 years since, it remains evident that the herpesviruses, particularly Epstein–Barr Virus (EBV, discussed in a separate chapter), cytomegalovirus (CMV), herpes-simplex virus (HSV), and human herpesvirus 6 and 8 (HHV-6, HHV-8) are clearly associated with triggering primary or secondary HLH (pHLH or sHLH, respectively). A causal association is often difficult to prove, however, particularly in light of the known propensity of latent herpesviruses to reactivate in settings of systemic stress. In this chapter, the clinical aspects of herpesvirus-associated CSS will be discussed, along with the current state of our understanding of the pathogenesis of—and specific genetic lesions predisposing to—herpesvirus-associated CSS.

Pathogenesis, Clinical Features, and Treatment

The specific molecular pathways that link non-EBV herpesvirus infection with CSS are yet to be elucidated. One defining feature of herpesviruses is their ability to modulate and evade the immune response and establish latent infection in the host. It stands to reason, therefore, that these same immunomodulatory features may, in certain situations or in susceptible hosts, lead to dysregulated inflammation and

D. Dulek · I. Thomsen (✉)
Division of Infectious Diseases, Department of Pediatrics, Vanderbilt University School of Medicine, Nashville, TN, USA
e-mail: isaac.thomsen@vanderbilt.edu

R. Q. Cron, E. M. Behrens (eds.), *Cytokine Storm Syndrome*,
https://doi.org/10.1007/978-3-030-22094-5_14

cytokine storm syndromes. Recent murine models suggest that chronic and excessive Toll-like receptor (TLR) stimulation, as may be present during chronic herpesvirus infection, is associated with the development of CSS [2]. In addition, the proven ability of the herpesviruses to interfere with natural killer cell or T-cell cytotoxic activity may predispose the host to CSS [3, 4]. Additional mechanisms such as molecular mimicry targeting cytokines and signaling pathways, as well as inhibition of apoptotic pathways, have also been proposed [5], and future work is needed to elucidate these complex pathways.

The clinical features of CSS are complex and variable, and diagnosis is frequently challenging (see chapters "Clinical Features of Cytokine Storm Syndrome" and "Criteria for Cytokine Storm Syndromes"). There are no specific or unique clinical features of CSS triggered by herpesviruses. The identification of active or recent herpesvirus infection in the setting of a clinical syndrome consistent with CSS, however, is often illuminating given the strong association of these viruses with CSS. While often useful in securing the diagnosis of CSS, the identification of herpesviral infection does not distinguish between primary or secondary CSS, as pCSS can be triggered by herpesvirus infection [6], particularly in infants [7].

While most commonly reported in the pediatric literature, herpesviruses-associated CSS is frequently reported in adults as well. In a recent cohort at a large medical center, 73 adult patients met full diagnostic criteria for HLH, of which 41% were associated with infections as a presumed trigger. Of these cases, herpesvirus infection was the most common infectious trigger identified (9/30 cases) [8]. It is overall estimated that non-EBV herpesviruses represent approximately 20% of HLH cases with an infectious trigger in adults, with CMV being the most frequently associated [9]. Identification of a triggering infectious agent, if present, is important for the management of CSS. Primary herpesvirus infection is most common in pediatric patients and immunocompetent hosts, whereas reactivation of chronic EBV or CMV infection is most common in immunocompromised adults. In the latter case, quantitative PCR from serum may be useful as serology is typically unhelpful in these cases.

No randomized controlled trials for the treatment of viral-triggered CSS exist. For cases proven or suspected to be triggered by a herpesvirus, most experts recommend appropriate antiviral therapy if available. Importantly, however, the treatment is not limited to pathogen-directed therapy. Immunomodulation and other CSS-specific supportive care (See chapter "Alternative Therapies for Cytokine Storm Syndromes") is critical, as cases typically do not resolve with antiviral therapy alone [10].

Intravenous immunoglobulin therapy has been investigated for use in cases of viral-triggered CSS. One small series showed apparent benefit in a case of CMV-triggered HLH [11], while a larger series of 17 patients with EBV- or HHV-6-associated HLH received IVIG or etoposide therapy [12]. Both modalities appeared effective in improving outcomes, and two of nine patients developed sustained complete response with IVIG therapy alone. Many experts recommend that for CSS triggered by herpesviruses, the potential benefits of IVIG therapy likely outweigh the risks [13].

Non-EBV Herpesviruses in CSS

Cytomegalovirus

Cytomegalovirus (CMV) is second only to EBV in frequency of reports as an infectious trigger of cytokine storm syndromes. The majority of humans have been infected by CMV by the age of six [14]. Similar to EBV, CMV infection persists for the life of the host (residing latent in a variety of organs such as the lungs, spleen, salivary glands, and especially within myeloid progenitor cells in bone marrow), and can reactivate in settings of immune suppression or settings of acute systemic stress such as critical illness [15, 16]. CMV has been associated with CSS in children and adults, as a trigger of both primary and secondary CSS, and following either primary infection or reactivation of latent disease. Identification of CMV in the blood (by culture, antigen detection, or PCR) is more suggestive of true disease (and, therefore, of a potential trigger for CSS), as the virus may be shed in the urine for months in the absence of clinical disease, potentially confounding the interpretation of CMV viruria/antigenuria [17, 18].

Many of the reported cases and series of CMV-associated CSS involve viral reactivation during immune suppressed states. A wide variety of immune-suppressing states have been implicated, including Inflammatory bowel disease [19], solid organ transplantation in both adults and children [20, 21], and rheumatologic disease [22].

The prognosis of CMV-associated CSS, as with most other cytokine storm syndromes, is traditionally poor, though recent reports indicate improved outcomes with modern supportive care and therapeutics. One series of infants with HLH in Japan included five children with CMV-associated HLH, four of whom did not survive [23]. Recently, there are reported cases of excellent outcomes with a combination of ganciclovir (an antiviral medication that inhibits viral DNA polymerase) and IVIG therapy for CMV-associated CSS, including in a very young child (5 months old) [24]. IVIG alone [25] or ganciclovir alone [19] has also been reported to lead to favorable outcomes. Foscarnet, a DNA polymerase inhibitor often used in acyclovir or ganciclovir-resistant infections, has also been used with success for CMV-associated HLH, in combination with corticosteroids or ganciclovir [26, 27].

Human Herpes Virus 8

Human herpesvirus-8 (HHV-8) is clearly implicated in numerous reported cases of CSS, though much less frequently than EBV or CMV. Following primary infection, HHV-8 establishes lifelong latency (primarily within CD19 B cells), and the virus is perhaps more notable for its propensity to cause lymphomas and other malignancies (e.g., Kaposi sarcoma). The majority of reported cases of HHV-8 associated CSS occurred in the setting of immunocompromise, most commonly HIV (especially with associated Kaposi sarcoma) or solid organ transplantation. Multicentric

Castleman disease, a lymphoproliferative disorder strongly associated with HHV-8 (especially in the setting of HIV coinfection) has also been associated with CSS. One recent report described a 61 year-old immunocompetent man who developed HHV-8 associated HLH in the setting of multicentric Castleman disease, which was unfortunately fatal. HHV-8 was also implicated in a case of female triplets, two of which developed HLH (one at age 4 months, the other at age 6 months) [28]. The two infants had identical mutations in the Perforin pathway, as well as proven HHV-8 infection triggering HLH. This case serves as an instructive example of herpesviral infection triggering primary CSS with a clear genetic susceptibility.

The overall outcomes in the reported cases of HHV-8 associated CSS have been fair. Several patients have recovered with an etoposide-containing regimen [29]; one reported patient recovered after treatment with foscarnet plus doxorubicin [30]. Current literature would suggest potential benefit from either an etoposide-containing or with antiviral therapy in addition to supportive care. The HLH-94 regimen (etoposide, corticosteroids, cyclosporin A, followed by bone marrow transplant) was employed in the case of triplets with HHV-8 associated HLH, with favorable outcome [31].

Herpes Simplex Virus

Like all members of the Herpesviridae family, Herpes simplex virus types 1 and 2 (HSV-1 and HSV-2) establish lifelong latency following primary infection, residing primarily in sensory neural ganglia. Much like HHV-8, the majority of cases of Herpes simplex virus (HSV)-associated CSS occur in immunocompromised hosts. Reported cases of CSS triggered by HSV involve underlying comorbidities such as autoimmune/rheumatologic diseases with concomitant immunomodulatory therapy [32, 33], bone marrow failure syndromes [34], or primary immunodeficiencies [35].

Cytokine storm syndromes are associated with both primary HSV infection as well as reactivated disease, even in immunocompetent hosts [36]. One high risk group appears to be pregnant women with genital herpetic lesions, in whom there are numerous reported cases of HSV-triggered CSS [37–39].

As with CSS triggered by the other herpesviruses, treatment typically involves CSS-specific supportive care, often with the addition of antiviral therapy, most commonly acyclovir.

Herpesviral Infection and CSS in Specific Clinical Scenarios

Neonatal HSV Infection and Possible HLH

One area of significant current controversy regarding herpesvirus-associated CSS is that of neonatal disseminated HSV infection and its possible role in triggering HLH. The large majority of putative cases of neonatal HSV-associated HLH reported in the literature do not involve the definitive diagnosis of HLH. In many cases, the diagnosis is presumed based on a markedly elevated ferritin level, and the association of high ferritin with HLH in pediatrics [40]. This study concluded that a ferritin level >10,000 μg/L was 96% specific for HLH, although only 10 of the 330 patients in this study were confirmed to have HLH and there were no cases of disseminated neonatal HSV in the comparator group. Further, there are no published studies of typical ferritin levels in cases of disseminated HSV with severe hepatic involvement.

Some reported cases of neonatal HSV-associated CSS involve a diagnosis that was based on suggestive findings such as hemophagocytosis on bone marrow aspirate or elevated circulating Interleukin-2 receptor alpha (IL-2R, CD25) levels. The presence of hemophagocytosis is neither sensitive nor specific for HLH [41], however, and is generally considered a weaker diagnostic criterion [13]. Elevated soluble CD25 levels are also associated simply with severe systemic inflammation and are not specific to HLH [42, 43]. It is also known that that fever, hypertriglyceridemia, and neutropenia were unreliable markers in newborns, thus making HLH diagnosis in this population even more challenging [44].

One recent case series reported three cases of neonatal HSV with massive hyperferritinemia which were initially diagnosed as presumed HSV-associated HLH consistent with previous literature [45]. In each of these cases, however, definitive diagnosis was never secured (e.g., no genetic lesions were identified), and the hyperferritinemia may have been due to hepatocyte destruction from disseminated HSV disease rather than extreme inflammation from concomitant HLH, as ferritin levels correlated closely with transaminase values for all three patients. Importantly, one of the subjects in this series died of an infectious complication of the immunomodulatory regimen that was initiated for presumed HLH. This emphasizes that the diagnosis of CSS should be very carefully considered in the neonate in general, and in the setting of disseminated HSV infection in particular. Early molecular diagnosis is recommended given the risks of immunosuppressive therapy, especially in the setting of systemic infection [13]. Overall, expert guidance indicates that immunosuppression should be used only with careful consideration in cases of disseminated HSV with possible associated CSS, as it remains unclear whether these cases represent HSV infection alone, or secondary HLH following HSV infection (for which treatment with antiviral therapy may be sufficient).

Herpesviral Infection and pHLH

The association between herpesviral infection and genetic predisposition to CSS (also referred to as pHLH) was first defined through investigation of severe EBV infection in a multigenerational kindred of males followed by subsequent genetic analysis [46, 47] (see chapter "Cytokine Storm Syndromes Associated with Epstein–Barr Virus"). To date, despite ten genetic defects having been shown to impart increased risk for HLH, virus-specific associations between non-EBV herpesviruses and specific genetic etiologies of pHLH have not been described. Several cases and case series of herpesviral infection associated with pHLH are reported in the literature (Table 1). Analysis of these cases and detection of future cases such as these holds potential for further understanding of herpesviral CSS/sHLH.

Three cases of HHV-8 infection have been detected in the setting of pHLH [28, 48] (Table 1). Two of these cases were in female infants of a triplet gestation in whom compound heterozygous mutations in *PRF1* were identified [28]. The first infant presented at 4 months of age with hepatosplenomegaly, anemia, leukopenia, thrombocytopenia, and elevated ferritin with hemophagocytosis noted in her bone marrow biopsy. HHV-8 PCR was positive from whole blood PCR but negative in cerebrospinal fluid (CSF). No other infectious triggers were identified. Genetic testing following illness onset revealed compound heterozygous mutations in *PRF1*. The second infant of this triplet gestation was found to have the same *PRF1* mutations as her sister but was asymptomatic until 6 months of age when she developed persistent fevers, hepatomegaly, neutropenia, thrombocytopenia, elevated ferritin, and bone marrow hemophagocytosis. HHV-8 PCR from her CSF was positive. In both infants, NK cell perforin expression was absent and NK cell target killing was significantly impaired. Both infants responded well to etoposide, dexamethasone, and cyclosporine therapy. There is no report of whether they required further therapy or underwent subsequent hematopoietic stem cell transplant (HSCT) [28].

In a separate case, a 2.5-year-old boy presented with night sweats and macular rash followed by persistent fever and diffuse lymphadenopathy [48] (Table 1). Hepatosplenomegaly, anemia, and thrombocytopenia were noted along with elevated ferritin. HHV-8 DNA was detected by PCR from peripheral blood mononuclear cells (PBMCs) and bone marrow. HHV-6 viremia was also documented with quantitative PCR showing 13,110 copies/mL in the plasma and 9685 copies/10^6 cells in PBMCs. The authors reached a diagnosis of X-linked lymphoproliferative

Table 1 Non-EBV Herpesvirus infections associated with pHLH

Year published (Reference)	Publication type	Virus	Age	Affected gene	Other infection history	Treatment/ outcome
2005 [26]	Case series	HHV-8	4m	*PRF1*	None	Tx: HLH-94
	$n = 2$		6m			Both survived
2012 [46]	Case report	HHV-8	2.5y	Absent SAP expression;	none	Tx: HLH-2004
	$n = 1$			No mutation identified		Survived
2014 [48]	Prospective cohort	HHV-6	2y	*SH2D1A*	Not reported	RIC-HCT
	$n = 1$					Outcome not reported
2016 [33]	Case report	HSV-1	18y	*GATA2*	*Blastomyces dermatitidis*	Death
	$n = 1$					

disorder in this patient by demonstrating absent SAP expression in activated T and NK cells. Interestingly, no mutations in *SH2D1A*, *BIRC4*, or *PRF1* were identified by "mutation analysis." This patient was treated with etoposide, dexamethasone, and cyclosporine with full clinical response.

There is one additional allusion to HHV-6 detection associated with lymphoproliferation in the setting of XLP due to SAP deficiency. Very little HHV-6 related information is given in this case as the study was mechanistic rather than descriptive [49]. A second case of HHV-6 associated HLH in an XLP patient noted in a separate study though this may be the same patient [50].

GATA2 deficiency has been associated with HSV-1-induced HLH in one reported case [35] (Table 1). This patient presented at 18 years of age with fevers and headache and developed pancytopenia, transaminase elevation, markedly elevated ferritin, and hypofibrinogenemia with hypotension. She progressed to cardiac arrest and death despite broad spectrum antimicrobial treatment. HSV-1 DNA was detected in her blood prior to death and disseminated HSV-1 infection involving lungs and liver was diagnosed on autopsy. Due to a prior history of lymphedema, NK-cell dysfunction, as well as pneumonia due to *Blastomyces dermatitidis*, GATA2 deficiency was suspected and genetic testing demonstrated a *GATA2* frameshift mutation [35].

To the best of our knowledge, the above cases represent the known literature for non-EBV herpesviral infection in the setting of known or subsequently diagnosed pHLH. Literature search using known pHLH-associated genes/gene products along with non-EBV herpesviruses did not reveal further cases. Attention to thorough evaluation for viral triggers, including herpesviruses, in the setting of known or potential pHLH is critical for identifying possible novel associations.

Herpesviral Infection and CSS in Immunocompromised Patients

As noted previously, CSS associated with non-EBV herpesvirus infection most frequently occurs in the setting of underlying immunocompromised states. Though some of this literature has been reviewed above, we will review here relevant English-language literature regarding non-EBV herpesvirus-associated CSS in the context of specific categories of immune-compromise with attention to treatment, outcomes, and potential viral associations. Recognizing the limitations of interpretations of accumulated case reports and small case series, we will point out potential trends and associations that merit further evaluation in future studies. Though little to no true incidence data is available for the frequency of CSS occurrence in each of these settings, the greatest number of cases is reported in two settings—underlying hematologic/oncologic/HSCT diagnosis (Table 2) and patients with iatrogenic immunosuppression due to autoimmune/autoinflammatory disease (Table 3). In this latter patient group, it is difficult to distinguish the relative contributions of immunosuppression and underlying disease in the predisposition to CSS.

Four recent single or multicenter studies have accumulated data related to HLH which provides insight into the relative incidence of specific etiologies of virus-associated HLH. In one recent single-institution series of 35 adult onset HLH patients, three-quarters of patients had an underlying hematologic diagnosis whereas 11% of patients had an underlying autoimmune/autoinflammatory disease [53]. In this series, EBV was the most frequently associated virus (28.6%) but was closely followed by CMV (22.9%) and HHV8 (17.1%). In a multicenter study of 162 adult HLH patients with infectious triggers, half of patients had a diagnosis of HIV and roughly one third of patients were immunocompromised for other reasons [52]. Of the patients with 17 patients with viral HLH triggers, 11 patients had CMV detected, three had EBV detection, and HHV-8 and HSV-2 were detected in two and one patient, respectively.

Non-EBV herpesvirus-associated CSS in patients with hematologic/oncologic/HSCT diagnoses is relatively frequently reported. Cases of each of the herpesviruses associated with CSS can be found in the literature (Table 2). No large prospective series exist for this patient population. Given the attention given to prophylaxis (HSV, VZV) and close monitoring/prophylaxis (CMV) in this patient population, it is likely that cases of non-EBV herpesvirus-associated CSS are prevented through these strategies. HSV-2 and VZV associated CSS are reported in three total patients; one patient with an undifferentiated bone marrow failure syndrome [34] and two patients with GVHD following HSCT for AML [58]. Two of these three patients did not survive CSS. CMV-associated CSS is reported in three patients with ALL [54, 59], two patients with aplastic anemia [60], one patient with multiple myeloma [60], and one patient with Hodgkin lymphoma [62]. HHV-6 and HHV-8 in patients with hematologic/oncologic/HSCT diagnoses occurred in patients with diagnoses of nephroblastomatosis, autoimmune hemolytic anemia, and multicentric Castleman disease (Table 2).

When the diagnosis of herpesviral-associated CSS was made antemortem, antiviral therapy was generally administered as part of the therapeutic approach. In several cases corticosteroids and/or IVIG were administered along with antivirals though data is too limited to comment on outcomes related to therapeutic approach. Outcomes following CSS diagnosis in these patients were poor with 50% mortality for reported in those patients for whom data was available. There was a potential trend towards improved survival in young pediatric patients compared to teenagers and adults (Table 2).

Patients with iatrogenic immunosuppression secondary to autoimmune or autoinflammatory diseases are also reported to develop CSS in association with non-EBV herpesvirus infection (Table 3). Underlying diagnoses range from inflammatory bowel disease, to neurologic autoimmune disease, to rheumatologic diagnoses. Notably, the majority of reported CSS cases in these patients were associated with CMV [19, 22, 27, 61, 63, 64, 66–68]. Identification of CMV in these cases led to antiviral treatment and outcomes were good with only one reported patient dying as a result of their CMV-associated CSS [67]. In contrast, one patient with HSV-2-associated CSS died, though HSV-2 was not identified until postmortem evaluation and antiviral therapy was not administered [33].

Table 2 Non-EBV Herpesvirus infections associated with CSS in the setting of hematologic/oncologic disorders or hematopoietic stem cell transplantation

Year published (Reference)	Virus	# patients	Age	Immunosuppression	Underlying diagnosis	Treatment/outcome
2006 [32]	HSV-2	1	33 years	Oxymetholone, corticosteroids	Bone marrow failure syndrome	Acyclovir, IVIG; pt died
2007 [51]	HHV-8	1	63 years	Corticosteroids	Autoimmune hemolytic anemia	IVIG, ganciclovir; pt died
2009 [52]	CMV; RSV	1	18 months	Consolidation chemotherapy	Pre-B cell ALL	Ganciclovir, IVIG, ribavirin; pt survived
2009 [53]	VZV	2	14 years	1: Corticosteroids	AML/HSCT with GVHD in both patients	1: Increased corticosteroid dose, cyclosporine; pt died
			17 years	2: Corticosteroids, mycophenolate, rituximab		2: Treatment NR; pt survived
2009 [54]	CMV	3	30 years, 18 years, 56 years	Atgam + Cyclophosphamide ×2 subjects; melphalan ×1 subject	Aplastic anemia ×2, multiple myeloma	NR
2010 [55]	HHV-6	1	14 months	Vincristine, actinomycin	Nephroblastomatosis	NR
2013 [56]	CMV	1	NR	NR	Schwachman–Diamond	NR
2016 [57]	HHV-8	1	61 years	None	Multicentric Castleman's	Vincristine, doxorubicin, dexamethasone; pt died
2017 [58]	CMV	2	7 years	Maintenance chemotherapy	ALL	1: Ganciclovir, foscarnet; survived
			6.5 years			2: Dexamethasone, ganciclovir; survived
2017 [59]	EBV, CMV	1	60 years	None	Hodgkin lymphoma	Dexamethasone, etoposide, IVIG, ganciclovir; survived

Table 3 Non-EBV Herpesvirus infections associated with CSS in the setting of autoimmune/autoinflammatory disease

Year published	Virus	# patients	Age	Immunosuppression	Underlying diagnosis	Treatment/ outcome
2006 [17]	CMV	1	22 years	Infliximab	Crohn disease	Ganciclovir, splenectomy; pt survived
				6-MP		
2011 [60]	CMV	1	32 years	Azathioprine	Crohn disease	Ganciclovir; pt survived
2011 [25]	CMV	2	32 years	Azathioprine (both)	Ulcerative colitis;	Ganciclovir (both);
					Crohn disease	Both survived
2013 [61]	CMV	1	52 years	None	Ulcerative colitis	Ganciclovir; survived
Neurologic autoimmune disease						
2014 [62]	CMV	1	47 years	Azathioprine, corticosteroids	Myasthenia gravis	Etoposide, dexamethasone, valganciclovir; survived
2016 [31]	HSV2	1	56 years	Fingolimod, corticosteroids	Multiple sclerosis	None; pt died
Rheumatologic disease						
2002 [63]	CMV	1	44y	Corticosteroids, cyclophosphamide	SLE	CMV-Ig, ganciclovir, corticosteroids; pt died
2005 [20]	CMV	1	80y	Corticosteroids	Adult-onset Still disease	Ganciclovir; pt survived
2008 [64]	CMV	1	58 years	Corticosteroids, azathioprine	SLE	Ganciclovir; survived
2010 [30]	HSV-1	1	57 years	Azathioprine, prednisone	Wegener/ GPA	Corticosteroids, etoposide, IVIG; survived
2015 [65]	VZV	1	5 years	Corticosteroids	HSP	Acyclovir, IVIG, plasma exchange; survived
2017 [66]	CMV	1	70 years	None	Sjogren syndrome	Acyclovir, corticosteroids, cyclosporine; survived

Fewer cases of CSS associated with non-EBV herpesviruses are reported in solid organ transplant patients (Table 4). Cases include patients with CMV, HHV-6, and HHV-8 [21, 69–73] with a notable absence of reported cases of HSV or VZV-associated CSS in this population. Outcomes were mixed with some patients responding to antiviral treatment and/or chemotherapy and others succumbing. Few conclusions can be drawn from this small series of cases.

Table 4 Non-EBV herpesvirus infections associated with CSS in the setting of solid organ transplantation

Year published	Virus	# patients	Age	Immunosuppression	Underlying diagnosis	Treatment/outcome
2002 [69]	HHV-8	1	61 years	Cyclosporine, corticosteroids	Kidney transplant, Kaposi sarcoma	IVIG, reduced immunosuppression, foscarnet, daunorubicin; pt survived
2006 [70]	CMV	2	46 years	Cyclosporine, azathioprine	Kidney transplant	1: Ganciclovir; pt died
			28 years			2: Ganciclovir, IVIG; pt died
2006 [21]	CMV	1	27 months	Tacrolimus, azathioprine, corticosteroids	Liver transplant	Ganciclovir; pt survived
2008 [71]	HHV-6	1	49 years	Basiliximab, cyclosporine, mycophenolate, corticosteroids	Liver/kidney transplant	No therapy; pt died
2016 [72]	HHV-8	1	66 years	Tacrolimus, mycophenolate, prednisone	Liver transplant	Immunosuppression decrease, ganciclovir, corticosteroids, rituximab, cyclophosphamide;
						pt died
2017 [73]	HHV-8	1	39 years	Tacrolimus	Liver/kidney transplant	Cidofovir; foscarnet+rituximab; corticosteroids+rituximab; foscarnet+rituximab; pt survived

Non-EBV herpesvirus-associated CSS in HIV patients is unique in that there is a clear association between HIV/AIDS and CSS due to HHV-8 (Table 5). The first such report was published in 1998 from a patient with AIDS and Kaposi Sarcoma (KS) who developed significant hepatosplenomegaly in association with cytopenias. HHV-8 DNA was detected by PCR of peripheral blood mononuclear cells, lymph node tissue, colonic tissue, and bronchoalveolar lavage fluid. Antiviral treatment, along with therapy for his underlying KS, resolved his HHV-8 viremia and antiretroviral therapy resulted in a good outcome [74]. Since that initial report, several other cases of HHV-8-associated CSS have been reported in the literature. Generally, these have been associated with high viral loads and marked CD4 T cell count suppression (Table 5). More recently, in two larger series of patients, this association between CSS and HHV-8 in HIV-positive patients has been described as "Kaposi sarcoma-associated herpesvirus inflammatory cytokine syndrome" (KICS)

Table 5 Non-EBV Herpesvirus infections associated with CSS in the setting of HIV/AIDS

Year published (Reference)	Virus	# patients	Age	CD4 count (cells/μL) HIV viral load (VL) (copies/mL)	Other concurrent infections or malignancy	Treatment/outcome
1998 [74]	HHV-8	1	28 years	CD4: 204 VL: NR	KS	Foscarnet, ART, vincristine, daunoxome; pt survived
2000 [75]	HHV-8	1	38 years	CD4: 94 VL: NR	PEL	Ganciclovir, vincristine, cyclophosphamide, corticosteroids; pt died
2003 [76]	HHV-8	5	NR	CD4: 165–234	KS, $n = 4$	Treatment NR
				VL: 50–25,000	MCD, $n = 3$	Survived, $n = 3$
2007 [77]	HHV-8, EBV[a]	1	45 years	CD4: 64 VL: $>1 \times 10^5$	KS, MCD	Corticosteroids, ganciclovir, cyclophosphamide; pt died
2009 [78]	HHV-8, CMV[b]	1	39 years	CD4: 16 VL: 1.3×10^6	KS	Foscarnet, ART, corticosteroids, doxorubicin; pt survived
2010 [79]	HHV-8	6	38 years[c]	CD4[c]: 255 VL[c]=4650	KS, $n = 3$	Treatment NR; $n = 3$ died
2013 [80]	CMV	1	29 years	CD4: 120 VL: 2×10^5	None	Foscarnet, ganciclovir; pt survived
2016 [81]	HHV-8	10	36 years[c]	CD4[c]: 88 VL[c]: 72	KS, $n = 6$	Treatment NR; $n = 6$ pts died

Abbreviations: *KS* Kaposi sarcoma, *ART* antiretroviral therapy, *MCD* multicentric Castleman disease, *NR* not reported, *PEL* primary effusion lymphoma
[a]EBV diagnosed by PCR of lymphoid tissue
[b]CMV viral load was at the lower limit of detection
[c]Values provided are medians

[79, 81]. Uldrich et al. first identified 10 HIV-positive patients with Kaposi Sarcoma or HHV-8 viremia but without evidence of multicentric Castleman disease (MCD). Patients with KICS had higher virally expressed serum IL-6 (vIL-6) levels than patients with HHV-8 associated MCD or patients with KS alone [79]. Subsequently, a similar syndrome was defined prospectively in a distinct cohort of 10 patients with KICS [81]. In this cohort, subjects with KICS had higher human IL-6 expression compared to controls with or without active HIV viremia. Unlike the prior study, differences in vIL-6 were not noted between KICS and control patients. Thus, HHV-8 associated CSS has a unique association with HIV-infection with or without associated HHV-8 tumors. Mortality in these patients is high and further study is warranted to better understand the pathogenesis of this association.

Conclusions

Herpesvirus infection plays a significant role in the etiology and pathogenesis of CSS. Despite the clear association of most members of the herpesvirus family with CSS, the specific factors responsible for this relationship remain unclear. All members of the herpesvirus family establish lifelong latent infection in the host, and the related balance of host control of a persistent pathogen may confer risk of immune dysregulation in the setting of reactivation. Unique aspects of non-EBV herpesvirus immunopathogenesis provide insight into the pathways of aberrant inflammation that contribute to CSS in both immune-competent and immunocompromised patients. Laboratory evaluation for evidence of primary or reactivated herpesviral infection in the setting of CSS provides opportunity for early initiation of specific antiviral therapy. Heightened clinical suspicion for non-EBV herpesvirus-associated CSS in immunocompromised patients is warranted given the apparent increased risk for CSS in these patients. Moreover, thorough immunophenotypic and genetic assessment of patients with non-EBV herpesvirus-associated CSS is critical for understanding of the pathogenesis and epidemiology of CSS. Multicenter prospective studies are needed to further define the epidemiology, optimal treatment, and outcomes of non-EBV herpesvirus-associated CSS.

References

1. Risdall, R. J., McKenna, R. W., Nesbit, M. E., Krivit, W., Balfour Jr., H. H., Simmons, R. L., et al. (1979). Virus-associated hemophagocytic syndrome: A benign histiocytic proliferation distinct from malignant histiocytosis. *Cancer, 44*, 993–1002.
2. Behrens, E. M., Canna, S. W., Slade, K., Rao, S., Kreiger, P. A., Paessler, M., et al. (2011). Repeated TLR9 stimulation results in macrophage activation syndrome-like disease in mice. *The Journal of Clinical Investigation, 121*, 2264–2277.
3. Orange, J. S., Fassett, M. S., Koopman, L. A., Boyson, J. E., & Strominger, J. L. (2002). Viral evasion of natural killer cells. *Nature Immunology, 3*, 1006–1012.
4. Odom, C. I., Gaston, D. C., Markert, J. M., & Cassady, K. A. (2012). Human herpesviridae methods of natural killer cell evasion. *Advances in Virology, 2012*, 359869.
5. Brisse, E., Wouters, C. H., Andrei, G., & Matthys, P. (2017). How viruses contribute to the pathogenesis of hemophagocytic lymphohistiocytosis. *Frontiers in Immunology, 8*, 1102.
6. Henter, J. I., Ehrnst, A., Andersson, J., & Elinder, G. (1993). Familial hemophagocytic lymphohistiocytosis and viral infections. *Acta Paediatrica, 82*, 369–372.
7. Henter, J. I., Elinder, G., Soder, O., & Ost, A. (1991). Incidence in Sweden and clinical features of familial hemophagocytic lymphohistiocytosis. *Acta Paediatrica Scandinavica, 80*, 428–435.
8. Otrock, Z. K., & Eby, C. S. (2015). Clinical characteristics, prognostic factors, and outcomes of adult patients with hemophagocytic lymphohistiocytosis. *American Journal of Hematology, 90*, 220–224.
9. Ramos-Casals, M., Brito-Zeron, P., Lopez-Guillermo, A., Khamashta, M. A., & Bosch, X. (2014). Adult haemophagocytic syndrome. *Lancet, 383*, 1503–1516.
10. Fisman, D. N. (2000). Hemophagocytic syndromes and infection. *Emerging Infectious Diseases, 6*, 601–608.

11. Chen, R. L., Lin, K. H., Lin, D. T., Su, I. J., Huang, L. M., Lee, P. I., et al. (1995). Immunomodulation treatment for childhood virus-associated haemophagocytic lymphohistiocytosis. *British Journal of Haematology, 89*, 282–290.
12. Gill, D. S., Spencer, A., & Cobcroft, R. G. (1994). High-dose gamma-globulin therapy in the reactive haemophagocytic syndrome. *British Journal of Haematology, 88*, 204–206.
13. Jordan, M. B., Allen, C. E., Weitzman, S., Filipovich, A. H., & McClain, K. L. (2011). How I treat hemophagocytic lymphohistiocytosis. *Blood, 118*, 4041–4052.
14. Staras, S. A., Dollard, S. C., Radford, K. W., Flanders, W. D., Pass, R. F., & Cannon, M. J. (2006). Seroprevalence of cytomegalovirus infection in the United States, 1988-1994. *Clinical Infectious Diseases, 43*, 1143–1151.
15. Heininger, A., Jahn, G., Engel, C., Notheisen, T., Unertl, K., & Hamprecht, K. (2001). Human cytomegalovirus infections in nonimmunosuppressed critically ill patients. *Critical Care Medicine, 29*, 541–547.
16. Hummel, M., & Abecassis, M. M. (2002). A model for reactivation of CMV from latency. *Journal of Clinical Virology, 25*(Suppl 2), S123–S136.
17. Amin, M. M., Bialek, S. R., Dollard, S. C., & Wang, C. (2018). Urinary cytomegalovirus shedding in the United States: The National Health and Nutrition Examination Surveys, 1999-2004. *Clinical Infectious Diseases, 67*(4), 587–592.
18. Cannon, M. J., Stowell, J. D., Clark, R., Dollard, P. R., Johnson, D., Mask, K., et al. (2014). Repeated measures study of weekly and daily cytomegalovirus shedding patterns in saliva and urine of healthy cytomegalovirus-seropositive children. *BMC Infectious Diseases, 14*, 569.
19. Kohara, M. M., & Blum, R. N. (2006). Cytomegalovirus ileitis and hemophagocytic syndrome associated with use of anti-tumor necrosis factor-alpha antibody. *Clinical Infectious Diseases, 42*, 733–734.
20. Bea Granell, S., Beneyto Castello, I., Ramos Escorihuela, D., Sanchez Plumed, J., Sanchez Perez, P., Hernandez-Jaras, J., et al. (2011). Cytomegalovirus-associated haemophagocytic syndrome in a kidney transplant patient. *Nefrología, 31*, 236–238.
21. Hardikar, W., Pang, K., Al-Hebbi, H., Curtis, N., & Couper, R. (2006). Successful treatment of cytomegalovirus-associated haemophagocytic syndrome following paediatric orthotopic liver transplantation. *Journal of Paediatrics and Child Health, 42*, 389–391.
22. Amenomori, M., Migita, K., Miyashita, T., Yoshida, S., Ito, M., Eguchi, K., et al. (2005). Cytomegalovirus-associated hemophagocytic syndrome in a patient with adult onset Still's disease. *Clinical and Experimental Rheumatology, 23*, 100–102.
23. Imashuku, S., Ueda, I., Teramura, T., Mori, K., Morimoto, A., Sako, M., et al. (2005). Occurrence of haemophagocytic lymphohistiocytosis at less than 1 year of age: Analysis of 96 patients. *European Journal of Pediatrics, 164*, 315–319.
24. Oloomi, Z., & Moayeri, H. (2006). Cytomegalovirus infection-associated hemophagocytic syndrome. *Archives of Iranian Medicine, 9*, 284–287.
25. Hot, A., Madoux, M. H., Viard, J. P., Coppere, B., & Ninet, J. (2008). Successful treatment of cytomegalovirus-associated hemophagocytic syndrome by intravenous immunoglobulins. *American Journal of Hematology, 83*, 159–162.
26. Knorr, B., Kessler, U., Poschl, J., Fickenscher, H., & Linderkamp, O. (2007). A haemophagocytic lymphohistiocytosis (HLH)-like picture following breastmilk transmitted cytomegalovirus infection in a preterm infant. *Scandinavian Journal of Infectious Diseases, 39*, 173–176.
27. van Langenberg, D. R., Morrison, G., Foley, A., Buttigieg, R. J., & Gibson, P. R. (2011). Cytomegalovirus disease, haemophagocytic syndrome, immunosuppression in patients with IBD: 'A cocktail best avoided, not stirred'. *Journal of Crohn's & Colitis, 5*, 469–472.
28. Grossman, W. J., Radhi, M., Schauer, D., Gerday, E., Grose, C., & Goldman, F. D. (2005). Development of hemophagocytic lymphohistiocytosis in triplets infected with HHV-8. *Blood, 106*, 1203–1206.
29. Li, C. F., Ye, H., Liu, H., Du, M. Q., & Chuang, S. S. (2006). Fatal HHV-8-associated hemophagocytic syndrome in an HIV-negative immunocompetent patient with plasmablastic variant of multicentric Castleman disease (plasmablastic microlymphoma). *The American Journal of Surgical Pathology, 30*, 123–127.

30. Bossini, N., Sandrini, S., Setti, G., Luppi, M., Maiorca, P., Maffei, C., et al. (2005). Successful treatment with liposomal doxorubicin and foscarnet in a patient with widespread Kaposi's sarcoma and human herpes virus 8-related, serious hemophagocytic syndrome, after renal transplantation. *Giornale Italiano di Nefrologia, 22*, 281–286.
31. Henter, J. I., Samuelsson-Horne, A., Arico, M., Egeler, R. M., Elinder, G., Filipovich, A. H., et al. (2002). Treatment of hemophagocytic lymphohistiocytosis with HLH-94 immunochemotherapy and bone marrow transplantation. *Blood, 100*, 2367–2373.
32. Cusini, A., Gunthard, H. F., Stussi, G., Schwarz, U., Fehr, T., Grueter, E., et al. (2010). Hemophagocytic syndrome caused by primary herpes simplex virus 1 infection: Report of a first case. *Infection, 38*, 423–426.
33. Ikumi, K., Ando, T., Katano, H., Katsuno, M., Sakai, Y., Yoshida, M., et al. (2016). HSV-2-related hemophagocytic lymphohistiocytosis in a fingolimod-treated patient with MS. *Neurol Neuroimmunol Neuroinflamm, 3*, e247.
34. Ramasamy, K., Lim, Z. Y., Savvas, M., Salisbury, J. R., Dokal, I., Mufti, G. J., et al. (2006). Disseminated herpes virus (HSV-2) infection with rhabdomyolysis and hemophagocytic lymphohistiocytosis in a patient with bone marrow failure syndrome. *Annals of Hematology, 85*, 629–630.
35. Spinner, M. A., Ker, J. P., Stoudenmire, C. J., Fadare, O., Mace, E. M., Orange, J. S., et al. (2016). GATA2 deficiency underlying severe blastomycosis and fatal herpes simplex virus-associated hemophagocytic lymphohistiocytosis. *The Journal of Allergy and Clinical Immunology, 137*, 638–640.
36. Mihalcea-Danciu, M., Ellero, B., Gandoin, M., Harlay, M. L., Schneider, F., & Bilbault, P. (2014). Herpes simplex hepatitis with macrophage activation syndrome in an immunocompetent patient. *La Revue de Médecine Interne, 35*, 823–826.
37. Desideri-Vaillant, C., Exbrayat, S., Sapin-Lory, J., Lambrechts, D., Rouxel, M., & Nicolas, X. (2012). Hemophagocytic syndrome due to Herpes simplex virus after hysteroscopy. *Journal de Gynécologie, Obstétrique et Biologie de la Reproduction, 41*, 672–675.
38. Goulding, E. A., & Barnden, K. R. (2014). Disseminated herpes simplex virus manifesting as pyrexia and cervicitis and leading to reactive hemophagocytic syndrome in pregnancy. *European Journal of Obstetrics, Gynecology, and Reproductive Biology, 180*, 198–199.
39. Yamaguchi, K., Yamamoto, A., Hisano, M., Natori, M., & Murashima, A. (2005). Herpes simplex virus 2-associated hemophagocytic lymphohistiocytosis in a pregnant patient. *Obstetrics and Gynecology, 105*, 1241–1244.
40. Allen, C. E., Yu, X., Kozinetz, C. A., & McClain, K. L. (2008). Highly elevated ferritin levels and the diagnosis of hemophagocytic lymphohistiocytosis. *Pediatric Blood & Cancer, 50*, 1227–1235.
41. Gupta, A., Weitzman, S., & Abdelhaleem, M. (2008). The role of hemophagocytosis in bone marrow aspirates in the diagnosis of hemophagocytic lymphohistiocytosis. *Pediatric Blood & Cancer, 50*, 192–194.
42. Holter, W., Goldman, C. K., Casabo, L., Nelson, D. L., Greene, W. C., & Waldmann, T. A. (1987). Expression of functional IL 2 receptors by lipopolysaccharide and interferon-gamma stimulated human monocytes. *Journal of Immunology, 138*, 2917–2922.
43. Seidler, S., Zimmermann, H. W., Weiskirchen, R., Trautwein, C., & Tacke, F. (2012). Elevated circulating soluble interleukin-2 receptor in patients with chronic liver diseases is associated with non-classical monocytes. *BMC Gastroenterology, 12*, 38.
44. Suzuki, N., Morimoto, A., Ohga, S., Kudo, K., Ishida, Y., Ishii, E., et al. (2009). Characteristics of hemophagocytic lymphohistiocytosis in neonates: A nationwide survey in Japan. *The Journal of Pediatrics, 155*, 235–238 e231.
45. Vladescu, I. A., Browning, W. L., & Thomsen, I. P. (2015). Massive ferritin elevation in neonatal herpes simplex virus infection: hemophagocytic lymphohistiocytosis or herpes simplex virus alone? *Journal of the Pediatric Infectious Diseases Society, 4*, e48–e52.
46. Purtilo, D. T., Cassel, C. K., Yang, J. P., & Harper, R. (1975). X-linked recessive progressive combined variable immunodeficiency (Duncan's disease). *Lancet, 1*, 935–940.

47. Sayos, J., Wu, C., Morra, M., Wang, N., Zhang, X., Allen, D., et al. (1998). The X-linked lymphoproliferative-disease gene product SAP regulates signals induced through the co-receptor SLAM. *Nature, 395*, 462–469.
48. Pasic, S., Cupic, M., & Lazarevic, I. (2012). HHV-8-related hemophagocytic lymphohistiocytosis in a boy with XLP phenotype. *Journal of Pediatric Hematology/Oncology, 34*, 467–471.
49. Snow, A. L., Marsh, R. A., Krummey, S. M., Roehrs, P., Young, L. R., Zhang, K., et al. (2009). Restimulation-induced apoptosis of T cells is impaired in patients with X-linked lymphoproliferative disease caused by SAP deficiency. *The Journal of Clinical Investigation, 119*, 2976–2989.
50. Marsh, R. A., Bleesing, J. J., Chandrakasan, S., Jordan, M. B., Davies, S. M., & Filipovich, A. H. (2014). Reduced-intensity conditioning hematopoietic cell transplantation is an effective treatment for patients with SLAM-associated protein deficiency/X-linked lymphoproliferative disease type 1. *Biology of Blood and Marrow Transplantation, 20*, 1641–1645.
51. Re, A., Facchetti, F., Borlenghi, E., Cattaneo, C., Capucci, M. A., Ungari, M., et al. (2007). Fatal hemophagocytic syndrome related to active human herpesvirus-8/Kaposi sarcoma-associated herpesvirus infection in human immunodeficiency virus-negative, non-transplant patients without related malignancies. *European Journal of Haematology, 78*, 361–364.
52. Lerolle, N., Laanani, M., Riviere, S., Galicier, L., Coppo, P., Meynard, J. L., et al. (2016). Diversity and combinations of infectious agents in 38 adults with an infection-triggered reactive haemophagocytic syndrome: A multicenter study. *Clinical Microbiology and Infection, 22*, 268.e261–268.e268.
53. Cattaneo, C., Oberti, M., Skert, C., Passi, A., Farina, M., Re, A., et al. (2016). Adult onset hemophagocytic lymphohistiocytosis prognosis is affected by underlying disease and coexisting viral infection: Analysis of a single institution series of 35 patients. *Hematological Oncology*.
54. Devecioglu, O., Anak, S., Atay, D., Aktan, P., Devecioglu, E., Ozalp, B., et al. (2009). Pediatric acute lymphoblastic leukemia complicated by secondary hemophagocytic lymphohistiocytosis. *Pediatric Blood & Cancer, 53*, 491–492.
55. Marabelle, A., Bergeron, C., Billaud, G., Mekki, Y., & Girard, S. (2010). Hemophagocytic syndrome revealing primary HHV-6 infection. *The Journal of Pediatrics, 157*, 511.
56. Schaballie, H., Renard, M., Vermylen, C., Scheers, I., Revencu, N., Regal, L., et al. (2013). Misdiagnosis as asphyxiating thoracic dystrophy and CMV-associated haemophagocytic lymphohistiocytosis in Shwachman-Diamond syndrome. *European Journal of Pediatrics, 172*, 613–622.
57. Zondag, T. C., Rokx, C., van Lom, K., van den Berg, A. R., Sonneveld, P., Dik, W. A., et al. (2016). Cytokine and viral load kinetics in human herpesvirus 8-associated multicentric Castleman's disease complicated by hemophagocytic lymphohistiocytosis. *International Journal of Hematology, 103*, 469–472.
58. van der Werff ten Bosch, J. E., Kollen, W. J., Ball, L. M., Brinkman, D. M., Vossen, A. C., Lankester, A. C., et al. (2009). Atypical varicella zoster infection associated with hemophagocytic lymphohistiocytosis. *Pediatric Blood & Cancer, 53*, 226–228.
59. Waddell, B., Belcher, C., & Willey, E. (2017). Cytomegalovirus induced hemophagocytic lymphocytic histiocytosis in two pediatric patients with acute lymphoblastic leukemia. *IDCases, 9*, 116–118.
60. Abdelkefi, A., Ben Jamil, W., Torjman, L., Ladeb, S., Ksouri, H., Lakhal, A., et al. (2009). Hemophagocytic syndrome after hematopoietic stem cell transplantation: a prospective observational study. *International Journal of Hematology, 89*, 368–373.
61. Frederiksen, J. K., & Ross, C. W. (2014). Cytomegalovirus-associated hemophagocytic lymphohistiocytosis in a patient with myasthenia gravis treated with azathioprine. *Blood, 123*, 2290.
62. Mustafa Ali, M., Ruano Mendez, A. L., & Carraway, H. E. (2017). Hemophagocytic lymphohistiocytosis in a patient with Hodgkin lymphoma and concurrent EBV, CMV, and

Candida infections. *Journal of Investigative Medicine High Impact Case Reports, 5*, 2324709616684514.

63. Garcia-Montoya, L., Saenz-Tenorio, C. N., Janta, I., Menarguez, J., Lopez-Longo, F. J., Monteagudo, I., et al. (2017). Hemophagocytic lymphohistiocytosis in a patient with Sjogren's syndrome: Case report and review. *Rheumatology International, 37*, 663–669.
64. Mun, J. I., Shin, S. J., Yu, B. H., Koo, J. H., Kim, D. H., Lee, K. M., et al. (2013). A case of hemophagocytic syndrome in a patient with fulminant ulcerative colitis superinfected by cytomegalovirus. *The Korean Journal of Internal Medicine, 28*, 352–355.
65. Gur, G., Cakar, N., Uncu, N., Ayar, G., Basaran, O., Taktak, A., et al. (2015). Hemophagocytic lymphohistiocytosis secondary to Varicella zoster infection in a child with Henoch-Schonlein purpura. *Pediatrics International, 57*, e37–e38.
66. Presti, M. A., Costantino, G., Della Torre, A., Belvedere, A., Cascio, A., & Fries, W. (2011). Severe CMV-related pneumonia complicated by the hemophagocytic lymphohistiocytic (HLH) syndrome in quiescent Crohn's colitis: Harmful cure? *Inflammatory Bowel Diseases, 17*, E145–E146.
67. Sakamoto, O., Ando, M., Yoshimatsu, S., Kohrogi, H., Suga, M., & Ando, M. (2002). Systemic lupus erythematosus complicated by cytomegalovirus-induced hemophagocytic syndrome and colitis. *Internal Medicine, 41*, 151–155.
68. Tanaka, Y., Seo, R., Nagai, Y., Mori, M., Togami, K., Fujita, H., et al. (2008). Systemic lupus erythematosus complicated by cytomegalovirus-induced hemophagocytic syndrome and pneumonia. *Nihon Rinshō Men'eki Gakkai Kaishi, 31*, 71–75.
69. Luppi, M., Barozzi, P., Rasini, V., Riva, G., Re, A., Rossi, G., et al. (2002). Severe pancytopenia and hemophagocytosis after HHV-8 primary infection in a renal transplant patient successfully treated with foscarnet. *Transplantation, 74*, 131–132.
70. Asci, G., Toz, H., Ozkahya, M., Cagirgan, S., Duman, S., Sezis, M., et al. (2006). High-dose immunoglobulin therapy in renal transplant recipients with hemophagocytic histiocytic syndrome. *Journal of Nephrology, 19*, 322–326.
71. Dharancy, S., Crombe, V., Copin, M. C., Boleslawski, E., Bocket, L., Declerck, N., et al. (2008). Fatal hemophagocytic syndrome related to human herpesvirus-6 reinfection following liver transplantation: A case report. *Transplantation Proceedings, 40*, 3791–3793.
72. Vijgen, S., Wyss, C., Meylan, P., Bisig, B., Letovanec, I., Manuel, O., et al. (2016). Fatal outcome of multiple clinical presentations of human herpesvirus 8-related disease after solid organ transplantation. *Transplantation, 100*, 134–140.
73. Mularoni, A., Gallo, A., Riva, G., Barozzi, P., Miele, M., Cardinale, G., et al. (2017). Successful treatment of kaposi sarcoma-associated herpesvirus inflammatory cytokine syndrome after kidney-liver transplant: Correlations with the human herpesvirus 8 miRNome and specific T cell response. *American Journal of Transplantation*.
74. Low, P., Neipel, F., Rascu, A., Steininger, H., Manger, B., Fleckenstein, B., et al. (1998). Suppression of HHV-8 viremia by foscarnet in an HIV-infected patient with Kaposi's sarcoma and HHV-8 associated hemophagocytic syndrome. *European Journal of Medical Research, 3*, 461–464.
75. Pastore, R. D., Chadburn, A., Kripas, C., & Schattner, E. J. (2000). Novel association of haemophagocytic syndrome with Kaposi's sarcoma-associated herpesvirus-related primary effusion lymphoma. *British Journal of Haematology, 111*, 1112–1115.
76. Fardet, L., Blum, L., Kerob, D., Agbalika, F., Galicier, L., Dupuy, A., et al. (2003). Human herpesvirus 8-associated hemophagocytic lymphohistiocytosis in human immunodeficiency virus-infected patients. *Clinical Infectious Diseases, 37*, 285–291.
77. Yates, J. A., Zakai, N. A., Griffith, R. C., Wing, E. J., & Schiffman, F. J. (2007). Multicentric Castleman disease, Kaposi sarcoma, hemophagocytic syndrome, and a novel HHV8-lymphoproliferative disorder. *The AIDS Reader, 17*, 596–598, 601.
78. Uneda, S., Murata, S., Sonoki, T., Matsuoka, H., & Nakakuma, H. (2009). Successful treatment with liposomal doxorubicin for widespread Kaposi's sarcoma and human herpesvirus-8 related severe hemophagocytic syndrome in a patient with acquired immunodeficiency syndrome. *International Journal of Hematology, 89*, 195–200.

79. Uldrick, T. S., Wang, V., O'Mahony, D., Aleman, K., Wyvill, K. M., Marshall, V., et al. (2010). An interleukin-6-related systemic inflammatory syndrome in patients co-infected with Kaposi sarcoma-associated herpesvirus and HIV but without Multicentric Castleman disease. *Clinical Infectious Diseases, 51*, 350–358.
80. Ohkuma, K., Saraya, T., Sada, M., & Kawai, S. (2013). Evidence for cytomegalovirus-induced haemophagocytic syndrome in a young patient with AIDS. *BML Case Reports, 2013*.
81. Polizzotto, M. N., Uldrick, T. S., Wyvill, K. M., Aleman, K., Marshall, V., Wang, V., et al. (2016). Clinical features and outcomes of patients with symptomatic Kaposi Sarcoma Herpesvirus (KSHV)-associated inflammation: Prospective characterization of KSHV Inflammatory Cytokine Syndrome (KICS). *Clinical Infectious Diseases, 62*, 730–738.

Cytokine Storm Syndromes Associated with Epstein–Barr Virus

Katherine C. Verbist and Kim E. Nichols

Introduction

The mammalian immune system incorporates an array of strategies to defend against infectious pathogens such as viruses. Under certain circumstances, however, the virus and host can adopt counter-regulatory defense strategies leading to a situation in which both exist in equilibrium. Epstein–Barr virus (EBV), which can establish lifelong infection in humans, is a perfect example of this relationship. Human beings are the only known natural host for EBV, and it is estimated that more than 90% of the human population is seropositive [1]. This observation highlights the strict dependency of the virus on humans for its existence and propagation.

Surveillance mediated by effector cells of the immune system is critically important in controlling EBV, with most individuals developing an asymptomatic infection in childhood and enduring a subclinical latent infection for life. This host–virus homeostasis requires a very delicate balance between the necessity of EBV to preserve the host and the host to protect itself from the virus. As might be expected, when this balance is perturbed, EBV-driven diseases can occur. Among these are included nonmalignant lymphoproliferative disorders such as acute infectious mononucleosis (IM), chronic active EBV infection (CAEBV), hemophagocytic lymphohistiocytosis (HLH) [1], and posttransplant lymphoproliferative disorder (PTLD) [2], as well as a wide spectrum of hematopoietic and solid malignancies [3–6] (Table 1).

In many of these diseases, feed-forward loops involving cells of the innate and adaptive immune systems and the cytokines that they produce drive development of associated cytokine storm syndromes (CSS), which at times may be associated with significant morbidity and mortality. In this chapter, we review the processes of EBV infection and

K. C. Verbist · K. E. Nichols (✉)
Department of Oncology, St. Jude Children's Research Hospital, Memphis, TN, USA
e-mail: kim.nichols@stjude.org

R. Q. Cron, E. M. Behrens (eds.), *Cytokine Storm Syndrome*,
https://doi.org/10.1007/978-3-030-22094-5_15

Table 1 EBV-driven disease states

	Reference
Nonmalignant	
Infectious mononucleosis (IM)	Epstein et al. [7]
Chronic active EBV (CAEBV)	Kimura et al. [8]
Hemophagocytic lymphohistiocytosis (HLH)	Sullivan et al. [9] and Imashuku et al. [10]
Posttransplant lymphoproliferative disorder (PTLD)	Paya et al. [11]
Malignant	
Hematopoietic: Burkitt Lymphoma, Hodgkin Lymphoma, Diffuse Large B cell Lymphoma, rare NK or T cell Lymphomas, and HIV-Associated Lymphomas	Weiss et al. [3] and Jones et al. [4]
Solid: Nasopharyngeal Carcinoma and EBV-Associated Gastric Cancer	Neparidze and Lacy [6]

the host anti-EBV immune response. We also discuss several EBV-induced CSS and describe the roles played by various immune cells and their cytokines in driving disease pathogenesis.

Epstein–Barr Virus Infection and the Anti-EBV Immune Response

EBV is a double-stranded DNA virus belonging to the lymphocryptovirus (LCV) genus of the γ-herpesvirus subfamily. Divergence in DNA sequence divides EBV into two variants referred to as type 1 and type 2 [12]. Both variants are found widely distributed throughout the world and exist in together in most populations. There appears to be no relationship between these different strains and specific EBV-related diseases [13]. The EBV genome codes for nearly 100 unique proteins that organize the virus into the characteristic nucleoid, icosahedral capsid, and protective envelope derived from host nuclear or cellular membrane [14]. It is these proteins that interact with and elicit responses from the host immune system.

Transmission of EBV occurs via saliva, where early infection occurs in cells of the oral cavity—probably in epithelial cells as well as B cells present at or just beneath mucosal surfaces (Fig. 1) [15, 16]. Mechanisms of viral entry into B cells and epithelial cells are distinct, though both depend on viral glycoproteins (gp) [16]. To infect B cells, EBV envelope proteins gp350 and 220 [17] bind with high affinity [18] to the complement receptor type 2 (CR2, also known as CD21) on the B cell surface [19], but other viral gp are also likely involved in this process [20]. The fusion between the virus and the host cell membrane is then mediated by the viral proteins gHgL, gB, and gp42 [21–23] which bind to B cell human leukocyte antigen (HLA) class II molecules. Either independently, or as part of a complex with the B cell surface proteins CD19 and CD35 (complement receptor 1; CR1), CR2 signaling activates nuclear factor (NF)-κB and induces production of interleukin (IL)-6 by the infected B cell, a cytokine that drives B cell survival, expansion, and maturation

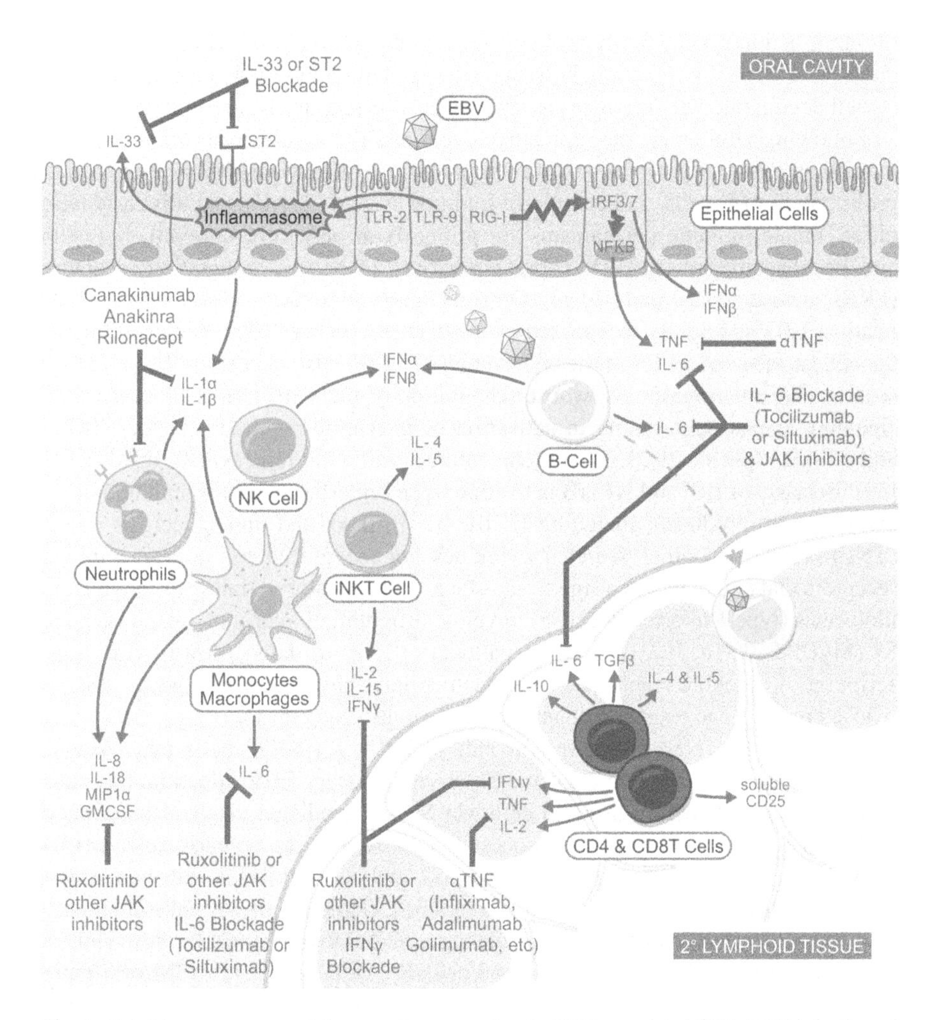

Fig. 1 Cytokine responses and therapeutic approaches to EBV-associated CSS. EBV infection of epithelial cells in the oral cavity is recognized by pattern recognition receptors and induces type I interferon production and NF-κB activation. Subsequent inflammasome activation results production of IL-33 and IL-1 family members. Viral infection moves to the B cell that induces a type I interferon antiviral state and IL-6 production. Innate cell recognition of infected cells causes IL-18 and chemokine production by neutrophils, IL-1 and IL-6 production by monocytes and macrophages, type 1 interferon production by NK cells, and Type I and Type II cytokine production by NK and iNKT cells. Infected B cells move into secondary lymphoid tissues and elicit cytokine production by CD4 and CD8 T cells, including copious amounts of IFNγ, TNF, IL-2 (and its receptor CD25), IL-6, and IL-10. There are numerous points of potential therapeutic intervention, including antibody-mediated neutralization of specific cytokines or their receptors, or inhibition of cytokine-induced intracellular signaling, as shown

[24–26]. It is not known whether cytokine production is initiated as a result of a gp42-HLA class II interaction; however, since signaling via HLA class II molecules is well documented under other circumstances, this remains a possibility [27].

Following viral entry, the replicative cycle of EBV results in the expression of more than 80 viral proteins involved in assembly of new virions, activation and proliferation of B cells, and evasion of immune system [28]. During this early lytic phase, innate immune mechanisms are primarily at play. Infected epithelial cells and B cells targeted by EBV express a variety of pattern recognition receptors (PRRs) whose downstream signaling results in an environment hostile to virus replication. EBV is initially recognized by toll-like receptors (TLRs), namely TLR2 at the cell surface and TLR9 at the endosome [29]. EBV also triggers RNA and DNA sensors and inflammasomes, whose recognition of the virus induces a cascade of signaling events culminating in activation of interferon-regulatory factors (IRFs) and NF-κB. Specifically, EBV proteins interact with (and repress) IRF3 and IRF 7 [30]. Because of IRF and NF-κB activation, type I interferons and proinflammatory cytokines are produced, including IL-1α, IL-1β, IL-6, and tumor necrosis factor (TNF). These and other secreted cytokines are recognized by surface cytokine receptors that signal through signal transducer and activator of transcription (STAT) molecules (type 1 interferons, IL-6), myeloid differentiation primary response gene 88 (MyD88; IL-1α, IL-1β) or other pathways (TNF) to induce downstream transcription of cytokine response genes, such as interferon-stimulated genes (ISGs), whose products can exert direct antiviral functions [28] (Fig. 1).

Natural killer (NK) and invariant natural killer T (iNKT) cells may also be important points of immune control early during primary EBV infection. NK cells recognize and kill virally infected cells via MHC-I mediated mechanisms though EBV proteins interfere with this process [31]. NK cells are potently activated by EBV-induced ligands on infected B cells [32], and NK and iNKT cells can inhibit EBV-mediated transformation of B cells [33–35]. Depletion of NK cells in humanized mouse models also exacerbates the signs of EBV infection [36, 37]. Thus, a failure of early viral control by NK and/or iNKT cells may lead to IM or a more fulminant primary viral infection. NK and iNKT cells respond to EBV infection rapidly and copiously produce cytokines such as IL-2, IL-15, and interferon (IFN) γ, which further drive NK cell activation and enhance the expansion, activation, and cytotoxic functions of CD8 T lymphocytes.

Once inside B cells, EBV quickly establishes a latent infection that is dependent on the expression of latent viral proteins. The EBV genome is replicated once and only once per cell cycle, leading to a "need" on the part of the virus to induce B cell immortalization for its continued replication. In this way, EBV DNA is maintained in latently infected B cells that undergo proliferation. EBNA2 and LMP1 are two viral proteins required for B cell immortalization [38, 39]. These two proteins, especially LMP1, activate NF-kB and c-Jun N-terminal kinase pathways, resulting in production of survival and growth factors, such as IL-6 and IL-8, by the B cells to stimulate differentiation and proliferation [39, 40]. The viral latency EBNA3 proteins regulate expression of certain cellular genes, including specific receptors such as CD28, CD19, CD21, CD23, and CD30; T cell costimulatory molecules

CD80/CD86; adhesion molecules such as intercellular adhesion molecule-1, leukocyte factor antigen-1, and leukocyte factor antigen-3; and a member of the src oncogene family, c-fgr [41]. Thus, as latent infection is established, B cells are simultaneously becoming lymphoblasts, functionally differentiating, secreting cytokines themselves, and disseminating into the circulation and secondary lymphoid organs (spleen, lymph nodes, and bone marrow).

Although the role for humoral immunity is less clear, cellular immunity is indisputably required for control of EBV. This is made evident by the fact that individuals with defects in T cell development or function exhibit greatly exacerbated disease upon infection [42, 43]. In acute IM, the absolute number of total peripheral blood T cells increases 5–10-fold above that of asymptomatic carriers, and during lytic infection up to 50% of the total CD8 T cell pool may be specific for EBV epitopes [42, 44]. During latent infection, there may be a tenfold reduction in the magnitude of this response, but the EBV-specific CD8 T cell response remains elevated [44]. EBV-specific CD4 T cells expand to a lesser degree, accounting for 0.1% (in latent infections) to 1% (in IM patients) of all circulating CD4 T cells [45, 46]. While CD4 T cell responses may be less pronounced than CD8 T cell responses, CD4 T cells exhibit specificity for a broader range of viral antigens and may be very important to EBV control, demonstrated by the observation that human immunodeficiency virus (HIV)+ individuals with low CD4 T cell counts have high EBV viral loads and are prone to develop EBV-associated B cell lymphomas [45, 47]. Consistently, improved outcomes have been noted in PTLD patients receiving both EBV-specific CD4 and CD8 T cells, as opposed to CD8 T cells alone [48, 49].

Normally, this acute expansion and activation of CD8 and CD4 T cells resolves, and a symbiosis between EBV and the infected host is reached. Remarkably, despite the robust T cell response, virus-infected B cells are never completely eradicated, even in immunocompetent individuals. Approximately 1 in 10,000 to 1 in 100,000 memory B cells remain infected throughout the life of the host. Under normal circumstances, this frequency of EBV-infected B cells is stable with an estimated total body load of 10^4–10^7 (mean 0.5×10^6) EBV-positive memory B cells [50–53]. It is from this pool of infected memory B cells that the virus periodically reactivates leading to the release of infectious virions and ultimately to reinfection of additional B cells.

Acute and Chronic Active EBV (CAEBV) Infection

For reasons that are not well understood, children infected with EBV do not generally develop symptoms. If infection occurs in an adolescent or adult, however, IM may develop, with peak incidences being between 17 and 25 years of age [7]. IM is traditionally characterized by fever, pharyngitis, generalized lymphadenopathy, and splenomegaly, with demonstration of an atypical lymphocytosis in the peripheral blood (defined as >50% lymphocytes and >10% atypical forms in the differential) and presence of a positive heterophile antibody titer. Most early symptoms are nonspecific and include myalgias, fatigue, and malaise. Fever occurs and may reach peaks of up to 40 ° C in up to a third of patients. While these are the most common and sometimes the only symptoms, other findings include cough, rashes, headaches, and ocular manifestations such as photophobia, ocular muscle pain, dry eyes, and conjunctivitis. The symptoms of IM develop gradually and usually resolve in 1–2 weeks; however, fatigue may linger longer. IM typically occurs only once in the lifetime of the host [7].

Rarely, EBV can cause CAEBV, a condition characterized by unrelenting active infection in apparently immunocompetent individuals. CAEBV is most common in older children or young adults and is characterized by persistent or recurring IM-like symptoms, often with hepatosplenomegaly and unusual anti-EBV antibody patterns. The diagnostic criteria consist of: (1) evidence of severe, progressive EBV illness beginning as a primary EBV infection and lasting more than 6 months with abnormally high anti-EBV titers (anti-VCA IgG > 1:5120 and anti EA >1:640), and/or abnormally low anti EBNA (<1:2); (2) histologic evidence of major organ involvement; and (3) increased amounts of EBV in peripheral blood and/or affected tissues. CAEBV may be lethal, with mortality rates around 30–50% within 5 years of diagnosis [8]. Most deaths are due to the development of an HLH-like syndrome, which is driven by hypercytokinemia (a CSS).

The etiology of CAEBV is unknown. Early studies suggested that the disease might be due to mutant strains of EBV that result only in lytic infection [54]. However, it was subsequently shown that the same lytic strain was present in control individuals [55]. Observations also favor a potential genetic cause. Towards this end, CAEBV is rare in the USA but relatively common in Asia and South America [56], indicating a possible ethnic bias. Second, CTL and NK cell activity are reduced when cells are obtained from CAEBV patients and their parents, suggesting a heritable nature to the cellular defect [57]. In support of this notion, Katano et al. recently identified biallelic inactivating mutations in the gene encoding perforin in a patient with CAEBV [57]. In CAEBV, high levels of EBV DNA are found in CD4 T cells and NK cells, but not in B cells [58], as is seen in typical EBV infection. The significance of this finding and the mechanisms underlying the preferential infection of T and NK cells remain to be elucidated. CAEBV clinically overlaps with EBV-HLH and further investigations are needed to differentiate between these disorders and determine the role of EBV in disease pathogenesis.

EBV–Associated Hemophagocytic Lymphohistiocytosis (EBV-HLH)

Sometimes referred to as fulminant infectious mononucleosis (FIM), EBV-associated HLH is a classic CSS. Hemophagocytic syndrome was first described in 1939 by Scott and Robb-Smith and represents a spectrum of diseases characterized by the aberrant activation of NK cells and macrophages resulting in hypercytokinemia, which in turn leads to cellular damage, organ dysfunction, and death [10, 59].

Classically, HLH has been categorized as "primary" (familial) or "secondary" (sporadic). In primary HLH, patients harbor germline mutations that disrupt the *PRF1*, *UNC13D*, *STX11*, *STXBP2*, or *RAB27A* genes. In contrast, patients with secondary HLH lack mutations in these genes. In either scenario, various infectious can trigger disease with EBV being a major culprit [9, 60]. Several additional primary immune deficiencies (PID) can lead to HLH, with many of these caused by mutations in genes associated with lymphocyte survival, activation, or cytotoxic function (see below, Primary Immune Deficiencies Associated with EBV-HLH) [61]. In adults, EBV-HLH often develops in association with NK or T cell lymphomas [62, 63].

In 1991 Henter et al., put forth diagnostic criteria for HLH that include fever, splenomegaly, cytopenia, hypertriglyceridemia, or hypofibrinogenemia, and hemophagocytosis in secondary lymphoid organs [64]. These criteria have been revised over time to also include low or absent NK cell activity, high serum ferritin levels, and high serum levels of soluble CD25 (the high affinity receptor for IL-2; also known as IL-2Ra) [64]. For EBV-HLH, the diagnostic criteria are the same, but the individual must also be EBV positive. As many patients with EBV-HLH are immunocompromised, serological responses to the virus may be greatly reduced or even absent. Accordingly, measurement of anti-EBV titers is not generally helpful in making the diagnosis [65]. Rather, EBV positivity is best determined by quantitation of EBV DNA via polymerase chain reaction of the blood or affected tissues.

The course of EBV-HLH can be variable, ranging from multiorgan failure developing over hours to persistent or recurring IM-like symptoms lasting for months. An atypical lymphocytosis is usually present during early stages, but patients subsequently develop cytopenias, hepatic dysfunction, central nervous system involvement, and ultimately multiorgan failure. In EBV-HLH, organs are extensively infiltrated by activated CD8 T cells and macrophages [66]. Early in the disease, the bone marrow may be hypercellular; however, the marrow later becomes hypocellular with increasing numbers of infiltrating histiocytes [67, 68] often with evidence of hemophagocytosis. Liver biopsies reveal portal infiltration with lymphocytes and histiocytes, with occasional erythrophagocytosis [69]. Perivascular mononuclear cell infiltrates of the brain, mononuclear cell myocarditis, and interstitial nephritis are also rarely observed [66–68, 70].

Primary Immune Deficiencies Associated with EBV-HLH

Several PID have been described in which EBV-induced lymphoproliferation is a common feature, and ruling out or confirming whether a patient has one of these PIDs is an important objective when EBV-HLH is suspected. PIDs are generally caused by monogenetic mutations that impair immune cell development, differentiation, proliferation, apoptosis, and/or function, leaving affected individuals more

Table 2 Primary immune deficiencies associated with EBV-HLH

Disease	EBV-Induced Lymphoproliferation	Gene (Location)	Deficient protein	Cellular phenotypes
XLP1	Increased susceptibility to EBV progressing to HLH (~55% of patients; lethal in ~65%) and lymphomas	*SH2D1A* (Xq25)	SAP SLAM-associated protein	• Loss of iNKT cells • Decrease in optimal T-B cell interactions and killing • Impaired T cell RICD • Th1-cytokine skewing
XLP2	Autoimmune manifestations and EBV-HLH in ~76% of patients without lymphomas	*BIRC4 / XIAP* (Xq25)	XIAP X-linked inhibitor of apoptosis protein	• Increased apoptosis of CD8 T cells • Increased cytokine production upon TNF stim. • Shift from apoptotic to necroptotic cell death
ITK deficiency	EBV-HLH-like symptoms with progression to Hodgkin lymphoma	*ITK* (5q33.3)	ITK IL-2 inducible T- cell kinase	• Compromised TCR activation • Reduced iNKT cells • Impaired Th2 with robust Th1 cytokine production • Impaired Th17 differentiation
XMEN disease	Indolent EBV-driven lymphoproliferation progressing to lymphoma with rare HLH	*MAGT1* (q21.1)	MAGT1 Magnesium transporter 1	• Decreased optimal TCR activation • Decreased NKG2D expression on NK and CD8 T cells

susceptible to EBV. Studies of these individuals provide important new insights into the molecular mechanisms regulating the interaction of EBV-infected B cells with T, NK, and other cells of the immune system (Table 2). Below, we discuss several PIDs in which EBV is responsible for inducing the most severe manifestations.

X-Linked Lymphoproliferative Disease Type 1 (XLP1; SAP Deficiency)

In 1974 and 1975, Purtilo described a family in which 6 of 18 boys died of a lymphoproliferative disease associated with EBV infection [71, 72]. Initially named Duncan's disease after this family, this condition is now more commonly referred to as XLP1. As the name implies, the genetic mutation responsible for disease is located on the X chromosome, and XLP1 affects approximately 1 out of every million male individuals [73]. Notably, affected boys and young men exhibit increased susceptibility to EBV but not to other childhood viruses. Three phenotypes are common and include [1] EBV-HLH, [2] malignant lymphoma, and [3] dysgammaglobulinemia. Less common manifestations include vasculitis, pulmonary lymphomatoid granulomatosis, and aplastic anemia [73, 74]. EBV-HLH affects 45–70% of XLP1 patients [73, 75, 76] and is the most lethal of the XLP phenotypes with up to 65% of patients dying, even with contemporary therapies [75]. The median age of EBV-HLH onset is 5 years, but the condition has occurred as late as 40 years [73]. The pathology and clinical findings are identical to those seen in patients with EBV-HLH but without XLP.

The gene responsible for XLP1 is *SH2D1A,* which encodes a small cytoplasmic adaptor protein known as signaling lymphocytic activation molecule (SLAM)-associated protein (SAP) [77–79]. The diagnosis of XLP1 can be strongly suspected based on reduced or lack of SAP protein expression by flow cytometric assessment of peripheral blood lymphocytes and confirmed by identification of inactivating mutations in *SH2D1A* [80]. SAP is an SH2-domain containing adaptor molecule that regulates intracellular signaling downstream of the SLAM family of receptors, which are broadly expressed on hematopoietic cells. Studies of SAP-deficient cells have revealed critical roles for this protein in the promotion of stable interactions between T and B cells, which are required for the recognition and killing of B cells by CD8 T cells and the induction of B cell dependent humoral responses by CD4 T follicular helper cells [78, 81, 82]. Attenuation of these processes in XLP1 patients presumably allows for the outgrowth of EBV-infected B cells and underlies the susceptibility to EBV that is observed in XLP1. Additionally, T cells from XLP patients exhibit impaired restimulation-induced cell death (RICD), which may lead to excessive accumulation of effector cells [83]. The exclusive susceptibility of these patients to EBV and not to other pathogens has revealed a potentially unique role for SAP in host defense against EBV.

Mice deficient for *Sh2d1a* have been generated, and studies of these animals have revealed a Th1 skewing in cytokine production upon infection with Lymphocytic choriomeningitis virus (LCMV) [84]. This abnormal response is not limited to viral infection, because exposure to *T. gondii* also results in elevated IFN-γ production from Sap-deficient T cells. Even lymphocytes from uninfected mice exhibit increased Th1 and decreased Th2 cytokine production (especially IL-4) [84]. iNKT cells, which play important immunoregulatory roles and produce copious amounts of cytokines [85], are also lacking in *Sap*-deficient mice [78]. Collectively, these observations suggest that poor control of EBV-infected B cells along with dysregulation of the cytokine milieu towards a more proinflammatory state, likely contribute to the pathogenesis of EBV-HLH in patients with XLP1.

X-Linked Lymphoproliferative Disease, Type 2 (XLP2; XIAP Deficiency)

In 2006, a second X-linked disorder associated with EBV-HLH was described as caused by inactivating mutations in the *BIRC4* gene (now known as *XIAP*) [86]. The similarity in EBV sensitivity between patients with SAP and XIAP deficiency initially led to the naming of this condition as XLP2. Over time, however, it has become apparent that XIAP deficient patients exhibit many differences in clinical manifestations compared to SAP deficient patients [87–89]. For example, XIAP deficient males are prone to developing colitis, uveitis, and other autoimmune manifestations, which are not observed in SAP-deficient patients. Conversely, SAP deficient patients develop lymphoma, a phenomenon not observed in XIAP deficient patients. A comparison of patients with SAP and XIAP deficiency showed that HLH was more common in XIAP (76%) versus SAP deficiency (55%), but more likely to be fatal in SAP versus XIAP deficiency [89]. Interestingly, HLH may be triggered in XIAP deficient patients by a broad array of infectious agents, not just by EBV. As with SAP deficiency, the diagnosis of XIAP deficiency is suspected based on reduced or lack of XIAP expression by flow cytometric assessment of lymphocytes and can be confirmed by identification of inactivating mutations in *XIAP* [80].

It remains unclear how loss of XIAP expression leads to the manifestations of XLP2. XIAP-deficient CD8 T cells exhibit normal cytotoxic activity so HLH is not likely to be caused by qualitative T cell defects. One function of XIAP is to inhibit apoptosis. Accordingly, T cells lacking XIAP exhibit increased apoptosis following T cell receptor engagement [86]. It has been suggested that this excessive death of cytotoxic T lymphocytes may lead to ineffective control of EBV infection and development of EBV-HLH. Studies in animals, though, have revealed that XIAP (and other IAP family members) repress inflammatory cytokine production. XIAP negatively regulates cytokine production, including TNF to disrupt myelopoiesis, and limits cytokine production in a dying cell by preventing RIP kinase activity [90]. Furthermore, treatment of mice with bivalent IAP antagonists results in a significant increase in IL-1β production by TNF-primed macrophages [91]. Deletion of XIAP in bone marrow derived macrophages also shifts their cell death pathways away from the anti-inflammatory apoptotic cell death towards the more proinflammatory necroptotic cell death pathways and primes cells for spontaneous inflammasome assembly [91, 92]. Thus, it remains possible that in XIAP deficiency, EBV infection drives B cells to undergo a more inflammatory cell death pathway, leading to excessive cytokine production by the cells that phagocytose them.

IL-2 Inducible T Cell Kinase (ITK) Deficiency

ITK encodes the IL-2 inducible T cell kinase, a non-receptor tyrosine kinase expressed by hematopoietic cells that is involved in proximal TCR signaling via its regulation of phospholipase C-γ phosphorylation [82]. Biallelic inactivating mutations in *ITK* cause an interesting autosomal recessive PID associated with development of EBV-positive Hodgkin lymphoma (HL). The disorder was originally identified in two female siblings from a consanguineous family who developed EBV-HLH-like symptoms with progression to HL [93]. Including this original report, eleven patients from eight unrelated families have been identified, 21 of whom presented with EBV viremia and lymphoproliferation that often progressed to HL [82]. Patients also exhibited CD4 T cell lymphopenia, reduced iNKT cells, and progressive hypogammaglobulinemia. Although most of the patients responded favorably to HL-directed chemotherapy, many experienced a relapse. Of the 13 reported cases, only 5 (42%) are alive at the time of reporting, with 2 having received allogeneic hematopoietic stem cell transplantation (HSCT) [82]. Collectively, these reports suggest that genetic defects that impair ITK function are an important cause of HL, especially when it occurs in young children and in the setting of EBV infection. Although the mechanisms underlying the sensitivity to EBV and development of HL remain to be determined, it is possible that ITK deficiency leads to disease by compromising TCR activation, which is further compounded by a reduction of iNKT cells, leading to an outgrowth of EBV-infected B cells.

Here, too, animal models of disease have yielded insights into how the loss of *ITK* gene function might contribute to disease [94]. ITK-deficient patients have high levels of EBV viremia, which may result from poor T cell activation and expansion and subsequent inability to control infection. However, analysis of T cells from *Itk−/−* mice reveals only a modest impact of Itk deficiency on naïve T cell activation [95]. Most strikingly, *Itk−/−* mice are resistant to the development of allergic asthma due to impaired secretion of Th2 and strong secretion of Th1 cytokines [96, 97]. T cells lacking ITK also display interesting selective alterations in cytokine production that highlight the complex nature of the signaling pathways that regulate T cell effector responses. One example is the role of Itk in the regulation of cytokine production by Th17 cells (a CD4 effector T cell lineage that differentiates in response to IL-6 and transforming growth factor (TGF)β and expresses the proinflammatory cytokines IL-17A, IL-17F, IL-21, and IL-22), as Itk deficient T cells show impaired differentiation into this lineage [98]. Although it is not known how or whether IL-17 plays a role in the CSS of EBV-HLH, these studies demonstrate the importance of ITK in regulating cytokine responses in general and highlight the impressive need for further research into the role of many of these cytokines in disease pathogenesis.

X-Linked Immunodeficiency with Magnesium Defect, EBV Infection and Neoplasia (XMEN) Disease

XMEN disease was discovered in 2011, and only seven cases have thus far been published [99] so the natural history of the disease is currently uncertain. As the inheritance is X-linked, all cases have been males, and prevalence of XMEN disease and the frequency of female carriers are unknown but expected to be uncommon [99]. Female carriers exhibit skewed lyonization, favoring expression of the wild-type X chromosome [100]. All patients thus far described exhibited persistent high levels of EBV, reduced CD4 T cell counts, and splenomegaly. B cell lymphoproliferation is the most common cause of morbidity and mortality with four post-pubertal males developing EBV-positive lymphomas [99, 100]. XMEN patients may also be more susceptible to sinopulmonary and ear infections, viral pneumonias, and other viral infections, but these are generally mild or infrequent. Indeed, all XMEN patients exhibit normal growth and development, and most did not even come to medical attention until an EBV-associated lymphoproliferative disorder developed.

XMEN is caused by loss of function mutations in the gene *MAGT1* (*Magnesium Transporter 1* located at Xq21.1), a ubiquitously expressed magnesium cation transporter. MAGT1 is a critical regulator of the levels of free (ionized) magnesium in cells, and although this pool is generally less that 5% of all intracellular magnesium [101], discovery of this disease has revealed an important and nonredundant role for free magnesium in T cells. A transient T cell receptor-mediated Mg^{2+} influx is a signaling step that enhances the efficiency of T cell activation [99]. Furthermore, influx of free magnesium is required for the upregulation of expression of the natural killer (NK) stimulatory receptor "natural-killer group 2, member D" (NKG2D) on NK and CD8 T cells, which is important for control of EBV, although exact mechanisms are not yet known. XMEN patients show defective expression of NKG2D, and remarkably supplementation with magnesium $_L$-threonate resulted in an increase of NKG2D on endogenous NK and CD8 T cells and enhanced control of EBV [102]. Treatment depends on the individual clinical features but is likely to include immunoglobulin replacement therapy and antibiotic prophylaxis (similar to other humoral PIDs), antiviral prophylaxis, chemotherapy for malignancy, rituximab, and allogeneic hematopoietic stem cell transplantation. Currently, there is no animal model for XMEN disease but generation of one would likely rapidly advance insights into the cellular mechanisms of disease and specific points of intervention. It remains unclear whether or not patients with XMEN, particularly those with EBV infection, are predisposed to CSS as are XLP patients.

Secondary Immune Deficiencies Associated with EBV-HLH

Individuals with acquired defects of the immune system, particularly those with uncontrolled human immunodeficiency virus (HIV) infection or those who have undergone solid organ or allogeneic HSCT, are at increased risk to develop EBV-associated cytokine storm and/or full blown HLH (as well as EBV-associated cancers). Similarly, elderly or malnourished individuals, or patients undergoing chemotherapy, may experience a CSS upon EBV infection. Therefore, a high index of suspicion is warranted in any immunosuppressed patient with signs and symptoms of EBV infection.

EBV Infection in Individuals with HIV

EBV infection is generally established prior to HIV infection, and EBV-infected B cells are kept in check by T cells. As HIV specifically infects and kills T cells, patients with HIV are at risk of losing control of EBV infection. It has been observed, for example, that patients with poorly controlled HIV infection have at least 10 times as many EBV-infected B cells in the circulation as healthy persons, and T cells from these HIV+ patients suppress EBV-infected B cells less effectively, leading to increased viral shedding of EBV in oropharyngeal tissue [103]. The onset of EBV-associated CSS in HIV-infected individuals generally begins with symptoms much like IM, including fever, pharyngitis, and lymphadenopathy. If left untreated, however, EBV infection can rapidly progress. Corticosteroids are often used to limit the detrimental effects of EBV-induced hyperinflammation, such as airway obstruction due to enlarged tonsils and lymph nodes, thrombocytopenia, and hemolytic anemia. Antiviral agents such as acyclovir and/or ganciclovir are also often employed to limit viral load and shedding [104, 105].

EBV-PTLD

Recipients of allogeneic HSCT and solid organ transplantation are susceptible to EBV infection or reactivation, which may occur naturally or be transmitted from the donor graft, and result in the development of nonmalignant or malignant PTLD. PTLD may present with a wide spectrum of clinical manifestations, including fever, lymphadenopathy, hepatitis, lymphoid interstitial pneumonitis, and meningoencephalitis, or it may present as an IM-like CSS [11]. Surgery, radiotherapy, or both are effective in curing localized disease, but this benefits only a small percentage of patients. Ultimately, successful treatment generally involves controlling EBV-driven B cell proliferation, usually by facilitating development of an appropriate EBV-specific cytotoxic T cell (EBV-CTL) response. The approach most widely used as initial therapy for patients post solid organ transplantation with PTLD is reduction of immunosuppression. In some cases, this is sufficient to control the disease, especially in localized, polymorphic cases or cases that present like IM. In contrast, reduction of immune suppression is rarely successful following allogeneic HSCT because the major issue in these patients is delayed CTL recovery, not suppression of CTL function. Therefore, provision of CTL is an alternative therapeutic intervention. In this approach, T cells are isolated from the patient with PTLD or an unrelated third party, expanded in vitro and activated by exposure to EBV-specific antigens. These cells are then reintroduced. A retrospective study of PID patients who received virus-specific CTL before or after HSCT showed that more than 70% of patients treated therapeutically for active EBV exhibited a complete or partial response to this therapy [106, 107].

Cytokines and Their Roles in EBV-HLH: Fueling the Fire

A hallmark of all forms of EBV-HLH is the cytokine storm. Failure to clear the infected B cells and/or the antigen presenting cells directing the T cell response leads to feed-forward loops that fuel immune activation and uncontrolled cytokine production. It has been observed that EBV-HLH patients whose disease is rapidly fatal often have extremely high serum cytokine levels. For example, IFNγ levels in these individuals can exceed 100 U/ml (normal levels are <1.0 U/ml), and sCD25 levels can exceed 10,000 U/ml (normal is <2000) [108, 109]. Serum levels of sCD25, IFNγ, IL-6, IL-10, and IL-18 (and TNF to a lesser extent) are elevated significantly above those of non-EBV-HLH patients [110], and these cytokines are not merely characteristic of the disease but are also central to the pathophysiology. Levels of IL-4, and IL-2 have also been observed to be significantly elevated in patients with EBV-HLH, but levels of these cytokines are not different from patients with sepsis. Moreover, levels of these cytokines did not change after treatment, so their relative importance is unclear [111].

Both immune effectors (mostly CD8 T cells but also CD4 T cells, NK/iNKT cells, monocytes, and neutrophils) and immune targets (mostly infected B cells but also epithelial cells) secrete copious amounts of cytokines in response to EBV (Fig. 1). Most immediately, signaling pathways downstream of PRRs culminate in the release of IL-1 family member cytokines and their receptors, such as IL-1α and β, IL-18, IL-33, and the IL-33 specific receptor chain ST2. In murine models of HLH, ST2 has been shown to greatly contribute to later T cell hyperactivation [112]. Also within the first 24 h of infection, type 1 interferons (IFNα and β) are released from NK cells and B cells [113]. EBV can also interact with neutrophils and monocytes. EBV binding of the surface of monocytes activates expression of IL-6 but inhibits that of TNF [114, 115]. EBV binding to macrophages also induces IL-8, macrophage inflammatory protein-1α (MIP-1α), and granulocyte macrophage colony stimulating factor (GM-CSF) [116]. During the innate phase of the antiviral response, NK cells and monocytes also produce IL-1α and β [113]. Neutrophils may also contribute to innate IL-1α and β production. Interestingly, EBV can modulate this response in favor of the production of IL-1 receptor antagonist (IL-1Rα) [117], and simultaneously EBV can induce neutrophil apoptosis [118]. Regardless, IL-1 levels are high early after infection, and this IL-1 together with rising levels of IL-2 from expanding T cells drive effector T cells to produce large quantities of IFNγ [113]. Accordingly, serum levels of IL-1α, IL-2, IL-6, and IFNγ are reported to be very high in symptomatic patients with acute or chronic EBV infection [119, 120]. Other reports based on gene expression data indicate that additional cytokines, such as IL-18 and monokine (in addition to IFNγ), are high in EBV-infected tonsils [121].

Circulating virus-specific CD8 and CD4 T cells produce numerous cytokines and are thought to be major contributors to the immune pathogenesis associated with EBV infection. Concomitant with the rise of T cell numbers is an increase in serum levels of the proinflammatory and immunoregulatory cytokines they produce, such as IFNγ, TNF, IL-6, IL-10, and TGFβ [82]. With high levels of circulating IFNγ, it stands to reason that T cells in EBV-infected individuals are highly Th1 polarized (strongly proinflammatory), and indeed there is evidence to this effect [122]. Conversely, Steigerwald-Mullen et al. showed that EBNA1-specific $CD4^+$ T cells preferentially produce Th2 type cytokines (IL-5) in response to antigenic stimulation [123]. These divergences in polarization could reflect differences in the time points in the immune response during which T cells were isolated or analyzed, or they could result from different contexts of in vivo T cell activation and/or in vitro stimulation. As several of the genes involved in PIDs associated with EBV-CSS disrupt iNKT development and/or function, it is tempting to speculate how these cells might influence the cytokine polarization towards Th1 and away from the more anti-inflammatory Th2 responses, but a great deal more research is needed in this area. Overall, the cytokine milieu is subject to constant change throughout the duration of the immune response and from site to site as some cytokines will be produced and/or consumed locally at sites of immune cell infiltration while others will be released systemically and lead to the pathology of CSS.

Treatment of EBV-Associated Disorders: Putting Out the Fire

Once EBV-HLH develops, therapy can be challenging. To be reductive, control of the CSS is largely dependent on immunosuppression and elimination of the trigger. Standard treatment often incorporates corticosteroids, most often administered with the chemotherapeutic agent etoposide [124, 125], while lymphomas have been treated with standard NHL therapy [8]. Cyclosporin A (CsA) with or without corticosteroids has also been shown beneficial [126]. For cases that do not respond to these treatments, T cell-depleting agents such as antithymocyte globulin (ATG) may be employed as a salvage therapy [127]. Because CD52 is broadly expressed on a variety of immune cells, alemtuzumab (anti-CD52) is profoundly immunosuppressive and may also be employed up front or as a salvage treatment [128, 129].

To eliminate the trigger in EBV-HLH, antiviral agents such as acyclovir and/or ganciclovir may be employed [130]; although, it remains unclear to what extent these agents lessen disease. More recently, drug screening has revealed that the use of EBNA1 inhibitors might provide a way to halt EBV infection [131, 132]. Eliminating the reservoir for the virus might also reduce viral load. The anti-CD20 monoclonal antibody rituximab has also been used to eliminate EBV-infected B cells. Rituximab is generally given in combination with chemotherapy or immunosuppressive medications (steroids, CsA), so successful outcomes are also dependent on the prevention and treatment of concurrent infections. EBV can, however, replicate in NK and T cells (as observed in individuals with CAEBV and in some patients with EBV-HLH); thus, lack of response or recurrence after rituximab treatment is a possibility [133]. The only curative therapy for patients with EBV-HLH and underlying HLH genetic mutations or those with a PID is allogeneic HSCT. Despite the use of this approach, the overall survival rate may only be as high as 50% [134, 135].

A more experimental approach focuses on targeting the cytokines that are elevated in EBV-HLH (Fig. 1). In this vein, clinical trials to block IFNγ are underway with the antibody NI-0501 (https://clinicaltrials.gov/ct2/show/NCT01818492). The IL-1 family of cytokines (IL-1β, IL-18, and IL-33) comprises a class of potential targets. Antibodies blocking/neutralizing IL-1, such as canakinumab and rilonacept, or proteins that block the IL-1 receptor, such as anakinra (recombinant IL-1Ra), have been shown to be helpful in other CSS such as macrophage activation syndrome [136], but to date these have not been explored in EBV-HLH. Similarly, in XIAP, blockade of IL-18 is in clinical trials and may also be a point of intervention in EBV-HLH (https://clinicaltrials.gov/ct2/show/NCT003113760). Blocking the IL-33 receptor ST2 is also an as yet unexplored treatment possibility, as this method has shown efficacy in mice [112]. Beyond the IL-1 family of cytokines, antibody-mediated blockade of IL-6 or TNF (using tocilizumab, siltuximab, infliximab, adalimumab, and golimumab) may be viable options for the treatment of EBV-HLH, as there have been successes reported with these approaches in patients with other CSS [133]. Many of the cytokines elevated in HLH or other EBV-driven CSS are downstream products of JAK-STAT signaling, so a final and potentially more attractive experimental approach with which to target numerous cytokines in HLH is the use of JAK inhibitors. Inhibitors of JAK signaling such as ruxolitinib, tofacitinib, and barocitinib have recently become available for the treatment of autoimmune diseases such as rheumatoid arthritis and ulcerative colitis, as well as myeloproliferative disorder [137]. As these inhibitors can target multiple cytokines by blocking the function of numerous JAK-dependent receptors, it is possible that unwanted side effects might result [133]. Nevertheless, a recent case report describes positive results in a single patient with refractory HLH following treatment with ruxolitinib [138].

Conclusion

EBV has a long history with its human host, and in most people the virus maintains a latent infection throughout life. This symbiosis between host and pathogen is dependent on immune surveillance, and loss of control of EBV infection by any number of mechanisms can lead to an overactive and detrimental immune response. EBV is a potent immune stimulus, and as such is associated with a variety of CSS, the most common of which are discussed above. Ultimately, treatment of these CSS depends on restoring the balance between the virus and the host, and many approaches towards the achievement of this goal have or are being explored. To date, treatment has focused on the use of antiviral and immunosuppressive agents. These treatments, however, are toxic and not always very effective. As we have come to appreciate the importance of cytokines in disease pathogenesis, we can now use targeted therapies to dampen hyperinflammation while minimizing unwanted side effects. In the future, cytokine-directed therapies might ultimately prove to be a safer and more effective treatment of EBV-driven CSS.

Acknowledgements We thank Brandon Stelter in the Biomedical Communications Department at St. Jude Children's Research Hospital for assisting with the creation of Fig. 1.

References

1. Kutok, J. L., & Wang, F. (2006). Spectrum of Epstein-Barr virus-associated diseases. *Annual Review of Pathology, 1*, 375–404.
2. Nagington, J., & Gray, J. (1980). Cyclosporin a immunosuppression, Epstein-Barr antibody, and lymphoma. *Lancet, 1*, 536–537.

3. Weiss, L. M., Strickler, J. G., Warnke, R. A., Purtilo, D. T., & Sklar, J. (1987). Epstein-Barr viral DNA in tissues of Hodgkin's disease. *The American Journal of Pathology, 129*, 86–91.
4. Jones, J. F., Shurin, S., Abramowsky, C., Tubbs, R. R., Sciotto, C. G., Wahl, R., et al. (1988). T-cell lymphomas containing Epstein-Barr viral DNA in patients with chronic Epstein-Barr virus infections. *The New England Journal of Medicine, 318*, 733–741.
5. Greenspan, J. S., Greenspan, D., Lennette, E. T., Abrams, D. I., Conant, M. A., Petersen, V., et al. (1985). Replication of Epstein-Barr virus within the epithelial cells of oral "hairy" leukoplakia, an AIDS-associated lesion. *The New England Journal of Medicine, 313*, 1564–1571.
6. Neparidze, N., & Lacy, J. (2014). Malignancies associated with epstein-barr virus: Pathobiology, clinical features, and evolving treatments. *Clinical Advances in Hematology & Oncology, 12*, 358–371.
7. Epstein, M. A., & Achong, B. G. (1977). Pathogenesis of infectious mononucleosis. *Lancet, 2*, 1270–1273.
8. Kimura, H., Hoshino, Y., Kanegane, H., Tsuge, I., Okamura, T., Kawa, K., et al. (2001). Clinical and virologic characteristics of chronic active Epstein-Barr virus infection. *Blood, 98*, 280–286.
9. Sullivan, J. L., Woda, B. A., Herrod, H. G., Koh, G., Rivara, F. P., & Mulder, C. (1985). Epstein-Barr virus-associated hemophagocytic syndrome: Virological and immunopathological studies. *Blood, 65*, 1097–1104.
10. Imashuku, S. (2002). Clinical features and treatment strategies of Epstein-Barr virus-associated hemophagocytic lymphohistiocytosis. *Critical Reviews in Oncology/Hematology, 44*, 259–272.
11. Paya, C. V., Fung, J. J., Nalesnik, M. A., Kieff, E., Green, M., Gores, G., et al. (1999). Epstein-Barr virus-induced posttransplant lymphoproliferative disorders. ASTS/ASTP EBV-PTLD task force and the mayo clinic organized international consensus development meeting. *Transplantation, 68*, 1517–1525.
12. Palser, A. L., Grayson, N. E., White, R. E., Corton, C., Correia, S., Ba Abdullah, M. M., et al. (2015). Genome diversity of Epstein-Barr virus from multiple tumor types and normal infection. *Journal of Virology, 89*, 5222–5237.
13. Neves, M., Marinho-Dias, J., Ribeiro, J., & Sousa, H. (2017). Epstein-Barr virus strains and variations: Geographic or disease-specific variants? *Journal of Medical Virology, 89*, 373–387.
14. Farrell, P. J. (2015). Epstein-Barr virus strain variation. *Current Topics in Microbiology and Immunology, 390*, 45–69.
15. Shannon-Lowe, C., & Rowe, M. (2011). Epstein-Barr virus infection of polarized epithelial cells via the basolateral surface by memory B cell-mediated transfer infection. *PLoS Pathogens, 7*, e1001338.
16. Hutt-Fletcher, L. M. (2007). Epstein-Barr virus entry. *Journal of Virology, 81*, 7825–7832.
17. Johannsen, E., Luftig, M., Chase, M. R., Weicksel, S., Cahir-McFarland, E., Illanes, D., et al. (2004). Proteins of purified Epstein-Barr virus. *Proceedings of the National Academy of Sciences of the United States of America, 101*, 16286–16291.
18. Moore, M. D., DiScipio, R. G., Cooper, N. R., & Nemerow, G. R. (1989). Hydrodynamic, electron microscopic, and ligand-binding analysis of the Epstein-Barr virus/C3dg receptor (CR2). *The Journal of Biological Chemistry, 264*, 20576–20582.
19. Fingeroth, J. D., Weis, J. J., Tedder, T. F., Strominger, J. L., Biro, P. A., & Fearon, D. T. (1984). Epstein-Barr virus receptor of human B lymphocytes is the C3d receptor CR2. *Proceedings of the National Academy of Sciences of the United States of America, 81*, 4510–4514.
20. Janz, A., Oezel, M., Kurzeder, C., Mautner, J., Pich, D., Kost, M., et al. (2000). Infectious Epstein-Barr virus lacking major glycoprotein BLLF1 (gp350/220) demonstrates the existence of additional viral ligands. *Journal of Virology, 74*, 10142–10152.
21. Haddad, R. S., & Hutt-Fletcher, L. M. (1989). Depletion of glycoprotein gp85 from virosomes made with Epstein-Barr virus proteins abolishes their ability to fuse with virus receptor-bearing cells. *Journal of Virology, 63*, 4998–5005.
22. Haan, K. M., Lee, S. K., & Longnecker, R. (2001). Different functional domains in the cytoplasmic tail of glycoprotein B are involved in Epstein-Barr virus-induced membrane fusion. *Virology, 290*, 106–114.

23. Miller, N., & Hutt-Fletcher, L. M. (1988). A monoclonal antibody to glycoprotein gp85 inhibits fusion but not attachment of Epstein-Barr virus. *Journal of Virology, 62*, 2366–2372.
24. Sugano, N., Chen, W., Roberts, M. L., & Cooper, N. R. (1997). Epstein-Barr virus binding to CD21 activates the initial viral promoter via NF-kappaB induction. *The Journal of Experimental Medicine, 186*, 731–737.
25. Tanner, J. E., Alfieri, C., Chatila, T. A., & Diaz-Mitoma, F. (1996). Induction of interleukin-6 after stimulation of human B-cell CD21 by Epstein-Barr virus glycoproteins gp350 and gp220. *Journal of Virology, 70*, 570–575.
26. Hunter, C. A., & Jones, S. A. (2015). IL-6 as a keystone cytokine in health and disease. *Nature Immunology, 16*, 448–457.
27. Leveille, C., Castaigne, J. G., Charron, D., & Al-Daccak, R. (2002). MHC class II isotype-specific signaling complex on human B cells. *European Journal of Immunology, 32*, 2282–2291.
28. Ressing, M. E., van Gent, M., Gram, A. M., Hooykaas, M. J., Piersma, S. J., & Wiertz, E. J. (2015). Immune evasion by Epstein-Barr virus. *Current Topics in Microbiology and Immunology, 391*, 355–381.
29. van Gent, M., Braem, S. G., de Jong, A., Delagic, N., Peeters, J. G., Boer, I. G., et al. (2014). Epstein-Barr virus large tegument protein BPLF1 contributes to innate immune evasion through interference with toll-like receptor signaling. *PLoS Pathogens, 10*, e1003960.
30. Bentz, G. L., Liu, R., Hahn, A. M., Shackelford, J., & Pagano, J. S. (2010). Epstein-Barr virus BRLF1 inhibits transcription of IRF3 and IRF7 and suppresses induction of interferon-beta. *Virology, 402*, 121–128.
31. Quinn, L. L., Zuo, J., Abbott, R. J., Shannon-Lowe, C., Tierney, R. J., Hislop, A. D., et al. (2014). Cooperation between Epstein-Barr virus immune evasion proteins spreads protection from CD8+ T cell recognition across all three phases of the lytic cycle. *PLoS Pathogens, 10*, e1004322.
32. Lanier, L. L. (2015). NKG2D receptor and its ligands in host defense. *Cancer Immunology Research, 3*, 575–582.
33. Strowig, T., Brilot, F., Arrey, F., Bougras, G., Thomas, D., Muller, W. A., et al. (2008). Tonsilar NK cells restrict B cell transformation by the Epstein-Barr virus via IFN-gamma. *PLoS Pathogens, 4*, e27.
34. Azzi, T., Lunemann, A., Murer, A., Ueda, S., Beziat, V., Malmberg, K. J., et al. (2014). Role for early-differentiated natural killer cells in infectious mononucleosis. *Blood, 124*, 2533–2543.
35. Chung, Y., Yamazaki, T., Kim, B. S., Zhang, Y., Reynolds, J. M., Martinez, G. J., et al. (2013). Epstein Barr virus-induced 3 (EBI3) together with IL-12 negatively regulates T helper 17-mediated immunity to Listeria monocytogenes infection. *PLoS Pathogens, 9*, e1003628.
36. Chijioke, O., Azzi, T., Nadal, D., & Munz, C. (2013). Innate immune responses against Epstein Barr virus infection. *Journal of Leukocyte Biology, 94*, 1185–1190.
37. Chijioke, O., Landtwing, V., & Munz, C. (2016). NK cell influence on the outcome of primary Epstein-Barr virus infection. *Frontiers in Immunology, 7*, 323.
38. Kempkes, B., & Ling, P. D. (2015). EBNA2 and its coactivator EBNA-LP. *Current Topics in Microbiology and Immunology, 391*, 35–59.
39. Kieser, A., & Sterz, K. R. (2015). The latent membrane protein 1 (LMP1). *Current Topics in Microbiology and Immunology, 391*, 119–149.
40. Eliopoulos, A. G., Gallagher, N. J., Blake, S. M., Dawson, C. W., & Young, L. S. (1999). Activation of the p38 mitogen-activated protein kinase pathway by Epstein-Barr virus-encoded latent membrane protein 1 coregulates interleukin-6 and interleukin-8 production. *The Journal of Biological Chemistry, 274*, 16085–16096.
41. Allday, M. J., Bazot, Q., & White, R. E. (2015). The EBNA3 family: Two oncoproteins and a tumour suppressor that are central to the biology of EBV in B cells. *Current Topics in Microbiology and Immunology, 391*, 61–117.
42. Taylor, G. S., Long, H. M., Brooks, J. M., Rickinson, A. B., & Hislop, A. D. (2015). The immunology of Epstein-Barr virus-induced disease. *Annual Review of Immunology, 33*, 787–821.
43. Palendira, U., & Rickinson, A. B. (2015). Primary immunodeficiencies and the control of Epstein-Barr virus infection. *Annals of the New York Academy of Sciences, 1356*, 22–44.

44. Callan, M. F., Fazou, C., Yang, H., Rostron, T., Poon, K., Hatton, C., et al. (2000). CD8(+) T-cell selection, function, and death in the primary immune response in vivo. *The Journal of Clinical Investigation, 106*, 1251–1261.
45. Amyes, E., Hatton, C., Montamat-Sicotte, D., Gudgeon, N., Rickinson, A. B., McMichael, A. J., et al. (2003). Characterization of the CD4+ T cell response to Epstein-Barr virus during primary and persistent infection. *The Journal of Experimental Medicine, 198*, 903–911.
46. Long, H. M., Chagoury, O. L., Leese, A. M., Ryan, G. B., James, E., Morton, L. T., et al. (2013). MHC II tetramers visualize human CD4+ T cell responses to Epstein-Barr virus infection and demonstrate atypical kinetics of the nuclear antigen EBNA1 response. *The Journal of Experimental Medicine, 210*, 933–949.
47. Petrara, M. R., Freguja, R., Gianesin, K., Zanchetta, M., & De Rossi, A. (2013). Epstein-Barr virus-driven lymphomagenesis in the context of human immunodeficiency virus type 1 infection. *Frontiers in Microbiology, 4*, 311.
48. Rooney, C. M., Smith, C. A., Ng, C. Y., Loftin, S. K., Sixbey, J. W., Gan, Y., et al. (1998). Infusion of cytotoxic T cells for the prevention and treatment of Epstein-Barr virus-induced lymphoma in allogeneic transplant recipients. *Blood, 92*, 1549–1555.
49. Haque, T., Wilkie, G. M., Jones, M. M., Higgins, C. D., Urquhart, G., Wingate, P., et al. (2007). Allogeneic cytotoxic T-cell therapy for EBV-positive posttransplantation lymphoproliferative disease: Results of a phase 2 multicenter clinical trial. *Blood, 110*, 1123–1131.
50. Hadinoto, V., Shapiro, M., Sun, C. C., & Thorley-Lawson, D. A. (2009). The dynamics of EBV shedding implicate a central role for epithelial cells in amplifying viral output. *PLoS Pathogens, 5*, e1000496.
51. Thorley-Lawson, D. A. (2015). EBV persistence–introducing the virus. *Current Topics in Microbiology and Immunology, 390*, 151–209.
52. Laichalk, L. L., Hochberg, D., Babcock, G. J., Freeman, R. B., & Thorley-Lawson, D. A. (2002). The dispersal of mucosal memory B cells: Evidence from persistent EBV infection. *Immunity, 16*, 745–754.
53. Callan, M. F., Tan, L., Annels, N., Ogg, G. S., Wilson, J. D., O'Callaghan, C. A., et al. (1998). Direct visualization of antigen-specific CD8+ T cells during the primary immune response to Epstein-Barr virus in vivo. *The Journal of Experimental Medicine, 187*, 1395–1402.
54. Alfieri, C., Ghibu, F., & Joncas, J. H. (1984). Lytic, nontransforming Epstein-Barr virus (EBV) from a patient with chronic active EBV infection. *Canadian Medical Association Journal, 131*, 1249–1252.
55. Alfieri, C., & Joncas, J. H. (1987). Biomolecular analysis of a defective nontransforming Epstein-Barr virus (EBV) from a patient with chronic active EBV infection. *Journal of Virology, 61*, 3306–3309.
56. Cohen, J. I. (2009). Optimal treatment for chronic active Epstein-Barr virus disease. *Pediatric Transplantation, 13*, 393–396.
57. Katano, H., Ali, M. A., Patera, A. C., Catalfamo, M., Jaffe, E. S., Kimura, H., et al. (2004). Chronic active Epstein-Barr virus infection associated with mutations in perforin that impair its maturation. *Blood, 103*, 1244–1252.
58. Kasahara, Y., Yachie, A., Takei, K., Kanegane, C., Okada, K., Ohta, K., et al. (2001). Differential cellular targets of Epstein-Barr virus (EBV) infection between acute EBV-associated hemophagocytic lymphohistiocytosis and chronic active EBV infection. *Blood, 98*, 1882–1888.
59. Scott, R. B. (1939). Leukopenic myelosis: (section of medicine). *Proceedings of the Royal Society of Medicine, 32*, 1429–1434.
60. Daum, G. S., Sullivan, J. L., Ansell, J., Mulder, C., & Woda, B. A. (1987). Virus-associated hemophagocytic syndrome: Identification of an immunoproliferative precursor lesion. *Human Pathology, 18*, 1071–1074.
61. Imashuku, S., Hlbi, S., & Todo, S. (1997). Hemophagocytic lymphohistiocytosis in infancy and childhood. *The Journal of Pediatrics, 130*, 352–357.
62. George, M. R. (2014). Hemophagocytic lymphohistiocytosis: Review of etiologies and management. *Journal of Blood Medicine, 5*, 69–86.

63. Takahashi, S., Oki, J., Miyamoto, A., Koyano, S., Ito, K., Azuma, H., et al. (1999). Encephalopathy associated with haemophagocytic lymphohistiocytosis following rotavirus infection. *European Journal of Pediatrics, 158*, 133–137.
64. Henter, J. I., Elinder, G., & Ost, A. (1991). Diagnostic guidelines for hemophagocytic lymphohistiocytosis. The FHL study Group of the Histiocyte Society. *Seminars in Oncology, 18*, 29–33.
65. Gartner, B. C., Kortmann, K., Schafer, M., Mueller-Lantzsch, N., Sester, U., Kaul, H., et al. (2000). No correlation in Epstein-Barr virus reactivation between serological parameters and viral load. *Journal of Clinical Microbiology, 38*, 2458.
66. Greiner, T. C., and T. G. Gross. 2004. Atypical immune lymphoproliferations. In Hematology: Basic principles and practice, 3rd ed. R. Hoffman, B. Furie, E. J. Benz, Jr., , and P. McGlave, eds. Churchill Livingston, Philadelphia. 1449.
67. Harrington, D. S., Weisenburger, D. D., & Purtilo, D. T. (1988). Epstein-Barr virus–associated lymphoproliferative lesions. *Clinics in Laboratory Medicine, 8*, 97–118.
68. Mroczek, E. C., Weisenburger, D. D., Grierson, H. L., Markin, R., & Purtilo, D. T. (1987). Fatal infectious mononucleosis and virus-associated hemophagocytic syndrome. *Archives of Pathology & Laboratory Medicine, 111*, 530–535.
69. Chen, J. H., Fleming, M. D., Pinkus, G. S., Pinkus, J. L., Nichols, K. E., Mo, J. Q., et al. (2010). Pathology of the liver in familial hemophagocytic lymphohistiocytosis. *The American Journal of Surgical Pathology, 34*, 852–867.
70. Seemayer, T. A., Gross, T. G., Hinrichs, S. H., & Egeler, R. M. (1994). Massive diffuse histiocytic myocardial infiltration in Epstein-Barr virus-associated hemophagocytic syndrome and fulminant infectious mononucleosis. *Cell Vision, 1*, 260.
71. Purtilo, D. T., Cassel, C., & Yang, J. P. (1974). Letter: Fatal infectious mononucleosis in familial lymphohistiocytosis. *The New England Journal of Medicine, 291*, 736.
72. Purtilo, D. T., Cassel, C. K., Yang, J. P., & Harper, R. (1975). X-linked recessive progressive combined variable immunodeficiency (Duncan's disease). *Lancet, 1*, 935–940.
73. Seemayer, T. A., Gross, T. G., Egeler, R. M., Pirruccello, S. J., Davis, J. R., Kelly, C. M., et al. (1995). X-linked lymphoproliferative disease: Twenty-five years after the discovery. *Pediatric Research, 38*, 471–478.
74. Talaat, K. R., Rothman, J. A., Cohen, J. I., Santi, M., Choi, J. K., Guzman, M., et al. (2009). Lymphocytic vasculitis involving the central nervous system occurs in patients with X-linked lymphoproliferative disease in the absence of Epstein-Barr virus infection. *Pediatric Blood & Cancer, 53*, 1120–1123.
75. Booth, C., Gilmour, K. C., Veys, P., Gennery, A. R., Slatter, M. A., Chapel, H., et al. (2011). X-linked lymphoproliferative disease due to SAP/SH2D1A deficiency: A multicenter study on the manifestations, management and outcome of the disease. *Blood, 117*, 53–62.
76. Sumegi, J., Huang, D., Lanyi, A., Davis, J. D., Seemayer, T. A., Maeda, A., et al. (2000). Correlation of mutations of the SH2D1A gene and epstein-barr virus infection with clinical phenotype and outcome in X-linked lymphoproliferative disease. *Blood, 96*, 3118–3125.
77. Nichols, K. E., Harkin, D. P., Levitz, S., Krainer, M., Kolquist, K. A., Genovese, C., et al. (1998). Inactivating mutations in an SH2 domain-encoding gene in X-linked lymphoproliferative syndrome. *Proceedings of the National Academy of Sciences of the United States of America, 95*, 13765–13770.
78. Nichols, K. E., Ma, C. S., Cannons, J. L., Schwartzberg, P. L., & Tangye, S. G. (2005). Molecular and cellular pathogenesis of X-linked lymphoproliferative disease. *Immunological Reviews, 203*, 180–199.
79. Coffey, A. J., Brooksbank, R. A., Brandau, O., Oohashi, T., Howell, G. R., Bye, J. M., et al. (1998). Host response to EBV infection in X-linked lymphoproliferative disease results from mutations in an SH2-domain encoding gene. *Nature Genetics, 20*, 129–135.
80. Gifford, C. E., Weingartner, E., Villanueva, J., Johnson, J., Zhang, K., Filipovich, A. H., et al. (2014). Clinical flow cytometric screening of SAP and XIAP expression accurately identifies patients with SH2D1A and XIAP/BIRC4 mutations. *Cytometry. Part B, Clinical Cytometry, 86*, 263–271.

81. Cohen, J. I. (2015). Primary immunodeficiencies associated with EBV disease. *Current Topics in Microbiology and Immunology, 390*, 241–265.
82. Tangye, S. G., Palendira, U., & Edwards, E. S. (2017). Human immunity against EBV-lessons from the clinic. *The Journal of Experimental Medicine, 214*, 269–283.
83. Ruffo, E., Malacarne, V., Larsen, S. E., Das, R., Patrussi, L., Wulfing, C., et al. (2016). Inhibition of diacylglycerol kinase alpha restores restimulation-induced cell death and reduces immunopathology in XLP-1. *Science Translational Medicine, 8*, 321ra327.
84. Czar, M. J., Kersh, E. N., Mijares, L. A., Lanier, G., Lewis, J., Yap, G., et al. (2001). Altered lymphocyte responses and cytokine production in mice deficient in the X-linked lymphoproliferative disease gene SH2D1A/DSHP/SAP. *Proceedings of the National Academy of Sciences of the United States of America, 98*, 7449–7454.
85. Coquet, J. M., Chakravarti, S., Kyparissoudis, K., McNab, F. W., Pitt, L. A., McKenzie, B. S., et al. (2008). Diverse cytokine production by NKT cell subsets and identification of an IL-17-producing CD4-NK1.1- NKT cell population. *Proceedings of the National Academy of Sciences of the United States of America, 105*, 11287–11292.
86. Rigaud, S., Fondaneche, M. C., Lambert, N., Pasquier, B., Mateo, V., Soulas, P., et al. (2006). XIAP deficiency in humans causes an X-linked lymphoproliferative syndrome. *Nature, 444*, 110–114.
87. Filipovich, A. H., Zhang, K., Snow, A. L., & Marsh, R. A. (2010). X-linked lymphoproliferative syndromes: Brothers or distant cousins? *Blood, 116*, 3398–3408.
88. Marsh, R. A., Madden, L., Kitchen, B. J., Mody, R., McClimon, B., Jordan, M. B., et al. (2010). XIAP deficiency: A unique primary immunodeficiency best classified as X-linked familial hemophagocytic lymphohistiocytosis and not as X-linked lymphoproliferative disease. *Blood, 116*, 1079–1082.
89. Pachlopnik Schmid, J., Canioni, D., Moshous, D., Touzot, F., Mahlaoui, N., Hauck, F., et al. (2011). Clinical similarities and differences of patients with X-linked lymphoproliferative syndrome type 1 (XLP-1/SAP deficiency) versus type 2 (XLP-2/XIAP deficiency). *Blood, 117*, 1522–1529.
90. Wong, W. W., Vince, J. E., Lalaoui, N., Lawlor, K. E., Chau, D., Bankovacki, A., et al. (2014). cIAPs and XIAP regulate myelopoiesis through cytokine production in an RIPK1- and RIPK3-dependent manner. *Blood, 123*, 2562–2572.
91. Lawlor, K. E., Khan, N., Mildenhall, A., Gerlic, M., Croker, B. A., D'Cruz, A. A., et al. (2015). RIPK3 promotes cell death and NLRP3 inflammasome activation in the absence of MLKL. *Nature Communications, 6*, 6282.
92. Lawlor, K. E., Feltham, R., Yabal, M., Conos, S. A., Chen, K. W., Ziehe, S., et al. (2017). XIAP loss triggers RIPK3- and caspase-8-driven IL-1beta activation and cell death as a consequence of TLR-MyD88-induced cIAP1-TRAF2 degradation. *Cell Reports, 20*, 668–682.
93. Huck, K., Feyen, O., Niehues, T., Ruschendorf, F., Hubner, N., Laws, H. J., et al. (2009). Girls homozygous for an IL-2-inducible T cell kinase mutation that leads to protein deficiency develop fatal EBV-associated lymphoproliferation. *The Journal of Clinical Investigation, 119*, 1350–1358.
94. Andreotti, A. H., Schwartzberg, P. L., Joseph, R. E., & Berg, L. J. (2010). T-cell signaling regulated by the Tec family kinase, Itk. *Cold Spring Harbor Perspectives in Biology, 2*, a002287.
95. Schaeffer, E. M., Debnath, J., Yap, G., McVicar, D., Liao, X. C., Littman, D. R., et al. (1999). Requirement for Tec kinases Rlk and Itk in T cell receptor signaling and immunity. *Science, 284*, 638–641.
96. Mueller, C., & August, A. (2003). Attenuation of immunological symptoms of allergic asthma in mice lacking the tyrosine kinase ITK. *Journal of Immunology, 170*, 5056–5063.
97. Fowell, D. J., Shinkai, K., Liao, X. C., Beebe, A. M., Coffman, R. L., Littman, D. R., et al. (1999). Impaired NFATc translocation and failure of Th2 development in Itk-deficient CD4+ T cells. *Immunity, 11*, 399–409.
98. Korn, T., Bettelli, E., Oukka, M., & Kuchroo, V. K. (2009). IL-17 and Th17 cells. *Annual Review of Immunology, 27*, 485–517.

99. Li, F. Y., Chaigne-Delalande, B., Su, H., Uzel, G., Matthews, H., & Lenardo, M. J. (2014). XMEN disease: A new primary immunodeficiency affecting Mg2+ regulation of immunity against Epstein-Barr virus. *Blood, 123*, 2148–2152.
100. Ravell, J., Chaigne-Delalande, B., & Lenardo, M. (2014). X-linked immunodeficiency with magnesium defect, Epstein-Barr virus infection, and neoplasia disease: A combined immune deficiency with magnesium defect. *Current Opinion in Pediatrics, 26*, 713–719.
101. Grubbs, R. D., & Maguire, M. E. (1987). Magnesium as a regulatory cation: Criteria and evaluation. *Magnesium, 6*, 113–127.
102. Chaigne-Delalande, B., Li, F. Y., O'Connor, G. M., Lukacs, M. J., Jiang, P., Zheng, L., et al. (2013). Mg2+ regulates cytotoxic functions of NK and CD8 T cells in chronic EBV infection through NKG2D. *Science, 341*, 186–191.
103. Cohen, J. I. (2000). Epstein-Barr virus infection. *The New England Journal of Medicine, 343*, 481–492.
104. McGowan Jr., J. E., Chesney, P. J., Crossley, K. B., & LaForce, F. M. (1992). Guidelines for the use of systemic glucocorticosteroids in the management of selected infections. Working group on steroid use, antimicrobial agents committee, infectious diseases society of America. *The Journal of Infectious Diseases, 165*, 1–13.
105. Pagano, J. S., Sixbey, J. W., & Lin, J. C. (1983). Acyclovir and Epstein-Barr virus infection. *The Journal of Antimicrobial Chemotherapy, 12*(Suppl B), 113–121.
106. Heslop, H. E. (2009). How I treat EBV lymphoproliferation. *Blood, 114*, 4002–4008.
107. Heslop, H. E., Slobod, K. S., Pule, M. A., Hale, G. A., Rousseau, A., Smith, C. A., et al. (2010). Long-term outcome of EBV-specific T-cell infusions to prevent or treat EBV-related lymphoproliferative disease in transplant recipients. *Blood, 115*, 925–935.
108. Imashuku, S., Hibi, S., Fujiwara, F., Ikushima, S., & Todo, S. (1994). Haemophagocytic lymphohistiocytosis, interferon-gamma-naemia and Epstein-Barr virus involvement. *British Journal of Haematology, 88*, 656–658.
109. Jordan, M. B., Allen, C. E., Weitzman, S., Filipovich, A. H., & McClain, K. L. (2011). How I treat hemophagocytic lymphohistiocytosis. *Blood, 118*, 4041–4052.
110. Imashuku, S., Hibi, S., Tabata, Y., Sako, M., Sekine, Y., Hirayama, K., et al. (1998). Biomarker and morphological characteristics of Epstein-Barr virus-related hemophagocytic lymphohistiocytosis. *Medical and Pediatric Oncology, 31*, 131–137.
111. Han, X. C., Ye, Q., Zhang, W. Y., Tang, Y. M., Xu, X. J., & Zhang, T. (2017). Cytokine profiles as novel diagnostic markers of Epstein-Barr virus-associated hemophagocytic lymphohistiocytosis in children. *Journal of Critical Care, 39*, 72–77.
112. Rood, J. E., Rao, S., Paessler, M., Kreiger, P. A., Chu, N., Stelekati, E., et al. (2016). ST2 contributes to T-cell hyperactivation and fatal hemophagocytic lymphohistiocytosis in mice. *Blood, 127*, 426–435.
113. Lotz, M., Tsoukas, C. D., Fong, S., Dinarello, C. A., Carson, D. A., & Vaughan, J. H. (1986). Release of lymphokines after Epstein Barr virus infection in vitro. I. Sources of and kinetics of production of interferons and interleukins in normal humans. *Journal of Immunology, 136*, 3636–3642.
114. Gosselin, J., Flamand, L., D'Addario, M., Hiscott, J., & Menezes, J. (1992). Infection of peripheral blood mononuclear cells by herpes simplex and Epstein-Barr viruses. Differential induction of interleukin 6 and tumor necrosis factor-alpha. *The Journal of Clinical Investigation, 89*, 1849–1856.
115. Gosselin, J., Menezes, J., D'Addario, M., Hiscott, J., Flamand, L., Lamoureux, G., et al. (1991). Inhibition of tumor necrosis factor-alpha transcription by Epstein-Barr virus. *European Journal of Immunology, 21*, 203–208.
116. Roberge, C. J., McColl, S. R., Larochelle, B., & Gosselin, J. (1998). Granulocyte-macrophage colony-stimulating factor enhances EBV-induced synthesis of chemotactic factors in human neutrophils. *Journal of Immunology, 160*, 2442–2448.
117. Roberge, C. J., Poubelle, P. E., Beaulieu, A. D., Heitz, D., & Gosselin, J. (1996). The IL-1 and IL-1 receptor antagonist (IL-1Ra) response of human neutrophils to EBV stimulation. Preponderance of IL-Ra detection. *Journal of Immunology, 156*, 4884–4891.

118. Larochelle, B., Flamand, L., Gourde, P., Beauchamp, D., & Gosselin, J. (1998). Epstein-Barr virus infects and induces apoptosis in human neutrophils. *Blood, 92*, 291–299.
119. Linde, A., Andersson, B., Svenson, S. B., Ahrne, H., Carlsson, M., Forsberg, P., et al. (1992). Serum levels of lymphokines and soluble cellular receptors in primary Epstein-Barr virus infection and in patients with chronic fatigue syndrome. *The Journal of Infectious Diseases, 165*, 994–1000.
120. Hornef, M. W., Wagner, H. J., Kruse, A., & Kirchner, H. (1995). Cytokine production in a whole-blood assay after Epstein-Barr virus infection in vivo. *Clinical and Diagnostic Laboratory Immunology, 2*, 209–213.
121. Setsuda, J., Teruya-Feldstein, J., Harris, N. L., Ferry, J. A., Sorbara, L., Gupta, G., et al. (1999). Interleukin-18, interferon-gamma, IP-10, and Mig expression in Epstein-Barr virus-induced infectious mononucleosis and posttransplant lymphoproliferative disease. *The American Journal of Pathology, 155*, 257–265.
122. Marshall, N. A., Culligan, D. J., Johnston, P. W., Millar, C., Barker, R. N., & Vickers, M. A. (2007). CD4(+) T-cell responses to Epstein-Barr virus (EBV) latent membrane protein 1 in infectious mononucleosis and EBV-associated non-Hodgkin lymphoma: Th1 in active disease but Tr1 in remission. *British Journal of Haematology, 139*, 81–89.
123. Steigerwald-Mullen, P., Kurilla, M. G., & Braciale, T. J. (2000). Type 2 cytokines predominate in the human CD4(+) T-lymphocyte response to Epstein-Barr virus nuclear antigen 1. *Journal of Virology, 74*, 6748–6759.
124. Imashuku, S., Kuriyama, K., Teramura, T., Ishii, E., Kinugawa, N., Kato, M., et al. (2001). Requirement for etoposide in the treatment of Epstein-Barr virus-associated hemophagocytic lymphohistiocytosis. *Journal of Clinical Oncology, 19*, 2665–2673.
125. Imashuku, S. (2011). Treatment of Epstein-Barr virus-related hemophagocytic lymphohistiocytosis (EBV-HLH); update 2010. *Journal of Pediatric Hematology/Oncology, 33*, 35–39.
126. Abella, E. M., Artrip, J., Schultz, K., & Ravindranath, Y. (1997). Treatment of familial erythrophagocytic lymphohistiocytosis with cyclosporine a. *The Journal of Pediatrics, 130*, 467–470.
127. Perel, Y., Alos, N., Ansoborlo, S., Carrere, A., & Guillard, J. M. (1997). Dramatic efficacy of antithymocyte globulins in childhood EBV-associated haemophagocytic syndrome. *Acta Paediatrica, 86*, 911.
128. Marsh, R. A., Allen, C. E., McClain, K. L., Weinstein, J. L., Kanter, J., Skiles, J., et al. (2013). Salvage therapy of refractory hemophagocytic lymphohistiocytosis with alemtuzumab. *Pediatric Blood & Cancer, 60*, 101–109.
129. Keith, M. P., Pitchford, C., & Bernstein, W. B. (2012). Treatment of hemophagocytic lymphohistiocytosis with alemtuzumab in systemic lupus erythematosus. *Journal of Clinical Rheumatology, 18*, 134–137.
130. Patton, L. L., Ramirez-Amador, V., Anaya-Saavedra, G., Nittayananta, W., Carrozzo, M., & Ranganathan, K. (2013). Urban legends series: Oral manifestations of HIV infection. *Oral Diseases, 19*, 533–550.
131. Gianti, E., Messick, T. E., Lieberman, P. M., & Zauhar, R. J. (2016). Computational analysis of EBNA1 "druggability" suggests novel insights for Epstein-Barr virus inhibitor design. *Journal of Computer-Aided Molecular Design, 30*, 285–303.
132. Thompson, S., Messick, T., Schultz, D. C., Reichman, M., & Lieberman, P. M. (2010). Development of a high-throughput screen for inhibitors of Epstein-Barr virus EBNA1. *Journal of Biomolecular Screening, 15*, 1107–1115.
133. Behrens, E. M., & Koretzky, G. A. (2017). Review: Cytokine storm syndrome: Looking toward the precision medicine era. *Arthritis & Rheumatology, 69*, 1135–1143.
134. Horne, A., Janka, G., Maarten Egeler, R., Gadner, H., Imashuku, S., Ladisch, S., et al. (2005). Haematopoietic stem cell transplantation in haemophagocytic lymphohistiocytosis. *British Journal of Haematology, 129*, 622–630.
135. Filipovich, A. H. (2005). Life-threatening hemophagocytic syndromes: Current outcomes with hematopoietic stem cell transplantation. *Pediatric Transplantation, 9*(Suppl 7), 87–91.

136. Miettunen, P. M., Narendran, A., Jayanthan, A., Behrens, E. M., & Cron, R. Q. (2011). Successful treatment of severe paediatric rheumatic disease-associated macrophage activation syndrome with interleukin-1 inhibition following conventional immunosuppressive therapy: Case series with 12 patients. *Rheumatology (Oxford), 50*, 417–419.
137. Baker, K. F., & Isaacs, J. D. (2017). Novel therapies for immune-mediated inflammatory diseases: What can we learn from their use in rheumatoid arthritis, spondyloarthritis, systemic lupus erythematosus, psoriasis, Crohn's disease and ulcerative colitis? *Annals of the Rheumatic Diseases, 77*(2), 175–187.
138. Broglie, L., Pommert, L., Rao, S., Thakar, M., Phelan, R., Margolis, D., et al. (2017). Ruxolitinib for treatment of refractory hemophagocytic lymphohistiocytosis. *Blood Advances, 1*, 1533–1536.

Cytokine Storm Syndrome Associated with Hemorrhagic Fever and Other Viruses

Ethan S. Sen and A. V. Ramanan

Introduction

Cytokine storm syndrome (CSS) results from the failure to regulate appropriately the immune response with particular dysfunction of cytotoxic T cells and NK cells [1]. A potent trigger for activation of these cells is infection, and particularly viral infections, in subjects both with and without genetic mutations associated with primary HLH (pHLH) [2]. The most common viral infection triggering CSS is Epstein–Barr virus (EBV) and was the cause in 74% of children in whom an infectious agent was identified from a cohort of 219 with infection-associated CSS [3]. CSS associated with EBV and other herpes viruses are discussed elsewhere and the focus of this chapter will be other viruses, including hemorrhagic fever viruses [4]. In order to diagnose virus-associated HLH, presence of the pathogen should be confirmed by serology (paired acute and convalescent samples) or specific viral polymerase chain reaction (PCR) testing of blood or tissue. Other infectious or noninfectious causes of secondary HLH would also need to be excluded, although in cases of severe or life-threatening CSS prompt initiation of immunosuppressive / immunomodulatory treatment without delay is more important than determination of the underlying etiology.

Viral Hemorrhagic Fevers

The viral hemorrhagic fevers (VHFs) are caused by viruses belonging to one of five families: *Arenaviridae*, *Bunyaviridae*, *Filoviridae*, *Falviviridae*, and *Togaviridae* [5, 6]. They are all RNA viruses and require a nonhuman vertebrate or insect host to

E. S. Sen · A. V. Ramanan (✉)
Pediatric Rheumatology, Bristol Royal Hospital for Children, Bristol, UK
e-mail: ethan.sen@doctors.org.uk

R. Q. Cron, E. M. Behrens (eds.), *Cytokine Storm Syndrome*,
https://doi.org/10.1007/978-3-030-22094-5_16

provide a natural reservoir. Initial infection occurs when humans come into contact with a host but subsequently human-to-human transmission of some of these viruses does occur. Typical presenting features include fever, dizziness, fatigue, muscle aches and weakness. Frequently VHFs occur as outbreaks or epidemics. The viruses and infections they cause are summarized in Table 1.

A systematic review of sHLH in zoonoses conducted in 2012 and covering studies published between 1950 and 2012 identified reports of HLH associated with Crimean-Congo hemorrhagic fever (CCHF) and hantaviruses [7]. Multiple cases of sHLH linked to dengue [8–39], Chikungunya [40], CCHF [41–43], hantavirus [44], and severe fever with thrombocytopenia syndrome (Bunyavirus) [45–47] have been described.

Dengue is a relatively common tropical infection [48] and may progress to more severe forms: dengue hemorrhagic fever (DHF) and dengue shock syndrome (DSS). DHF is defined as dengue infection accompanied by fever lasting 2–7 days, hemorrhagic tendencies, thrombocytopenia and evidence of plasma leakage due to increased vascular permeability [49]. DSS is classified as DHF plus evidence of circulatory failure. The primary aspects of management are supportive with fluids and blood

Table 1 Viral hemorrhagic fever viruses and their associated diseases

Virus family	Host	Examples	Disease
Arenaviridae	Rodents—Spread through contact with excrement	Junin virus	Argentine hemorrhagic fever
		Lassa virus	Lassa fever
		Lujo virus	Lujo hemorrhagic fever
Bunyaviridae	Arthropods (ticks, mosquitoes, sand flies) and rodents	*Phlebovirus*—Rift valley fever virus	Rift valley fever
	Ticks	*Nairovirus*—Crimean-Congo hemorrhagic fever virus	Crimean-Congo hemorrhagic fever
	Rodents	*Hantavirus*—Hantaan virus Sin Nombre virus	Hemorrhagic fever with renal syndrome (HFRS) Hantavirus pulmonary syndrome (HPS)
Filoviridae	Fruit bats, *Rousettus* bats	Marburgvirus	Marburg hemorrhagic fever
	Fruit bats, primates	Ebolavirus	Ebola hemorrhagic fever
Flaviviridae	Mosquitoes	Yellow fever virus	Yellow fever
	Mosquitoes	Dengue virus	Dengue fever, dengue hemorrhagic fever
	Mosquitoes	Japanese encephalitis virus	Japanese encephalitis
	Mosquitoes	West Nile virus	West Nile fever
	Mosquitoes	Zika virus	Zika
	Ticks	Tick-borne encephalitis virus	Tick-borne encephalitis
Togaviridae		Chikungunya virus	Chikungunya

Adapted from [5, 6]

products as required [50]. Corticosteroids have been used; however, there has been uncertainty among clinicians about the efficacy and safety of corticosteroids in treatment of dengue with some considering them harmful [51]. An RCT of oral prednisolone (0.5 mg/kg or 2 mg/kg daily for 3 days) versus placebo, which included 225 patients with early dengue infection, found no prolongation of viremia or other adverse events in the steroid recipients [52]. It was not powered to assess efficacy; however, there appeared to be no reduction in the development of shock or other dengue-related complications. A Cochrane review published in 2014 which included 8 RCTs or quasi-randomized studies found the evidence to be of low or very low quality and insufficient to conclude whether corticosteroids are of benefit in dengue at an early stage or DSS [53]. Others, however, have suggested that the timing of steroids and patient selection may be critical [54]. A non-randomized retrospective study of adults with DSS with the most severe disease found those given a single dose of methylprednisolone (1 g intravenously) as a rescue treatment had a lower mortality (13%, 3/13 patients) than those not receiving steroids (47%, 15/32 patients) [55]. In these most severely affected cases it may be that patients had developed CSS.

Dengue has frequently been the trigger for HLH [56]. A large study in India of 212 patients with dengue identified 31 (14.6%) who developed CSS, including 23 with evidence of bone marrow hemophagocytosis [29]. In this group of 23 patients, 19 received IVIg and all recovered. Another study from Kolkata, India, reported 8 patients (2.2%) with HLH of 358 with dengue during the outbreak in 2012 [27]. The 8 cases received supportive therapy, blood component transfusions as required and parenteral dexamethasone (10 mg/m^2 in 3–4 divided doses/day) until hemodynamically stable before switching to oral tapering treatment for 21 days. IVIg (1 g/kg) was used in one patient as rescue therapy after failing to respond to 48 h of steroids. The patients with HLH were distinguished by persistence of fever for more than 7 days together with prolonged or progressive cytopenias, organomegaly, and sterile cultures. In a series of 33 children with HLH from Chennai, India, an infectious etiology was identified in 14 and specifically dengue in 5 [57].

Using a case-control design comparing patients with dengue who developed HLH (cases, $n = 22$) with patients with dengue without HLH (controls, $n = 88$), one study found cases had a younger age (median 1 vs. 13 years, $p < 0.01$), more frequent coinfection (18.2% vs. 4.5%, $p = 0.04$), and longer duration of fever (7 vs. 5 days, $p < 0.01$) [15]. Several studies have suggested testing for laboratory markers of HLH in cases of dengue, in particular ferritin, sCD163, and sCD25 [9, 10]. In a cohort of 208 patients with dengue, ferritin and sCD163 were significantly increased in patients with severe dengue. A report including patients with dengue during an outbreak on Aruba in the Caribbean found that levels of ferritin were significantly higher in patients with dengue compared with other febrile illnesses [58]. In another cohort of dengue-infected patients in Brazil, hyperferritinemia was associated with disease severity and a pro-inflammatory cytokine profile [58].

The features of severe dengue infection, as seen in dengue hemorrhagic fever or dengue shock syndrome, overlap with HLH suggesting a similar pathogenesis involving overactivation of the immune system leading to a hypercytokinemia [58, 59]. These features also seem to be shared in some cases of CCHF [41, 43]. One reported a 14-year-old boy from Turkey with CCHF associated with leukopenia, thrombocytopenia, hypertriglyceridemia, hyperferritinemia and bone marrow hemophagocytosis. The pathogenesis of viral hemorrhagic fevers may overlap with CSS. A report of 5 patients with CCHF treated with high dose intravenous methylprednisolone (IVMP) suggested resolution of fever, increase in leukocyte and platelet counts and clinical improvement within 5 days of treatment [60]. Another study reported outcomes in 12 patients with CCHF treated with IVMP (up to 30 mg/kg/day), fresh frozen plasma (FFP), and intravenous immunoglobulin (IVIg) [61]. The treatment appeared to be successful with reduction of fever within 2 days, white cell count above 4500/μL in 4 days and platelets above 150,000/μL in 9 days. Finally, it has been suggested that Ebola outbreaks share many features of CSS [62–64].

While the VHFs represent an important group of viruses associated with CSS, a host of other viruses more commonly seen in North America and Europe can also trigger hemophagocytic syndromes.

Nonhemorrhagic Fever Viruses

A wide range of viruses other than those discussed above have been associated with CSS. These are summarized in Table 2 and reviewed in more detail below.

Influenza and Parainfluenza

CSS has been identified in association with seasonal influenza [65, 86–89], influenza A (H5N1, "avian flu") [90, 91], and pandemic influenza A (H1N1, 2009 "swine flu") infection [92–100]. In some cases, patients were immunocompromised or had additional risk factors such as leukemia [86, 97, 98], post-bone marrow transplantation for lymphoma [93], genetic predisposition [101], or cystic fibrosis [99]. However, CSS developed in previously healthy individuals following influenza H1N1 and H5N1 leading to death in both adults [95] and children [91, 100, 102]. A case of HLH following influenza vaccination in a patient with aplastic anemia undergoing allogeneic bone marrow transplantation has also been documented [103].

During the 2009 influenza H1N1 pandemic, a center in Germany conducted a prospective observational study of 25 critically ill patients with the infection [66]. All developed severe acute respiratory distress syndrome and hypoxemia and were mechanically ventilated. HLH was diagnosed based on the presence of three of four major criteria (fever, cytopenia, hepatitis, or splenomegaly) and at least one minor criterion (evidence of hemophagocytosis in bone marrow samples or increase in serum level of sIL-2Rα or ferritin, respectively). Nine (36%) of 25 patients met these criteria and eight (89%) of them died, compared with 4 (25%) of 16 patients without HLH. Six of the patients with HLH were treated (four with etoposide and dexamethasone, two with steroids alone) but the other three were moribund at the

Table 2 Nonhemorrhagic fever viruses associated with cytokine storm syndromes

Virus	Clinical associations	Outcomes	References
Influenza	Pneumonia, myocarditis, encephalitis, myositis	Mechanical ventilation, recovery	[65–67]
Parainfluenza	Croup, bronchiolitis, pneumonia	Recovery	[68]
Adenovirus	Upper respiratory tract infection, pneumonia, conjunctivitis, gastroenteritis, hepatitis, myocarditis, encephalitis	Recovery, more severe manifestations in immunocompromised hosts	[69, 70]
Parvovirus	"Slapped cheek syndrome"/ fifth disease, aplastic crisis, arthropathy, hepatitis, myocarditis	Recovery, pure red cell aplasia, chronic arthritis, hydrops fetalis, chronic fatigue syndrome	[71]
Hepatitis viruses	Hepatitis, arthritis, leukocytoclastic vasculitis	Recovery (hepatitis A), chronic hepatitis, cirrhosis, hepatocellular carcinoma	[72–74]
Measles	Interstitial pneumonia, encephalitis, thrombocytopenic purpura	Recovery, subacute sclerosing panencephalitis	[75–77]
Mumps	Parotitis, pancreatitis, orchitis, meningitis, encephalitis	Recovery, deafness, sterility rarely after orchitis	[78, 79]
Rubella	Arthralgia, arthritis, encephalitis, congenital rubella syndrome	Deafness, developmental delay, cardiovascular and ocular defects in congenital rubella syndrome	[80]
Enterovirus	Respiratory and gastrointestinal infections, pancreatitis, meningitis, encephalitis, neonatal sepsis,	Neurological impairment in some children after meningitis	[81]
Parechovirus	Sepsis-like illness, meningitis, encephalitis, hepatitis	Neurological sequelae in some young infants	[82, 83]
Rotavirus	Gastroenteritis, seizures, encephalopathy/encephalitis	Recovery, rarely intussusception	[84]
Human T-lymphotropic virus	Adult T cell leukemia/ lymphoma, demyelinating disease, autoimmune diseases	Tropical spastic paraparesis, systemic lupus erythematosus, Sjögren's syndrome	[85]

time of diagnosis with HLH and were not considered suitable for treatment. The study suggests that CSS/HLH may have been a significant contributor to multiorgan failure and death in critically ill patients during the influenza A H1N1 pandemic.

Reports have indicated that avian influenza A (H5N1) can lead to severe and widely disseminated infection outside the respiratory system. In one case, a previously healthy nine-year-old Vietnamese girl died following encephalitis and coma with virus detected in rectal swabs, serum and cerebrospinal fluid [102]. Hemophagocytosis was detected in bone marrow from several patients with H5N1 infection [90, 91, 104–106]. In vitro studies indicated that recombinant hemagglutinin (H5) from H5N1 influenza suppressed perforin expression and reduced cytotoxicity of human $CD8^+$ T cells to kill H5-bearing cells [107]. This failure of clearance of infected cells could promote lymphoproliferation and hypercytokinemia as seen in CSS. Another study measured cytokine levels in a familial group of patients with H5N1 influenza in Hong Kong and found particularly high serum concentrations of interferon induced protein-10 (also known as CXCL10) and monokine induced by interferon γ (CXCL9) [108]. Taken together with other in vitro work, the authors suggest this hypercytokinemia may contribute to pathogenesis in fatal influenza infection. How might these cases be treated?

Drawing parallels with EBV-associated HLH, Henter et al. proposed a modified HLH-94 treatment protocol for influenza A (H5N1)-associated HLH in addition to antiviral and supportive therapy [109, 110]. For children, they suggested intravenous etoposide 150 mg/m^2 once per week and dexamethasone initially 10 mg/m^2 once daily. They recommended reduced doses of both drugs in patients aged 15 years or older and advised against upfront use of cyclosporin A (CsA) in all cases due to the relative frequency of renal complications in H5N1 infection.

Influenza B has been suggested as the trigger for HLH in a 24-year-old man with systemic lupus erythematosus (SLE) [111]. Several weeks after his initial diagnosis of SLE when he presented with a pericardial effusion, he developed fever, erythematous rash, splenomegaly, hypertriglyceridemia, and ferritin of 95,703 ng/mL. Influenza virus B was detected in the patient's nasal lavage sample and no other infective triggers for HLH were apparent. His pericardial effusion recurred with progression to cardiac tamponade, and he was managed with ventilation and pericardiocentesis. His condition initially improved after steroid pulse treatment, but he was started on colchicine 1 mg/day after reaccumulation of the pericardial fluid. He remained stable on low-dose prednisolone and colchicine. The relative contributions of SLE and influenza infection to pathogenesis of CSS in this case are unclear. While many reports have associated influenza with HLH, parainfluenza has been documented once.

A case of CSS in a 33-year-old Chilean man characterized by fever, evanescent rash, hepatosplenomegaly, anemia, thrombocytopenia, hyperferritinemia, and hemophagocytosis on bone marrow biopsy was attributed to parainfluenza virus-2 infection detected by polymerase chain reaction (PCR) testing of respiratory and enteric samples [112]. He was treated with etoposide, dexamethasone, and CsA following the HLH-94 protocol, and he made a complete recovery remaining in remission after 2 years of follow-up.

Adenovirus

Adenovirus infections are frequent in childhood presenting with respiratory, gastrointestinal, or ocular manifestations [70]. These are usually mild in immunocompetent hosts but may lead to more severe disease including pneumonia, hepatitis, and encephalitis in the immunocompromised.

CSS has been described in adults and children secondary to adenovirus infection [113–124]. It has occurred in previously healthy children with the initial presentation of pneumonia [114, 115, 120]. One of the patients was treated with dexamethasone and CsA and the other two with IVIg, and all recovered. In a large single-center study of HLH in children under 1 year of age, three of four infants with adenovirus-associated HLH survived [116]. A case in an adult while receiving chemotherapy for a solid tumor has been reported [121]. A young child being treated with chemotherapy for Langerhans cell histiocytosis developed recurrent viral-associated HLH, in one instance caused by adenovirus [117]. Several cases of adenovirus-associated HLH have been reported post-hematopoietic stem cell transplantation [113], associated with sJIA [119], and in brothers with X-linked agammaglobulinemia [124].

Parvovirus

Parvovirus B19 is the cause of erythema infectiosum (fifth disease or "slapped cheek syndrome"), which is common in childhood [71]. In addition to a rash, adults more frequently develop arthralgia and myalgia.

CSS has been reported in adults and children in association with parvovirus B19 infection [125–132]. One report detailed five previously healthy adults whose disease resolved spontaneously [133]. In other cases, the most frequent underlying disease was hereditary spherocytosis [4, 128]. Cases of CSS triggered by parvovirus B19 in patients with another underlying condition have been reported: post-renal transplant [134], post kidney-pancreas transplant [135], B-cell acute lymphoblastic leukemia [136], autoimmune hemolytic anemia [137], Evans syndrome (autoimmune hemolytic anemia and autoimmune thrombocytopenia) [138], and pregnancy [139].

Coinfection with parvovirus B19 and another pathogen has been reported in association with HLH in several cases, including EBV [126, 140] and Klebsiella [141].

Among 28 cases of parvovirus-associated HLH, the majority were woman over 15 years of age, and 22 survived despite 16 of them having no specific treatment [142]. This suggests a better prognosis than other forms of viral-associated CSS [4], although fatalities and serious complications such as acute myocarditis have been reported [143, 144]. Case reports have detailed use of glucocorticoids, IVIg, CsA, and anakinra (IL-1 receptor antagonist) in treatment of parvovirus B19-associated HLH [145].

In some of the cases of HLH attributed to parvovirus B19, viral nucleic acid was detected in blood or tissues by PCR. It is known that virus may persist for weeks or months, and therefore detection of viral DNA in tissues does not definitively confirm acute infection [71]. The most reliable marker of this is detection of virus-specific IgM and a fourfold increase or seroconversion of IgG in paired serum samples. However, care should be taken in interpretation of serology samples in patients after treatment with IVIg.

Hepatitis Viruses

Various hepatitis viruses have been detected in association with CSS. Hepatitis A virus is the most-frequently reported [146–157]. Features of fulminant acute viral hepatitis may be similar to CSS. Hepatitis A-triggered MAS has been reported in several patients with underlying systemic JIA or Still disease [148, 154, 157]. Successful treatments for hepatitis A-triggered HLH have included glucocorticoids, IVIg [150, 156], and the HLH-2004 etoposide-based protocol [154]. In addition to hepatitis A, other hepatitis viruses have also been found to cause HLH.

Hepatitis B virus has been reported as a potential trigger for CSS [158, 159]. The first reported case did not respond to steroids, IVIg, or CsA, but the patient did respond to etoposide, although subsequently succumbed to fulminant infection [158]. A fatal case of CSS was reported in a patient with the combination of chronic active hepatitis B and acute hepatitis C infection despite intensive immunosuppressive (intravenous methylprednisolone, intravenous CsA, granulocyte-colony stimulating factor, IVIg, and anti-thymocyte globulin), and supportive treatment [160].

CSS was identified in a 60-year-old woman with chronic hepatitis C infection [161]. In this case, it was speculated whether the more acute triggers for development of sHLH were the interferon and ribavirin used as treatment for hepatitis C virus which were started 3 months before the characteristic features of CSS: fever, splenomegaly, coagulopathy, anemia, and thrombocytopenia.

Hepatitis E is typically a self-limited illness with average duration of 4–6 weeks and presenting with fever, nausea, vomiting, abdominal pain, anorexia, hepatomegaly, and jaundice [162]. CSS associated with hepatitis E infection has been reported in a small number of cases [163–165]. In one patient with rheumatoid arthritis, MAS developed within 24 h of her fourth tocilizumab infusion. Investigations for infections revealed positive serology, and hepatitis E virus RNA was detected in blood and stool by PCR [164]. In a second patient, sHLH appeared to be triggered by coinfection with hepatitis A and hepatitis E [165].

Measles, Mumps, and Rubella Viruses

CSS has occurred in a small number of cases following measles infection [18, 75, 76, 166]. In eight of the cases patients developed interstitial pneumonia. The reported therapies included supportive treatment alone, intravenous methylprednisolone, or the HLH-2004 protocol. One case of sHLH following measles vaccination has been reported in a 19-month-old girl who developed persistent fever, hepatosplenomegaly, pancytopenia, liver dysfunction, and hemophagocytosis 1 week after vaccination [167]. A second case of suspected sHLH was reported in a previously healthy 14-month-old girl following the combined measles, mumps and rubella (MMR) vaccination and the authors speculate whether there may have been an underlying genetic predisposition [168].

Mumps infection has rarely been associated with CSS [78, 169]. In one case from China, a previously healthy 21-year-old male had persistent painful parotid gland swelling for 30 days and fever for 15 days together with typical features of CSS, including hemophagocytic macrophage infiltration on bone marrow biopsy. Anti-mumps virus IgM was positive but testing for bacteria and other viruses in peripheral blood was negative. Although there was initial response to high doses of methylprednisolone, IVIg, and etoposide, he succumbed 4 weeks after admission. In the second case, a 39-year-old female with parotitis and acute pancreatitis followed by features of CSS achieved complete response with corticosteroids.

Rubella virus-associated HLH has been reported in patients with ages ranging from young infants to adults [170–173]. In one case of a 26-year-old woman, serology was positive for both varicella-zoster virus (VZV) IgM and rubella virus IgM suggesting dual infection.

Enterovirus

Enteroviruses, which include the Coxsackieviruses and echoviruses, can cause a wide range of clinical presentations from mild respiratory and gastrointestinal infections, and hand-foot-and-mouth disease, to more severe conditions such as pancreatitis, meningitis, encephalitis, and neonatal sepsis [81]. Over 10 cases of enterovirus-associated CSS have been reported in the literature [116, 174–181]. One case occurred in an adult infected with Coxsackie virus A9 [174]. However, a significant proportion were in neonates or infants [116, 175, 179–181]. A case of vertical transmission of Coxsackie virus B1 leading to HLH in a 4-day-old neonate was reported with virus isolated in the throat and stool, and exclusion of inherited disease with normal perforin and CD107a expression [180]. The baby was successfully treated with corticosteroids, CsA, and etoposide. A case of fatal HLH was described in a 4-month-old infant with liver dysfunction (AST 626 IU/L, ALT 121 IU/L), high ferritin (1100 ng/mL), and hemophagocytosis in a liver biopsy [176].

Parechovirus

Virus of the *Parechovirus* genus are within the family *Picornaviridae* and were previously considered to be within the *Enterovirus* genus [81]. Human parechovirus-3 (HPeV-3) in neonates and infants can manifest with a sepsis-like presentation, and one publication has reported young infants with fever, rash, leukopenia, thrombocytopenia, and hyperferritinemia [181]. This and other studies have suggested that HPeV-3 can cause an HLH-like illness, although the reason why other types of HPeV do not seem to cause a similar febrile illness is not clear [82].

Rotavirus

Rotavirus is one of the leading causes of infectious, dehydrating gastroenteritis in children globally with over 200,000 deaths reported annually [84]. CSS associated with rotavirus infection has been described in a small number of cases in the published literature [182–184]. In two of the cases there were significant underlying conditions. In the first, a 67-year-old man developed rotavirus enteritis 1 month after live-unrelated renal transplantation, and he developed fever, pancytopenia, altered consciousness, elevated liver enzymes, hypofibrinogenemia, and hyperferritinemia [182]. He also had herpes zoster infection and varicella zoster virus DNA was detected in his CSF. Bone marrow analysis confirmed hemophagocytosis, and he responded to acyclovir and pulsed methylprednisolone therapy. The second case was a 3-year-old child approximately 30 months after allogeneic hematopoietic stem cell transplantation for familial HLH [183]. She developed fever, diarrhea and lethargy, and she progressed to multiorgan failure meeting six of the HLH criteria on day 2 of admission. At this stage, rotavirus antigen was detected in stool and all other bacterial, viral, and fungal testing was negative. She received high dose corticosteroids and IVIg but developed invasive fungal infection and succumbed 4 weeks later.

Human T-Lymphotropic Virus

Human T-lymphotropic viruses (HTLVs) belong to the family of retroviruses. HTLV type-1 (HTLV-1) is associated with adult T cell leukemia/lymphoma, demyelinating disease, and autoimmune conditions [85]. A case has been reported of a woman who was a carrier of HTLV-1 who was diagnosed as having adult T-cell leukemia/lymphoma and B cell lymphoma-associated hemophagocytic syndrome [185]. HTLV type-3, the cause of AIDS, and its association with CSS is covered in a separate chapter in this textbook.

Conclusion

Viruses are the single most common infectious trigger for the final common pathway resulting in CSS. Identification of the pathogen usually requires serological or PCR testing, although caution is required in interpretation of these investigations in relation to the timing of the acute infection. In many of the reported cases, development of CSS appeared to be multifactorial involving more than a single infectious agent sometimes on a background of genetic predisposition, malignancy, immune suppression, or rheumatological disease. In the acute setting of life-threatening CSS, determining the etiology is subsidiary to management with multiorgan supportive therapy and high-dose glucocorticoids. Other than a modified HLH-94 treatment protocol for influenza A (H5N1)-associated HLH, there is no evidence from controlled trials for a particular treatment regime based on the triggering virus.

Author Contributions E. S. Sen performed a literature review and wrote the chapter. A. V. Ramanan made contributions to reviewing and editing of the manuscript before submission.
Competing Interests Statement
E. S. Sen declares no competing interests.
A. V. Ramanan has received speaker fees from SOBI.

References

1. Sen, E. S., Clarke, S. L., & Ramanan, A. V. (2016). Macrophage activation syndrome. *Indian Journal of Pediatrics, 83*, 248–253.
2. Sen, E. S., Steward, C. G., & Ramanan, A. V. (2017). Diagnosing haemophagocytic syndrome. *Archives of Disease in Childhood, 102*, 279–284.
3. Janka, G., Imashuku, S., Elinder, G., Schneider, M., & Henter, J. I. (1998). Infection- and malignancy-associated hemophagocytic syndromes. Secondary hemophagocytic lymphohistiocytosis. *Hematology/Oncology Clinics of North America, 12*, 435–444.
4. Maakaroun, N. R., Moanna, A., Jacob, J. T., & Albrecht, H. (2010). Viral infections associated with haemophagocytic syndrome. *Reviews in Medical Virology, 20*, 93–105.
5. World Health Organisation. (2017). Haemorrhagic fevers, viral. http://www.who.int/topics/haemorrhagic_fevers_viral/en/.
6. Centers for Disease Control and Prevention. (2014). Viral hemorrhagic fevers (VHFs). https://www.cdc.gov/vhf/.
7. Cascio, A., Pernice, L. M., Barberi, G., Delfino, D., Biondo, C., Beninati, C., et al. (2012). Secondary hemophagocytic lymphohistiocytosis in zoonoses. A systematic review. *European Review for Medical and Pharmacological Sciences, 16*, 1324–1337.
8. Waxman, M. A. (2014). Update on emerging infections from the Centers for Disease Control and Prevention. Fatal hemophagocytic lymphohistiocytosis associated with locally acquired dengue virus infection-New Mexico and Texas, 2012. *Annals of Emergency Medicine, 64*, 55–57. discussion 58.
9. Ab-Rahman, H. A., Wong, P. F., Rahim, H., Abd-Jamil, J., Tan, K. K., Sulaiman, S., et al. (2015). Dengue death with evidence of hemophagocytic syndrome and dengue virus infection in the bone marrow. *Springerplus, 4*, 665.
10. Ab-Rahman, H. A., Rahim, H., AbuBakar, S., & Wong, P. F. (2016). Macrophage activation syndrome-associated markers in severe dengue. *International Journal of Medical Sciences, 13*, 179–186.
11. Arshad, U., Ahmad, S. Q., & Khan, F. (2015). Hemophagocytic lymphohistiocytosis in a patient with dengue infection. *Hematology/Oncology and Stem Cell Therapy, 8*, 189–190.
12. Chung, S. M., Song, J. Y., Kim, W., Choi, M. J., Jeon, J. H., Kang, S., et al. (2017). Dengue-associated hemophagocytic lymphohistiocytosis in an adult: A case report and literature review. *Medicine, 96*, e6159.
13. De Koninck, A. S., Dierick, J., Steyaert, S., & Taelman, P. (2014). Hemophagocytic lymphohistiocytosis and dengue infection: Rare case report. *Acta Clinica Belgica, 69*, 210–213.
14. Ray, S., Kundu, S., Saha, M., & Chakrabarti, P. (2011). Hemophagocytic syndrome in classic dengue fever. *Journal of Global Infectious Diseases, 3*, 399–401.
15. Ellis, E. M., Sharp, T. M., Perez-Padilla, J., Gonzalez, L., Poole-Smith, B. K., Lebo, E., et al. (2016). Incidence and risk factors for developing dengue-associated hemophagocytic lymphohistiocytosis in Puerto Rico, 2008–2013. *PLoS Neglected Tropical Diseases, 10*, e0004939.

16. Jain, D., & Singh, T. (2008). Dengue virus related hemophagocytosis: A rare case report. *Hematology (Amsterdam, Netherlands), 13*, 286–288.
17. Jasmine, Y. S., Lee, S. L., & Kan, F. K. (2017). Infection associated haemophagocytic syndrome in severe dengue infection - a case series in a district hospital. *The Medical Journal of Malaysia, 72*, 62–64.
18. Joshi, R., Phatarpekar, A., Currimbhoy, Z., & Desai, M. (2011). Haemophagocytic lymphohistiocytosis: A case series from Mumbai. *Annals of Tropical Paediatrics, 31*, 135–140.
19. Kapdi, M., & Shah, I. (2012). Dengue and haemophagocytic lymphohistiocytosis. *Scandinavian Journal of Infectious Diseases, 44*, 708–709.
20. Kobayashi, K., Hikone, M., Sakamoto, N., Iwabuchi, S., Kashiura, M., Takasaki, T., et al. (2015). Dengue-associated hemophagocytic syndrome in a Japanese traveler: A case report. *Journal of Travel Medicine, 22*, 64–66.
21. Koshy, M., Mishra, A. K., Agrawal, B., Kurup, A. R., & Hansdak, S. G. (2016). Dengue fever complicated by hemophagocytosis. *Oxford Medical Case Reports, 2016*, 121–124.
22. Krithika, M. V., Amboiram, P., Latha, S. M., Ninan, B., Suman, F. R., & Scott, J. (2017). Neonate with haemophagocytic lymphohistiocytosis secondary to dengue infection: A case report. *Tropical Doctor, 47*, 253–255.
23. Lakhotia, M., Pahadiya, H. R., Gandhi, R., Prajapati, G. R., & Choudhary, A. (2016). Stuck with pancytopenia in dengue fever: Evoke for hemophagocytic syndrome. *Indian journal of critical care medicine : peer-reviewed, official publication of Indian Society of. Critical Care Medicine, 20*, 55–56.
24. Lu, P. L., Hsiao, H. H., Tsai, J. J., Chen, T. C., Feng, M. C., Chen, T. P., et al. (2005). Dengue virus-associated hemophagocytic syndrome and dyserythropoiesis: A case report. *The Kaohsiung Journal of Medical Sciences, 21*, 34–39.
25. Mitra, S., & Bhattacharyya, R. (2014). Hemophagocytic syndrome in severe dengue fever: A rare presentation. *Indian Journal of Hematology & Blood Transfusion, 30*, 97–100.
26. My, L. T., Lien le, B., Hsieh, W. C., Imamura, T., Anh, T. N., Anh, P. N., et al. (2010). Comprehensive analyses and characterization of haemophagocytic lymphohistiocytosis in Vietnamese children. *British Journal of Haematology, 148*, 301–310.
27. Pal, P., Giri, P. P., & Ramanan, A. V. (2014). Dengue associated hemophagocytic lymphohistiocytosis: A case series. *Indian Pediatrics, 51*, 496–497.
28. Pongtanakul, B., Narkbunnam, N., Veerakul, G., Sanpakit, K., Viprakasit, V., Tanphaichitr, V. T., et al. (2005). Dengue hemorrhagic fever in patients with thalassemia. *Journal of the Medical Association of Thailand = Chotmaihet Thangphaet, 88*(Suppl 8), S80–S85.
29. Raju, S., Kalyanaraman, S., Swaminathan, K., Nisha, A., & Praisid, S. (2014). Hemophagocytic lymphohistiocytosis syndrome in dengue hemorrhagic fever. *Indian Journal of Pediatrics, 81*, 1381–1383.
30. Ramanathan, M., & Duraisamy, G. (1991). Haemophagocytosis in dengue haemorrhagic fever: A case report. *Annals of the Academy of Medicine, Singapore, 20*, 803–804.
31. Ray, U., Dutta, S., Mondal, S., & Bandyopadhyay, S. (2017). Severe dengue due to secondary hemophagocytic lymphohistiocytosis: A case study. *IDCases, 8*, 50–53.
32. Ribeiro, E., Kassab, S., Pistone, T., Receveur, M. C., Fialon, P., & Malvy, D. (2014). Primary dengue fever associated with hemophagocytic syndrome: A report of three imported cases, Bordeaux France. *Internal Medicine, 53*, 899–902.
33. Sharp, T. M., Gaul, L., Muehlenbachs, A., Hunsperger, E., Bhatnagar, J., Lueptow, R., et al. (2014). Fatal hemophagocytic lymphohistiocytosis associated with locally acquired dengue virus infection - New Mexico and Texas, 2012. *MMWR. Morbidity and Mortality Weekly Report, 63*, 49–54.
34. Soler Rosario, Y., Garcia, R., & Fernandez Sein, A. (2010). Dengue virus associated hemophagocytic syndrome in children: A case report. *Boletín de la Asociación Médica de Puerto Rico, 102*, 49–54.
35. Srichaikul, T., Punyagupta, S., Kanchanapoom, T., Chanokovat, C., Likittanasombat, K., & Leelasiri, A. (2008). Hemophagocytic syndrome in dengue hemorrhagic fever with severe

multiorgan complications. *Journal of the Medical Association of Thailand = Chotmaihet Thangphaet, 91*, 104–109.
36. Tan, L. H., Lum, L. C., Omar, S. F., & Kan, F. K. (2012). Hemophagocytosis in dengue: Comprehensive report of six cases. *Journal of Clinical Virology, 55*, 79–82.
37. Wan Jamaludin, W. F., Periyasamy, P., Wan Mat, W. R., & Abdul Wahid, S. F. (2015). Dengue infection associated hemophagocytic syndrome: Therapeutic interventions and outcome. *Journal of Clinical Virology, 69*, 91–95.
38. Yoshifuji, K., Oshina, T., Sonokawa, S., Noguchi, Y., Suzuki, S., Tanaka, K., et al. (2016). Domestic dengue infection with hemophagocytic lymphohistiocytosis successfully treated by early steroid therapy. *Rinsho ketsueki, 57*, 864–868.
39. Hein, N., Bergara, G. H., Moura, N. B., Cardoso, D. M., Hirose, M., Ferronato, A. E., et al. (2015). Dengue fever as a cause of hemophagocytic lymphohistiocytosis. *Autopsy & Case Reports, 5*, 33–36.
40. Betancur, J. F., Navarro, E. P., Echeverry, A., Moncada, P. A., Canas, C. A., & Tobon, G. J. (2015). Hyperferritinemic syndrome: Still's disease and catastrophic antiphospholipid syndrome triggered by fulminant Chikungunya infection: A case report of two patients. *Clinical Rheumatology, 34*, 1989–1992.
41. Bicakci, Z., Tavil, B., Tezer, H., & Olcay, L. (2013). Hemophagocytosis in a case with Crimean-Congo hemorrhagic fever and an overview of possible pathogenesis with current evidence. *The Turkish Journal of Pediatrics, 55*, 344–348.
42. Erduran, E., & Cakir, M. (2010). Reactive hemophagocytic lymphohistiocytosis and Crimean-Congo hemorrhagic fever. In *International journal of infectious diseases. IJID : official publication of the International Society for Infectious Diseases*, Canada. e349; author reply e350.
43. Tasdelen Fisgin, N., Fisgin, T., Tanyel, E., Doganci, L., Tulek, N., Guler, N., et al. (2008). Crimean-Congo hemorrhagic fever: Five patients with hemophagocytic syndrome. *American Journal of Hematology, 83*, 73–76.
44. Lee, J. J., Chung, I. J., Shin, D. H., Cho, S. H., Cho, D., Ryang, D. W., et al. (2002). Hemorrhagic fever with renal syndrome presenting with hemophagocytic lymphohistiocytosis. *Emerging Infectious Diseases, 8*, 209–210.
45. Lee, J., Jeong, G., Lim, J. H., Kim, H., Park, S. W., Lee, W. J., et al. (2016). Severe fever with thrombocytopenia syndrome presenting with hemophagocytic lymphohistiocytosis. *Infect Chemother, 48*(4), 338–341.
46. Lin, L., Xu, Y. Z., Wu, X. M., Ge, H. F., Feng, J. X., Chen, M. F., et al. (2016). A rare fatal case of a novel bunyavirus-associated hemophagocytic lymphohistiocytosis. *Journal of Infection in Developing Countries, 10*, 533–536.
47. Nakano, A., Ogawa, H., Nakanishi, Y., Fujita, H., Mahara, F., Shiogama, K., et al. (2017). Hemophagocytic lymphohistiocytosis in a fatal case of severe fever with thrombocytopenia syndrome. *Internal Medicine, 56*, 1597–1602.
48. Wiwanitkit, V. (2010). Dengue fever: Diagnosis and treatment. *Expert Review of Anti-Infective Therapy, 8*, 841–845.
49. Hasan, S., Jamdar, S. F., Alalowi, M., Al Ageel, S. M., & Beaiji, A. (2016). Dengue virus: A global human threat: Review of literature. *Journal of International Society of Preventive and Community Dentistry, 6*, 1–6.
50. Kularatne, S. A. (2015). Dengue fever. *BMJ, 351*, h4661.
51. Rajapakse, S., Ranasinghe, C., & Rodrigo, C. (2010). Corticosteroid therapy in dengue infection- opinions of junior doctors. *Journal of Global Infectious Diseases, 2*, 199–200.
52. Tam, D. T., Ngoc, T. V., Tien, N. T., Kieu, N. T., Thuy, T. T., Thanh, L. T., et al. (2012). Effects of short-course oral corticosteroid therapy in early dengue infection in Vietnamese patients: A randomized, placebo-controlled trial. *Clinical Infectious Diseases, 55*, 1216–1224.
53. Zhang, F., & Kramer, C. V. (2014). Corticosteroids for dengue infection. *Cochrane Database of Systematic Reviews, 7*, CD003488.

54. Rajapakse, S., Rodrigo, C., Maduranga, S., & Rajapakse, A. C. (2014). Corticosteroids in the treatment of dengue shock syndrome. *Infection and Drug Resistance, 7*, 137–143.
55. Premaratna, R., Jayasinghe, K. G., Liyanaarachchi, E. W., Weerasinghe, O. M., Pathmeswaran, A., & de Silva, H. J. (2011). Effect of a single dose of methyl prednisolone as rescue medication for patients who develop hypotensive dengue shock syndrome during the febrile phase: A retrospective observational study. *International Journal of Infectious Diseases, 15*, e433–e434.
56. Wiwanitkit, V. (2015). Haemophagocytic lymphohistiocytosis and dengue. *Acta Clinica Belgica, 70*, 72.
57. Ramachandran, B., Balasubramanian, S., Abhishek, N., Ravikumar, K. G., & Ramanan, A. V. (2011). Profile of hemophagocytic lymphohistiocytosis in children in a tertiary care hospital in India. *Indian Pediatrics, 48*, 31–35.
58. van de Weg, C. A., Huits, R. M., Pannuti, C. S., Brouns, R. M., van den Berg, R. W., van den Ham, H. J., et al. (2014). Hyperferritinaemia in dengue virus infected patients is associated with immune activation and coagulation disturbances. *PLoS Neglected Tropical Diseases, 8*, e3214.
59. Martina, B. E., Koraka, P., & Osterhaus, A. D. (2009). Dengue virus pathogenesis: An integrated view. *Clinical Microbiology Reviews, 22*, 564–581.
60. Dilber, E., Cakir, M., Erduran, E., Koksal, I., Bahat, E., Mutlu, M., et al. (2010). High-dose methylprednisolone in children with Crimean-Congo haemorrhagic fever. *Tropical Doctor, 40*, 27–30.
61. Erduran, E., Bahadir, A., Palanci, N., & Gedik, Y. (2013). The treatment of crimean-Congo hemorrhagic fever with high-dose methylprednisolone, intravenous immunoglobulin, and fresh frozen plasma. *Journal of Pediatric Hematology/Oncology, 35*, e19–e24.
62. McElroy, A. K., Erickson, B. R., Flietstra, T. D., Rollin, P. E., Nichol, S. T., Towner, J. S., et al. (2014). Ebola hemorrhagic fever: Novel biomarker correlates of clinical outcome. *The Journal of Infectious Diseases, 210*, 558–566.
63. van der Ven, A. J., Netea, M. G., van der Meer, J. W., & de Mast, Q. (2015). Ebola virus disease has features of hemophagocytic lymphohistiocytosis syndrome. *Frontiers in Medicine, 2*, 4.
64. Cron, R. Q., Behrens, E. M., Shakoory, B., Ramanan, A. V., & Chatham, W. W. (2015). Does viral hemorrhagic fever represent reactive Hemophagocytic syndrome? *The Journal of Rheumatology, 42*, 1078–1080.
65. Samransamruajkit, R., Hiranrat, T., Chieochansin, T., Sritippayawan, S., Deerojanawong, J., Prapphal, N., et al. (2008). Prevalence, clinical presentations and complications among hospitalized children with influenza pneumonia. *Japanese Journal of Infectious Diseases, 61*, 446–449.
66. Beutel, G., Wiesner, O., Eder, M., Hafer, C., Schneider, A. S., Kielstein, J. T., et al. (2011). Virus-associated hemophagocytic syndrome as a major contributor to death in patients with 2009 influenza A (H1N1) infection. *Critical Care (London, England), 15*, R80.
67. Centers for Disease Control and Prevention. (2018). Influenza (Flu). https://www.cdc.gov/flu.
68. Centers for Disease Control and Prevention. (2017). Human Parainfluenza Viruses (HPIVs). https://www.cdc.gov/parainfluenza.
69. Centers for Disease Control and Prevention. (2018). Adenoviruses. https://www.cdc.gov/adenovirus.
70. Echavarría, M. (2008). Adenoviruses in immunocompromised hosts. *Clinical Microbiology Reviews, 21*, 704–715.
71. Qiu, J., Söderlund-Venermo, M., & Young, N. S. (2017). Human parvoviruses. *Clinical Microbiology Reviews, 30*, 43–113.
72. Centers for Disease Control and Prevention. (2018). Viral Hepatitis. https://www.cdc.gov/hepatitis.
73. Inman, R. D., Hodge, M., Johnston, M. E., Wright, J., & Heathcote, J. (1986). Arthritis, vasculitis, and cryoglobulinemia associated with relapsing hepatitis a virus infection. *Annals of Internal Medicine, 105*, 700–703.

74. Schiff, E. R. (1992). Atypical clinical manifestations of hepatitis a. *Vaccine, 10*(Suppl 1), S18–S20.
75. Iaria, C., Leonardi, M. S., Buda, A., Toro, M. L., & Cascio, A. (2012). Measles and secondary hemophagocytic lymphohistiocytosis. *Emerging Infectious Diseases, 18*, 1529. author reply 1529-1530.
76. Komatsuda, A., Chubachi, A., & Miura, A. B. (1995). Virus-associated hemophagocytic syndrome due to measles accompanied by acute respiratory failure. *Internal Medicine, 34*, 203–206.
77. Centers for Disease Control and Prevention. (2018). Measles (Rubeola). https://www.cdc.gov/measles.
78. Hiraiwa, K., Obara, K., & Sato, A. (2005). Mumps virus-associated hemophagocytic syndrome. *Emerging Infectious Diseases, 11*, 343.
79. Centers for Disease Control and Prevention. (2018). Mumps. https://www.cdc.gov/mumps.
80. Centers for Disease Control and Prevention. (2017). Rubella (German measles, three day measles). https://www.cdc.gov/rubella.
81. de Crom, S. C., Rossen, J. W., van Furth, A. M., & Obihara, C. C. (2016). Enterovirus and parechovirus infection in children: A brief overview. *European Journal of Pediatrics, 175*, 1023–1029.
82. Yuzurihara, S. S., Ao, K., Hara, T., Tanaka, F., Mori, M., Kikuchi, N., et al. (2013). Human parechovirus-3 infection in nine neonates and infants presenting symptoms of hemophagocytic lymphohistiocytosis. *Journal of Infection and Chemotherapy, 19*, 144–148.
83. Vergnano, S., Kadambari, S., Whalley, K., Menson, E. N., Martinez-Alier, N., Cooper, M., et al. (2015). Characteristics and outcomes of human parechovirus infection in infants (2008-2012). *European Journal of Pediatrics, 174*, 919–924.
84. Crawford, S. E., Ramani, S., Tate, J. E., Parashar, U. D., Svensson, L., Hagbom, M., et al. (2017). Rotavirus infection. *Nature Reviews. Disease Primers, 3*, 17083.
85. Quaresma, J. A., Yoshikawa, G. T., Koyama, R. V., Dias, G. A., Fujihara, S., & Fuzii, H. T. (2015). HTLV-1, immune response and autoimmunity. *Viruses, 8*, E5.
86. Potter, M. N., Foot, A. B., & Oakhill, A. (1991). Influenza A and the virus associated haemophagocytic syndrome: Cluster of three cases in children with acute leukaemia. *Journal of Clinical Pathology, 44*, 297–299.
87. Ando, M., Miyazaki, E., Hiroshige, S., Ashihara, Y., Okubo, T., Ueo, M., et al. (2006). Virus associated hemophagocytic syndrome accompanied by acute respiratory failure caused by influenza A (H3N2). *Internal Medicine, 45*, 1183–1186.
88. Watanabe, T., Okazaki, E., & Shibuya, H. (2003). Influenza A virus-associated encephalopathy with haemophagocytic syndrome. *European Journal of Pediatrics, 162*, 799–800.
89. Mou, S. S., Nakagawa, T. A., Riemer, E. C., McLean, T. W., Hines, M. H., & Shetty, A. K. (2006). Hemophagocytic lymphohistiocytosis complicating influenza A infection. *Pediatrics, 118*, e216–e219.
90. Zhang, Z., Zhang, J., Huang, K., Li, K. S., Yuen, K. Y., Guan, Y., et al. (2009). Systemic infection of avian influenza A virus H5N1 subtype in humans. *Human Pathology, 40*, 735–739.
91. To, K. F., Chan, P. K., Chan, K. F., Lee, W. K., Lam, W. Y., Wong, K. F., et al. (2001). Pathology of fatal human infection associated with avian influenza A H5N1 virus. *Journal of Medical Virology, 63*, 242–246.
92. Harms, P. W., Schmidt, L. A., Smith, L. B., Newton, D. W., Pletneva, M. A., Walters, L. L., et al. (2010). Autopsy findings in eight patients with fatal H1N1 influenza. *American Journal of Clinical Pathology, 134*, 27–35.
93. Katsumi, A., Nishida, T., Murata, M., Terakura, S., Shimada, K., Saito, S., et al. (2011). Virus-associated hemophagocytic syndrome caused by pandemic swine-origin influenza A (H1N1) in a patient after unrelated bone marrow transplantation. *Journal of Clinical and Experimental Hematopathology, 51*, 63–65.

94. Ozdemir, H., Ciftci, E., Ince, E. U., Ertem, M., Ince, E., & Dogru, U. (2011). Hemophagocytic lymphohistiocytosis associated with 2009 pandemic influenza A (H1N1) virus infection. *Journal of Pediatric Hematology/Oncology, 33*, 135–137.
95. Willekens, C., Cornelius, A., Guerry, M. J., Wacrenier, A., & Fourrier, F. (2011). Fulminant hemophagocytic lymphohistiocytosis induced by pandemic A (H1N1) influenza: A case report. *Journal of Medical Case Reports, 5*, 280.
96. Zhang, X. Y., Ye, X. W., Feng, D. X., Han, J., Li, D., & Zhang, C. (2011). Hemophagocytic lymphohistiocytosis induced by severe pandemic influenza A (H1N1) 2009 virus infection: A case report. *Case Reports in Medicine, 2011*, 951910.
97. Lai, S., Merritt, B. Y., Chen, L., Zhou, X., & Green, L. K. (2012). Hemophagocytic lymphohistiocytosis associated with influenza A (H1N1) infection in a patient with chronic lymphocytic leukemia: An autopsy case report and review of the literature. *Annals of Diagnostic Pathology, 16*, 477–484.
98. Yontem, Y., Ilker, D., Yesim, O., Aysen, T., Gulcihan, O., Ozgur, C., et al. (2013). Analysis of fatal cases of pandemic influenza A (H1N1) virus infections in pediatric patients with leukemia. *Pediatric Hematology and Oncology, 30*, 437–444.
99. Casciaro, R., Cresta, F., Favilli, F., Naselli, A., De Alessandri, A., & Minicucci, L. (2014). Macrophage activation syndrome induced by a/H1N1 influenza in cystic fibrosis. *Pediatric Pulmonology, 49*, E10–E12.
100. Shrestha, B., Omran, A., Rong, P., & Wang, W. (2015). Report of a fatal pediatric case of hemophagocytic lymphohistiocytosis associated with pandemic influenza A (H1N1) infection in 2009. *Pediatrics and Neonatology, 56*, 189–192.
101. Schulert, G. S., Zhang, M., Fall, N., Husami, A., Kissell, D., Hanosh, A., et al. (2016). Whole-exome sequencing reveals mutations in genes linked to hemophagocytic lymphohistiocytosis and macrophage activation syndrome in fatal cases of H1N1 influenza. *The Journal of Infectious Diseases, 213*, 1180–1188.
102. de Jong, M. D., Bach, V. C., Phan, T. Q., Vo, M. H., Tran, T. T., Nguyen, B. H., et al. (2005). Fatal avian influenza A (H5N1) in a child presenting with diarrhea followed by coma. *The New England Journal of Medicine, 352*, 686–691.
103. Ikebe, T., Takata, H., Sasaki, H., Miyazaki, Y., Ohtsuka, E., Saburi, Y., et al. (2017). Hemophagocytic lymphohistiocytosis following influenza vaccination in a patient with aplastic anemia undergoing allogeneic bone marrow stem cell transplantation. *International Journal of Hematology, 105*, 389–391.
104. Chokephaibulkit, K., Uiprasertkul, M., Puthavathana, P., Chearskul, P., Auewarakul, P., Dowell, S. F., et al. (2005). A child with avian influenza A (H5N1) infection. *The Pediatric Infectious Disease Journal, 24*, 162–166.
105. Chan, P. K. (2002). Outbreak of avian influenza A (H5N1) virus infection in Hong Kong in 1997. *Clinical Infectious Diseases, 34*(Suppl 2), S58–S64.
106. Ng, W. F., To, K. F., Lam, W. W., Ng, T. K., & Lee, K. C. (2006). The comparative pathology of severe acute respiratory syndrome and avian influenza A subtype H5N1--a review. *Human Pathology, 37*, 381–390.
107. Hsieh, S. M., & Chang, S. C. (2006). Insufficient perforin expression in CD8+ T cells in response to hemagglutinin from avian influenza (H5N1) virus. *Journal of Immunology, 176*, 4530–4533.
108. Peiris, J. S., Yu, W. C., Leung, C. W., Cheung, C. Y., Ng, W. F., Nicholls, J. M., et al. (2004). Re-emergence of fatal human influenza A subtype H5N1 disease. *Lancet, 363*, 617–619.
109. Henter, J. I., Chow, C. B., Leung, C. W., & Lau, Y. L. (2006). Cytotoxic therapy for severe avian influenza A (H5N1) infection. *Lancet, 367*, 870–873.
110. Henter, J. I., Samuelsson-Horne, A., Aricò, M., Egeler, R. M., Elinder, G., Filipovich, A. H., et al. (2002). Treatment of hemophagocytic lymphohistiocytosis with HLH-94 immunochemotherapy and bone marrow transplantation. *Blood, 100*, 2367–2373.

111. Horai, Y., Miyamura, T., Takahama, S., Sonomoto, K., Nakamura, M., Ando, H., et al. (2010). Influenza virus B-associated hemophagocytic syndrome and recurrent pericarditis in a patient with systemic lupus erythematosus. *Modern Rheumatology, 20*, 178–182.
112. Beffermann, N., Pilcante, J., & Sarmiento, M. (2015). Acquired hemophagocytic syndrome related to parainfluenza virus infection: Case report. *Journal of Medical Case Reports, 9*, 78.
113. Levy, J., Wodell, R. A., August, C. S., & Bayever, E. (1990). Adenovirus-related hemophagocytic syndrome after bone marrow transplantation. *Bone Marrow Transplantation, 6*, 349–352.
114. Seidel, M. G., Kastner, U., Minkov, M., & Gadner, H. (2003). IVIG treatment of adenovirus infection-associated macrophage activation syndrome in a two-year-old boy: Case report and review of the literature. *Pediatric Hematology and Oncology, 20*, 445–451.
115. Morimoto, A., Teramura, T., Asazuma, Y., Mukoyama, A., & Imashuku, S. (2003). Hemophagocytic syndrome associated with severe adenoviral pneumonia: Usefulness of real-time polymerase chain reaction for diagnosis. *International Journal of Hematology, 77*, 295–298.
116. Imashuku, S., Ueda, I., Teramura, T., Mori, K., Morimoto, A., Sako, M., et al. (2005). Occurrence of haemophagocytic lymphohistiocytosis at less than 1 year of age: Analysis of 96 patients. *European Journal of Pediatrics, 164*, 315–319.
117. Klein, A., Corazza, F., Demulder, A., Van Beers, D., & Ferster, A. (1999). Recurrent viral associated hemophagocytic syndrome in a child with Langerhans cell histiocytosis. *Journal of Pediatric Hematology/Oncology, 21*, 554–556.
118. Iyama, S., Matsunaga, T., Fujimi, A., Murase, K., Kuribayasi, K., Sato, T., et al. (2005). Successful treatment with oral ribavirin of adenovirus-associated hemophagocytic syndrome in a stem cell transplantation recipient. *Rinshō Ketsueki, 46*, 363–367.
119. Lin, C. I., Yu, H. H., Lee, J. H., Wang, L. C., Lin, Y. T., Yang, Y. H., et al. (2012). Clinical analysis of macrophage activation syndrome in pediatric patients with autoimmune diseases. *Clinical Rheumatology, 31*, 1223–1230.
120. Hosnut, F. O., Ozcay, F., Malbora, B., Hizli, S., & Ozbek, N. (2014). Severe adenovirus infection associated with hemophagocytic lymphohistiocytosis. *Turkish Journal of Haematology, 31*, 103–105.
121. Mellon, G., Henry, B., Aoun, O., Boutolleau, D., Laparra, A., Mayaux, J., et al. (2016). Adenovirus related lymphohistiocytic hemophagocytosis: Case report and literature review. *Journal of Clinical Virology, 78*, 53–56.
122. Reardon, D. A., Roskos, R., Hanson, C. A., & Castle, V. (1991). Virus-associated hemophagocytic syndrome following bone marrow transplantation. *The American Journal of Pediatric Hematology/Oncology, 13*, 305–309.
123. Risdall, R. J., McKenna, R. W., Nesbit, M. E., Krivit, W., Balfour, H. H., Simmons, R. L., et al. (1979). Virus-associated hemophagocytic syndrome: A benign histiocytic proliferation distinct from malignant histiocytosis. *Cancer, 44*, 993–1002.
124. Schultz, K. A., Neglia, J. P., Smith, A. R., Ochs, H. D., Torgerson, T. R., & Kumar, A. (2008). Familial hemophagocytic lymphohistiocytosis in two brothers with X-linked agammaglobulinemia. *Pediatric Blood & Cancer, 51*, 293–295.
125. Kaya, Z., Ozturk, G., Gursel, T., & Bozdayi, G. (2005). Spontaneous resolution of hemophagocytic syndrome and disseminated intravascular coagulation associated with parvovirus b19 infection in a previously healthy child. *Japanese Journal of Infectious Diseases, 58*, 149–151.
126. Larroche, C., Scieux, C., Honderlick, P., Piette, A. M., Morinet, F., & Bletry, O. (2002). Spontaneous resolution of hemophagocytic syndrome associated with acute parvovirus B19 infection and concomitant Epstein-Barr virus reactivation in an otherwise healthy adult. *European Journal of Clinical Microbiology & Infectious Diseases, 21*, 739–742.
127. Dutta, U., Mittal, S., Ratho, R. K., & Das, A. (2005). Acute liver failure and severe hemophagocytosis secondary to parvovirus B19 infection. *Indian Journal of Gastroenterology, 24*, 118–119.

128. Yilmaz, S., Oren, H., Demircioglu, F., Firinci, F., Korkmaz, A., & Irken, G. (2006). Parvovirus B19: A cause for aplastic crisis and hemophagocytic lymphohistiocytosis. *Pediatric Blood & Cancer, 47*, 861.
129. Hermann, J., Steinbach, D., Lengemann, J., & Zintl, F. (2003). Parvovirus B 19 associated hemophagocytic syndrome in a patient with hereditary sperocytosis. *Klinische Pädiatrie, 215*, 270–274.
130. Syruckova, Z., Stary, J., Sedlacek, P., Smisek, P., Vavrinec, J., Komrska, V., et al. (1996). Infection-associated hemophagocytic syndrome complicated by infectious lymphoproliferation: A case report. *Pediatric Hematology and Oncology, 13*, 143–150.
131. Yuan, C., Asad-Ur-Rahman, F., & Abusaada, K. (2016). A rare case of Hemophagocytic Lymphohistiocytosis associated with parvovirus B19 infection. *Cureus, 8*, e897.
132. Yufu, Y., Matsumoto, M., Miyamura, T., Nishimura, J., Nawata, H., & Ohshima, K. (1997). Parvovirus B19-associated haemophagocytic syndrome with lymphadenopathy resembling histiocytic necrotizing lymphadenitis (Kikuchi's disease). *British Journal of Haematology, 96*, 868–871.
133. Shirono, K., & Tsuda, H. (1995). Parvovirus B19-associated haemophagocytic syndrome in healthy adults. *British Journal of Haematology, 89*, 923–926.
134. Ardalan, M. R., Shoja, M. M., Tubbs, R. S., Esmaili, H., & Keyvani, H. (2008). Postrenal transplant hemophagocytic lymphohistiocytosis and thrombotic microangiopathy associated with parvovirus b19 infection. *American Journal of Transplantation : Official Journal of the American Society of Transplantation and the American Society of Transplant Surgeons, 8*, 1340–1344.
135. Tavera, M., Petroni, J., Leon, L., Minue, E., & Casadei, D. (2012). Reactive haemophagocytic syndrome associated with parvovirus B19 in a kidney-pancreas transplant patient. *Nefrología, 32*, 125–126.
136. Matsubara, K., Uchida, Y., Wada, T., Iwata, A., Yura, K., Kamimura, K., et al. (2011). Parvovirus B19-associated hemophagocytic lymphohistiocytosis in a child with precursor B-cell acute lymphoblastic leukemia under maintenance chemotherapy. *Journal of Pediatric Hematology/Oncology, 33*, 565–569.
137. Sekiguchi, Y., Shimada, A., Imai, H., Wakabayashi, M., Sugimoto, K., Nakamura, N., et al. (2014). A case of recurrent autoimmune hemolytic anemia during remission associated with acute pure red cell aplasia and hemophagocytic syndrome due to human parvovirus B19 infection successfully treated by steroid pulse therapy with a review of the literature. *International Journal of Clinical and Experimental Pathology, 7*, 2624–2635.
138. Uike, N., Miyamura, T., Obama, K., Takahira, H., Sato, H., & Kozuru, M. (1993). Parvovirus B19-associated haemophagocytosis in Evans syndrome: Aplastic crisis accompanied by severe thrombocytopenia. *British Journal of Haematology, 84*, 530–532.
139. Mayama, M., Yoshihara, M., Kokabu, T., & Oguchi, H. (2014). Hemophagocytic lymphohistiocytosis associated with a parvovirus B19 infection during pregnancy. *Obstetrics and Gynecology, 124*, 438–441.
140. Kishore, J., & Kishore, D. (2014). Fatal missed case of hemophagocytic lymphohistiocytosis co-infected with parvovirus B19 and Epstein-Barr virus in an infant: Test hyperferritinaemia early. *Indian Journal of Medical Microbiology, 32*, 181–183.
141. Sood, N., & Yadav, P. (2012). Hemophagocytic syndrome associated with concomitant Klebsiella and parvovirus B-19 infection. *Indian Journal of Pathology & Microbiology, 55*, 124–125.
142. Rouphael, N. G., Talati, N. J., Vaughan, C., Cunningham, K., Moreira, R., & Gould, C. (2007). Infections associated with haemophagocytic syndrome. *The Lancet Infectious Diseases, 7*, 814–822.
143. Ramachandra, G., Shields, L., Brown, K., & Ramnarayan, P. (2010). The challenges of prompt identification and resuscitation in children with acute fulminant myocarditis: Case series and review of the literature. *Journal of Paediatrics and Child Health, 46*, 579–582.

144. Tsuda, H., Maeda, Y., Nakagawa, K., Nakayama, M., Nishimura, H., Ishihara, A., et al. (1994). Parvovirus B19-associated haemophagocytic syndrome with prominent neutrophilia. *British Journal of Haematology, 86*, 413–414.
145. Butin, M., Mekki, Y., Phan, A., Billaud, G., Di Filippo, S., Javouhey, E., et al. (2013). Successful immunotherapy in life-threatening parvovirus B19 infection in a child. *The Pediatric Infectious Disease Journal, 32*, 789–792.
146. Watanabe, M., Shibuya, A., Okuno, J., Maeda, T., Tamama, S., & Saigenji, K. (2002). Hepatitis a virus infection associated with hemophagocytic syndrome: Report of two cases. *Internal Medicine, 41*, 1188–1192.
147. Tuon, F. F., Gomes, V. S., Amato, V. S., Graf, M. E., Fonseca, G. H., Lazari, C., et al. (2008). Hemophagocytic syndrome associated with hepatitis a: Case report and literature review. *Revista do Instituto de Medicina Tropical de São Paulo, 50*, 123–127.
148. Russo, R. A., Rosenzweig, S. D., & Katsicas, M. M. (2008). Hepatitis A-associated macrophage activation syndrome in children with systemic juvenile idiopathic arthritis: Report of 2 cases. *The Journal of Rheumatology, 35*, 166–168.
149. Bay, A., Bosnak, V., Leblebisatan, G., Yavuz, S., Yilmaz, F., & Hizli, S. (2012). Hemophagocytic lymphohistiocytosis in 2 pediatric patients secondary to hepatitis a virus infection. *Pediatric Hematology and Oncology, 29*, 211–214.
150. Canoz, P. Y., Afat, E., Temiz, F., Azizoglu, N. O., Citilcioglu, H. B., Tumgor, G., et al. (2014). Reactive hemophagocytic lymphohistiocytosis after hepatitis a infection. *Indian Journal of Hematology & Blood Transfusion, 30*, 46–48.
151. Cho, E., Cha, I., Yoon, K., Yang, H. N., Kim, H. W., Kim, M. G., et al. (2010). Hemophagocytic syndrome in a patient with acute tubulointerstitial nephritis secondary to hepatitis a virus infection. *Journal of Korean Medical Science, 25*, 1529–1531.
152. Ishii, H., Yamagishi, Y., Okamoto, S., Saito, H., Kikuchi, H., & Kodama, T. (2003). Hemophagocytic syndrome associated with fulminant hepatitis a: A case report. *The Keio Journal of Medicine, 52*, 38–51.
153. Kyoda, K., Nakamura, S., Machi, T., Kitagawa, S., Ohtake, S., & Matsuda, T. (1998). Acute hepatitis a virus infection-associated hemophagocytic syndrome. *The American Journal of Gastroenterology, 93*, 1187–1188.
154. Navamani, K., Natarajan, M. M., Lionel, A. P., & Kumar, S. (2014). Hepatitis a virus infection-associated hemophagocytic lymphohistiocytosis in two children. *Indian Journal of Hematology & Blood Transfusion, 30*, 239–242.
155. Onaga, M., Hayashi, K., Nishimagi, T., Nagata, K., Uto, H., Kubuki, Y., et al. (2000). A case of acute hepatitis a with marked hemophagocytosis in bone marrow. *Hepatology Research : The Official Journal of the Japan Society of Hepatology, 17*, 205–211.
156. Tai, C. M., Liu, C. J., & Yao, M. (2005). Successful treatment of acute hepatitis A-associated hemophagocytic syndrome by intravenous immunoglobulin. *Journal of the Formosan Medical Association, 104*, 507–510.
157. McPeake, J. R., Hirst, W. J., Brind, A. M., & Williams, R. (1993). Hepatitis a causing a second episode of virus-associated haemophagocytic lymphohistiocytosis in a patient with Still's disease. *Journal of Medical Virology, 39*, 173–175.
158. Aleem, A., Al Amoudi, S., Al-Mashhadani, S., & Siddiqui, N. (2005). Haemophagocytic syndrome associated with hepatitis-B virus infection responding to etoposide. *Clinical and Laboratory Haematology, 27*, 395–398.
159. Yu, M. G., & Chua, J. (2016). Virus-associated haemophagocytic lymphohistiocytosis in a youn-g Filipino man. *BMJ Case Reports, 2016*, bcr2016214655.
160. Faurschou, M., Nielsen, O. J., Hansen, P. B., Juhl, B. R., & Hasselbalch, H. (1999). Fatal virus-associated hemophagocytic syndrome associated with coexistent chronic active hepatitis B and acute hepatitis C virus infection. *American Journal of Hematology, 61*, 135–138.
161. Tierney, L. M., Thabet, A., & Nishino, H. (2011). Case records of the Massachusetts General Hospital. Case 10-2011. A woman with fever, confusion, liver failure, anemia, and thrombocytopenia. *The New England Journal of Medicine, 364*, 1259–1270.

162. Guerra, J. A. A. A., Kampa, K. C., Morsoletto, D. G. B., Junior, A. P., & Ivantes, C. A. P. (2017). Hepatitis E: A literature review. *Journal of Clinical and Translational Hepatology, 5*, 376–383.
163. Kamihira, T., Yano, K., Tamada, Y., Matsumoto, T., Miyazato, M., Nagaoka, S., et al. (2008). Case of domestically infected hepatitis E with marked thrombocytopenia. *Nihon Shokakibyo Gakkai Zasshi, 105*, 841–846.
164. Leroy, M., Coiffier, G., Pronier, C., Triquet, L., Perdriger, A., & Guggenbuhl, P. (2015). Macrophage activation syndrome with acute hepatitis E during tocilizumab treatment for rheumatoid arthritis. *Joint, Bone, Spine, 82*, 278–279.
165. Choudhary, S. K., Agarwal, A., Mandal, R. N., & Grover, R. (2016). Hemophagocytic lymphohistiocytosis associated with hepatitis a and hepatitis E co-infection. *Indian Journal of Pediatrics, 83*, 607–608.
166. Tsunemi, Y., Matsushita, T., Nagayama, T., Takahashi, T., & Tamaki, T. (2001). A case of virus-associated hemophagocytic syndrome due to measles. *The Japanese Journal of Dermatology, 111*, 1591–1596.
167. Otagiri, T., Mitsui, T., Kawakami, T., Katsuura, M., Maeda, K., Ikegami, T., et al. (2002). Haemophagocytic lymphohistiocytosis following measles vaccination. *European Journal of Pediatrics, 161*, 494–496.
168. Buda, P., Gietka, P., Wieteska-Klimczak, A., Smorczewska-Kiljan, A., & Książyk, J. (2016). Autoimmune/inflammatory syndrome leading to macrophage activation syndrome: An example of autoinflammatory spectrum disorder? *The Israel Medical Association Journal, 18*, 571.
169. Xing, Q., & Xing, P. (2013). Mumps caused hemophagocytic syndrome: A rare case report. *The American Journal of Emergency Medicine, 31*, 1000.e1001–1000.e1002.
170. Takenaka, H., Kishimoto, S., Ichikawa, R., Shibagaki, R., Kubota, Y., Yamagata, N., et al. (1998). Virus-associated haemophagocytic syndrome caused by rubella in an adult. *The British Journal of Dermatology, 139*, 877–880.
171. Takeoka, Y., Hino, M., Oiso, N., Nishi, S., Koh, K. R., Yamane, T., et al. (2001). Virus-associated hemophagocytic syndrome due to rubella virus and varicella-zoster virus dual infection in patient with adult idiopathic thrombocytopenic purpura. *Annals of Hematology, 80*, 361–364.
172. Baykan, A., Akcakus, M., & Deniz, K. (2005). Rubella-associated hemophagocytic syndrome in an infant. *Journal of Pediatric Hematology/Oncology, 27*, 430–431.
173. Koubaa, M., Marrakchi, C., Maaloul, I., Makni, S., Berrajah, L., Elloumi, M., et al. (2012). Rubella associated with hemophagocytic syndrome. First report in a male and review of the literature. *Mediterranean Journal of Hematology and Infectious Diseases, 4*, e2012050.
174. Guerin, C., Pozzetto, B., & Berthoux, F. (1989). Hemophagocytic syndrome associated with coxsackie virus a 9 infection in a non-immunosuppressed adult. *Intensive Care Medicine, 15*, 547–548.
175. Barre, V., Marret, S., Mendel, I., Lesesve, J. F., & Fessard, C. I. (1998). Enterovirus-associated haemophagocytic syndrome in a neonate. *Acta Paediatrica, 87*, 469–471.
176. Kashiwagi, Y., Kawashima, H., Sato, S., Ioi, H., Amaha, M., Takekuma, K., et al. (2007). Virological and immunological characteristics of fatal virus-associated haemophagocytic syndrome (VAHS). *Microbiology and Immunology, 51*, 53–62.
177. Letsas, K. P., Filippatos, G. S., Delimpasi, S., Spanakis, N., Kounas, S. P., Efremidis, M., et al. (2007). Enterovirus-induced fulminant myocarditis and hemophagocytic syndrome. *The Journal of Infection, 54*, e75–e77.
178. Katsibardi, K., Moschovi, M. A., Theodoridou, M., Spanakis, N., Kalabalikis, P., Tsakris, A., et al. (2008). Enterovirus-associated hemophagocytic syndrome in children with malignancy: Report of three cases and review of the literature. *European Journal of Pediatrics, 167*, 97–102.
179. Lindamood, K. E., Fleck, P., Narla, A., Vergilio, J. A., Degar, B. A., Baldwin, M., et al. (2011). Neonatal enteroviral sepsis/meningoencephalitis and hemophagocytic lymphohistiocytosis: Diagnostic challenges. *American Journal of Perinatology, 28*, 337–346.

180. Fukazawa, M., Hoshina, T., Nanishi, E., Nishio, H., Doi, T., Ohga, S., et al. (2013). Neonatal hemophagocytic lymphohistiocytosis associated with a vertical transmission of coxsackievirus B1. *Journal of Infection and Chemotherapy, 19*, 1210–1213.
181. Hara, S., Kawada, J., Kawano, Y., Yamashita, T., Minagawa, H., Okumura, N., et al. (2014). Hyperferritinemia in neonatal and infantile human parechovirus-3 infection in comparison with other infectious diseases. *Journal of Infection and Chemotherapy, 20*, 15–19.
182. Nanmoku, K., Yamamoto, T., Tsujita, M., Hiramitsu, T., Goto, N., Katayama, A., et al. (2015). Virus-associated hemophagocytic syndrome in renal transplant recipients: Report of 2 cases from a single center. *Case Reports in Hematology, 2015*, 876301.
183. Park, M., Yun, Y. J., Woo, S. I., Lee, J. W., Chung, N. G., & Cho, B. (2015). Rotavirus-associated hemophagocytic lymphohistiocytosis (HLH) after hematopoietic stem cell transplantation for familial HLH. *Pediatrics International, 57*, e77–e80.
184. Takahashi, S., Oki, J., Miyamoto, A., Koyano, S., Ito, K., Azuma, H., et al. (1999). Encephalopathy associated with haemophagocytic lymphohistiocytosis following rotavirus infection. *European Journal of Pediatrics, 158*, 133–137.
185. Nagao, T., Takahashi, N., Saitoh, H., Noguchi, S., Guo, Y. M., Ito, M., et al. (2012). Adult T-cell leukemia-lymphoma developed from an HTLV-1 carrier during treatment of B-cell lymphoma-associated hemophagocytic syndrome. *Rinshō Ketsueki, 53*, 2008–2012.

Cytokine Storm Syndrome as a Manifestation of Primary HIV Infection

Nathan Erdmann and Sonya L. Heath

Introduction

Cytokine storm syndrome (CSS) has a recognized association with HIV/AIDS, where profound immune suppression and dysregulation lead to opportunistic infections that sometimes trigger fulminant and often fatal inflammation [1, 2]. In these late stage presentations, HIV is not the primary instigator of CSS, but rather provides a host susceptible to a pathogen capable of precipitating CSS. This is in contrast to CSS associated with primary HIV. Primary HIV infection has been recognized as a cause of CSS, with reports dating from 1992 [3]. As detailed in this chapter, primary HIV infection is capable of instigating a profound inflammatory cascade that often meets the criteria for macrophage activation syndrome. However, unlike many other manifestations of CSS, presentations associated with primary HIV have excellent outcomes when recognized and treated urgently with HIV therapeutics.

Primary HIV Infection

Primary HIV infection is characterized by very high viral loads (100,000- > 10 million copies of HIV RNA/mL) peaking 10–14 days after transmission. The host immune response eventually controls plasma viremia to lower levels (usually 10,000 to 100,000 HIV RNA copies/mL), coinciding with the emergence of HIV-specific T cells. While not all patients infected with HIV have symptoms during this acute phase, it is estimated that 50–80% of patients have some form of acute retroviral syndrome (ARS). ARS is characterized by a nonspecific flu-like syndrome

N. Erdmann · S. L. Heath (✉)
Department of Medicine, University of Alabama at Birmingham, Birmingham, AL, USA
e-mail: slheath@uabmc.edu

R. Q. Cron, E. M. Behrens (eds.), *Cytokine Storm Syndrome*,
https://doi.org/10.1007/978-3-030-22094-5_17

including high fever, lymphadenopathy, acute pharyngitis, myalgia, and diffuse maculopapular rash. Some patients also develop gastrointestinal symptoms including diarrhea, or central nervous symptoms including headache, meningismus, and altered mental status. Clinically, acute retroviral syndrome manifest with cytopenias including leukopenia with lymphopenia, anemia, and thrombocytopenia, as well as transaminitis, elevated creatinine kinase, elevated lactate dehydrogenase, and other laboratory anomalies seen with a generalized inflammatory response. The acute retroviral syndrome is self-limited, typically lasting 2–4 weeks, and resolves with decline in HIV plasma viral load. However, given the spectrum of the syndrome, some patients develop fulminant disease processes with multiorgan system involvement. These patients have cytokine storm syndromes secondary to high levels of circulating virus and the immune response generated to control this.

Suspicion for acute retroviral syndrome must be high in order to make the correct diagnosis. Standard third-generation HIV assays rely on an HIV antibody response that typically occurs 4–12 weeks after initial infection. Given that most patients are symptomatic within 2–4 weeks of infection, many are pre-seroconversion (i.e., HIV antibody negative) and in the "window period." When the clinical suspicion for primary HIV infection is high, plasma viral load testing will confirm the diagnosis of HIV during the window period of pre-seroconversion. Newer, fourth-generation combination antigen–antibody HIV assays can also confirm primary HIV, given that these assays detect p24 HIV antigens, which are present within days of transmission, as well as HIV antibodies. Fourth-generation HIV assays are the currently recommended HIV test based on CDC guidelines; however, they are not available in all testing settings including some hospitals. Thus, providers evaluating patients with cytokine storm syndrome should not only consider primary HIV infection as an etiology but also recognize the optimal test (either HIV viral load or p24 antigen via fourth-generation combination Ag/Ab HIV assay) to accurately make the diagnosis.

Cytokine Storm Syndrome and Primary HIV Infection

CSS as a clinical syndrome has been associated with viral infections dating back to Risdall et al. in 1979 [4], and causes an aggressive and often fatal course. The syndrome is a consequence of profound cytokine secretion, macrophage activation and proliferation, leading to excessive phagocytosis. Case definitions have evolved since the initial description, and are characterized by fever, cytopenias (of at least 2 types), hypertriglyceridemia and/or hypofibrinogenemia, hyperferritinemia, hemophagocytosis, elevated CD25, decreased NK cell activity, splenomegaly, and hemophagocytosis in BM, LN, or spleen. Not coincidentally, there is significant overlap in the syndromes of acute retroviral syndrome and cytokine storm syndrome (Table 1). Fever, cytopenias, and splenomegaly/lymphadenopathy are cardinal features of ARS, and it is likely that CSS in the setting of primary HIV infection is underrecognized. Histological examination in CSS should reveal infiltration by

Table 1 Cardinal features of ARS and CSS

Acute retroviral syndrome	Cytokine storm syndrome
Fever	Fever
Splenomegaly/lymphadenopathy	Splenomegaly/lymphadenopathy
Cytopenias	Cytopenias
Rash	Hemophagocytosis
Sore throat	Decreased NK cell function
Drop in CD4 count	Triglycerides
High degree of viremia	Hyperferritinemia

histiocytes and lymphocytes with the presence of hemophagocytosis. Biopsy is not a routine component of clinical workup for ARS, although as documented below, when completed, biopsy was consistent with CSS in all reported cases [3, 5–13]. ARS is thought to be clinically evident in approximately 70% of individuals acutely infected with HIV. Gathering sufficient data for formal diagnosis of CSS is often difficult, particularly given the pace of presentation; thus, it is very likely that manifestations in HIV are underappreciated and may approach 20% and may be less severe in many cases.

A majority of cases of primary HIV infection result in a clinical syndrome, most frequently associated with fever, sore throat, cytopenias, and lymphadenopathy/splenomegaly. However, what differentiates individuals that develop more severe presentations from individuals that have an asymptomatic course is unclear. There does not appear to be an obvious association between how HIV was acquired and development of MAS, nor is there evidence for underlying genetic predisposition such as mutations in primary HLH genes, although this certainly cannot be ruled out based upon limited number of cases. Various studies have revealed progression of HIV disease is influenced by host and viral genetics [14], but there is no available evidence suggesting this drives ARS.

There is a body of case reports available in the literature of more severe cases of acute retroviral syndrome that are consistent with CSS [3, 5–13]. To what degree these cases represent a unique subset versus the extremes on a continuum is unclear. When available, the observed ferritin is frequently elevated above 15,000 ng/mL. The presence of fever, cytopenias, and lymphadenopathy/splenomegaly is all but uniform, and other criteria are routinely observed when checked in these cases. HIV viral load is exceptionally high, reaching $1 \times 10^{\wedge 6}$–$1 \times 10^{\wedge 8}$ copies/mL, measurements typical of ARS. Although these cases are by definition early in infection, CD4 count is often depressed. While absolute CD4 count is often a labile measurement in HIV infection, the degree of CD4 T cell deficiency in cases with CSS is notable from the reports available in the literature with a median of 138 cells/$mm^3$and average of 183 cells/mm^3. Further, the presence of opportunistic infection, thrush specifically, was noted in 5 of the 13 reported cases. Together, these data suggest that manifestation of CSS in ARS represents the more severe cases of primary HIV infection. These clinical characteristics and diagnostic criteria are summarized in Table 2 [3, 5–13].

Table 2 Clinical characteristics and diagnostic criteria for ARS and CSS cases

Age/gender	Viral load	Cd4 count	Ferritin	Fever	Splenomegaly/lymphadenopathy	Cytopenias	Biopsy	TG	OI
48, M	3 M	90	34,000	✓	X	✓	✓	✓	✓
18, M	522,105	63	17,000	✓	✓	✓	✓	N/A	X
27, F	384,000	13	2095	✓	N/A	✓	✓	✓	X
31, M	N/A	300	N/A	✓	✓	✓	✓	N/A	✓
45, M	>7 M	137	69,000	✓	✓	✓	✓	✓	✓
27, M	27 M	138	N/A	✓	✓	✓	✓	✓	✓
44, M	>10 M	157	>30,000	✓	✓	✓	✓	✓	✓
27, M	N/A	500	N/A	✓	✓	✓	✓	N/A	N/A
28, F	N/A	N/A	N/A	✓	✓	✓	✓	N/A	N/A
31, M	>700,000	324	4625	✓	✓	✓	✓	182	N/A
23, M	>700,000	194	2227	✓	✓	✓	✓	168	N/A
25, M	>700,000	101	29,000	✓	✓	✓	✓	99	N/A

Viral load = copies/mL
M = million
CD4 count = cells/mm
Ferritan = ng/mL
Triglycerides = mg/dL

Table 3 Treatment and outcomes for ARS and CSS cases

	Immunomodulatory agents	ART	OI	Patient outcomes
48y/o M	None	Yes	✓	Survived
18y/o M	IVIG	Yes	X	Survived
27y/o F	None	Yes	X	Survived
31y/o M	IVIG, steroids	No	✓	Survived
45y/o M	None	Yes	✓	Survived
27y/o M	None	Yes	✓	Survived
44y/o M	None	Yes	✓	Survived
27y/o M	NA	No	N/A	Survived
31y/o M	IVIG	No	✓	Survived
23y/o M	IVIG	No	✓	Survived
25y/o M	None	No	✓	Survived

Treatment and Outcomes

Treatment guidelines for HIV have changed over recent decades, and it is now recognized that treatment should occur for all patients infected with HIV regardless of CD4 count or viral load. Since the beginning of the AIDS epidemic, numerous cases of primary HIV infection have been reported with significant CD4 T-cell declines below 200 cells/mm^3. Since ARS produces a profound inflammatory response and high viremia, it is perhaps not surprising that opportunistic infections such as thrush, *Pneumocystis jirovecii* pneumonia, and others are sometime observed. In primary HIV infection, rapid control of the virus with antiretroviral therapy not only improves patient symptoms but may also restore immune function and reconstitutes CD4 T cells to normal or near-normal levels. As is generally the case in patients with CSS, treatment should be directed at the primary condition instigating the inflammatory cascade driving the cytokine storm. Thus, in the setting of primary HIV infection, treatment should first be directed at HIV with antiretroviral therapy, and regimens containing integrase inhibitors, which results in more rapid control of viremia, are recommended. After therapy is initiated, there is reliable, predictable, and rapid decline in circulating viremia and subsequent increases in CD4 T cells. While some case reports of primary HIV infection have included treatment of CSS with immune modulating agents including steroids, intravenous immunoglobulin, and biologics it is unclear if this is necessary or of added benefit. The mortality from cytokine syndrome storm resulting from primary HIV infection is near zero when it is recognized and treated appropriately [3, 5–13] (Table 3).

Chronic HIV and CSS

While this chapter has focused on CSS occurring in the setting of primary HIV infection, it is important to recognize the presence of CSS in advanced HIV infection and the differences in driving factors, treatments, and outcomes. In an all cause of death autopsy study of 56 AIDS patients, histopathological evidence of hemophagocytosis was found in 20% of cases [15]. These data suggest that similar to CSS in primary HIV infection, the presence of CSS in advanced HIV is likely underrecognized by clinicians. HIV infection clearly disrupts normal host immunity, establishing a pro-inflammatory environment. In later stages of HIV/AIDS, there is profound immune dysregulation affecting both innate and adaptive immunity resulting in increased host susceptibility to opportunistic infections associated with CSS. Here the acute trigger could be a variety of viral infections such as CMV, EBV, VZV; bacterial or fungal sepsis, or malignancy, (58 cases reviewed in Fardet et al.) [1]. In these cases, the degree of immunodeficiency is more profound, and thus comorbid viral and bacterial infections driving CSS are more difficult to manage. Here, use of immune modulators and steroids is reasonable, although supporting data is limited. Rarely, in an AIDS patient with CSS is there not another active infectious process. In such reported cases, ART treatment may be sufficient to control the syndrome [16]. Despite more aggressive treatment of HIV and any recognized coinfections, CSS outcomes are more typical of CSS overall, with mortality reports as high as 80% [17–22].

While ART in general is greatly beneficial to controlling HIV or HIV-related CSS, there are occasions where effective treatment can precipitate cytokine storm. As HIV leads to profound immunosuppression, abruptly restricting viral replication can lead to an immune reconstitution inflammatory response, or IRIS. This paradoxical immune process is particularly common in the setting of advanced disease where the immune system becomes capable of responding to chronic infections such as *Mycobacterium avium* or *Cryptococcus neoformans*. A rare, but recognized process is IRIS-mediated CSS, where immune recovery leads to a burst of cytokine production which then escalates to CSS [23–27]. This process is typically more responsive to treatment interventions than other CSS manifestations. Again, directing specific treatment to the underlying infectious causes, HIV and the opportunistic infection, is usually adequate.

Conclusion

Cytokine storm syndrome can occur at multiple stages of HIV including primary HIV infection, and end stage AIDS as well as during immune reconstitution after starting antiretroviral therapy in a patient with profound immunosuppression and comorbid opportunistic infections. Therapy should always be directed at the underlying cause, most notably antiretroviral therapy and when present, therapy directed at comorbid opportunistic infections. The outcomes of patients with CSS varies with near 100% survival in primary HIV infection and CSS and high mortality (~80%) in end stage AIDS with CSS. Therapy to dampen CSS may be of benefit in some patients; however, there are no controlled trials to support this. Early recognition of primary HIV is imperative to initiating appropriate antiretroviral therapy to halt CSS associated with the acute retroviral syndrome. Physicians must recognize the importance of primary HIV infection as a driver of CSS and know the optimal laboratory tests (p24 antigen and/or HIV viral load) to ensure the diagnosis early after infection.

References

1. Fardet, L., Lambotte, O., Meynard, J. L., Kamouh, W., Galicier, L., Marzac, C., et al. (2010). Reactive haemophagocytic syndrome in 58 HIV-1-infected patients: Clinical features, underlying diseases and prognosis. *AIDS, 24*, 1299–1306.
2. Doyle, T., Bhagani, S., & Cwynarski, K. (2009). Haemophagocytic syndrome and HIV. *Current Opinion in Infectious Diseases, 22*, 1–6.
3. Pellegrin, J. L., Merlio, J. P., Lacoste, D., Barbeau, P., Brossard, G., Beylot, J., et al. (1992). Syndrome of macrophagic activation with hemophagocytosis in human immunodeficiency virus infection. *La Revue de Médecine Interne, 13*, 438–440.
4. Risdall, R. J., McKenna, R. W., Nesbit, M. E., Krivit, W., Balfour Jr., H. H., Simmons, R. L., et al. (1979). Virus-associated hemophagocytic syndrome: A benign histiocytic proliferation distinct from malignant histiocytosis. *Cancer, 44*, 993–1002.
5. Adachi, E., Koibuchi, T., Imai, K., Kikuchi, T., Shimizu, S., Koga, M., et al. (2013). Hemophagocytic syndrome in an acute human immunodeficiency virus infection. *Internal Medicine, 52*, 629–632.
6. Castilletti, C., Preziosi, R., Bernardini, G., Caterini, A., Gomes, V., Calcaterra, S., et al. (2004). Hemophagocytic syndrome in a patient with acute human immunodeficiency virus infection. *Clinical Infectious Diseases, 38*, 1792–1793.
7. Chen, T. L., Wong, W. W., & Chiou, T. J. (2003). Hemophagocytic syndrome: An unusual manifestation of acute human immunodeficiency virus infection. *International Journal of Hematology, 78*, 450–452.
8. Ferraz, R. V., Carvalho, A. C., Araujo, F., Koch, C., Abreu, C., & Sarmento, A. (2016). Acute HIV infection presenting as hemophagocytic syndrome with an unusual serological and virological response to ART. *BMC Infectious Diseases, 16*, 619.
9. Manji, F., Wilson, E., Mahe, E., Gill, J., & Conly, J. (2017). Acute HIV infection presenting as hemophagocytic lymphohistiocytosis: Case report and review of the literature. *BMC Infectious Diseases, 17*, 633.
10. Martinez-Escribano, J. A., Pedro, F., Sabater, V., Quecedo, E., Navarro, V., & Aliaga, A. (1996). Acute exanthem and pancreatic panniculitis in a patient with primary HIV infection and haemophagocytic syndrome. *The British Journal of Dermatology, 134*, 804–807.
11. Park, K. H., Yu, H. S., Jung, S. I., Shin, D. H., & Shin, J. H. (2008). Acute human immunodeficiency virus syndrome presenting with hemophagocytic lymphohistiocytosis. *Yonsei Medical Journal, 49*, 325–328.
12. Pontes, J., Mateo, O., Gaspar, G., & Vascones, S. (1995). Hemophagocytosis syndrome associated with acute HIV infection. *Enfermedades Infecciosas y Microbiología Clínica, 13*, 441–442.
13. Sun, H. Y., Chen, M. Y., Fang, C. T., Hsieh, S. M., Hung, C. C., & Chang, S. C. (2004). Hemophagocytic lymphohistiocytosis: An unusual initial presentation of acute HIV infection. *Journal of Acquired Immune Deficiency Syndromes, 37*, 1539–1540.
14. Carlson, J. M., Du, V. Y., Pfeifer, N., Bansal, A., Tan, V. Y., Power, K., et al. (2016). Impact of pre-adapted HIV transmission. *Nature Medicine, 22*, 606–613.
15. Niedt, G. W., & Schinella, R. A. (1985). Acquired immunodeficiency syndrome. Clinicopathologic study of 56 autopsies. *Archives of Pathology & Laboratory Medicine, 109*, 727–734.

16. Fitzgerald, B. P., Wojciechowski, A. L., & Bajwa, R. P. S. (2017). Efficacy of prompt initiation of antiretroviral therapy in the treatment of hemophagocytic lymphohistiocytosis triggered by uncontrolled human immunodeficiency virus. *Case Reports in Critical Care, 2017*, 8630609.
17. Bourquelot, P., Oksenhendler, E., Wolff, M., Fegueux, S., Piketty, C., D'Agay, M. F., et al. (1993). Hemophagocytic syndrome in HIV infection. *Presse Médicale, 22*, 1217–1220.
18. Grateau, G., Bachmeyer, C., Blanche, P., Jouanne, M., Tulliez, M., Galland, C., et al. (1997). Haemophagocytic syndrome in patients infected with the human immunodeficiency virus: Nine cases and a review. *The Journal of Infection, 34*, 219–225.
19. Nie, Y., Zhang, Z., Wu, H., & Wan, L. (2017). Hemophagocytic lymphohistiocytosis in a patient with human immunodeficiency virus infection: A case report. *Experimental and Therapeutic Medicine, 13*, 2480–2482.
20. Tiab, M., Mechinaud Lacroix, F., Hamidou, M., Gaillard, F., & Raffi, F. (1996). Reactive haemophagocytic syndrome in AIDS. *AIDS, 10*, 108–111.
21. Uemura, M., Huynh, R., Kuo, A., Antelo, F., Deiss, R., & Yeh, J. (2013). Hemophagocytic lymphohistiocytosis complicating T-cell lymphoma in a patient with HIV infection. *Case Reports in Hematology, 2013*, 687260.
22. Tsuda, H. (1997). Hemophagocytic syndrome (HPS) in children and adults. *International Journal of Hematology, 65*, 215–226.
23. Cuttelod, M., Pascual, A., Baur Chaubert, A. S., Cometta, A., Osih, R., Duchosal, M. A., et al. (2008). Hemophagocytic syndrome after highly active antiretroviral therapy initiation: A life-threatening event related to immune restoration inflammatory syndrome? *AIDS, 22*, 549–551.
24. De Lavaissiere, M., Manceron, V., Bouree, P., Garcon, L., Bisaro, F., Delfraissy, J. F., et al. (2009). Reconstitution inflammatory syndrome related to histoplasmosis, with a hemophagocytic syndrome in HIV infection. *The Journal of Infection, 58*, 245–247.
25. Huang, D. B., Wu, J. J., & Hamill, R. J. (2004). Reactive hemophagocytosis associated with the initiation of highly active antiretroviral therapy (HAART) in a patient with AIDS. *Scandinavian Journal of Infectious Diseases, 36*, 516–519.
26. Shelburne 3rd, S. A., Hamill, R. J., Rodriguez-Barradas, M. C., Greenberg, S. B., Atmar, R. L., Musher, D. W., et al. (2002). Immune reconstitution inflammatory syndrome: Emergence of a unique syndrome during highly active antiretroviral therapy. *Medicine (Baltimore), 81*, 213–227.
27. Zorzou, M. P., Chini, M., Lioni, A., Tsekes, G., Nitsotolis, T., Tierris, I., et al. (2016). Successful treatment of immune reconstitution inflammatory syndrome-related Hemophagocytic syndrome in an HIV patient with primary effusion lymphoma. *Hematology Reports, 8*, 6581.

Bacteria-Associated Cytokine Storm Syndrome

Esraa M. Eloseily and Randy Q. Cron

Macrophage activation syndrome (MAS) or hemophagocytic lymphohistiocytosis (HLH) are life threatening conditions that are described among febrile hospitalized patients. They present commonly with unremitting fever, and a shock-like multiorgan dysfunction picture. Laboratory studies show pancytopenia, elevated liver enzymes, and elevated ferritin and triglycerides, among others. Of note, hemophagocytosis in bone marrow is identified in only 60% of cases and is often absent during the early stages of MAS. Thought to be caused by a dysregulation of the immune response, a continuous activation and expansion of T lymphocytes and macrophages leads to a cytokine storm, ultimately resulting in multiorgan failure [1–4].

Hemophagocytic syndromes are divided into primary and secondary forms. Primary cases commonly present in the first year of life. They include familial forms that have specific genetic mutations [5–8] and certain immunodeficiency syndromes, such as Chédiak–Higashi syndrome, type II Hermansky–Pudlak syndrome, and Griscelli syndrome [9]. Secondary forms are usually associated with conditions that lead to chronic immune dysregulation such as rheumatologic diseases, with sJIA being the most commonly described, and certain malignancies.

As mentioned above, MAS develops as a "cytokine storm" which is often precipitated by infections, rheumatologic conditions or malignancies [10]. This storm develops due to an imbalance between the pro-inflammatory and the anti-inflammatory arms of the cytokines cascade. The pro-inflammatory cytokines associated with MAS likely include interferon-γ, IL-1, IL-6, IL-12, IL-18, and TNF [2, 11]. Elevated levels of IL-27, megakaryocyte colony-stimulating factor (M-CSF) and granulocyte-macrophage colony-stimulating factor (GM)-CSF have also been

E. M. Eloseily
Assiut University Children's Hospital, Assiut, Egypt

R. Q. Cron (✉)
UAB School of Medicine, University of Alabama at Birmingham, Birmingham, AL, USA
e-mail: rcron@peds.uab.edu

R. Q. Cron, E. M. Behrens (eds.), *Cytokine Storm Syndrome*,
https://doi.org/10.1007/978-3-030-22094-5_18

reported [12–14]. Furthermore, chemokines, such as IL-8/CXCL8, MIG/CXCL9, IP10/CXCL10, I-TAC/CXCL11, MCP-1/CCL2, MIP-1α/CCL3, and MIP-1β/CCL4, have been reported to be increased [11, 15–17]. Elevation of the cytokines and chemokines activates the immune system, perpetuating the on-going cytokine storm. On the other arm, elevated levels of anti-inflammatory cytokines, such as IL-10 and IL-18-binding protein (IL-18BP), might not be sufficient to check the ongoing inflammation [18, 19]. Abnormally high levels of free IL-18 is considered to be the result of the discrepancy between the increase in IL-18 and its antagonist IL-18BP [18].

It is proposed that MAS is due to a combination of genetic predisposition and a hyper-inflammatory state reducing cytolytic function, put into action by a trigger (e.g., infection, cancer, immunodeficiency, autoimmunity, and autoinflammation) [20–22].

Either biallelic genetic defects in the familial forms of CSS that usually present in infancy, or single copy gene mutations [22–24] in older children and adults, involving the perforin-mediated cytolytic pathway used by natural killer (NK) cells and CD8 T lymphocytes [25, 26] leads to the inability to clear the antigenic stimulus and thus turn off the inflammatory response culminating in hyper-cytokinemia [27]. Either the inadequate levels of perforin itself or improper granule exocytosis leads to impaired apoptosis of the target cell, improper removal of the stimulating antigen, and ultimately ongoing inflammation.

Like viruses, intracellular bacteria have commonly been the precipitating agents for CSS. This is probably related to the high levels of activating cytokines produced by the host lymphocytes and monocytes, and is also likely due to defective NK and cytotoxic T cell function [28]. A model of secondary HLH utilizing the intracellular bacteria *Salmonella enterica* to trigger the disease suggested that viruses and intracellular bacteria might exploit a common immunologic weakness in driving CSS [29].

While infections are a common inciting trigger for CSS as mentioned earlier, it is also expected that they become acquired during treatment with immune suppression which is an essential part of most CSS treatment protocols. It is probable that opportunistic bacteria are more likely to be acquired during treatment than to be the triggering agent in an otherwise healthy individual. Careful monitoring for symptoms and signs of acquiring an infection is crucial during the treatment process, since many of the clinical features are shared with the CSS picture. Moreover, clinical and laboratory parameters are commonly altered either due to the CSS diagnosis or as a result of the immune suppressive or the biologic treatment. Therefore, it is essential that along with the CSS treatment protocols, prompt treatment with broad spectrum antimicrobials should be initiated if infection is suspected, and appropriately tailored when a pathogen is confirmed, to achieve successful control of the disease. Antimicrobial prophylaxis is also usually added alongside the use of immune suppressive medications. For example, the HLH-2004 protocol suggested the use of cotrimoxazole, an oral antifungal during the initial phase, considering the use of antivirals in patients with ongoing viral infections, and IVIG once every 4 weeks (during the initial and continuation therapy) [30].

There is a broad range of bacterial triggers of CSS. Listed below are some of the bacterial infections believed to be potential triggers of CSS in a variety of clinical settings, summarized in Table 1.

Table 1 An alphabetical list of common bacterial triggers associated with CSS

Bacteria	Comments
Abiotrophia defectiva [31]	Endocarditis
Acinetobacter baumannii [32]	Nosocomial infection
Aeromonas hydrophila [33]	Taiwanese adult cohort
Bartonella sp. [34, 35]	Renal transplant patients and one case of underlying DOC8 mutation (unpublished)
Borrelia sp. [36]	Lyme disease
Brucella sp. [37–51]	Several reports of children and adult cases with brucellosis and HLH
Campylobacter sp. [52]	*Campylobacter fetus* infection in HIV patient
Capnocytophaga sp. [53]	Sudden onset hearing loss with HLH
Chlamydia sp. [54]	*Chlamydia pneumoniae* with HLH and acute encephalitis and poliomyelitis-like flaccid paralysis.
Clostridium sp. [55, 56]	Recurrent HLH in HIV patient, pancreatic carcinoma patient
Coxiella burnetti [57–59]	Q fever
Ehrlichia sp. [60–63]	Children and adult cases in the USA
Escherichia coli [64]	Nephrotic syndrome
Fusobacterium sp. [65, 66]	Lemierre disease and 19-year-old man
Intravesical BCG [67, 68]	Following installation for urothelial carcinomas
Klebsiella pneumonia [33]	Taiwanese adult cohort
Legionella sp. [69]	Chronic lymphocytic leukemia
Leptospira sp. [70, 71]	4-year-old boy and neglected case of ARF
Listeria sp. [72]	Bone marrow transplant recipient
Mycobacterium avium [73, 74]	HIV & SLE
Mycobacterium bovis—weakened form (Bacillus Calmette–Guérin) [75, 76]	Disseminated cutaneous eruption after BCG vaccination, BCG lymphadenitis in neonates with pHLH
Mycobacterium tuberculosis [77, 78]	Disseminated TB, perinatal TB
Mycobacterium leprae [79]	Leprosy
Mycoplasma [80, 81]	Retrospective analysis of 4 pediatric cases, 2 pediatric cases
Rickettsia sp. [82–84]	*Rickettsia conorii*, murine typhus and MSF
Salmonella typhi [85]	Typhoid fever with HLH and rhabdomyolysis
Salmonella sp. (other than typhi) [86, 87]	Child with CGD and *Salmonella typhimurium* septicemia
Staphylococcus aureus [88, 89]	Toxic epidermal necrolysis and HLH & 3 months old girl with sepsis

HLH hemophagocytic lymphohistiocytosis, *CGD* chronic granulomatous disease, *ARF* acute renal failure, *MSF* Mediterranean spotted fever

Acinetobacter baumannii

Acinetobacter baumannii septicemia following urinary tract infection was reported to trigger CSS in a 3-year-old child. The case showed complete recovery using only repeated transfusions and multiple doses of granulocyte colony stimulating factor [32].

Legionella Species

Nguyen et al. reported a 50-year-old male with a history of chronic lymphocytic leukemia who presented with *Legionella pneumonia*, persistent fevers despite levofloxacin treatment and was found to have hemophagocytosis on bone marrow biopsy. The patient died in spite of treatment with corticosteroids [69].

Leptospira Species

Leptospirosis has been described with CSS and has required treatment with corticosteroids, intravenous immunoglobulin or etoposide in addition to antibiotic treatment. Niller and colleagues suggested that an insufficient, dysfunctional, or misdirected immune response to Leptospira may culminate in myelodysplastic syndrome (MDS) in cases not initially recognized as Leptospira-triggered CSS [90]. Also a 4-year-old boy with pallor and hepatosplenomegaly as the initial presentation was reported to have CSS triggered by Leptospirosis.

Mycobacterium tuberculosis

There are several reports of *Mycobacterium tuberculosis* triggering CSS. It can occur in otherwise healthy individuals [91], or in patients with end-stage renal disease either receiving hemodialysis [92] or had undergone renal transplantation [93], malignancy [94], AIDS [95], and sarcoidosis [96]. In a review of 36 cases of CSS triggered by *Mycobacterium tuberculosis* by Brastianos et al., 83% of cases had evidence of extrapulmonary tuberculosis. The mortality rate was approximately 50% touting the poor outcome of TB-CSS. However, anti-tuberculous and immunomodulatory therapy (consisting of high-dose corticosteroids, intravenous immunoglobulin, anti-thymocyte globulin, cyclosporine A, and epipodophyllotoxin, or plasma exchange) may lead to a better outcome [77]. Early diagnosis and timely administration of anti-tuberculous treatment is crucial in these patients.

Bacillus Calmette–Guérin vaccination was also reported to trigger CSS in one pediatric case [75].

Mycoplasma pneumoniae

Mycoplasma pneumoniae has been identified in a few reports of pediatric cases of HLH. Yoshiyama et al. reported 4 cases, aged 1–11 years, of mycoplasma related HLH, one of them also had concurrent rubella infection. All the patients had typical radiologic picture of *M. pneumoniae* pneumonia, and one patient also developed encephalopathy. All the children had unrelenting fever, mild hepatosplenomegaly, cytopenia, and elevated serum ferritin levels and urinary beta-2-microglobulin. Bone marrow hemophagocytosis was found in all cases. Of note, cytopenia and hepatosplenomegaly were relatively mild as compared to other cases of infection-associated HLH. Two cases promptly responded to corticosteroids, high-dose intravenous immunoglobulin (IVIG) achieved a complete response in another child, while spontaneous recovery with symptomatic treatment and antibiotics alone was observed in one case [80].

Staphylococcus aureus

Staphylococcus aureus, a relatively common bacterial pathogen, was also described to trigger CSS. Sniderman et al. reported a 17-month-old boy that grew methicillin-sensitive *Staphylococcus aureus* from endotracheal cultures while on ventilator support due to laryngotracheitis. The boy presented 9 days after discharge with a history of spreading rash and high fevers that progressed to full desquamation and be diagnosed with biopsy to be toxic epidermal necrolysis (TEN). During progression of his illness he developed all eight criteria of HLH-2004 and achieved full recovery after treatment with dexamethasone and etoposide [88]. Additionally, Dube et al. reported a 3-month-old girl that was diagnosed with Staphylococcal pneumonia and multiorgan failure and was found to have hemophagocytosis on bone marrow biopsy. She died 6 days after admission in spite of treatment with antibiotics, corticosteroids, and IVIG [89].

Brucella, Ehrlichia, and *Rickettsia* as triggers of CSS are discussed in the chapter of zoonotic bacterial infections triggering CSS.

In conclusion, bacterial triggers are not only common triggering agents for CSS, but are also a serious complication of its various treatment protocols. Early identification of such infections is crucial to the outcome. Treating bacterial infections whether they trigger CSS or they develop during the disease course is a cornerstone in achieving successful control of the disease along with controlling the cytokine storm with immune suppressive medicine and/or biologic therapies.

References

1. Davì, S., Minoia, F., Pistorio, A., Horne, A., Consolaro, A., Rosina, S., et al. (2014). Performance of current guidelines for diagnosis of macrophage activation syndrome complicating systemic juvenile idiopathic arthritis. *Arthritis & Rheumatology, 66*(10), 2871–2880.
2. Ravelli, A., Grom, A. A., Behrens, E. M., & Cron, R. Q. (2012). Macrophage activation syndrome as part of systemic juvenile idiopathic arthritis: Diagnosis, genetics, pathophysiology and treatment. *Genes and Immunity, 13*(4), 289–298.
3. Ravelli, A., Minoia, F., Davi, S., Horne, A., Bovis, F., Pistorio, A., et al. (2016). Expert consensus on dynamics of laboratory tests for diagnosis of macrophage activation syndrome complicating systemic juvenile idiopathic arthritis. *RMD Open, 2*(1), e000161.
4. Ravelli, A., Minoia, F., Davi, S., Horne, A., Bovis, F., Pistorio, A., et al. (2016). 2016 classification criteria for macrophage activation syndrome complicating systemic juvenile idiopathic arthritis: A European league against rheumatism/American college of rheumatology/paediatric rheumatology international trials organisation collaborative initiative. *Arthritis & Rheumatology, 68*(3), 566–576.
5. Feldmann, J., Callebaut, I., Raposo, G., Certain, S., Bacq, D., Dumont, C., et al. (2003). Munc13-4 is essential for cytolytic granules fusion and is mutated in a form of familial hemophagocytic lymphohistiocytosis (FHL3). *Cell, 115*(4), 461–473.
6. zur Stadt, U., Rohr, J., Seifert, W., Koch, F., Grieve, S., Pagel, J., et al. (2009). Familial hemophagocytic lymphohistiocytosis type 5 (FHL-5) is caused by mutations in Munc18-2 and impaired binding to syntaxin 11. *The American Journal of Human Genetics, 85*(4), 482–492.
7. zur Stadt, U., Schmidt, S., Kasper, B., Beutel, K., Diler, A. S., Henter, J.-I., et al. (2005). Linkage of familial hemophagocytic lymphohistiocytosis (FHL) type-4 to chromosome 6q24 and identification of mutations in syntaxin 11. *Human Molecular Genetics, 14*(6), 827–834.
8. Stepp, S. E., Dufourcq-Lagelouse, R., Le Deist, F., Bhawan, S., Certain, S., Mathew, P. A., et al. (1999). Perforin gene defects in familial hemophagocytic lymphohistiocytosis. *Science, 286*(5446), 1957–1959.
9. Emmenegger, U., Schaer, D., Larroche, C., & Neftel, K. (2005). Haemophagocytic syndromes in adults: Current concepts and challenges ahead. *Swiss Medical Weekly, 135*(21–22), 299–314.
10. Grom, A. A., & Mellins, E. D. (2010). Macrophage activation syndrome: Advances towards understanding pathogenesis. *Current Opinion in Rheumatology, 22*(5), 561.
11. Put, K., Avau, A., Brisse, E., Mitera, T., Put, S., Proost, P., et al. (2015). Cytokines in systemic juvenile idiopathic arthritis and haemophagocytic lymphohistiocytosis: Tipping the balance between interleukin-18 and interferon-γ. *Rheumatology, 54*(8), 1507–1517.
12. Akashi, K., Hayashi, S., Gondo, H., Mizuno, S. I., Harada, M., Tamura, K., et al. (1994). Involvement of interferon-γ and macrophage colony-stimulating factor in pathogenesis of haemophagocytic lymphohistiocytosis in adults. *British Journal of Haematology, 87*(2), 243–250.
13. Kuriyama, T., Takenaka, K., Kohno, K., Yamauchi, T., Daitoku, S., Yoshimoto, G., et al. (2012). Engulfment of hematopoietic stem cells caused by down-regulation of CD47 is critical in the pathogenesis of hemophagocytic lymphohistiocytosis. *Blood, 120*(19), 4058–4067.
14. Nold-Petry, C. A., Lehrnbecher, T., Jarisch, A., Schwabe, D., Pfeilschifter, J. M., Muhl, H., et al. (2010). Failure of interferon γ to induce the anti-inflammatory interleukin 18 binding protein in familial hemophagocytosis. *PLoS One, 5*(1), e8663.
15. Bracaglia, C., Marafon, D. P., Caiello, I., de Graaf, K., Guilhot, F., Ferlin, W., et al. (2015). High levels of interferon-gamma (IFNγ) in macrophage activation syndrome (MAS) and CXCL9 levels as a biomarker for IFNγ production in MAS. *Pediatric Rheumatology, 13*(1), 1.
16. Tamura, K., Kanazawa, T., Tsukada, S., Kobayashi, T., Kawamura, M., & Morikawa, A. (2008). Increased serum monocyte chemoattractant protein-1, macrophage inflammatory protein-1β, and interleukin-8 concentrations in hemophagocytic lymphohistiocytosis. *Pediatric Blood & Cancer, 51*(5), 662–668.

17. Teruya-Feldstein, J., Setsuda, J., Yao, X., Kingma, D. W., Straus, S., Tosato, G., et al. (1999). MIP-1alpha expression in tissues from patients with hemophagocytic syndrome. *Laboratory Investigation, 79*(12), 1583–1590.
18. Mazodier, K., Marin, V., Novick, D., Farnarier, C., Robitail, S., Schleinitz, N., et al. (2005). Severe imbalance of IL-18/IL-18BP in patients with secondary hemophagocytic syndrome. *Blood, 106*(10), 3483–3489.
19. Osugi, Y., Hara, J., Tagawa, S., Takai, K., Hosoi, G., Matsuda, Y., et al. (1997). Cytokine production regulating Th1 and Th2 cytokines in hemophagocytic lymphohistiocytosis. *Blood, 89*(11), 4100–4103.
20. Schulert, G. S., & Grom, A. A. (2014). Macrophage activation syndrome and cytokine-directed therapies. *Best Practice & Research Clinical Rheumatology, 28*(2), 277–292.
21. Weaver, L. K., & Behrens, E. M. (2014). Hyperinflammation, rather than hemophagocytosis, is the common link between macrophage activation syndrome and hemophagocytic lymphohistiocytosis. *Current Opinion in Rheumatology, 26*(5), 562.
22. Zhang, M., Behrens, E. M., Atkinson, T. P., Shakoory, B., Grom, A. A., & Cron, R. Q. (2014). Genetic defects in cytolysis in macrophage activation syndrome. *Current Rheumatology Reports, 16*(9), 1–8.
23. Saltzman, R. W., Monaco-Shawver, L., Zhang, K., Sullivan, K. E., Filipovich, A. H., & Orange, J. S. (2012). Novel mutation in syntaxin-binding protein 2 (STXBP2) prevents IL-2-induced natural killer cell cytotoxicity. *The Journal of Allergy and Clinical Immunology, 129*(6), 1666.
24. Spessott, W. A., Sanmillan, M. L., McCormick, M. E., Patel, N., Villanueva, J., Zhang, K., et al. (2015). Hemophagocytic lymphohistiocytosis caused by dominant-negative mutations in STXBP2 that inhibit SNARE-mediated membrane fusion. *Blood, 125*(10), 1566–1577.
25. Jenkins, M. R., Rudd-Schmidt, J. A., Lopez, J. A., Ramsbottom, K. M., Mannering, S. I., Andrews, D. M., et al. (2015). Failed CTL/NK cell killing and cytokine hypersecretion are directly linked through prolonged synapse time. *The Journal of Experimental Medicine, 212*(3), 307–317.
26. Janka, G. (2009). Hemophagocytic lymphohistiocytosis: When the immune system runs amok. *Klinische Pädiatrie, 221*(05), 278–285.
27. Grom, A. A. (2004). Natural killer cell dysfunction: A common pathway in systemic-onset juvenile rheumatoid arthritis, macrophage activation syndrome, and hemophagocytic lymphohistiocytosis? *Arthritis and Rheumatism, 50*(3), 689–698.
28. Janka, G. (2012). Familial and acquired hemophagocytic lymphohistiocytosis. *Annual Review of Medicine, 63*, 233–246.
29. Brown, D. E., McCoy, M. W., Pilonieta, M. C., Nix, R. N., & Detweiler, C. S. (2010). Chronic murine typhoid fever is a natural model of secondary hemophagocytic lymphohistiocytosis. *PLoS One, 5*(2), e9441.
30. Henter, J. I., Horne, A., Arico, M., Egeler, R. M., Filipovich, A. H., Imashuku, S., et al. (2007). HLH-2004: Diagnostic and therapeutic guidelines for hemophagocytic lymphohistiocytosis. *Pediatric Blood & Cancer, 48*(2), 124–131.
31. Kiernan, T. J., O'Flaherty, N., Gilmore, R., Ho, E., Hickey, M., Tolan, M., et al. (2008). Abiotrophia defectiva endocarditis and associated hemophagocytic syndrome—A first case report and review of the literature. *International Journal of Infectious Diseases, 12*(5), 478–482.
32. Gosh, J., Roy, M., & Bala, A. (2009). Infection associated with hemophagocytic lymphohistiocytosis triggered by nosocomial infection. *Oman Medical Journal, 24*(3), 223–225.
33. Tseng, Y.-T., Sheng, W.-H., Lin, B.-H., Lin, C.-W., Wang, J.-T., Chen, Y.-C., et al. (2011). Causes, clinical symptoms, and outcomes of infectious diseases associated with hemophagocytic lymphohistiocytosis in Taiwanese adults. *Journal of Microbiology, Immunology, and Infection, 44*(3), 191–197.
34. Karras, A., Thervet, E., & Legendre, C. (2004). Groupe Cooperatif de transplantation d'Ile de F. Hemophagocytic syndrome in renal transplant recipients: Report of 17 cases and review of literature. *Transplantation, 77*(2), 238–243.

35. Poudel, A., Lew, J., Slayton, W., & Dharnidharka, V. R. (2014). Bartonella henselae infection inducing hemophagocytic lymphohistiocytosis in a kidney transplant recipient. *Pediatric Transplantation, 18*(3), E83–E87.
36. Cantero-Hinojosa, J., D'iez-Ruiz, A., Santos-Perez, J., Aguilar-Martinez, J., & Ramos-Jimenez, A. (1993). Lyme disease associated with hemophagocytic syndrome. *Journal of Molecular Medicine, 71*(8), 620.
37. Akbayram, S., Dogan, M., Akgun, C., Peker, E., Parlak, M., Caksen, H., et al. (2011). An analysis of children with brucellosis associated with pancytopenia. *Pediatric Hematology and Oncology, 28*(3), 203–208.
38. Erduran, E., Makuloglu, M., & Mutlu, M. (2010). A rare hematological manifestation of brucellosis: Reactive hemophagocytic syndrome. *Journal of Microbiology, Immunology, and Infection, 43*(2), 159–162.
39. Meneses, A., Epaulard, O., Maurin, M., Gressin, R., Pavese, P., Brion, J. P., et al. (2010). Brucella bacteremia reactivation 70 years after the primary infection. *Médecine et Maladies Infectieuses, 40*(4), 238–240.
40. Sari, I., Altuntas, F., Hacioglu, S., Kocyigit, I., Sevinc, A., Sacar, S., et al. (2008). A multicenter retrospective study defining the clinical and hematological manifestations of brucellosis and pancytopenia in a large series: Hematological malignancies, the unusual cause of pancytopenia in patients with brucellosis. *American Journal of Hematology, 83*(4), 334–339.
41. Karakukcu, M., Patiroglu, T., Ozdemir, M. A., Gunes, T., Gumus, H., & Karakukcu, C. (2004). Pancytopenia, a rare hematologic manifestation of brucellosis in children. *Journal of Pediatric Hematology/Oncology, 26*(12), 803–806.
42. Yildirmak, Y., Palanduz, A., Telhan, L., Arapoglu, M., & Kayaalp, N. (2003). Bone marrow hypoplasia during Brucella infection. *Journal of Pediatric Hematology/Oncology, 25*(1), 63–64.
43. al-Eissa, Y., & al-Nasser, M. (1993). Haematological manifestations of childhood brucellosis. *Infection, 21*(1), 23–26.
44. Andreo, J. A., Vidal, J. B., Hernandez, J. E., Serrano, P., Lopez, V. M., & Soriano, J. (1988). Hemophagocytic syndrome associated with brucellosis. *Medicina Clínica (Barcelona), 90*(12), 502–505.
45. Lopez-Gomez, M., Hernandez, J., Sampalo, A., Biedma, A., Alcala, A., & Mateas, F. (1994). Reactive hemophagocytic syndrome with disseminated intravascular coagulation secondary to acute brucellosis. *Enfermedades Infecciosas Y Microbiologia Clinica, 12*(10), 519–520.
46. Martin-Moreno, S., Soto-Guzman, O., Bernaldo-de-Quiros, J., Reverte-Cejudo, D., & Bascones-Casas, C. (1983). Pancytopenia due to hemophagocytosis in patients with brucellosis: A report of four cases. *The Journal of Infectious Diseases, 147*(3), 445–449.
47. Mondal, N., Suresh, R., Acharya, N. S., Praharaj, I., Harish, B. N., & Mahadevan, S. (2010). Hemophagocytic syndrome in a child with brucellosis. *Indian Journal of Pediatrics, 77*(12), 1434–1436.
48. Ullrich, C. H., Fader, R., Fahner, J. B., & Barbour, S. D. (1993). Brucellosis presenting as prolonged fever and hemophagocytosis. *American Journal of Diseases of Children, 147*(10), 1037–1038.
49. Zuazu, J. P., Duran, J. W., & Julia, A. F. (1979). Hemophagocytosis in acute brucellos. *The New England Journal of Medicine, 301*(21), 1185–1186.
50. Demir, C., Karahocagil, M. K., Esen, R., Atmaca, M., Gonullu, H., & Akdeniz, H. (2012). Bone marrow biopsy findings in brucellosis patients with hematologic abnormalities. *Chinese Medical Journal, 125*(11), 1871–1876.
51. Heydari, A. A., Ahmadi, F., Sarvghad, M. R., Safari, H., Bajouri, A., & Saeidpour, M. (2007). Hemophagocytosis and pulmonary involvement in brucellosis. *International Journal of Infectious Diseases, 11*(1), 89–90.
52. Anstead, G. M., Jorgensen, J. H., Craig, F. E., Blaser, M. J., & Patterson, T. F. (2001). Thermophilic multidrug-resistant campylobacter fetus infection with hypersplenism and

histiocytic phagocytosis in a patient with acquired immunodeficiency syndrome. *Clinical Infectious Diseases, 32*(2), 295–296.
53. Tamura, A., Matsunobu, T., Kurita, A., & Shiotani, A. (2012). Hemophagocytic syndrome in the course of sudden sensorineural hearing loss. *ORL: Journal for Oto-rhino-laryngology and Its Related Specialties, 74*(4), 211–214.
54. Yagi, K., Kano, G., Shibata, M., Sakamoto, I., Matsui, H., & Imashuku, S. (2011). Chlamydia pneumoniae infection-related hemophagocytic lymphohistiocytosis and acute encephalitis and poliomyelitis-like flaccid paralysis. *Pediatric Blood & Cancer, 56*(5), 853–855.
55. Ramon, I., Libert, M., Guillaume, M.-P., Corazza, F., & Karmali, R. (2010). Recurrent haemophagocytic syndrome in an HIV-infected patient. *Acta Clinica Belgica, 65*(4), 276–278.
56. Chinen, K., Ohkura, Y., Matsubara, O., & Tsuchiya, E. (2004). Hemophagocytic syndrome associated with clostridial infection in a pancreatic carcinoma patient. *Pathology, Research and Practice, 200*(3), 241–245.
57. Estrov, Z., Bruck, R., Shtalrid, M., Berrebi, A., & Resnitzky, P. (1984). Histiocytic hemophagocytosis in Q fever. *Archives of Pathology & Laboratory Medicine, 108*(1), 7.
58. Harris, P., Dixit, R., & Norton, R. (2011). Coxiella burnetii causing haemophagocytic syndrome: A rare complication of an unusual pathogen. *Infection, 39*(6), 579–582.
59. Hufnagel, M., Niemeyer, C., Zimmerhackl, L. B., Tüchelmann, T., Sauter, S., & Brandis, M. (1995). Hemophagocytosis: A complication of acute Q fever in a child. *Clinical Infectious Diseases, 21*(4), 1029–1031.
60. Abbott, K. C., Vukelja, S. J., Smith, C. E., McAllister, C. K., Konkol, K. A., O'rourke, T. J., et al. (1991). Hemophagocytic syndrome: A cause of pancytopenia in human ehrlichiosis. *American Journal of Hematology, 38*(3), 230–234.
61. Burns, S., Saylors, R., & Mian, A. (2010). Hemophagocytic lymphohistiocytosis secondary to Ehrlichia chaffeensis infection: A case report. *Journal of Pediatric Hematology/Oncology, 32*(4), e142–e1e3.
62. Hanson, D., Walter, A. W., & Powell, J. (2011). Ehrlichia-induced hemophagocytic lymphohistiocytosis in two children. *Pediatric Blood & Cancer, 56*(4), 661–663.
63. Marty, A. M., Dumler, J. S., Imes, G., Brusman, H. P., Smrkovski, L. L., & Frisman, D. M. (1995). Ehrlichiosis mimicking thrombotic thrombocytopenic purpura. Case report and pathological correlation. *Human Pathology, 26*(8), 920–925.
64. Chang, C.-C., Hsiao, P.-J., Chiu, C.-C., Chen, Y.-C., Lin, S.-H., Wu, C.-C., et al. (2015). Catastrophic hemophagocytic lymphohistiocytosis in a young man with nephrotic syndrome. *Clinica Chimica Acta, 439*, 168–171.
65. Mohyuddin, G. R., & Male, H. J. (2016). A rare cause of hemophagocytic lymphohistiocytosis: Fusobacterium infection-a case report and review of the literature. *Case Reports in Hematology, 2016*, 4839146.
66. Ellis, G. R., Gozzard, D. I., Looker, D. N., & Green, G. J. (1998). Postanginal septicaemia (Lemmiere's disease) complicated by haemophagocytosis. *The Journal of Infection, 36*(3), 340–341.
67. Schleinitz, N., Bernit, E., & Harle, J.-R. (2002). Severe hemophagocytic syndrome after intravesical BCG instillation. *The American Journal of Medicine, 112*(7), 593–594.
68. Gonzalez, M. J., Franco, A. G., & Alvaro, C. G. (2008). Hemophagocytic lymphohistiocytosis secondary to Calmette-Guerin bacilli infection. *European Journal of Internal Medicine, 19*(2), 150.
69. Nguyen, L., Ebaee, A., Pham, L., Ghani, H., French, S. W., & Qing, X. (2017). Hemophagocytic lymphohistiocytosis in a patient with legionella infection and bone marrow monotypic plasma cells in the setting of chronic lymphocytic leukemia. *The FASEB Journal, 31*(1 Supplement), 807.19–807.19.
70. Krishnamurthy, S., Mahadevan, S., Mandal, J., & Basu, D. (2013). Leptospirosis in association with hemophagocytic syndrome: A rare presentation. *Indian Journal of Pediatrics, 80*(6), 524–525.

71. Yang, C. W., Pan, M. J., Wu, M. S., Chen, Y. M., Tsen, Y. T., Lin, C. L., et al. (1997). Leptospirosis: An ignored cause of acute renal failure in Taiwan. *American Journal of Kidney Diseases, 30*(6), 840–845.
72. Lambotte, O., Fihman, V., Poyart, C., Buzyn, A., Berche, P., & Soumelis, V. (2005). Listeria monocytogenes skin infection with cerebritis and haemophagocytosis syndrome in a bone marrow transplant recipient. *Journal of Infection, 50*(4), 356–358.
73. Pellegrin, J., Merlio, J., Lacoste, D., Barbeau, P., Brossard, G., Beylot, J., et al. (1992). Syndrome of macrophagic activation with hemophagocytosis in human immunodeficiency virus infection. *La Revue de Medecine Interne, 13*(6), 438–440.
74. Yang, W., Fu, L., Lan, J., Shen, G., Chou, G., Tseng, C., et al. (2003). Mycobacterium avium complex-associated hemophagocytic syndrome in systemic lupus erythematosus patient: Report of one case. *Lupus, 12*(4), 312–316.
75. Rositto, A., Molinaro, L., Larralde, M., Ranalletta, M., & Drut, R. (1996). Disseminated cutaneous eruption after BCG vaccination. *Pediatric Dermatology, 13*(6), 451–454.
76. Wali, Y., & Beshlawi, I. (2012). BCG lymphadenitis in neonates with familial hemophagocytic lymphohistiocytosis. *The Pediatric Infectious Disease Journal, 31*(3), 324.
77. Brastianos, P. K., Swanson, J. W., Torbenson, M., Sperati, J., & Karakousis, P. C. (2006). Tuberculosis-associated haemophagocytic syndrome. *The Lancet Infectious Diseases, 6*(7), 447–454.
78. Maheshwari, P., Chhabra, R., & Yadav, P. (2012). Perinatal tuberculosis associated hemophagocytic lymphohistiocytosis. *The Indian Journal of Pediatrics, 79*(9), 1228–1229.
79. Saidi, W., Gammoudi, R., Korbi, M., Aounallah, A., Boussofara, L., Ghariani, N., et al. (2015). Hemophagocytic lymphohistiocytosis: An unusual complication of leprosy. *International Journal of Dermatology, 54*(9), 1054–1059.
80. Yoshiyama, M., Kounami, S., Nakayama, K., Aoyagi, N., & Yoshikawa, N. (2008). Clinical assessment of mycoplasma pneumoniae-associated hemophagocytic lymphohistiocytosis. *Pediatrics International, 50*(4), 432–435.
81. Ishida, Y., Hiroi, K., Tauchi, H., Oto, Y., Tokuda, K., & Kida, K. (2004). Hemophagocytic lymphohistiocytosis secondary to mycoplasma pneumoniae infection. *Pediatrics International, 46*(2), 174–177.
82. Cascio, A., Giordano, S., Dones, P., Venezia, S., Iaria, C., & Ziino, O. (2011). Haemophagocytic syndrome and rickettsial diseases. *Journal of Medical Microbiology, 60*(4), 537–542.
83. Premaratna, R., Williams, H. S., Chandrasena, T. G., Rajapakse, R. P., Kularatna, S. A., & de Silva, H. J. (2009). Unusual pancytopenia secondary to haemophagocytosis syndrome in rickettsioses. *Transactions of the Royal Society of Tropical Medicine and Hygiene, 103*(9), 961–963.
84. Walter, G., Botelho-Nevers, E., Socolovschi, C., Raoult, D., & Parola, P. (2012). Murine typhus in returned travelers: A report of thirty-two cases. *The American Journal of Tropical Medicine and Hygiene, 86*(6), 1049–1053.
85. Non, L. R., Patel, R., Esmaeeli, A., & Despotovic, V. (2015). Typhoid fever complicated by hemophagocytic lymphohistiocytosis and rhabdomyolysis. *The American Journal of Tropical Medicine and Hygiene, 93*(5), 1068–1069.
86. Benz-Lemoine, E., Bordigoni, P., Schaack, J. C., Briquel, E., Chiclet, A. M., & Olive, D. (1983). Systemic reactive histiocytosis with hemophagocytosis and hemostasis disorders associated with septic granulomatosis. *Archives Françaises de Pédiatrie, 40*(3), 179–182.
87. Gutierrez-Rave Pecero, V., Luque Marquez, R., Ayerza Lerchundi, M., Canavate Illescas, M., & Prados Madrona, D. (1990). Reactive hemophagocytic syndrome: Analysis of a series of 7 cases. *Medicina Clínica (Barcelona), 94*(4), 130–134.
88. Sniderman, J. D., Cuvelier, G. D., Veroukis, S., & Hansen, G. (2015). Toxic epidermal necrolysis and hemophagocytic lymphohistiocytosis: A case report and literature review. *Clinical Case Reports, 3*(2), 121–125.
89. Dube, R., Kar, S. S., Mahapatro, S., & Ray, R. (2013). Infection associated Hemophagocytic Lymphohistiocytosis: A case report. *Indian Journal of Clinical Practice, 24*(2), 163–165.

90. Niller, H. (2010). Myelodysplastic syndrome (MDS) as a late stage of subclinical hemophagocytic lymphohistiocytosis (HLH): A putative role for Leptospira infection. A hypothesis. *Acta Microbiologica et Immunologica Hungarica, 57*(3), 181–189.
91. Verma, T., & Aggarwal, S. (2012). Childhood tuberculosis presenting with haemophagocytic syndrome. *Indian Journal of Hematology and Blood Transfusion, 28*(3), 178–180.
92. Yang, C., Lee, J., Kim, Y., Kim, P., Lee, S., Kim, B., et al. (1996). Tuberculosis-associated hemophagocytic syndrome in a hemodialysis patient: Case report and review of the literature. *Nephron, 72*(4), 690–692.
93. Karras, A., Thervet, E., & Legendre, C. (2004). Hemophagocytic syndrome in renal transplant recipients: Report of 17 cases and review of literature. *Transplantation, 77*(2), 238–243.
94. Ruiz-argüelles, G. J., Arizpe-Bravo, D., Garces-Eisele, J., Sanchez-Sosa, S., Ruiz-argüelles, A., & Ponce-de-Leon, S. (1998). Tuberculosis-associated fatal hemophagocytic syndrome in a patient with lymphoma treated with fludarabine. *Leukemia & Lymphoma, 28*(5–6), 599–602.
95. Baraldes, M. A., Domingo, P., Gonzalez, M. J., Aventin, A., & Coll, P. (1998). Tuberculosis-associated hemophagocytic syndrome in patients with acquired immunodeficiency syndrome. *Archives of Internal Medicine, 158*(2), 194–195.
96. Lam, K., Ng, W., & Chan, A. (1994). Miliary tuberculosis with splenic rupture: A fatal case with hemophagocytic syndrome and possible association with long standing sarcoidosis. *Pathology, 26*(4), 493–496.

Zoonotic Bacterial Infections Triggering Cytokine Storm Syndrome

Zaher K. Otrock and Charles S. Eby

Introduction

Zoonotic infections transmitted by tick bites or directly from infected animals can produce life-threatening complications which can manifest with hemophagocytic lymphohistiocytosis (HLH)/cytokine storm syndrome (CSS). While the supporting literature consists of isolated case reports and small case series, increased awareness that HLH can be a complication of a diverse group of zoonotic infections may be increasing diagnostic testing, and in some settings, empiric antibiotic treatment for these pathogens.

Zoonotic infections represent a potential trigger for HLH and this is evident in the increasing number of reported cases in the literature. Cascio and colleagues reviewed published cases of HLH triggered by zoonotic diseases from January 1950 until August 2012 [1]. Their search revealed that HLH can be associated with many zoonotic infections including bacterial diseases and viral, protozoal, and fungal infections. More reports on zoonotic infections triggering HLH have been recently published which might reflect increased awareness to this entity among health care professional [1–3]. Table 1 summarizes reported bacterial zoonotic infections causing HLH.

Bacterial infections may constitute the largest group of reported cases associated with HLH among other zoonotic infections [1]. The reported bacterial infections included *Brucella* spp., *Rickettsia* spp., *Ehrlichia*, *Coxiella burnetii*, *Mycobacterium* spp., *Leptospira* spp., and *Salmonella* spp. [1]. The majority of cases are associated with intracellular organisms causing splenomegaly and leukopenia [1, 2]. We hereby review the most commonly reported zoonotic bacterial infections triggering HLH/CSS.

Z. K. Otrock (✉) · C. S. Eby
Department of Pathology and Laboratory Medicine, Henry Ford Hospital, Detroit, MI, USA

Department of Pathology and Immunology, Washington University School of Medicine, St. Louis, MO, USA
e-mail: zotrock1@hfhs.org

R. Q. Cron, E. M. Behrens (eds.), *Cytokine Storm Syndrome*,
https://doi.org/10.1007/978-3-030-22094-5_19

Table 1 Summary of reported bacterial zoonotic infections causing HLH/CSS

Infection	Reservoir	Vector of human transmission	Symptoms	Treatment
Brucella	Infected animals; unpasteurized milk and dairy products from infected animals	NA	Fever, sweats, arthralgia, myalgia, fatigue, weight loss, hepatomegaly, splenomegaly	Doxycycline, rifampin
Rickettsia	Cattle, goats, sheep	Arthropod host (tick, louse, mite, flea, or other insect)	Fever, localized lymphadenopathy, neutropenia, thrombocytopenia	Doxycycline
Ehrlichia	The white-tailed deer *Odocoileus virginianus*	Lone Star tick (*Amblyomma americanum*)	Fever, myalgia, headache, cough, chills, maculopapular rash, leukopenia, thrombocytopenia, elevated liver enzymes	Doxycycline, rifampin
Coxiella burnetii	Birth products (i.e., placenta, amniotic fluid), urine, feces, and milk of infected animals	NA	Can be asymptomatic; febrile illness associated with signs of pneumonia, increased transaminases and thrombocytopenia	Doxycycline
Mycobacterium avium	Domestic animals; aerosols from infected subjects, soil	NA	Fever, weight loss, cough, cytopenias, hepatomegaly, splenomegaly	Isoniazid, rifampin, pyrazinamide, ethambutol
Clostridium	Cows, pigs	NA	Diarrhea, abdominal pain, cramping, low grade fever, leukocytosis	Metronidazole
Leptospira	Feral and domestic animals	NA	High fever, chills, vomiting, headache, myalgia, jaundice	*Doxycycline*, ampicillin, amoxicillin, IV penicillin

NA not applicable

Brucella

Brucella spp. are common gram-negative bacteria zoonoses among wildlife and domestic animals. Brucellosis is rare in the USA while endemic in the eastern Mediterranean Basin, Middle East, the Arabian Peninsula, Mexico, Central and South America, Central Asia, Southern Europe, and the Indian subcontinent [4]. Transmission to humans occurs via contact with infected animal tissues or ingestion of derived food products. The disease occurs among the general population almost equally among children and adults [5]. The incubation period is variable but usually ranges between 1 and 4 weeks. The disease onset is insidious and can present with a diverse range of nonspecific clinical findings, such as fever, sweats, arthralgia, myalgia, fatigue, weight loss, hepatomegaly, and splenomegaly. Thus, the diagnosis of brucellosis can be a dilemma and may be delayed in some cases [6].

Brucella has been reported to be associated with HLH especially in children. The largest series came from a single institution in Turkey of seven pediatric cases with brucellosis confirmed by standard tube agglutination test [7]. The average age of patients was 10.2 years (range 4–14 years); none of the patients had a history of any hematologic disorder. Blood cultures were positive for *Brucella melitensis* in 3 patients, and bone marrow cultures were positive for *B. melitensis* in 4 patients. All patients fulfilled the HLH-2004 diagnostic criteria, and hemophagocytosis was documented in bone marrow examinations of 5 children. All patients recovered completely after antibiotic treatment of brucellosis [7]. The majority of reported cases of Brucella-associated HLH occurred in patients who had consumed unpasteurized dairy products or had contact with animals [7–10]. Early treatment for Brucella-associated HLH with appropriate antibiotics often results in complete recovery, contrary to the generally poor outcome of secondary HLH [7, 11].

Rickettsia

Rickettsia spp. are transmitted to humans by an arthropod host (tick, louse, mite, flea, or other insect). Major clinical and pathological findings in rickettsial infections include fever, localized lymphadenopathy, neutropenia, thrombocytopenia and moderate increases in transaminases [12, 13]. There have been many reports of HLH in patients with rickettsial infections including *Rickettsia typhi* [14, 15], *Rickettsia conorii* [16, 17], *Orientia tsutsugamushi* [18, 19], and *Rickettsia japonica* [20]. Most of the reports recommend considering a diagnosis of HLH in severe cases of rickettsial disease especially if associated with pancytopenia [16]. The prognosis of rickettsial infection-associated HLH is unknown. Outcome may depend on many factors, such as the *Rickettsia* spp. involved, host factors, comorbidities, and the prompt initiation of antibiotic therapy [16].

Ehrlichia

Ehrlichiosis is the term used to refer to acute febrile tick-borne infections caused by obligate intracellular bacteria. The majority of these infections that affect humans are caused by three distinct species: *Ehrlichia chaffeensis*, *Ehrlichia ewingii*, and *Anaplasma phagocytophilum* [12, 21]. *Ehrlichia chaffeensis* infects monocytes and causes human monocytotropic ehrlichiosis (HME). The primary tick vector for HME is the Lone Star tick (*Amblyomma americanum*) which occurs across the south-central and southeastern states [22]. The primary host of this tick is the white-tailed deer *Odocoileus virginianus*. *Ehrlichia ewingii* is serologically similar to *Ehrlichia chaffeensis* and is also transmitted by *Amblyomma americanum*. *Anaplasma phagocytophilum* infects granulocytes and causes human granulocytotropic anaplasmosis (HGA; previously known as human granulocytotropic ehrlichiosis or HGE). *Anaplasma phagocytophilum* is transmitted by *Ixodes scapularis*, which also transmits agents that cause Lyme disease and babesiosis [23]. The infection areas of endemicity in the USA include northeastern and mid-Atlantic, Upper Midwest, and Pacific Northwest states.

Patients usually present with fever, myalgia, headache, cough, and chills. A nonspecific maculopapular rash has been described in children [24]. In addition, patients commonly have leukopenia, thrombocytopenia, and elevated liver enzymes [25]. Ehrlichiosis is a relatively severe disease with 49% of patients requiring hospitalization, and a case-fatality rate of 1.0–1.9% [26, 27]. Prognosis is better if treatment is initiated early. The suspicion for this diagnosis is heightened in a patient exposed in a tick-endemic area during seasons of increased tick activity. Doxycycline is the treatment of choice and the recommended dose is 100 mg administered twice daily.

Tissue damage in HME is more likely a result of poorly controlled macrophage activation and release of effector molecules, including nitric oxide and reactive oxygen species [28]. Investigations of *Ehrlichia* infection in a murine model of fatal monocytotropic ehrlichiosis highlight the immune response and the role of tumor necrosis factor (TNF) and interleukin-10 (IL-10). These findings include focal hepatic necrosis and apoptosis, leukopenia and lymphopenia, and CD4+ T cell apoptosis [29, 30]. Excessive cytokine production is induced with *E. chaffeensis* infection and this is believed to contribute to the septic shock-like presentation seen in many HME cases [29].

Ehrlichia is a rarely reported trigger for HLH. We reported a series of five cases of *Ehrlichia*-induced HLH treated at Washington University Medical Center in St. Louis, Missouri [3]. These cases were identified among 76 HLH cases reviewed *between October 2003 and June 2014*. All *Ehrlichia*-induced HLH patients presented with fever, cytopenias and hyperferritinemia, and two of them had CNS involvement. Treatment with doxycycline was effective with no recurrence of HLH. We recently reviewed our experience with *Ehrlichia* infections at the same institution; over 10 years we identified 157 cases of ehrlichiosis [31]. Ten patients (10/157, 6.37%) fulfilled the HLH-2004 diagnostic criteria (5 of these cases were reported by our group previously [3]). Table 2 summarizes all reported cases of *Ehrlichia*-induced HLH in the literature [32–41]. All patients were immunocompetent except two patients (one had HIV [34] and another had bilateral lung transplant [3]). Although HLH often has a dismal prognosis even with treatment with an overall mortality of 58–75% [41–43], the prognosis of *Ehrlichia*-induced HLH appears to be excellent with early recognition and initiation of treatment. Some of the reported patients were treated, in addition to doxycycline, with etoposide, dexamethasone, or HLH-2004 protocol to suppress the inflammatory response and the exaggerated proliferation of macrophages that characterize HLH. Interestingly, one of the patients responded to doxycycline, steroids, and anakinra, which is a soluble interleukin-1 receptor antagonist [33], supporting the presence of inflammatory cytokines that are the basis of HLH/CSS.

Table 2 Summary of published cases of *Ehrlichia*-induced HLH

Characteristic	[3] [b]	[3] [b]	[3]	[3]	[3]	[32]	[33]
Age (years)	52	47	59	16	62	7	63
Immunocompromised	N	N	N	N	N	N	N
Gender	F	F	F	F	M	F	M
Fever	**Y**	**Y**	**Y**	**Y**	**Y**	**Y**	**Y**
Splenomegaly	N	N	N	N	**Y**	NA	**Y**
ANC (×10³/μL)	**0.1**	12.1	**0.6**	**0.4**	3.9	**Present**	1.33
Hemoglobin (g/dL)	**7.9**	**8.6**	10.6	11.9	**8.8**	**Present**	**8**
Platelets (×10³/μL)	**25**	**65**	**41**	**37**	**20**	**Present**	**19**
Triglycerides (mg/dL)	**650**	**710**	**307**	**319**	**516**	NA	**436**
Fibrinogen (mg/dL)	173	178	NA	187	312	**84**	**90**
Ferritin (μg/L)	**47,290**	**10,002**	**2863**	**85,517**	**84,676**	**8750**	**70,097**
Hemophagocytosis	**Y (BM)**	N	**Y (BM)**	**Y (BM)**	NA	**Y (BM)**	NA
Soluble IL-2 receptor (U/mL)[c]	**>6500**	**5873**	NA	NA	NA	NA	NA
Low/absent NK cell activity	Failed	NA	NA	NA	NA	NA	NA
Molecular testing	NA	NA	NA	Negative	NA	Negative	NA
Treatment	Doxycycline, rifampin, and dexamethasone	Doxycycline and methylprednisolone	Doxycycline	Doxycycline and dexamethasone	Doxycycline and dexamethasone	Doxycycline and corticosteroids	Doxycycline, steroids, and anakinra[a]
Outcome	Recovered	Recovered	Recovered	Recovered	Recovered	Recovered	Doing well after 2 months

(continued)

Table 2 (continued)

Characteristic	[34]	[35]	[36]	[37]	[38]	[39]	[39] [b]	[40]
Age (years)	66	9	41	7	74	10	13	10
Immunocompromised	Y	N	N	N	N	N	N	N
Gender	F	M	F	M	M	F	M	M
Fever	**Y**	**Y**	**Y**	**Y**	**Y**	**Y**	**Y**	**Y**
Splenomegaly	N	N	N	**Y**	N	N	NA	**Y**
ANC ($\times 10^3/\mu L$)	2.9	2.7	NA	1.62	**Present**	1.3	**0.56**	**Present**
Hemoglobin (g/dL)	12.2	**8.2**	**8.4**	10.1	NA	9.1	**7.9**	10.2
Platelets ($\times 10^3/\mu L$)	**22**	**68**	**27**	109	**16**	**38**	**57**	**50**
Triglycerides (mg/dL)	**358**	161	**829**	147	**387**	**327**	**605**	**287**
Fibrinogen (mg/dL)	225	**138**	156	**71**	337	**88**	**118**	**93**
Ferritin (μg/L)	**>40,000**	**>40,000**	**13,257**	**5306**	**12,369**	**3517**	**31,022**	**>10,000**
Hemophagocytosis	**Y**	**Y**	**Y (BM)**	**Y (BM)**	**Y (BM)**	**Y (BM)**	**Y (BM)**	**Y (BM)**
Soluble IL-2 receptor (U/mL)[c]	NA	**3022**	NA	**10,650**	NA	**4692**	**4454**	NA
Low/absent NK cell activity	**Y**	NA	**Y**	Normal	NA	Normal	Normal	Normal
Molecular testing	NA	Negative	NA	**Heterozygous for MUNC and perforin genes**	NA	Negative	Negative	Negative
Treatment	HLH-2004 protocol	Doxycycline, etoposide, and dexamethasone	Doxycycline, prednisone, and IVIG	Doxycycline (initially started on dexamethasone)	Doxycycline, vancomycin, and imipenem	Doxycycline (initially started on steroids)	Doxycycline plus HLH-2004 protocol	HLH-2004 protocol then doxycycline
Outcome	Died	Recovered	Recovered	Asymptomatic at 1 year	Recovered	Recovered	Recovered with 2-month follow up	Asymptomatic at 1 year

Characteristic	[31]	[31]	[31]	[31]	[31]
Age (years)	9	7	77	11	7
Immunocompromised	N	N	N	N	N
Gender	F	F	M	M	F
Fever	**Y**	**Y**	**Y**	**Y**	**Y**
Splenomegaly	N	**Y**	N	N	N
ANC (×103/μL)	**0.5**	NA	1.7	**0.8**	NA
Hemoglobin (g/dL)	10.2	**8.4**	**7.55**	9.9	**7.9**
Platelets (×103/μL)	**28**	**51**	**40**	**15**	**44**
Triglycerides (mg/dL)	**314**	**401**	**700**	**340**	126
Fibrinogen (mg/dL)	**138**	253	192	159	**93**
Ferritin (μg/L)	**36,282**	**3183**	**79,101**	**21,187**	**44,095**
Hemophagocytosis	**Y**	NA	NA	NA	NA
Soluble IL-2 receptor (U/mL)	**4336**	**10,143**	**7443**	**13,505**	**11,072**
Low/absent NK cell activity	**Low**	Normal	NA	Normal	Normal
Molecular testing	Negative	NA	NA	NA	NA
Treatment	Doxycycline and dexamethasone	Doxycycline	Doxycycline and dexamethasone	Doxycycline	Doxycycline
Outcome	Recovered	Recovered	Died at 15 days of septic shock	Recovered	Recovered

Parameters that fulfill the HLH-2004 diagnostic criteria are indicated in **bold**

F female, *M* male, *Y* yes, *N* no, *NA* not available, *ANC* absolute neutrophil count, *BM* bone marrow, *NK* natural killer

[a]Anakinra is a soluble interleukin-1 receptor antagonist

[b]CNS involvement with *Ehrlichia*

[c]Normal soluble IL-2 receptor <2400 U/mL

Coxiella burnetii

Q fever is the clinical manifestation of symptomatic *Coxiella burnetii* infection. *Coxiella burnetii* is an obligate intracellular bacterium most commonly found in ruminants including cattle, sheep, and goats. Numerous tick species either harbor or transmit *Coxiella burnetii*; however, tick transmission is not considered a major route of transmission to humans [44]. Q fever is a worldwide infection with endemic regions in the Mediterranean countries especially Spain and France [45]. Transmission to humans most often occurs through inhalation of aerosolized animal wastes or contaminated soil. Infection may be asymptomatic, acute, or chronic and most commonly presents as a nonspecific febrile illness (acute Q fever) associated with signs of pneumonia, increased transaminases, and thrombocytopenia [45]. It is an intracellular pathogen that preferentially infects monocytes and macrophages; the majority of infected macrophages will be polarized towards an inflammatory response stimulating the release of inflammatory cytokines [46]. HLH is a rarely reported complication of *Coxiella burnetii* [47–49]. However, it is recommended that profound and persistent hematological abnormalities especially thrombocytopenia in the context of Q fever should raise suspicion for HLH as a diagnosis. Conversely, Q fever should be investigated by serology and PCR in endemic areas as a potential trigger for an established HLH diagnosis.

Mycobacterium avium

Mycobacterial infections have a diverse range of clinical manifestations. Mycobacteria, whether tuberculous or non-tuberculous, have been associated with HLH. Tuberculosis-associated HLH cases were reviewed in a recent paper by Padhi and colleagues [50]. A high proportion (41/63, 65%) of patients had underlying comorbidities such as end-stage renal disease, type 2 diabetes mellitus, past history of malignancies, and autoimmune diseases. The median duration of symptoms before diagnosis of tuberculosis-associated HLH was 45 days. Fever was present in all cases. Hepatosplenomegaly was observed in 43 of 61 cases (70.5%), and 6 (9.8%) cases had isolated splenomegaly. A higher proportion (32/59, 54.2%) of patients had pancytopenia, and 22/59 (37.2%) had bicytopenia. Hemophagocytosis was detected on bone marrow examination in 58/63 (92%) cases. Non-tuberculous mycobacterial infection-associated HLH cases were also reviewed recently [51]. Six of the seven published cases occurred among patients with underlying immune disorders.

Mycobacterial zoonotic infections associated with HLH are very uncommon but should be considered in patients with fever and underlying immunosuppressive conditions. Among the zoonotic mycobacterial infections, HLH was reported in patients with *Mycobacterium avium* and HIV, systemic lupus erythematosus, and sickle cell disease [52–54]. Although early recognition of the etiology and optimal treatment of the underlying mycobacterial infection might improve the outcome of HLH, these cases are often fatal [54].

Conclusions

In summary, zoonotic bacterial infections, particularly intracellular pathogens, are potential triggers for HLH/CSS. These two syndromes are characterized by hypercytokinemia which manifests with the release of many inflammatory mediators including cytokines. The list of zoonotic bacterial infections associated with HLH is long and includes Brucella, *Rickettsia* spp., *Ehrlichia*, *Coxiella burnetii*, and *Mycobacterium* spp., among others. The institution of antimicrobial therapy is of paramount importance, and in some cases, immunosuppressive therapy has been beneficial in treating these infections. A high index of suspicion for one of these infections should be maintained in seriously ill patients presenting with clinical findings suggestive of HLH/CSS especially in areas endemic with these zoonoses.

References

1. Cascio, A., Pernice, L. M., Barberi, G., Delfino, D., Biondo, C., Beninati, C., et al. (2012). Secondary hemophagocytic lymphohistiocytosis in zoonoses. A systematic review. *European Review for Medical and Pharmacological Sciences, 16*, 1324–1337.
2. Lecronier, M., Prendki, V., Gerin, M., Schneerson, M., Renvoisé, A., Larroche, C., et al. (2013). Q fever and Mediterranean spotted fever associated with hemophagocytic syndrome: Case study and literature review. *International Journal of Infectious Diseases, 17*, e629–e633.
3. Otrock, Z. K., Gonzalez, M. D., & Eby, C. S. (2015). Ehrlichia-induced hemophagocytic lymphohistiocytosis: A case series and review of literature. *Blood Cells, Molecules & Diseases, 55*, 191–193.
4. Pappas, G., Papadimitriou, P., Akritidis, N., Christou, L., & Tsianos, E. V. (2006). The new global map of human brucellosis. *The Lancet Infectious Diseases, 6*, 91–99.
5. Dean, A. S., Crump, L., Greter, H., Schelling, E., & Zinsstag, J. (2012). Global burden of human brucellosis: A systematic review of disease frequency. *PLoS Neglected Tropical Diseases, 6*, e1865.
6. Dean, A. S., Crump, L., Greter, H., Hattendorf, J., Schelling, E., & Zinsstag, J. (2012). Clinical manifestations of human brucellosis: A systematic review and meta-analysis. *PLoS Neglected Tropical Diseases, 6*, e1929.
7. Karaman, K., Akbayram, S., Kaba, S., Karaman, S., Garipardiç, M., Aydin, I., et al. (2016). An analysis of children with brucellosis associated with haemophagocytic lymphohistiocytosis. *Le Infezioni in Medicina, 24*, 123–130.
8. Yaman, Y., Gözmen, S., Özkaya, A. K., Oymak, Y., Apa, H., Vergin, C., et al. (2015). Secondary hemophagocytic lymphohistiocytosis in children with brucellosis: Report of three cases. *Journal of Infection in Developing Countries, 9*, 1172–1176.
9. Erduran, E., Makuloglu, M., & Mutlu, M. (2010). A rare hematological manifestation of brucellosis: Reactive hemophagocytic syndrome. *Journal of Microbiology, Immunology, and Infection, 43*, 159–162.
10. Mondal, N., Suresh, R., Acharya, N. S., Praharaj, I., Harish, B. N., & Mahadevan, S. (2010). Hemophagocytic syndrome in a child with brucellosis. *Indian Journal of Pediatrics, 77*, 1434–1436.
11. David, A., Iaria, C., Giordano, S., Iaria, M., & Cascio, A. (2012). Secondary hemophagocytic lymphohistiocytosis: Forget me not! *Transplant Infectious Disease, 14*, E121–E123.
12. Dumler, J. S., Madigan, J. E., Pusterla, N., & Bakken, J. S. (2007). Ehrlichioses in humans: Epidemiology, clinical presentation, diagnosis, and treatment. *Clinical Infectious Diseases, 45*(Suppl 1), S45–S51.

13. Parola, P., Paddock, C. D., & Raoult, D. (2005). Tick-borne rickettsioses around the world: Emerging diseases challenging old concepts. *Clinical Microbiology Reviews, 18*, 719–756.
14. Jayakrishnan, M. P., Veny, J., & Feroze, M. (2011). Rickettsial infection with hemophagocytosis. *Tropical Doctor, 41*, 111–112.
15. Walter, G., Botelho-Nevers, E., Socolovschi, C., Raoult, D., & Parola, P. (2012). Murine typhus in returned travelers: A report of thirty-two cases. *The American Journal of Tropical Medicine and Hygiene, 86*, 1049–1053.
16. Cascio, A., Giordano, S., Dones, P., Venezia, S., Iaria, C., & Ziino, O. (2011). Haemophagocytic syndrome and rickettsial diseases. *Journal of Medical Microbiology, 60*, 537–542.
17. Premaratna, R., Williams, H. S., Chandrasena, T. G., Rajapakse, R. P., Kularatna, S. A., & de Silva, H. J. (2009). Unusual pancytopenia secondary to haemophagocytosis syndrome in rickettsioses. *Transactions of the Royal Society of Tropical Medicine and Hygiene, 103*, 961–963.
18. Basheer, A., Padhi, S., Boopathy, V., Mallick, S., Nair, S., Varghese, R. G., et al. (2015 Jan 1). Hemophagocytic Lymphohistiocytosis: an Unusual Complication of Orientia tsutsugamushi Disease (Scrub Typhus). *Mediterr J Hematol Infect Dis., 7*(1), e2015008.
19. Kwon, H. J., Yoo, I. H., Lee, J. W., Chung, N. G., Cho, B., Kim, H. K., et al. (2013 Feb). Life-threatening scrub typhus with hemophagocytosis and acute respiratory distress syndrome in an infant. *J Trop Pediatr., 59*(1), 67–69.
20. Otsuki, S., Iwamoto, S., Azuma, E., Nashida, Y., Akachi, S., Taniguchi, K., et al. (2015). Hemophagocytic Lymphohistiocytosis due to rickettsia japonica in a 3-month-old infant. *Journal of Pediatric Hematology/Oncology, 37*, 627–628.
21. Dumler, J. S. (2005). Anaplasma and Ehrlichia infection. *Annals of the New York Academy of Sciences, 1063*, 361–373.
22. Paddock, C. D., & Childs, J. E. (2003). Ehrlichia chaffeensis: A prototypical emerging pathogen. *Clinical Microbiology Reviews, 16*, 37–64.
23. Dumler, J. S., Choi, K. S., Garcia-Garcia, J. C., Barat, N. S., Scorpio, D. G., Garyu, J. W., et al. (2005). Human granulocytic anaplasmosis and Anaplasma phagocytophilum. *Emerging Infectious Diseases, 11*, 1828–1834.
24. Lantos, P. M., & Krause, P. J. (2002). Ehrlichiosis in children. *Seminars in Pediatric Infectious Diseases, 13*, 249–256.
25. Pujalte, G. G., & Chua, J. V. (2013). Tick-borne infections in the United States. *Primary Care, 40*, 619–635.
26. Dahlgren, F. S., Heitman, K. N., & Behravesh, C. B. (2016). Undetermined human ehrlichiosis and anaplasmosis in the United States, 2008–2012: A catch-all for passive surveillance. *The American Journal of Tropical Medicine and Hygiene, 94*, 299–301.
27. Dahlgren, F. S., Mandel, E. J., Krebs, J. W., Massung, R. F., & McQuiston, J. H. (2011). Increasing incidence of Ehrlichia chaffeensis and Anaplasma phagocytophilum in the United States, 2000–2007. *The American Journal of Tropical Medicine and Hygiene, 85*, 124–131.
28. Ismail, N., Olano, J. P., Feng, H. M., & Walker, D. H. (2002). Current status of immune mechanisms of killing of intracellular microorganisms. *FEMS Microbiology Letters, 207*, 111–120.
29. Ismail, N., Soong, L., McBride, J. W., Valbuena, G., Olano, J. P., Feng, H. M., et al. (2004). Overproduction of TNF-alpha by CD8+ type 1 cells and down-regulation of IFN-gamma production by CD4+ Th1 cells contribute to toxic shock-like syndrome in an animal model of fatal monocytotropic ehrlichiosis. *Journal of Immunology, 172*, 1786–1800.
30. Ismail, N., Stevenson, H. L., & Walker, D. H. (2006). Role of tumor necrosis factor alpha (TNF-alpha) and interleukin-10 in the pathogenesis of severe murine monocytotropic ehrlichiosis: Increased resistance of TNF receptor p55- and p75-deficient mice to fatal ehrlichial infection. *Infection and Immunity, 74*, 1846–1856.
31. Otrock, Z. K., Eby, C. S., & Burnham, C. D. (2019). Human ehrlichiosis at a tertiary-care academic medical center: Clinical associations and outcomes of transplant patients and patients with hemophagocytic lymphohistiocytosis. *Blood Cells Mol Dis., 77*, 17–22.
32. Statler, V. A., & Marshall, G. S. (2015). Hemophagocytic lymphohistiocytosis induced by monocytic ehrlichiosis. *The Journal of Pediatrics, 166*, 499–99.e1.

33. Kumar, N., Goyal, J., Goel, A., Shakoory, B., & Chatham, W. (2014). Macrophage activation syndrome secondary to human monocytic ehrlichiosis. *Indian Journal of Hematology and Blood Transfusion, 30*(Suppl 1), 145–147.
34. Naqash, A. R., Yogarajah, M., Vallangeon, B. D., Hafiz, M., Patel, D., Kolychev, E., et al. (2017). Hemophagocytic lymphohistiocytosis (HLH) secondary to Ehrlichia chaffeensis with bone marrow involvement. *Annals of Hematology, 96*, 1755–1758.
35. Cheng, A., Williams, F., Fortenberry, J., Preissig, C., Salinas, S., & Kamat, P. (2016). Use of extracorporeal support in Hemophagocytic Lymphohistiocytosis secondary to Ehrlichiosis. *Pediatrics, 138*(4), e20154176.
36. Kaplan, R. M., Swat, S. A., & Singer, B. D. (2016). Human monocytic ehrlichiosis complicated by hemophagocytic lymphohistiocytosis and multi-organ dysfunction syndrome. *Diagnostic Microbiology and Infectious Disease, 86*, 327–328.
37. Vijayan, V., Thambundit, A., & Sukumaran, S. (2015). Hemophagocytic lymphohistiocytosis secondary to ehrlichiosis in a child. *Clinical Pediatrics (Phila), 54*, 84–86.
38. Pandey, R., Kochar, R., Kemp, S., Rotaru, D., & Shah, S. V. (2013). Ehrlichiosis presenting with toxic shock-like syndrome and secondary hemophagocytic lymphohistiocytosis. *The Journal of the Arkansas Medical Society, 109*, 280–282.
39. Hanson, D., Walter, A. W., & Powell, J. (2011). Ehrlichia-induced hemophagocytic lymphohistiocytosis in two children. *Pediatric Blood & Cancer, 56*, 661–663.
40. Burns, S., Saylors, R., & Mian, A. (2010). Hemophagocytic lymphohistiocytosis secondary to Ehrlichia chaffeensis infection: A case report. *Journal of Pediatric Hematology/Oncology, 32*, e142–e143.
41. Otrock, Z. K., & Eby, C. S. (2015). Clinical characteristics, prognostic factors, and outcomes of adult patients with hemophagocytic lymphohistiocytosis. *American Journal of Hematology, 90*, 220–224.
42. Parikh, S. A., Kapoor, P., Letendre, L., Kumar, S., & Wolanskyj, A. P. (2014). Prognostic factors and outcomes of adults with hemophagocytic lymphohistiocytosis. *Mayo Clinic Proceedings, 89*, 484–492.
43. Shabbir, M., Lucas, J., Lazarchick, J., & Shirai, K. (2011). Secondary hemophagocytic syndrome in adults: A case series of 18 patients in a single institution and a review of literature. *Hematological Oncology, 29*, 100–106.
44. Anderson, A., Bijlmer, H., Fournier, P. E., Graves, S., Hartzell, J., Kersh, G. J., et al. (2013). Diagnosis and management of Q fever–United States, 2013: Recommendations from CDC and the Q fever working group. *MMWR - Recommendations and Reports, 62*, 1–30.
45. Million, M., & Raoult, D. (2015). Recent advances in the study of Q fever epidemiology, diagnosis and management. *The Journal of Infection, 71*(Suppl 1), S2–S9.
46. Amara, A. B., Bechah, Y., & Mege, J. L. (2012). Immune response and Coxiella burnetii invasion. *Advances in Experimental Medicine and Biology, 984*, 287–298.
47. Chen, T. C., Chang, K., Lu, P. L., Liu, Y. C., Chen, Y. H., Hsieh, H. C., et al. (2006). Acute Q fever with hemophagocytic syndrome: Case report and literature review. *Scandinavian Journal of Infectious Diseases, 38*, 1119–1122.
48. Harris, P., Dixit, R., & Norton, R. (2011). Coxiella burnetii causing haemophagocytic syndrome: A rare complication of an unusual pathogen. *Infection, 39*, 579–582.
49. Paine, A., Miya, T., & Webb, B. J. (2015). Coxiella burnetii infection with severe hyperferritinemia in an asplenic patient. *Open Forum Infectious Diseases, 2*, ofv125.
50. Padhi, S., Ravichandran, K., Sahoo, J., Varghese, R. G., & Basheer, A. (2015). Hemophagocytic lymphohistiocytosis: An unusual complication in disseminated Mycobacterium tuberculosis. *Lung India, 32*, 593–601.
51. Grandjean Lapierre, S., Toro, A., & Drancourt, M. (2017). Mycobacterium iranicum bacteremia and hemophagocytic lymphohistiocytosis: A case report. *BMC Research Notes, 10*, 372.
52. Castelli, A. A., Rosenthal, D. G., Bender Ignacio, R., & Chu, H. Y. (2015). Hemophagocytic Lymphohistiocytosis secondary to human immunodeficiency virus-associated Histoplasmosis. *Open Forum Infectious Diseases, 2*, ofv140.

53. Yang, W. K., Fu, L. S., Lan, J. L., Shen, G. H., Chou, G., Tseng, C. F., et al. (2003). Mycobacterium avium complex-associated hemophagocytic syndrome in systemic lupus erythematosus patient: Report of one case. *Lupus, 12*, 312–316.
54. Chamsi-Pasha, M. A., Alraies, M. C., Alraiyes, A. H., & Hsi, E. D. (2013). Mycobacterium avium complex-associated hemophagocytic lymphohistiocytosis in a sickle cell patient: An unusual fatal association. *Case Reports in Hematology, 2013*, 291518.

Parasitic and Fungal Triggers

Gary Sterba and Yonit Sterba

Introduction

CSS occurs associated with infections, or is triggered by them. Their coexistence often hides one another. A high index of suspicion will help to recognize a CSS and the inciting organism, thus enabling a diagnosis and the administration of an appropriate treatment to reduce morbidity and mortality. Patients with a CSS that live in or have traveled to an endemic area of parasites or fungi should always be thoroughly studied. Patients with underlying inflammatory, autoimmune, immune-deficient diseases, or cancer are prone to have parasitic or fungal infections as triggers of a CSS [1]. The first CSS triggered by Histoplasma was in a patient with inflammatory bowel disease (IBD) [2]. Since then, many parasites and fungi have been associated with IBD and CSS 2. CSS needs aggressive therapy, in many instances, treating the infection may resolve the CSS [3].

Three main classes of parasites cause disease in humans: protozoa, helminths, and ectoparasites. In the tropics, visceral leishmaniasis, rickettsia, malaria, Histoplasma, enteric fever, and tuberculosis cause 50% of CSS. Viral agents like Epstein–Barr virus and Parvovirus B19 contribute to another 30% [4]. Three hundred species of fungi can cause disease; the most frequent are aspergillosis, blastomycosis, candidiasis, coccidioidomycosis, cryptococcosis, histoplasmosis, mucormycosis, and pneumocystis pneumonia.

The clinical picture of the CSS secondary to parasites or fungi is like those due to other triggers. It is difficult to identify these infectious agents in patients with a CSS. No specific CSS diagnostic criteria for these situations have been established further complicating diagnosis. Parasites and fungi should always be looked for in

G. Sterba (✉)
South Miami Hospital, Baptist system, Miami, FL, USA

Y. Sterba
Nicklaus Children's Hospital, Miami, FL, USA

R. Q. Cron, E. M. Behrens (eds.), *Cytokine Storm Syndrome*,
https://doi.org/10.1007/978-3-030-22094-5_20

patients on biological therapies that develop CSS. In a review of 30 patients on biologics, three had Histoplasma, and one had Leishmania [5]. In another study, inpatients with HIV infection and a CSS 39 patients described, two had candidiasis and six disseminated histoplasmosis [6].

Parasites

Protozoan infections include malaria (*Plasmodium (P) vivax, P. malariae*), babesiosis (*Babesia microti*), gastroenteritis (*Entamoeba histolytica, Giardia lamblia*), leishmaniasis (*Leishmania (L)* sp.), sleeping sickness (*Trypanosoma brucei gambiense, T. b. rhodiense*), and vaginal infections (*Trichomonas vaginalis*). Also, *Cryptosporidium parvum* and *Toxoplasma gondii* are seen in patients who are immunosuppressed by disease or treatment.

Plasmodium species *(sp.)*

Malaria causes 650,000 deaths yearly due to *P. falciparum, P. vivax, P. malariae, P. ovale,* and *P. knowlesi. P. vivax* and *P. falciparum* are the most common. *P. falciparum* is the deadliest in South-East Asia. Malaria occurs in sub-Sahara, Africa, Asia, Latin America, the Middle East, and parts of Europe.

Polymerase chain reaction (PCR) to detect parasite DNA and blood microscopy search for the parasite are diagnostic tools. *Plasmodium* infections generate high serum cytokine levels that are associated with cerebral involvement, hypoglycemia, and death [7]; the role of cytokines in malaria has been studied [8]. Many reports describe the association of *Plasmodium* with a CSS (Table 1): in fifty-two patients with a CSS, three had malaria [9], a patient with Langerhans cell histiocytosis and malaria and developed a CSS [10], and a patient with *Mycoplasma* and *P. falciparum* developed CSS [11]. In one case, Antimalarial treatment resolved the CSS. HLH, can be a fatal complication of plasmodium vivax , as it has been described [12], Specific CSS treatment may be required for the CSS [13, 81]. Many r reports show *Plasmodium* as the trigger for CSS [14–17].

Babesiosis

Human babesiosis, or piroplasmosis, is transmitted by *Ixodes. Babesia (Bb) microti* is endemic on the northeastern coastline and the upper mid-west of the USA, and *B. divergens* and *B. venatorum* infections are found in the Mediterranean coastlines, the travel history is very important when considering *Babesia* [18]. In dogs, *Babesia*

Table 1 Parasites associated with CSS: protozoa and helminths

Organism	References and comment
Plasmodium sp.	52 children from Thailand with HLH, (1989–1998), 15 related to infection, three had malaria [9] A case report of a CSS induced by malaria with Langerhans histiocytosis [10] *Falciparum* co-infection with *Mycoplasma* [11] Case report with a bad outcome [12] HLH secondary to *Plasmodium vivax* and difficult therapeutic response [13] CSS required corticosteroid treatment [14] Case report, resolution of CSS, after eradication of parasite [14] Case report, *Falciparum* with CSS, despite eradication of the parasite [15] Case report, *Plasmodium vivax with CSS* [16] Case report of *Plasmodium vivax* and a review of the literature [17]
Leishmania sp.	A review of 30 patients, one had leishmania [5] 14 cases, nine with leishmania [18] Case reports [19–40] Case report in association with another virus [41–43] Two children with hepatic failure that responded to treatment [44] A case report of an adult with a review of the literature [45, 46] Case where CSS responded when the parasite was treated [47] Case treated with an anti-TNF agent that developed CSS with leishmania [48] A report where cyclosporine was substituted in the therapy of a transplanted patient [49] A case report with tocilizumab was added to the CSS treatment [50] A report with three cases of chronic granulomatous disease and CSS [51] In a ten-year review, 13 patients with HLH, two had leishmania [52] Four cases reported from China (full text in Chinese) [53] A report with three cases from India [54] A report of a patient with HIV and visceral leishmaniasis [55] Case reports [51–66] Fatal case report with hepatitis [63] Cerebrospinal findings in a case report [64] In a 7-year period review in Turkey, 18 children had HLH, and two were due to leishmania [65]
Toxoplasma gondii	Case report triggered by disseminated infection, no response to therapy [67] Case reports in immunocompetent individuals [68, 69] Case report of a patient diagnosed at autopsy after bone marrow transplant [70] An immunocompetent child with *Toxoplasma* infection [71] Case repot after kidney transplant [72, 73] Case report in an HIV-infected patient [74] Case report immunodeficient patient with toxoplasma [75]
Entamoeba histolytica	A case report, together with *Endolimax nana* [76]
Babesia sp.	An European traveler that had CSS from *Babesia* he had acquired in the USA [18] Case report [77] Case report of a transplanted patient [78, 79]
Ascaris lumbricoides	A child that resolved CSS after expulsion of the helminth [80]

is responsible for a CSS 1 [82]. Babesiosis and malaria have the same clinical features, and babesiosis and Lyme disease share the transmitting tick , thus the three diseases are considered in the differential diagnosis [83]. *Babesia* is searched for on blood smears, and PCR confirms the diagnosis. Eradication of *Babesia* resolves the CSS [77–79].

Entamoeba

Entamoeba histolytica causes worldwide extra- and intra-intestinal infections. Male homosexuals, travelers, recent immigrants and, institutionalized people are high-risk groups. The infection is generally localized to the gastrointestinal tract [84], inflammatory markers are local in contrast to systemic disease when or if CSS developes. One such case has been reported (Table 1) [76].

Leishmaniasis

The sand-fly in Africa, Europe, and Asia, and *Lutzomyia* ticks in America transmit 300,000 infections yearly of leishmaniasis. Visceral (kala-azar), cutaneous, and the mucocutaneous are the typical presentations [85]. In patients with a CSS from endemic areas, visceral leishmaniasis must be considered, but the diagnosis is difficult. Amastocytes are found in bone marrow and spleen, and serology is also used for diagnosis [86]. Many cases associated with a CSS have been reported (Table 1) [19–40]. Some cases were resistant to therapy [87], and some presented with other infections, such as viruses [41–43]. In a patient with severe hepatic failure, making the diagnosis prevented a fatal outcome [44, 63]. CSS associated with leishmaniasis can affect immunocompetent individuals [45–47] or patients that are immunocompromised, under immunosuppressive therapy [51, 66], or receiving antitumor necrosis factor (TNF) treatment [48]. Patients with CSS not responding to therapy, had *Leishmania* identified worsening the CSS [33, 44]. Resolving the infection has resolved the CSS [47].

Cyclosporine has been used when there was no response to corticosteroids, anakinra (recombinant IL-1 receptor antagonist), and etoposide [49]. Tocilizumab (anti-IL-6 receptor antibody) has been added for a better response [50]. Many other cases have been reported [51–66].

Toxoplasmosis

Toxoplasma gondii is present worldwide with a 65% prevalence of serum anti-toxoplasma antibody. The infection ranges from asymptomatic primary infections to mononucleosis-like syndromes [88]. In a case report, MAS was documented after

Toxoplasma resolution; the patient developed the typical features of a CSS, but did not respond to typical therapy [67]. A CSS associated with toxoplasmosis is described in immunocompetent individuals [68, 69] , in patients following bone marrow transplant [70], and in patients with primary infection that became immune-deficient [71], other presentations have been described [72–75] (Table 1). The characteristics of the CSS are like those triggered by other infections.

Helminths

Ascaris lumbricoides is an infection in the intestine that affects millions of people. A case was reported of a 5-year-old child with typical features of a CSS that underwent multiple studies to rule out a trigger for his disease, all negative until an abdominal ultrasound demonstrated tubular structures. He was treated with IVIG and received mebendazole. After the third day of therapy, he passed the *Ascaris* in stools, and the CSS resolved with no more fever and complete recovery of all the typical features that were present for a CSS (Table 1) [80].

Fungus

Candidiasis

Candida albicans is the most common fungus in the intestinal tract, mucosae, and skin. Overgrowth causes oral–pharyngeal–esophageal, vaginal, or disseminated candidiasis. A review of 39 patients with HIV and a related CSS describes an acutely ill patient with an ovarian tube abscess due to *Candida* (Table 2) [6]. Severe pancytopenia, elevated serum ferritin, LDH, triglycerides, prothrombin time, and a bone marrow with hemophagocytosis led to the CSS diagnosis. Methamine silver staining and culture were used to diagnose *Candida*. Ninety percent of all HIV patients with a CSS had fever, 100% anemia, 78% leukopenia, and 80% thrombocytopenia. Serum ferritin and triglycerides were very high in most, 50% had hepatosplenomegaly, and two of the patients had candidiasis [107]. Another infant with a CSS and disseminated *Candida lusitaniae* sepsis was also found to have chronic granulomatous disease (CGD). Others with a CSS associated with *Candida* have had lymphoma or were concurrently infected with other organisms [108, 109]. In a kidney-pancreas transplant recipient with a CSS due to *Candida*, treatment with cyclosporine was useful [110].

Table 2 Fungi associated with CSS

Cryptococcus neoformans	Report of a child with cryptococcal meningoencephalitis [89] Case report of secondary CSS in an immunocompromised patient [90]
Histoplasma capsulatum	Review of 30 patients with CCS, three had *Histoplasma* [5] Report on an AIDS patient [91] CSS complicating H. Capsulatum infection in immunocompetent patients [92–94] Report in an immunocompetent individual [95] Report in an HIV-infected patient [100–102] Cases where CSS responded after fungus was eradicated [96] Case report diagnosed at autopsy [97] Case report of a surgical patient that responded to therapy for CSS and *Histoplasma* [98] Case report in a chronic lymphocytic leukemia patient [99, 100] Report of a patient with chronic arthritis on anti-TNF agent with disseminated *Histoplasma* [101] Case report in a kidney transplant recipient [98, 102] Case report in a heart transplant patient [98] Case report of a patient with sarcoidosis on corticosteroids [103]
Penicillium marneffei	52 children with HLH (1989–1998), 15 infections, one with *Penicillium marneffei* [9] Case reports in patients with HIV infection [104, 105] Review of 18 published cases and a report of four Chinese cases [104] Case report, a woman with Sjogren's syndrome [106]
Candida sp.	Case report and a review of the literature 38 patients with HLH, two had *Candida* [6] An association with HIV and a review of 39 cases [107] Case report of a patient with chronic granulomatous disease [108] Case associated with other viruses (EBV, CMV) and lymphoma [109] A kidney-pancreas transplant patient that was treated with cyclosporine and steroids [110]
Pneumocystis sp.	Case report in a patient with lymphoma [111] Case report of a patient on biologic therapy for a rheumatic disease [112] Report of a renal transplant patient [113]

Histoplasmosis

Histoplasma fungal spores enter the body through the lungs, and typically do not cause illness. However, some individuals develop fever, cough, and fatigue that improve without medication. Nevertheless, immunocompromised patients can develop a CSS associated with histoplasmosis. *Histoplasma* is found worldwide and in the USA, around the Ohio and Mississippi River valleys [114]. In Latin America, *Histoplasma* is the most common opportunistic infection among people with HIV, and 30% of those will die [115]. The fungus is detected with computed tomography scanning of the chest and methenamine silver staining of the bone marrow [91, 92]. Disseminated histoplasmosis with CSS occurs in immunocompetent [93–95] and immunocompromised patients [5, 116–120] and may respond to fungus treatment

(Table 2) [96, 121]. In some cases, the diagnosis was made at autopsy [97]. A retrospective review of all inpatients with HLH and *Histoplasma* in a ten-year period revealed 11 cases. Nine were males. Nine had HIV. One was a renal transplant patient, and one was immunocompetent. Seven died within 90 days in this study. Thus, histoplasmosis-associated HLH is a highly lethal disease [122]. Many other case reports have been described [98–101, 112, 123–128].

Cryptococcus

Cryptococcus neoformans is a fungus that infects via inhalation; however, most exposures do not result in illness. *C. neoformans* infections are extremely rare in people who are healthy; CSS occurs in immunocompromised patients [89, 90], although they have also been described in immunocompetent patients (Table 2).

Pneumocystis sp.

Previously known as *Pneumocystis (P) carinii, P. jirovecii*, a yeast-like fungus is an important human pathogen. Mild pulmonary infection is prevalent in half of the general adult population [129]. In the immunocompromised patient, *Pneumocystis* may reactivate leading to disease. It is extracellular, found in the lungs, and is identified with methamine staining. CSS have been reported in patients with *Pneumocystis* and lymphoma [111], with biologic therapy for rheumatic disease , and renal transplantation 2 [125]. As in other infections, CSS typically resolve after the infection has been controlled.

Penicillinosis

Penicillium marneffei is a fungus that has been reported in Thailand and Southern China [102, 130]. In the late 1980s, the appearance of HIV in Southeast Asia made penicillinosis one of the acquired immunodeficiency syndrome (AIDS) -defining illnesses and was listed as a HIV clinical stage 4, clinical condition [131, 132]. It has been reported in different parts of the world as infected travelers return to their home countries from endemic areas [133]. The disseminated form of *P. marneffei* infection can be fatal with an in-hospital mortality rate at about 20% [129].

Penicillinosis can trigger a CSS, the first case of a CSS caused by penicillinosis was a HIV-infected child who presented with cytopenia and developed an Immune reconstitution inflammatory syndrome and recovered promptly after antifungal and intravenous immunoglobulin therapy [104–106].

Conclusion

Human pathogenic parasites and fungi are found worldwide, and they can infect all mucous membranes. They can cause disease in immunocompetent and immunodeficient individuals, often becoming a serious complication. They have been shown to trigger CSS. It is of great importance to consider these organisms as responsible in these situations since their treatment may lead to control of infection and resolution of the otherwise fatal CSS.

References

1. Reiner, A. P., & Spivak, J. L. (1988). Hematophagic histiocytosis. A report of 23 new patients and a review of the literature. *Medicine (Baltimore), 67*, 369–388.
2. Fries, W., Cottone, M., & Cascio, A. (2013). Systematic review: Macrophage activation syndrome in inflammatory bowel disease. *Alimentary Pharmacology & Therapeutics, 37*(11), 1033–1045.
3. Singh, Z. N., Rakheja, A., Yadav, T. P., & Shome, J. (2005). Infection-associated haemophagocytosis: The tropical spectrum. *Clinical and Laboratory Haematology, 27*(5), 312–315.
4. Rajagopala, S., & Singh, N. (2012). Diagnosing and treating haemophagocytic Lymphohistiocytosis in the tropics: Systematic review from the Indian subcontinent. *Acta Medica Academica, 41*, 161–174.
5. Brito-Zerón, P., Bosch, X., Pérez-de-Lis, M., Pérez-Álvarez, R., Fraile, G., Gheitasi, H., et al. (2016). Infection is the major trigger of hemophagocytic syndrome in adult patients treated with biological therapies. *Seminars in Arthritis and Rheumatism, 45*(4), 391–399.
6. Shailender, B., Bauer, F., & Bilgrami, S. A. (2003). Candidiasis-associated Hemophagocytic Lymphohistiocytosis in a patient infected with human immunodeficiency virus. *Clinical Infectious Diseases, 37*(11), 161–166.
7. Grau, G. E., Taylor, T. T., Molyneux, M. E., Wirima, J. J., Vassalli, P., Hommel, M., et al. (1989). Tumor necrosis factor and disease severity in children with falciparum malaria. *The New England Journal of Medicine, 320*, 1586.
8. Ioannidis, L. J., Nie, C. Q., & Hansen, D. S. (2014). The role of chemokines in severe malaria: More than meets the eye. *Parasitology, 141*(5), 602–613.
9. Veerakul, G., Sanpakit, K., Tanphaichitr, V. S., Mahasandana, C., & Jirarattanasopa, N. (2002). Secondary hemophagocytic lymphohistiocytosis in children: An analysis of etiology and outcome. *Journal of the Medical Association of Thailand, 85*(Suppl 2), S530–S541.
10. Saribeyoglu, E. T., Anak, S., Agaoglu, L., Boral, O., Unuvar, A., & Devecioglu, O. (2004). Secondary hemophagocytic lymphohistiocytosis induced by malaria infection in a child with Langerhans cell histiocytosis. *Pediatric Hematology and Oncology, 21*(3), 267–272.
11. Weeratunga, P., Rathnayake, G., Sivashangar, A., Karunanayake, P., Gnanathasan, A., & Chang, T. (2016). Plasmodium falciparum and mycoplasma pneumoniae co-infection presenting with cerebral malaria manifesting orofacial dyskinesia and Haemophagocytic lymphohistiocytosis. *Malaria Journal, 15*, 461.
12. Ullah, W., Abdullah, H. M., Qadir, S., & Shahzad, M. A. (2016). Haemophagocytic lymphohistiocytosis (HLH): A rare but potentially fatal association with plasmodium vivax malaria. *British Medical Journal Case Reports, 2016*, bcr2016215366.
13. Bhagat, M., Kanhere, S., Kadakia, P., Phadke, V., George, R., & Chaudhari, K. (2015). Haemophagocytic lymphohistiocytosis: a cause of unresponsive malaria in a 5-year-old girl. *Paediatrics and International Child Health, 35*(4), 333–336.

14. Ohno, T., Shirasaka, A., Sugiyama, T., & Furukawa, H. (1996). Hemophagocytic syndrome induced by plasmodium falciparum malaria infection. *International Journal of Hematology, 64*(3–4), 263–266.
15. Ohnishi, K., Mitsui, K., Komiya, N., Iwasaki, N., Akashi, A., & Hamabe, Y. (2007). Clinical case report: Falciparum malaria with hemophagocytic syndrome. *The American Journal of Tropical Medicine and Hygiene, 76*, 1016–1018.
16. Bae, E., Jang, S., Park, C. J., & Chi, H. S. (2011). Plasmodium vivax malaria-associated hemophagocytic lymphohistiocytosis in a young man with pancytopenia and fever. *Annals of Hematology, 90*(4), 491–492. Epub 2010 Aug 19.
17. Sung, P. S., Kim, I. H., Lee, J. H., & Park, J. W. (2011). Hemophagocytic lymphohistiocytosis (HLH) associated with plasmodium vivax infection: Case report and review of the literature. *Chonnam Medical Journal, 47*(3), 173–176. Epub 2011 Dec 26.
18. Poisnel, E., Ebbo, M., Berda-Haddad, Y., Faucher, B., Bernit, E., Carcy, B., et al. (2013). Babesia microti: An unusual travel-related disease. *BMC Infectious Diseases, 13*, 99.
19. Visentin, S., Baudesson de Chanville, A., Loosveld, M., Chambost, H., & Barlogis, V. (2013). Infantile visceral leishmaniasis, an etiology of easily curable hemophagocytic lymphohistiocytosis syndrome. *Archives de Pédiatrie, 20*(11), 1225–1229. Epub 2013 Sep 26.
20. Marom, D., Offer, I., Tamary, H., Jaffe, C. L., & Garty, B. (2001). Hemophagocytic Lymphohistiocytosis associated with visceral leishmaniasis. *Pediatric Hematology and Oncology, 18*(1), 65.
21. Cancado, G. G., Freitas, G. G., Faria, F. H., de Macedo, A. V., & Nobre, V. (2013). Hemophagocytic lymphohistiocytosis associated with visceral leishmaniasis in late adulthood. *The American Journal of Tropical Medicine and Hygiene, 88*(3), 575–577.
22. Bode, S. F., Bogdan, C., Beutel, K., Behnisch, W., Greiner, J., Henning, S., et al. (2014). Hemophagocytic lymphohistiocytosis in imported pediatric visceral leishmaniasis in a nonendemic area. *The Journal of Pediatrics, 165*(1), 147–153. Epub 2014 May 3.
23. Watkins, E. R., Shamasunder, S., Cascino, T., White, K. L., Katrak, S., Bern, C., et al. (2014). Visceral leishmaniasis-associated hemophagocytic lymphohistiocytosis in a traveler returning from a pilgrimage to the Camino de Santiago. *Journal of Travel Medicine, 21*(6), 429–432. Epub 2014 Aug 21.
24. Scalzone, M., Ruggiero, A., Mastrangelo, S., Trombatore, G., Ridola, V., Maurizi, P., et al. (2016). Hemophagocytic lymphohistiocytosis and visceral leishmaniasis in children: Case report and systematic review of literature. *Journal of Infection in Developing Countries, 10*(1), 103–108.
25. Hernández-Jiménez, P., Díaz-Pedroche, C., Laureiro, J., Madrid, O., Martín, E., & Lumbreras, C. (2016). Hemophagocytic lymphohistiocytosis: Analysis of 18 cases. [article in Spanish]. *Medicina Clínica (Barcelona), 147*(11), 495–498. Epub 2016 Oct 7.
26. Aydın Teke, T., Metin Timur, Ö., Gayretli Aydın, Z. G., Öz, N., Bayhan, G. İ., Yılmaz, N., et al. (2015). Three pediatric cases of leishmaniasis with different clinical forms and treatment regimens. *Türkiye Parazitolojii Dergisi, 39*(2), 147–150. in Turkish.
27. Ranjan, P., Kumar, V., Ganguly, S., Sukumar, M., Sharma, S., Singh, N., et al. (2016). Hemophagocytic lymphohistiocytosis associated with visceral leishmaniasis: Varied presentation. *Indian Journal of Hematology and Blood Transfusion, 32*(Suppl 1), 351–354. https://doi.org/10.1007/s12288-015-0541-2. Epub 2015 Apr 28.
28. Michel, G., Simonin, G., Perrimond, H., & Coignet, J. (1988). Syndrome of activation of the mononuclear phagocyte system. Initial manifestation of visceral leishmaniasis. *Archives Françaises de Pédiatrie, 45*, 45–46.
29. Kilani, B., Ammari, L., Kanoun, F., Ben Chaabane, T., Abdellatif, S., & Chaker, E. (2006). Hemophagocytic syndrome associated with visceral leishmaniasis. *International Journal of Infectious Diseases, 10*, 85–86.
30. Levy, L., Nasereddin, A., Rav-Acha, M., Kedmi, M., Rund, D., & Gatt, M. E. (2009). Prolonged fever, Hepatosplenomegaly, and pancytopenia in a 46-year-old woman. *PLoS Medicine, 6*(4), e1000053. https://doi.org/10.1371/journal.pmed.1000053

31. Thabet, F., Tabarki, B., Fehem, R., Yacoub, M., Selmi, H., & Essoussi, A. S. (1999). Syndrome of inappropriate macrophage activation associated with infantile visceral leishmaniasis. *La Tunisie Médicale, 77*, 648–650.
32. Nadrid, A., Pousse, H., Laradi-Chebil, S., Khelif, A., Bejaoui, M., Besbes, A., et al. (1996). Infantile visceral leishmaniasis: Difficult diagnosis in cases complicated by hemophagocytosis. *Archives de Pédiatrie, 3*, 881–883.
33. Kram, M. K., Bjering, S., Hermansen, N. O., Dini, L., & Hellebostad, M. (2011). A 15-month-old girl with fever and pancytopenia. *Tidsskrift for den Norske Lægeforening, 131*, 2482–2486.
34. Sukova, M., Stary, J., Houskova, J., & Nohynkova, E. (2002). Hemophagocytic lymphohistiocytosis as a manifestation of visceral leishmaniasis. *Casopís Lékařů Českých, 141*, 581–584.
35. Sotoca Fernandez, J. V., Garcia Villaescusa, L., Lillo Lillo, M., Garcia Mialdea, O., Carrascosa Romero, M. C., & Tebar Gil, R. (2008). Hemophagocytic syndrome secondary to visceral leishmaniasis. *Anales de Pediatría (Barcelona, Spain), 69*, 46–48.
36. Santamaria, M., Carrillo, J., Perez-Navero, J., Mateos, E., Ibarra, I., Fernandez, S., et al. (2008). Leishmaniasis and concurrent hemophagocytosis with or without transient perforin expression perturbation. *Pediatric Blood & Cancer, 51*, 310.
37. Rodriguez-Cuartero, A., Salas-Galan, A., Perezgalvez, M. N., & Perez-Blanco, F. J. (1991). Haemophagocytic visceral kala azar. *Infection, 19*(184), 1333.
38. Lazanas, M., Perronne, C., Lomverdos, D., Galariotis, C., Arapakis, G., & Vilde, J. L. (1990). Visceral leishmaniasis disclosed by histiocytosis with erythrophagocytosis. *Presse Médicale, 19*, 765.
39. Rajagopala, S., Dutta, U., Chandra, K. S., Bhatia, P., Varma, N., & Kochhar, R. (2008). Visceral leishmaniasis associated hemophagocitic lymphohisticytosis. A case report and systematic review. *Journal of Infection, 56*(5), 381–388.
40. Higel, L., Froehlich, C., Pages, M. P., Dupont, D., Collardeau-Frachon, S., Dijoud, F., et al. (2015). Macrophage activation syndrome and autoimmunity due to visceral leishmaniasis [article in French]. *Archives de Pédiatrie, 22*(4), 397–400. Epub 2015 Jan 22.
41. Ay, Y., Yildiz, B., Unver, H., Karapinar, D. Y., & Vardar, F. (2012). Hemophagocytic lymphohistiocytosis associated with H1N1 virus infection and visceral leishmaniasis in a 4.5-month-old infant. *Revista da Sociedade Brasileira de Medicina Tropical, 45*(3), 407–409.
42. Gaifer, Z., & Boulassel, M. R. (2016). Leishmania infantum and epstein-barr virus co-infection in a patient with hemophagocytosis. *Infectious DiseaseReports, 8*(4), 6545. eCollection 2016 Dec 31.
43. Koliou, M. G., Soteriades, E. S., Ephros, M., Mazeris, A., Antoniou, M., Elia, A., et al. (2008). Hemophagocytic lymphohistiocytosis associated with Epstein Barr virus and Leishmania donovani coinfection in a child from Cyprus. *Journal of Pediatric Hematology/Oncology, 30*, 704–707.
44. Chaudary, S. R., & Arpit, M. K. (2013). Fulminant hepatic failure in kala azar, hemophagocytic lymphohistiocytosis? *Pediatric Infectious Disease, 5*(3), 127–129.
45. Koubaa, M., Maaloul, I., Marrakchi, C., et al. (2012). Hemophagocytic syndrome associated with visceral leishmaniasis in an immunocompetent adult-case report and review of the literature. *Annals of Hematology, 91*(7), 1143–1145.
46. Cort, P., Cardefiosa, N., & Muñoz, C. (1997). Case report Hemophagocytic syndrome associated with visceral Leishmaniasis in an Immunocompetent patient. *Clinical Microbiology Newsletter, 19*(2), 14–16.
47. Zyurek, E., Ozcay, F., Yilmaz, B., & Ozbek, N. (2005). Hemophagocytic Lymphohistiocytosis associated with visceral leishmaniasis a case report. *Pediatric Hematology and Oncology, 22*, 409–414.
48. Moltó, A., Mateo, L., Loveras, N., Olivé, A., & Minguez, S. (2010). Visceral Leishmaniasis and macrophage activation syndrome in a patient with rheumatoid arthritis under treatment with adalimumab. *Joint, Bone, Spine, 77*(3), 271–273.

49. Merelli, M., Quartuccio, L., Bassetti, M., Pecori, D., Gandolfo, S., Avellini, C., et al. (2015). Efficacy of intravenous cyclosporine in a case of cytophagic histiocytic panniculitis complicated by haemophagocytic syndrome after visceral leishmania infection. *Clinical and Experimental Rheumatology, 33*(6), 906–909.
50. Rios-Fernández, R., Callejas-Rubio, J. L., García-Rodríguez, S., Sancho, J., Zubiaur, M., & Ortego-Centeno, N. (2016). Tocilizumab as an adjuvant therapy for hemophagocytic lymphohistiocytosis associated with visceral leishmaniasis. *American Journal of Therapeutics, 23*(5), e1193–e1196.
51. Martin, A., Marques, L., Soler-Palacin, P., et al. (2009). Visceral leishmaniasis associated hemophagocytic syndrome in patients with chronic granulomatous disease. *The Pediatric Infectious Disease Journal, 28*(8), 753–754.
52. Krivokapic-Dokmanovic, L., Krstovski, N., Jankovic, S., Lazic, J., Radlovic, N., & Janic, D. (2012). Clinical characteristics and disease course in children with haemophagocytic lymphohistiocytosis treated at the university Children's Hospital in Belgrade. *Srpski Arhiv za Celokupno Lekarstvo, 140*, 191–197.
53. Guo, X., Chen, N., Wang, T. Y., Zhou, C. Y., Li, Q., & Gao, J. (2011). Visceral leishmaniasis associated hemophagocytic lymphohistiocytosis: Report of four childhood cases. *Zhonghua Er Ke Za Zhi, 49*, 550–553.
54. Rasad, R., Muthusami, S., Pandey, N., Tilak, V., Shukla, J., & Mishra, O. P. (2009). Unusual presentations of visceral leishmaniasis. *Indian Journal of Pediatrics, 76*, 843–845.
55. Patel, K. K., Patel, A. K., Sarda, P., Shah, B. A., & Ranjan, R. (2009). Immune reconstitution visceral leishmaniasis presented as hemophagocytic syndrome in a patient with AIDS from a nonendemic area: A case report. *Journal of the International Association of Physicians in AIDS Care (Chicago, Ill.), 8*, 217–220.
56. Bhutani, V., Dutta, U., Das, R., & Singh, K. (2002). Hemophagocytic syndrome as the presenting manifestation of visceral leishmaniasis. *The Journal of the Association of Physicians of India, 50*, 838–839.
57. Celik, U., Alabaz, D., Alhan, E., Bayram, I., & Celik, T. (2007). Diagnostic dilemma in an adolescent boy: Hemophagocytic syndrome in association with kala azar. *The American Journal of the Medical Sciences, 334*, 139–141.
58. Kocak, N., Eren, M., Yuce, A., & Gumruk, F. (2004). Hemophagocytic syndrome associated with visceral leishmaniasis. *Indian Pediatrics, 41*, 605–607.
59. Tapisiz, A., Belet, N., Ciftci, E., Ince, E., & Dogru, U. (2007). Hemophagocytic lymphohistiocytosis associated with visceral leishmaniasis. *Journal of Tropical Pediatrics, 53*, 359–361.
60. Tunc, B., & Ayata, A. (2001). Hemophagocytic syndrome: A rare life-threatening complication of visceral leishmaniasis in a young boy. *Pediatric Hematology and Oncology, 18*, 531–536.
61. Ozyurek, E., Ozcay, F., Yilmaz, B., & Ozbek, N. (2005). Hemophagocytic lymphohistiocytosis associated with visceral leishmaniasis: A case report. *Pediatric Hematology and Oncology, 22*, 409–414.
62. Kontopoulou, T., Tsaousis, G., Vaidakis, E., Fanourgiakis, P., Michalakeas, E., Trigoni, E., et al. (2002). Hemophagocytic syndrome in association with visceral leishmaniasis. *The American Journal of Medicine, 113*, 439–440.
63. Mathur, P., Samantaray, J. C., & Samanta, P. (2007). Fatal haemophagocytic syndrome and hepatitis associated with visceral leishmaniasis. *Indian Journal of Medical Microbiology, 25*, 416–418.
64. Fathalla, M., Hashim, J., Alkindy, H., & Wali, Y. (2007). Cerebrospinal fluid involvement in a case of visceral leishmaniasis associated with hemophagocytic lymphohistiocytosis. *Sultan Qaboos University Medical Journal, 7*, 253–256.
65. Gurgey, A., Secmeer, G., Tavil, B., Ceyhan, M., Kuskonmaz, B., Cengiz, B., et al. (2005). Secondary hemophagocytic lymphohistiocytosis in Turkish children. *The Pediatric Infectious Disease Journal, 24*, 1116–1117.
66. Sipahi, T., Tavil, B., & Oksal, A. (2005). Visceral leishmaniasis and pseudomonas septicemia associated with hemophagocytic syndrome and myelodysplasia in a Turkish child. *The Turkish Journal of Pediatrics, 47*, 191–194.

67. Arslan, F., Batirel, A., Ramazan, M., Ozer, S., & Mert, A. (2012). Macrophage activation syndrome triggered by primary disseminated toxoplasmosis. *Scandinavian Journal of Infectious Diseases, 44*(12), 1001–1004. Epub 2012 Jul 17.
68. Yan, K., Yang, F., Zuo, W., Hu, J., Esch, G. W., & Zuo, Y. X. (2013). The American society of tropical medicine and hygiene short report: Hemophagocytic syndrome as uncommon presentation of disseminated toxoplasmosis in an immunocompetent adult from Chinese. *The American Journal of Tropical Medicine and Hygiene, 88*(6), 1209–1211.
69. Undseth, Ø., Gerlyng, P., Goplen, A. K., Holter, E. S., von der Lippe, E., & Dunlop, O. (2014). Primary toxoplasmosis with critical illness and multi-organ failure in an immunocompetent young man. *Scandinavian Journal of Infectious Diseases, 46*(1), 58–62.
70. Duband, S., Cornillon, J., Tavernier, E., Dumollard, J. M., & Guyotat, D. (2008). Pe'oc'h M,.Toxoplasmosis with hemophagocytic syndrome after bone marrow transplantation: Diagnosis at autopsy. *Transplant Infectious Disease, 10*, 372–374.
71. Briand, P. Y., Gangneux, J. P., Favaretto, G., Ly-Sunnaram, B., Godard, M., Robert-Gangneux, F., et al. (2008). Hemophagocytic syndrome and toxoplasmic primo-infection. *Annales de Biologie Clinique, 66*(2), 199–205.
72. Hebraud, B., Kamar, N., Borde, J. S., Bessieres, M. H., & Galinier, M. (2008). Rostaing l. unusual presentation of primary toxoplasmosis infection in a kidney transplant patient complicated by an acute left ventricular failure. *NDT Plus, 1*, 429–432.
73. Segall, L., Moal, M. C., Doucet, L., Kergoat, N., & Bourbigot, B. (2006). Toxoplasmosis-associated hemophagocytic syndrome in renal transplantation. *Transplant International, 19*, 78–80.
74. Guillaume, M. P., Driessens, N., Libert, M., De Bels, D., Corazza, F., & Karmali, R. (2006). Hemophagocytic syndrome associated with extracerebral toxoplasmosis in an HIV-infected patient. *European Journal of Internal Medicine, 17*, 503–504.
75. Blanche, P., Robert, F., Dupouy-Camet, J., & Sicard, D. (1994). Toxoplasmosis-associated hemophagocytic syndrome in a patient with AIDS: Diagnosis by the polymerase chain reaction. *Clinical Infectious Diseases, 19*, 989–990.
76. D'Errico, M. M., Cuoco, F., Biancardi, C., Torcoletti, M., Filocamo, G., & Corona, F. (2014). Macrophage activation syndrome: The role of infectious triggers. *Pediatric Rheumatology Online Journal, 12*(Sup. 1), Y5.
77. Auerbach, M., Haubenstock, A., & Soloman, G. (1986). Systemic babesiosis. Another cause of the hemophagocytic syndrome. *The American Journal of Medicine, 80*, 301–303.
78. Slovut, D. P., Benedetti, E., & Matas, A. J. (1996). Babesiosis and hemophagocytic syndrome in an asplenic renal transplant recipient. *Transplantation, 62*, 537–539.
79. Gupta, P., Hurley, R. W., Helseth, P. H., Goodman, J. L., & Hammerschmidt, D. E. (1995). Pancytopenia due to hemophagocytic syndrome as a presenting manifestation of babesiosis. *American Journal of Hematology, 50*, 60–62.
80. Bayhan, G. İ., Çenesiz, F., Tanır, G., Taylan Özkan, A., & Çınar, G. (2015). First case of Ascaris lumbricoides infestation complicated with Hemophagocytic Lymphohistiocytosis. *Turkish Journal of Parasitology (Türkiye Parazitol Derg), 39*, 2.
81. Trapani, S., Canessa, C., Fedi, A., Giusti, G., Barni, S., Montagnani, C., et al. (2013). Macrophage activation syndrome in a child affected by malaria: The choice of steroid. *International Journal of Immunopathology and Pharmacology, 26*(2), 535–539.
82. Sudhakara, R., Sivajothi, S., Varaprasad, R., & Solmon, R. (2016). Clinical and laboratory findings of *Babesia* infection in dogs. *Journal of Parasitic Diseases, 40*(2), 268–272.
83. Vannier, E., & Krause, P. J. (2012). Human babesiosis. *The New England Journal of Medicine, 366*, 2397–2407.
84. Zavala, G. A., García, O. P., Camacho, M., Ronquillo, D., Campos-Ponce, M., Doak, C., et al. (2018). Intestinal parasites: Associations with intestinal and systemic inflammation. *Parasite Immunology, 40*(4), e12518.
85. Pigott, D. M., Golding, N., Messina, J. P., Battle, K. E., Duda, K. A., Balard, Y., et al. (2014). Global database of leishmaniasis occurrence locations, 1960–20. *Scientific Data, 1*, 140036.

86. Boalaert, M., Verdonck, K., Menten, J., Sunyoto, T., Griensven, J. V., Chappuis, F., et al. (2014). Rapid test for the diagnosis of visceral leishmaniasis in patients with suspected disease. *Cochrane Database of Systematic Reviews, 6*, 1–119.
87. Bouguila, J., Chabchoub, I., Moncef, Y., Mlika, A., Saghrouni, F., Boughamoura, L., et al. (2010). Treatment of severe hemophagocytic syndrome associated with visceral leishmaniasis. *Archives de Pédiatrie, 17*, 1566–1570.
88. Chaichan, P., Mercier, A., Galal, L., Mahittikorn, A., Ariey, F., Morand, S., et al. (2017). Geographical distribution of toxoplasma gondii genotypes in Asia: A link with neighboring continents. *Infection, Genetics and Evolution, 53*, 227–238. Epub 2017 Jun 3.
89. Numata, K., Tsutsumi, H., Wakai, S., Tachi, N., & Chiba, S. (1998). A child case of haemophagocytic syndrome associated with cryptococcal meningoencephalitis. *The Journal of Infection, 36*(1), 118–119.
90. Powel, M. S., Alizadeh, A. A., Budvytiene, I., Schaenman, J. M., & Banaei, N. (2012). First isolation of *Cryptococcus uzbekistanensis* from an Immunocompromised patient with lymphoma. *Journal of Clinical Microbiology, 50*(3), 1125–1127.
91. Chandra, H., Chandra, S., & Sharma, A. (2012). Histoplasmosis on bone marrow aspirate: Cytological examination associated with haemophagocytosis and pancytopenia in an AIDS patient. *The Korean journal of hematology, 47*, 77–79.
92. Kumar, N., Jain, S., & Singh, Z. N. (2000). Disseminated histoplasmosis with reactive haemophagocytosis: Aspiration cytology findings in two cases. *Diagnostic Cytopathology, 23*, 422–424.
93. Sonawane, P. B., Chandak, S. V., & Rathi, P. M. (2016). Disseminated histoplasmosis with haemophagocytic lymphohistiocytosis in an immunocompetent host. *Journal of Clinical and Diagnostic Research, 10*(3), OD03–OD05.
94. Saluja, S., Sunita, Bhasin, S., Gupta, D. K., Gupta, B., Kataria, S. P., et al. (2005). Disseminated histoplasmosis with reactive haemophagocytosis presenting as PUO in an immunocompetent host. *The Journal of the Association of Physicians of India, 53*, 906–908.
95. De, D., & Nath, U. K. (2015). Disseminated Histoplasmosis in Immunocompetent individuals- not a so rare entity, in India. *Mediterranean Journal of Hematology and Infectious Diseases, 7*(1), 45.
96. Phillips, J., Staszewski, H., & Garrison, M. (2008). Successful treatment of secondary hemophagocytic lymphohistiocytosis in a patient with disseminated histoplasmosis. *Hematology, 13*(5), 282–285.
97. Wang, Z., Duarte, A. G., & Schnadig, V. J. (2007). Fatal reactive hemophagocytosis related to disseminated histoplasmosis with endocarditis: An unusual case diagnosed at autopsy. *Southern Medical Journal, 100*(2), 208–211.
98. Masri, K., Mahon, N., Rosario, A., Mirza, L., Keys, T. F., Ratliff, N. B., et al. (2003). Reactive hemophagocytic syndrome associated with disseminated histoplasmosis in a heart transplant recipient. *The Journal of Heart and Lung Transplantation, 22*, 487–491.
99. Van Koeveringe, M. P., & Brouwer, R. E. (2010). Histoplasma capsulatum reactivation with haemophagocytic syndrome in a patient with chronic lymphocytic leukaemia. *The Netherlands Journal of Medicine, 68*, 418–421.
100. Rao, R. D., Morice, W. G., & Phyliky, R. L. (2002). Hemophagocytosis in a patient with chronic lymphocytic leukemia and histoplasmosis. *Mayo Clinic Proceedings, 77*, 287–290.
101. Majluf-Cruz, A., Hurtado Monroy, R., Souto-Meirino, C., Del Rio Chiriboga, C., & Simon, J. (1993). Hemophagocytic syndrome associated with histoplasmosis in the acquired immunodeficiency syndrome: Description of 3 cases and review of the literature. *Sangre (Barc), 38*, 51–55.
102. Supparatpinyo, K., Khamwan, C., Baosoung, V., Nelson, K. E., & Sirisanthana, T. (1994). Disseminated *P. marneffei* infection in Southeast Asia. *Lancet, 344*, 110–113.
103. Abughanimeh, O., Qasrawi, A., & Abu Ghanimeh, M. (2018). Hemophagocytic Lymphohistiocytosis Complicating Systemic Sarcoidosis. *Cureus, 10*(6), e2838.
104. Chokephaibulkit, K., Veerakul, G., Vanprapar, N., Chaiprasert, A., Tanphaichitr, V., & Chearskul, S. (2001). Penicilliosis-associated hemophagocytic syndrome in a human

immunodeficiency virus-infected child: The first case report in children. *Journal of the Medical Association of Thailand, 84*(3), 426–429.
105. Pei, S. N., Lee, C. H., & Liu, J. W. (2008). Hemophagocytic syndrome in a patient with acquired immunodeficiency syndrome and acute disseminated penicilliosis. *The American Journal of Tropical Medicine and Hygiene, 78*, 11–13.
106. Chim, C. S., Fong, C. Y., Ma, S. K., Wong, S. S., & Yuen, K. Y. (1998). Reactive hemophagocytic syndrome associated with Penicillium marneffei infection. *The American Journal of Medicine, 104*(2), 196–197.
107. Bhatia, S., Bauer, F., & Bilgrami, S. A. (2003). Candidiasis-associated hemophagocytic lymphohistiocytosis in a patient infected with human immunodeficiency virus. *Clinical Infectious Diseases, 37*(11), e161–e166.
108. Valentine, G., Thomas, T. A., Nguyen, T., & Lai, Y. C. (2014). Chronic granulomatous disease presenting as hemophagocytic lymphohistiocytosis: A case report. *Pediatrics, 134*(6), e1727–e1730.
109. Mustafa Ali, M., Ruano Mendez, A. L., & Carraway, H. E. (2017). Hemophagocytic Lymphohistiocytosis in a patient with Hodgkin lymphoma and concurrent EBV, CMV, and *Candida* infections. *Journal of Investigative Medicine High Impact Case Reports, 5*(1), 2324709616684514.
110. González-Posada, J. M., Hernández, D., Martin, A., Raya, J. M., Pitti, S., Bonilla, A., et al. (2008). Hemophagocytic lymphohistiocytosis in a pancreas-kidney transplant recipient: Response to dexamethasone and cyclosporine. *Clinical Nephrology, 70*(1), 82–86.
111. Machaczka, M., & Vaktnas, J. (2007). Haemophagocytic syndrome associated with Hodgkin lymphoma and pneumocystis jirovecii pneumonitis. *British Journal of Haematology, 138*(6), 672.
112. Watanabe, E., Sugawara, H., Yamashita, T., Ishii, A., Oda, A., & Tera, C. H. (2016). Successful tocilizumab therapy for macrophage activation syndrome associated with adult-onset still's disease: A case-based review. *Case Reports in Medicine, 2016*, 5656320.
113. Karras, A., Thervet, E., & Legendre, C. (2004). Hemophagocytic syndrome in renal transplant recipients: Report of 17 cases and review of literature. *Transplantation, 77*(2), 238–243.
114. Baddley, J. W., Winthrop, K. L., Patkar, N. M., Delzell, E., Beukelman, T., & Xie, F. (2011). Geographic distribution of endemic fungal infections among older persons, United States. *Emerging Infectious Diseases, 17*(9), 1664–1669.
115. Colombo, A. L., Tobon, A., Restrepo, A., Queiroz-Telles, F., & Nucci, M. (2011). Epidemiology of endemic systemic fungal infections in Latin America. *Medical Mycology, 49*(8), 785–798.
116. Vaid, N., & Patel, P. (2011). A case of haemophagocytic syndrome in HIV associated disseminated histoplasmosis. *Acute Medicine, 10*, 142–144.
117. Chemlal, K., Andrieu-Bautru, V., & Couvelard, A. (1997). Hemophagocytic syndrome during Histoplasma capsulatum infection. *Haematologica, 82*(6), 726.
118. Koduri, P. R., Chundi, V., DeMarais, P., Mizock, B. A., Patel, A. R., & Weinstein, R. A. (1995). Reactive hemophagocytic syndrome: A new presentation of disseminated histoplasmosis in patients with AIDS. *Clinical Infectious Diseases, 21*, 1463–1465.
119. Mukherjee, T., & Basu, A. (2015). Disseminated histoplasmosis presenting as a case of erythema nodosum and haemophagocytic Lymphohistiocytosis. *Medical Journal, Armed Forces India, 71*(Suppl 2), S598–S600.
120. Sanchez, A., Celaya, A. K., & Victorio, A. (2007). Histoplasmosis-associated hemophagocytic syndrome: A case report. *The AIDS Reader, 17*, 496–499.
121. Nieto, J. F., Gómez, S. M., Moncada, D. C., Serna, L. M., & Hidrón, A. I. (2016). Successful treatment of hemophagocytic lymphohistiocytosis and disseminated intravascular coagulation secondary to histoplasmosis in a patient with HIV/AIDS. *Biomédica, 36*(0), 9–14.
122. Townsend, J. L., Shanbhag, S., Hancock, J., Bowman, K., & Nijhawan, A. E. (2015). Histoplasmosis-induced hemophagocytic syndrome: A case series and review of the literature. *Open Forum Infectious Diseases, 2*(2), ofv055.

123. Agarwal, S., Moodley, J., Ajani Goel, G., Theil, K. S., Mahmood, S. S., & Lang, R. S. (2009). A rare trigger for macrophage activation syndrome. *Rheumatology International, 31*, 405–407.
124. Lo, M. M., Mo, J. Q., Dixon, B. P., & Czech, K. A. (2010). Disseminated histoplasmosis associated with hemophagocytic lymphohistiocytosis in kidney transplant recipients. *American Journal of Transplantation, 10*, 687–691.
125. Keller, F. G., & Kurtzberg, J. (1994). Disseminated histoplasmosis: A cause of infection-associated hemophagocytic syndrome. *The American Journal of Pediatric Hematology/Oncology, 16*, 368–371.
126. Guiot, H. M., Bertran-Pasarell, J., Tormos, L. M., Gonzalez-Keelan, C., Procop, G. W., Fradera, J., et al. (2007). Ileal perforation and reactive hemophagocytic syndrome in a patient with disseminated histoplasmosis: The role of the real-time polymerase chain reaction in the diagnosis and successful treatment with amphotericin B lipid complex. *Diagnostic Microbiology and Infectious Disease, 57*, 429–433.
127. Gil-Brusola, A., Peman, J., Santos, M., Salavert, M., Lacruz, J., & Gobernado, M. (2007). Disseminated histoplasmosis with hemophagocytic syndrome in a patient with AIDS: Description of one case and review of the Spanish literature. *Revista Iberoamericana de Micología, 24*, 312–316.
128. Mehta, B. M., Hashkes, P. J., Avery, R., & Deal, C. L. (2010). A 21-year-old man with Still's disease with fever, rash, and pancytopenia. *Arthritis Care & Research (Hoboken), 62*, 575–579.
129. Ponce, C. A., Gallo, M., Bustamante, R., & Vargas, S. L. (2010). Pneumocystis colonization is highly prevalent in the autopsied lungs of the general population. *Clinical Infectious Diseases, 50*(3), 347–353.
130. Deng, Z., Ribas, J. L., Gibson, D. W., & Connor, D. H. (1988). Infections caused by *Penicillium marneffei* in China and Southeast Asia: Review of eighteen published cases and report of four more Chinese cases. *Reviews of Infectious Diseases, 10*, 640–652.
131. World Health Organization. (2007). *WHO case definitions of HIV for surveillance and revised clinical staging and immunological classification of HIV-related disease in adults and children*. WHO press.
132. Le, T., Wolbers, M., Chi, N. H., et al. (2011). Epidemiology, seasonality, and predictors of outcome of AIDS-associated Penicillium marneffei infection in Ho Chi Minh City, Viet Nam. *Clinical Infectious Diseases, 52*, 945–952.
133. Antinori, S., Gianelli, E., Bonaccorso, C., et al. (2006). Disseminated Penicillium marneffei infection in an HIV-positive Italian patient and a review of cases reported outside endemic regions. *Journal of Travel Medicine, 13*, 181–188.

Part V
Rheumatic Triggers of Cytokine Storm Syndromes

Cytokine Storm Syndrome Associated with Systemic Juvenile Idiopathic Arthritis

Rayfel Schneider, Susan P. Canny, and Elizabeth D. Mellins

The cytokine storm syndrome (CSS) associated with systemic juvenile idiopathic arthritis (sJIA) has widely been referred to as macrophage activation syndrome (MAS). In this chapter, we use the term sJIA-associated CSS (sJIA-CSS) when referring to this syndrome and will use the term MAS when referencing specific publications that report on sJIA-associated MAS.

The frequency, the potential severity, and the imperative to promptly initiate appropriate treatment of this hyperinflammatory state place the diagnosis of CSS at the forefront of the clinician's assessment of all febrile children with sJIA. Even in the current era, CSS is associated with a high mortality rate of 8% [1].

Epidemiology

Overt sJIA-CSS occurs in 7–17% of children with sJIA at some point during the disease course [2, 3]. Subclinical or occult CSS is seen even more commonly in >30% of children with new-onset sJIA or with a flare of systemic symptoms of the disease [4, 5]. Classification criteria for sJIA-CSS were developed in 2016 (Table 1)

R. Schneider
Department of Paediatrics, University of Toronto and The Hospital for Sick Children, Toronto, Canada
e-mail: rayfel.schneider@sickkids.ca

S. P. Canny
Department of Pediatrics, University of Washington, Seattle, WA, USA
e-mail: susan.canny@seattlechildrens.org

E. D. Mellins (✉)
Department of Pediatrics and Program in Immunology, Stanford University, Stanford, CA, USA
e-mail: mellins@stanford.edu

R. Q. Cron, E. M. Behrens (eds.), *Cytokine Storm Syndrome*,
https://doi.org/10.1007/978-3-030-22094-5_21

Table 1 Classification of MAS in sJIA [22]

A *febrile* patient with known or suspected systemic JIA is classified as having MAS if the patient has:
Ferritin > 684 ng/L
AND
At least two of the following laboratory abnormalities:
Platelets ≤ 181 × 10^9 /mL
AST > 48 U/L
Triglycerides > 1.76 mmol/L (156 mg/dL)
Fibrinogen ≤ 3.6 g/L (360 mg/mL)

and only very recently validated in one study [6] and it is therefore unsurprising that there is no consensus about the definition of occult CSS in sJIA. Patients with active sJIA often have laboratory features suggestive of CSS, without meeting the proposed classification criteria. These features can include rising levels of serum ferritin, transaminases, D-dimers, triglycerides, and LDH, and lower platelet, white blood cell, and neutrophil counts and fibrinogen levels than would be anticipated for the degree of systemic inflammation. Activation of T-lymphocytes and macrophages reflected by high levels of sCD25 and sCD163, respectively, may also be seen.

The largest reported cohort of sJIA-CSS included 362 international children (57.5% female) with onset of sJIA at a median age of 5.3 years. CSS was diagnosed after a median disease duration of 3.5 months, but developed at the onset of sJIA in >20% of patients [1].

Clinical Features

The hallmark clinical feature of CSS is fever, which is characteristically high and sustained. Children are often acutely unwell with a "septic" appearance, which may rapidly progress to shock and multiorgan failure. A change in a child's fever pattern from the typical intermittent, quotidian fever of active sJIA to a more persistent, non-remitting fever should alert the clinician to the potential evolution of CSS. Fewer than 5% of children do not have fever with sJIA-CSS, but this proportion may be higher in children being treated with IL-6 inhibitors [7].

Additional clinical features seen in >50% of patients (Table 2) include hepatomegaly, splenomegaly and diffuse lymphadenopathy. Jaundice occurs in some patients. Central nervous system involvement occurs in approximately one-third of patients with symptoms that range from headache, irritability, lethargy, and mood changes to seizures, confusion, and coma. Cardiac involvement may include pericardial disease, arrhythmia, and myocardial dysfunction with cardiac failure and shock. Coagulopathy may initially manifest as a petechial or purpuric skin rash but

Table 2 Major clinical features of sJIA-CSS [1]

Major clinical features of CSS-associated sJIA	Frequency (%)
Fever	96
Hepatomegaly	70
Splenomegaly	58
Lymphadenopathy	51
CNS involvement	35
Cardiac	26
Pulmonary	22
Hemorrhagic manifestations	20
Renal	15

may progress to frank gastrointestinal, upper respiratory tract, or pulmonary bleeding. Pulmonary infiltrates may be seen and progressive respiratory failure may ensue. The development of interstitial lung disease with or without pulmonary vascular involvement has been reported with increasing frequency in patients with sJIA and may be associated with sJIA-CSS episodes [8] Renal involvement may be mild with hematuria and proteinuria, but renal function can also rapidly deteriorate.

Differential Diagnosis

CSS should be considered in any child with sJIA or strongly suspected sJIA, who has a febrile illness. The most important confounding conditions are active sJIA and infections. Diagnosis of CSS at the onset of sJIA may be particularly difficult, as the diagnosis of sJIA may not yet be firmly established. In order to meet the widely accepted International League of Associations for Rheumatology (ILAR) criteria for the classification of sJIA [9] children must have at least a 2-week history of fever, that is quotidian in character for at least 3 days, and persistent arthritis, in at least one joint, for at least 6 weeks. Although most children with sJIA have arthralgia at presentation, overt arthritis may only develop weeks or sometimes months after the onset of the disease. Children with new-onset sJIA-CSS require urgent treatment that often cannot be deferred until the ILAR criteria sJIA are fulfilled. Applying the Yamaguchi criteria for the diagnosis of adult-onset Still's disease (AOSD) [10] may be particularly helpful in this circumstance, as AOSD can be considered part of the spectrum of sJIA [11]. The Yamaguchi criteria, which require fever for only 1 week and arthralgia, rather than overt arthritis for 2 weeks, may be more sensitive than the ILAR criteria for the diagnosis of new-onset sJIA [12]. The challenges in making a timely diagnosis of new-onset sJIA have also been addressed by the inclusion criteria developed for the North American Childhood Arthritis and Rheumatology Research Alliance (CARRA) consensus treatment protocols for newly diagnosed sJIA [13]. These criteria allow inclusion of patients within 2 weeks of the onset of fever.

When a definitive diagnosis of sJIA cannot be made and CSS features are prominent, infections, sepsis with disseminated intravascular coagulation, malignancy, drug reactions, thrombotic thrombocytopenia purpura, systemic lupus erythematosus (SLE), and Kawasaki disease must be considered. SLE and Kawasaki disease are complicated by CSS in 9% [14] and 1–2% [15, 16] of cases, respectively (see chapters "Systemic Lupus Erythematosus and Cytokine Storm" and "Kawasaki Disease-Associated Cytokine Storm Syndrome"). CSS associated with malignancy must always be excluded, particularly at the initial presentation. As the onset of sJIA under 1 year of age is uncommon [17], genetic CSS syndromes (see chapter "Cytokine Storm Syndrome Associated with Systemic Juvenile Idiopathic Arthritis") and autoinflammatory syndromes with CSS, such as the NLRC4 inflammasome mutation (see chapter "The Intersections of Autoinflammation and Cytokine Storm"), must be considered in this age group.

The possibility of a coexisting infection must always be considered, particularly in patients with sJIA on treatment with biologic other immunosuppressive agents, because CSS is associated with infection in more than one-third of cases [1, 7] and see *Pathophysiology* below. To differentiate infections from CSS, serum ferritin and sCD25 levels are typically higher in association with CSS than in infections and sepsis [18].

Diagnosis of sJIA-CSS

Guidelines for the diagnosis of primary or genetic CSS [19] are well established and have been applied to secondary CSS. However, these guidelines have low sensitivity for the diagnosis of sJIA-CSS for several reasons: fever, hepatosplenomegaly, lymphadenopathy, and elevated levels of ferritin and d-dimers are commonly seen in active sJIA; on the other hand, white cell counts, platelet counts, and fibrinogen levels may be high in sJIA and may not fall as rapidly or as profoundly in sJIA-CSS. In addition, some laboratory test results such as NK cell number and function and sCD25 and sCD163 levels may not be available quickly enough to impact timely clinical decision-making for the treatment of sJIA-CSS. Finally, tissue hemophagocytosis, most often evaluated by bone marrow aspirate and biopsy is only detected in 30–60% of cases [1].

Over the past 10–15 years, there have been a number of efforts to develop more sensitive and specific guidelines for the diagnosis of sJIA-CSS (see chapter "Alternative Therapies for Cytokine Storm Syndromes"). In 2005, Ravelli et al. published preliminary diagnostic guidelines for MAS complicating sJIA (Table 3) [20]. These guidelines were based on a retrospective study designed to differentiate sJIA-CSS from active sJIA. These criteria emphasized many of the clinical and laboratory features (Tables 4 and 5), which differentiate sJIA-CSS from active sJIA, including neurological abnormalities and hemorrhages [20]. Interestingly, this study did not evaluate serum ferritin as a diagnostic test, but perhaps most saliently, this study highlighted the importance of a *relatively* low platelet count, taking into account the thrombocytosis that typically accompanies active sJIA. These criteria

Table 3 Preliminary diagnostic guidelines for MAS complicating sJIA [20]

• *Clinical criteria*
– CNS dysfunction
– Hemorrhages
– Hepatomegaly
• *Laboratory criteria*
– Platelets $\leq 262 \times 10^9$/L
– AST > 59 U/L
– WBC $\leq 4.0 \times 10^9$/L
– Fibrinogen ≤2.5 g/L
• Diagnosis requires the presence of any two or more laboratory criteria, or any two or three or more clinical and/or lab criteria
• *Bone marrow only required in "doubtful cases"*

Table 4 Clinical features of CSS-associated sJIA compared to active sJIA [20]

Feature	CSS-sJIA	Active sJIA
Fever	Non-remitting	Remitting, quotidien
Rash	Petechiae, purpura	Pink, maculopapular
Diffuse lymphadenopathy	++	+
Jaundice	+	–
Hepatomegaly	++	+
Splenomegaly	++	+
Serositis	±	+
Arthritis	±	++
Encephalopathy, seizures	+	–
Hemorrhages	+	–
Tachycardia	+	±
Hypotension	±	–

have been shown to be more sensitive for the diagnosis of sJIA-CSS than the HLH-2004 criteria [21].

Although there are still no validated diagnostic criteria for sJIA-CSS, robust international collaborative efforts, involving pediatric rheumatologists and hematologists, have made substantial progress by developing classification criteria for sJIA-CSS. These criteria were established in an intensive multistep process that included consensus of an international expert panel and analysis of real patient data of over 400 patients who had been diagnosed with sJIA-CSS or a confusable condition (active sJIA without CSS or infection) (Table 1) [22].

Paying close attention to the changes in relevant laboratory values over time may be even more important than defining absolute threshold values in making an early diagnosis. A fall in platelet count, an increase in ferritin and AST levels were considered most important, followed by changes in white cell count, neutrophil count, fibrinogen, and ESR [23, 24]. However, the composite pattern of changes in laboratory values is still most important; for example, a falling ESR, associated with a rising CRP may suggest evolving CSS. Extreme hyperferritinemia with ferritin levels

Table 5 Laboratory features of CSS-associated sJIA compared to active sJIA [20]

Feature	CSS	Active sJIA
White blood cell count	↑/Normal/↓ or decreasing	↑↑
Neutrophil count	↑/Normal/↓ or decreasing	↑↑
Platelet count	↓/Decreasing	↑
Hemoglobin	↓/Decreasing	Normal/↓
CRP	↑	↑
ESR	Normal/decreasing	↑↑
AST/ALT	↑/↑↑	Normal/↑
Bilirubin	Normal/↑	Normal
LDH	↑↑	Normal/↑
Ferritin	↑↑	↑
Triglycerides	↑	Normal
Fibrinogen	↓	↑
D-dimer	↑↑	↑
PT/PTT	Normal/↑	Normal
Soluble IL-2Rα	↑↑	Normal/↑
Soluble CD163	↑↑	Normal/↑
CXCL-9/CXCL-10	↑↑	Normal

Table 6 The MAS/HLH score, differentiating primary HLH and sJIA-CSS [25]

Feature	Points for scoring
Age at onset, y	0 (>1.6); 37 (≤1.6)
Neutrophil count, × 10^9/L	0 (>1.4); 37 (≤1.4)
Fibrinogen, mg/dL	0 (>131); 15 (≤131)
Splenomegaly	0 (no); 12 (yes)
Platelet count, × 10^9/L	0 (>78); 11 (≤78)
Hemoglobin, g/dL	0 (>8.3); 11 (≤8.3)

MH score ≥60 (score range 0–123) proved best in discriminating
sJIA-CSS and primary HLH, with sensitivity 91% and specificity 93%

>10,000 μg/L is strongly associated with CSS [24] but still needs to be interpreted in the context of other clinical and laboratory features.

The differentiation of primary HLH and sJIA-CSS is important because there are significant differences in their treatment and expected outcomes. Recently the MAS/HLH (MH) score, was developed to discriminate between these diagnoses (Table 6, [25]), using the following discriminating variables: age at onset, neutrophil count, fibrinogen, platelet count and hemoglobin. Each variable was assigned a score based on its statistical weight. An MH score ≥60 (score range 0–123) proved best in discriminating sJIA-CSS and primary HLH, with a sensitivity of 91% and a specificity of 93%. The authors report that the MH score performed similarly well in differentiating primary CSS from a subset of patients with CSS at the onset of sJIA.

CSS on Biologic Agents

The proportions of sJIA patients who developed CSS in clinical trials of tocilizumab and canakinumab [26] were reported to be approximately 3% and 5% respectively, compared with 7–17% of sJIA patients reported to develop CSS in the literature. A somewhat more meaningful comparison is the MAS rate per 100 patient years, which was 1.8 in the tocilizumab trials and 2.8 in the canakinumab trials, compared to a rate of 4–6 in a large pediatric center [27]. However, these trials were not appropriately designed or powered to accurately detect true rates of MAS. Although there are reports of sJIA-CSS occurring in patients treated with anakinra, the MAS rate in sJIA treated with anakinra has not been studied [28, 29]. Overall, the rate of MAS does not appear to be substantially altered in patients treated with standard doses of IL-1 and IL-6 inhibitors.

A systematic review of 84 patients with sJIA-CSS occurring while on treatment with canakinumab ($n = 35$) and tocilizumab ($n = 49$) [7] has helped to elucidate the clinical and laboratory features of sJIA-CSS on these biologic agents. These features were compared to those of the large historical cohort of 362 patients with sJIA-CSS [1]. Patients who developed sJIA-CSS on background treatment with canakinumab or tocilizumab had lower median white blood cell counts. A lower proportion of patients on tocilizumab had fever (74% vs 96%) and hepatosplenomegaly; platelet counts, fibrinogen and LDH were lower and AST was higher. Only 53.4% of the patients on background tocilizumab met the 2016 classification criteria for sJIA-associated MAS, compared with 77.1% and 78.5% of patients in the canakinumab-treated and historical cohorts, respectively. The relatively low sensitivity could be attributed to absence of fever or low ferritin levels. These data suggest that the new classification criteria may not be sufficiently sensitive for the diagnosis of sJIA-CSS in all sJIA patients who are treated with biologic agents. Another study showed that tocilizumab-treated patients with sJIA-CSS had significantly lower CRP levels than CRP levels in the historical cohort mentioned above [30].

Predictors of Poor Outcome

Multiorgan failure and central nervous system involvement tend to predict a severe course of sJIA-CSS [31].

Pathophysiology

Data from genetically driven hemophagocytic lymphohistiocytosis (HLH) (see "Genetics of Cytokine Storm Syndromes") support several mechanistic models for the inherited disorders, depending on whether the genetic lesion affects cytotoxicity, immunodeficiency, or the inflammasome. In sJIA-CSS, similar complexity may underlie genetic predisposition to developing CSS, with further mechanistic heterogeneity arising due to the various triggers that can incite this hyper-inflammatory state. Despite this upstream heterogeneity, data from mouse models of CSS and sJIA-CSS, together with data from patients with sJIA-CSS, suggest a pathophysiologic convergence with interferon gamma (IFNγ) as a key effector cytokine in at least one prevalent scenario. In this section, we summarize information supporting this view and mention some exceptions that leave open the possibility of other mechanistic models. We emphasize results from studies of patients with sJIA-CSS and cite some data from relevant animal models. We highlight open questions and note some particular areas warranting further investigation.

Triggers of sJIA-CSS

Uncontrolled sJIA is often the presumed trigger in sJIA-CSS episodes that occur at disease onset or during disease flares [32–34]. However, active sJIA is frequently associated with recent or concurrent viral infection. Thus, the role of sJIA versus infection in CSS episodes in the context of active sJIA is difficult to resolve. In a multinational, multicenter study of 362 patients with sJIA-CSS, co-occurrence of infection was documented in one third of the cases [1]. Notably, sJIA-CSS can erupt during well-controlled sJIA, including in sJIA patients on biologic immunosuppressive agents. In this case, infections are more strongly implicated as triggers. In a compelling example, a study of sJIA-CSS in canakinumab-treated patients identified CSS episodes in inactive patients and, overall, infections were the most common apparent trigger [27]. In infection-associated CSS, herpes viruses, Epstein–Barr virus (EBV), and cytomegalovirus, are frequently culprits ([35] and see chapters "Infectious Triggers of Cytokine Storm Syndromes: Herpes Virus Family (Non-EBV)" and "Cytokine Storm Syndromes Associated with Epstein-Barr Virus"). Less often, drugs, including aspirin, NSAIDs, methotrexate and biologic agents, are implicated as possible triggers of sJIA-CSS [33, 36, 37], although it is difficult to prove these medications are causative. Gold injections and sulfasalazine have long been considered contraindicated in sJIA, because of their reported association with CSS [36].

Table 7 Pathological implications of clinical and laboratory features of sJIA-CSS

Feature	Candidate mechanism
Fever	Increased pyrogenic cytokines (IL-1, IL-6)
Splenomegaly	Splenic infiltration by lymphocytes and macrophages
Cytopenias	Suppression by cytokines and ferritin
Hyperferritinemia	Macrophage activation
Hyperfibrinogenemia	Plasminogen-activator induced macrophages
Hypertriglyceridemia	Inhibition of lipoprotein lipase by cytokines
Elevated soluble CD25 (IL-2Rα)	T cell activation
Elevated soluble CD163	Macrophage activation
Reduced NK cell cytotoxicity	Genetic defect or transient dysfunction
Elevated transaminases and bilirubin	Liver infiltration by lymphocytes and macrophages
Elevated LDH	Increased cell death
Elevated D-dimer	Hyperfibrinolysis
Elevated CSF cells or protein	CNS exposure to cytokines and infiltration by immune cells

Clinical Markers of sJIA-CSS: Implications for Pathophysiology

Clinical markers used to diagnose sJIA-CSS provide clues to cell types contributing to its pathophysiology (Table 7). The high levels of serum ferritin implicate CD163 (hemoglobin-haptoglobin scavenger receptor)-bearing macrophages, which are major producers of ferritin [38]. CD163 is expressed only by monocytes and macrophages [39], and preferentially on cells undergoing alternate activation and differentiation associated with enhanced phagocytic capacity [40]. It has been argued that serum ferritin derives primarily from cell damage [41], but there is evidence for secretion from activated murine macrophages by a nonclassical secretory pathway [42]. Current data are conflicting as to whether serum ferritin is solely a biomarker of inflammation in CSS, contributes to inflammation and immune dysregulation through effects on cells, or induces immunosuppression as part of a negative feedback loop [41, 43, 44].

Soluble CD163 (sCD163), generated by metalloproteinase-mediated shedding of CD163, is also associated with sJIA-CSS. sCD163 indicates monocyte/macrophage activation [45, 46], although its specific function is unknown. Other sJIA-CSS clinical markers reflecting macrophage activation include low levels of fibrinogen, which likely reflects plasminogen activator release from macrophages [47] and phagocytosis of progenitor and mature blood elements in bone marrow, liver or other organs. Phagocytosis of hematopoietic stem cells, in particular, suggests cytokine-driven downregulation of surface CD47 on these cells [48]. CD47 interacts with signal regulatory protein α (SIRPA) on macrophages and inhibits their phagocytic function.

Soluble CD25 (sCD25) is also elevated in sJIA-CSS serum. CD25, the alpha subunit of the IL-2 receptor, confers high affinity to the IL-2 receptor complex, by combining with CD122 (IL-2rβ) and CD132 (γ_c). CD25 is expressed on effector T (Teff) cells, T regulatory (Treg) cells, immature B cells, $CD56^{hi}$ natural killer (NK) cells (an NK cell subset), natural killer T (NKT) cells, and dendritic cells (DCs) [49]. sCD25 results from shedding of the receptor from activated cells; in sJIA-CSS, Teff likely make a prominent contribution to sCD25 levels. Published reports suggest various functions for sCD25, including antagonism of IL-2, enhancement of T cell activation, and immunoregulation of non-T immune cells [50, 51].

In summary, elevations of ferritin, sCD163, and sCD25 implicate macrophages and T cells in sJIA-CSS. Additional clinical features of sJIA-CSS with mechanistic implications are summarized on Table 7 (modified from [52]).

sJIA-CSS: Inducing Conditions and Effector Phase

sJIA-CSS is thought to be a process in which a predisposing, dysfunctional immune state associated with sJIA is triggered to escalate to a hyper-inflammatory cytokine storm. Emerging data, described below, suggest that a subset of sJIA children may be at highest risk for this dangerous complication, due to a genetic predisposition to dysregulated immune processes, but more investigation of genetic risk is needed. An alternative hypothesis is that encounters with triggering stimuli lead to sJIA-CSS, and that all children with sJIA are at equal risk. However, the same sJIA patients get multiple episodes of CSS, whereas some patients, despite a lot of inflammation, never develop CSS. To discuss the pathophysiology, we consider a current paradigm (Fig. 1) and describe the CSS process chronologically, in terms of inductive conditions and effector mechanisms.

Inductive Phase of sJIA-CSS

Impaired Cytotoxic Function: Genetic

The investigation of impaired cytotoxic function in sJIA-CSS was motivated by the clinical similarity of sJIA-CSS to a group of monogenic disorders, referred to as primary HLH (pHLH), in which mutations affect the perforin-mediated cytolytic pathway (see chapter “Murine Models of Familial Cytokine Storm Syndromes”). Evidence from animal models with disruptions of the pHLH genes ([53] and chapter “Murine Models of Familial Cytokine Storm Syndromes”) argues that reduced function of key cytolytic lymphocytes (especially CD8 T cells of the adaptive immune system) leads to persistence of an ongoing immune response, particularly due to lack of killing of stimulatory antigen presenting dendritic cells.

The strongest evidence for a possible contribution of impaired cytotoxicity to sJIA-CSS is genetic. (For a comprehensive discussion of the genetics of sJIACSS,

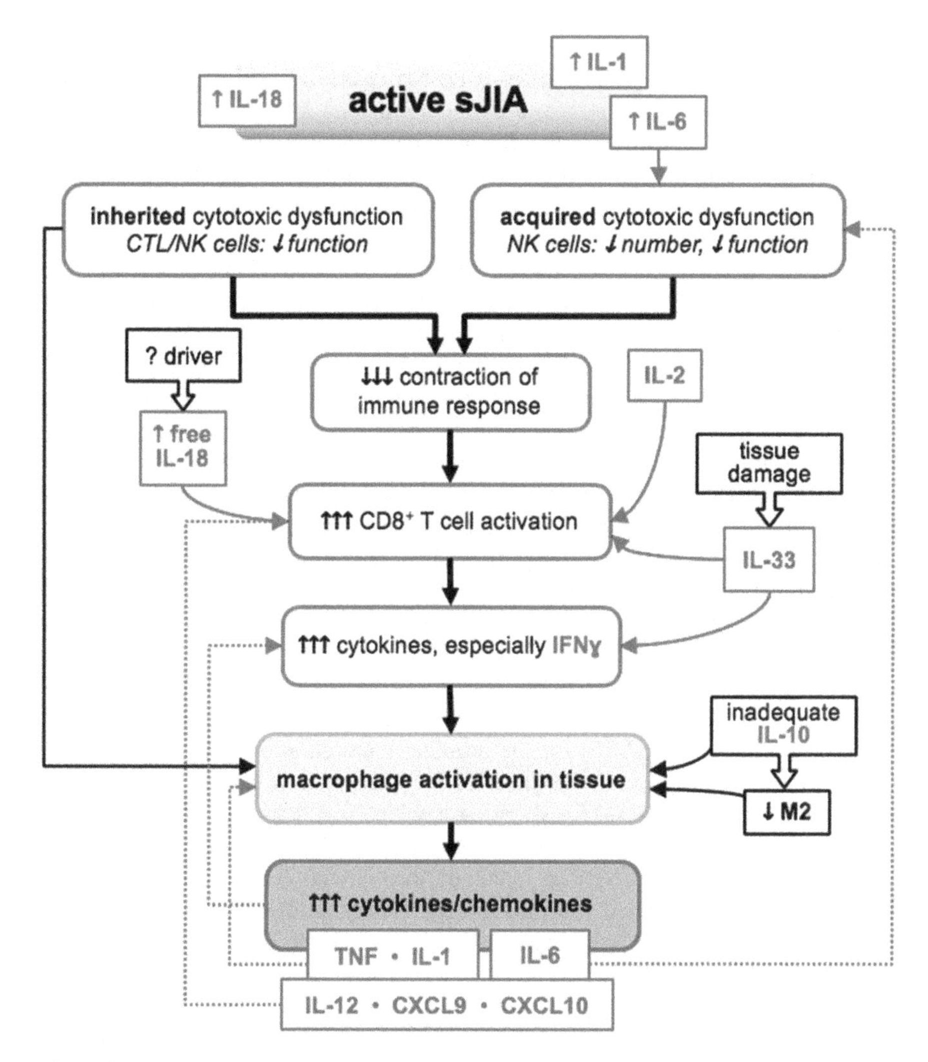

Fig. 1 Schematic diagram of pathophysiology of sJIA-CSS. Feed-forward processes shown in solid black or red arrows, the latter for cytokine effects. Feedback processes shown in dotted arrows. Cytokines and chemokines shown in red. Chemokines CXCL9 and CXCL10 recruit CD8 T cells to tissue sites; IL-12 and IL-18 synergize to activate CD8 T cells

see chapter "Genetics of Macrophage Activation Syndrome in Systemic Juvenile Idiopathic Arthritis"). Studies of patients with sJIA-CSS identify a subset (~35%) with heterozygous mutations in causative genes for pHLH [54–61]. Mutations in pHLH genes in sJIA patients include those previously observed in pHLH patients, as well as novel rare variants [56, 57, 60, 62]. One third to one half of pHLH cases are caused by inactivating mutations in *PRF1,* which encodes perforin. Perforin is stored in secretory granules of cytotoxic CD8 T lymphocytes (CTL) and NK cells. It is released, together with granzyme proteases, into the cleft of an immune synapse

(IS) formed with a target cell. Perforin oligomerizes to form a pore in the target cell membrane. This allows granzymes to enter the target cell and activate apoptotic pathways. pHLH can also be caused by mutations in *UNC13D, RAB27A, STX11, STXBP2,* which encode proteins that facilitate delivery of perforin to the IS through effects on cytolytic granules. The protein encoded by *UNC13D,* Munc13-4, is a Ca^{2+}-dependent SNARE (soluble *N*-ethylmaleimide-sensitive factor attachment protein receptor) and phospholipid-binding protein that localizes to and primes granules for membrane fusion and exocytosis [54, 59]. Munc13-4 interacts with the small GTPase, Rab27a, an endosome-associated protein, to regulate exocytosis of lytic granules in CTL [63] and NK cells [64]. Munc13-4 also interacts with Rab11a [65], which marks recycling endosomes whose fusion with the target cell plasma membrane at the IS is essential for cytotoxicity [66]. *STX11* and *STXBP2* encode syntaxin 11 and syntaxin binding protein 2, respectively; these interacting proteins are required for fusion of Rab11a + endosomes with the plasma membrane [67].

Mutations in pHLH genes have been investigated in patients with sJIA and sJIA-CSS. An analysis of 31 Italian sJIA-CSS patients for mutations in *PRF1*, *UNC13D*, *STX11*, *STXBP2*, and *RAB27A* found monoallelic mutations in 11 (35.5%); clinical features of the children with mutations were not distinguishable from those without [57]. A recent whole-exome sequencing (WES) study of sJIA patients identified rare protein-altering variants in *UNC13D*, *STXBP2*, and *LYST* in 5/14 (36%) sJIA-CSS patients versus 4 variants in 4 of 29 (14%) sJIA patients without sJIA-CSS by the time of study [56]. *LYST* (lysosomal trafficking regulator) is involved in trafficking of Munc13-4 and Rab27a for the terminal maturation of perforin-containing vesicles into secretory cytotoxic granules for exocytosis [68]. *LYST* mutation is responsible for the Chediak–Higashi disease, which is sporadically associated with CSS [69]. In the WES study, 22 rare variants were found in at least two patients, and pathway analysis suggested that many of the variant genes encoded proteins involved in vesicle-mediated transport and cellular organization [56]. Investigations of the effects on cytotoxic cell function of new rare variants in pHLH genes or new candidate genes are on-going [62, 70]. Taken together, these genetic studies support the hypothesis that inherited alterations affecting granule-mediated cytolytic function increase susceptibility to sJIA-CSS in a subset of sJIA patients, although more work is needed to definitively link these mutations to altered cytolytic function in patients and to sJIA-CSS predisposition.

Impaired Cytotoxic Function: Acquired

The investigations of cytotoxic cell dysfunction in sJIA began before there was evidence that a subset of sJIA patients have genetic alterations in the pathway. Initial data suggesting a cytotoxicity defect emerged primarily from studies of NK cells, cytotoxic lymphocytes of the innate immune system. These cells, like CD8 T cells, are important in host defense and in homeostatic immune mechanisms required to curtail immune responses. Evidence (though conflicting, see below) has been presented suggesting sJIA NK cells can be defective in sJIA and more frequently in sJIA-CSS. The most consistent observation is reversibly lowered numbers during sJIA-CSS; other phenotypes variably reported for sJIA NK cells include reversible decreases in NK activity, associated with low perforin and granzyme B, and defective degranulation [57, 71–73]. Circulating CD56^{hi} NK cells isolated from active sJIA patients show reduced levels of granzyme K [74], an enzyme implicated in killing of autologous activated T cells by this subset. This specific killing defect may contribute to expansion of activated CD8 T cells in sJIA-CSS.

The reversibility of all the NK defects reported to date implicates environmental influence, and indeed, NK cell development, number and function are modulated by cytokines (reviewed in [75]). For example, after ligation of activating receptors, NK cells require inflammatory cytokine stimulation for optimal effector function. Two cytokines elevated in serum during active sJIA are among those that regulate NK cells: interleukin (IL)-6 and IL-18. Transcriptional analysis of circulating NK cells from active sJIA patients revealed effects of in vivo exposure to IL-6 [74]. In an IL-6 transgenic mouse that develops CCS features after exposure to innate stimuli, elevated IL-6 led to reversible reductions in perforin and granzyme B in NK cells, without effects on degranulation or IFNγ production, and data from four sJIA patients suggested similar changes that improved after treatment with IL-6 inhibition [76]. In the case of IL-18, a known stimulator of NK function, the initial finding of high circulating IL-18 and low NK function in sJIA-CSS seemed hard to reconcile; however, a negative correlation between IL-18 serum levels and NK cell number is observed in studies of several diseases (e.g., [77]). In vitro evidence suggests that sJIA NK cells isolated from blood are IL-18-insensitive, as reflected in reduced downstream signaling [74, 78]. Interestingly, in one study, patient cells showed higher baseline activation of downstream IL-18 mediators (ERK1/2), raising the possibility that in vivo exposure to high IL-18 might drive receptor insensitivity tested in vitro. Decreased IL-18 responsiveness of NK cells, in particular reduced production of IFNγ, was proposed as a possible feedback mechanism to inhibit sJIA-CSS [74]. [see more on IL-18 below].

Other Candidate CSS-Predisposing Immune States in sJIA

One consequence of reduced cytolytic activity in mouse models of pHLH type CSS is the expansion and activation of poorly cytolytic CD8 T cells in tissue. This is likely due, at least in part, to their continued interaction with antigen presenting cells that are not killed. sJIA-CSS liver biopsies show extensive infiltration of IFNγ+ CD8 T cells [79]; the proposed role(s) of these cells in sJIA-CSS is discussed below (*sJIA-CSS effector phase*). Recent intriguing evidence from a perforin-null mouse model of CSS adds to this mechanism; a synergistic effect on IFNγ+ CD8 T cell expansion was observed, mediated by the alarmin IL-33, which is released from necrotic cells and signals through an innate immune receptor [80]. However, genetic causes of sporadic or persistent CSS without cytotoxic defects are also known [81].

Interleukin-1β (IL-1) and IL-6 are strongly implicated in sJIA pathogenesis, most notably by the clinical efficacy of their inhibitors [82, 83]. Neither appears directly linked to CD8 T cell expansion (except as mediated indirectly by IL-6 effects on NK cells as discussed above). Monogenetic autoinflammatory diseases caused by IL-1 excess are not known to be accompanied by CD8 T cell expansion [84]. Serum levels of IL-1 and IL-6 are usually noted to be similar between sJIA patients with and without CSS [85]. Although successful treatment of SJIA-CSS with inhibitors of these cytokines, particularly with the IL-1 receptor antagonist, anakinra, has been observed in the effector phase (reviewed in [86]), these therapies do not fully protect against CSS [7, 26, 87, 88] as might be expected if either IL-1 or IL-6 were required or were solely sufficient to drive an early step in CSS pathogenesis.

Notably, evidence is building for a possible predisposing role of interleukin 18 (IL-18) in sJIA-CSS, although many open questions remain. IL-18 is a pleotropic cytokine, whose earliest known effects were promotion of T_H1 cell differentiation and IFNγ secretion by NK and T cells [89, 90]. The inactive IL-18 precursor is constitutively expressed in nearly all cells. Processing and secretion of mature IL-18 requires activation of the NLRP3 inflammasome, an intracellular complex of proteins that forms after innate stimuli. IL-18 activity is regulated by its endogenous inhibitor, IL-18 binding protein (IL-18BP), which is induced by IFNγ. Recently, Canna et al. identified an NLRC4 (inflammasome) activating mutation resulting in excessive free IL-18 and CSS-associated autoinflammation. Successful treatment of NLRC4-CSS with IL-18BP indicated the pathogenic role of IL-18. Cytotoxic defects are not observed in this disorder, implying a distinct route to CSS [91].

Importantly, high serum IL-18 is a biomarker of CSS-risk in sJIA patients [92, 93]. Levels in sJIA patients can be elevated to over 30X the level found in healthy controls. High levels persist during active sJIA-CSS and slowly decrease with the administration of immunosuppressive therapy [94]. The cellular origins of IL-18 in sJIA-CSS have not been elucidated fully. However, there is evidence for bone marrow macrophages [95] and gut epithelia [96]. The mechanism linking IL-18 to CSS risk is unknown. Effects of IL-18 on sJIA NK cells purified from circulation have been investigated (discussed above), but data on effects of high IL-18 on tissue T cells are lacking and this is an important avenue for future study.

To assess IL-18 effects, there is general agreement that free IL-18 is the relevant metric that correlates with clinical status in CSS [97]. There is some controversy over how best to measure the free form. Another issue is that IL-18 has different effects depending on the overall cytokine milieu. In the presence of IL-12, IL-18 drives T cell differentiation to T_H1 cells and IFNγ secretion by NK and CD8 T cells [89]. However, in the absence of sufficient IL-12, IL-18 can promote T cell differentiation to T_H2 cells [90]. Although IL-12 is not identified as a signature serum cytokine in sJIA, increases are observed in serum and/or CSS-relevant tissues, such as liver and spleen, in animal models that resemble sJIA-CSS phenotypically [98, 99].

Effector Phase of sJIA-CSS

Interferon Gamma and Insufficient Counter-Regulation by Interleukin-10

Expanded and activated CD8 T cells appear to be key drivers of the effector phase of sJIA-CSS. No therapeutic interventions currently target activated CD8 T cells specifically, but the efficacy of cyclosporine A in sJIA-CSS [33] is consistent with a key role for T cells [100]. In the inductive pathway characterized by cytotoxic defects, the duration of the CD8 T cell/dendritic cell IS is likely extended, increasing cytokine production by cytotoxic cells, in particular, IFNγ [101]. Based on mouse models, other factors, including innate stimuli and cytokine milieu, likely contribute to IFNγ production by tissue CD8 T and by other cells [80, 98, 99].

Regardless of mechanism leading to its expression, IFNγ has been shown to be a critical effector in conditions relevant to sJIA-CSS, including pHLH disorders, animal models based on pHLH (reviewed in [102]) and other animal models resembling sJIA-CSS [98, 99]. Antibody-mediated blockade of IFNγ is sufficient to improve survival in the animal models [53, 99] and a pilot phase 2 study of anti-IFNγ antibody in children with pHLH also showed efficacy [103].

Two studies found significantly elevated serum levels of IFNγ and IFNγ-induced proteins (e.g., CXCL9, CXCL10, IL-18BP) in sJIA-CSS, compared to levels in active or inactive sJIA patients or in healthy controls [85, 104]. These levels decreased with sJIA-CSS resolution [85]. The chemokines, CXCL9 and CXCL10, reflect tissue activity of the IFNγ pathway and act to recruit CD8 T cells [105], and T_H1 and NK cells to tissue, amplifying the potential for persistent IFNγ production. Different cells produce IFNγ in various animal models of CSS [53, 98, 106, 107], and although CD8 T cells are strong candidates for this function in the current sJIA-CSS paradigm, identification of all IFNγ-producing cell(s) in sJIA-CSS requires more investigation.

IL-10 is a regulatory/immunosuppressive cytokine, known to counter-regulate IFNγ [107, 108]. In the repeated TLR9 stimulation mouse model of CSS, IL-10 receptor blockade greatly increases disease [107]. Transcriptional profiles of children with sJIA-CSS show signatures consistent with activity of IL-10, suggesting an effort at suppression of the on-going inflammation [109]. Notably, sJIA-linked IL-10 polymorphisms are associated with decreased IL-10 activity [110], which could predispose to a more aggressive effector phase in sJIA-CSS. Less IL-10 activity may limit, for example, IL-10-driven, regulatory NK cells, shown to arise in systemic infection/inflammation in mice [111] or B-regulatory cells [112]. The range of cellular sources of IL-10 in sJIA-CSS has not been elucidated.

A substantial body of evidence argues for an etiologic role for IFNγ in sJIA-CSS. However, a CSS condition has been described in two children who lack IFNγ receptors, implying other pathways [113]. In a mouse model of CSS based on repeated TLR9 stimulation, features of CSS, except for anemia, are independent of IFNγ in the absence of IL-10 [107], and in a mouse model of IL-17-driven inflammation with clinical similarity to sJIA, without ferritinemia, IFNγ derived from NK cells provides an ameliorating effect [114, 115]. Thus, IFNγ may not be essential to disease pathophysiology in all types of CSS.

Monocytes/Macrophages

A central player in the effector phase of sJIA-CSS appears to be the macrophage, which is a well-studied target of IFNγ activation and a robust producer of cytokines. Macrophages are highly sensitive to their microenvironment and highly plastic in response, leading to a spectrum of polarized states from pro-inflammatory to regulatory/repair [116]. At the inflammatory end of the spectrum, driven by IFNγ, are M1 or classically activated macrophages. A number of lines of evidence implicate M1 macrophages in the sJIA-CSS effector phase in tissues like liver and spleen, where the IFN pathway is activated [99]. For example, in a study that included patients with sJIA-CSS, neopterin, a product of IFNγ-stimulated macrophages, was 70% sensitive and 95% specific for CSS [117]. IFNγ drives macrophage production of IL-6 and TNF, the cytokines observed in macrophages in sJIA-CSS liver biopsies [79]. Other products of M1 macrophages include IL-12, CXCL9, and CXCL10, which amplify the IFNγ effect. Polymorphisms in IRF5 (interferon regulatory factor 5), are associated with sJIA-CSS in Japanese patients [118]. IRF5 is a transcription factor, downstream of toll-like receptors (TLR) 7,8,9 and other stimuli, that also drives the M1 macrophage phenotype [119]. Analyses of IRF5 polymorphisms in other sJIA cohorts are needed, but, notably, a murine model of CSS driven by repeated TLR9 stimulation also argues for a contribution of macrophages to sJIA-CSS pathology [107]. Cytotoxic cells participate in removal of activated macrophages (reviewed in [86]). Therefore, the defects described as predisposing influences for sJIA-CSS development likely also contribute to the uncontrolled macrophage activation of the effector phase of sJIA-CSS.

IFNγ is also implicated as the direct or indirect driver of hemophagocyte development [120]. The latter may be more likely as hemophagocytic macrophages in tissues bear CD163, a key marker of alternatively activated (M2) macrophages, which are associated with regulation of inflammation, tissue repair and phagocytosis. Indeed, no direct evidence indicates a pathogenic role for hemophagocytic macrophages, although they are often described as pathognomonic of sJIA-CSS [5, 121]. These cells are neither sensitive (only 60%) nor specific for sJIA-CSS, as they also occur in other conditions.

The appearance of alternatively activated macrophages in sJIA-CSS brings to mind several studies of circulating mononuclear cells in sJIA. Transcriptional profiling of these cells in sJIA patients with highly elevated ferritin (consistent with a more sJIA-CSS like-phenotype) compared to those without hyper-ferritinemia showed enrichment for genes that downregulate TLR/interleukin-1 receptor-triggered inflammation and for genes encoding markers of M2 macrophage differentiation [122]. MicroRNA analysis of monocytes from active sJIA patients showed a significant increase in miR-125a-5p, which plays a role in M2 polarization [123]. In active sJIA, circulating CD14+ monocytes (representing ~85% of all blood monocytes) with M1 phenotype are reduced and monocytes with mixed M1/M2 phenotype, associated with myeloid derived suppressor cells that downregulate inflammatory responses are increased [124]. Serum markers of M2 monocytes, namely, heme-oxygenase 1 (HO-1), sCD163, and IL-10, are elevated in sJIA-CSS [125]. These signatures in blood appear to reflect a response to inflammatory activity in tissues. Notably, steroids, an effective therapy for sJIA-CSS, promote monocyte differentiation into M2 macrophages [126]. In clinically inactive disease, circulating M2 monocytes tend to increase as a proportion of monocytes [124]. Serum concentrations of HO-1, sCD163, and IL-10, as well as IL-18, remain elevated, whereas clinical parameters and other pro-inflammatory cytokines normalize [125], apparently reflecting a state of compensated inflammation. Thus, circulating immunosuppressive monocytes likely are a response to hyper-inflammation in the tissues.

Treatment of sJIA-CSS

General Principles

Infections, particularly viruses, may trigger CSS and potential organisms should be diligently sought and treated, if possible. Given the potential for the development of multiorgan failure in CSS, appropriate supportive care, including an intensive care unit, should be readily accessible.

Initial Approach

There is widespread consensus among pediatric rheumatologists that full-blown sJIA-CSS should initially be treated with high dose systemic glucocorticoids. Intravenous pulse methylprednisolone (30 mg/kg up to 1000 mg) daily for 3–5 days is commonly used, followed by prednisone 1–3 mg/kg/day, although higher doses may be required if there is not a satisfactory and prompt response. Dexamethasone may be preferred to oral prednisone if there is significant central nervous system involvement.

Since sJIA-CSS often occurs in a pro-inflammatory context associated with high systemic disease activity, treatment is best directed at controlling both active sJIA and CSS. Although ongoing treatment with IL-1 inhibitors in sJIA does not prevent the development of CSS, there are several reports that suggest anakinra is effective in treating sJIA-CSS [28, 127–130].

Higher doses of anakinra than are typically used to treat active sJIA, even exceeding 8–10 mg/kg/day or 400 mg daily in AOSD, have been used [131, 132]. Clinical trials in patients with septic shock suggest that much higher doses of anakinra of

1–2 mg/kg/hour by intravenous infusion for 3 days are well tolerated [133]. Moreover, patients with sepsis syndrome and features of CCS, defined as hepatobiliary dysfunction and disseminated intravascular coagulation, have been reported to have improved survival with treatment with these very high doses of anakinra [134]. However, there are currently no formal studies to support this approach in severe sJIA-CSS.

Our preferred approach is to start treatment of new-onset sJIA-CSS with the combination of intravenous pulse methylprednisolone and anakinra. A prompt and substantial improvement in CSS may enable accelerated weaning of systemic glucocorticoids, particularly if anakinra also proves to be effective in controlling the underlying systemic disease. In the absence of a rapid response to high-dose systemic glucocorticoids or if there is significant multiorgan involvement, adding cyclosporine (2–7 mg/kg daily) can induce rapid improvement in clinical and laboratory features of CSS, often within 24–48 h [100, 135–137]. Dose adjustment to reach a trough serum level of cyclosporine of 150–200 may be necessary to achieve an optimal response.

Intravenous immunoglobulin is sometimes used in combination with glucocorticoids or cyclosporine. Although IVIg has been reported to be effective as initial, isolated treatment for CSS [138], there is little evidence for its efficacy as an isolated treatment for sJIA-CSS. If there is an imperative to initiate treatment for CSS before completing the diagnostic workup such as bone marrow examination or tissue biopsy, IVIg or anakinra may be considered, as these agents will not confound the subsequent diagnosis of a malignancy and IVIg will not exacerbate an associated infection.

Severe or Refractory sJIA-CSS

For severe CSS that is unresponsive to the combination of high dose systemic glucocorticoids, anakinra and cyclosporine, there is no consensus about the optimal approach to treatment. In the large international series of patients with sJIA-CSS, almost all (98%) received systemic glucocorticoids, administered intravenously in 89% [1]. The other commonly administered medications were cyclosporine (62%) and intravenous immunoglobulin (36%). Biologic agents, most commonly anakinra, were used in only 15% and etoposide in 12% of patients and other immunosuppressive agents, including methotrexate and cyclophosphamide, in 7% of patients. Plasma exchange was performed in 4%.

Treatments used for primary HLH may be considered, including those used for refractory primary HLH. Consultation with a hematologist experienced in the management of CSS may be helpful. Etoposide can be very effective, but its use must be balanced with the risks of serious infections and bone marrow suppression [139]. To minimize the side effects of etoposide, the dose should be reduced if there is impairment of renal function or liver function. It may be possible to retain the efficacy of etoposide while limiting its potential toxicity by using lower doses of etoposide, administered less frequently and for shorter duration than typically used for the treatment of primary HLH. In one case series, etoposide 50–120 mg/m^2 was administered once weekly for 4–7 weeks, considerably less than suggested in the primary HLH protocols (150 mg/m^2 twice weekly) [140]. A recent consensus statement by the HLH Steering Committee of the Histiocyte Society similarly supports consideration of etoposide for severe, refractory secondary CSS at a dose of 50–100 mg/m^2 [141].

Anti-thymocyte globulin, which depletes CD4 and CD8 T cells, has been used to treat secondary CSS refractory to high dose glucocorticoids and cyclosporine [142] and may be considered especially when sJIA-CSS is associated with severe liver or renal impairment [102].

Alemtuzumab, a monoclonal antibody which effectively and rapidly depletes CD52 expressing B and T lymphocytes, monocytes and macrophages, has been used to treat refractory primary HLH [143]. Alemtuzumab has also been used as a component of a hematopoietic stem cell transplant (HSCT) conditioning regimen for JIA patients, including some with CSS and has been purported to be less toxic than the standard regimen [144]. Thus far, however, there are only anecdotal reports of alemtuzumab treatment for severe sJIA-CSS who do not require HSCT and its use may be complicated by severe viral infections.

Plasma exchange may be considered as a temporizing measure for severe manifestations of CSS or for patients with CSS that is refractory to standard therapy [145–148].

Other Biologic Agents

Although there are some reports of improvement of sJIA-CSS with treatment with etanercept and infliximab [149–151], etanercept has been associated both with initiation of CSS in sJIA [37] and worsening of CSS in AOSD [152]. Thus, there is no clear role for TNF inhibitors in the treatment of sJIA-CSS.

There are several reasons to consider treatment of sJIA-CSS with tocilizumab, an anti-IL6 receptor antibody: it is a very effective for treating active sJIA [82]; it is used to the treat the CSS associated with chimeric antigen receptor-modified T cell therapy in patients with leukemia [153], and there are reports of its efficacy in treating AOSD-associated CSS [154–156]. However, evidence for this approach in sJIA-CSS is lacking and therefore the role of tocilizumab in sJIA-CSS is unclear.

Patients with sJIA who develop CSS associated with an Epstein-Barr virus infection may benefit from treatment with rituximab to eliminate EBV-infected B cells and to increase viral clearance, in conjunction with immunosuppressive therapy [157] (see chapter "Cytokine Storm Syndromes Associated with Epstein-Barr Virus"). However, there are no safety data to support the use of rituximab in patients already on a longer-acting biologic agent, such as canakinumab, tocilizumab, or a TNF inhibitor. Although antiviral agents may be considered, their role remains unclear, particularly considering their potential myelotoxicity [158].

Treatment of sJIA-CSS While on Treatment with Biologic Agents

When CSS occurs in sJIA patients on treatment with biologic agents, standard treatment can be initiated with high dose systemic glucocorticoids, with or without, cyclosporine, IVIg and other primary HLH treatments added as needed. However, the optimal role of biologic agents is not clear. In the major clinical trials of IL-1 and IL-6 inhibitors in sJIA, these biologics have been discontinued when the diagnosis of CSS has been made. However, withdrawal of IL-1 or IL-6 inhibition should be done with caution and vigilance, as this may amplify the inflammatory milieu and potentially worsen features of MAS. One approach to patients who develop CSS while already treated with anakinra is to increase the dose of anakinra in accordance with doses used for CSS (see above and [131]). The safety of adding anakinra to other biologic agents, particularly those targeting a different cytokine pathway, has not been established, although it has been used with apparent safety together with abatacept in a small case series [159]. Although there are currently no clear guidelines for restarting biologics after an episode of sJIA-CSS has been controlled, Yokota et al. [87] reported restarting tocilizumab in sJIA patients who developed CSS while on this biologic agent, without recurrence of CSS.

Novel Treatment Approaches Under Active Investigation

Interferon-γ (IFNγ) Neutralization

IFNγ has been identified as a pivotal mediator in murine models of primary HLH [53] and marked elevations of IFNγ and IFNγ-induced cytokines, especially CXCL9 and CXCL10, are seen in sJIA-CSS but not in active sJIA [98]. Moreover, IFNγ, CXCL9, and CXCL10 levels correlate with laboratory features of CSS, providing a strong rationale for a trial of IFNγ blockade in sJIA-CSS. NI-0501 (emapalumab) is a fully human IgG1 monoclonal antibody that binds free and receptor-bound IFNγ and is being studied in a clinical trial of children with active primary HLH [85]. Analysis of preliminary data from this pilot phase 2 study suggested that emapalumab was well tolerated, even in patients who had concurrent infections, and that there was an improvement in HLH features with reduction in systemic glucocorticoid doses in responders [103].

Targeting IL-18

IL-18 levels are not only increased in sJIA, but they are also predictive of the development of sJIA-CSS [93]. Moreover, IL-18 levels may be even higher during sJIA-CSS, with free IL-18 levels being specifically elevated [96, 160]. These observations make IL-18 blockade an attractive option for sJIA-CSS. This approach is supported by a mouse model in which IL-18 blockade with a monoclonal antibody, reduced the severity of CSS in IL-18BP knock-out mice [161] and by another murine HLH model, in which treatment with IL-18BP reduced hemophagocytosis and organ damage [162]. Also, of interest, the preliminary results of an open-label, phase II clinical trial of tadekinig alfa (recombinant IL-18BP) in patients with active AOSD suggest that this treatment has a favorable safety profile and may show early signs of efficacy [163].

JAK Inhibitors

Janus tyrosine kinases (JAK), JAK 1 and 2, are critical to the downstream signaling of IFNγ. In mouse models of both primary and secondary HLH, inhibition JAK function has been shown to improve clinical and laboratory features of CSS [164]. A recent report of the successful treatment of a child with refractory, secondary CSS with the JAK 1/2 inhibitor, ruxolitinib [165], suggests that JAK inhibition might be another therapeutic avenue that merits further investigation.

Allogeneic HSCT

With improved outcomes likely associated with a less toxic conditioning regimen that includes alemtuzumab, allogeneic HSCT may be considered for severe, recurrent or refractory sJIA-CSS. Of 16 JIA patients recently reported to have allogeneic HSCT using this regimen, 5 had sJIA and CSS. One patient died and 4 were in clinical remission at the time of the report [144].

Treatment Summary

The key to successful treatment of sJIA-CSS is early recognition and diagnosis of CSS. The prompt administration of high dose systemic glucocorticoids, with or without high dose anakinra, often results in clinical improvement. If there is an inadequate response to treatment within 24–48 h, clinical and laboratory deterioration or severe multisystemic involvement, adding cyclosporine can be very effective. For sJIA-CSS refractory to these treatments, the clinician should consider etoposide or other treatments used for primary HLH. Further studies of novel, promising treatments with potentially less toxicity, particularly those targeting IFNγ and IL-18, will help to define their roles in treating sJIA-CSS.

References

1. Minoia, F., Davi, S., Horne, A., Demirkaya, E., Bovis, F., Li, C., et al. (2014). Clinical features, treatment, and outcome of macrophage activation syndrome complicating systemic juvenile idiopathic arthritis: A multinational, multicenter study of 362 patients. *Arthritis & Rhematology, 66*, 3160–3169.
2. Sawhney, S., Woo, P., & Murray, K. J. (2001). Macrophage activation syndrome: A potentially fatal complication of rheumatic disorders. *Archives of Disease in Childhood, 85*, 421–426.
3. Moradinejad, M. H., & Ziaee, V. (2011). The incidence of macrophage activation syndrome in children with rheumatic disorders. *Minerva Pediatrica, 63*, 459–466.
4. Bleesing, J., Prada, A., Siegel, D. M., Villanueva, J., Olson, J., Ilowite, N. T., et al. (2007). The diagnostic significance of soluble CD163 and soluble interleukin-2 receptor alpha-chain in macrophage activation syndrome and untreated new-onset systemic juvenile idiopathic arthritis. *Arthritis and Rheumatism, 56*, 965–971.
5. Behrens, E. M., Beukelman, T., Paessler, M., & Cron, R. Q. (2007). Occult macrophage activation syndrome in patients with systemic juvenile idiopathic arthritis. *The Journal of Rheumatology, 34*, 1133–1138.
6. Shimizu, M., Mizuta, M., Yasumi, T., Iwata, N., Okura, Y., Kinjo, N., et al. (2018). Validation of classification criteria of macrophage activation syndrome in Japanese patients with systemic juvenile idiopathic arthritis. *Arthritis Care and Research (Hoboken), 70*, 1412–1415.
7. Schulert, G. S., Minoia, F., Bohnsack, J., Cron, R. Q., Hashad, S., Kone-Paut, I., et al. (2018). Effect of biologic therapy on clinical and laboratory features of macrophage activation syndrome associated with systemic juvenile idiopathic arthritis. *Arthritis Care and Research (Hoboken), 70*, 409–419.
8. Kimura, Y., Weiss, J. E., Haroldson, K. L., Lee, T., Punaro, M., Oliveira, S., et al. (2013). Pulmonary hypertension and other potentially fatal pulmonary complications in systemic juvenile idiopathic arthritis. *Arthritis Care and Research (Hoboken), 65*, 745–752.
9. Petty, R. E., Southwood, T. R., Manners, P., Baum, J., Glass, D. N., Goldenberg, J., et al. (2004). International league of associations for rheumatology classification of juvenile idiopathic arthritis: Second revision, Edmonton, 2001. *The Journal of Rheumatology, 31*, 390–392.
10. Yamaguchi, M., Ohta, A., Tsunematsu, T., Kasukawa, R., Mizushima, Y., Kashiwagi, H., et al. (1992). Preliminary criteria for classification of adult Still's disease. *The Journal of Rheumatology, 19*, 424–430.
11. Nigrovic, P. A., Raychaudhuri, S., & Thompson, S. D. (2018). Review: Genetics and the classification of arthritis in adults and children. *Arthritis & Rhematology, 70*, 7–17.
12. Kumar, S., Kunhiraman, D. S., & Rajam, L. (2012). Application of the Yamaguchi criteria for classification of "suspected" systemic juvenile idiopathic arthritis (sJIA). *Pediatric Rheumatology Online Journal, 10*, 40.
13. DeWitt, E. M., Kimura, Y., Beukelman, T., Nigrovic, P. A., Onel, K., Prahalad, S., et al. (2012). Consensus treatment plans for new-onset systemic juvenile idiopathic arthritis. *Arthritis Care and Research (Hoboken), 64*, 1001–1010.
14. Borgia, R. E., Gerstein, M., Levy, D. M., Silverman, E. D., & Hiraki, L. T. (2018). Features, treatment, and outcomes of macrophage activation syndrome in childhood-onset systemic lupus erythematosus. *Arthritis & Rhematology, 70*, 616–624.

15. Wang, W., Gong, F., Zhu, W., Fu, S., & Zhang, Q. (2015). Macrophage activation syndrome in Kawasaki disease: More common than we thought? *Seminars in Arthritis and Rheumatism, 44*, 405–410.
16. Latino, G. A., Manlhiot, C., Yeung, R. S., Chahal, N., & McCrindle, B. W. (2010). Macrophage activation syndrome in the acute phase of Kawasaki disease. *Journal of Pediatric Hematology/Oncology, 32*, 527–531.
17. Behrens, E. M., Beukelman, T., Gallo, L., Spangler, J., Rosenkranz, M., Arkachaisri, T., et al. (2008). Evaluation of the presentation of systemic onset juvenile rheumatoid arthritis: Data from the Pennsylvania systemic onset juvenile arthritis registry (PASOJAR). *The Journal of Rheumatology, 35*, 343–348.
18. Machowicz, R., Janka, G., & Wiktor-Jedrzejczak, W. (2017). Similar but not the same: Differential diagnosis of HLH and sepsis. *Critical Reviews in Oncology/Hematology, 114*, 1–12.
19. Henter, J. I., Horne, A., Arico, M., Egeler, R. M., Filipovich, A. H., Imashuku, S., et al. (2007). HLH-2004: Diagnostic and therapeutic guidelines for hemophagocytic lymphohistiocytosis. *Pediatric Blood & Cancer, 48*, 124–131.
20. Ravelli, A., Magni-Manzoni, S., Pistorio, A., Besana, C., Foti, T., Ruperto, N., et al. (2005). Preliminary diagnostic guidelines for macrophage activation syndrome complicating systemic juvenile idiopathic arthritis. *The Journal of Pediatrics, 146*, 598–604.
21. Davi, S., Minoia, F., Pistorio, A., Horne, A., Consolaro, A., Rosina, S., et al. (2014). Performance of current guidelines for diagnosis of macrophage activation syndrome complicating systemic juvenile idiopathic arthritis. *Arthritis & Rhematology, 66*, 2871–2880.
22. Ravelli, A., Minoia, F., Davi, S., Horne, A., Bovis, F., Pistorio, A., et al. (2016). Classification criteria for macrophage activation syndrome complicating systemic juvenile idiopathic arthritis: A European league against rheumatism/American college of rheumatology/paediatric rheumatology international trials organisation collaborative initiative. *Annals of the Rheumatic Diseases, 75*, 481–489.
23. Ravelli, A., Minoia, F., Davi, S., Horne, A., Bovis, F., Pistorio, A., et al. (2016). Expert consensus on dynamics of laboratory tests for diagnosis of macrophage activation syndrome complicating systemic juvenile idiopathic arthritis. *RMD Open, 2*, e000161.
24. Allen, C. E., Yu, X., Kozinetz, C. A., & McClain, K. L. (2008). Highly elevated ferritin levels and the diagnosis of hemophagocytic lymphohistiocytosis. *Pediatric Blood & Cancer, 50*, 1227–1235.
25. Minoia, F., Bovis, F., Davi, S., Insalaco, A., Lehmberg, K., Shenoi, S., et al. (2017). Development and initial validation of the macrophage activation syndrome/primary hemophagocytic lymphohistiocytosis score, a diagnostic tool that differentiates primary hemophagocytic lymphohistiocytosis from macrophage activation syndrome. *The Journal of Pediatrics, 189*, 72–78 e73.
26. Grom, A. A., Horne, A., & De Benedetti, F. (2016). Macrophage activation syndrome in the era of biologic therapy. *Nature Reviews Rheumatology, 12*, 259–268.
27. Grom, A. A., Ilowite, N. T., Pascual, V., Brunner, H. I., Martini, A., Lovell, D., et al. (2016). Rate and clinical presentation of macrophage activation syndrome in patients with systemic juvenile idiopathic arthritis treated with canakinumab. *Arthritis & Rhematology, 68*, 218–228.
28. Nigrovic, P. A., Mannion, M., Prince, F. H., Zeft, A., Rabinovich, C. E., van Rossum, M. A., et al. (2011). Anakinra as first-line disease-modifying therapy in systemic juvenile idiopathic arthritis: Report of forty-six patients from an international multicenter series. *Arthritis and Rheumatism, 63*, 545–555.
29. Zeft, A., Hollister, R., LaFleur, B., Sampath, P., Soep, J., McNally, B., et al. (2009). Anakinra for systemic juvenile arthritis: The Rocky Mountain experience. *Journal of Clinical Rheumatology, 15*, 161–164.
30. Ravelli, A., Schneider, R., Weitzman, S., Devlin, C., Daimaru, K., Yokota, S., et al. (2014). Macrophage activation syndrome in patients with systemic juvenile idiopathic arthritistreated with tocilizumab. *Arthritis & Rhematology, 66*, S83–S84.

31. Minoia, F., Davi, S., Horne, A., Bovis, F., Demirkaya, E., Akikusa, J., et al. (2015). Dissecting the heterogeneity of macrophage activation syndrome complicating systemic juvenile idiopathic arthritis. *The Journal of Rheumatology, 42*, 994–1001.
32. Avcin, T., Tse, S. M., Schneider, R., Ngan, B., & Silverman, E. D. (2006). Macrophage activation syndrome as the presenting manifestation of rheumatic diseases in childhood. *The Journal of Pediatrics, 148*, 683–686.
33. Stephan, J. L., Kone-Paut, I., Galambrun, C., Mouy, R., Bader-Meunier, B., & Prieur, A. M. (2001). Reactive haemophagocytic syndrome in children with inflammatory disorders. A retrospective study of 24 patients. *Rheumatology (Oxford, England), 40*, 1285–1292.
34. Bracaglia, C., Prencipe, G., & De Benedetti, F. (2017). Macrophage activation syndrome: Different mechanisms leading to a one clinical syndrome. *Pediatric Rheumatology Online Journal, 15*, 5.
35. Hashemi-Sadraei, N., Vejpongsa, P., Baljevic, M., Chen, L., & Idowu, M. (2015). Epstein-barr virus-related hemophagocytic lymphohistiocytosis: Hematologic emergency in the critical care setting. *Case Reports in Hematology, 2015*, 491567.
36. Ravelli, A. (2002). Macrophage activation syndrome. *Current Opinion in Rheumatology, 14*, 548–552.
37. Ramanan, A. V., & Schneider, R. (2003). Macrophage activation syndrome following initiation of etanercept in a child with systemic onset juvenile rheumatoid arthritis. *The Journal of Rheumatology, 30*, 401–403.
38. Dinkla, S., van Eijk, L. T., Fuchs, B., Schiller, J., Joosten, I., Brock, R., et al. (2016). Inflammation-associated changes in lipid composition and the organization of the erythrocyte membrane. *BBA Clinical, 5*, 186–192.
39. Schaer, D. J., Schleiffenbaum, B., Kurrer, M., Imhof, A., Bachli, E., Fehr, J., et al. (2005). Soluble hemoglobin-haptoglobin scavenger receptor CD163 as a lineage-specific marker in the reactive hemophagocytic syndrome. *European Journal of Haematology, 74*, 6–10.
40. Mendoza-Coronel, E., & Ortega, E. (2017). Macrophage polarization modulates fcgammar- and cd13-mediated phagocytosis and reactive oxygen species production, independently of receptor membrane expression. *Frontiers in Immunology, 8*, 303.
41. Kell, D. B., & Pretorius, E. (2014). Serum ferritin is an important inflammatory disease marker, as it is mainly a leakage product from damaged cells. *Metallomics, 6*, 748–773.
42. Cohen, L. A., Gutierrez, L., Weiss, A., Leichtmann-Bardoogo, Y., Zhang, D. L., Crooks, D. R., et al. (2010). Serum ferritin is derived primarily from macrophages through a nonclassical secretory pathway. *Blood, 116*, 1574–1584.
43. Rosario, C., Zandman-Goddard, G., Meyron-Holtz, E. G., D'Cruz, D. P., & Shoenfeld, Y. (2013). The hyperferritinemic syndrome: Macrophage activation syndrome, Still's disease, septic shock and catastrophic antiphospholipid syndrome. *BMC Medicine, 11*, 185.
44. Kernan, K., & Carcillo, J. A. (2017). Hyperferritinemia and inflammation. *International Immunology, 29*, 401–409.
45. Moller, H. J., Aerts, H., Gronbaek, H., Peterslund, N. A., Hyltoft Petersen, P., Hornung, N., et al. (2002). Soluble CD163: A marker molecule for monocyte/macrophage activity in disease. *Scandinavian Journal of Clinical and Laboratory Investigation. Supplementum, 237*, 29–33.
46. Moller, H. J. (2012). Soluble CD163. *Scandinavian Journal of Clinical and Laboratory Investigation, 72*, 1–13.
47. Loscalzo, J. (1996). The macrophage and fibrinolysis. *Seminars in Thrombosis and Hemostasis, 22*, 503–506.
48. Kuriyama, T., Takenaka, K., Kohno, K., Yamauchi, T., Daitoku, S., Yoshimoto, G., et al. (2012). Engulfment of hematopoietic stem cells caused by down-regulation of CD47 is critical in the pathogenesis of hemophagocytic lymphohistiocytosis. *Blood, 120*, 4058–4067.
49. Boyman, O., & Sprent, J. (2012). The role of interleukin-2 during homeostasis and activation of the immune system. *Nature Reviews. Immunology, 12*, 180–190.

50. Buhelt, S., Ratzer, R. L., Christensen, J. R., Bornsen, L., Sellebjerg, F., & Sondergaard, H. B. (2017). Relationship between soluble CD25 and gene expression in healthy individuals and patients with multiple sclerosis. *Cytokine, 93*, 15–25.
51. Maier, L. M., Anderson, D. E., Severson, C. A., Baecher-Allan, C., Healy, B., Liu, D. V., et al. (2009). Soluble IL-2RA levels in multiple sclerosis subjects and the effect of soluble IL-2RA on immune responses. *Journal of Immunology (Baltimore, Md. : 1950), 182*, 1541–1547.
52. Janka, G. E., & Lehmberg, K. (2014). Hemophagocytic syndromes–an update. *Blood Reviews, 28*, 135–142.
53. Jordan, M. B., Hildeman, D., Kappler, J., & Marrack, P. (2004). An animal model of hemophagocytic lymphohistiocytosis (HLH): CD8+ T cells and interferon gamma are essential for the disorder. *Blood, 104*, 735–743.
54. Feldmann, J., Callebaut, I., Raposo, G., Certain, S., Bacq, D., Dumont, C., et al. (2003). Munc13-4 is essential for cytolytic granules fusion and is mutated in a form of familial hemophagocytic lymphohistiocytosis (FHL3). *Cell, 115*, 461–473.
55. Hazen, M. M., Woodward, A. L., Hofmann, I., Degar, B. A., Grom, A., Filipovich, A. H., et al. (2008). Mutations of the hemophagocytic lymphohistiocytosis-associated gene UNC13D in a patient with systemic juvenile idiopathic arthritis. *Arthritis and Rheumatism, 58*, 567–570.
56. Kaufman, K. M., Linghu, B., Szustakowski, J. D., Husami, A., Yang, F., Zhang, K., et al. (2014). Whole-exome sequencing reveals overlap between macrophage activation syndrome in systemic juvenile idiopathic arthritis and familial hemophagocytic lymphohistiocytosis. *Arthritis & Rhematology, 66*, 3486–3495.
57. Bracaglia, C., Sieni, E., Da Ros, M., De Fusco, C., Micalizzi, C., Cetica, V., et al. (2014). Mutations of familial hemophagocytic lymphohistiocytosis (FHL) related genes and abnormalities of cytotoxicity function tests in patients with macrophage activation syndrome (MAS) occurring in systemic juvenile idiopathic arthritis (sJIA). *Pediatric Rheumatology, 12*, P53.
58. Vastert, S. J., van Wijk, R., D'Urbano, L. E., de Vooght, K. M., de Jager, W., Ravelli, A., et al. (2010). Mutations in the perforin gene can be linked to macrophage activation syndrome in patients with systemic onset juvenile idiopathic arthritis. *Rheumatology (Oxford, England), 49*, 441–449.
59. Woo, S. S., James, D. J., & Martin, T. F. (2017). Munc13-4 functions as a Ca2+ sensor for homotypic secretory granule fusion to generate endosomal exocytic vacuoles. *Molecular Biology of the Cell, 28*, 792–808.
60. Zhang, K., Biroschak, J., Glass, D. N., Thompson, S. D., Finkel, T., Passo, M. H., et al. (2008). Macrophage activation syndrome in patients with systemic juvenile idiopathic arthritis is associated with MUNC13-4 polymorphisms. *Arthritis and Rheumatism, 58*, 2892–2896.
61. Zhang, M., Behrens, E. M., Atkinson, T. P., Shakoory, B., Grom, A. A., & Cron, R. Q. (2014). Genetic defects in cytolysis in macrophage activation syndrome. *Current Rheumatology Reports, 16*, 439.
62. Schulert, G. S., M, Z., Husami, A., Fall, N., Brunner, H., Zhang, K., et al. (2018). Novel UNC13D intronic variant disrupting a NFkB enhancer in a patient with recurrent macrophage activation syndrome and systemic juvenile idiopathic arthritis. *Arthritis and Rheumatism, 70*, 963–970.
63. Menager, M. M., Menasche, G., Romao, M., Knapnougel, P., Ho, C. H., Garfa, M., et al. (2007). Secretory cytotoxic granule maturation and exocytosis require the effector protein hMunc13-4. *Nature Immunology, 8*, 257–267.
64. Wood, S. M., Meeths, M., Chiang, S. C., Bechensteen, A. G., Boelens, J. J., Heilmann, C., et al. (2009). Different NK cell-activating receptors preferentially recruit Rab27a or Munc13-4 to perforin-containing granules for cytotoxicity. *Blood, 114*, 4117–4127.
65. Johnson, J. L., He, J., Ramadass, M., Pestonjamasp, K., Kiosses, W. B., Zhang, J., et al. (2016). Munc13-4 is a Rab11-binding protein that regulates rab11-positive vesicle trafficking and docking at the plasma membrane. *The Journal of Biological Chemistry, 291*, 3423–3438.

66. Marshall, M. R., Pattu, V., Halimani, M., Maier-Peuschel, M., Muller, M. L., Becherer, U., et al. (2015). VAMP8-dependent fusion of recycling endosomes with the plasma membrane facilitates T lymphocyte cytotoxicity. *The Journal of Cell Biology, 210*, 135–151.
67. Cote, M., Menager, M. M., Burgess, A., Mahlaoui, N., Picard, C., Schaffner, C., et al. (2009). Munc18-2 deficiency causes familial hemophagocytic lymphohistiocytosis type 5 and impairs cytotoxic granule exocytosis in patient NK cells. *The Journal of Clinical Investigation, 119*, 3765–3773.
68. Sepulveda, F. E., Burgess, A., Heiligenstein, X., Goudin, N., Menager, M. M., Romao, M., et al. (2015). LYST controls the biogenesis of the endosomal compartment required for secretory lysosome function. *Traffic (Copenhagen, Denmark), 16*, 191–203.
69. Faigle, W., Raposo, G., Tenza, D., Pinet, V., Vogt, A. B., Kropshofer, H., et al. (1998). Deficient peptide loading and MHC class II endosomal sorting in a human genetic immunodeficiency disease: The Chediak-Higashi syndrome. *The Journal of Cell Biology, 141*, 1121–1134.
70. Zhang, M., Bracaglia, C., Prencipe, G., Bemrich-Stolz, C. J., Beukelman, T., Dimmitt, R. A., et al. (2016). A heterozygous RAB27A mutation associated with delayed cytolytic granule polarization and hemophagocytic lymphohistiocytosis. *Journal of Immunology (Baltimore, Md. : 1950), 196*, 2492–2503.
71. Wulffraat, N. M., Rijkers, G. T., Elst, E., Brooimans, R., & Kuis, W. (2003). Reduced perforin expression in systemic juvenile idiopathic arthritis is restored by autologous stem-cell transplantation. *Rheumatology (Oxford, England), 42*, 375–379.
72. Grom, A. A., Villanueva, J., Lee, S., Goldmuntz, E. A., Passo, M. H., & Filipovich, A. (2003). Natural killer cell dysfunction in patients with systemic-onset juvenile rheumatoid arthritis and macrophage activation syndrome. *The Journal of Pediatrics, 142*, 292–296.
73. Villanueva, J., Lee, S., Giannini, E. H., Graham, T. B., Passo, M. H., Filipovich, A., et al. (2005). Natural killer cell dysfunction is a distinguishing feature of systemic onset juvenile rheumatoid arthritis and macrophage activation syndrome. *Arthritis Research & Therapy, 7*, R30–R37.
74. Put, K., Vandenhaute, J., Avau, A., van Nieuwenhuijze, A., Brisse, E., Dierckx, T., et al. (2017). Inflammatory gene expression profile and defective interferon-gamma and granzyme K in natural killer cells from systemic juvenile idiopathic arthritis patients. *Arthritis & Rhematology, 69*, 213–224.
75. Avau, A., Put, K., Wouters, C. H., & Matthys, P. (2015). Cytokine balance and cytokine-driven natural killer cell dysfunction in systemic juvenile idiopathic arthritis. *Cytokine & Growth Factor Reviews, 26*, 35–45.
76. Cifaldi, L., Prencipe, G., Caiello, I., Bracaglia, C., Locatelli, F., De Benedetti, F., et al. (2015). Inhibition of natural killer cell cytotoxicity by interleukin-6: Implications for the pathogenesis of macrophage activation syndrome. *Arthritis & Rhematology, 67*, 3037–3046.
77. Shibatomi, K., Ida, H., Yamasaki, S., Nakashima, T., Origuchi, T., Kawakami, A., et al. (2001). A novel role for interleukin-18 in human natural killer cell death: High serum levels and low natural killer cell numbers in patients with systemic autoimmune diseases. *Arthritis and Rheumatism, 44*, 884–892.
78. de Jager, W., Vastert, S. J., Beekman, J. M., Wulffraat, N. M., Kuis, W., Coffer, P. J., et al. (2009). Defective phosphorylation of interleukin-18 receptor beta causes impaired natural killer cell function in systemic-onset juvenile idiopathic arthritis. *Arthritis and Rheumatism, 60*, 2782–2793.
79. Billiau, A. D., Roskams, T., Van Damme-Lombaerts, R., Matthys, P., & Wouters, C. (2005). Macrophage activation syndrome: Characteristic findings on liver biopsy illustrating the key role of activated, IFN-gamma-producing lymphocytes and IL-6- and TNF-alpha-producing macrophages. *Blood, 105*, 1648–1651.
80. Rood, J. E., Rao, S., Paessler, M., Kreiger, P. A., Chu, N., Stelekati, E., et al. (2016). ST2 contributes to T-cell hyperactivation and fatal hemophagocytic lymphohistiocytosis in mice. *Blood, 127*, 426–435.

81. Tsoukas, P., & Canna, S. W. (2017). No shortcuts: New findings reinforce why nuance is the rule in genetic autoinflammatory syndromes. *Current Opinion in Rheumatology, 29*, 506–515.
82. De Benedetti, F., Brunner, H. I., Ruperto, N., Kenwright, A., Wright, S., Calvo, I., et al. (2012). Randomized trial of tocilizumab in systemic juvenile idiopathic arthritis. *The New England Journal of Medicine, 367*, 2385–2395.
83. Ruperto, N., Brunner, H. I., Quartier, P., Constantin, T., Wulffraat, N., Horneff, G., et al. (2012). Two randomized trials of canakinumab in systemic juvenile idiopathic arthritis. *The New England Journal of Medicine, 367*, 2396–2406.
84. de Jesus, A. A., Canna, S. W., Liu, Y., & Goldbach-Mansky, R. (2015). Molecular mechanisms in genetically defined autoinflammatory diseases: Disorders of amplified danger signaling. *Annual Review of Immunology, 33*, 823–874.
85. Bracaglia, C., de Graaf, K., Pires Marafon, D., Guilhot, F., Ferlin, W., Prencipe, G., et al. (2017). Elevated circulating levels of interferon-gamma and interferon-gamma-induced chemokines characterise patients with macrophage activation syndrome complicating systemic juvenile idiopathic arthritis. *Annals of the Rheumatic Diseases, 76*, 166–172.
86. Ravelli, A., Grom, A. A., Behrens, E. M., & Cron, R. Q. (2012). Macrophage activation syndrome as part of systemic juvenile idiopathic arthritis: Diagnosis, genetics, pathophysiology and treatment. *Genes and Immunity, 13*, 289–298.
87. Yokota, S., Itoh, Y., Morio, T., Sumitomo, N., Daimaru, K., & Minota, S. (2015). Macrophage activation syndrome in patients with systemic juvenile idiopathic arthritis under treatment with tocilizumab. *The Journal of Rheumatology, 42*, 712–722.
88. Hedrich, C. M., Bruck, N., Fiebig, B., & Gahr, M. (2012). Anakinra: A safe and effective first-line treatment in systemic onset juvenile idiopathic arthritis (SoJIA). *Rheumatology International, 32*, 3525–3530.
89. Dinarello, C. A. (2007). Interleukin-18 and the pathogenesis of inflammatory diseases. *Seminars in Nephrology, 27*, 98–114.
90. Dinarello, C. A., Novick, D., Kim, S., & Kaplanski, G. (2013). Interleukin-18 and IL-18 binding protein. *Frontiers in Immunology, 4*, 289.
91. Canna, S. W., Girard, C., Malle, L., de Jesus, A., Romberg, N., Kelsen, J., et al. (2017). Life-threatening NLRC4-associated hyperinflammation successfully treated with IL-18 inhibition. *The Journal of Allergy and Clinical Immunology, 139*, 1698–1701.
92. Shimizu, M., Nakagishi, Y., & Yachie, A. (2013). Distinct subsets of patients with systemic juvenile idiopathic arthritis based on their cytokine profiles. *Cytokine, 61*, 345–348.
93. Shimizu, M., Nakagishi, Y., Inoue, N., Mizuta, M., Ko, G., Saikawa, Y., et al. (2015). Interleukin-18 for predicting the development of macrophage activation syndrome in systemic juvenile idiopathic arthritis. *Clinical Immunology, 160*, 277–281.
94. Shimizu, M., Yokoyama, T., Yamada, K., Kaneda, H., Wada, H., Wada, T., et al. (2010). Distinct cytokine profiles of systemic-onset juvenile idiopathic arthritis-associated macrophage activation syndrome with particular emphasis on the role of interleukin-18 in its pathogenesis. *Rheumatology (Oxford, England), 49*, 1645–1653.
95. Maeno, N., Takei, S., Imanaka, H., Yamamoto, K., Kuriwaki, K., Kawano, Y., et al. (2004). Increased interleukin-18 expression in bone marrow of a patient with systemic juvenile idiopathic arthritis and unrecognized macrophage-activation syndrome. *Arthritis and Rheumatism, 50*, 1935–1938.
96. Weiss, E. S., Girard-Guyonvarc'h, C., Holzinger, D., de Jesus, A. A., Tariq, Z., Picarsic, J., et al. (2018). Interleukin-18 diagnostically distinguishes and pathogenically promotes human and murine macrophage activation syndrome. *Blood, 131*, 1442–1455.
97. Mazodier, K., Marin, V., Novick, D., Farnarier, C., Robitail, S., Schleinitz, N., et al. (2005). Severe imbalance of IL-18/IL-18BP in patients with secondary hemophagocytic syndrome. *Blood, 106*, 3483–3489.
98. Behrens, E. M., Canna, S. W., Slade, K., Rao, S., Kreiger, P. A., Paessler, M., et al. (2011). Repeated TLR9 stimulation results in macrophage activation syndrome-like disease in mice. *The Journal of Clinical Investigation, 121*, 2264–2277.

99. Prencipe, G., Caiello, I., Pascarella, A., Grom, A. A., Bracaglia, C., Chatel, L., et al. (2018). Neutralization of interferon-gamma reverts clinical and laboratory features in a mouse model of macrophage activation syndrome. *The Journal of Allergy and Clinical Immunology, 141*, 1439–1449.
100. Mouy, R., Stephan, J. L., Pillet, P., Haddad, E., Hubert, P., & Prieur, A. M. (1996). Efficacy of cyclosporine a in the treatment of macrophage activation syndrome in juvenile arthritis: Report of five cases. *The Journal of Pediatrics, 129*, 750–754.
101. Jenkins, M. R., Rudd-Schmidt, J. A., Lopez, J. A., Ramsbottom, K. M., Mannering, S. I., Andrews, D. M., et al. (2015). Failed CTL/NK cell killing and cytokine hypersecretion are directly linked through prolonged synapse time. *The Journal of Experimental Medicine, 212*, 307–317.
102. Schulert, G. S., & Grom, A. A. (2015). Pathogenesis of macrophage activation syndrome and potential for cytokine- directed therapies. *Annual Review of Medicine, 66*, 145–159.
103. Jordan, M. B., Locattelli, F., Allen, C., De Benedetti, F., Grom, A. A., Ballabio, M., et al. (2015). A novel targeted approach to the treatment of hemophagocytic lymphohistiocytosis (HLH) with an anti-interferon gamma (IFNγ) monoclonal antibody (mAb), NI-0501: First results from a pilot phase 2 study in children with primary HLH. *Blood, 126*, LBA–LB3.
104. Put, K., Avau, A., Brisse, E., Mitera, T., Put, S., Proost, P., et al. (2015). Cytokines in systemic juvenile idiopathic arthritis and haemophagocytic lymphohistiocytosis: Tipping the balance between interleukin-18 and interferon-gamma. *Rheumatology (Oxford, England), 54*, 1507–1517.
105. Kohanbash, G., Carrera, D. A., Shrivastav, S., Ahn, B. J., Jahan, N., Mazor, T., et al. (2017). Isocitrate dehydrogenase mutations suppress STAT1 and CD8+ T cell accumulation in gliomas. *The Journal of Clinical Investigation, 127*, 1425–1437.
106. Pachlopnik Schmid, J., Ho, C. H., Chretien, F., Lefebvre, J. M., Pivert, G., Kosco-Vilbois, M., et al. (2009). Neutralization of IFNgamma defeats haemophagocytosis in LCMV-infected perforin- and Rab27a-deficient mice. *EMBO Molecular Medicine, 1*, 112–124.
107. Canna, S. W., Wrobel, J., Chu, N., Kreiger, P. A., Paessler, M., & Behrens, E. M. (2013). Interferon-gamma mediates anemia but is dispensable for fulminant toll-like receptor 9-induced macrophage activation syndrome and hemophagocytosis in mice. *Arthritis and Rheumatism, 65*, 1764–1775.
108. Moller, J. C., Paul, D., Ganser, G., Range, U., Gahr, M., Kelsch, R., et al. (2010). IL10 promoter polymorphisms are associated with systemic onset juvenile idiopathic arthritis (SoJIA). *Clinical and Experimental Rheumatology, 28*, 912–918.
109. Sumegi, J., Barnes, M. G., Nestheide, S. V., Molleran-Lee, S., Villanueva, J., Zhang, K., et al. (2011). Gene expression profiling of peripheral blood mononuclear cells from children with active hemophagocytic lymphohistiocytosis. *Blood, 117*, e151–e160.
110. Hersh, A. O., & Prahalad, S. (2017). Genetics of juvenile idiopathic arthritis. *Rheumatic Diseases Clinics of North America, 43*, 435–448.
111. Perona-Wright, G., Mohrs, K., Szaba, F. M., Kummer, L. W., Madan, R., Karp, C. L., et al. (2009). Systemic but not local infections elicit immunosuppressive IL-10 production by natural killer cells. *Cell Host & Microbe, 6*, 503–512.
112. Lykken, J. M., Candando, K. M., & Tedder, T. F. (2015). Regulatory B10 cell development and function. *International Immunology, 27*, 471–477.
113. Tesi, B., Sieni, E., Neves, C., Romano, F., Cetica, V., Cordeiro, A. I., et al. (2015). Hemophagocytic lymphohistiocytosis in 2 patients with underlying IFN-gamma receptor deficiency. *The Journal of Allergy and Clinical Immunology, 135*, 1638–1641.
114. Avau, A., & Matthys, P. (2015). Therapeutic potential of interferon-gamma and its antagonists in autoinflammation: Lessons from murine models of systemic juvenile idiopathic arthritis and macrophage activation syndrome. *Pharmaceuticals (Basel, Switzerland), 8*, 793–815.
115. Avau, A., Mitera, T., Put, S., Put, K., Brisse, E., Filtjens, J., et al. (2014). Systemic juvenile idiopathic arthritis-like syndrome in mice following stimulation of the immune system with

Freund's complete adjuvant: Regulation by interferon-gamma. *Arthritis & Rhematology, 66*, 1340–1351.
116. Murray, P. J. (2017). Macrophage polarization. *Annual Review of Physiology, 79*, 541–566.
117. Ibarra, M. F., Klein-Gitelman, M., Morgan, E., Proytcheva, M., Sullivan, C., Morgan, G., et al. (2011). Serum neopterin levels as a diagnostic marker of hemophagocytic lymphohistiocytosis syndrome. *Clinical and Vaccine Immunology, 18*, 609–614.
118. Yanagimachi, M., Naruto, T., Miyamae, T., Hara, T., Kikuchi, M., Hara, R., et al. (2011). Association of IRF5 polymorphisms with susceptibility to macrophage activation syndrome in patients with juvenile idiopathic arthritis. *The Journal of Rheumatology, 38*, 769–774.
119. Krausgruber, T., Blazek, K., Smallie, T., Alzabin, S., Lockstone, H., Sahgal, N., et al. (2011). IRF5 promotes inflammatory macrophage polarization and TH1-TH17 responses. *Nature Immunology, 12*, 231–238.
120. Zoller, E. E., Lykens, J. E., Terrell, C. E., Aliberti, J., Filipovich, A. H., Henson, P. M., et al. (2011). Hemophagocytosis causes a consumptive anemia of inflammation. *The Journal of Experimental Medicine, 208*, 1203–1214.
121. Behrens, E. M. (2008). Macrophage activation syndrome in rheumatic disease: What is the role of the antigen presenting cell? *Autoimmunity Reviews, 7*, 305–308.
122. Fall, N., Barnes, M., Thornton, S., Luyrink, L., Olson, J., Ilowite, N. T., et al. (2007). Gene expression profiling of peripheral blood from patients with untreated new-onset systemic juvenile idiopathic arthritis reveals molecular heterogeneity that may predict macrophage activation syndrome. *Arthritis and Rheumatism, 56*, 3793–3804.
123. Schulert, G. S., Fall, N., Harley, J. B., Shen, N., Lovell, D. J., Thornton, S., et al. (2016). Monocyte microRNA expression in active systemic juvenile idiopathic arthritis implicates MicroRNA-125a-5p in polarized monocyte phenotypes. *Arthritis & Rhematology, 68*, 2300–2313.
124. Macaubas, C., Nguyen, K. D., Peck, A., Buckingham, J., Deshpande, C., Wong, E., et al. (2012). Alternative activation in systemic juvenile idiopathic arthritis monocytes. *Clinical Immunology, 142*, 362–372.
125. Shimizu, M., & Yachie, A. (2012). Compensated inflammation in systemic juvenile idiopathic arthritis: Role of alternatively activated macrophages. *Cytokine, 60*, 226–232.
126. Martinez, F. O., Sica, A., Mantovani, A., & Locati, M. (2008). Macrophage activation and polarization. *Frontiers in Bioscience, 13*, 453–461.
127. Kelly, A., & Ramanan, A. V. (2008). A case of macrophage activation syndrome successfully treated with anakinra. *Nature Clinical Practice. Rheumatology, 4*, 615–620.
128. Bruck, N., Suttorp, M., Kabus, M., Heubner, G., Gahr, M., & Pessler, F. (2011). Rapid and sustained remission of systemic juvenile idiopathic arthritis-associated macrophage activation syndrome through treatment with anakinra and corticosteroids. *Journal of Clinical Rheumatology, 17*, 23–27.
129. Miettunen, P. M., Narendran, A., Jayanthan, A., Behrens, E. M., & Cron, R. Q. (2011). Successful treatment of severe paediatric rheumatic disease-associated macrophage activation syndrome with interleukin-1 inhibition following conventional immunosuppressive therapy: Case series with 12 patients. *Rheumatology (Oxford, England), 50*, 417–419.
130. Durand, M., Troyanov, Y., Laflamme, P., & Gregoire, G. (2010). Macrophage activation syndrome treated with anakinra. *The Journal of Rheumatology, 37*, 879–880.
131. Kahn, P. J., & Cron, R. Q. (2013). Higher-dose Anakinra is effective in a case of medically refractory macrophage activation syndrome. *The Journal of Rheumatology, 40*, 743–744.
132. Parisi, F., Paglionico, A., Varriano, V., Ferraccioli, G., & Gremese, E. (2017). Refractory adult-onset still disease complicated by macrophage activation syndrome and acute myocarditis: A case report treated with high doses (8 mg/kg/d) of anakinra. *Medicine, 96*, e6656.
133. Fisher Jr., C. J., Dhainaut, J. F., Opal, S. M., Pribble, J. P., Balk, R. A., Slotman, G. J., et al. (1994). Recombinant human interleukin 1 receptor antagonist in the treatment of patients with sepsis syndrome. Results from a randomized, double-blind, placebo-controlled trial. Phase III rhIL-1ra sepsis syndrome study group. *JAMA, 271*, 1836–1843.

134. Shakoory, B., Carcillo, J.A., Chatham, W.W., Amdur, R.L., Zhao. H., Dinarello, C. A., et. al. (2016). Interleukin-1 receptor blockade is associated with reduced mortality in sepsis patients with features of macrophage activation syndrome: Reanalysis of a prior phase III trial. *Critical Care Medicine, 44*, 275–281.
135. Stephan, J. L., Zeller, J., Hubert, P., Herbelin, C., Dayer, J. M., & Prieur, A. M. (1993). Macrophage activation syndrome and rheumatic disease in childhood: A report of four new cases. *Clinical and Experimental Rheumatology, 11*, 451–456.
136. Ravelli, A., De Benedetti, F., Viola, S., & Martini, A. (1996). Macrophage activation syndrome in systemic juvenile rheumatoid arthritis successfully treated with cyclosporine. *The Journal of Pediatrics, 128*, 275–278.
137. Quesnel, B., Catteau, B., Aznar, V., Bauters, F., & Fenaux, P. (1997). Successful treatment of juvenile rheumatoid arthritis associated haemophagocytic syndrome by cyclosporin a with transient exacerbation by conventional-dose G-CSF. *British Journal of Haematology, 97*, 508–510.
138. Gupta, A. A., Tyrrell, P., Valani, R., Benseler, S., Abdelhaleem, M., & Weitzman, S. (2009). Experience with hemophagocytic lymphohistiocytosis/macrophage activation syndrome at a single institution. *Journal of Pediatric Hematology/Oncology, 31*, 81–84.
139. Sung, L., King, S. M., Carcao, M., Trebo, M., & Weitzman, S. S. (2002). Adverse outcomes in primary hemophagocytic lymphohistiocytosis. *Journal of Pediatric Hematology/Oncology, 24*, 550–554.
140. Palmblad, K., Schierbeck, H., Sundberg, E., Horne, A. C., Harris, H. E., Henter, J. I., et al. (2015). High systemic levels of the cytokine-inducing HMGB1 isoform secreted in severe macrophage activation syndrome. *Molecular Medicine, 20*, 538–547.
141. Ehl, S., Astigarraga, I., von Bahr Greenwood, T., Hines, M., Horne, A., Ishii, E., et al. (2018). Recommendations for the use of etoposide-based therapy and bone marrow transplantation for the treatment of HLH: Consensus statements by the HLH steering committee of the histiocyte society. *The Journal of Allergy and Clinical Immunology. In Practice, 6*, 1508–1517.
142. Coca, A., Bundy, K. W., Marston, B., Huggins, J., & Looney, R. J. (2009). Macrophage activation syndrome: Serological markers and treatment with anti-thymocyte globulin. *Clinical Immunology, 132*, 10–18.
143. Marsh, R. A., Allen, C. E., McClain, K. L., Weinstein, J. L., Kanter, J., Skiles, J., et al. (2013). Salvage therapy of refractory hemophagocytic lymphohistiocytosis with alemtuzumab. *Pediatric Blood & Cancer, 60*, 101–109.
144. Silva, J. M. F., Ladomenou, F., Carpenter, B., Chandra, S., Sedlacek, P., Formankova, R., et al. (2018). Allogeneic hematopoietic stem cell transplantation for severe, refractory juvenile idiopathic arthritis. *Blood Advances, 2*, 777–786.
145. Bosnak, M., Erdogan, S., Aktekin, E. H., & Bay, A. (2016). Therapeutic plasma exchange in primary hemophagocytic lymphohistiocytosis: Reports of two cases and a review of the literature. *Transfusion and Apheresis Science, 55*, 353–356.
146. Nusshag, C., Morath, C., Zeier, M., Weigand, M. A., Merle, U., & Brenner, T. (2017). Hemophagocytic lymphohistiocytosis in an adult kidney transplant recipient successfully treated by plasmapheresis: A case report and review of the literature. *Medicine, 96*, e9283.
147. Nakakura, H., Ashida, A., Matsumura, H., Murata, T., Nagatoya, K., Shibahara, N., et al. (2009). A case report of successful treatment with plasma exchange for hemophagocytic syndrome associated with severe systemic juvenile idiopathic arthritis in an infant girl. *Therapeutic Apheresis and Dialysis, 13*, 71–76.
148. Demirkol, D., Yildizdas, D., Bayrakci, B., Karapinar, B., Kendirli, T., Koroglu, T. F., et al. (2012). Hyperferritinemia in the critically ill child with secondary hemophagocytic lymphohistiocytosis/sepsis/multiple organ dysfunction syndrome/macrophage activation syndrome: What is the treatment? *Critical Care, 16*, R52.
149. Makay, B., Yilmaz, S., Turkyilmaz, Z., Unal, N., Oren, H., & Unsal, E. (2008). Etanercept for therapy-resistant macrophage activation syndrome. *Pediatric Blood & Cancer, 50*, 419–421.
150. Prahalad, S., Bove, K. E., Dickens, D., Lovell, D. J., & Grom, A. A. (2001). Etanercept in the treatment of macrophage activation syndrome. *The Journal of Rheumatology, 28*, 2120–2124.

151. Aeberli, D., Oertle, S., Mauron, H., Reichenbach, S., Jordi, B., & Villiger, P. M. (2002). Inhibition of the TNF-pathway: Use of infliximab and etanercept as remission-inducing agents in cases of therapy-resistant chronic inflammatory disorders. *Swiss Medical Weekly, 132*, 414–422.
152. Stern, A., Riley, R., & Buckley, L. (2001). Worsening of macrophage activation syndrome in a patient with adult onset Still's disease after initiation of etanercept therapy. *Journal of Clinical Rheumatology, 7*, 252–256.
153. Maude, S. L., Barrett, D., Teachey, D. T., & Grupp, S. A. (2014). Managing cytokine release syndrome associated with novel T cell-engaging therapies. *Cancer Journal, 20*, 119–122.
154. Savage, E., Wazir, T., Drake, M., Cuthbert, R., & Wright, G. (2014). Fulminant myocarditis and macrophage activation syndrome secondary to adult-onset Still's disease successfully treated with tocilizumab. *Rheumatology (Oxford, England), 53*, 1352–1353.
155. Kobayashi, D., Ito, S., Murasawa, A., Narita, I., & Nakazono, K. (2015). Two cases of adult-onset Still's disease treated with tocilizumab that achieved tocilizumab-free remission. *Internal Medicine, 54*, 2675–2679.
156. Watanabe, E., Sugawara, H., Yamashita, T., Ishii, A., Oda, A., & Terai, C. (2016). Successful tocilizumab therapy for macrophage activation syndrome associated with adult-onset still's disease: A case-based review. *Case Reports in Medicine, 2016*, 5656320.
157. Chellapandian, D., Das, R., Zelley, K., Wiener, S. J., Zhao, H., Teachey, D. T., et al. (2013). Treatment of Epstein Barr virus-induced haemophagocytic lymphohistiocytosis with rituximab-containing chemo-immunotherapeutic regimens. *British Journal of Haematology, 162*, 376–382.
158. Stefanou, C., Tzortzi, C., Georgiou, F., & Timiliotou, C. (2016). Combining an antiviral with rituximab in EBV-related haemophagocytic lymphohistiocytosis led to rapid viral clearance; and a comprehensive review. *BMJ Case Reports, 2016*, bcr2016216488.
159. Record, J. L., Beukelman, T., & Cron, R. Q. (2011). Combination therapy of abatacept and anakinra in children with refractory systemic juvenile idiopathic arthritis: A retrospective case series. *The Journal of Rheumatology, 38*, 180–181.
160. Yasin, S., & Schulert, G. S. (2018). Systemic juvenile idiopathic arthritis and macrophage activation syndrome: Update on pathogenesis and treatment. *Current Opinion in Rheumatology, 30*, 514–520.
161. Girard-Guyonvarc'h, C., Palomo, J., Martin, P., Rodriguez, E., Troccaz, S., Palmer, G., et al. (2018). Unopposed IL-18 signaling leads to severe TLR9-induced macrophage activation syndrome in mice. *Blood, 131*, 1430–1441.
162. Chiossone, L., Audonnet, S., Chetaille, B., Chasson, L., Farnarier, C., Berda-Haddad, Y., et al. (2012). Protection from inflammatory organ damage in a murine model of hemophagocytic lymphohistiocytosis using treatment with IL-18 binding protein. *Frontiers in Immunology, 3*, 239.
163. Gabay, C., Fautrel, B., Rech, J., Spertini, F., Feist, E., Kotter, I., et al. (2018). Open-label, multicentre, dose-escalating phase II clinical trial on the safety and efficacy of tadekinig alfa (IL-18BP) in adult-onset Still's disease. *Annals of the Rheumatic Diseases, 77*, 840–847.
164. Das, R., Guan, P., Sprague, L., Verbist, K., Tedrick, P., An, Q. A., et al. (2016). Janus kinase inhibition lessens inflammation and ameliorates disease in murine models of hemophagocytic lymphohistiocytosis. *Blood, 127*, 1666–1675.
165. Broglie, L., Pommert, L., Rao, S., Thakar, M., Phelan, R., Margolis, D., et al. (2017). Ruxolitinib for treatment of refractory hemophagocytic lymphohistiocytosis. *Blood Advances, 1*, 1533–1536.

Systemic Lupus Erythematosus and Cytokine Storm

Roberto Caricchio

Introduction

The concept of autoimmune-associated hemophagocytic syndrome in Systemic Lupus Erythematosus (SLE) was first introduced in the early 1990s, when six cases of hemophagocytosis were described as complication of SLE [1, 2]. SLE is the prototype of autoimmune diseases and can manifest with a plethora of clinical signs and symptoms associated with a myriad of laboratory abnormalities [3]. SLE is characterized by the production of autoantibodies mainly directed toward nuclear components, named anti-nuclear antibodies (ANA) [4]. It primarily affects females with a 10:1 female-to-male ratio, and with a US incidence and prevalence of ~73 and ~5 per 100,000 persons, respectively [5]. Genetics, sex, and environment are the three major factors that influence the development of SLE, that is, the "lupus troika" [6, 7]. The last two decades have witnessed tremendous advancement in both research and treatment; nevertheless, lupus remains a disease of unknown etiology [8]. An infrequent but potentially lethal complication of SLE is macrophage activation syndrome (MAS). MAS is generally considered part of the spectrum of hemophagocytic lymphohystiocytosis (HLH), and it can potentially occur in any autoimmune condition although adult-onset Still disease (AOSD) and SLE are the most frequent [9]. The diagnosis of MAS in SLE can be very challenging due to similarities in presentation of both flares and infections [10, 11]. And while MAS classification criteria have been used and in part validated for pediatric SLE [12], there is no consensus for adult SLE, leaving the latter at increased risk of poor outcome. In this chapter, we discuss several aspects of MAS in the context of SLE. In particular, we discuss the pathogenesis of MAS in SLE, how MAS presents in pediatric versus adult SLE (Table 1), and, finally, MAS treatment in SLE and future directions.

R. Caricchio, MD (✉)
Lewis Katz School of Medicine, Temple University, Philadelphia, PA, USA
e-mail: roc@temple.edu

R. Q. Cron, E. M. Behrens (eds.), *Cytokine Storm Syndrome*,
https://doi.org/10.1007/978-3-030-22094-5_22

Table 1 Clinical characteristics of MAS in SLE

Variables	cSLE	aSLE
Fever	+++	+++
Cytopenias	+++	+++
Hyperferritinemia	+++	+++
Increased liver function tests	+++	+++
Acute pancreatitis	+++	+
With lupus onset	+++	+
With lupus disease flare	+	+++
With infections	++	+
Recurrence	−	+++
Ravelli lupus MAS criteria	+++	++
HLH-2004 criteria	+	++
2006 sJIA criteria	+++	++
High dose corticosteroids	+++	+++
Anakinra/cyclosporine	+++	+
IVIg	+++	+
Cyclophosphamide	+	+++

SLE and MAS Pathogenesis

Hemophagocytic lymphohystiocytosis (HLH) is a spectrum of hyperinflammatory conditions which can be inherited or acquired [13]. When HLH occurs in the context of autoimmune conditions, it is termed macrophage activation syndrome, or MAS [14]. Patients with SLE are at significantly increased risk of developing MAS, in both children and adults [13]. An aggravating factor for MAS in lupus is that the two conditions share several manifestations such as fever, lymphadenopathy, splenomegaly, and cytopenias [15]. Very often, it is difficult to distinguish them; indeed, lupus flares can resemble MAS. Hence, early diagnosis is extremely important, as MAS in lupus can be fatal [15].

The pathogenesis of SLE is considered multifactorial [16]. The major players are autoreactive B and T cells along with dysregulated dendritic cells and cytokines, such as interferon alpha [7, 17, 18]. A paramount feature is the production of autoantibodies directed toward nuclear components (ANA), but the true cause of the so-called "horror autotoxicus" (the horror of self-toxicity) remains elusive [4, 19]. Genetic studies in the last two decades have pointed to an extraordinary number of single nucleotide polymorphisms in three major areas of the immune system: dysregulation of T and B cells signaling, delayed immune complexes clearance, and intrinsic activation of the Innate immune system [20]. The tremendous knowledge acquired with these studies has allowed for specifically targeting T and B cells functions and even the innate immune activation of interferon alpha, the latter considered a major player in the activation of the autoimmune process [8, 21]. Nevertheless, most large international trials have been quite disappointing and have not met their primary end point, demonstrating how challenging the treatment for this complex disease can be [22].

Although the genetic mutations of familial (*f*)HLH generally do not always occur in SLE [23], both MAS and SLE share a number of immune molecular pathways. This is perhaps why lupus is the autoimmune disease in which MAS most commonly occurs.

MAS pathogenesis is in part considered driven by a series of events that involve Toll Like Receptor (TLR) 9 engagement by CpGs, production of interferon gamma which via JAK1 and JAK2 and STAT1 phosphorylation activate the transcription of interferon-stimulated genes [24]. These gene products eventually culminate in the ignition of the "cytokine storm" characterized by the unusually high production of proinflammatory cytokines, such as IL-6, TNF, IL-12, and IL-1 (please see previous chapters for more extensive and comprehensive description of MAS pathogenesis). Interestingly, TLR triggering by CpGs is thought to be an important activator of the lupus autoimmune system [25], and viral infections have been one of the first environmental factors identified as capable of predisposing to lupus or inducing flares [26]. Attempts at blocking TLR9 in clinical trials have yet to demonstrate efficacy. Nevertheless, hydroxychloroquine, which is widely used in SLE, is thought to reduce TLR activation [27], and recently has been shown to decreases the incidence of MAS in SLE [28]. Another common pathway of activation and actively targeted in SLE is JAK1/2 phosphorylation [29]. The latter is specifically important in the signaling of cytokines critical in the pathogenesis of SLE such as interferon alpha which would in turn activate production of IL-6, but also IL-12 and IL-23 [30, 31]. A phase II clinical trial with baricitinib, a JAK1/2 inhibitor, and a phase II clinical trial with ustekinumab, a monoclonal antibody targeting IL-12/IL-23, have demonstrated promising efficacy in moderate lupus and are undergoing phase III clinical trials [32, 33]. Finally, both IL-10 and IL-6 are considered important participants to the inflammatory damage and B cell activation in lupus, and are currently targeted in Phase I and II clinical trials to demonstrate that blocking their activity would also impair the autoimmune response in lupus [21]. It will be interesting to determine if they will show efficacy in lupus disease activity but also decrease the occurrence of MAS as shown by the use of hydroxychloroquine [28]. For many commonalities among the two pathogeneses, SLE and MAS also differ. For example, IL-1 production is considered a fundamental step in activating the cytokine storm [23]. Indeed, anakinra, an IL-1 receptor antagonist (IL-1ra), has shown dramatic therapeutic effects [34, 35]. Surprisingly, canakinumab, another IL1ra, although has shown important effects on sJIA, it did not prevent MAS [36]. Nevertheless, IL-1 tends not to increase during lupus flares and IL-1ra has yielded conflicting results [37–39]. Hence, IL-1ra therapy in lupus has yet to be fully tested [34, 40–43]. Finally, NK cells and cytotoxic CD8 T cell dysregulation are both important in the pathogenesis of MAS, while a definitive role in SLE has yet to be demonstrated [24].

MAS in Childhood-Onset SLE

Lupus disease is a complex autoimmune condition which can occur in both children and adults. Childhood-onset SLE (cSLE) is rare in infants but its incidence and prevalence in children is estimated to be 2.22 and 9.73 per 100,000 respectively, which is ten time less than the adult-onset (a)SLE [44–46]. Interestingly the female-to-male ratio in cSLE is lower in young children but approaches the adult ratio by teenage onset [44–46]. The difference is thought to be due to the hormonal environment in the adult life which would predispose females to lupus [44–46]. Although, a definitive demonstration has yet to be achieved. Major clinical differences between cSLE and aSLE are the increased incidence of neurological and renal organ involvement and the higher use of steroids in the former [47, 48]. Nevertheless, large cSLE cohorts have not demonstrated a significant increase in accrual damage in cSLE when compared to aSLE. One worrisome exception is the significant increased mortality in cSLE during renal-replacement therapy [49]. Finally, mid- to long-term studies in cSLE are lacking, primarily due to the scarcity of dedicated lupus adolescent transition clinics [50]. Some of the clinical differences in cSLE compared to aSLE also account for the increased difficulties in diagnosing MAS in cSLE. One of the most challenging aspects of MAS in cSLE is the capacity to discriminate lupus flares, and/or infections, from the occurrence of MAS. And although aSLE is a borderline rare disease, the prevalence of MAS in cSLE is estimated to be up to 9%, and mortality near 20–25%, hence a particularly dangerous and troublesome manifestation [51].

The latest MAS classification criteria for Systemic juvenile idiopathic arthritis (sJIA) were developed in 2016 and have not been validated for cSLE [52]. A major issue for cSLE is that the disease itself is characterized by cytopenias even during periods of remission, rendering the classification criteria particularly difficult to apply [15]. In 2009, Parodi and colleagues developed the preliminary diagnostic guideline specifically for MAS as complication of cSLE [12]. These criteria yielded a sensitivity of 92.1% and a specificity of 90.9%. A recently published large cSLE cohort from a single pediatric tertiary center reported 38 patients with concomitant MAS [53]. In this particular cohort, all cSLE patients diagnosed with MAS met the preliminary diagnostic criteria providing the support for further validation [53]. Multiple studies have reported that cSLE patients do not carry gene mutations in known fHLH genes involved in the granule-dependent cytolytic secretory pathway and rarely have heterozygosity of variants of uncertain significance [53]. These findings, albeit limited by the lack of systematic gene sequencing in cohorts of cSLE patients, suggest the multifactorial nature of MAS in cSLE and that the former is not driven by the same genetic predispositions.

A consistent observation across multiple case series and large cSLE cohorts is that the vast majority of MAS cases are diagnosed concomitantly with the onset of cSLE, interestingly when the lupus disease is the most active [12, 42, 51, 53, 54]. Since cSLE and MAS may not share genetic predisposition, one can speculate that the activation of the innate immune system in newly diagnosed cSLE, where the overactivation of type I IFN was first described [55], is a potential driving force, possibly induced by yet unknown viral infections that, along with the lupus disease activity, could facilitate the occurrence of MAS. It is therefore predictable, based on common observations among published cohorts, and contrary to sJIA, that MAS in cSLE almost never reoccurs [12, 42, 51, 53, 54], as to indicate that by controlling lupus activity, MAS can also be controlled. Indeed, the minority of MAS diagnosed after lupus onset, occur during lupus flares [12, 42, 51, 53, 54] and Table 1.

cSLE with or without MAS do not seem to be different in terms of clinical manifestations and laboratory findings, with the exception in some cohorts of increased hematological manifestations, such as lymphopenias, lymphadenopathy, and increased renal disease [12]. One common and worrisome observation is that MAS in cSLE tends to be associated with higher incidence of CNS involvement [53]. The latter is associated with increased incidence of multiorgan failure and unfortunately increased mortality [53]. Indeed, among the autoimmune diseases, MAS in cSLE has the highest mortality rate [54, 56, 57]. The CNS involvement is particularly problematic as it occurs as part of lupus manifestations but also MAS, therefor prompting significant immune suppression regardless of the cause [58].

One of the most challenging aspects of cSLE-associated MAS is to differentiate it from a lupus flare and/or an infection [15]. Hepato-splenomegaly, hemorrhages, and CNS dysfunction are able to discriminate MAS from lupus active disease but not in all published cohorts [12, 15, 42, 51, 53, 54]. An interesting retrospective cohort study reported that the incidence of acute pancreatitis (AP) in cSLE with MAS was significantly higher than without MAS (55% and 0.6%, respectively) [51]. The presence of AP was also associated with increased lupus disease activity; the authors concluded that the screening for MAS should be initiated when a lupus patient presents with AP [51] and Table 1. A number of retrospective cohort studies have reported that MAS in cSLE very often presents with clinical findings such as non-remitting fever and cytopenia (often of more than one lineage), along with elevated ferritin and liver enzymes levels [12, 15, 42, 51, 53, 54]. Therefore, when a cSLE patients presents with a similar clinical picture, the investigation of MAS should be considered. Interestingly, a comparison between sJIA and cSLE associated MAS, demonstrated a delay in obtaining ferritin levels in cSLE patients, suggesting perhaps that MAS is yet to be fully part of the differential diagnosis in active cSLE [54]. Another helpful laboratory marker is the level of C-reactive protein (CRP), which in cSLE generally does not significantly increase during flares but does so during infections [10, 59, 60]. It has been found that in cSLE-associated MAS, CRP is significantly elevated. Nevertheless, the infectious work-up is mandatory, which includes a bone marrow aspirate to rule out Leishmaniasis, a notoriously strong trigger of MAS [61]. Interestingly, the absence of macrophage hemophagocytosis in the bone marrow should not rule out the diagnosis of cSLE-associate MAS, as it is only found in a minority of patients [12, 15, 42, 51, 53, 54].

MAS in Adult-Onset SLE

Lupus is a systemic autoimmune disease with a US prevalence of 20–150 and incidence of 1–25 per 100,000 individuals [5, 46]. Although lupus can manifest in virtually any organ of the body and with a myriad of autoantibody specificities, most adults with lupus at a given time have a mild to moderate disease, possibly characterized by arthritis/arthralgia and cutaneous rash [3]. Nevertheless, the course of the disease can greatly vary, and it is often unpredictable; indeed, there is no biomarker able to anticipate a lupus flare, or more importantly, its severity [62, 63]. Interestingly, lupus is one of the most frequently reported autoimmune conditions associated with MAS [64]. In adult-onset SLE (aSLE), MAS has been reported with a frequency of up to 5%, but the accuracy is still unclear and the consensus is that lupus-associated MAS is significantly underdiagnosed [15, 34]. MAS mortality in aSLE is dramatic with reported fatalities of up to 50% [65]. Race plays an important role in the pathogenesis of SLE; in particular, African-Americans in the USA have three- to fivefold increased risk of developing lupus [46]. Although race and the incidence of SLE-associated MAS has not been investigated due to the design of the studies, it is interesting that in a recent retrospective cohort of a large academic center that spanned over 45 years, half of the cases where identified as African-Americans, perhaps suggesting that the known increased severity of disease in African-Americans can also increase the risk of MAS [28]. An important characteristic of lupus-associated MAS is that it often presents during severe flares or with the very onset of disease [66] (Table 1); these common findings among adult cohorts in the USA, Europe, and Asia suggest that while active lupus autoimmune mechanisms can trigger it, when they do, it becomes very challenging to promptly diagnose MAS because it can be confused with the flare itself [28, 34, 64–67]. On the other hand, when a patient presents with MAS, the diagnosis of infections and/or malignancies must be excluded, but SLE should also be taken into consideration. To complicate the MAS picture in SLE, in retrospective cohorts it was found to be frequently triggered by infections, mostly viral [28, 34, 64–67] (Table 1).

There have been several attempts to develop diagnostic criteria based on the validated ones for sJIA [51]. Parodi and colleagues in 2009 developed preliminary criteria for cSLE, the "Ravelli" criteria [12], which have been subsequently validated in other cSLE cohorts with high sensitivity and specificity and across different ethnicities [28]. In aSLE, the "Ravelli" and the HLH-2004 criteria have shown mixed results, where the first has yielded good while the latter inconsistent results [28, 67] (Table 1); generally these criteria performed best when they were applied to hospitalized patients, hence either during very severe manifestations or infections, and especially for predicting in-hospital mortality [34, 64, 67]. In these circumstances, often both criteria were met with 100% success rate. Nevertheless, almost invariably, the cohorts were retrospective [34]. Often patients were selected based on very high levels of ferritin, which might have biased the selection of patients toward the criteria itself. Therefore, at present, the diagnosis of MAS in aSLE remains a combination of clinical judgment and supportive criteria.

Because of the severity of MAS, lupus patients in cohorts are identified from hospital admissions. Therefore, lupus manifestations may vary from the outpatient settings. There is consensus in the literature that if a lupus patient presents on hospital admission with unexplained high fevers, cytopenia of at least two cell lines, and elevation of liver function tests (LFTs), then the determination of ferritin levels and the screening for MAS should not be delayed [12, 15, 42, 51, 53, 54]. In this setting, almost invariably, the levels of ferritin are very high. Interestingly, in two Asian cohorts (from China and Korea), thrombocytopenia discriminated SLE-associated MAS from SLE flares [64, 65] (Table 1). The screening should be preferably done with the "Ravelli" or the 2016 sJIA criteria [28, 34, 64–67]. Importantly, a negative bone marrow aspirate for hemophagocytosis should not rule out the diagnosis of MAS, as this is not present in all the patients evaluated [34, 66, 68]. Of note is that MAS has been shown to be much harder to diagnose when it develops after hospital admission, perhaps because the focus is mostly on controlling lupus activity or on infections due to the immune suppression. Early diagnosis and intervention are paramount to mitigate the often poor outcome of MAS [67, 69]. In fact, delayed diagnosis dramatically increases the risk of ICU admission and mortality irrespective of the presence of SLE flare [64]. Among reported cases and large cohorts from the USA, Europe, and Asia, there are few commonalities and limited patterns as to how SLE presents associated with MAS. French and other European cohorts have reported higher frequencies of cutaneous vasculitis, polyarthritis, pericarditis, and renal disease at onset and when MAS was diagnosed [34, 66, 67]. Importantly, in European cohorts and contrary to cSLE [53], MAS in aSLE could reoccur, especially in a subset of patients who tend to have a more severe lupus with multiple flares and prolonged immunosuppression [34, 66] (Table 1). Korean and Chinese cohorts have reported no particular lupus manifestation associated with MAS onset or occurrence, except for increased renal involvement [64, 65]. Finally, a recent large US cohort reported an increased frequency of lupus nephritis flares at presentation of MAS, but the cohort had 70% overall frequency of renal disease [28]. It is important to note that the SLE disease activity index (SLEDAI), a measure commonly used in investigational lupus studies [70], although at times elevated, was not predictive of in-hospital mortality in most cohorts [47, 51, 57, 64]. This suggests that although MAS occurs in a subsets of patients with more aggressive lupus, MAS is the driving force of the high mortality.

In conclusion, in the setting of a hospital admission, aSLE with unexplained high fevers, elevation of LFTs, and serum ferritin should prompt, not only the evaluation of infections and possibly malignancies, but also the investigation of MAS.

MAS Treatment in SLE

Clinically, once the diagnosis of MAS is made, strong immune suppression is required. The cornerstone of initial treatment in SLE-associated MAS remains high dose corticosteroids, preferentially intravenously (IV) [12, 15, 28, 34, 42, 51, 53,

54, 64–67, 69]. Several approaches have been used after the initial steroids, or as steroid-sparing agents, to control the cytokine storm. SLE-associated MAS often occurs during lupus flares, hence immunosuppression frequently used in lupus, such as cyclophosphamide, mycophenolate mofetil, or azathioprine, are also often reported in retrospective cohorts, especially in aSLE [28, 34, 64–67] (Table 1). In cases where SLE-associated MAS also presents with a superimposed infection, several series have reported successful remission with IV immunoglobulins (IVIg) or B cell depletion with rituximab so that further immune suppression is avoided [15, 51, 54]. A noticeable difference in the literature between the aSLE and the cSLE is the more frequently reported successful use of anakinra and cyclosporine in the latter [12, 15, 28, 34, 42, 51, 53, 54, 64–67, 69], possibly due to the "extrapolation" of the therapy from sJIA, where both have been used extensively [36, 71]. Anakinra and cyclosporine have been used in cSLE, even when MAS presents along with infections or lupus flares [68] (Table 1). This suggests, perhaps, that such an approach should be successfully used in aSLE as well. Etoposide is primarily used for secondary HLH during malignancies, but it has been reported with some success in aSLE-associated MAS, albeit in a very limited number of cases with significant morbidity [64, 68]. An interesting retrospective observation was that in aSLE the use of hydroxychloroquine and the presence of arthritis as lupus manifestation was associated with a significantly reduced risk of MAS [28], adding yet one more benefit of hydroxychloroquine to SLE [72]. Successful use of tocilizumab and etanercept has been also reported in small case series and definitive clinical trials from sJIA will be soon available.

There are a number of clinical trials currently enrolling for sJIA-associated MAS or MAS regardless of the association. Promising approaches are with tocilizumab (NCT02007239), anti-interferon gamma (NCT03311854), low-dose IL-2 (NCT02569463), and ruxolitinib, a Janus kinase 1/2 (JAK1/2) inhibitor [73] (NCT03795909), while validating studies are ongoing for anakinra [67] (NCT02780583). These clinical trials will help solidify and expand the therapeutic arsenal to help patients with SLE-associated MAS, which at the moment remains invariably lethal if untreated and still has high mortality rate even with treatment.

References

1. Wong, K. F., Hui, P. K., Chan, J. K., Chan, Y. W., & Ha, S. Y. (1991). The acute lupus hemophagocytic syndrome. *Annals of Internal Medicine, 114*, 387–390.
2. Ruperto, N., Brunner, H. I., Quartier, P., Constantin, T., Wulffraat, N. M., Horneff, G., et al. (2018). Canakinumab in patients with systemic juvenile idiopathic arthritis and active systemic features: Results from the 5-year long-term extension of the phase III pivotal trials. *Annals of the Rheumatic Diseases, 77*, 1710–1719.
3. Fava, A., & Petri, M. (2019). Systemic lupus erythematosus: Diagnosis and clinical management. *Journal of Autoimmunity, 96*, 1–13.
4. Pisetsky, D. S. (2017). Antinuclear antibody testing—Misunderstood or misbegotten? *Nature Reviews Rheumatology, 13*, 495–502.

5. Carter, E. E., Barr, S. G., & Clarke, A. E. (2016). The global burden of SLE: Prevalence, health disparities and socioeconomic impact. *Nature Reviews Rheumatology, 12*, 605–620.
6. Alarcon-Segovia, D. (1984). The pathogenesis of immune dysregulation in systemic lupus erythematosus. A troika. *The Journal of Rheumatology, 11*, 588–590.
7. Tsokos, G. C., Lo, M. S., Costa Reis, P., & Sullivan, K. E. (2016). New insights into the immunopathogenesis of systemic lupus erythematosus. *Nature Reviews Rheumatology, 12*, 716–730.
8. Vukelic, M., Li, Y., & Kyttaris, V. C. (2018). Novel treatments in lupus. *Frontiers in Immunology, 9*, 2658.
9. Ravelli, A., Davi, S., Minoia, F., Martini, A., & Cron, R. Q. (2015). Macrophage activation syndrome. *Hematology/Oncology Clinics of North America, 29*, 927–941.
10. Jung, J. Y., & Suh, C. H. (2017). Infection in systemic lupus erythematosus, similarities, and differences with lupus flare. *The Korean Journal of Internal Medicine, 32*, 429–438.
11. Dima, A., Opris, D., Jurcut, C., & Baicus, C. (2016). Is there still a place for erythrocyte sedimentation rate and C-reactive protein in systemic lupus erythematosus? *Lupus, 25*, 1173–1179.
12. Parodi, A., Davi, S., Pringe, A. B., Pistorio, A., Ruperto, N., Magni-Manzoni, S., et al. (2009). Macrophage activation syndrome in juvenile systemic lupus erythematosus: A multinational multicenter study of thirty-eight patients. *Arthritis and Rheumatism, 60*, 3388–3399.
13. Ramos-Casals, M., Brito-Zeron, P., Lopez-Guillermo, A., Khamashta, M. A., & Bosch, X. (2014). Adult haemophagocytic syndrome. *Lancet, 383*, 1503–1516.
14. Cron, R. Q., Davi, S., Minoia, F., & Ravelli, A. (2015). Clinical features and correct diagnosis of macrophage activation syndrome. *Expert Review of Clinical Immunology, 11*, 1043–1053.
15. Pringe, A., Trail, L., Ruperto, N., Buoncompagni, A., Loy, A., Breda, L., et al. (2007). Macrophage activation syndrome in juvenile systemic lupus erythematosus: An underrecognized complication? *Lupus, 16*, 587–592.
16. Lisnevskaia, L., Murphy, G., & Isenberg, D. (2014). Systemic lupus erythematosus. *Lancet, 384*, 1878–1888.
17. Muskardin, T. L. W., & Niewold, T. B. (2018). Type I interferon in rheumatic diseases. *Nature Reviews Rheumatology, 14*, 214–228.
18. Tipton, C. M., Hom, J. R., Fucile, C. F., Rosenberg, A. F., & Sanz, I. (2018). Understanding B-cell activation and autoantibody repertoire selection in systemic lupus erythematosus: A B-cell immunomics approach. *Immunological Reviews, 284*, 120–131.
19. Silverstein, A. M. (2001). Autoimmunity versus horror autotoxicus: The struggle for recognition. *Nature Immunology, 2*, 279–281.
20. Deng, Y., & Tsao, B. P. (2017). Updates in lupus genetics. *Current Rheumatology Reports, 19*, 68.
21. Felten, R., Dervovic, E., Chasset, F., Gottenberg, J. E., Sibilia, J., Scher, F., et al. (2018). The 2018 pipeline of targeted therapies under clinical development for systemic lupus erythematosus: A systematic review of trials. *Autoimmunity Reviews, 17*, 781–790.
22. Thanou, A., & Merrill, J. T. (2018). New trials in lupus and where are we going. *Current Rheumatology Reports, 20*, 34.
23. Zhang, M., Behrens, E. M., Atkinson, T. P., Shakoory, B., Grom, A. A., & Cron, R. Q. (2014). Genetic defects in cytolysis in macrophage activation syndrome. *Current Rheumatology Reports, 16*, 439.
24. Crayne, C. B., Albeituni, S., Nichols, K. E., & Cron, R. Q. (2019). The immunology of macrophage activation syndrome. *Frontiers in Immunology, 10*, 119.
25. Elkon, K. B. (2018). Review: Cell death, nucleic acids, and immunity: Inflammation beyond the grave. *Arthritis & Rhematology, 70*, 805–816.
26. James, J. A., & Robertson, J. M. (2012). Lupus and Epstein-Barr. *Current Opinion in Rheumatology, 24*, 383–388.
27. Torigoe, M., Sakata, K., Ishii, A., Iwata, S., Nakayamada, S., & Tanaka, Y. (2018). Hydroxychloroquine efficiently suppresses inflammatory responses of human class-switched memory B cells via Toll-like receptor 9 inhibition. *Clinical Immunology, 195*, 1–7.

28. Cohen, E. M., D'Silva, K., Kreps, D., Son, M. B., & Costenbader, K. H. (2018). Arthritis and use of hydroxychloroquine associated with a decreased risk of macrophage activation syndrome among adult patients hospitalized with systemic lupus erythematosus. *Lupus, 27*, 1065–1071.
29. Mok, C. C. (2019). The Jakinibs in systemic lupus erythematosus: Progress and prospects. *Expert Opinion on Investigational Drugs, 28*, 85–92.
30. Liu, E., & Perl, A. (2019). Pathogenesis and treatment of autoimmune rheumatic diseases. *Current Opinion in Rheumatology, 31*, 307–315.
31. Grammer, A. C., Ryals, M. M., Heuer, S. E., Robl, R. D., Madamanchi, S., Davis, L. S., et al. (2016). Drug repositioning in SLE: Crowd-sourcing, literature-mining and Big Data analysis. *Lupus, 25*, 1150–1170.
32. Costedoat-Chalumeau, N., & Houssiau, F. A. (2018). Ustekinumab: A promising new drug for SLE? *Lancet, 392*, 1284–1286.
33. Wallace, D. J., Furie, R. A., Tanaka, Y., Kalunian, K. C., Mosca, M., Petri, M. A., et al. (2018). Baricitinib for systemic lupus erythematosus: A double-blind, randomised, placebo-controlled, phase 2 trial. *Lancet, 392*, 222–231.
34. Gavand, P. E., Serio, I., Arnaud, L., Costedoat-Chalumeau, N., Carvelli, J., Dossier, A., et al. (2017). Clinical spectrum and therapeutic management of systemic lupus erythematosus-associated macrophage activation syndrome: A study of 103 episodes in 89 adult patients. *Autoimmunity Reviews, 16*, 743–749.
35. Chamseddin, B., Marks, E., Dominguez, A., Wysocki, C., & Vandergriff, T. (2019). Refractory macrophage activation syndrome in the setting of adult onset Still's disease with hemophagocytic lymphohistiocytosis detected on skin biopsy treated with canakinumab and tacrolimus. *Journal of Cutaneous Pathology*. https://doi.org/10.1111/cup.13466
36. Grom, A. A., Horne, A., & De Benedetti, F. (2016). Macrophage activation syndrome in the era of biologic therapy. *Nature Reviews Rheumatology, 12*, 259–268.
37. Brugos, B., Kiss, E., Dul, C., Gubisch, W., Szegedi, G., Sipka, S., et al. (2010). Measurement of interleukin-1 receptor antagonist in patients with systemic lupus erythematosus could predict renal manifestation of the disease. *Human Immunology, 71*, 874–877.
38. Chang, D. M. (1997). Interleukin-1 and interleukin-1 receptor antagonist in systemic lupus erythematosus. *Immunological Investigations, 26*, 649–659.
39. Capper, E. R., Maskill, J. K., Gordon, C., & Blakemore, A. I. (2004). Interleukin (IL)-10, IL-1ra and IL-12 profiles in active and quiescent systemic lupus erythematosus: Could longitudinal studies reveal patient subgroups of differing pathology? *Clinical and Experimental Immunology, 138*, 348–356.
40. Karakike, E., & Giamarellos-Bourboulis, E. J. (2019). Macrophage activation-like syndrome: A distinct entity leading to early death in sepsis. *Frontiers in Immunology, 10*, 55.
41. Toplak, N., Blazina, S., & Avcin, T. (2018). The role of IL-1 inhibition in systemic juvenile idiopathic arthritis: Current status and future perspectives. *Drug Design, Development and Therapy, 12*, 1633–1643.
42. Aytac, S., Batu, E. D., Unal, S., Bilginer, Y., Cetin, M., Tuncer, M., et al. (2016). Macrophage activation syndrome in children with systemic juvenile idiopathic arthritis and systemic lupus erythematosus. *Rheumatology International, 36*, 1421–1429.
43. Tayer-Shifman, O. E., & Ben-Chetrit, E. (2015). Refractory macrophage activation syndrome in a patient with SLE and APLA syndrome—Successful use of PET-CT and Anakinra in its diagnosis and treatment. *Modern Rheumatology, 25*, 954–957.
44. Kim, H., Levy, D. M., Silverman, E. D., Hitchon, C., Bernatsky, S., Pineau, C., et al. (2019). A comparison between childhood and adult onset systemic lupus erythematosus adjusted for ethnicity from the 1000 Canadian Faces of Lupus Cohort. *Rheumatology (Oxford)*. https://doi.org/10.1093/rheumatology/kez006
45. Mina, R., & Brunner, H. I. (2013). Update on differences between childhood-onset and adult-onset systemic lupus erythematosus. *Arthritis Research & Therapy, 15*, 218.

46. Somers, E. C., Marder, W., Cagnoli, P., Lewis, E. E., DeGuire, P., Gordon, C., et al. (2014). Population-based incidence and prevalence of systemic lupus erythematosus: The Michigan Lupus Epidemiology and Surveillance program. *Arthritis & Rhematology, 66*, 369–378.
47. Hiraki, L. T., Benseler, S. M., Tyrrell, P. N., Hebert, D., Harvey, E., & Silverman, E. D. (2008). Clinical and laboratory characteristics and long-term outcome of pediatric systemic lupus erythematosus: A longitudinal study. *The Journal of Pediatrics, 152*, 550–556.
48. Papadimitraki, E. D., & Isenberg, D. A. (2009). Childhood- and adult-onset lupus: An update of similarities and differences. *Expert Review of Clinical Immunology, 5*, 391–403.
49. Bundhun, P. K., Kumari, A., & Huang, F. (2017). Differences in clinical features observed between childhood-onset versus adult-onset systemic lupus erythematosus: A systematic review and meta-analysis. *Medicine (Baltimore), 96*, e8086.
50. Son, M. B., Sergeyenko, Y., Guan, H., & Costenbader, K. H. (2016). Disease activity and transition outcomes in a childhood-onset systemic lupus erythematosus cohort. *Lupus, 25*, 1431–1439.
51. Gormezano, N. W., Otsuzi, C. I., Barros, D. L., da Silva, M. A., Pereira, R. M., Campos, L. M., et al. (2016). Macrophage activation syndrome: A severe and frequent manifestation of acute pancreatitis in 362 childhood-onset compared to 1830 adult-onset systemic lupus erythematosus patients. *Seminars in Arthritis and Rheumatism, 45*, 706–710.
52. Ravelli, A., Minoia, F., Davi, S., Horne, A., Bovis, F., Pistorio, A., et al. (2016). 2016 Classification criteria for macrophage activation syndrome complicating systemic juvenile idiopathic arthritis: A European League Against Rheumatism/American College of Rheumatology/Paediatric Rheumatology International Trials Organisation Collaborative Initiative. *Annals of the Rheumatic Diseases, 75*, 481–489.
53. Borgia, R. E., Gerstein, M., Levy, D. M., Silverman, E. D., & Hiraki, L. T. (2018). Features, treatment, and outcomes of macrophage activation syndrome in childhood-onset systemic lupus erythematosus. *Arthritis & Rhematology, 70*, 616–624.
54. Bennett, T. D., Fluchel, M., Hersh, A. O., Hayward, K. N., Hersh, A. L., Brogan, T. V., et al. (2012). Macrophage activation syndrome in children with systemic lupus erythematosus and children with juvenile idiopathic arthritis. *Arthritis and Rheumatism, 64*, 4135–4142.
55. Bennett, L., Palucka, A. K., Arce, E., Cantrell, V., Borvak, J., Banchereau, J., et al. (2003). Interferon and granulopoiesis signatures in systemic lupus erythematosus blood. *The Journal of Experimental Medicine, 197*, 711–723.
56. Kumakura, S., & Murakawa, Y. (2014). Clinical characteristics and treatment outcomes of autoimmune-associated hemophagocytic syndrome in adults. *Arthritis & Rhematology, 66*, 2297–2307.
57. Fukaya, S., Yasuda, S., Hashimoto, T., Oku, K., Kataoka, H., Horita, T., et al. (2008). Clinical features of haemophagocytic syndrome in patients with systemic autoimmune diseases: Analysis of 30 cases. *Rheumatology (Oxford), 47*, 1686–1691.
58. Papo, T., Andre, M. H., Amoura, Z., Lortholary, O., Tribout, B., Guillevin, L., et al. (1999). The spectrum of reactive hemophagocytic syndrome in systemic lupus erythematosus. *The Journal of Rheumatology, 26*, 927–930.
59. Littlejohn, E., Marder, W., Lewis, E., Francis, S., Jackish, J., McCune, W. J., et al. (2018). The ratio of erythrocyte sedimentation rate to C-reactive protein is useful in distinguishing infection from flare in systemic lupus erythematosus patients presenting with fever. *Lupus, 27*, 1123–1129.
60. Gaitonde, S., Samols, D., & Kushner, I. (2008). C-reactive protein and systemic lupus erythematosus. *Arthritis and Rheumatism, 59*, 1814–1820.
61. Singh, G., Shabani-Rad, M. T., Vanderkooi, O. G., Vayalumkal, J. V., Kuhn, S. M., Guilcher, G. M., et al. (2013). Leishmania in HLH: A rare finding with significant treatment implications. *Journal of Pediatric Hematology/Oncology, 35*, e127–e129.
62. Jog, N. R., & James, J. A. (2017). Biomarkers in connective tissue diseases. *The Journal of Allergy and Clinical Immunology, 140*, 1473–1483.

63. Felten, R., Sagez, F., Gavand, P. E., Martin, T., Korganow, A. S., Sordet, C., et al. (2019). 10 most important contemporary challenges in the management of SLE. *Lupus Science and Medicine, 6*, e000303.
64. Ahn, S. S., Yoo, B. W., Jung, S. M., Lee, S. W., Park, Y. B., & Song, J. J. (2017). In-hospital mortality in febrile lupus patients based on 2016 EULAR/ACR/PRINTO classification criteria for macrophage activation syndrome. *Seminars in Arthritis and Rheumatism, 47*, 216–221.
65. Liu, A. C., Yang, Y., Li, M. T., Jia, Y., Chen, S., Ye, S., et al. (2018). Macrophage activation syndrome in systemic lupus erythematosus: A multicenter, case-control study in China. *Clinical Rheumatology, 37*, 93–100.
66. Lambotte, O., Khellaf, M., Harmouche, H., Bader-Meunier, B., Manceron, V., Goujard, C., et al. (2006). Characteristics and long-term outcome of 15 episodes of systemic lupus erythematosus-associated hemophagocytic syndrome. *Medicine (Baltimore), 85*, 169–182.
67. Ruscitti, P., Cipriani, P., Ciccia, F., Masedu, F., Liakouli, V., Carubbi, F., et al. (2017). Prognostic factors of macrophage activation syndrome, at the time of diagnosis, in adult patients affected by autoimmune disease: Analysis of 41 cases collected in 2 rheumatologic centers. *Autoimmunity Reviews, 16*, 16–21.
68. Dhote, R., Simon, J., Papo, T., Detournay, B., Sailler, L., Andre, M. H., et al. (2003). Reactive hemophagocytic syndrome in adult systemic disease: Report of twenty-six cases and literature review. *Arthritis and Rheumatism, 49*, 633–639.
69. Lerkvaleekul, B., & Vilaiyuk, S. (2018). Macrophage activation syndrome: Early diagnosis is key. *Open access Rheumatology: Research and Reviews, 10*, 117–128.
70. Kalunian, K. C., Urowitz, M. B., Isenberg, D., Merrill, J. T., Petri, M., Furie, R. A., et al. (2018). Clinical trial parameters that influence outcomes in lupus trials that use the systemic lupus erythematosus responder index. *Rheumatology (Oxford), 57*, 125–133.
71. Grevich, S., & Shenoi, S. (2017). Update on the management of systemic juvenile idiopathic arthritis and role of IL-1 and IL-6 inhibition. *Adolescent Health, Medicine and Therapeutics, 8*, 125–135.
72. Durcan, L., & Petri, M. (2016). Immunomodulators in SLE: Clinical evidence and immunologic actions. *Journal of Autoimmunity, 74*, 73–84.
73. Maschalidi, S., Sepulveda, F. E., Garrigue, A., Fischer, A., & de Saint Basile, G. (2016). Therapeutic effect of JAK1/2 blockade on the manifestations of hemophagocytic lymphohistiocytosis in mice. *Blood, 128*, 60–71.

Kawasaki Disease-Associated Cytokine Storm Syndrome

Rolando Ulloa-Gutierrez, Martin Prince Alphonse, Anita Dhanranjani, and Rae S. M. Yeung

Kawasaki Disease

Kawasaki disease (KD) is an acute self-limited febrile vasculitis of childhood and currently the leading cause of acquired cardiac disease in children from developed countries [1]. Although it can occur at any age, around 80% of cases occur during the first 5 years of life and 50% in the first 24 months. Diagnosis is more difficult at extreme age groups [2] and most patients have been seen more than once prior to establishing the final diagnosis [3]. In patients less than 6 months of age, the diagnosis is often difficult and late, resistance to treatment is higher, and there is a higher probability of acute complications [1, 4]. New onset presentations are seen but

R. Ulloa-Gutierrez
Division of Cell Biology, The Hospital for Sick Children, Department of Paediatrics, University of Toronto, Toronto, ON, Canada

M. P. Alphonse
Division of Cell Biology, The Hospital for Sick Children, Department of Paediatrics, University of Toronto, Toronto, ON, Canada

Department of Immunology, University of Toronto, Toronto, ON, Canada

A. Dhanranjani
Division of Rheumatology, The Hospital for Sick Children, Department of Paediatrics, University of Toronto, Toronto, ON, Canada

R. S. M. Yeung (✉)
Division of Cell Biology, The Hospital for Sick Children, Department of Paediatrics, University of Toronto, Toronto, ON, Canada

Department of Immunology, University of Toronto, Toronto, ON, Canada

Division of Rheumatology, The Hospital for Sick Children, Department of Paediatrics, University of Toronto, Toronto, ON, Canada
e-mail: rae.yeung@sickkids.ca

R. Q. Cron, E. M. Behrens (eds.), *Cytokine Storm Syndrome*,
https://doi.org/10.1007/978-3-030-22094-5_23

uncommon in adolescents and adults, in whom it is more common to see late-sequelae of childhood KD.

KD affects predominantly medium-size arteries and systemic inflammation also occurs in many organs and tissues [1, 4]. The majority of vascular compromise occurs in the coronary arteries; however, extra coronary complications can occur. The immunobiology of KD revolves around massive systemic immune activation leading to localized inflammation at the blood vessels, resulting in coronary arteritis and aneurysm formation. Therefore, the main goal of therapy is to reduce inflammation as much and as early as possible to prevent coronary artery damage.

Although the precise cause of the disease has remained a mystery for more than six decades since the original description by Dr. Tomisaku Kawasaki in Japan [5], significant advances have been made over the last decades to characterize the pathobiology. Many agents and triggering factors have been implicated, with epidemiological, climatic and genetic factors in susceptible hosts playing a major role in modulating disease risk and outcome. If left untreated, up to 25% of children will develop coronary artery aneurysms. Standard first line treatment includes intravenous immunoglobulin (IVIG) at high immunosuppressive doses (2 g/kg) in a single dose together with acetyl salicylic acid (ASA) [4, 6]. Antiplatelet doses of ASA at 3–5 mg/kg/day once daily is continued until standard laboratory measures of inflammation are resolved (usually 4–6 weeks) and the patient has no evidence of coronary changes on follow up echocardiography. Resistance to first dose of IVIG varies but in general is around 20%.

Following Dr. Kawasaki's first descriptions, clinical criteria have been defined for classic and incomplete KD by the American Heart Association [4], the Japanese Society for Paediatric Cardiology [6], and different guidelines have been published in other countries [7, 8]. Classic (complete or typical) Kawasaki disease is defined as the presence of fever of ≥5 days plus at least ≥4 of the following diagnostic criteria: (1) oral mucosal changes including erythema and cracking of the lips, strawberry tongue, and/or erythema of oral and pharyngeal mucosa; (2) non-suppurative conjunctival injection; (3) polymorphous skin rash; (4) peripheral changes, including erythema and/or edema of hands and feet; and (5) cervical lymphadenopathy ≥1.5 cm in diameter and usually unilateral. Children with incomplete or atypical Kawasaki disease do not fulfill the classic criteria and have less than four of the diagnostic criteria. It is more common in infants younger than 6 months of age, in whom the diagnosis is more difficult and often late [9], and in this age group, coronary artery lesions (CALs) are more common [10].

Pathobiology of KD and CSS

Cytokine storm syndrome (CSS)/macrophage activation syndrome (MAS) or secondary hemophagocytic lymphohistiocytosis (sHLH), is an inflammatory reaction produced by an excessive cytokine production and release, macrophage activation, and hemophagocytosis [11] and is detailed in previous chapters. Despite the diverse etiologies leading to CSS, the clinical features remain similar, leading to the speculation that CSS may represent the severe end of a spectrum of inflammatory conditions, rather than a discrete entity. High grade sustained fever is almost unequivocally present in CSS. An unremitting fever pattern is common to both KD and CSS. In fact, a low platelet count is one of the predictive factors of poor outcome in KD and may in fact represent patients who have CSS at presentation of KD. Additionally, KD patients with recalcitrant fever despite treatment with IVIG should also be investigated for the possibility of CSS. The key pathological feature of CSS is attributed to extensive activation of the immune response, including both adaptive and innate immune cells such as T-cells and macrophages [12–14] and soluble mediators of inflammation creating the classic "cytokine storm" [15] detailed in previous chapters. Similarly, both the adaptive and innate arms of the immune system are implicated in the acute phase of KD [16–19]. An infectious trigger is thought to initiate the immune response in KD, with activation of T-cells and macrophages by antigens and danger signals, respectively. The presence of superantigens (SAg) has been identified during some outbreaks of KD, thought to be one of the mechanism underlying massive activation of T-cells [20]. Superantigens are produced by some bacteria and viruses, and are able to stimulate up to 30% of all T-cells by binding to a common element of the T-cell receptor outside of the antigen binding cleft [21]. Superantigenic activity leads to massive T-cell activation and release of pro-inflammatory cytokines IL-2, TNF, and IFN-ɣ, which among their many pleiotropic activities, can mediate activation of monocytes/macrophages [22].

The innate immune system is also implicated in KD, with evidence of inflammasome activation and production of IL-1β and IL-18 by monocytes and macrophages. Inflammasomes are multimeric protein complexes consisting of NOD-like receptors (NLRs), which are cytosolic sensors for danger signals including endogenous and exogenous insults, such as microbes. Once the inflammasome protein complexes have assembled, the inflammasomes activate caspase-1, which proteolytically cleaves the precursor forms of the pro-inflammatory cytokines IL-1β and IL-18 into their active proteins. Intracellular calcium signaling plays a central role in mediating the processes leading to inflammasome activation. Genome-wide association studies (GWAS) have identified inositol triphosphate 3-kinase C (ITPKC), an enzyme in Ca^{2+} signaling pathway, to be associated with KD. The KD-associated genetic polymorphism in *ITPKC* has important functional consequences, governing ITPKC expression levels and intracellular calcium levels, which in turn regulates NLRP3 expression and production of IL-1β and IL-18. Treatment failure in those with the "high-risk" *ITPKC*-genotype is associated with the highest intracellular calcium levels and with increased production of IL-1β and IL-18 by macrophages and higher circulating levels of both cytokines [23]. These findings provide the mechanism behind the observed efficacy of rescue therapy with IL-1 blockade in children with recalcitrant KD and highlight the common pathobiology underlying KD and CSS.

Animal Models of HLH/CSS and KD

Common themes in immunobiology link KD and CSS. Animal models of HLH/CSS such as perforin-deficient [24], Munc13-4 deficient and Rab27a [25] deficient mouse models infected with lymphocytic choriomeningitis virus provide some clues into the complex pathology of CSS (described in previous chapters). These animals develop CSS-like clinical features in an IFNγ dependent manner [24–26]. Similarly, in patients with CSS, evidence point to IFNγ producing CD8+ T-cells [13] with liver biopsies from patients with CSS showing extensive periportal infiltration with hemophagocytic macrophages and IFN-γ producing CD8+ T-cells [27]. Similarly, in the *Lactobacillus casei* cell wall extract (LCWE) mouse model of KD, IFN-γ plays an important role and is present both at mRNA and protein levels in affected vessels. However, IFN-γ appears to play a regulatory role and its presence does not indicate its necessity for induction of coronary arteritis in the KD animal model [28]. The LCWE animal model has also been shown to require T-cell activation for development of KD [17], with CSS parallels pointing to excessive T-cell activation [29].

TLR signaling leading to inflammasome activation is also a shared pathogenic pathway between CSS and KD. Repeated innate immune system stimulation via TLR9 signaling in a mouse leads to MAS features, including hepatic dysfunction and cytopenia [30]. Similarly, TLR2 stimulation and inflammasome (NLRP3) activation are essential for development and severity of disease in the LCWE animal model of KD [23, 31, 32], with production of IL-1β and IL-18 present in both preclinical models. Aneurysm formation in the LCWE mouse model of KD is mediated by both IL-1α and IL-1β and is successfully treated with anakinra [recombinant IL-1 receptor antagonist (IL-1RA)] [33], in accord with clinical observations of treatment success in children with recalcitrant KD [34]. Interestingly, genetic polymorphisms in IL-18 and serum levels of IL-18 have also been associated with risks and outcomes in children with KD [35–37]. IL-18 is unique in the IL-1 family of cytokines, able to induce production of IFN-γ by natural killer (NK) cells and T-cells along with TNF and chemokine secretion by macrophages [38]. Serum IL-18 levels are elevated out of proportion when compared with other cytokines in MAS/CSS and are correlated with clinical measures of CSS, including serum levels of soluble IL-2 receptor-α chain and IFN-γ [39]. Serum IL-18 binding protein (IL-18BP) levels appear to be disproportionately elevated compared to the marked increase in IL-18 levels in CSS, resulting in high levels of biologically active IL-18 [40]. Interestingly, children with KD also show increased secretion of IL-18 and IL-18BP during the acute phase of KD [23]. A severe IL-18/IL-18BP imbalance may contribute to excessive T-cell and macrophage activation in patients with underlying inflammatory disease, creating a favorable milieu for development of MAS/CSS [15].

CSS in Children with Kawasaki Disease

Although KD was described more than six decades ago and many aspects of the disease have been elucidated, MAS/CSS has been underrecognized and underdiagnosed. Many of the clinical and laboratory findings consistent with MAS/CSS have been considered by KD clinicians and researchers as part of the clinical picture of the KD itself, including persistence of fever, increased liver enzymes (seen in over 30% of children in some studies), and thrombocytopenia, among others [41, 42]. Duration of fever is an indirect measure of the severity of the underlying vasculitis and the best predictor of poor coronary outcome. In previous studies it has been found that age, fever duration, platelet count, and albumin level are highly predictive of coronary abnormalities [43]. Platelet count has been considered in many studies and risk scoring algorithms as an important predictor of coronary outcome, but these have failed to agree on whether increased or decreased platelet count is the high risk factor [44–46]. Both thrombocytopenia and thrombocytosis reflect massive immune activation, with the former suggesting a consumption process triggered by and consistent with cytokine storm, and the latter reflecting a pro-inflammatory process mediated by IL-6 [43].

The first report in the English language literature of MAS/CSS in KD was made by Ohga et al. in 1995 [47] in a 32-month-old Japanese boy who was diagnosed and treated for KD but had persistence of fever, recurrent rash, hepatomegaly and other clinical and laboratory findings that led further investigations and the diagnosis of MAS/CSS. Since that initial report more than two decades ago, CSS has been described in KD infants and children from different parts of the world. The majority of these publications have been single case reports, small series or retrospective studies, and usually from single-site institutions [48–57], (Table 1), including a review of the literature in one of these publications [54]. In most of these publications, the common remarkable findings have been prolonged fever, lack of clinical suspicion early in the course of the disease, IVIG-refractory KD, and late diagnosis of CSS.

A large case series of MAS in acute KD patients during the last two decades was published [58]. This retrospective study at the Hospital for Sick Children in Toronto analyzed all KD patients from January 2001 to March 2008. Among 638 identified patients with KD, 12 (1.9%) had a co diagnosis of MAS and KD, with a median age at diagnosis of 7 years (8 months to 14 years). In this series, 9/12 patients fulfilled criteria for complete KD, 7/12 were males more than 5 years of age. Eleven of 12 patients manifested hepatosplenomegaly, and all patients had prolonged fever beyond initial intravenous immunoglobulin treatment. Despite receiving treatment with IVIG, all patients were persistently febrile for at least 48 h and prompted further laboratory workup and additional cardiac imaging studies and other therapeutic interventions. Remarkable laboratory findings included hyperferritinemia (≥500 μg/L) in all patients and thrombocytopenia in 83%, elevated D-dimers in

Table 1 Clinical characteristics of patients with KD and HLH reported in selected publications

Author	Number of patients	Sex	Age or age ranges	Serum ferritin	Bone marrow with hemophagocytosis	Treatment	Outcome	Reference
Muise	1	M	9 years	1156 μg/L	Yes	IVIG (2 doses) ASA Methylprednisolone Prednisone	Survived	[48]
Titze	1	M	7 weeks	447 μg/L	Yes	No IVIG No ASA Dexamethasone Cyclosporine A Etoposide	Died (KD diagnosed on autopsy)	[49]
Hendricks	1	F	6 years	10,000 μg/L	Yes	IVIG ASA Dexamethasone	Survived	[50]
Simonini	1	F	20 months	Not specified	Not performed	IVIG (2 doses)	Survived	[51]
Kim	5	3M 2F	4–14 years	Range 768–43,216 ng/mL	Yes	HLH-2004 IVIG ASA	2 deaths (40%)	[52]
Mukherjee	1	M	4 years	15,716 ng/dL	Yes	IVIG ASA Methylprednisolone	Survived	[53]
Ogawa	1	F	9 years	823 mg/dL	Yes	IVIG Flurbiprofen	Survived	[54]
Wang	8	8M 0F	18 months–12 years	785–>1500 ng/mL	Yes in 3/7 (42.9%)	IVIG ASA Methylprednisolone Dexamethasone VP16 CSA	1 death (12.5%)	[55]
Dogan	1	F	4 months	386 ng/mL	BM Not performed	IVIG Heparin	Died	[57]

M male, *F* female, *ASA* acetyl salicylic acid

92% and 33% with biopsy-proven evidence of hemophagocytosis. Acute cardiac complications were seen in 50% of patients, predominantly coronary dilations; however, of interest, none had residual coronary abnormalities during follow up visits. Prolonged hospitalizations occurred in this group (median 19 days, range 9–49 days). Features associated with CSS in this case series were older age at diagnosis of KD, male sex, non-responsiveness to initial IVIG and thrombocytopenia, consistent with other reports in the literature.

More recent updated (unpublished) information from our group extended our previous study [59] and analyzed 1021 patients with KD from January 2001 to February 2014. Fifty-eight patitents (5.7%) were evaluated for possible MAS/CSS with 23 patients (2.3%) diagnosed and treated for it. In this series, 47.8% patients fulfilled criteria for complete KD, and 67% were males more than 5 years of age. In accord with the literature and our previous observations, older boys with longer fever dominated, with mean age at diagnosis of 6.22 years and mean duration of fever prior to IVIG treatment of 11.5 days. All 23 patients had prolonged fever with the following KD diagnostic features: rash 22 (95.7%) patients, cervical lymphadenopathy 14 (60.9%), erythema and/or edema of extremities in 16 (69.6%), oral mucositis in 13 (56.5%), bilateral conjunctivitis 13 (56.5%). MAS/CSS features included persistent fever 21/22 (95.5%), splenomegaly 12/23 (52.2%), cytopenias 20/23 (87%), ferritin >500 μg/L in 22/23 (95.7%), hemophagocytosis 6/23 (26.1%), low or absent KN activity 1/7 (14.3%), and increased serum sIL-2Rα (CD25) 5/6 (83.3%). Total duration of fever was 17.3 days (mean) with all 23 patients receiving standard IVIG treatment for KD. Additional treatment included: A second and third dose of IVIG in 14/23 (60.9%) and 2 (8.7%) respectively; oral and IV steroids in 15 (65.2%) and 17 (73.9%), respectively; and cyclosporine in 3 (13%) patients. Supportive therapy including transfusion of red blood cells/platelets in 7/23 (30.4%), fresh frozen plasma in 2/21 (9.5%) and cryoprecipitates in 1/21 (4.8%) was required, with three (13%) patients requiring admission to the pediatric intensive care unit (PICU). Interestingly, increased ferritin levels has been reported to be useful to predict refractory Kawasaki disease patients [60], highlighting the fact that this subgroup of patients may actually have features of CSS. To address detection and early recognition of CSS complicating KD, all children with recalcitrant KD have a serum ferritin added to their standard KD laboratory workup at our institution. In those with evidence of CSS, therapeutic management includes use of systemic steroids and IL-1 blockade with anakinra.

A systematic review of the literature on MAS/CSS in KD patients was published by García-Pavón et al. in 2017 [61]. The authors performed a literature review of published cases since Dr. Kawasaki's first report and until September 2016, which included 69 reported cases. The mean age of diagnosis of 5.6 years, with a range of 7 weeks to 17 years of age; 68% were boys and 75% had an Asian ethnicity. They found that CSS diagnosis was made before the diagnosis of KD in 6%, simultaneous diagnosis in 21%, and in 73% after KD was diagnosed. Tissue demonstration of hemophagocytosis was found in 88% (51/58) of patients. The reported mortality was 13% (9/69). All patients received IVIG and not surprisingly, 90% of them required two or more doses. Steroids were given in 87%, cyclosporine in 49%, etoposide in 39%, anti-TNF in 5.7%, and IL-1 receptor antagonist in 4.3%. Fatalities occurred in 9 (13%) patients with the cause of death including acute myocardial infarction, pneumonia, and sepsis among others. All studies except one, included in the systematic review used the HLH-2004 criteria to diagnose MAS/CSS. Seventy-eight percent (54/69) of the patients fulfilled HLH-2004 criteria and 74% (51/69) fulfilled the 2016 consensus criteria for MAS-SJIA, the classification criteria for MAS/CSS complicating systemic JIA (sJIA) proposed by Ravelli et al., based on a combination of expert consensus, available evidence from the medical literature and analysis of real patient data [11, 62].

The evidence for use of various criteria for MAS/CSS in KD is mostly extrapolated from studies in sJIA and other inflammatory conditions, with sparse literature on the specific use of these diagnostic criteria in KD. In a retrospective chart review of 719 patients with KD by Wang et al. [55], 8 patients (1.11%) were identified as MAS with KD according to Ravelli criteria, but only 3 patients were independently identified by the HLH-2009 criteria. In this series of 8 patients, aspartate aminotransferase was significantly elevated in all cases, whereas alanine aminotransferase, lactate dehydrogenase and serum ferritin were abnormal in 7/8 cases. Cytopenia and hypertriglyceridemia were relatively common, seen in 6/8 and 5/8 cases respectively, whereas hypofibrinogenemia was noted only in one case. Three patients had histopathological evidence of hemophagocytosis, but only one fulfilled the HLH-2009 criteria.

CSS, Kawasaki Disease, and sJIA

CSS, KD, and sJIA share many common clinical and laboratory features. Coronary artery abnormalities are not exclusive of KD and can be seen in sJIA. sJIA with CSS can be initially diagnosed as KD, and a review of the literature in 2013 [63] found that most sJIA patients with coronary artery dilations were classified initially as classic or incomplete KD, and treated with multiple doses of IVIG. Other investigators have suggested that serum IL-18 and ferritin levels can help clinicians distinguishing between KD and sJIA [64, 65]; however, IL-18 is also elevated in acute KD and may be predictive of those at high risk for treatment failure [23].

We reviewed patients with combined diagnosis of KD and sJIA (KD/sJIA) [66]. Children diagnosed with either KD (n = 1765) or sJIA (n = 112) between January 1990 and December 2011 were analyzed. Among over 20 patients with the co-diagnosis of KD and sJIA, only those who fulfilled both the American Heart Association guidelines for KD and the International League of Associations for Rheumatology (ILAR) classification criteria for sJIA were included (n = 8) for further analysis. In the KD cohort 0.5% (8/1765) children had a co-diagnosis of sJIA; whereas KD preceded sJIA in 7% (8/112) of the cases.

A report by Dong et al. in 2015 [67] utilized the Pediatric Health Information System (PHIS) to estimate the incidence and to characterize those patients initially treated for KD but later received the diagnosis of sJIA. This retrospective study analyzed only patients diagnosed with sJIA within the initial 6-month period after the first hospitalization for KD. Among 176 patients at their institution who were treated for KD with IVIG during the study period, 2 (1.1%) were diagnosed with sJIA after KD treatment. When they analyzed the PHIS database, the authors identified 10/6475 cases (0.2%) between 2009 and 2013. MAS was more common in patients with KD and subsequent diagnosis of sJIA than in patients with KD alone (30% and 0.3%, respectively; $p < 0.001$).

Given the common clinical and laboratory features and the shared IL-1β and IL-18 immunobiology, we propose that KD, sJIA, and CSS are parts of the same disease spectrum with the distinguishing features being the intensity and duration of the immune response (Fig. 1). KD having the shortest disease duration and sJIA the longest with CSS having the most dramatic immune response. There are many reports of successful use of anakinra, a recombinant IL-1 receptor antagonist in CSS associated with sJIA [68–70]. Refractory KD has also been shown in case reports to have a good response to anakinra [34]. A single case report of high-dose anakinra (9 mg/kg/day) in a neonate with refractory KD, suspected to have CSS, abated the inflammatory response, highlighting the role of anti-IL-1 therapy in patients with KD and CSS [71]. Considering the high mortality of this condition, and the difficulty in establishing a definite co-diagnosis of CSS, the early use of anakinra in refractory KD may prove to be a life-saving measure and has been used successfully as such in our institution in the past several years with good effect.

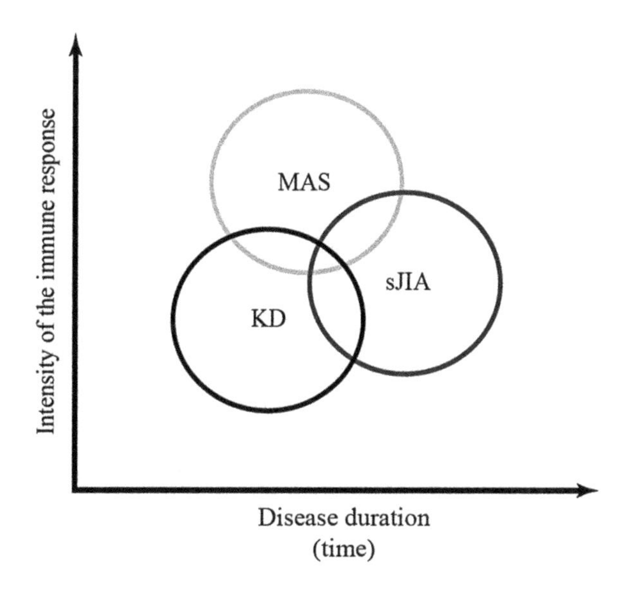

Fig. 1 KD, sJIA and MAS/CSS share many common clinical and immunobiologic features and may be part of the same disease spectrum with the distinguishing features being the intensity and duration of the immune response. On the x-axis the disease duration is represented and on the y-axis the intensity of the immune response

Conclusion

CSS occurs in at least 2% of children with KD. It is undoubtedly underrecognized and underdiagnosed with the two syndromes sharing many overlapping of features and each lacking specific diagnostic tests. Although no disease-specific criteria for diagnosing CSS in KD is available, the classification criteria for CSS in sJIA have been shown to be useful for CSS in KD. Among the laboratory criteria, hyperferritinemia (>500 μg/L) is the most practical addition to standard KD laboratory tests, but the usefulness of serial laboratory measurements to detect trends in transaminases, white blood cell count, platelet count, and fibrinogen levels cannot be overemphasized. Early recognition and prompt institution of immunomodulatory treatment can substantially reduce the mortality and morbidity of MAS/CSS in KD. Given the known pathogenetic role of IL-1β in both syndromes, the early use of IL-1 blockers in refractory KD with CSS deserves consideration.

Disclosures R.U.: Nothing to disclose.

M.A.: Nothing to disclose.

A.D.: Nothing to disclose.

R.S.M.Y.: Research grant funding from the Canadian Institutes of Health Research, Genome Canada, ZonMw, Reumafonds, Arthritis Foundation, The Arthritis Society, CARRA, and the Heart and Stroke Foundation of Canada.

Financial support for this work: None.

References

1. Newburger, J. W., Takahashi, M., & Burns, J. C. (2016). Kawasaki disease. *Journal of the American College of Cardiology, 67*, 1738–1749.
2. Manlhiot, C., Yeung, R. S. M., Clarizia, N. A., Chahal, N., & McCrindle, B. W. (2009). Kawasaki disease at the extremes of the age spectrum. *Pediatrics, 124*, e410–e415.
3. Anderson, M. S., Todd, J. K., & Glodé, M. P. (2005). Delayed diagnosis of Kawasaki syndrome: An analysis of the problem. *Pediatrics, 115*, e428–e433.
4. McCrindle, B. W., Rowley, A. H., Newburger, J. W., Burns, J. C., Bolger, A. F., Gewitz, M., et al. (2017). Diagnosis, treatment, and long-term management of Kawasaki disease: A scientific statement for health professionals from the American Heart Association. *Circulation, 135*, e927–e999.
5. Kawasaki, T. (1967). Acute febrile mucocutaneous syndrome with lymphoid involvement with specific desquamation of the fingers and toes in children. *Arerugi Allergy, 16*, 178–222.
6. Research Committee of the Japanese Society of Pediatric Cardiology, and Cardiac Surgery Committee for Development of Guidelines for Medical Treatment of Acute Kawasaki Disease. (2014). Guidelines for medical treatment of acute Kawasaki disease: Report of the Research Committee of the Japanese Society of Pediatric Cardiology and Cardiac Surgery (2012 revised version). *Pediatrics International, 56*, 135–158.
7. Marchesi, A., Tarissi de Jacobis, I., Rigante, D., Rimini, A., Malorni, W., Corsello, G., et al. (2018). Kawasaki disease: Guidelines of the Italian Society of Pediatrics, Part I—Definition, epidemiology, etiopathogenesis, clinical expression and management of the acute phase. *Italian Journal of Pediatrics, 44*, 102.

8. Marchesi, A., Tarissi de Jacobis, I., Rigante, D., Rimini, A., Malorni, W., Corsello, G., et al. (2018). Kawasaki disease: Guidelines of Italian Society of Pediatrics, Part II—Treatment of resistant forms and cardiovascular complications, follow-up, lifestyle and prevention of cardiovascular risks. *Italian Journal of Pediatrics, 44*, 103.
9. Manlhiot, C., Christie, E., McCrindle, B. W., Rosenberg, H., Chahal, N., & Yeung, R. S. M. (2012). Complete and incomplete Kawasaki disease: Two sides of the same coin. *European Journal of Pediatrics, 171*, 657–662.
10. Salgado, A. P., Ashouri, N., Berry, E. K., Sun, X., Jain, S., Burns, J. C., et al. (2017). High risk of coronary artery aneurysms in infants younger than 6 months of age with Kawasaki disease. *The Journal of Pediatrics, 185*, 112–116.e1.
11. Ravelli, A., Minoia, F., Davì, S., Horne, A., Bovis, F., Pistorio, A., et al. (2016). 2016 Classification criteria for macrophage activation syndrome complicating systemic juvenile idiopathic arthritis: A European League Against Rheumatism/American College of Rheumatology/Paediatric Rheumatology International Trials Organisation Collaborative Initiative. *Annals of the Rheumatic Diseases, 75*, 481–489.
12. Hadchouel, M., Prieur, A. M., & Griscelli, C. (1985). Acute hemorrhagic, hepatic, and neurologic manifestations in juvenile rheumatoid arthritis: Possible relationship to drugs or infection. *The Journal of Pediatrics, 106*, 561–566.
13. Stéphan, J. L., Zeller, J., Hubert, P., Herbelin, C., Dayer, J. M., & Prieur, A. M. (1993). Macrophage activation syndrome and rheumatic disease in childhood: A report of four new cases. *Clinical and Experimental Rheumatology, 11*, 451–456.
14. Stéphan, J. L., Koné-Paut, I., Galambrun, C., Mouy, R., Bader-Meunier, B., & Prieur, A. M. (2001). Reactive haemophagocytic syndrome in children with inflammatory disorders. A retrospective study of 24 patients. *Rheumatology, 40*, 1285–1292.
15. Schulert, G. S., & Grom, A. A. (2015). Pathogenesis of macrophage activation syndrome and potential for cytokine-directed therapies. *Annual Review of Medicine, 66*, 145–159.
16. Moolani, Y. (2010). The role of costimulation in the persistence of the immune response in Kawasaki disease. https://tspace.library.utoronto.ca/handle/1807/18898
17. Schulte, D. J., Yilmaz, A., Shimada, K., Fishbein, M. C., Lowe, E. L., Chen, S., et al. (2009). Involvement of innate and adaptive immunity in a murine model of coronary arteritis mimicking Kawasaki disease. *Journal of Immunology, 1950*(183), 5311–5318.
18. Lau, A. C., Duong, T. T., Ito, S., & Yeung, R. S. M. (2008). Matrix metalloproteinase 9 activity leads to elastin breakdown in an animal model of Kawasaki disease. *Arthritis and Rheumatism, 58*, 854–863.
19. Lau, A. C., Rosenberg, H., Duong, T. T., McCrindle, B. W., & Yeung, R. S. M. (2007). Elastolytic matrix metalloproteinases and coronary outcome in children with Kawasaki disease. *Pediatric Research, 61*, 710–715.
20. Morita, A., Imada, Y., Igarashi, H., & Yutsudo, T. (1997). Serologic evidence that streptococcal superantigens are not involved in the pathogenesis of Kawasaki disease. *Microbiology and Immunology, 41*, 895–900.
21. Herman, A., Kappler, J. W., Marrack, P., & Pullen, A. M. (1991). Superantigens: Mechanism of T-cell stimulation and role in immune responses. *Annual Review of Immunology, 9*, 745–772.
22. Duong, T. T., & Yeung, R. S. M. (2017). Update on pathogenesis: Lessons learned from animal models of disease. In B. T. Saji, J. W. Newburger, J. C. Burns, & M. Takahashi (Eds.), *Kawasaki disease: Current understanding of the mechanism and evidence-based treatment* (pp. 45–51). Tokyo: Springer.
23. Alphonse, M. P., Duong, T. T., Shumitzu, C., Hoang, T. L., McCrindle, B. W., Franco, A., et al. (2016). Inositol-triphosphate 3-kinase C mediates inflammasome activation and treatment response in Kawasaki disease. *Journal of Immunology, 1950*(197), 3481–3489.
24. Pachlopnik Schmid, J., Ho, C.-H., Chrétien, F., Lefebvre, J. M., Pivert, G., Kosco-Vilbois, M., et al. (2009). Neutralization of IFNgamma defeats haemophagocytosis in LCMV-infected perforin- and Rab27a-deficient mice. *EMBO Molecular Medicine, 1*, 112–124.

25. Jordan, M. B., Hildeman, D., Kappler, J., & Marrack, P. (2004). An animal model of hemophagocytic lymphohistiocytosis (HLH): CD8+ T-cells and interferon gamma are essential for the disorder. *Blood, 104*, 735–743.
26. Grom, A. A., Horne, A., & De Benedetti, F. (2016). Macrophage activation syndrome in the era of biologic therapy. *Nature Reviews Rheumatology, 12*, 259–268.
27. Billiau, A. D., Roskams, T., Van Damme-Lombaerts, R., Matthys, P., & Wouters, C. (2005). Macrophage activation syndrome: Characteristic findings on liver biopsy illustrating the key role of activated, IFN-gamma-producing lymphocytes and IL-6- and TNF-alpha-producing macrophages. *Blood, 105*, 1648–1651.
28. Chan, W. C., Duong, T. T., & Yeung, R. S. M. (2004). Presence of IFN-gamma does not indicate its necessity for induction of coronary arteritis in an animal model of Kawasaki disease. *Journal of Immunology, 1950*(173), 3492–3503.
29. Bleesing, J., Prada, A., Siegel, D. M., Villanueva, J., Olson, J., Ilowite, N. T., et al. (2007). The diagnostic significance of soluble CD163 and soluble interleukin-2 receptor alpha-chain in macrophage activation syndrome and untreated new-onset systemic juvenile idiopathic arthritis. *Arthritis and Rheumatism, 56*, 965–971.
30. Behrens, E. M., Canna, S. W., Slade, K., Rao, S., Kreiger, P. A., Paessler, M., et al. (2011). Repeated TLR9 stimulation results in macrophage activation syndrome-like disease in mice. *The Journal of Clinical Investigation, 121*, 2264–2277.
31. Rosenkranz, M. E., Schulte, D. J., Agle, L. M. A., Wong, M. H., Zhang, W., Ivashkiv, L., et al. (2005). TLR2 and MyD88 contribute to Lactobacillus casei extract-induced focal coronary arteritis in a mouse model of Kawasaki disease. *Circulation, 112*, 2966–2973.
32. Lee, Y., Schulte, D. J., Shimada, K., Chen, S., Crother, T. R., Chiba, N., et al. (2012). Interleukin-1β is crucial for the induction of coronary artery inflammation in a mouse model of Kawasaki disease. *Circulation, 125*, 1542–1550.
33. Wakita, D., Kurashima, Y., Crother, T. R., Noval Rivas, M., Lee, Y., Chen, S., et al. (2016). Role of interleukin-1 signaling in a mouse model of Kawasaki disease-associated abdominal aortic aneurysm. *Arteriosclerosis, Thrombosis, and Vascular Biology, 36*, 886–897.
34. Cohen, S., Tacke, C. E., Straver, B., Meijer, N., Kuipers, I. M., & Kuijpers, T. W. (2012). A child with severe relapsing Kawasaki disease rescued by IL-1 receptor blockade and extracorporeal membrane oxygenation. *Annals of the Rheumatic Diseases, 71*, 2059–2061.
35. Hsueh, K.-C., Lin, Y.-J., Chang, J.-S., Wan, L., Tsai, Y.-H., Tsai, C.-H., et al. (2008). Influence of interleukin 18 promoter polymorphisms in susceptibility to Kawasaki disease in Taiwan. *The Journal of Rheumatology, 35*, 1408–1413.
36. Biezeveld, M. H., Kuipers, I. M., Geissler, J., Lam, J., Ottenkamp, J. J., Hack, C. E., et al. (2003). Association of mannose-binding lectin genotype with cardiovascular abnormalities in Kawasaki disease. *Lancet, 361*, 1268–1270.
37. Weng, K.-P., Hsieh, K.-S., Huang, S.-H., Ou, S.-F., Lai, T.-J., Tang, C.-W., et al. (2013). Interleukin-18 and coronary artery lesions in patients with Kawasaki disease. *Journal of the Chinese Medical Association, 76*, 438–445.
38. Akira, S. (2000). The role of IL-18 in innate immunity. *Current Opinion in Immunology, 12*, 59–63.
39. Takada, H., Takahata, Y., Nomura, A., Ohga, S., Mizuno, Y., & Hara, T. (2003). Increased serum levels of interferon-gamma-inducible protein 10 and monokine induced by gamma interferon in patients with haemophagocytic lymphohistiocytosis. *Clinical and Experimental Immunology, 133*, 448–453.
40. Novick, D., Schwartsburd, B., Pinkus, R., Suissa, D., Belzer, I., Sthoeger, Z., et al. (2001). A novel IL-18BP ELISA shows elevated serum IL-18BP in sepsis and extensive decrease of free IL-18. *Cytokine, 14*, 334–342.
41. Nofech-Mozes, Y., & Garty, B.-Z. (2003). Thrombocytopenia in Kawasaki disease: A risk factor for the development of coronary artery aneurysms. *Pediatric Hematology and Oncology, 20*, 597–601.

42. Eladawy, M., Dominguez, S. R., Anderson, M. S., & Glodé, M. P. (2011). Abnormal liver panel in acute Kawasaki disease. *The Pediatric Infectious Disease Journal, 30*, 141–144.
43. Yeung, R. S. M. (2007). Phenotype and coronary outcome in Kawasaki's disease. *Lancet, 369*, 85–87.
44. Harada, K. (1991). Intravenous gamma-globulin treatment in Kawasaki disease. *Acta paediatrica Japonica: Overseas Edition, 33*, 805–810.
45. Kobayashi, T., Inoue, Y., Takeuchi, K., Okada, Y., Tamura, K., Tomomasa, T., et al. (2006). Prediction of intravenous immunoglobulin unresponsiveness in patients with Kawasaki disease. *Circulation, 113*, 2606–2612.
46. Beiser, A. S., Takahashi, M., Baker, A. L., Sundel, R. P., & Newburger, J. W. (1998). A predictive instrument for coronary artery aneurysms in Kawasaki disease. US Multicenter Kawasaki Disease Study Group. *The American Journal of Cardiology, 81*, 1116–1120.
47. Ohga, S., Ooshima, A., Fukushige, J., & Ueda, K. (1995). Histiocytic haemophagocytosis in a patient with Kawasaki disease: Changes in the hypercytokinaemic state. *European Journal of Pediatrics, 154*, 539–541.
48. Muise, A., Tallett, S. E., & Silverman, E. D. (2003). Are children with Kawasaki disease and prolonged fever at risk for macrophage activation syndrome? *Pediatrics, 112*, e495.
49. Titze, U., Janka, G., Schneider, E. M., Prall, F., Haffner, D., & Classen, C. F. (2009). Hemophagocytic lymphohistiocytosis and Kawasaki disease: Combined manifestation and differential diagnosis. *Pediatric Blood & Cancer, 53*, 493–495.
50. Hendricks, M., Pillay, S., Davidson, A., De Decker, R., & Lawrenson, J. (2010). Kawasaki disease preceding haemophagocytic lymphohistiocytosis: Challenges for developing world practitioners. *Pediatric Blood & Cancer, 54*, 1023–1025.
51. Simonini, G., Pagnini, I., Innocenti, L., Calabri, G. B., De Martino, M., & Cimaz, R. (2010). Macrophage activation syndrome/hemophagocytic lymphohistiocytosis and Kawasaki disease. *Pediatric Blood & Cancer, 55*, 592.
52. Kim, H. K., Kim, H. G., Cho, S. J., Hong, Y. M., Sohn, S., Yoo, E.-S., et al. (2011). Clinical characteristics of hemophagocytic lymphohistiocytosis related to Kawasaki disease. *Pediatric Hematology and Oncology, 28*, 230–236.
53. Mukherjee, D., Pal, P., Kundu, R., & Niyogi, P. (2014). Macrophage activation syndrome in Kawasaki disease. *Indian Pediatrics, 51*, 148–149.
54. Ogawa, M., & Hoshina, T. (2015). Hemophagocytic lymphohistiocytosis prior to the diagnosis of Kawasaki disease. *Indian Pediatrics, 52*, 78.
55. Wang, W., Gong, F., Zhu, W., Fu, S., & Zhang, Q. (2015). Macrophage activation syndrome in Kawasaki disease: More common than we thought? *Seminars in Arthritis and Rheumatism, 44*, 405–410.
56. Choi, U.-Y., Han, S.-B., Lee, S.-Y., & Jeong, D.-C. (2017). Should refractory Kawasaki disease be considered occult macrophage activation syndrome? *Seminars in Arthritis and Rheumatism, 46*, e17.
57. Doğan, V., Karaaslan, E., Özer, S., Gümüşer, R., & Yılmaz, R. (2016). Hemophagocytosis in the acute phase of fatal Kawasaki disease in a 4 month-old girl. *Balkan Medical Journal, 33*, 470–472.
58. Latino, G. A., Manlhiot, C., Yeung, R. S. M., Chahal, N., & McCrindle, B. W. (2010). Macrophage activation syndrome in the acute phase of Kawasaki disease. *Journal of Pediatric Hematology/Oncology, 32*, 527–531.
59. Latino Giuseppe, A., Cedric, M., McCrindle Brian, W., Nita, C., & Yeung Rae, S. (2015). Abstract O.28: Macrophage activation syndrome associated with Kawasaki disease. *Circulation, 131*, AO28.
60. Yamamoto, N., Sato, K., Hoshina, T., Kojiro, M., & Kusuhara, K. (2015). Utility of ferritin as a predictor of the patients with Kawasaki disease refractory to intravenous immunoglobulin therapy. *Modern Rheumatology, 25*, 898–902.

61. García-Pavón, S., Yamazaki-Nakashimada, M. A., Báez, M., Borjas-Aguilar, K. L., & Murata, C. (2017). Kawasaki disease complicated with macrophage activation syndrome: A systematic review. *Journal of Pediatric Hematology/Oncology, 39*, 445–451.
62. Ravelli, A., Magni-Manzoni, S., Pistorio, A., Besana, C., Foti, T., Ruperto, N., et al. (2005). Preliminary diagnostic guidelines for macrophage activation syndrome complicating systemic juvenile idiopathic arthritis. *The Journal of Pediatrics, 146*, 598–604.
63. Kumar, S., Vaidyanathan, B., Gayathri, S., & Rajam, L. (2013). Systemic onset juvenile idiopathic arthritis with macrophage activation syndrome misdiagnosed as Kawasaki disease: Case report and literature review. *Rheumatology International, 33*, 1065–1069.
64. Takahara, T., Shimizu, M., Nakagishi, Y., Kinjo, N., & Yachie, A. (2015). Serum IL-18 as a potential specific marker for differentiating systemic juvenile idiopathic arthritis from incomplete Kawasaki disease. *Rheumatology International, 35*, 81–84.
65. Mizuta, M., Shimizu, M., Inoue, N., Kasai, K., Nakagishi, Y., Takahara, T., et al. (2016). Serum ferritin levels as a useful diagnostic marker for the distinction of systemic juvenile idiopathic arthritis and Kawasaki disease. *Modern Rheumatology, 26*, 929–932.
66. Yeung, R. S., van Veenendaal, M., Manlhiot, C., Schneider, R., & McCrindle, B. W. (2015). Abstract O.36: Kawasaki disease and systemic juvenile idiopathic arthritis—Two ends of the same spectrum. *Circulation, 131*, AO36.
67. Dong, S., Bout-Tabaku, S., Texter, K., & Jaggi, P. (2015). Diagnosis of systemic-onset juvenile idiopathic arthritis after treatment for presumed Kawasaki disease. *The Journal of Pediatrics, 166*, 1283–1288.
68. Kelly, A., & Ramanan, A. V. (2008). A case of macrophage activation syndrome successfully treated with anakinra. *Nature Clinical Practice. Rheumatology, 4*, 615–620.
69. Miettunen, P. M., Narendran, A., Jayanthan, A., Behrens, E. M., & Cron, R. Q. (2011). Successful treatment of severe paediatric rheumatic disease-associated macrophage activation syndrome with interleukin-1 inhibition following conventional immunosuppressive therapy: Case series with 12 patients. *Rheumatology, 50*, 417–419.
70. Durand, M., Troyanov, Y., Laflamme, P., & Gregoire, G. (2010). Macrophage activation syndrome treated with anakinra. *The Journal of Rheumatology, 37*, 879–880.
71. Shafferman, A., Birmingham, J. D., & Cron, R. Q. (2014). High dose Anakinra for treatment of severe neonatal Kawasaki disease: A case report. *Pediatric Rheumatology Online Journal, 12*, 26.

The Intersections of Autoinflammation and Cytokine Storm

Scott W. Canna

Introduction

The inflammatory nature of sepsis was formally recognized well over a decade ago, when investigators coined the term systemic inflammatory response syndrome (SIRS) [1]. Despite great preclinical promise, trials of targeted anti-inflammatory therapies were initially found to be failures in sepsis (see chapter "Cytokine Storm and Sepsis-Induced Multiple Organ Dysfunction Syndrome"). Just as these setbacks in understanding the immunopathology of sepsis were occurring, a new family of diseases caused by primary defects in innate inflammatory pathways was growing: the autoinflammatory disorders (AIDs). This chapter briefly summarizes the ways in which insights from the study of AIDs have advanced the understanding of human inflammation in general and then focuses on the limited, though very instructive, circumstances in which autoinflammation and SIRS overlap.

Defining Autoinflammation

Autoinflammation is the term ascribed to persistent organ-specific and/or systemic inflammation not explained by infection, malignancy, or the effects of autoantibodies or antigen-specific T-cells. The term does not specify acute versus chronic inflammation, nor is it restricted to any set of tissues or organs. Importantly, its meaning is enmeshed with its history.

The concept of autoinflammation arose in the context of genetic research into the periodic fever syndromes. It has blossomed as next-generation sequencing

S. W. Canna (✉)
RK Mellon Institute for Pediatric Research/Pediatric Rheumatology, University of Pittsburgh/UPMC Children's Hospital of Pittsburgh, Pittsburgh, PA, USA
e-mail: scott.canna@chp.edu

R. Q. Cron, E. M. Behrens (eds.), *Cytokine Storm Syndrome*,
https://doi.org/10.1007/978-3-030-22094-5_24

techniques revealed the genetic origins for a startling number of these disorders. The first genetic association was made with familial Mediterranean fever (FMF), a disorder of recurrent attacks of fever, serositis, and erisypelas-like rash that can result in amyloidosis and renal failure. The gene for FMF was identified by positional cloning and ultimately mapped to a gene dubbed *MEFV*, encoding the protein Pyrin [2]. The mechanisms by which Pyrin mutations resulted in autoinflammation, and an explanation for the high carrier frequency of such mutations in Mediterranean populations, would take another 25 years to understand. However, soon after the FMF/*MEFV* association, another hereditary periodic fever syndrome was associated with mutations in the tumor necrosis factor receptor (TNF-R). The disease was renamed TNF-R Associated Periodic Syndrome (TRAPS), and the authors coined the term "autoinflammatory" [3]. Quickly thereafter, three independent groups made a pivotal discovery: gain-of-function mutations in the *NLRP3* gene resulted in hyperactivation of a newly described innate immune complex called the inflammasome. Different mutations in *NLRP3* resulted in a spectrum of diseases, eventually called cryopyrin-associated periodic syndromes (CAPS), ranging from familial cold-induced autoinflammatory syndrome (FCAS) to neonatal-onset multisystem inflammatory disease (NOMID: urticaria, sensorineural hearing loss, bony arthropathy, and sterile meningitis) [4–7].

The inflammasome is a potentially massive innate immune complex strongly linked to the concept of autoinflammation [8]. Inflammasomes form in response to the sensing of cytosolic danger- or pathogen-associated molecular patterns (DAMPs/PAMPs). The rate-limiting step of inflammasome formation seems to be the homo-oligomerization of several proteins capable of "nucleating" an inflammasome. Four proteins have been well-characterized to nucleate inflammasomes in mice and humans: Pyrin, NLRP3, NLRC4, and AIM2. Of these, only AIM2 has not (yet) been associated with a monogenic autoinflammatory disorder. Activated and oligomerized inflammasome nucleators then trigger an exponential cascade wherein thousands of molecules of an adaptor protein (ASC) recruit many thousands of an effector protease called Caspase-1. Once itself proteolytically activated, Caspase-1 proteolytically activates three highly inflammatory effector molecules: IL-1β, IL-18, and Gasdermin-D. IL-1β and IL-18 signal through their respective IL-1 family receptors, whereas Gasdermin-D mediates an inflammatory cell death known as pyroptosis [9]. Only recently was the link identified between the Pyrin-inflammasome, pathogen-sensing, and gain-of-function *MEFV* mutations in FMF [10]. By contrast, the role of NLRC4 as a detector of bacterial proteins in the cytosol was known many years before the first identification of gain-of-function *NLRC4* mutations with autoinflammation [11, 12]. These diseases caused by mutations that directly result in inflammasome activity are collectively known as inflammasomopathies.

The association of these diseases with the inflammasome not only ignited mechanistic research, it indicated a specific therapeutic intervention: blocking IL-1 signaling in the NLRP3-mediated CAPS was overwhelmingly successful [13, 14] and paved the way for the efficacious use of this strategy in a host of other autoinflammatory (including TRAPS, FMF, hyper-IgD syndrome (HIDS), deficiency of IL-1 receptor antagonist (DIRA)) and rheumatic (systemic juvenile idiopathic arthritis (SJIA), adult-onset Still's disease (AOSD), gout) diseases [7]. IL-1 was initially blocked with a recombinant IL-1 Receptor Antagonist (IL-1RA, anakinra) that inhibited signaling by both IL-1α and IL-1β, but similar efficacy has been observed with strategies specific to IL-1β.

For many years, the association of autoinflammation with IL-1 defined the field. More recently, however, diseases with little/no response to IL-1 blockade have been discovered. As opposed to triggering the inflammasome, many of these diseases are associated with abnormal production of Interferon (IFN). These "interferonopathies" are associated with genetic defects causing dysregulation of cytosolic nucleic acid sensing/handling or proteasomal protein degradation. A staggering array of autoinflammatory disorders have also been identified through whole-exome sequencing. Their mechanisms are not so easily categorized, and their precise description is beyond the scope of this chapter but has been recently and extensively reviewed [7, 15, 16].

Problems with Defining Autoinflammation

As with any classification system in the complex and evolving landscape of human immunology, the concept of autoinflammation is imperfect and at times arbitrary. The two major limitations of the autoinflammatory paradigm most relevant to the subject of cytokine storm syndromes (CSS) are (1) diseases with features that overlap between autoinflammation and immunodeficiency or autoimmunity, and (2) diseases with phenotypic features suggestive of autoinflammation, but lacking a known genetic cause to support such a categorization.

Genetic impairment of cellular cytotoxicity represents an important overlap between the concepts of autoinflammation and immunodeficiency. This mechanism is discussed at length throughout this book as an important cause of primary hemophagocytic lymphohistiocytosis (pHLH) and a potential contributor to secondary HLH (sHLH) and macrophage activation syndrome (MAS). The hyperinflammation and CSS associated with pHLH is canonically triggered by viral infection, associated with persistent viremia, and caused by antigen-specific T-cell activation. Thus, deficiency of Perforin, Munc13-4, etc. is listed among the immunodeficiencies. However, mechanistic work has shown that cytotoxicity is an immunoregulatory mechanism important both for preventing excessive antigen-dependent T- and NK-cell activation as well as clearing virus. Likewise, damage in pHLH is best prevented by blocking the inflammatory response rather than targeting viral cytopathic effects [17–19]. Genetic defects resulting in inflammation in the context of other immunodeficiencies, such as NEMO or HOIL1 deficiency, reinforce the notion that severe infection can drive CSS, and are discussed in detail in chapter "Primary Immunodeficiencies and Cytokine Storm Syndromes".

Several monogenic disorders, particularly in the category of Interferonopathies, blur the line between autoinflammation and autoimmunity [7], but are beyond the scope of this chapter. The overlap between autoimmunity and autoinflammation in CSS is perhaps best explored in the context of systemic lupus erythematosus-associated MAS, as detailed in chapter "Systemic Lupus Erythematosus and Cytokine Storm".

Several rheumatic diseases, most notably SJIA, AOSD, Kawasaki disease, and sarcoidosis, meet the definition of autoinflammation, but are not routinely categorized as such (in part) owing to the absence of a single genetic cause. These are discussed in detail in chapter "Cytokine Storm Syndrome Associated with Systemic Juvenile Idiopathic Arthritis", chapter "Kawasaki Disease-Associated Cytokine Storm Syndrome", and chapter "Macrophage Activation Syndrome in the Setting of Rheumatic Diseases". Thus, this chapter will focus on the ways in which monogenic, canonically autoinflammatory disorders inform our understanding of CSS.

The Rare Concordance of Autoinflammation and Cytokine Storm

If by CSS we mean a pattern of systemic inflammation, typified by HLH and MAS, that can result in shock, multiorgan dysfunction syndrome (MODS), and death, then the CSS phenotype is notably absent in most autoinflammation. Notwithstanding a few case reports, MAS/HLH, shock, and MODS are not listed among the cardinal features of most autoinflammatory diseases [7, 16] (Table 1). Many of these cohorts, particularly in FMF and CAPS, include reasonably large numbers of patients followed longitudinally over many years. The more recent broadening of the autoinflammatory phenotype to include disorders of excessive IFN production, ubiquitylation disorders, and causes of vasculitis (Table 1) has not resulted in the inclusion of CSS as a feature. Thus, it would seem that systemic inflammation, even in its (perhaps) purest forms, is not sufficient in degree and/or direction to induce a CSS.

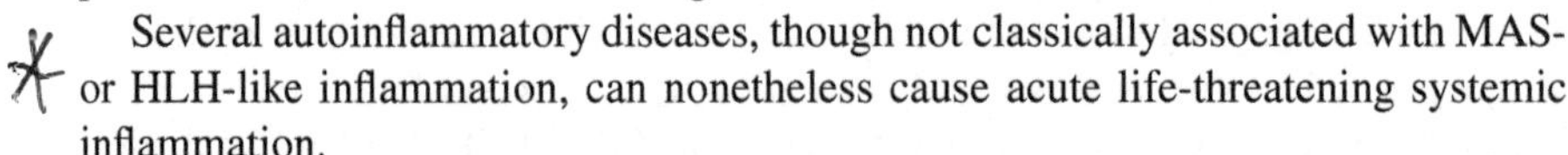

Several autoinflammatory diseases, though not classically associated with MAS- or HLH-like inflammation, can nonetheless cause acute life-threatening systemic inflammation.

Aicardi–Goutieres Syndrome (AGS): AGS is classically a neonatal-onset, IFN-mediated autoinflammatory condition that mimics in utero viral infection [20]. There are at least seven monogenic causes associated with AGS (Table 1), but of these the earliest-onset and most severe phenotypes have been associated with deficiency of the nucleases *TREX1*, *RNASEH2A*, and, *RNASEH2C*. Severe AGS presenting in infancy can manifest as livedoid skin rash, symptomatic thrombocytopenia, hepatosplenomegaly, and occasionally fever [7]. The most consistent and devastating feature, however, is subacute encephalomyelitis that can be associated with rapid neurologic decline. Mortality with these more severe mutations can be over 30% [20]. Severe AGS mutations all result in an impairment in the ability degrade (usually endogenous) cytosolic nucleic acids. The build-up of these nucleic acids triggers cytosolic viral-sensing mechanisms and an immense type I IFN response detectable in both blood and CSF. Stress and infection can increase the production/accumulation of these nucleic acids, as well. Strategies to limit endogenous nucleic

Table 1 Overview of canonical autoinflammatory disorders

Syndrome	Acronym	Gene	Protein	Primary manifestations
IL-1/inflammasome mediated disorders				
Familial Mediterranean fever	FMF	*MEFV*	PYRIN	Periodic fever, serositis, neutrophilic rash, renal amyloidosis
Familial cold-induced autoinflammatory syndrome	FCAS[a]	*NLRP3*	NLRP3/ Cryopyrin	Cold-induced fever, malaise, neutrophilic urticaria
Muckle–Wells syndrome	MWS[a]	*NLRP3*	NLRP3/ Cryopyrin	Fever, malaise, neutrophilic urticaria, sensorineural hearing loss
Neonatal onset multisystem inflammatory disease	NOMID[a]	*NLRP3*	NLRP3/ Cryopyrin	As above + sterile meningitis and bony
Hyper-IgD syndrome	HIDS	*MVK*	Mevalonate kinase	Fever, rash, arthralgia, GI upset, adenopathy, hepatosplenomegaly
TNF-receptor associated periodic syndromes	TRAPS	*TNFRSF1A*	TNFR1A	Myalgia/arthralgia, periorbital edema, serositis
Deficiency of IL-1 receptor antagonist	DIRA	*IL1RN*	IL-1RA	Sterile osteomyelitis, pustular rash, thrombosis
NLRC4-associated autoinflammatory diseases	N4AID	*NLRC4*	NLRC4	Spectrum including lymphohistiocytic rash to MAS, infantile enterocolitis
IFN-mediated disorders				
STING-associated vasculopathy of infancy	SAVI	*TMEM173*	STING	Fever, purpura, acral necrosis, interstitial lung disease, cytopenias
Aicardi-Goutieres Syndrome 1	AGS1	*TREX1*	TREX1	Livedo reticularis, H/ Smegaly, basal ganglia calficification, developmental delay, dystonia, thrombocytopenia
Aicardi-Goutieres Syndrome 2	AGS2	*RNASEH2B*	RNH2B	
Aicardi-Goutieres Syndrome 3	AGS3	*RNASEH2C*	RNH2C	
Aicardi-Goutieres Syndrome 4	AGS4	*RNASEH2A*	RNH2A	
Aicardi-Goutieres Syndrome 5	AGS5	*SAMHD1*	*SAMH1*	
Aicardi-Goutieres Syndrome 6	AGS6	*ADAR*	DSRAD	
Aicardi-Goutieres Syndrome 7	AGS7	*IFIH1*	MDA5	
Chronic atypical neutrophilic dermatosis lipodystrophy elevated temperature	CANDLE	*PSMB8, etc.*	PSB8, etc.	Nodular panniculitis, lipodystrophy, myositis, H/ Smegaly
NF-κB/ubiquitin-mediated disorders				

(continued)

Table 1 (continued)

Syndrome	Acronym	Gene	Protein	Primary manifestations
Pediatric granulomatous arthritis	PGA/ Blau	*NOD2*	NOD2	Fever, papular rash, arthritis, uveitis, adenopathy
Haploinsufficiency of A20	HA20	*TNFAIP3*	A20	Oral/genital ulcers, uveitis, arthritis, colitis, erythema nodosum
Otulopenia	–	*OTULIN*	OTULIN	Fevers, neutrophilic dermatosis, panniculitis, failure to thrive, infections
XIAP—deficiency	XLP2	*XIAP/ BIRC4*	XIAP	Variable from IBD, uveitis, arthritis to (often EBV-associated) MAS
Unclassified disorders				
Pyogenic arthritis, pyoderma gangrenosum, acne	PAPA	*PSTPIP1*	PPIP1	Pyogenic arthritis, pyoderma gangrenosum, acne
Deficiency of ADA2	DADA2	*CECR1*	ADA2	Cutaneous, visceral, and CNS vasculitis/stroke

H/Smegaly = hepatosplenomegaly
[a]Part of the continuum known as Cryopyrin-Associated Periodic Syndromes (CAPS)

acid production or inhibit IFN signaling are under investigation in AGS [21] (NCT02363452, NCT01724580).

Deficiency of IL-1 Receptor Antagonist (DIRA): After the successful use of recombinant IL-1 Receptor Antagonist (IL-1RA, anakinra) to treat NLRP3-mediated diseases, investigators found that a syndrome of severe neonatal inflammation, pustulosis, and osteomyelitis mapped to a region of chromosome 2 containing the gene for endogenous IL-1RA. Patients with biallelic, loss-of-function mutations resulting in deficiency of IL-1RA [22] often presented within the first days of life with fetal distress, pustular skin rash, and oral mucosal lesions, and were found to have characteristic sterile osteomyelitis (often including the ribs). At least two children expired due to SIRS. Administration of anakinra (recombinant IL-1RA) usually results in dramatic cessation of ongoing inflammation, but can trigger anaphylaxis [22, 23].

Autoinflammatory Mutations and Sepsis

Despite the apparent failure of cytokine blockade in sepsis, the search for proinflammatory defects in sepsis patients has continued. Several investigators hypothesized that, though the rate of sepsis in canonical autoinflammatory diseases may not be substantially elevated, patients with severe sepsis may have contributory mutations in autoinflammatory genes. For example, the carrier frequency of pathogenic

MEFV mutations in critically ill Turkish adults (30%) appeared significantly higher than the general population carrier rate (10%) [24]. Despite this, multiple GWAS studies in sepsis have failed to link inflammasome-associated genes with sepsis outcomes [25]. However, typical autoinflammatory mutations occur at too low a population frequency to reach significance in most GWAS studies. As discussed elsewhere in this text, heterozygous mutations in genes affecting perforin-mediated cytotoxicity may predispose to viral sepsis [26]. Ongoing genome-wide investigations (including whole exome sequencing) in individual sepsis patients should help elucidate the strength of this connection.

Innate Immunity in HLH

Notwithstanding the debate about whether impairment of cytotoxicity causes HLH through immunodeficiency or autoinflammation, there is considerable evidence for a crucial role of innate immunity in murine models of HLH. Unsurprisingly, innate immune signaling is a necessary and critical precondition for the CD8 activation that drives experimental HLH. In particular, MyD88 [27] and IL-33 [18] have been shown to be essential. IL-33 is a widely expressed alarmin that, like most IL-1 family cytokines and Toll-like Receptors, requires MyD88 for its signaling. Furthermore, innate immune cells appear to be necessary effectors for HLH immunopathology. IFN-γ signaling in murine macrophages was essential for anemia and hemophagocytosis [28], and a xenograft model of HLH was amenable to treatment aimed at depleting myeloid cells [29]. Finally, HLH was far less severe and associated with notably less myeloid cell infiltration when NK cells, but not T-cells, retained cytotoxic function in murine pHLH [30]. This was despite equal levels of viremia and IFN-γ. These murine data support a critical role of innate immunity both in enabling and executing immunopathology in an infectious, antigen-specific model of HLH.

NLRC4, XIAP, and MAS Under the Umbrella of Autoinflammation

MAS is most canonically associated with systemic juvenile idiopathic arthritis (SJIA) and adult onset Still's disease (AOSD), and it is in the context of these diseases that it is best studied (see chapter "Cytokine Storm Syndrome Associated with Systemic Juvenile Idiopathic Arthritis" for details). Though there is notable disagreement about their classification [31], SJIA and AOSD are commonly categorized as autoinflammatory diseases. Several observations drive this presumptive classification, including the response to IL-1 and/or IL-6 inhibition [32, 33], the genetic distinction from other forms of JIA [34], and the absence of known autoantigens. Perhaps most uniquely and importantly, increasing evidence has associated extreme elevations of the inflammasome-activated cytokine IL-18 with MAS in SJIA and

AOSD [35, 36]. Early clinical trial data suggest IL-18 blockade is efficacious in AOSD [37]. Interestingly, the degree of IL-18 elevation seen in SJIA/AOSD and MAS has not been observed in classical autoinflammatory diseases [11, 38].

In 2014, two groups discovered that gain-of-function mutations in *NLRC4* resulted in a syndrome of infantile enterocolitis and recurrent, life-threatening MAS [11, 12]. As with the *NLRP3* and *MEFV* inflammasomopathies, these *NLRC4* mutations gave rise to spontaneous/increased macrophage inflammasome formation, cell death, IL-1β and IL-18 secretion, and a disease potentially responsive to IL-1 inhibition [11]. However, as previously stated, both MAS and enterocolitis are extremely rare in *NLRP3*- or *MEFV*-mediated diseases (CAPS and FMF, respectively). Furthermore, levels of peripheral IL-18 were consistently and extremely elevated in all affected *NLRC4* patients, but not in CAPS patients [11]. Reinforcing the potentially violent potency of NLRC4 activation, an infant born premature due to placental thrombotic disease and fetal edema soon succumbed to MAS and enterocolitis. She was posthumously found to have somatic mosaicism (not present in all cells) for a pathogenic *NLRC4* mutation [39]. Two other patients with activating NLRC4 mutations and extremely elevated serum IL-18 succumbed to MAS without severe intestinal disease [40]. In 2016, an infant was identified with life-threatening MAS and enterocolitis due to *NLRC4* mutation, and confirmed both extremely elevated total IL-18 as well as elevated free IL-18 [41]. Free IL-18 is unbound by the endogenous, soluble inhibitor known as IL-18 Binding Protein (IL-18BP), and is thought to be the biologically active IL-18 fraction [36]. After a minimal response to aggressive MAS treatment, including IL-1 inhibition, this patient showed a dramatic and acute response to IL-18 inhibition in the form of recombinant IL-18BP. Notably, the relative amounts of IL-18 and IL-18BP in serum versus stool (diarrheal) during active disease suggested IL-18 was regulated differently in the inflamed gut versus peripheral blood [41]. Consistent with IL-18 as a driver of IFN-γ production, one patient with infantile NLRC4-MAS (without overt enterocolitis) responded dramatically to IFN-γ inhibition, a strategy now approved for use in refractory pHLH [42]. Likewise, biomarkers of IFN-γ activity associate with active disease in both MAS and HLH [43]. Indeed, extremely elevated IL-18 appears to distinguish patients at risk for MAS both from other causes of hyperferritinemic CSS as well as other autoinflammatory diseases, and exacerbates experimental models of MAS [38, 44]. The expression of NLRC4 in intestinal epithelia, a rich source of proIL-18, offers one potential explanation for the association of extreme IL-18 elevation with NLRC4, but not NLRP3 or MEFV inflammasomopathies [38].

XIAP (aka *BIRC4*) deficiency is classically associated with an X-linked syndrome of EBV-associated HLH [45, 46]. Unlike pHLH, it is not associated with cytotoxic impairment; and unlike SAP deficiency (another cause of EBV-HLH), XIAP deficiency is not associated with lymphoma risk. XIAP is a widely expressed protein whose multiple functions likely all revolve around its ability to stabilize particular NF-κB signaling complexes. It inhibits apoptosis downstream of many in vitro stimuli, it is required for NOD2 signaling, and it prevents the inflammatory effects of RIPK3 and Caspase-8 [47–49]. Importantly, HLH associated with XIAP deficiency has also been associated with extreme and chronic IL-18 elevation [50]. More recently, MAS and chronic IL-18 elevation were also associated with specific mutations in the C-terminus of the Rho-GTPase CDC42 [51].

Phenotypes in patients with activating *NLRC4* mutations or XIAP deficiency are not restricted to CSS. XIAP deficiency is a major cause of monogenic inflammatory bowel disease (possibly relating to its role in NOD2 signaling), but has been associated with phenotypes like periodic fevers, recurrent skin infections, and arthritis [52]. Though recently described, *NLRC4* mutations have been associated with an extremely wide spectrum of disease severity, including mild cold-induced urticaria, erythematous nodules, colitis, and hearing loss/urticaria/CNS inflammation reminiscent of severe CAPS [53–55]. IL-18 levels have not been broadly assessed in these milder XIAP- and NLRC4-associated phenotypes, but such patients do not appear to be at higher risk for MAS. Nonetheless, the effects of IL-18 inhibition in NLRC4- and XIAP-related inflammation are currently under investigation (NCT03113760).

IL-18 at the Intersection of Autoinflammation and CSS

The correlation of extremely elevated IL-18 levels and MAS in SJIA/AOSD, NLRC4, and XIAP deficiency is striking. IL-18 is somewhat unique among inflammatory cytokines in that its mRNA expression is relatively stable in the face of varied stimuli, and it is expressed in both myeloid and epithelial cells. Its effects have been best studied on cytotoxic lymphocytes, where it strongly synergizes with cytokines like IL-12, IL-2, and IL-15 to drive IFN-γ production and cytotoxicity. The IL-18 receptor is constitutively expressed on Natural Killer cells, and induced upon activation in most T-cells. SJIA patients' NK cells appear to become insensitive to IL-18 through downregulation of cytosolic signaling, but otherwise retain cytotoxicity and cytokine production [56, 57]. In contrast to its origins in and effects on immune cells, IL-18 has an increasingly appreciated role at mucosal barrier sites. Murine data show a complex role for IL-18 in the gut, where it both promotes barrier immunity and gut immunopathology [58–61]. Thus insights into the biology of IL-18 in MAS appear poised to inform our understanding of the complex interplay between barrier and systemic inflammation.

Synthesis

Advances in the field of autoinflammation have largely followed watershed genetic discoveries. These discoveries have flowed into downstream mechanistic insights, ultimately validated by responses to targeted cytokine blockade (Table 2) [7, 16].

Table 2 Targeted treatment of monogenic autoinflammatory cytokine storm syndromes

Syndrome	Gene	Drug	Category	Effect	Reference/trial ID
AGS2	RNASEH2B	Ruxolitinib	Jak inhibitor	Given after CSS, improvement of IFN score	Tungler et al., Ann Rheum Dis [21]
AGS	Several	Baracitinib	Jak inhibitor	Unknown	NCT01724580 https://www.ncbi.nlm.nih.gov/pubmed/29649002
DIRA	IL1RN	Anakinra	IL-1 inhibitor	Rapid improvement, risk of anaphylaxis	Aksentijevich et al., NEJM [22]
					Mendonca et al., J Clin Immunol [23]
MAS	NLRC4	Anakinra	IL-1 inhibitor	Prevention of MAS flares	Canna et al., Nat Gen [11]
		Tadekinig-α	IL-18 inhibitor	Given with IL-1 inhibitor, MAS resolution/flare prevention	Canna et al., JACI [41]
Auto-inflammation	NLRC4/ XIAP	Tadekinig-α	IL-18 inhibitor	Unknown	NCT03113760
MAS	NLRC4	Emapalumab	IFNγ inhibitor	Improvement of MAS	Bracaglia et al., Pediatr Blood & Ca [42] (abstract)
MAS	N/A	Emapalumab	IFNγ inhibitor	Unknown	NCT03311854

Exemplifying the power of this pattern, the safety and utility of IL-1 blockade in monogenic autoinflammation prompted its use in genetically indistinct diseases like SJIA, and ultimately the review of IL-1 responses in MAS-like sepsis subsets [62]. Still, the lack of clinical overlap between canonical autoinflammation and CSS syndromes is itself instructive. The isolated effects of excess IL-1β or type I IFN appear insufficient to drive SIRS. However, biomarkers suggest that MAS and HLH are united by evidence for pathogenic IFN-γ activity. That NLRC4 hyperactivity and XIAP deficiency are both associated with MAS/HLH as well as extreme and chronic IL-18 suggests one potential mechanism of convergence. Taken together, these data suggest that the CSS/SIRS phenotype may rely on simultaneous excess of both IFN and IL-1 family members. As we improve our ability to immunophenotype CSS patients, including their responses to an ever-widening array of targeted therapies, the near future looks bright for precision medicine in CSS.

References

1. Goldstein, B., Giroir, B., & Randolph, A. (2005). International pediatric sepsis consensus conference: Definitions for sepsis and organ dysfunction in pediatrics. *Pediatric Critical Care Medicine, 6*(1), 2–8. https://doi.org/10.1097/01.PCC.0000149131.72248.E6
2. Pras, E., Aksentijevich, I., Gruberg, L., Balow Jr., J. E., Prosen, L., Dean, M., et al. (1992). Mapping of a gene causing familial Mediterranean fever to the short arm of chromosome 16. *The New England Journal of Medicine, 326*(23), 1509–1513. https://doi.org/10.1056/NEJM199206043262301
3. McDermott, M. F., Aksentijevich, I., Galon, J., McDermott, E. M., Ogunkolade, B. W., Centola, M., et al. (1999). Germline mutations in the extracellular domains of the 55 kDa TNF receptor, TNFR1, define a family of dominantly inherited autoinflammatory syndromes. *Cell, 97*(1), 133–144.
4. Hoffman, H. M., Mueller, J. L., Broide, D. H., Wanderer, A. A., & Kolodner, R. D. (2001). Mutation of a new gene encoding a putative pyrin-like protein causes familial cold autoinflammatory syndrome and Muckle-Wells syndrome. *Nature Genetics, 29*(3), 301–305. https://doi.org/10.1038/ng756
5. Feldmann, J., Prieur, A. M., Quartier, P., Berquin, P., Certain, S., Cortis, E., et al. (2002). Chronic infantile neurological cutaneous and articular syndrome is caused by mutations in CIAS1, a gene highly expressed in polymorphonuclear cells and chondrocytes. *American Journal of Human Genetics, 71*(1), 198–203.
6. Aksentijevich, I., Nowak, M., Mallah, M., Chae, J. J., Watford, W. T., Hofmann, S. R., et al. (2002). De novo CIAS1 mutations, cytokine activation, and evidence for genetic heterogeneity in patients with neonatal-onset multisystem inflammatory disease (NOMID): A new member of the expanding family of pyrin-associated autoinflammatory diseases. *Arthritis and Rheumatism, 46*(12), 3340–3348. https://doi.org/10.1002/art.10688
7. de Jesus, A. A., Canna, S. W., Liu, Y., & Goldbach-Mansky, R. (2015). Molecular mechanisms in genetically defined autoinflammatory diseases: Disorders of amplified danger signaling. *Annual Review of Immunology, 33*, 823–874. https://doi.org/10.1146/annurev-immunol-032414-112227
8. Agostini, L., Martinon, F., Burns, K., McDermott, M. F., Hawkins, P. N., & Tschopp, J. (2004). NALP3 forms an IL-1beta-processing inflammasome with increased activity in Muckle-Wells autoinflammatory disorder. *Immunity, 20*(3), 319–325.
9. Broz, P., & Dixit, V. M. (2016). Inflammasomes: Mechanism of assembly, regulation and signalling. *Nature Reviews. Immunology, 16*(7), 407–420. https://doi.org/10.1038/nri.2016.58
10. Park, Y. H., Wood, G., Kastner, D. L., & Chae, J. J. (2016). Pyrin inflammasome activation and RhoA signaling in the autoinflammatory diseases FMF and HIDS. *Nature Immunology, 17*(8), 914–921. https://doi.org/10.1038/ni.3457
11. Canna, S. W., de Jesus, A. A., Gouni, S., Brooks, S. R., Marrero, B., Liu, Y., et al. (2014). An activating NLRC4 inflammasome mutation causes autoinflammation with recurrent macrophage activation syndrome. *Nature Genetics, 46*(10), 1140–1146. https://doi.org/10.1038/ng.3089
12. Romberg, N., Al Moussawi, K., Nelson-Williams, C., Stiegler, A. L., Loring, E., Choi, M., et al. (2014). Mutation of NLRC4 causes a syndrome of enterocolitis and autoinflammation. *Nature Genetics, 46*(10), 1135–1139. https://doi.org/10.1038/ng.3066
13. Goldbach-Mansky, R., Dailey, N. J., Canna, S. W., Gelabert, A., Jones, J., Rubin, B. I., et al. (2006). Neonatal-onset multisystem inflammatory disease responsive to interleukin-1beta inhibition. *The New England Journal of Medicine, 355*(6), 581–592. https://doi.org/10.1056/NEJMoa055137
14. Hawkins, P. N., Lachmann, H. J., Aganna, E., & McDermott, M. F. (2004). Spectrum of clinical features in Muckle-Wells syndrome and response to anakinra. *Arthritis and Rheumatism, 50*(2), 607–612. https://doi.org/10.1002/art.20033

15. Tsoukas, P., & Canna, S. W. (2017). No shortcuts: New findings reinforce why nuance is the rule in genetic autoinflammatory syndromes. *Current Opinion in Rheumatology, 29*(5), 506–515. https://doi.org/10.1097/BOR.0000000000000422
16. Manthiram, K., Zhou, Q., Aksentijevich, I., & Kastner, D. L. (2017). The monogenic autoinflammatory diseases define new pathways in human innate immunity and inflammation. *Nature Immunology, 18*(8), 832–842. https://doi.org/10.1038/ni.3777
17. Lykens, J. E., Terrell, C. E., Zoller, E. E., Risma, K., & Jordan, M. B. (2011). Perforin is a critical physiologic regulator of T-cell activation. *Blood*. https://doi.org/10.1182/blood-2010-12-324533
18. Rood, J. E., Rao, S., Paessler, M., Kreiger, P. A., Chu, N., Stelekati, E., et al. (2016). ST2 contributes to T-cell hyperactivation and fatal hemophagocytic lymphohistiocytosis in mice. *Blood, 127*(4), 426–435. https://doi.org/10.1182/blood-2015-07-659813
19. Jordan, M. B., Locatelli, F., Allen, C., de Benedetti, F., Grom, A., Ballabio, M., et al. (2015). Abstract: A novel targeted approach to the treatment of hemophagocytic lymphohistiocytosis (HLH) with an anti-interferon gamma (IFNγ) monoclonal antibody (mAb), NI-0501: First results from a pilot phase 2 study in children with primary HLH. *Blood, 126*(23), LBA-3.
20. Rice, G., Patrick, T., Parmar, R., Taylor, C. F., Aeby, A., Aicardi, J., et al. (2007). Clinical and molecular phenotype of Aicardi-Goutieres syndrome. *American Journal of Human Genetics, 81*(4), 713–725. https://doi.org/10.1086/521373
21. Tungler, V., Konig, N., Gunther, C., Engel, K., Fiehn, C., Smitka, M., et al. (2016). Response to: 'JAK inhibition in STING-associated interferonopathy' by Crow et al. *Annals of the Rheumatic Diseases, 75*(12), e76. https://doi.org/10.1136/annrheumdis-2016-210565
22. Aksentijevich, I., Masters, S. L., Ferguson, P. J., Dancey, P., Frenkel, J., van Royen-Kerkhoff, A., et al. (2009). An autoinflammatory disease with deficiency of the interleukin-1-receptor antagonist. *The New England Journal of Medicine, 360*(23), 2426–2437. https://doi.org/10.1056/NEJMoa0807865
23. Mendonca, L. O., Malle, L., Donovan, F. X., Chandrasekharappa, S. C., Montealegre Sanchez, G. A., Garg, M., et al. (2017). Deficiency of interleukin-1 receptor antagonist (DIRA): Report of the first Indian patient and a novel deletion affecting IL1RN. *Journal of Clinical Immunology, 37*(5), 445–451. https://doi.org/10.1007/s10875-017-0399-1
24. Koc, B., Oktenli, C., Bulucu, F., Karadurmus, N., Sanisoglu, S. Y., & Gul, D. (2007). The rate of pyrin mutations in critically ill patients with systemic inflammatory response syndrome and sepsis: A pilot study. *The Journal of Rheumatology, 34*(10), 2070–2075.
25. Rodrigue-Gervais, I. G., & Saleh, M. (2010). Genetics of inflammasome-associated disorders: A lesson in the guiding principals of inflammasome function. *European Journal of Immunology, 40*(3), 643–648. https://doi.org/10.1002/eji.200940225
26. Schulert, G. S., Zhang, M., Fall, N., Husami, A., Kissell, D., Hanosh, A., et al. (2016). Whole-exome sequencing reveals mutations in genes linked to hemophagocytic lymphohistiocytosis and macrophage activation syndrome in fatal cases of H1N1 influenza. *The Journal of Infectious Diseases, 213*(7), 1180–1188. https://doi.org/10.1093/infdis/jiv550
27. Krebs, P., Crozat, K., Popkin, D., Oldstone, M. B., & Beutler, B. (2011). Disruption of MyD88 signaling suppresses hemophagocytic lymphohistiocytosis in mice. *Blood, 117*(24), 6582–6588. https://doi.org/10.1182/blood-2011-01-329607
28. Zoller, E. E., Lykens, J. E., Terrell, C. E., Aliberti, J., Filipovich, A. H., Henson, P. M., et al. (2011). Hemophagocytosis causes a consumptive anemia of inflammation. *The Journal of Experimental Medicine, 208*(6), 1203–1214. https://doi.org/10.1084/jem.20102538
29. Wunderlich, M., Stockman, C., Devarajan, M., Ravishankar, N., Sexton, C., Kumar, A. R., et al. (2016). A xenograft model of macrophage activation syndrome amenable to anti-CD33 and anti-IL-6R treatment. *JCI Insight, 1*(15), e88181. https://doi.org/10.1172/jci.insight.88181
30. Sepulveda, F. E., Maschalidi, S., Vosshenrich, C. A., Garrigue, A., Kurowska, M., Menasche, G., et al. (2014). A novel immunoregulatory role for NK cell cytotoxicity in protection from HLH-like immunopathology in mice. *Blood, 125*(9), 1427–1434. https://doi.org/10.1182/blood-2014-09-602946

31. Ombrello, M. J., Remmers, E. F., Tachmazidou, I., Grom, A., Foell, D., Haas, J. P., et al. (2015). HLA-DRB1*11 and variants of the MHC class II locus are strong risk factors for systemic juvenile idiopathic arthritis. *Proceedings of the National Academy of Sciences of the United States of America, 112*(52), 15970–15975. https://doi.org/10.1073/pnas.1520779112
32. De Benedetti, F., Brunner, H. I., Ruperto, N., Kenwright, A., Wright, S., Calvo, I., et al. (2012). Randomized trial of tocilizumab in systemic juvenile idiopathic arthritis. *The New England Journal of Medicine, 367*(25), 2385–2395. https://doi.org/10.1056/NEJMoa1112802
33. Ruperto, N., Brunner, H. I., Quartier, P., Constantin, T., Wulffraat, N., Horneff, G., et al. (2012). Two randomized trials of canakinumab in systemic juvenile idiopathic arthritis. *The New England Journal of Medicine, 367*(25), 2396–2406. https://doi.org/10.1056/NEJMoa1205099
34. Ombrello, M. J., Arthur, V. L., Remmers, E. F., Hinks, A., Tachmazidou, I., Grom, A. A., et al. (2016). Genetic architecture distinguishes systemic juvenile idiopathic arthritis from other forms of juvenile idiopathic arthritis: Clinical and therapeutic implications. *Annals of the Rheumatic Diseases, 76*(5), 906–913. https://doi.org/10.1136/annrheumdis-2016-210324
35. Shimizu, M., Nakagishi, Y., Inoue, N., Mizuta, M., Ko, G., Saikawa, Y., et al. (2015). Interleukin-18 for predicting the development of macrophage activation syndrome in systemic juvenile idiopathic arthritis. *Clinical Immunology, 160*(2), 277–281. https://doi.org/10.1016/j.clim.2015.06.005
36. Girard, C., Rech, J., Brown, M., Allali, D., Roux-Lombard, P., Spertini, F., et al. (2016). Elevated serum levels of free interleukin-18 in adult-onset Still's disease. *Rheumatology (Oxford), 55*(12), 2237–2247. https://doi.org/10.1093/rheumatology/kew300
37. Gabay, C., Fautrel, B., Rech, J., Spertini, F., Feist, E., Kotter, I., et al. (2018). Open-label, multicentre, dose-escalating phase II clinical trial on the safety and efficacy of tadekinig alfa (IL-18BP) in adult-onset Still's disease. *Annals of the Rheumatic Diseases*. https://doi.org/10.1136/annrheumdis-2017-212608
38. Weiss, E. S., Girard-Guyonvarc'h, C., Holzinger, D., de Jesus, A. A., Tariq, Z., Picarsic, J., et al. (2018). Interleukin-18 diagnostically distinguishes and pathogenically promotes human and murine macrophage activation syndrome. *Blood, 131*(13), 1442–1455. https://doi.org/10.1182/blood-2017-12-820852
39. Liang, J., Alfano, D. N., Squires, J. E., Riley, M. M., Parks, W. T., Kofler, J., et al. (2017). Novel NLRC4 mutation causes a syndrome of perinatal autoinflammation with hemophagocytic lymphohistiocytosis, hepatosplenomegaly, fetal thrombotic vasculopathy, and congenital anemia and ascites. *Pediatric and Developmental Pathology, 20*(6), 498–505.
40. Moghaddas, F., Zeng, P., Zhang, Y., Schutzle, H., Brenner, S., Hofmann, S. R., et al. (2018). Autoinflammatory mutation in NLRC4 reveals an LRR-LRR oligomerization interface. *The Journal of Allergy and Clinical Immunology*. https://doi.org/10.1016/j.jaci.2018.04.033
41. Canna, S. W., Girard, C., Malle, L., de Jesus, A., Romberg, N., Kelsen, J., et al. (2016). Life-threatening NLRC4-associated hyperinflammation successfully treated with Interleukin-18 inhibition. *The Journal of Allergy and Clinical Immunology, 139*(5), 1698–1701. https://doi.org/10.1016/j.jaci.2016.10.022
42. Bracaglia, C., Prencipe, G., Gatto, A., Pardeo, M., Lapeyre, G., Raganelli, L., et al. (2015). Anti interferon-gamma (IFN gamma) monoclonal antibody treatment in a child with NLRC4-related disease and severe hemophagocytic lymphohistiocytosis (HLH). *Pediatric Blood & Cancer, 62*, S123–S123.
43. Bracaglia, C., de Graaf, K., Pires Marafon, D., Guilhot, F., Ferlin, W., Prencipe, G., et al. (2017). Elevated circulating levels of interferon-gamma and interferon-gamma-induced chemokines characterise patients with macrophage activation syndrome complicating systemic juvenile idiopathic arthritis. *Annals of the Rheumatic Diseases, 76*(1), 166–172. https://doi.org/10.1136/annrheumdis-2015-209020
44. Girard-Guyonvarc'h, C., Palomo, J., Martin, P., Rodriguez, E., Troccaz, S., Palmer, G., et al. (2018). Unopposed IL-18 signaling leads to severe TLR9-induced macrophage activation syndrome in mice. *Blood, 131*(13), 1430–1441. https://doi.org/10.1182/blood-2017-06-789552

45. Rigaud, S., Fondaneche, M. C., Lambert, N., Pasquier, B., Mateo, V., Soulas, P., et al. (2006). XIAP deficiency in humans causes an X-linked lymphoproliferative syndrome. *Nature, 444*(7115), 110–114. https://doi.org/10.1038/nature05257
46. Marsh, R. A., Madden, L., Kitchen, B. J., Mody, R., McClimon, B., Jordan, M. B., et al. (2010). XIAP deficiency: A unique primary immunodeficiency best classified as X-linked familial hemophagocytic lymphohistiocytosis and not as X-linked lymphoproliferative disease. *Blood, 116*(7), 1079–1082. https://doi.org/10.1182/blood-2010-01-256099
47. Lawlor, K. E., Feltham, R., Yabal, M., Conos, S. A., Chen, K. W., Ziehe, S., et al. (2017). XIAP loss triggers RIPK3- and caspase-8-driven IL-1beta activation and cell death as a consequence of TLR-MyD88-induced cIAP1-TRAF2 degradation. *Cell Reports, 20*(3), 668–682. https://doi.org/10.1016/j.celrep.2017.06.073
48. Kenneth, N. S., & Duckett, C. S. (2012). IAP proteins: Regulators of cell migration and development. *Current Opinion in Cell Biology, 24*(6), 871–875. https://doi.org/10.1016/j.ceb.2012.11.004
49. Yabal, M., Muller, N., Adler, H., Knies, N., Gross, C. J., Damgaard, R. B., et al. (2014). XIAP restricts TNF- and RIP3-dependent cell death and inflammasome activation. *Cell Reports, 7*(6), 1796–1808. https://doi.org/10.1016/j.celrep.2014.05.008
50. Wada, T., Kanegane, H., Ohta, K., Katoh, F., Imamura, T., Nakazawa, Y., et al. (2014). Sustained elevation of serum interleukin-18 and its association with hemophagocytic lymphohistiocytosis in XIAP deficiency. *Cytokine, 65*(1), 74–78. https://doi.org/10.1016/j.cyto.2013.09.007
51. Gernez, Y., de Jesus, A. A., Alsaleem, H., Macaubas, C., Roy, A., Lovell, D., et al. (2019). Severe autoinflammation in 4 patients with C-terminal variants in cell division control protein 42 (CDC42) successfully treated with IL-1beta inhibition. The Journal of Allergy and Clinical Immunology, epub Jul 2. https://doi.org/10.1016/j.jaci.2019.06.017
52. Speckmann, C., Lehmberg, K., Albert, M. H., Damgaard, R. B., Fritsch, M., Gyrd-Hansen, M., et al. (2013). X-linked inhibitor of apoptosis (XIAP) deficiency: The spectrum of presenting manifestations beyond hemophagocytic lymphohistiocytosis. *Clinical Immunology, 149*(1), 133–141. https://doi.org/10.1016/j.clim.2013.07.004
53. Kitamura, A., Sasaki, Y., Abe, T., Kano, H., & Yasutomo, K. (2014). An inherited mutation in NLRC4 causes autoinflammation in human and mice. *The Journal of Experimental Medicine, 211*(12), 2385–2396. https://doi.org/10.1084/jem.20141091
54. Volker-Touw, C. M., de Koning, H. D., Giltay, J., de Kovel, C., van Kempen, T. S., Oberndorff, K., et al. (2016). Erythematous nodes, urticarial rash and arthralgias in a large pedigree with NLRC4-related autoinflammatory disease, expansion of the phenotype. *The British Journal of Dermatology, 176*(1), 244–248. https://doi.org/10.1111/bjd.14757
55. Kawasaki, Y., Oda, H., Ito, J., Niwa, A., Tanaka, T., Hijikata, A., et al. (2017). Identification of a high-frequency somatic NLRC4 mutation as a cause of autoinflammation by pluripotent cell-based phenotype dissection. *Arthritis & Rheumatology, 69*(2), 447–459. https://doi.org/10.1002/art.39960
56. de Jager, W., Vastert, S. J., Beekman, J. M., Wulffraat, N. M., Kuis, W., Coffer, P. J., et al. (2009). Defective phosphorylation of interleukin-18 receptor beta causes impaired natural killer cell function in systemic-onset juvenile idiopathic arthritis. *Arthritis and Rheumatism, 60*(9), 2782–2793. https://doi.org/10.1002/art.24750
57. Put, K., Vandenhaute, J., Avau, A., van Nieuwenhuijze, A., Brisse, E., Dierckx, T., et al. (2017). Inflammatory gene expression profile and defective interferon-gamma and granzyme K in natural killer cells from systemic juvenile idiopathic arthritis patients. *Arthritis & Rheumatology, 69*(1), 213–224. https://doi.org/10.1002/art.39933
58. Munoz, M., Eidenschenk, C., Ota, N., Wong, K., Lohmann, U., Kuhl, A. A., et al. (2015). Interleukin-22 induces interleukin-18 expression from epithelial cells during intestinal infection. *Immunity, 42*(2), 321–331. https://doi.org/10.1016/j.immuni.2015.01.011
59. Nowarski, R., Jackson, R., Gagliani, N., de Zoete, M. R., Palm, N. W., Bailis, W., et al. (2015). Epithelial IL-18 equilibrium controls barrier function in colitis. *Cell, 163*(6), 1444–1456. https://doi.org/10.1016/j.cell.2015.10.072

60. Rauch, I., Deets, K. A., Ji, D. X., von Moltke, J., Tenthorey, J. L., Lee, A. Y., et al. (2017). NAIP-NLRC4 inflammasomes coordinate intestinal epithelial cell expulsion with eicosanoid and IL-18 release via activation of caspase-1 and -8. *Immunity, 46*(4), 649–659. https://doi.org/10.1016/j.immuni.2017.03.016
61. Chudnovskiy, A., Mortha, A., Kana, V., Kennard, A., Ramirez, J. D., Rahman, A., et al. (2016). Host-protozoan interactions protect from mucosal infections through activation of the inflammasome. *Cell, 167*(2), 444–456.e414. https://doi.org/10.1016/j.cell.2016.08.076
62. Shakoory, B., Carcillo, J. A., Chatham, W. W., Amdur, R. L., Zhao, H., Dinarello, C. A., et al. (2016). Interleukin-1 receptor blockade is associated with reduced mortality in sepsis patients with features of macrophage activation syndrome: Reanalysis of a prior phase III trial. *Critical Care Medicine, 44*(2), 275–281. https://doi.org/10.1097/CCM.0000000000001402

Macrophage Activation Syndrome in the Setting of Rheumatic Diseases

W. Winn Chatham

Macrophage activation syndrome (MAS) is a hemophagocytic cytokine storm syndrome (CSS) that may occur in patients with established rheumatic disorders. Development of MAS may be due to activity of the underlying disease lowering thresholds for activation of immune cells involved in MAS pathogenesis (one example being systemic lupus erythematosus, see chapter on Systemkic Lupus). In other cases, shared genetic factors favoring development of the underlying rheumatic disease may also predispose to the development of MAS (as may be the case in patients with adult Still disease). Alternatively, in many cases the factors leading to development of MAS and those causing development of the underlying disorder may be unrelated. Treatment for rheumatic disorders may increase the risk for acquired infections known to serve as triggers for MAS in genetically susceptible individuals. Indeed, with the notable exceptions of lupus and adult Still disease, in the majority of reported cases of MAS occurring in patients with underlying rheumatic disease an infectious trigger (most commonly herpes virus infection such as CMV or EBV) has been implicated, with biologic therapy being a potential risk factor [1]. Nonetheless, as the genetic basis of rheumatic disorders as well as MAS become further elucidated, pathophysiologic links between a number of rheumatic disorders and MAS risk may become increasingly apparent.

Rheumatoid Arthritis

There are case reports of hemophagocytic syndrome noted in adult as well as in pediatric patients with underlying seropositive rheumatoid arthritis (RA) [1–8]. In most reported cases, identifying triggers were not identified but it is unclear as to

W. W. Chatham (✉)
University of Alabama at Birmingham, Birmingham, AL, USA
e-mail: wchatham@uabmc.edu

R. Q. Cron, E. M. Behrens (eds.), *Cytokine Storm Syndrome*,
https://doi.org/10.1007/978-3-030-22094-5_25

Table 1 MAS in established rheumatic disease: identified infection triggers

Rheumatic disease	Identified infectious triggers
Rheumatoid arthritis	*EBV*, *CMV* [4] *Hepatitis E* [6] *E. coli* [7] *Leishmaniasis* [8]
Spondyloarthropathy/inflammatory bowel disease	*Mycobacterium tuberculosis* [11, 12]
Polyarteritis nodosa	*EBV* [13, 14]
Granulomatosis with polyangiitis	*HSV* [15]
Eosinophilic granulomatosis with polyangiitis	*Aspergillosis* [16]
Dermatomyositis	*CMV* [17]
Sarcoidosis	*EBV* [18] *Histoplasmosis* [18, 19]

CMV cytomegalovirus, *EBV* Epstein–Barr virus, *HSV* herpes simplex virus

what extent systematic approaches to identify such were undertaken. Given the relatively high prevalence of RA, it is not surprising there would be reported cases of MAS in this population, but in reported case series examining rheumatic disease associations with MAS, RA is a seldom reported association [9], and relative to disorders with a much lower prevalence there have been significantly fewer case reports of MAS occurring in the setting of RA [10]. It is therefore likely that MAS arising in the context of RA is likely attributable to acquired infections and/or genetic and epigenetic factors unrelated to those responsible for RA pathogenesis.

In several reported cases, the syndrome developed in the context of changes in disease-modifying therapy and in cases where infectious triggers were identified (Table 1), patients were on biologic and/or immunosuppressive therapy for their RA [4–8]. Prescribed therapies for RA may therefore increase risk for acquired infections known to trigger MAS in susceptible individuals. Conversely, biologics such as anakinra and tocilizumab that are used in the management of RA have reported efficacy in managing MAS, and their use may confer a protective role by suppressing the development of MAS that may otherwise develop in a susceptible patient with RA exposed to a known MAS trigger.

Seronegative Spondyloarthropathies

MAS has been reported to occur in the setting of ankylosing spondylitis and psoriatic arthritis, and in patients with Crohn disease or ulcerative colitis with or without enteropathic arthritis [10–12, 20–24]. Of interest, in several reports development of MAS antedated the onset of manifestations of spondyloarthropathy [21–23]. In some of these reported cases, genetic testing confirmed the presence of dominant negative mutations in perforin pathway genes, including *PRF1* [22], *UNC13D* [23], and *RAB27A* [23] that have been linked to MAS. Whether perforin pathway genes increase risk for spondyloarthropathy has not been systematically evaluated.

As has been observed in patients with RA, MAS has been reported in patients with spondylitis in the setting of acquired infection while on therapy with biologic modifiers [11, 12]. Acute infections with Salmonella and *Clostridium difficile* have been implicated as infectious triggers of MAS in patients with or without underlying rheumatic disease [25, 26]. Despite the association of acquired infection with reactive arthritis, MAS has not been reported in this setting.

Treatment of MAS that develops in the context of any of the spondyloarthopathies should be directed toward eliminating infections identified as likely triggers and appropriate immunosuppressive therapy to minimize the production of cytokines by activated T cells and macrophages. In addition to initiating high doses of corticosteroids, early use of anti-cytokine therapy targeting IL-1 (anakinra) or IL-6 (tocilizumab) are of reported benefit in effecting resolution of the MAS with less corticosteroid exposure. Biologics targeting tumor necrosis-alpha, IL-12/IL-23, or IL-17 that are commonly used in the management of inflammatory bowel disease or spondyloarthopathies may not be as efficacious in suppressing MAS features as those that target IL-1 or IL-6. To minimize additional infection risk, it may therefore be prudent to withhold these background treatment biologic therapies when employing anti-IL-1 or anti-IL-6 therapies to treat supervening MAS.

Vasculitis

There are several reported cases of MAS occurring in patients with *polyarteritis nodosa* (*PAN*). In some reported cases, MAS was attributed to underlying viral infection (EBV) thought to trigger both the vasculitis and the observed MAS [13, 14]. In others, MAS appears to have occurred as a complication of secondarily acquired opportunistic infections complicating prescribed immunosuppressive therapy [13]. In a number of case reports, active hepatitis B virus infection has been implicated as the likely trigger for MAS, although associated arteritis was not a reported clinical feature [27–29]. MAS has been reported in patients with *HCV associated cryoglobulin syndromes*, but usually in the context of other acquired infection during the course of prescribed treatment [30].

There are rare case reports of MAS occurring in the setting of *anti-neutrophil cytoplasmic antibody* (*ANCA*) *associated vasculitis* syndromes including granulomatosis with polyangiitis, microscopic polyangiitis, eosinophilic granulomatosis with polyangiitis, or anti-MPO associated Goodpasture syndrome [15, 16, 31, 32]. However, it is important to recognize that complicating infections have been identified as the likely inciting trigger for MAS observed in the reported cases, and careful evaluation for known infectious triggers (Table 1) should be undertaken when MAS develops in this setting [10, 15].

Similarly, MAS has not been reported as a presenting or complicating feature of *giant cell arteritis* or *Takayasu arteritis* [10]. As is applicable to the other aforementioned vasculitides, the occurrence of MAS in a patient with either of these disorders should likewise prompt investigation for infections known to be associated with the development of MAS.

Since the development of MAS in patients with vasculitis most commonly occurs in the setting of acquired (often opportunistic) infection, treatment should be directed toward elimination of identified infectious triggers as well as immunotherapy directed toward minimizing organ damage from inflammatory cytokines. In addition to escalating doses of prescribed corticosteroids, high dose intravenous immunoglobulin (IVIG) (2 g/kg every 3–4 weeks) may be of particular utility in managing MAS in this setting as it has been shown to be effective in suppressing vascular inflammation in polyarteritis as well as ANCA-associated vasculitides, and may also help clear viruses (notably CMV) or other MAS-triggering microbial pathogens. Caution should be undertaken when using IVIG in patients with vasculitis due to cryoglobulins containing rheumatoid factors as doing so may exacerbate the vasculitis. Rituximab or other B-cell-depleting monoclonal reagents is an effective adjunct in the treatment of EBV-associated MAS and may have additional remission attributes for patients with underlying ANCA-associated vasculitis who develop MAS due to EBV infection.

Kawasaki Disease

MAS has been increasingly recognized as a complicating feature of Kawasaki disease (KD), occurring in at least 1–2% of cases in one reported series [33, 34]. In both murine models of Kawasaki disease as well as human immunophenotyping studies in affected patients, IL-1β has been shown to be central to the development of vascular lesions [35, 36]. Indeed, these observations as well as a number of shared clinical and laboratory features with systemic juvenile idiopathic arthritis (sJIA) have led to the consideration of Kawasaki disease as an IL-1β mediated autoinflammatory disorder, with several clinical trials underway to determine the efficacy of anakinra or canakinumab in managing KD [37, 38]. Hyperferritinemia is not uncommonly observed in patients with KD, but when observed levels of ferritin exceed 5000 ng/ml and are accompanied by leukopenia/thrombocytopenia, evolving MAS should be considered. In such cases, therapy should be escalated beyond treatment with IVIG and corticosteroids, incorporating anti-cytokine therapy with anakinra or canakinumab. If there is still failure to respond, further escalation of treatment with calcineurin inhibitors can be undertaken.

Adult-Onset Still Disease

Patients presenting with adult-onset Still disease (AOSD) also have clinical and laboratory features very much in common with those associated with sJIA, many of which are also characteristic of MAS (Table 2). These shared features include

Table 2 Clinical and laboratory features of AOSD, sJIA, and MAS

	AOSD	sJIA	MAS
Hyperferritinemia	+	+	++
Fever	+	+	+
AST, ALT elevation	+	+/−	+
High CRP	+	+	+
Hepatomegaly	+	+	+
Splenomegaly	+	+	+
High triglycerides	−	−	+
Cytopenias	−	−	+
Low fibrinogen Elevated d-dimer	−	−	+

significant hyperferritinemia, fever, and elevated serum transaminase levels. The underlying cytokine profiles are also similar, with elevations in IL-1β and IL-18 characteristic of both disorders [39]. Just as patients with active flares of sJIA or KD should be monitored for development of MAS, such is also the case for patients who present with AOSD. The initial clinical presentation of AOSD may well be that of MAS [40, 41]. While hyperferritinemia and elevation of liver transaminase levels are commonly seen in patients with active AOSD, the development of leukopenia, thrombocytopenia, and/or evidence of coagulopathy in a patient with AOSD portend developing MAS. Given the increased mortality associated with AOSD complicated by MAS [39], it is therefore prudent to monitor serial blood counts as well as serum levels of lactate dehydrogenase (LDH), plasma levels of d-dimer, and serum levels of fibrinogen in patients with active AOSD.

As has been observed in patients with sJIA and KD, excess production of IL-1β appears to play a significant role in the pathogenesis of AOSD [39]. For patients with AOSD who have developed features of MAS, initiation of anti-IL-1β therapy with glucocorticoids (if not yet administered) is recommended as the majority of patients will respond rapidly to this combined treatment. Anakinra has been used most commonly, usually in doses of 100 mg administered subcutaneously every 12 h, and may lessen overall corticosteroid exposure [42]. If needed, the dose can safely be escalated to every 6 h for critically ill patients [43, 44]. For patients failing to respond within 48 h to maximum doses of anakinra, addition of cyclosporine at 2.5 mg/kg/day may effect clinical resolution of MAS [41]. Alternatively, an 8 mg/kg dose of tocilizumab can be administered in patients failing to respond or who may develop intolerable significant injection site reactions to anakinra [45].

Additional reported complications seen in AOSD patients who have developed MAS include myocarditis, pulmonary hypertension, and interstitial lung disease with nonspecific interstitial pneumonitis (NSIP) [43, 46]. The combined approach of anti-IL-1 therapy with anakinra and cyclosporine or use of tocilizumab has been used with success in patients with these complications [43, 46, 47].

Once the MAS flare has abated with resolution of cytopenias and any noted elevations in serum LDH and d-dimer levels, therapy can be sequentially attenuated starting with tapering of corticosteroids, then discontinuation of cyclosporin or other employed calcineurin inhibitor. Since AOSD is most effectively managed with anti-IL-1β therapy or tocilizumab, biologic therapy should be continued to avoid severe relapses of AOSD and associated MAS. In most patients, anakinra can be decreased to once daily dosing; when tocilizumab is employed, either weekly subcutaneous (100 mg) or monthly intravenous dosing (4–8 mg/kg) can be employed to maintain disease remission and avoid relapse. With prolonged remission it may be reasonable to discontinue therapy with careful subsequent observation for relapse.

Behcet and Autoinflammatory Disease

Due to associated excessive production of inflammatory cytokines capable of promoting activation of macrophages, monogenic autoinflammatory disorders as well as Behcet disease (BD) may be associated with the development of MAS [10, 48]. MAS has been reported to be the presenting manifestation in patients with familial Mediterranean fever [49] and TNF (tumor necrosis factor) receptor associated periodic syndrome [50]. MAS was reported to have occurred during one of many sequential febrile attacks in a patient with hyperimmunoglobulin D syndrome [51]. Patients with cryopyrin associated periodic syndrome (CAPS) have also been reported to develop MAS [52], whether due to gain of function mutations in NACHT, LRR, and PYD domain-containing protein 3 (NLRP3) or the NLR-family CARD-containing protein 4 (NLRC4) with attendant excessive production of IL-18 [53]. Defects in cytotoxicity have been identified in some of the reported cases, but have not been a consistent finding in patients who have developed MAS on this background [48]. Regardless of similarities or differences in canonical IL-1β and IL-18 driven pathways of inflammation, it is important to be mindful of possible MAS development in patients with the monogenic inflammatory disorders, as escalation of therapy to manage complicating MAS may be required in the setting of hyperferritinemia with cytopenias and coagulopathy.

Studies in patients with BD have traditionally demonstrated evidence of significant monocyte and macrophage activation [54, 55] as well as increased numbers and activation of circulating gamma-delta T cell subsets and natural killer (NK) cells [56]. However, reports of MAS in patients with BD are uncommon. Reported cases have been in association with intercurrent herpes virus infection such as EBV or CMV that potentially triggered observed MAS [57].

Polymyositis/Dermatomyositis

MAS has been reported in patients with idiopathic inflammatory myopathy, with the preponderance of reported cases associated with dermatomyositis (DM), particularly, juvenile-onset DM. MAS may be the presenting feature of dermatomyositis [58], and may be associated with pulmonary [59], as well as CNS [60] complications. Favorable responses to treatment with calcineurin inhibitors and corticosteroids, with or without plasma exchange, have been reported in this clinical setting [61, 62]. As has been noted to occur in other rheumatic disorders, complicating infection with herpes viruses such as CMV may trigger MAS in the setting of DM with fatal consequences despite CMV and MAS-directed therapy [17].

Systemic Sclerosis

MAS has rarely been reported in the setting of systemic sclerosis [10, 13, 63, 64]. The treatment of MAS in this setting may be challenging as high-dose corticosteroids carries an increased risk of precipitating scleroderma renal crisis, and use of calcineurin inhibitors may exacerbate underlying renal arteriopathy prevalent in patients with this disorder. Early intervention with an IL-1β inhibitor such as anakinra may therefore be a preferred initial intervention. Etoposide has been used with success in this setting but may require support with G-CSF [65].

Sarcoidosis

MAS is rarely reported in patients with established sarcoidosis, occurring either in the presence or absence of identified infectious triggers [10, 18]. In the majority of reported cases infectious triggers including EBV and histoplasmosis I have been identified [18]. Histoplasmosis is a well-established trigger for MAS, and the acute presentations of sarcoidosis and histoplasmosis share both clinical and pathologic features [19]. It is therefore particularly prudent to perform diagnostic studies including serum and urine antigen studies, blood smears, and fungal cultures for histoplasmosis in the setting of MAS and granulomatous disease. Moreover, since disseminated granulomatous lesions may be characteristic of patients with common variable immune deficiency (CVID), it is furthermore prudent to assess immunoglobulin levels in this setting to determine whether MAS is occurring in the context of CVID rather than sarcoidosis. This is of particular relevance in the context of reported EBV triggered MAS in patients with sarcoidosis, as EBV as well as CMV infections have been implicated in the majority of case reports of MAS associated with CVID [66, 67]. As is recommended for MAS occurring in other rheumatic entities, treatment is best directed toward identified triggering infections as well as combinations of high-dose corticosteroids, calcineurin inhibitors, and biologic therapies targeting IL-1 and/or IL-6 to suppress the associated cytokine storm.

Summary

MAS has been reported with variable frequency in the context of most established inflammatory rheumatic and autoimmune disorders. It is most commonly seen in disorders whereby by the underlying innate inflammation defects (AOSD, AID) or autoimmune pathways (SLE) underlying the disorder might be predicted to give rise to MAS during severe flares. In other disorders, MAS may arise in the context of mutations in the perforin pathway known to be associated with MAS and inciting epigenetic infectious triggers, most commonly herpes virus family infection. As such, for these patients presenting with MAS, it is important to assess for the presence of infectious triggers as well as genetic studies targeting mutations in perforin pathway proteins to ascertain future risk.

Therapy directed toward treating severe flares of underlying rheumatic disease may abate developing MAS, as is not uncommonly observed in patients with AOSD or systemic lupus. However, resolution of MAS in these disorders and more often than not in other rheumatic disorders may require escalation of therapy directed specifically toward decreasing the production of cytokines driving MAS. Successful reported outcomes employ a combined approach with initial high dose corticosteroids, inhibitors of IL-1β and/or IL-6, and the use of calcineurin inhibitors targeting production of T cell-derived cytokines. Careful dosing with etoposide may be used in refractory cases with vigilance for and appropriate interventions for attendant neutropenia. The role of antibodies to interferon-gamma, now available for use in the management of patients with primary/familial hemophagocytic lymphohistiocytosis (fHLH), in the treatment of secondary forms of MAS/HLH remains to be determined. Caution should be exercised in extrapolating the experiences with this and other established protocols for familial HLH to therapeutic use in secondary/acquired forms of macrophage activation syndrome.

References

1. Brito-Zerón, P., Bosch, X., Pérez-de-Lis, M., Pérez-Álvarez, R., Fraile, G., Gheitasi, H., et al. (2016). Infection is the major trigger of hemophagocytic syndrome in adult patients treated with biological therapies. *Seminars in Arthritis and Rheumatism, 45*(4), 391–399.
2. Collamer, A., & Arroyo, R. (2010). Bone marrow hemophagocytosis complicating rheumatoid arthritis. *Journal of Clinical Rheumatology, 16*(3), 151.
3. Niang, A., Diallo, S., Ka, M. M., Pouye, A., Diop, S., Ndongo, S., et al. (2004). Hemophagocytic syndrome complicating adult's seropositive rheumatoid arthritis. *La Revue de Médecine Interne, 25*(11), 826–828.
4. Ricci, M., Rossi, P., De Marco, G., Varisco, V., & Marchesoni, A. (2010). Macrophage activation syndrome after leflunomide treatment in an adult rheumatoid arthritis patient. *Rheumatology (Oxford, England), 49*(10), 2001.
5. Sandhu, C., Chesney, A., Piliotis, E., Buckstein, R., & Koren, S. (2007). Macrophage activation syndrome after etanercept treatment. *The Journal of Rheumatology, 34*(1), 241–242.
6. Leroy, M., Coiffier, G., Pronier, C., Triquet, L., Perdriger, A., & Guggenbuhl, P. (2015). Macrophage activation syndrome with acute hepatitis E during tocilizumab treatment for rheumatoid arthritis. *Joint, Bone, Spine, 82*(4), 278–279.

7. Sibilia, J., Javier, R. M., Albert, A., Cazenave, J. P., & Kuntz, J. L. (1998). Pancytopenia secondary to hemophagocytic syndrome in rheumatoid arthritis treated with methotrexate and sulfasalazine. *The Journal of Rheumatology, 25*(6), 1218–1220.
8. Moltó, A., Mateo, L., Lloveras, N., Olivé, A., & Minguez, S. (2010). Visceral leishmaniasis and macrophagic activation syndrome in a patient with rheumatoid arthritis under treatment with adalimumab. *Joint, Bone, Spine, 77*(3), 271–273.
9. Deane, S., Selmi, C., Teuber, S. S., & Gershwin, M. E. (2010). Macrophage activation syndrome in autoimmune disease. *International Archives of Allergy and Immunology, 153*(2), 109–120.
10. Atteritano, M., David, A., Bagnato, G., Beninati, C., Frisina, A., Iaria, C., et al. (2012). Haemophagocytic syndrome in rheumatic patients. A systematic review. *European Review for Medical and Pharmacological Sciences, 16*(10), 1414–1424.
11. André, V., Liddell, C., Guimard, T., Tanguy, G., & Cormier, G. (2013). Macrophage activation syndrome revealing disseminated tuberculosis in a patient on infliximab. *Joint, Bone, Spine, 80*(1), 109–110.
12. Troncoso Mariño, A., Campelo Sánchez, E., Martínez López de Castro, N., & Inaraja Bobo, M. T. (2010). Haemophagocytic syndrome and paradoxical reaction to tuberculostatics after treatment with infliximab. *Pharmacy World & Science, 32*(2), 117–119.
13. Dhote, R., Simon, J., Papo, T., et al. (2003). Reactive hemophagocytic syndrome in adult systemic disease: Report of twenty-six cases and literature review. *Arthritis and Rheumatism, 49*(5), 633–639.
14. Hayakawa, I., Shirasaki, F., Ikeda, H., Oishi, N., Hasegawa, M., Sato, S., et al. (2006). Reactive hemophagocytic syndrome in a patient with polyarteritis nodosa associated with Epstein-Barr virus reactivation. *Rheumatology International, 26*(6), 573–576.
15. Cusini, A., Günthard, H. F., Stussi, G., Schwarz, U., Fehr, T., Grueter, E., et al. (2010). Hemophagocytic syndrome caused by primary herpes simplex virus 1 infection: Report of a first case. *Infection, 38*, 423–426.
16. García Escudero, A., Benítez Moya, J. M., & Lag Asturiano, E. (2000). Hemophagocytic syndrome and invasive aspergillosis in a patient with Churg-Strauss vasculitis. *Medicina Clínica (Barcelona), 115*(15), 598.
17. Lange, A. V., Kazi, S., Chen, W., & Barnes, A. (2018). Fatal case of macrophage activation syndrome (MAS) in a patient with dermatomyositis and cytomegalovirus (CMV) viraemia. *BML Case Reports*, pii: bcr-2018-225231.
18. Abughanimeh, O., Qasrawi, A., & Abu Ghanimeh, M. (2018). Hemophagocytic lymphohistiocytosis complicating systemic sarcoidosis. *Cureus, 10*(6), e2838.
19. Schulze, A. B., Heptner, B., Kessler, T., Baumgarten, B., Stoica, V., Mohr, M., et al. (2017). Progressive histoplasmosis with hemophagocytic lymphohistiocytosis and epithelioid cell granulomatosis: A case report and review of the literature. *European Journal of Haematology, 99*(1), 91–100.
20. Lou, Y. J., Jin, J., & Mai, W. Y. (2007). Ankylosing spondylitis presenting with macrophage activation syndrome. *Clinical Rheumatology, 26*(11), 1929–1930.
21. Park, J. H., Seo, Y. M., Han, S. B., Kim, K. H., Rhim, J. W., Chung, N. G., et al. (2016). Recurrent macrophage activation syndrome since toddler age in an adolescent boy with HLA B27 positive juvenile ankylosing spondylitis. *Korean Journal of Pediatrics, 59*(10), 421–424.
22. Filocamo, G., Petaccia, A., Torcoletti, M., Sieni, E., Ravelli, A., & Corona, F. (2016). Recurrent macrophage activation syndrome in spondyloarthritis and monoallelic missense mutations in PRF1: A description of one paediatric case. *Clinical and Experimental Rheumatology, 34*(4), 719.
23. Cron, R. Q., & Chatham, W. W. (2016). Development of spondyloarthropathy following episodes of macrophage activation syndrome in children with heterozygous mutations in haemophagocytic lymphohistiocytosis-associated genes. *Clinical and Experimental Rheumatology, 34*(5), 953.
24. Fries, W., Cottone, M., & Cascio, A. (2013). Systematic review: Macrophage activation syndrome in inflammatory bowel disease. *Alimentary Pharmacology & Therapeutics, 37*(11), 1033–1045.

25. Sánchez-Moreno, P., Olbrich, P., Falcón-Neyra, L., Lucena, J. M., Aznar, J., & Neth, O. (2018). Typhoid fever causing haemophagocytic lymphohistiocytosis in a non-endemic country—First case report and review of the current literature. *Enfermedades Infecciosas y Microbiología Clínica, 18*, 30186–30181.
26. Ramon, I., Libert, M., Guillaume, M. P., Corazza, F., & Karmali, R. (2010). Recurrent haemophagocytic syndrome in an HIV-infected patient. *Acta Clinica Belgica, 65*(4), 276–278.
27. Aleem, A., Al Amoudi, S., Al-Mashhadani, S., & Siddiqui, N. (2005). Haemophagocytic syndrome associated with hepatitis-B virus infection responding to etoposide. *Clinical and Laboratory Haematology, 27*(6), 395–398.
28. Yu, M. G., & Chua, J. (2016). Virus-associated haemophagocytic lymphohistiocytosis in a young Filipino man. *BMJ Case Reports*. https://doi.org/10.1136/bcr-2016-214655
29. Faurschou, M., Nielsen, O. J., Hansen, P. B., Juhl, B. R., & Hasselbalch, H. (1999). Fatal virus-associated hemophagocytic syndrome associated with coexistent chronic active hepatitis B and acute hepatitis C virus infection. *American Journal of Hematology, 61*(2), 135–138.
30. Wang, Z., Duarte, A. G., & Schnadig, V. J. (2007). Fatal reactive hemophagocytosis related to disseminated histoplasmosis with endocarditis: An unusual case diagnosed at autopsy. *Southern Medical Journal, 100*(2), 208–211.
31. Amlani, A., Bromley, A., & Fifi-Mah, A. (2018). ANCA vasculitis and hemophagocytic lymphohistiocytosis following a fecal microbiota transplant. *Case Reports in Rheumatology, 2018*, 9263537.
32. Basnet, A., & Cholankeril, M. R. (2014). Hemophagocytic lymphohistiocytosis in a patient with Goodpasture's syndrome: A rare clinical association. *The American Journal of Case Reports, 15*, 431–436.
33. García-Pavón, S., Yamazaki-Nakashimada, M. A., Báez, M., Borjas-Aguilar, K. L., & Murata, C. (2017). Kawasaki disease complicated with macrophage activation syndrome: A systematic review. *Journal of Pediatric Hematology/Oncology, 39*(6), 445–451.
34. Wang, W., Gong, F., Zhu, W., Fu, S., & Zhang, Q. (2015). Macrophage activation syndrome in Kawasaki disease: More common than we thought? *Seminars in Arthritis and Rheumatism, 44*(4), 405–410.
35. Lee, Y., Wakita, D., Dagvadorj, J., Shimada, K., Chen, S., Huang, G., et al. (2015). IL-1 signaling is critically required in stromal cells in Kawasaki disease vasculitis mouse model: Role of both IL-1α and IL-1β. *Arteriosclerosis, Thrombosis, and Vascular Biology, 35*(12), 2605–2616.
36. Lee, Y., Schulte, D. J., Shimada, K., Chen, S., Crother, T. R., & Chiba, N. (2012). Interleukin-1β is crucial for the induction of coronary artery inflammation in a mouse model of Kawasaki disease. *Circulation, 125*(12), 1542–1550.
37. Dusser, P., & Koné-Paut, I. (2017). IL-1 inhibition may have an important role in treating refractory Kawasaki disease. *Frontiers in Pharmacology, 8*, 163.
38. Burns, J. C., Kone-Paut, I., & Kuijpers, T. (2017). Review: Found in translation: International initiatives pursuing interleukin-1 blockade for treatment of acute Kawasaki disease. *Arthritis & Rheumatology, 69*(2), 268–276.
39. Inoue, N., Shimizu, M., Tsunoda, S., Kawano, M., Matsumura, M., & Yachie, A. (2016). Cytokine profile in adult-onset Still's disease: Comparison with systemic juvenile idiopathic arthritis. *Clinical Immunology, 169*, 8–13.
40. Ruscitti, P., Rago, C., Breda, L., Cipriani, P., Liakouli, V., Berardicurti, O., et al. (2017). Macrophage activation syndrome in Still's disease: Analysis of clinical characteristics and survival in paediatric and adult patients. *Clinical Rheumatology, 36*(12), 2839–2845.
41. Lenert, A., & Yao, Q. (2016). Macrophage activation syndrome complicating adult onset Still's disease: A single center case series and comparison with literature. *Seminars in Arthritis and Rheumatism, 45*(6), 711–716.
42. Yoo, D. H. (2017). Treatment of adult-onset Still's disease: Up to date. *Expert Review of Clinical Immunology, 13*(9), 849–866.
43. Parisi, F., Paglionico, A., Varriano, V., Ferraccioli, G., & Gremese, E. (2017). Refractory adult-onset Still disease complicated by macrophage activation syndrome and acute myocarditis:

A case report treated with high doses (8mg/kg/d) of anakinra. *Medicine (Baltimore), 96*(24), e6656.
44. Kahn, P. J., & Cron, R. Q. (2013). Higher-dose Anakinra is effective in a case of medically refractory macrophage activation syndrome. *The Journal of Rheumatology, 40*(5), 743–744.
45. Watanabe, E., Sugawara, H., Yamashita, T., Ishii, A., Oda, A., & Terai, C. (2016). Successful tocilizumab therapy for macrophage activation syndrome associated with adult-onset Still's disease: A case-based review. *Case Reports in Medicine, 2016*, 5656320.
46. Mehta, M. V., Manson, D. K., Horn, E. M., & Haythe, J. (2016). An atypical presentation of adult-onset Still's disease complicated by pulmonary hypertension and macrophage activation syndrome treated with immunosuppression: A case-based review of the literature. *Pulmonary Circulation, 6*(1), 136–142.
47. Savage, E., Wazir, T., Drake, M., Cuthbert, R., & Wright, G. (2014). Fulminant myocarditis and macrophage activation syndrome secondary to adult-onset Still's disease successfully treated with tocilizumab. *Rheumatology (Oxford, England), 53*(7), 1352–1353.
48. Rigante, D., Emmi, G., Fastiggi, M., Silvestri, E., & Cantarini, L. (2015). Macrophage activation syndrome in the course of monogenic autoinflammatory disorders. *Clinical Rheumatology, 34*(8), 1333–1339.
49. Rossi-Semerano, L., Hermeziu, B., Fabre, M., & Koné-Paut, I. (2011). Macrophage activation syndrome revealing familial Mediterranean fever. *Arthritis Care and Research (Hoboken), 63*(5), 780–783.
50. Horneff, G., Rhouma, A., Weber, C., & Lohse, P. (2013). Macrophage activation syndrome as the initial manifestation of tumour necrosis factor receptor 1-associated periodic syndrome (TRAPS). *Clinical and Experimental Rheumatology, 31*(3 Suppl 77), 99–102.
51. Rigante, D., Capoluongo, E., & Bertoni, B. (2007). First report of macrophage activation syndrome in hyperimmunoglobulinemia D with periodic fever syndrome. *Arthritis and Rheumatism, 56*(2), 658–661.
52. Eroglu, F. K., Kasapcopur, O., & Beşbaş, N. (2016). Genetic and clinical features of cryopyrin-associated periodic syndromes in Turkish children. *Clinical and Experimental Rheumatology, 34*(6 Suppl 102), S115–S120.
53. Canna, S. W., de Jesus, A. A., Gouni, S., Brooks, S. R., Marrero, B., Liu, Y., et al. (2014). An activating NLRC4 inflammasome mutation causes autoinflammation with recurrent macrophage activation syndrome. *Nature Genetics, 46*(10), 1140–1146.
54. Sahin, S., Lawrence, R., Direskeneli, H., Hamuryudan, V., Yazici, H., & Akoğlu, T. (1996). Monocyte activity in Behçet's disease. *British Journal of Rheumatology, 35*(5), 424–429.
55. Alpsoy, E., Kodelja, V., Goerdt, S., Orfanos, C. E., & Zouboulis, C. C. (2003). Serum of patients with Behçet's disease induces classical (pro-inflammatory) activation of human macrophages in vitro. *Dermatology, 206*(3), 225–232.
56. Yamashita, N. (1997). Hyperreactivity of neutrophils and abnormal T cell homeostasis: A new insight for pathogenesis of Behçet's disease. *International Reviews of Immunology, 14*(1), 11–19.
57. Lee, S. H., Kim, S. D., Kim, S. H., Kim, H. R., Oh, E. J., Yoon, C. H., et al. (2005). EBV-associated haemophagocytic syndrome in a patient with Behcet's disease. *Scandinavian Journal of Rheumatology, 34*, 320–323.
58. Poddighe, D., Cavagna, L., Brazzelli, V., Bruni, P., & Marseglia, G. L. (2014). A hyper-ferritinemia syndrome evolving in recurrent macrophage activation syndrome, as an onset of amyopathic juvenile dermatomyositis: A challenging clinical case in light of the current diagnostic criteria. *Autoimmunity Reviews, 13*(11), 1142–1148.
59. Wakiguchi, H., Hasegawa, S., Hirano, R., Kaneyasu, H., Wakabayashi-Takahara, M., & Ohga, S. (2015). Successful control of juvenile dermatomyositis-associated macrophage activation syndrome and interstitial pneumonia: Distinct kinetics of interleukin-6 and -18 levels. *Pediatric Rheumatology Online Journal, 13*, 49.
60. Lilleby, V., Haydon, J., Sanner, H., Krossness, B. K., Ringstad, G., & Flatø, B. (2014). Severe macrophage activation syndrome and central nervous system involvement in juvenile dermatomyositis. *Scandinavian Journal of Rheumatology, 43*(2), 171–173.

61. Kaieda, S., Yoshida, N., Yamashita, F., Okamoto, M., Ida, H., Hoshino, T., et al. (2015). Successful treatment of macrophage activation syndrome in a patient with dermatomyositis by combination with immunosuppressive therapy and plasmapheresis. *Modern Rheumatology, 25*(6), 962–966.
62. Bustos, B. R., Carrasco, A. C., & Toledo, R. C. (2012). Plasmapheresis for macrophage activation syndrome and multiorgan failure as first presentation of juvenile dermatomyositis. *Anales de Pediatría (Barcelona, Spain), 77*(1), 47–50.
63. Hone, N., Donnelly, C., Houk, J. B., & Mina, R. (2017). Macrophage activation syndrome in scleroderma. *Journal of Clinical Rheumatology, 23*(2), 120–121.
64. Tochimoto, A., Nishimagi, E., Kawaguchi, Y., Kobashigawa, T., Okamoto, H., Harigai, M., et al. (2001). A case of recurrent hemophagocytic syndrome complicated with systemic sclerosis: Relationship between disease activity and serum level of IL-18. *Ryūmachi, 41*(3), 659–664.
65. Katsumata, Y., Okamoto, H., Harigai, M., Ota, S., Uesato, M., Tochimoto, A., et al. (2002). Etoposide ameliorated refractory hemophagocytic syndrome in a patient with systemic sclerosis. *Ryūmachi, 42*(5), 820–826.
66. Bajaj, P., Clement, J., Bayerl, M. G., Kalra, N., Craig, T. J., & Ishmael, F. T. (2014). High-grade fever and pancytopenia in an adult patient with common variable immune deficiency. *Allergy and Asthma Proceedings, 35*(1), 78–82.
67. Aghamohammadi, A., Abolhassani, H., Hirbod-Mobarakeh, A., Ghassemi, F., Shahinpour, S., & Behniafard, N. (2012). The uncommon combination of common variable immunodeficiency, macrophage activation syndrome, and cytomegalovirus retinitis. *Viral Immunology, 25*(2), 161–165.

Part VI
Other Triggers of Cytokine Storm Syndromes

Hemophagocytic Lymphohistiocytosis in the Context of Hematological Malignancies and Solid Tumors

Kai Lehmberg

Abbreviations

CMV	Cytomegalovirus
CT	Computed tomography
DLBCL	Diffuse large B cell lymphoma
EBNA	Epstein–Barr nuclear antigen
EBV	Epstein–Barr virus
HIV	Human immune deficiency virus
HLH	Hemophagocytic lymphohistiocytosis
IFNγ	Interferon gamma
IL-6	Interleukin-6
ITK	Inducible T cell kinase
LCH	Langerhans cell histiocytosis
MAGT1	Magnesium transporter 1
MRI	Magnetic resonance imaging
PET	Positron emission tomography
sCD25	Soluble IL2 receptor
XLP	x-Linked lymphoproliferative disease

K. Lehmberg (✉)
Division of Stem Cell Transplantation and Immunology, University Medical Center Hamburg Eppendorf, Hamburg, Germany
e-mail: k.lehmberg@uke.de

R. Q. Cron, E. M. Behrens (eds.), *Cytokine Storm Syndrome*,
https://doi.org/10.1007/978-3-030-22094-5_26

Two Types of HLH in the Context of Malignancies

Malignancies constitute a major underlying condition of the hyperinflammatory syndrome hemophagocytic lymphohistiocytosis, a subset of cytokine storm syndromes (CSS). The number of reported cases is increasing steadily [1]. The diagnosis is challenging as the overlap of HLH features and features of neoplastic disease is substantial. HLH in the context of malignant conditions can occur in the two different settings [2]:

In the first setting, the hyperinflammatory nature of HLH is directly driven by the neoplasm. In this chapter, this is referred to as malignancy-triggered HLH. Manifestation is usually at presentation or relapse of the malignancy. Lymphoma cell lines secret inflammatory cytokines such as interferon-γ (IFN-γ) and interleukin-6 (IL-6) that are key players in different types of cytokine storm syndromes [3, 4]. Additionally, certain markers of HLH and lymphoma show relevant overlap. Elevated soluble interleukin-2 receptor alpha chain (sCD25) is a feature of both HLH and of non-Hodgkin lymphoma [5, 6]. In some lymphomas, herpes viruses, particularly Epstein–Barr virus (EBV), can further aggravate the hyperinflammation [7].

In the second setting, HLH can occur related to the profound immune suppression conferred by the cytostatic treatment. This increases the likelihood of infection or reactivations and consequent HLH [8]. Indeed, infectious triggers are frequently found, which renders HLH in this setting more similar to infection-associated secondary HLH. The microbiological spectrum extends from viral triggers (e.g., EBV, cytomegalovirus (CMV), BK virus, human herpes virus-6) to invasive fungi and bacteria [9–12]. HLH in this setting tends to occur several months into treatment. Patients may already be in remission of the neoplasm. It is sometimes difficult to clearly differentiate between malignancy-triggered HLH and HLH in the context of chemotherapy. Coexistence is possible when infectious agents enhance malignancy-triggered HLH. Management, however, is different, and it should thus always be attempted to distinguish the two subtypes, which is not always the case in the available literature.

Frequent Malignant Conditions in Malignancy-Triggered HLH

Lymphomas make up for 75–80% of malignant conditions associated with HLH in children, adolescents, and adults [1, 13]. Table 1 lists the most pertinent malignancies. The most frequent malignancies in adults are T and NK cell lymphomas (35%), B cell lymphomas (32%), leukemias (6%), Hodgkin lymphomas (6%), and other hematologic cancers (14%). Solid tumors (3%) and other malignancies (3%) are rare. The distribution of entities appears to differ in different global regions. B cell lymphomas are predominantly reported from Western countries and Japan [14, 15], while cohorts from China and Korea mainly include T cell malignancies [16–18]. In children and adolescents, T cell malignancies predominate [13, 19].

Table 1 Malignancies associated with HLH

Type of malignancy	Sub-entities with particular risk of HLH	Prevalence in adults with malignancy-associated HLH (%) [1]
T and NK cell lymphoma		35
	Subcutaneous panniculitis-like T cell lymphoma	
	Primary cutaneous γδ-T cell lymphoma NK cell lymphoma	
	Anaplastic large cell lymphoma	
B cell lymphoma		32
	Diffuse large B cell lymphoma	
	Intravascular large B cell lymphoma	
Hodgkin lymphoma		6
Leukemia		6
Other hematologic malignancies		14
Solid tumors		3
	Mediastinal germ cell tumors	
	Langerhans cell histiocytosis	
Others and related disorders		
	Cytophagic histiocytic panniculitis	
	EBV+ T cell and NK cell lymphoproliferative disorders	
	Multicentric Castleman disease with HIV infection	

Among the T cell neoplasms, mature subtypes are more prone to elicit HLH. This includes subcutaneous panniculitis-like T cell lymphoma, primary cutaneous γδ-T cell lymphoma, and anaplastic large cell lymphoma [20]. Lymphoblastic T cell lymphomas and leukemias have less frequently been reported [13, 19]. Diffuse large B cell lymphoma (DLBCL) is the main entity among the B cell neoplasms. The prevalence of HLH in patients with intravascular large B cell lymphoma, especially in Far East Asia, is high [21]. In contrast, B precursor malignancies are infrequent triggers of HLH [22]. Among the rarely occurring solid tumors, a few have a particular propensity to elicit HLH, namely, mediastinal germ-cell tumors [23] and Langerhans cell histiocytosis (LCH) [24], for which increasing evidence indicates it is a malignant condition. Occasional reports described HLH with embryonal tumors [10]. Some lymphomas have a strong correlation with EBV as a co-trigger. This applies to Hodgkin lymphoma, where the prevalence of EBV reaches 90% if HLH occurs [7, 25], and peripheral T cell lymphoma (30%) [15, 17]. In DLBCL, however, a viral co-trigger is rarely found [26, 27].

Differential diagnoses for HLH associated hematopoietic disorders include cytophagic histiocytic panniculitis [28], EBV-driven T and NK cell lymphoproliferative disorders (particularly in Far East Asia) [29–31], and multicentric Castleman disease with human immune deficiency virus (HIV) infection [32].

Treatment Regimens Associated with HLH in the Context of Chemotherapy

The prevalence of documented infectious triggers in this cohort of malignancy associated HLH is 75–100%. This includes viruses (reactivations and primary infections) and fungi [9, 10, 13, 22, 33, 34]. As bacterial septicemia per se shares many features with HLH [35], it is controversial if these cases should be regarded as HLH or rather as a "mimic". The intensive cytostatic drug schemes for leukemias and lymphomas have a strong correlation with the occurrence of HLH [9]. It is, however, noteworthy that HLH not only occurs during the intensive induction and consolidation schemes, but HLH on maintenance therapy can be equally severe and life-threatening [13]. T cell engaging treatments with chimeric antigen receptor modified T cells and bispecific T cell engaging antibodies can elicit cytokine release syndromes. In this iatrogenic subtype of HLH, the forced recruitment of T cells triggers the cytokine storm [36].

Diagnostic Workup

Different scenarios lead to different questions in the diagnostic workup: (1) A patient has confirmed HLH. Workup must address the question if there is a malignant driver of the hyperinflammation; (2) A patient has confirmed malignancy and HLH is suspected; or (3) A patient is on chemotherapy treatment and HLH is suspected. In the latter two cases, workup must provide evidence for or against HLH, to allow for therapeutic consequences to be taken.

HLH is a condition that can only be diagnosed on the grounds of an entire set of characteristics, including clinical features and laboratory parameters. The most frequently used criteria for clinical and scientific purposes were devised for the HLH-2004 treatment protocol [37]. These criteria include fever, splenomegaly, decreased blood counts and fibrinogen, elevated ferritin, triglycerides, and sCD25, decreased natural killer cell function, and the finding of hemophagocytosis, typically in the bone marrow. (PLEASE REFER TO DIAGNOTIC CHAPTER WHERE HLH-2004 CRITERIA ARE DISPLAYED IN A TABLE.) Elevated lactate dehydrogenase, transaminases, and d-dimers, and decreased albumin may support the diagnosis. In the context of malignant diseases, there are substantial imperfections of the criteria, mainly related to the fact that several findings per se may be present in neoplastic disease or during cytostatic treatment ("B symptoms," cytopenia, organomegaly). Different groups of adult physicians have attempted to come up with other criteria, reviewed in [38]. This includes a scoring system based on a cohort in which almost half of patients had malignancy-associated HLH [39]. Others have described typical parameters of HLH during therapy for acute myeloid leukemia [9]. The acceptance of the different sets of criteria however varies. Even though hemophagocytosis is eponymous in this condition, it is neither very specific nor sensitive [40]. It is a diagnostic error to diagnose HLH solely based on the finding of hemophagocytosis. Soluble CD25 (sCD25) is typically more pronouncedly elevated in HLH with underlying lymphoma or Langerhans cell histiocytosis (LCH) than in other forms of HLH, which is why a high sCD25–ferritin ratio should raise suspicion of hidden lymphoma or LCH [41].

Despite the importance of formal criteria for clinical and scientific purposes, a general judgement is usually helpful if the combination, the extent, and the progression of parameters is unexpected, unusual, and otherwise unexplained [2]. There is no single pathognomonic feature of HLH. However, the combination of features makes the condition unique and guides the way to the diagnosis. The extent of laboratory abnormalities tends to be extreme, which should raise suspicion. The dynamics of HLH are frequently fulminant. To stop the condition in time, this progression warrants early detection.

The diagnosis of HLH must always prompt a search for a potential malignant disease, which in most cases is lymphoma. The higher the age of the patient the more likely is a neoplasm [15]. Basic diagnostic procedures should be performed (cytology of blood and marrow, ultrasound of abdomen and lymph nodes, chest X-ray). If no other plausible cause is found (infection, hereditary disease, autoimmune or autoinflammatory condition), magnetic resonance imaging (MRI), computed tomography (CT), positron emission tomography CT (PET-CT) scans, and lymph node biopsies should be considered. Organ biopsies (e.g., liver) in patients with HLH carry a substantial risk of bleeding complications in the context of thrombocytopenia and consumptive hypofibrinogenemia. Risks and benefits must be carefully considered. Other conditions may be present, concomitantly with a neoplasm. EBV and HIV are potent triggers of HLH and may be associated with hematological malignancies [42]. A cerebral MRI and lumbar tap should be considered to exclude central nervous system involvement, particularly in patients presenting with neurological symptoms.

Monitoring of disease progression and response to treatment is based on the HLH parameters. Platelets quickly indicate improvement or worsening of disease activity. A challenge, however, is the differentiation of myelotoxicity of treatment and persistence of inflammation, as both the former and the latter result in cytopenia. A bone marrow aspirate may sometimes be helpful in this situation. Normalization of ferritin may take several weeks or months [43] and repeated red blood cell transfusions may constitute a confounder.

Cytotoxicity Defects and Malignancy

Most pertinent hereditary defects not only predispose to HLH but to malignancy as well. The degree of predisposition varies. In a large retrospective study on patients with X-linked lymphoproliferative syndrome type (XLP) 1, most individuals presented with EBV-associated HLH (approx. 40%). However, B cell lymphoma was the first manifestation in 14% and occurred at any time in 25% of patients [44]. A history of EBV-associated HLH and B cell lymphoma is thus suggestive of XLP1 in male patients. No patients with lymphoma have however been reported for XIAP deficiency (XLP2) [45]. Several EBV-susceptibility syndromes additionally predispose to lymphoma, including deficiencies of the magnesium transporter 1 (MAGT1), inducible T cell kinase (ITK), and CD27 [46].

Familial HLH mouse models reveal an increased incidence of lymphoma [47]. In perforin and degranulation defects (see chapter "Murine Models of Familial Cytokine Storm Syndromes"), malignancy usually occurs in patients with hypomorphic defects who have survived to adolescence or adulthood without stem cell transplant (SCT). Patients with complete defects usually experience HLH and receive an SCT before malignancy can develop. In humans, perforin mutations are associated with hematologic malignancies [48, 49]. The role of the prevalent hypomorphic mutation p.A91V in hematological malignancies is debated [50, 51]. Several case reports ascribe the development of Hodgkin lymphoma to inherited cytotoxicity defects [52–55], with or without EBV. Heterozygous familial HLH mutations do not seem to be associated with hematological malignancies [56]. Somatic loss of heterozygosity of STX11, the gene mutated in familial HLH 4, has been found in adult peripheral T cell lymphoma [57]. An increased incidence of gynecological tumors was found for heterozygous carriers of mutations conferring familial HLH [56]. Colorectal and ovarian carcinoma patients are not more likely to have perforin mutations [58].

Whether a predisposing hereditary defect should be excluded in a patient with malignancy-associated HLH must be decided case-by-case. Young age, previous episodes of full or partial HLH, and a positive family history render a hereditary background more likely. A positive EBNA IgG as proof of a previous infection without a prior episode of HLH decreases the likelihood of a genetic defect, as EBV is considered the most potent trigger of HLH in cytotoxicity defects and is likely to have elicited HLH on primary infection. If required, screening for the relevant defects may be performed by flow cytometry (protein stains and degranulation assays) [59, 60] (REFER TO RESPECTIVE CHAPTER), followed by genetic analysis in the case of abnormal findings. Next-generation sequencing may be indicated in special cases.

Treatment and Prognosis of Malignancy-Triggered HLH

Treatment of HLH must not be delayed as the course of disease may be fulminant. It is not known whether primarily HLH-directed treatment or malignancy-directed treatment is most effective for overall outcome. However, there is substantial overlap in the agents used for treatment of malignancies and HLH. Decisions must be taken on a case-by-case basis. Etoposide and glucocorticosteroids, in particular, are used for both conditions. Etoposide selectively ablates activated T cells [61] and in that way dampens inflammation. Further chemotherapeutic agents with efficacy in murine models of primary HLH include cyclophosphamide and methotrexate [61]. Malignancy-directed protocols containing dexamethasone, etoposide, or cyclophosphamide may thus constitute the preferred treatment option to address HLH when it occurs in the context of a neoplasm. In patients where initially an HLH-directed approach is pursued, a regimen addressing the neoplasm must follow once HLH parameters have stabilized or resolved.

Initial HLH-directed immunosuppressive treatment may be an option particularly in patients with poor general condition and thusly prepares the patient for more intensive subsequent malignancy-directed therapy. A prospective study of patients with lymphoma-triggered HLH and failure of first-line treatment analyzed a regimen with liposomal doxorubicin, methylprednisolone, and etoposide. Complete response was achieved in 17%, partial response (i.e., moderate improvement of parameters) in 59%, and no response in 24% of patients [62]. Some retrospective studies and case series have indicated better survival if etoposide was administered; substantial limitations of these analyses however preclude generalization [63, 64]. In cytophagic histiocytic panniculitis and subcutaneous panniculitis-like T cell lymphoma, cyclosporine A and anakinra have shown beneficial effects [28, 65, 66]. Ruxolitinib, a janus kinase inhibitor, was effective in mouse models of primary and secondary HLH [67, 68]. Its role for the treatment of HLH in the context of malignancies remains to be determined. Stem cell transplantation is usually warranted if a hereditary defect predisposing to HLH is found.

Extensive anti-infectious treatment of viruses, bacteria, and fungi, as well as anti-infectious prophylaxis, including *Pneumocystis jirovecii*, and frequent screening for fungi and viruses (EBV, CMV, adenovirus) is essential to fight triggering agents. In highly replicative EBV infection, rituximab should be considered to address this strong co-trigger by elimination of B cells [69]. An additional antitumor effect can be expected, if the neoplasm is CD20 positive. Adjustments of treatment doses and renal replacement therapy are required in patients with acute kidney failure, which is more frequent in adults [70].

The interpretation of outcome data may prove difficult because it is often not possible to distinguish whether HLH or the underlying malignancy or both are the major causes of death. The substantial mortality of underlying neoplasms contributes to the poorer outcome of HLH in the context of malignancy when comparing to other subtypes of HLH. Prognosis of HLH in T cell malignancies is inferior as compared to B cell lymphomas. In general, the occurrence of HLH in a patient with malignancy is associated with a poorer outcome. The 30-day survival of the acute phase of HLH in adults is reported to be 56–70% (depending on the subtype of the neoplasm), the median overall survival 36–230 days, and the 3-year survival

18–55% [15, 17, 18, 62–64, 71–74]. Children have a better prognosis, surviving the acute phase of HLH in 56–67% of cases and a median overall survival of approximately 1 year [13, 19].

Therapy of HLH in the Context of Chemotherapy

Evidence regarding treatment for HLH in the context of chemotherapy is limited. Pathogen-specific therapy is crucial if an infectious trigger is found, including antiviral, antibacterial, and antifungal treatment. Rituximab has been proven beneficial in EBV-driven HLH [69] and is worth considering in patients with substantial EB-viremia. In general, antimicrobial prophylaxis against other viruses, fungi, and bacteria is advisable, as patients with HLH in the context of chemotherapy are usually profoundly neutropenic and lymphopenic and are at risk for further aggravating infections. The antifungal spectrum must include *Aspergilli.*

Further chemotherapeutic courses should be postponed or maintenance medication interrupted until HLH is controlled. Administration of immunoglobulins may be attempted as an immune modulatory approach. Glucocorticosteroids can be used; however, particularly if the infectious trigger is fungal, the period should be kept as short as possible. It is debatable if more profound immune suppression with etoposide [13] is beneficial or counterproductive. This decision must thus be discussed case-by-case. In cytokine-release syndrome in the context of T cell engaging therapies, the anti-IL-6 monoclonal antibody tocilizumab has been shown to be efficacious [36]. Overall survival was significantly lower in adult AML patients where features of HLH occurred during treatment [9]. In a small pediatric cohort of HLH in the context of chemotherapy mainly for leukemia, overall survival was 0.9 years [13].

Conclusions for HLH Associated with Malignancies

Malignancy-triggered HLH and HLH in the context of chemotherapy constitute a major challenge in hematology with poorer outcomes than other forms of HLH. The criteria used for the definition of HLH in malignant conditions need refinement. However, awareness of the condition may facilitate timely initiation of therapy. Since it is unknown if initial HLH-directed or malignancy-directed treatment is superior, therapy must be tailored on a case-by-case basis.

References

1. Ramos-Casals, M., Brito-Zeron, P., Lopez-Guillermo, A., Khamashta, M. A., & Bosch, X. (2014). Adult haemophagocytic syndrome. *Lancet, 383*, 1503–1516.
2. Lehmberg, K., Nichols, K. E., Henter, J. I., Girschikofsky, M., Greenwood, T., Jordan, M., et al. (2015). Consensus recommendations for the diagnosis and management of hemophagocytic lymphohistiocytosis associated with malignancies. *Haematologica, 100*, 997–1004.
3. Al-Hashmi, I., Decoteau, J., Gruss, H. J., Zielenska, M., Thorner, P., Poon, A., et al. (2001). Establishment of a cytokine-producing anaplastic large-cell lymphoma cell line containing the t(2;5) translocation: Potential role of cytokines in clinical manifestations. *Leukemia & Lymphoma, 40*, 599–611.
4. Siebert, S., Amos, N., Williams, B. D., & Lawson, T. M. (2007). Cytokine production by hepatic anaplastic large-cell lymphoma presenting as a rheumatic syndrome. *Seminars in Arthritis and Rheumatism, 37*, 63–67.
5. Perez-Encinas, M., Villamayor, M., Campos, A., Gonzalez, S., & Bello, J. L. (1998). Tumor burden and serum level of soluble CD25, CD8, CD23, CD54 and CD44 in non-Hodgkin's lymphoma. *Haematologica, 83*, 752–754.
6. Janik, J. E., Morris, J. C., Pittaluga, S., McDonald, K., Raffeld, M., Jaffe, E. S., et al. (2004). Elevated serum-soluble interleukin-2 receptor levels in patients with anaplastic large cell lymphoma. *Blood, 104*, 3355–3357.
7. Menard, F., Besson, C., Rince, P., Lambotte, O., Lazure, T., Canioni, D., et al. (2008). Hodgkin lymphoma-associated hemophagocytic syndrome: A disorder strongly correlated with Epstein-Barr virus. *Clinical Infectious Diseases, 47*, 531–534.
8. Risdall, R. J., McKenna, R. W., Nesbit, M. E., Krivit, W., Balfour Jr., H. H., Simmons, R. L., et al. (1979). Virus-associated hemophagocytic syndrome: A benign histiocytic proliferation distinct from malignant histiocytosis. *Cancer, 44*, 993–1002.
9. Delavigne, K., Berard, E., Bertoli, S., Corre, J., Duchayne, E., Demur, C., et al. (2014). Hemophagocytic syndrome in patients with acute myeloid leukemia undergoing intensive chemotherapy. *Haematologica, 99*, 474–480.
10. Celkan, T., Berrak, S., Kazanci, E., Ozyurek, E., Unal, S., Ucar, C., et al. (2009). Malignancy-associated hemophagocytic lymphohistiocytosis in pediatric cases: A multicenter study from Turkey. *The Turkish Journal of Pediatrics, 51*, 207–213.
11. Lehmberg, K., Sprekels, B., Nichols, K. E., Woessmann, W., Müller, I., Suttorp, M., et al. (2015). Malignancy-associated haemophagocytic lymphohistiocytosis in children and adolescents. *British Journal of Haematology, 170*, 539.
12. Strenger, V., Merth, G., Lackner, H., Aberle, S. W., Kessler, H. H., Seidel, M. G., et al. (2018). Malignancy and chemotherapy induced haemophagocytic lymphohistiocytosis in children and adolescents-a single centre experience of 20 years. *Annals of Hematology, 97*, 989–998.
13. Lehmberg, K., Sprekels, B., Nichols, K. E., Woessmann, W., Muller, I., Suttorp, M., et al. (2015). Malignancy-associated haemophagocytic lymphohistiocytosis in children and adolescents. *British Journal of Haematology, 170*, 539.
14. Riviere, S., Galicier, L., Coppo, P., Marzac, C., Aumont, C., Lambotte, O., et al. (2014). Reactive hemophagocytic syndrome in adults: A multicenter retrospective analysis of 162 patients. *The American Journal of Medicine, 127*(11), 1118–1125.
15. Ishii, E., Ohga, S., Imashuku, S., Yasukawa, M., Tsuda, H., Miura, I., et al. (2007). Nationwide survey of hemophagocytic lymphohistiocytosis in Japan. *International Journal of Hematology, 86*, 58–65.
16. Li, J., Wang, Q., Zheng, W., Ma, J., Zhang, W., Wang, W., et al. (2014). Hemophagocytic lymphohistiocytosis: Clinical analysis of 103 adult patients. *Medicine, 93*, 100–105.
17. Yu, J. T., Wang, C. Y., Yang, Y., Wang, R. C., Chang, K. H., Hwang, W. L., et al. (2013). Lymphoma-associated hemophagocytic lymphohistiocytosis: Experience in adults from a single institution. *Annals of Hematology, 92*, 1529–1536.

18. Han, A. R., Lee, H. R., Park, B. B., Hwang, I. G., Park, S., Lee, S. C., et al. (2007). Lymphoma-associated hemophagocytic syndrome: Clinical features and treatment outcome. *Annals of Hematology, 86*, 493–498.
19. Veerakul, G., Sanpakit, K., Tanphaichitr, V. S., Mahasandana, C., & Jirarattanasopa, N. (2002). Secondary hemophagocytic lymphohistiocytosis in children: An analysis of etiology and outcome. *Journal of the Medical Association of Thailand, 85*(Suppl 2), S530–S541.
20. Go, R. S., & Wester, S. M. (2004). Immunophenotypic and molecular features, clinical outcomes, treatments, and prognostic factors associated with subcutaneous panniculitis-like T-cell lymphoma: A systematic analysis of 156 patients reported in the literature. *Cancer, 101*, 1404–1413.
21. Ferreri, A. J., Dognini, G. P., Campo, E., Willemze, R., Seymour, J. F., Bairey, O., et al. (2007). Variations in clinical presentation, frequency of hemophagocytosis and clinical behavior of intravascular lymphoma diagnosed in different geographical regions. *Haematologica, 92*, 486–492.
22. Kelly, C., Salvi, S., McClain, K., & Hayani, A. (2011). Hemophagocytic lymphohistiocytosis associated with precursor B acute lymphoblastic leukemia. *Pediatric Blood & Cancer, 56*, 658–660.
23. Nichols, C. R., Roth, B. J., Heerema, N., Griep, J., & Tricot, G. (1990). Hematologic neoplasia associated with primary mediastinal germ-cell tumors. *The New England Journal of Medicine, 322*, 1425–1429.
24. Chellapandian, D., Zhang, R., Jeng, M., van den Bos, C., Santa-María López, V., Lehmberg, K., et al. (2016). Hemophagocytic lymphohistiocytosis in Langerhans cell histiocytosis: A multicenter retrospective descriptional study. In *32nd Annual Meeting of the Histiocyte Society, Dublin*.
25. Chang, Y. H., Lu, P. J., Lu, M. Y., Wang, J. S., Tung, C. L., & Shaw, C. F. (2009). Sequential transplants for respective relapse of Hodgkin disease and hemophagocytic lymphohistiocytosis: A treatment dilemma. *Journal of Pediatric Hematology/Oncology, 31*, 778–781.
26. Shimazaki, C., Inaba, T., & Nakagawa, M. (2000). B-cell lymphoma-associated hemophagocytic syndrome. *Leukemia & Lymphoma, 38*, 121–130.
27. Murase, T., Nakamura, S., Kawauchi, K., Matsuzaki, H., Sakai, C., Inaba, T., et al. (2000). An Asian variant of intravascular large B-cell lymphoma: Clinical, pathological and cytogenetic approaches to diffuse large B-cell lymphoma associated with haemophagocytic syndrome. *British Journal of Haematology, 111*, 826–834.
28. Aronson, I. K., & Worobec, S. M. (2010). Cytophagic histiocytic panniculitis and hemophagocytic lymphohistiocytosis: An overview. *Dermatologic Therapy, 23*, 389–402.
29. Hong, M., Ko, Y. H., Yoo, K. H., Koo, H. H., Kim, S. J., Kim, W. S., et al. (2013). EBV-positive T/NK-cell lymphoproliferative disease of childhood. *Korean Journal of Pathology, 47*, 137–147.
30. Kimura, H., Ito, Y., Kawabe, S., Gotoh, K., Takahashi, Y., Kojima, S., et al. (2012). EBV-associated T/NK-cell lymphoproliferative diseases in nonimmunocompromised hosts: Prospective analysis of 108 cases. *Blood, 119*, 673–686.
31. Paik, J. H., Choe, J. Y., Kim, H., Lee, J. O., Kang, H. J., Shin, H. Y., et al. (2016). Clinicopathological categorization of Epstein-Barr virus-positive T/NK-cell lymphoproliferative disease: An analysis of 42 cases with an emphasis on prognostic implications. *Leukemia & Lymphoma, 58*, 53–63.
32. Stebbing, J., Ngan, S., Ibrahim, H., Charles, P., Nelson, M., Kelleher, P., et al. (2008). The successful treatment of haemophagocytic syndrome in patients with human immunodeficiency virus-associated multi-centric Castleman's disease. *Clinical and Experimental Immunology, 154*, 399–405.
33. Trebo, M. M., Attarbaschi, A., Mann, G., Minkov, M., Kornmuller, R., & Gadner, H. (2005). Histiocytosis following T-acute lymphoblastic leukemia: A BFM study. *Leukemia & Lymphoma, 46*, 1735–1741.

34. Lackner, H., Urban, C., Sovinz, P., Benesch, M., Moser, A., & Schwinger, W. (2008). Hemophagocytic lymphohistiocytosis as severe adverse event of antineoplastic treatment in children. *Haematologica, 93*, 291–294.
35. Machowicz, R., Janka, G., & Wiktor-Jedrzejczak, W. (2017). Similar but not the same: Differential diagnosis of HLH and sepsis. *Critical Reviews in Oncology/Hematology, 114*, 1–12.
36. Teachey, D. T., Rheingold, S. R., Maude, S. L., Zugmaier, G., Barrett, D. M., Seif, A. E., et al. (2013). Cytokine release syndrome after blinatumomab treatment related to abnormal macrophage activation and ameliorated with cytokine-directed therapy. *Blood, 121*, 5154–5157.
37. Henter, J. I., Horne, A., Arico, M., Egeler, R. M., Filipovich, A. H., Imashuku, S., et al. (2007). HLH-2004: Diagnostic and therapeutic guidelines for hemophagocytic lymphohistiocytosis. *Pediatric Blood & Cancer, 48*, 124–131.
38. Hayden, A., Park, S., Giustini, D., Lee, A. Y., & Chen, L. Y. (2016). Hemophagocytic syndromes (HPSs) including hemophagocytic lymphohistiocytosis (HLH) in adults: A systematic scoping review. *Blood Reviews, 30*, 411.
39. Fardet, L., Galicier, L., Lambotte, O., Marzac, C., Aumont, C., Chahwan, D., et al. (2014). Development and validation of the HScore, a score for the diagnosis of reactive hemophagocytic syndrome. *Arthritis & Rhematology, 66*, 2613–2620.
40. Gupta, A., Tyrrell, P., Valani, R., Benseler, S., Weitzman, S., & Abdelhaleem, M. (2008). The role of the initial bone marrow aspirate in the diagnosis of hemophagocytic lymphohistiocytosis. *Pediatric Blood & Cancer, 51*, 402–404.
41. Tsuji, T., Hirano, T., Yamasaki, H., Tsuji, M., & Tsuda, H. (2014). A high sIL-2R/ferritin ratio is a useful marker for the diagnosis of lymphoma-associated hemophagocytic syndrome. *Annals of Hematology, 93*, 821–826.
42. Fardet, L., Lambotte, O., Meynard, J. L., Kamouh, W., Galicier, L., Marzac, C., et al. (2010). Reactive haemophagocytic syndrome in 58 HIV-1-infected patients: Clinical features, underlying diseases and prognosis. *AIDS (London, England), 24*, 1299–1306.
43. Lehmberg, K., & Ehl, S. (2012). Diagnostic evaluation of patients with suspected haemophagocytic lymphohistiocytosis. *British Journal of Haematology, 160*, 275–287.
44. Booth, C., Gilmour, K. C., Veys, P., Gennery, A. R., Slatter, M. A., Chapel, H., et al. (2011). X-linked lymphoproliferative disease due to SAP/SH2D1A deficiency: A multicenter study on the manifestations, management and outcome of the disease. *Blood, 117*, 53–62.
45. Speckmann, C., Lehmberg, K., Albert, M. H., Damgaard, R. B., Fritsch, M., Gyrd-Hansen, M., et al. (2013). X-linked inhibitor of apoptosis (XIAP) deficiency: The spectrum of presenting manifestations beyond hemophagocytic lymphohistiocytosis. *Clinical Immunology, 149*, 133–141.
46. Cohen, J. I. (2015). Primary immunodeficiencies associated with EBV disease. *Current Topics in Microbiology and Immunology, 390*, 241–265.
47. Smyth, M. J., Thia, K. Y., Street, S. E., MacGregor, D., Godfrey, D. I., & Trapani, J. A. (2000). Perforin-mediated cytotoxicity is critical for surveillance of spontaneous lymphoma. *The Journal of Experimental Medicine, 192*, 755–760.
48. Chia, J., Yeo, K. P., Whisstock, J. C., Dunstone, M. A., Trapani, J. A., & Voskoboinik, I. (2009). Temperature sensitivity of human perforin mutants unmasks subtotal loss of cytotoxicity, delayed FHL, and a predisposition to cancer. *Proceedings of the National Academy of Sciences of the United States of America, 106*, 9809–9814.
49. Clementi, R., Locatelli, F., Dupre, L., Garaventa, A., Emmi, L., Bregni, M., et al. (2005). A proportion of patients with lymphoma may harbor mutations of the perforin gene. *Blood, 105*, 4424–4428.
50. Santoro, A., Cannella, S., Trizzino, A., Lo Nigro, L., Corsello, G., & Arico, M. (2005). A single amino acid change A91V in perforin: A novel, frequent predisposing factor to childhood acute lymphoblastic leukemia. *Haematologica, 90*, 697–698.
51. Mehta, P. A., Davies, S. M., Kumar, A., Devidas, M., Lee, S., Zamzow, T., et al. (2006). Perforin polymorphism A91V and susceptibility to B-precursor childhood acute lymphoblastic leukemia: A report from the Children's Oncology Group. *Leukemia, 20*, 1539–1541.

52. Machaczka, M., Klimkowska, M., Chiang, S. C., Meeths, M., Muller, M. L., Gustafsson, B., et al. (2013). Development of classical Hodgkin's lymphoma in an adult with biallelic STXBP2 mutations. *Haematologica, 98*, 760–764.
53. Lorenzi, L., Tabellini, G., Vermi, W., Moratto, D., Porta, F., Notarangelo, L. D., et al. (2013). Occurrence of nodular lymphocyte-predominant Hodgkin lymphoma in Hermansky-Pudlak type 2 syndrome is associated to natural killer and natural killer T cell defects. *PLoS One, 8*, e80131.
54. Nagai, K., Ochi, F., Terui, K., Maeda, M., Ohga, S., Kanegane, H., et al. (2013). Clinical characteristics and outcomes of Chediak-Higashi syndrome: A nationwide survey of Japan. *Pediatric Blood & Cancer, 60*, 1582–1586.
55. Pagel, J., Beutel, K., Lehmberg, K., Koch, F., Maul-Pavicic, A., Rohlfs, A. K., et al. (2012). Distinct mutations in STXBP2 are associated with variable clinical presentations in patients with familial hemophagocytic lymphohistiocytosis type 5 (FHL5). *Blood, 119*, 6016–6024.
56. Lofstedt, A., Chiang, S. C., Onelov, E., Bryceson, Y. T., Meeths, M., & Henter, J. I. (2015). Cancer risk in relatives of patients with a primary disorder of lymphocyte cytotoxicity: A retrospective cohort study. *The Lancet. Haematology, 2*, e536–e542.
57. Yoshida, N., Tsuzuki, S., Karube, K., Takahara, T., Suguro, M., Miyoshi, H., et al. (2015). STX11 functions as a novel tumor suppressor gene in peripheral T-cell lymphomas. *Cancer Science, 106*, 1455–1462.
58. Trapani, J. A., Thia, K. Y., Andrews, M., Davis, I. D., Gedye, C., Parente, P., et al. (2013). Human perforin mutations and susceptibility to multiple primary cancers. *Oncoimmunology, 2*, e24185.
59. Bryceson, Y. T., Pende, D., Maul-Pavicic, A., Gilmour, K. C., Ufheil, H., Vraetz, T., et al. (2012). A prospective evaluation of degranulation assays in the rapid diagnosis of familial hemophagocytic syndromes. *Blood, 119*, 2754–2763.
60. Marsh, R. A., Bleesing, J. J., & Filipovich, A. H. (2013). Flow cytometric measurement of SLAM-associated protein and X-linked inhibitor of apoptosis. *Methods in Molecular Biology, 979*, 189–197.
61. Johnson, T. S., Terrell, C. E., Millen, S. H., Katz, J. D., Hildeman, D. A., & Jordan, M. B. (2014). Etoposide selectively ablates activated T cells to control the immunoregulatory disorder hemophagocytic lymphohistiocytosis. *Journal of Immunology, 192*, 84–91.
62. Wang, Y., Huang, W., Hu, L., Cen, X., Li, L., Wang, J., et al. (2015). Multicenter study of combination DEP regimen as a salvage therapy for adult refractory hemophagocytic lymphohistiocytosis. *Blood, 126*, 2186–2192.
63. Arca, M., Fardet, L., Galicier, L., Riviere, S., Marzac, C., Aumont, C., et al. (2014). Prognostic factors of early death in a cohort of 162 adult haemophagocytic syndrome: Impact of triggering disease and early treatment with etoposide. *British Journal of Haematology, 168*(1), 63–68.
64. Schram, A. M., Comstock, P., Campo, M., Gorovets, D., Mullally, A., Bodio, K., et al. (2016). Haemophagocytic lymphohistiocytosis in adults: A multicentre case series over 7 years. *British Journal of Haematology, 172*, 412–419.
65. Mizutani, S., Kuroda, J., Shimura, Y., Kobayashi, T., Tsutsumi, Y., Yamashita, M., et al. (2011). Cyclosporine A for chemotherapy-resistant subcutaneous panniculitis-like T cell lymphoma with hemophagocytic syndrome. *Acta Haematologica, 126*, 8–12.
66. Behrens, E. M., Kreiger, P. A., Cherian, S., & Cron, R. Q. (2006). Interleukin 1 receptor antagonist to treat cytophagic histiocytic panniculitis with secondary hemophagocytic lymphohistiocytosis. *The Journal of Rheumatology, 33*, 2081–2084.
67. Maschalidi, S., Sepulveda, F. E., Garrigue, A., Fischer, A., & de Saint Basile, G. (2016). Therapeutic effect of JAK1/2 blockade on the manifestations of hemophagocytic lymphohistiocytosis in mice. *Blood, 128*, 60–71.
68. Das, R., Guan, P., Sprague, L., Verbist, K., Tedrick, P., An, Q. A., et al. (2016). Janus kinase inhibition lessens inflammation and ameliorates disease in murine models of hemophagocytic lymphohistiocytosis. *Blood, 127*, 1666–1675.

69. Chellapandian, D., Das, R., Zelley, K., Wiener, S. J., Zhao, H., Teachey, D. T., et al. (2013). Treatment of Epstein Barr virus-induced haemophagocytic lymphohistiocytosis with rituximab-containing chemo-immunotherapeutic regimens. *British Journal of Haematology, 162*, 376–382.
70. Aulagnon, F., Lapidus, N., Canet, E., Galicier, L., Boutboul, D., Peraldi, M. N., et al. (2015). Acute kidney injury in adults with hemophagocytic lymphohistiocytosis. *American Journal of Kidney Diseases, 65*, 851–859.
71. Takahashi, N., Chubachi, A., Kume, M., Hatano, Y., Komatsuda, A., Kawabata, Y., et al. (2001). A clinical analysis of 52 adult patients with hemophagocytic syndrome: The prognostic significance of the underlying diseases. *International Journal of Hematology, 74*, 209–213.
72. Parikh, S. A., Kapoor, P., Letendre, L., Kumar, S., & Wolanskyj, A. P. (2014). Prognostic factors and outcomes of adults with hemophagocytic lymphohistiocytosis. *Mayo Clinic Proceedings, 89*, 484–492.
73. Tong, H., Ren, Y., Liu, H., Xiao, F., Mai, W., Meng, H., et al. (2008). Clinical characteristics of T-cell lymphoma associated with hemophagocytic syndrome: Comparison of T-cell lymphoma with and without hemophagocytic syndrome. *Leukemia & Lymphoma, 49*, 81–87.
74. Buyse, S., Teixeira, L., Galicier, L., Mariotte, E., Lemiale, V., Seguin, A., et al. (2010). Critical care management of patients with hemophagocytic lymphohistiocytosis. *Intensive Care Medicine, 36*, 1695–1702.

Cytokine Storm and Sepsis-Induced Multiple Organ Dysfunction Syndrome

Joseph A. Carcillo and Bita Shakoory

Because this chapter overlaps with many others in this text book, we have included four figures to help the reader appreciate similarities and differences between sepsis-induced MODS-associated MAS or MALS (macrophage activation like syndrome) and rheumatologic disease-associated MAS, and familial HLH. Importantly there are no distinguishing clinical characteristics between the three; however, there are distinctive differences in sepsis cytokine patterns (Fig. 1, Table 1), NK cell number and functional response (Fig. 2, Table 1), T cell numbers and response (Fig. 3, Table 1), and putative therapies (Fig. 4, Table 1) compared to rheumatologic disease-associated MAS, and familial HLH. This chapter explores these differences.

Mortality in sepsis is associated with a cytokine storm known as MARS or mixed anti-inflammatory response syndrome. This cytokine storm, said to be a result of immunologic dissonance, is characterized by high circulating IL-6 and IL-10 levels, and decreased ability of monocytes, macrophages, and dendritic cells to produce TNF alpha or interferon gamma (IFN-γ) in response to endotoxin. Hotchkiss and colleagues reviewed autopsies in adult patients who died after 7 days of sepsis and found profound circulating T cell, B cell, dendritic cell, and NK cell lymphopenia along with lymphoid depletion related to apoptosis in the thymus, lymph nodes, spleen, and bone marrow [1]. Felmet et al. reported similar findings in children who died of lymphopenic sepsis-induced multiple organ failure (MOF) [2], and Gurevich et al. also reported this phenomenon in neonates who died with sepsis [3]. In addition to lymphocyte numbers being depressed, monocyte and macrophage functions were severely depressed in these patients who had a condition known as "immunoparalysis" characterized by inability to produce IFN-γ, to clear infection, and to survive sepsis.

J. A. Carcillo (✉)
Children's Hospital of Pittsburgh, Pittsburgh, PA, USA
e-mail: carcilloja@ccm.upmc.edu

B. Shakoory
National Institutes of Health, NIAID, Bethesda, MD, USA

R. Q. Cron, E. M. Behrens (eds.), *Cytokine Storm Syndrome*,
https://doi.org/10.1007/978-3-030-22094-5_27

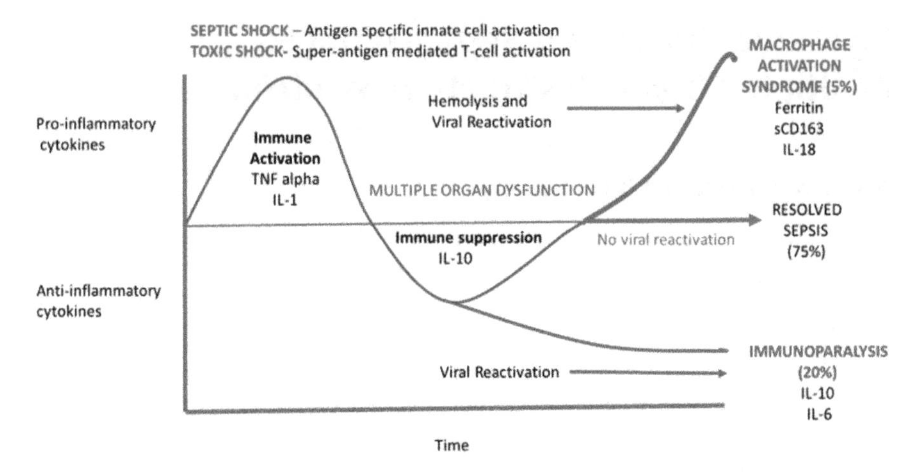

Fig. 1 Cytokine response patterns in septic shock, toxic shock, multiple organ dysfunction syndrome, immunoparalysis, and macrophage activation syndrome

Table 1 Differences between sepsis-related MAS/MALS, rheumatology-associated MAS, and familial HLH

	Rheumatologic MAS	Familial HLH	Sepsis-induced MODS	Sepsis-associated MAS/MALS
Cytokine pattern	Hyperferritinemia Very high IL-18 Increased interferon γ	Extreme hyperferritinemia High interferon γ Increased IL-18	Increased ferritin Absent interferon γ	Hyperferritinemia Somewhat increased IL-18 Low interferon γ
NK cell numbers/ function	Normal numbers, decreased cytolytic function	Normal numbers, absent cytolytic function	Low numbers, normal cytolytic function per cell	Less than 10% of normal numbers, normal cytolytic function per cell
T cell numbers/ function	T-cell activation/no proliferation	T cell activation/ proliferation	Decreased T cell numbers	Profound lymphopenia and T-cell exhaustion
Putative therapies	Corticosteroids IVIG Anakinra Interferon γ Ab?	Etoposide Dexamethasone Interferon γ Ab	Remove source of infection GM-CSF? Checkpoint inhibitors? IL-7?	Remove source of infection IVIG Methylprednisone, Anakinra?

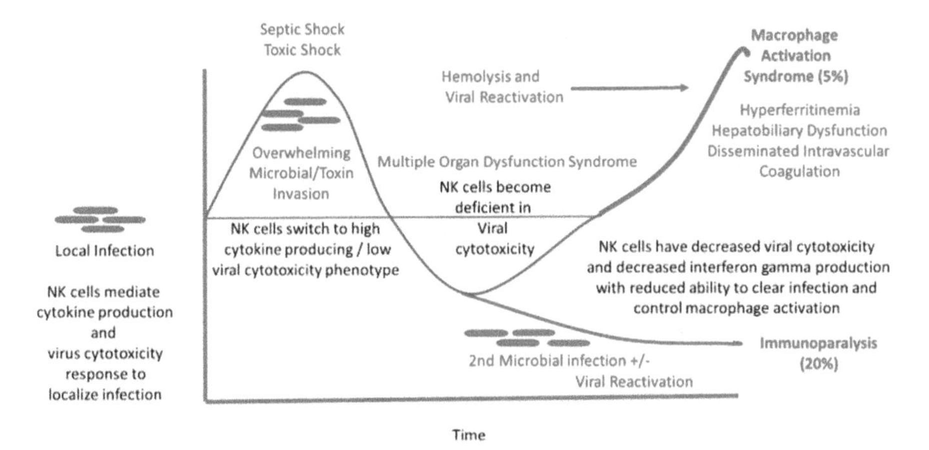

Fig. 2 Natural killer cell function response during infection, septic shock, toxic shock, multiple organ dysfunction syndrome, immunoparalysis, and macrophage activation syndrome

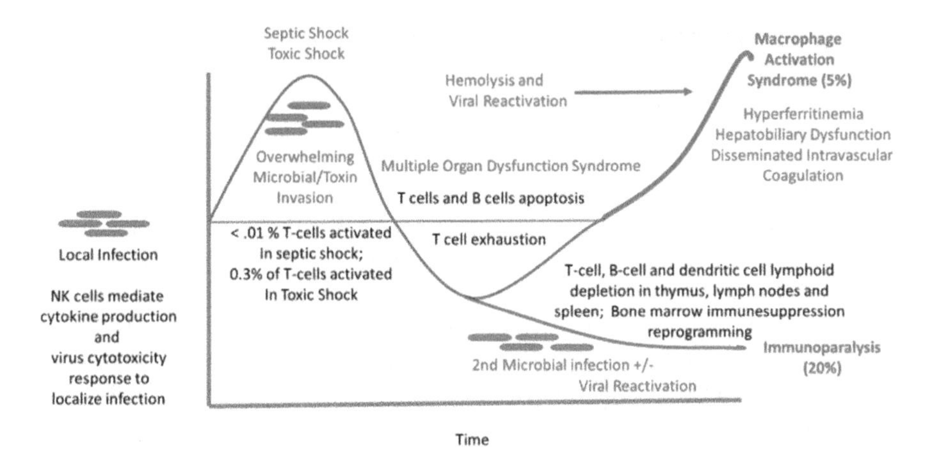

Fig. 3 T cell response in local infection, septic shock, toxic shock, multiple organ dysfunction, immunoparalysis, and macrophage activation syndrome

Hotchkiss and colleagues investigated sepsis immunology further using the cecal ligation and perforation (CLP) experimental model of mixed bacterial sepsis. Puncture with an increasing gauge needle increases mortality rate as the bacterial load leaking into the peritoneum increases with this maneuver. Initially, approximately 50% of the rodents who die do so from shock followed by the other 50% who die later from multiple organ failure. Antibiotics and fluids can markedly reduce mortality when a lower gauge puncture is used but are less effective when a larger gauge puncture is used on the cecum. In a series of elegant experiments Hotchkiss studied the influence of apoptotic lymphocytes on development of immunoparalysis and subsequent death by employing heterologous transfer of

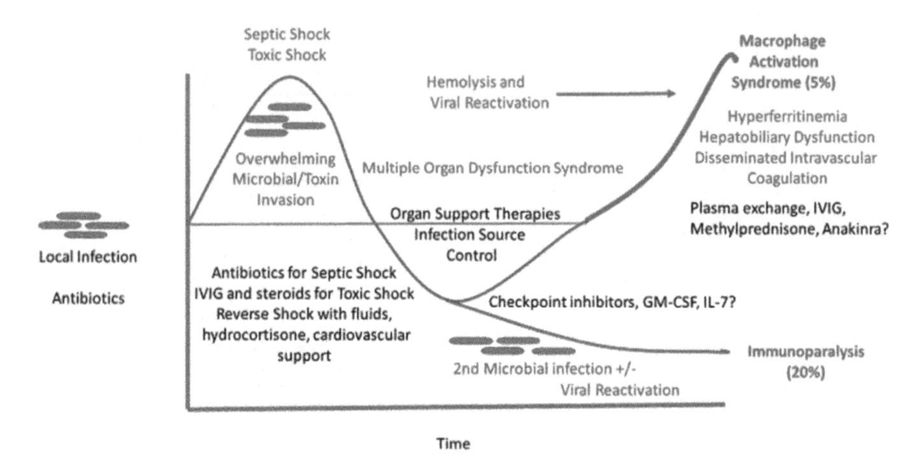

Fig. 4 Therapies used and being studied for local infection, septic shock, toxic shock, multiple organ dysfunction, immunoparalysis, and macrophage activation syndrome

sham treated, freeze thawed (necrotic cells), and irradiated (apoptotic cells) splenocytes from a control rodent into the blood stream of sham and cecal ligation puncture treated animals. Compared to animals who received untreated heterologous splenocytes, those who received necrotic cells (induces dendritic cell and macrophage activation) were protected from cecal ligation sepsis. They exhibited higher IFN-γ production, lower circulating bacterial colony forming units, and improved survival. By contrast, compared to animals who received untreated heterologous cells, those who received apoptotic heterologous cells (deactivates dendritic cells and macrophages) were more susceptible to cecal ligation and puncture induced sepsis. They exhibited decreased IFN-γ production, increased circulating bacterial colony forming units, and increased mortality [4]. Chiche and colleagues similarly showed that cytokine storm is related to NK cell and T cell apoptosis in late experimental sepsis [5]. Hotchkiss and colleagues have now further demonstrated that check point inhibitors, including the PDL1 and PD1 antagonists that prevent lymphocyte apoptosis, as well as the lymphocyte growth factor IL-7, improve survival in the cecal ligation and perforation model by decreasing bacterial colony forming unit counts [6–8]. These investigators are now studying PDL1 and PD1 checkpoint inhibitors and IL-7 in human sepsis clinical trials.

Because sepsis does not induce T lymphocyte derived IFN-γ production, there is no role for etoposide therapy directed at CD8 T cells, nor for IFN-γ inhibitors [9–15]. Both of these therapies can harm sepsis patients rather than help them because recovery of T cell and NK cell numbers is necessary for recovery of IFN-γ production that is needed to reduce bacterial cell counts and resolve infection. The anti-cytokine strategy that has been successful in improving survival in endotoxin models and cecal ligation puncture sepsis models has been combined treatment with IL-1 and IL-18 antagonists [16]. Interleukin-1 and IL-18 are produced in a feedforward fashion during inflammasome activation which is the likely the target of these anti-cytokine therapies.

Toxic Shock and the Cytokine Storm

Although most sepsis is caused by bacteria that stimulate a specific immune response that requires antigen presentation and specific T-cell activation; bacteria that shed toxins that are super antigens induce amplified promiscuous T cell activation and a more pronounced cytokine storm. These bacteria include *Streptococcus pyogenes* which sheds the TSST toxin and *Staphylococcus aureus* which sheds the lipoteichoic acid toxin. Toxin-mediated cytokine storms can be treated with intravenous gamma globulin which neutralizes the toxin, clindamycin which prevents further toxin production, and methylprednisolone which calms the T cell response [17–20].

Hyperferritinemic Sepsis and the Cytokine Storm

The greatest cytokine storm is observed in patients with hyperferritinemic sepsis. We have previously published a literature review derived working model that proposes a role for ferritin as a mediator of feedforward inflammation in sepsis. Two IFN-γ independent inducers of macrophage activation related hyperferritinemia in patients with sepsis-induced MOF, are free hemoglobin and DNA viremia coinfection [21]. Endotheliopathy (a disease that affects the endothelium) in sepsis leads to hemolysis, particularly in patients with gene variants related to atypical HUS and low to absent inhibitory complement production. The released free hemoglobin complexes with haptoglobin and binds to the macrophage haptoglobin receptor, CD163, which is internalized leading to production and release of extracellular ferritin. Extracellular ferritin activates liver stellate cells causing proinflammatory cytokine mediated liver injury. Ferritin also increases Toll like receptor (TLR) expression including TLR9 on innate immune cells, while inhibiting the adaptive immune response by preventing lymphopoiesis and increasing IL-10 production by dendritic cells. DNA viremia, ether by primary infection or by reactivation, allowed through sepsis-induced lymphopenia and immunoparalysis complexes with the macrophage TLR9 receptor resulting in inflammasome activation and production of IL-1, IL-18, and more extracellular ferritin. This results in a feedforward positive feedback inflammation loop with more liver injury, innate immune cell inflammation, and adaptive immune cell depression. Plasma exchange can be used to remove free hemoglobin and extracellular ferritin as well as to replace inhibitory complement. In patients with atypical HUS gene variants related to reduced or absent inhibitory complement, the C5A monoclonal antibody eculizumab can also be considered but care must be taken to not consider this in patients with meningococcemia. Intravenous immunoglobulin (IVIG) is given to neutralize DNA viremia and to block further TLR9 stimulation. Interestingly, IL-1 receptor antagonist protein is effective in reversing both ferritin and TLR9 induced liver injury while inducing a delayed type 1 interferon response to combat DNA viremia [21–25]. It goes without saying that removal of the inciting source of infection is paramount to survival in sepsis.

Macrophage Activation and Hyperferritinemic MOF in the Intensive Care Unit

Autopsy studies have provided evidence linking sepsis and blood transfusion exposure to the development of macrophage activation and hyperferritinemic MOF during critical illness. In 1988, Suster and colleagues reviewed bone marrow, lymph node, and spleen histology from 230 consecutive intensive care unit adult autopsies to identify cases of histiocytic hyperplasia with hemophagocytosis (HHH) [26]. They reported moderate to severe HHH in 102/230 bone marrows (44%), 79/191 lymph nodes (41%), and 16/209 spleens (8%). There was a strong dose related relationship to both blood transfusion (adjusted odds ratio 59.9 for ≥5 transfusions compared to 0 transfusions) and bacterial sepsis (adjusted risk ratio 4.10). In 2004, Strauss and colleagues evaluated 107 consecutive medical intensive care unit (MICU) patient autopsies and found mild to severe HHH in 69 (64.5%) [27]. The authors similarly found HHH to be associated with sepsis and blood transfusions. Patients with HHH were less likely to have died due to cardiovascular causes (HHH 22/69, 32% vs. no HHH 28/38, 74%; $p < 0.01$) and more likely to die due to multiple organ failure (MOF) (HHH 27/69, 39% vs. no HHH 7/38, 18%; $p < 0.05$) with a characteristic organ failure pattern of elevated bilirubin, liver enzymes, and disseminated intravascular coagulation. Patients with HHH required more catecholamine infusions, mechanical ventilation, and continuous renal replacement therapy (CRRT). Severe HHH had more red blood cell siderosis suggesting iron overload, and CD8 T cells in bone marrow suggesting T cell activation, hence the clinical descriptor "macrophage activation syndrome (MAS)/Secondary (2°) hemophagocytic lymphohistiocytosis (HLH)/sepsis-induced MOF" [28]. These two autopsy studies associated HHH/MAS/2° HLH with more severe disease but could not determine if it was a novel and clinically relevant process or a secondary phenomenon.

Experimental models provide evidence that supports macrophage activation as an important pathway to MOF. Steinberg et al. developed the sterile model of zymosan (Saccharomycoses cerevisiae) plus mineral oil intraperitoneal injection-induced MOF in rodents. This model results in initial hypovolemic shock followed by persistent macrophage activation [29]. Injection of either zymosan or mineral oil alone does not induce MOF, suggesting the need for both TLR stimulation and unremitting particulate irritation to induce persistent macrophage activation. In another rodent model, Behrens et al. reported that repeated TLR9 stimulation with

CpG oligodeoxynucleotides without TL4 stimulation led to MAS with cytopenias, splenomegaly, hyperferritinemia, and hepatitis [30]. Similarly, in a third murine model of cecal ligation and perforation induced sepsis, additional CpG injection induced cytokine production by macrophages and hepatic mononuclear cells with subsequent development of liver injury and MODS induced mortality [31]. The MAS/2° HLH phenotype elicited with repeated TLR9 stimulation is exacerbated in knockout mice deficient in the type 1 interferon receptor and ameliorated by interferon-α induced production of IL-1 receptor antagonist protein or with direct administration of recombinant IL-1 receptor antagonist protein (anakinra) in vivo. Liver dysfunction in this model appears to be related in part to IL-1 mediated inflammation [23, 32]. In a knockout for IL-18 binding protein (the natural inhibitor of IL-18), TLR-9-induced MAS is further exacerbated. Together these experimental models support a potential therapeutic role for Type 1 interferons, Interleukin 1 receptor antagonist protein, and IL-18 as specific approaches for controlling MAS [16, 23, 31–33].

There are some clinical data supporting nonspecific approaches to treatment of MAS/2° HLH defined by the presence of five of eight clinical criteria that include ferritin > 500 ng/mL, two line cytopenia, splenomegaly, hypertriglyceridemia, hypofibrinogenemia, elevated sCD25, absent NK cytotoxic activity, and hemophagocytosis. Demirkol et al. evaluated therapies for Turkish children who met these criteria for hyperferritinemic MAS/2° HLH/sepsis-induced MOF in a cohort study [34]. They first excluded children who were under 2 years, had a history of consanguineous parenting, or had a previous young family member dying from fever because they were considered to be at high risk for having familial HLH and were referred for treatment by hematologists with etoposide and bone marrow transplantation and subsequently experienced a 50% mortality rate. All of the remaining children without these HLH risk factors had 5–6 organ failure hyperferritinemic MOF with hepatobiliary dysfunction and DIC. The centers in one treatment cohort administered the HLH protocol of dexamethasone and/or etoposide along with daily plasma exchange and observed a 50% morality rate, whereas centers in the other treatment cohort administered the less immune suppressive regimen of methylprednisolone and/or intravenous immune globulin (IVIG) with daily plasma exchange and observed a 0% mortality rate. MOF resolved as ferritin levels decreased to normal values. Similar to the excluded patients at high risk for familial HLH, in the etoposide and dexamethasone cohort without these risk factors, patients who did not resolve hyperferritinemic MOF died with uneradicated infection. The authors thought that these children likely died from too much immune suppression related to dexamethasone and etoposide.

Because three of the clinical eight criteria (sCD25 levels, NK cytotoxicity, and hemophagocytosis) used to identify patients with hyperferritinemic HLH/MAS/sepsis-induced MODS are not easily accessible tests, rheumatologists have sought to redefine MAS/2° HLH using organ dysfunction patterns that use more readily available laboratory tests. Ravelli and colleagues have provided a consensus statement defining MAS in any child with known systemic juvenile arthritis (sJIA) who presents with *fever* and *a ferritin > 684 ng/mL* with any two of the following: *plate-*

let count < 181K, ALT > 48 IU/L, triglycerides > 156 mg/dL, and *fibrinogen < 360 mg/dL* [35]. The authors suggest that these criteria should allow patients to be eligible for early immunomodulatory clinical trials.

In this regard, Shakoory and colleagues considered the combination *Hepatobiliary dysfunction and Disseminated Intravascular Coagulation (DIC)* as representative of features of MAS in adults with severe sepsis [36]. They hypothesized that if the combination of these two organ dysfunctions represents MAS/2° HLH then a specific treatment approach with IL-1 receptor antagonist protein should improve sepsis-related MAS as it does in the previously mentioned experimental model and in children with sJIA related MAS. In a secondary analysis of an adult severe sepsis IL-1 receptor antagonist protein trial they compared patients with combined *Hepatobiliary dysfunction (HBD) and DIC (HBD + DIC)* to those without this combination (non-HBD + DIC) and found the following: (1) Only 5.6% of severe sepsis patients had these features of MAS (HBD + DIC), (2) patients with these features of MAS had higher incidence of shock (HBD + DIC = 95% vs. non-HBD + DIC = 79%), and acute kidney injury (HBD + DIC = 61% vs. non-HBD + DIC = 29%), but not acute respiratory distress syndrome (HBD + DIC = 21% vs. non-HBD + DIC = 26%), and (3) although IL-1 receptor antagonist protein had no effect on non-HBD + DIC 28 day survival (IL-1 blockade = 71% vs. placebo 71%) it significantly increased HBD + DIC 28 day survival (IL-1 blockade = 65.4% vs. placebo 35%). If HBD + DIC is indeed reflective of MAS/2° HLH, then we can conclude that though the condition is rare and associated with a high incidence of MODS and death, it is remediable in part by IL-1 receptor antagonist protein. More recently Hellenic and Swedish investigators have defined these adult patients as having sepsis with macrophage activation like syndrome (MALS) with the degree of hyperferritinemia (>4400 ng/mL) being most predictive of mortality [37]. They concurred that *hyperferritinemia, hepatobiliary dysfunction*, and *disseminated intravascular coagulation* can feasibly identify patients with MAS/2° HLH/sepsis-induced MOF for "early clinical trials" of inflammation modulating therapies.

Varied NK Cell Pathobiology Leading to Macrophage Activation

The eight clinical criteria used to describe the constellation of symptoms and signs indicative of these syndromes are considered to be biomarkers of a state of uncontrolled inflammation. The natural killer (NK) cell is among the most important "cellular" controllers of inflammation. Uncontrolled inflammation can be due in part to ineffective NK cell and CD8 T cell cytolytic function. Defects or deficiencies in the ability of the NK cell to kill viruses and cancer cells, and to turn off the host reticuloendothelial system, and macrophage, dendritic cell, and lymphocyte activation can be related to one of three conditions: (1) absent NK cell cytolytic activity unrelated to numbers of NK cells present (HLH treated by hematologists), (2) reduced

NK cell cytolytic activity unrelated to numbers of NK cells present (MAS treated by rheumatologists), and (3) normal NK cell cytolytic activity per cell but NK cell cytopenia (sepsis-induced hyperferritinemic MOF treated by intensivists and infectious disease specialists) [38, 39].

Familial HLH can be considered as a group of inherited homozygous gene variant primary immune deficiency diseases characterized by absence of crucial components of the perforin-granzyme pathway needed for NK cells and CD8 T cells to kill virus infected cells, and induce apoptosis in cancer cells and host inflammatory cells. A perforin gene knockout model for this pathway results in murine HLH and death after infection with LCMV infection at a dose which is innocuous in wild-type mice. Etoposide cures these mice by eliminating activated host T cells. In the wild-type mouse exposed to LCMV infection, host NK and T cells respond to control the virus. In the perforin knockout mouse LCMV infection results in T cell proliferation and activation which overproduces IFN-γ reflected by very high levels of CXCL9 (the monokine induced by gamma interferon). IFN-γ induces macrophage activation and organ injury. Etoposide destruction of proliferating activated T cells reduces IFN-γ production, prevents undue macrophage activation, and reverses organ failure. The US FDA has approved IFN-γ monoclonal antibody therapy for this condition in humans.

Rheumatology and drug related MAS patients are less likely driven by gene variant conditions. Nevertheless, some sJIA patients are hypomorphic or heterozygotes for the perforin-granzyme pathway gene variants. As heterozygotes these patients have some decreased NK cell cytolytic activity. For the most part these patients respond nicely to anti-inflammatory therapies such as corticosteroids or anakinra (IL-1 receptor antagonist protein). A group of rheumatologic MAS patients with cryopyrin-associated periodic syndromes has autosomal gene variants in NLRP leading to increased inflammasome activation that generates positive feedback IL-1 and IL-18 production. The IL-1 receptor antagonist protein, anakinra is FDA approved for these inflammasome driven conditions. Because IL-18 is the IFN-γ inducible factor there is optimism that IL18-BP (neutralizes IL-18) as well as IFN-γ monoclonal antibody could help these patients.

By contrast, hyperferritinemic sepsis-induced MOF is related to a reduction in NK cell numbers, not cytolytic activity per cell [38, 39]. When NK cell numbers recover, then inflammation and MODS resolve. During bacterial infection, NK cells switch from an overall low cytokine producing high cytolytic activity phenotype to a high cytokine producing low cytolytic activity phenotype. In sepsis-induced MOF, lymphopenia with associated NK cell cytopenia and T cell cytopenia occurs with reticuloendothelial system activation but decreased to absent IFN-γ production. When this reduction in NK cell and T cell numbers is below 10% of normal there is a significant decrease in host ability to kill viruses and cancer cells as well as to induce apoptosis in activated macrophages. Compared to HLH patients, these patients with sepsis-induced MOF have lower production of CXCL9 because of lymphopenia. Because T cells and IFN-γ production are already low to absent, etoposide is unlikely to be of benefit in reducing macrophage activation in these children. Indeed, etoposide may worsen outcomes in sepsis patients by preventing recovery of lymphocyte counts needed to resolve infection. Compared to rheumatologic MAS patients, patients with sepsis-induced MOF have a much lower increase in IL-18 production (IFN-γ inducing cytokine). The hoped for promise of IFN-γ monoclonal antibodies in treating HLH and rheumatologic MAS is less likely to be realized in hyperferritinemic sepsis-induced MOF. Indeed, inability to produce IFN-γ is associated with increased mortality in experimental models of sepsis [4, 9–15].

Management of Hyperferritinemic Sepsis

The clinician is not uncommonly faced with bedside controversy over how to manage patients with cytokine storm and hyperferritinemic MOF, a condition associated with increased mortality in critically ill children. In a review article in *Pediatric Critical Care Medicine* [40] Dr. Castillo and Dr. Carcillo showed that the alternate diagnoses of familial hemophagocytic lymphohistiocytosis (HLH) by hematologists, rheumatological disease or drug-associated macrophage activation syndrome (MAS) by rheumatologists, or hyperferritinemic sepsis-induced MOF by intensivists and infectious disease specialists are given to critically ill children who show at least five of eight inflammatory clinical criteria that include fever, hyperferritinemia (>500 ng/mL), hypertriglyceridemia/hypofibrinogenemia, two cell line cytopenia, splenomegaly, hemophagocytosis, elevated sCD25 (lymphocyte marker), and NK cell activity <10%.

Each of these specialist groups uses different therapeutic strategies to treat these different syndromes. The hematologist generally prescribes etoposide and dexamethasone for first episodes followed by interferon γ monoclonal antibody and then if ineffective bone marrow transplantation for recurrent episodes. The rheumatologist most commonly uses corticosteroids, and biologics such as anakinra (IL-1 receptor antagonist protein) or tocilizumab (IL-6 antibody) among others. The intensivists and infectious disease specialists generally use source control as well as methylprednisolone, IVIG, plasma exchange, and IL-1 receptor antagonist protein [41]. In addition, rituximab (anti-CD20 monoclonal antibody) is used when the process is driven by EBV infection to reduce the B-lymphocyte reservoir for the DNA viral infection.

In terms of monitoring systemic inflammation, macrophage activation syndromes can be followed at the bedside by tracking fever and measuring C-reactive protein and ferritin levels at least twice weekly. C-reactive protein (CRP) is a pattern recognition receptor made by the liver in response to bacterial infection or necrotic tissue. C-reactive protein binds to C-components of microbes or the externalized phosphatidyl choline moiety of necrotic cells, complexes with complement, and attaches to the CRP receptor on the macrophage for internalization, degradation, and presentation to the adaptive immune system. Ferritin is released by macrophages in response to free hemoglobin and to DNA viremia. Hyperferritinemia occurs in iron overload states and can also be released by dying cells during necrosis. Mortality risk increases as the circulating levels of CRP and ferritin increase. The goal of therapy is to normalize CRP and ferritin levels [42, 43]. If CRP levels increase while ferritin levels decrease this is a harbinger of new or worsening infection. This warrants attention to better source control and reduction in immune suppression (if being used). If ferritin does not come down or increases, then this suggest ongoing iron overload and or macrophage activation possibly associated with DNA viremia. Attention should be given to reducing hemolysis, quelling macrophage activation, and neutralizing DNA viremia. If, however, CRP rises and the ESR falls (a result of consumptive coagulopathy consuming fibrinogen—a driver of increased ESR), then this often indicates worsening MAS coagulopathy.

The approach to use for hyperferritinemic sepsis-induced MODS is first and foremost source control and organ support. Empiric antibiotics are started within 1 h. In previously healthy children and children with erythroderma, one must consider toxic shock from Group A streptococcus or *Staphylococcus aureus*. Antibiotics should include clindamycin to prevent toxin production and an antimicrobial drug which kills MRSA as well as Group A streptococcus. It is advisable to avoid linezolid because hepatobiliary dysfunction reduces the safety window for this drug which becomes a host mitochondrial toxin when blood levels increase due to poor clearance. For immune suppression, corticosteroids (30 mg/kg per day of methylprednisolone × 3 days) are given along with IVIG (2 g/kg over 1–4 days). If the patients have AKI, thrombocytopenia and elevated LDH, consider daily plasma exchange 1½ volume followed by 1 volume daily until ferritin levels decrease to less than 500 ng/mL. In addition, anakinra (2.5 mg/kg every 6 h IV/SC to maximum of 100 mg q 6 h) for 3 days is also helpful [43].

Conflict of Interest Statement On behalf of the authors there are no conflicts of interest.

References

1. Hotchkiss, R. S., Monneret, G., & Payen, D. (2013). Immunosuppression in sepsis: A novel understanding of the disorder and a new therapeutic approach. *The Lancet Infectious Diseases, 13*(3), 260–268.
2. Felmet, K. A., Hall, M. W., Clark, R. S., Jaffe, R., & Carcillo, J. A. (2005). Prolonged lymphopenia, lymphoid depletion, and hypoprolactinemia in children with nosocomial sepsis and multiple organ failure. *Journal of Immunology, 174*(6), 3765–3772.
3. Gurevich, P., Ben-Hur, H., Czernobilsky, B., Nyska, A., Zuckerman, A., & Zusman, I. (1995). Pathology of lymphoid organs in low birth weight infants subjected to antigen-related diseases: A morphological and morphometric study. *Pathology, 27*(2), 121–126.
4. Hotchkiss, R. S., Chang, K. C., Grayson, M. H., Tinsley, K. W., Dunne, B. S., Davis, C. G., et al. (2003). Adoptive transfer of apoptotic splenocytes worsens survival, whereas adoptive transfer of necrotic splenocytes improves survival in sepsis. *Proceedings of the National Academy of Sciences of the United States of America, 100*(11), 6724–6729.
5. Chiche, L., Forel, J. M., Thomas, G., Farnarier, C., Vely, F., Bléry, M., et al. (2011). The role of natural killer cells in sepsis. *Journal of Biomedicine & Biotechnology, 2011*, 986491.

6. Brahmamdam, P., Inoue, S., Unsinger, J., Chang, K. C., McDunn, J. E., & Hotchkiss, R. S. (2010). Delayed administration of anti-PD-1 antibody reverses immune dysfunction and improves survival during sepsis. *Journal of Leukocyte Biology, 88*(2), 233–240.
7. Chang, K., Svabek, C., Vazquez-Guillamet, C., Sato, B., Rasche, D., Wilson, S., et al. (2014). Targeting the programmed cell death 1: Programmed cell death ligand 1 pathway reverses T cell exhaustion in patients with sepsis. *Critical Care, 18*(1), R3.
8. Shindo, Y., Unsinger, J., Burnham, C. A., Green, J. M., & Hotchkiss, R. S. (2015). Interleukin-7 and anti-programmed cell death 1 antibody have differing effects to reverse sepsis-induced immunosuppression. *Shock, 43*(4), 334–343.
9. Poujol, F., Monneret, G., Gallet-Gorius, E., Pachot, A., Textoris, J., & Venet, F. (2018). Ex vivo stimulation of lymphocytes with IL-10 mimics sepsis-induced intrinsic T-cell alterations. *Immunological Investigations, 47*(2), 154–168.
10. Boomer, J. S., Shuherk-Shaffer, J., Hotchkiss, R. S., & Green, J. M. (2012). A prospective analysis of lymphocyte phenotype and function over the course of acute sepsis. *Critical Care, 16*(3), R112.
11. Hotchkiss, R. S., Chang, K. C., Swanson, P. E., Tinsley, K. W., Hui, J. J., Klender, P., et al. (2000). Caspase inhibitors improve survival in sepsis: A critical role of the lymphocyte. *Nature Immunology, 1*(6), 496–501.
12. Ronit, A., Plovsing, R. R., Gaardbo, J. C., Berg, R. M., Hartling, H. J., Ullum, H., et al. (2017). Inflammation-induced changes in circulating T-cell subsets and cytokine production during human endotoxemia. *Journal of Intensive Care Medicine, 32*(1), 77–85.
13. Condotta, S. A., Khan, S. H., Rai, D., Griffith, T. S., & Badovinac, V. P. (2015). Polymicrobial sepsis increases susceptibility to chronic viral infection and exacerbates CD8+ T cell exhaustion. *Journal of Immunology, 195*(1), 116–125.
14. Jensen, I. J., Sjaastad, F. V., Griffith, T. S., & Badovinac, V. P. (2018). Sepsis-induced T cell immunoparalysis: The ins and outs of impaired T cell immunity. *Journal of Immunology, 200*(5), 1543–1553.
15. Souza-Fonseca-Guimaraes, F., Parlato, M., Philippart, F., Misset, B., Cavaillon, J. M., Adib-Conquy, M., et al. (2012). Toll-like receptors expression and interferon-γ production by NK cells in human sepsis. *Critical Care, 16*(5), R206.
16. Vanden Berghe, T., Demon, D., Bogaert, P., Vandendriessche, B., Goethals, A., Depuydt, B., et al. (2014). Simultaneous targeting of IL-1 and IL-18 is required for protection against inflammatory and septic shock. *American Journal of Respiratory and Critical Care Medicine, 189*(3), 282–291.
17. Miethke, T., Duschek, K., Wahl, C., Heeg, K., & Wagner, H. (1993). Pathogenesis of the toxic shock syndrome: T cell mediated lethal shock caused by the superantigen TSST-1. *European Journal of Immunology, 23*(7), 1494–1500.
18. Low, D. E. (2013). Toxic shock syndrome: Major advances in pathogenesis, but not treatment. *Critical Care Clinics, 29*(3), 651–675.
19. Chuang, Y. Y., Huang, Y. C., & Lin, T. Y. (2005). Toxic shock syndrome in children: Epidemiology, pathogenesis, and management. *Paediatric Drugs, 7*(1), 11–25.
20. Parks, T., Wilson, C., Curtis, N., Norrby-Teglund, A., & Sriskandan, S. (2018). Polyspecific intravenous immunoglobulin in clindamycin-treated patients with streptococcal toxic shock syndrome: A systematic review and meta-analysis. *Clinical Infectious Diseases, 67*(9), 1434–1436.
21. Kernan, K. F., & Carcillo, J. A. (2017). Hyperferritinemia and inflammation. *International Immunology, 29*(9), 401–409.
22. Conrad, E., Resch, T. K., Gogesch, P., Kalinke, U., Bechmann, I., Bogdan, C., et al. (2014). Protection against RNA-induced liver damage by myeloid cells requires type I interferon and IL-1 receptor antagonist in mice. *Hepatology, 59*(4), 1555–1563.
23. Petrasek, J., Dolganiuc, A., Csak, T., Kurt-Jones, E. A., & Szabo, G. (2011). Type I interferons protect from Toll-like receptor 9-associated liver injury and regulate IL-1 receptor antagonist in mice. *Gastroenterology, 140*(2), 697–708.

24. Quartier, P., Allantaz, F., Cimaz, R., Pillet, P., Messiaen, C., Bardin, C., et al. (2011). A multicentre, randomised, double-blind, placebo-controlled trial with the interleukin-1 receptor antagonist anakinra in patients with systemic-onset juvenile idiopathic arthritis (ANAJIS trial). *Annals of the Rheumatic Diseases, 70*(5), 747–754.
25. Roh, Y. S., Park, S., Kim, J. W., Lim, C. W., Seki, E., & Kim, B. (2014). Toll-like receptor 7-mediated type I interferon signaling prevents cholestasis- and hepatotoxin-induced liver fibrosis. *Hepatology, 60*(1), 237–249.
26. Suster, S., Hilsenbeck, S., & Rywlin, A. M. (1988). Reactive histiocytic hyperplasia with hemophagocytosis in hematopoietic organs: A reevaluation of the benign hemophagocytic proliferations. *Human Pathology, 19*, 705–712.
27. Strauss, R., Neureiter, D., Westenburger, B., Wehler, M., Kirchner, T., & Hahn, E. G. (2004). Multifactorial risk analysis of bone marrow histiocytic hyperplasia with hemophagocytosis in critically ill medical patients—A postmortem clinicopathologic analysis. *Critical Care Medicine, 32*(6), 1316–1321.
28. Inai, K., Noriki, S., Iwasaki, H., & Naiki, H. (2014). Risk factor analysis for bone marrow histiocytic hyperplasia with hemophagocytosis: An autopsy study. *Virchows Archives, 465*, 109–118.
29. Steinberg, S., Flynn, W., Kelley, K., Bitzer, L., Sharma, P., Gutierrez, C., et al. (1989). Development of a bacteria-independent model of the multiple organ failure syndrome. *Archives Surgery, 124*, 1390–1395.
30. Behrens, E. M., Canna, S. W., Slade, K., Rao, S., Kreiger, P. A., Paessler, M., et al. (2011). Repeated TLR9 stimulation results in macrophage activation syndrome-like disease in mice. *Journal of Clinical Investigtaion, 121*, 2264–2277.
31. Tsujimoto, H., Ono, S., Matsumoto, A., Kawabata, T., Kinoshita, M., Majima, T., et al. (2006). A critical role of CpG motifs in a murine peritonitis model by their binding to highly expressed toll-like receptor-9 on liver NKT cells. *Journal of Hepatology, 45*, 836–843.
32. Szabo, G., & Petrasek, J. (2015). Inflammasome activation and function in liver disease. *Nature Reviews. Gastroenterology and Hepatology, 12*, 387–400.
33. Girard-Guyonvarc'h, S., Palomo, J., Martin, P., Rodriguez, E., Troccaz, S., Palmer, G., et al. (2018). Unopposed IL-18 signaling leads to severe TLR9-induced macrophage activation syndrome in mice. *Blood, 131*(13), 1430–1441.
34. Demirkol, D., Yildizdas, D., Bayrakci, B., Karapinar, B., Kendirli, T., Koroglu, T. F., et al. (2012). Hyperferritinemia in the critically ill child with secondary hemophagocytic lymphohistiocytosis/sepsis/multiple organ dysfunction syndrome/macrophage activation syndrome: What is the treatment? *Critical Care, 16*, R52.
35. Ravelli, A., Minoia, F., Davì, S., Horne, A., Bovis, F., Pistorio, A., et al. (2016). 2016 Classification criteria for macrophage activation syndrome complicating systemic juvenile idiopathic arthritis: A European League Against Rheumatism/American College of Rheumatology/Paediatric Rheumatology International Trials Organisation Collaborative Initiative. *Arthritis Rheumatology, 68*, 566–576.
36. Shakoory, B., Carcillo, J. A., Chatham, W. W., Amdur, R. L., Zhao, H., Dinarello, C. A., et al. (2016). Interleukin-1 receptor blockade is associated with reduced mortality in sepsis patients with features of macrophage activation syndrome: Reanalysis of a prior phase III trial. *Critical Care Medicine, 44*, 275–281.
37. Kyriazopoulou, E., Leventogiannis, K., Norrby-Teglund, A., Dimopoulos, G., Pantazi, A., Orfanos, S. E., et al. (2017). Macrophage activation-like syndrome: An immunological entity associated with rapid progression to death in sepsis. *BMC Medicine, 15*(1), 172.
38. Halstead, E. S., Carcillo, J. A., Schilling, B., Greiner, R. J., & Whiteside, T. L. (2013). Reduced frequency of CD56 dim CD16 pos natural killer cells in pediatric systemic inflammatory response syndrome/sepsis patients. *Pediatric Research, 74*, 427–432.
39. Carcillo, J. A., Podd, B., & Simon, D. W. (2017). From febrile pancytopenia to hemophagocytic lymphohistiocytosis-associated organ dysfunction. *Intensive Care Medicine, 43*(12), 1853–1855.

40. Castillo, L., & Carcillo, J. (2009). Secondary hemophagocytic lymphohistiocytosis and severe sepsis/systemic inflammatory response syndrome/multiorgan dysfunction syndrome/macrophage activation syndrome share common intermediate phenotypes on a spectrum of inflammation. *Pediatric Critical Care Medicine, 10*, 387–392.
41. Carcillo, J. A., Halstead, E. S., Hall, M. W., Nguyen, T. C., Reeder, R., Aneja, R., et al. (2017). Three hypothetical inflammation pathobiology phenotypes and pediatric sepsis-induced multiple organ failure outcome. *Pediatric Critical Care Medicine, 18*(6), 513–523.
42. Carcillo, J. A., Sward, K., Halstead, E. S., Telford, R., Jimenez-Bacardi, A., Shakoory, B., et al. (2017). A systemic inflammation mortality risk assessment contingency table for severe sepsis. *Pediatric Critical Care Medicine, 18*(2), 143–150.
43. Rajasekaran, S., Kruse, K., Kovey, K., Davis, A. T., Hassan, N. E., Ndika, A. N., et al. (2014). Therapeutic role of anakinra, an interleukin-1 receptor antagonist, in the management of secondary hemophagocytic lymphohistiocytosis/sepsis/multiple organ dysfunction/macrophage activating syndrome in critically ill children. *Pediatric Critical Care Medicine, 15*, 401–408.

Part VII
Murine Models of Cytokine Storm Syndromes

Murine Models of Familial Cytokine Storm Syndromes

Benjamin Volkmer, Peter Aichele, and Jana Pachlopnik Schmid

Introduction

Animal models are often essential for identifying and characterizing the mechanisms that underlie systemic diseases. Indeed, in vitro techniques cannot fully replace animal models. For hemophagocytic lymphohistiocytosis (HLH) and many other diseases caused by immune system defects, the mouse is the animal model of choice. On the one hand, the biology of laboratory mice is well known, an overwhelming number of experimental tools are available, and resources such as the Mouse Genome Database provide detailed information on genetically modified strains used in research [1]. On the other hand, the use of inbred mice, inter-strain differences in the genetic background, and the specific settings of mouse experiments such as differences in water or microbiome between facilities must be borne in mind [2] when interpreting the results and when considering translational aspects of this type of research.

Given that (1) primary HLH (pHLH) is a hereditary disease and (2) the underlying disease-causing gene mutations are known for most monogenic forms of pHLH (with the exception of familial HLH subtype 1, FHL1), murine models with defects in the corresponding genes are of great interest. There are several different strategies

B. Volkmer
Division of Immunology, University Children's Hospital Zurich, Zurich, Switzerland
e-mail: Benjamin.Volkmer@kispi.uzh.ch

P. Aichele
Department of Immunology, Institute for Medical Microbiology and Hygiene, University of Freiburg, Freiburg, Germany

J. Pachlopnik Schmid (✉)
Division of Immunology, University Children's Hospital Zurich, Zurich, Switzerland

University of Zurich, Zurich, Switzerland
e-mail: Jana.Pachlopnik@kispi.uzh.ch

R. Q. Cron, E. M. Behrens (eds.), *Cytokine Storm Syndrome*,
https://doi.org/10.1007/978-3-030-22094-5_28

for generating mice with a defect in a particular gene. Historically, mice with spontaneous mutations and interesting clinical phenotypes were inbred to generate disease models [3]. With the advent of sequencing techniques, the disease-causing mutations in murine models and patients could be identified. For mutations in the same gene, this enabled direct comparisons of disease induction, clinical symptoms and disease progression in humans vs. mice. Today, the mouse genome can be edited so as to introduce almost any imaginable genomic feature [4, 5]. The most common changes are selective gene modifications that prevent the expression of a given protein (i.e., complete, inducible or conditional gene knockout). Inducible knockouts are triggered by administration of an external stimulus, while conditional knockouts combine gene deletion with an endogenous promoter that is specific for a certain cell type and/or developmental time point. The introduction of a gene mutation identified in patients into the corresponding locus in mice is another option for studying the pathogenesis of a hereditary disease. After the assessment of protein expression in vitro, this strategy is often chosen for cases in which the mutation allows the expression of a protein with modified functional or binding properties. Even the detailed characterization of gene mutations and their consequences on protein expression cannot replace an in vivo analysis of physiological effects in a complex biological system like the mouse.

Murine orthologs are available for all known pHLH-causing genes in humans. Accordingly, the corresponding knockout mouse strains are available (Table 1). Although the human and murine immune systems are similar in many respects, one cannot expect a mouse to perfectly replicate a human disease [6]. In mice and humans, the genes coding for the perforin-dependent cytolytic machinery of T cells and NK cells (i.e., proteins involved in vesicle loading, vesicle transport, vesicle fusion, and effector functions) are quite homologous. However, one striking difference

Table 1 Murine models with pHLH-causing gene mutations

Disease	Causative gene	Commonly used strain name(s)	Main background	Number of variants available
FHL1	Unknown	–	–	–
FHL2	*PRF1*	Prf1$^{-/-}$	C57BL/6	14
FHL3	*UNC13D*	Jinx	C57BL/6	2
FHL4	*STX11*	Stx11$^{-/-}$	C57BL/6	2
FHL5	*STXBP2*	Stxbp2$^{-/-}$	C57BL/6	2
GS2	*RAB27A*	Ashen	C3H/HeSn, C57BL/6	9
CHS	*LYST*	Beige/souris	C57BL/6	24
HPS2	*AP3B1*	Pearl	B6.C3	18
XLP1	*SH2D1A*	SAP$^{-/-}$	129S6	18
XLP2	*XIAP*	XIAP$^{-/-}$	B6.129S1	4

The columns indicate the type of pHLH, the causative gene, the name of the most commonly used mutant mouse strain, the main genetic background, and the number of distinct mouse variants available. Source: the Mouse Genome Database [1]

between mice and humans with regard to the development of HLH is the need for an infectious trigger in mice. In humans, the need for an external trigger is subject to debate. Once HLH has been induced by infection in the mouse, the clinical and diagnostic hallmarks of the disease [7] are well recapitulated—as will be shown in the following sections. Ultimately, murine models of HLH are used to (1) confirm the disease-causing nature of mutations found in patients, (2) understand the disease's mechanisms and pathogenesis, and (3) test novel therapeutic approaches. Of course, ethical considerations of using animals for research must always to be considered, and the possible development of treatment approaches in humans needs to be weighed against the animals' suffering. However, given the complexity of HLH, the use of in vivo disease models is essential for making substantial progress.

Genetic Defects in the Cytolytic Machinery: Murine Models

For most genes encoding proteins critically involved in the cytolytic mechanism, several mouse strains with distinct genetic modifications are available. In this context, it is important to consider the strain's genetic background because it may significantly influence the animal's immune response [8, 9]. When compared with wild mice, inbred laboratory mice differ immunologically in terms of both low genetic variability and reduced exposure to antigens [2]. When deciding on which particular mouse strain to use in a given experiment, the Mouse Genome Database [1] and the International Mouse Strain Resource [10] often provide very helpful information, including links to the associated literature. In the following sections, the available murine models for pHLH will be presented for each known causative gene (Table 1). The HLH-like symptoms observed in each murine model are summarized in Table 2 [11].

The Prf1$^{-/-}$ Mouse (A Model of FHL2)

The first report (in 1994) on this murine model (the perforin-deficient mouse, (C57BL/6-Prf1^{tm1Sdz} or Prf1$^{-/-}$) reported that lymphocytic choriomeningitis virus (LCMV)-triggered immune disease led to death 13–16 days post-infection [12, 13]. The first description of perforin-mutant humans with pHLH confirmed that perforin-based effector systems are involved not only in the lysis of abnormal cells but also in the downregulation of immune activation in humans [14]. A careful, detailed description of the model's clinical and histopathological features showed how well this murine disease corresponds to human pHLH [15]. Prf1$^{-/-}$ mice harbor a targeted mutation in the exon 3 of *Prf1*, leading to a lack of protein expression and thus impaired T-cell cytotoxicity [12]. The mice are fertile, develop normally, and are not distinguishable from wild-type mice (other than for the drastically impaired cytotoxic functions of T cells and NK cells). However, after LCMV infection

Table 2 Overview of the HLH-2004 criteria met (or not) in different murine models [7]

	HLH-2004 diagnostic criteria											
Mouse strain and virus	Number of criteria met (out of 8)	Fever	Splenomegaly	Anemia	Thrombo-cytopenia	Neutropenia	Hypertriglyceri-demia	Hypofibrino-genemia	Hemophago-cytosis	Low NK-cell activity	High ferritin	High soluble CD25
				≥2 of 3			1≥ of 2					
FHL2												
Prf1$^{-/-}$ + LCMV	8	x	x	x	x	x	x	x	x	x	x	x
Prf1$^{-/-}$ + MCMV	4		x		x				x	x		
FHL3												
Jinx + LCMV	4		x	x	x	↑			x	x		
FHL4												
Stx11$^{-/-}$ + LCMV	7	↓	x	x	x	n	x		x	x	x	x
GS2												
Ashen + LCMV	8	x	x	x	x	x	x		x	x	x	x
CHS												
Souris + LCMV	6	↓	x	x	x		n		x	x	x	x
HPS2												
Pearl + LCMV	4	n	x	n	x		n		x	x	x	n
XLP1												
SAP$^{-/-}$ + LCMV	1		↓							x		
SAP$^{-/-}$ + MHV-68	3		x						x	x		
XLP2												
XIAP$^{-/-}$ + MHV-68	0											

The table was adapted from Brisse et al. [11]. x = criteria fulfilled, n = criteria not fulfilled, ↑ = increase, ↓ = reduction, empty box = no data available

$Prf1^{-/-}$ mice develop overt HLH, fulfilling all the diagnostic HLH-criteria [7] around day 10 and succumbing before 14 days post infection [15]. But this is not the case for $Prf1^{-/-}$ mice with a BALB/C background, where the mice do not die, but LCMV clearance is delayed beyond 100 days post-infection [16]. The LCMV-infected $Prf1^{-/-}$ mouse has been extensively studied and could now be described as the "standard" murine model of pHLH.

The Discovery of Perforin-Dependent Cytotoxicity

The study of perforin-deficient mouse strains has led to the discovery of several features of perforin-dependent cytotoxicity. In particular, it was found that (1) perforin was required in cytotoxic T lymphocytes (CTLs), and (2) NK cell cytotoxicity was not mediated by Fas [12, 13, 17, 18]. Interestingly, $Prf1^{-/-}$ $Fas^{-/-}$ double knockout mice die early from macrophage expansion and severe pancreatitis; this is due to the T and NK cells' inability to restrict immune responses by killing antigen-presenting cells (APCs) in a negative feedback loop [19]. Furthermore, tumor surveillance is limited in $Prf1^{-/-}$ mice [20], whereas $Prf1^{-/-}$ nonobese diabetic mice display a reduced incidence and delayed onset of diabetes [21].

Infections in $Prf1^{-/-}$ Mice

$Prf1^{-/-}$ mice show altered responses to other infections, in addition to LCMV. In fact, these mice are more susceptible to the natural pathogen ectromelia virus but less susceptible to cowpox virus [22]. The pathogenesis of corneal inflammation induced by herpes simplex virus type 1 is less severe in $Prf1^{-/-}$ mice [23]. Although MCMV infection is lethal in $Prf1^{-/-}$ mice, this is not the case in granzyme A/B--deficient mice [24]. During prolonged infections with *Listeria monocytogenes*, perforin was seen to be important for the contraction of antigen-specific T cells [25]. In mice, CD8-mediated protection against Ebola virus infection was perforin-dependent [26]. Lastly, in an acute Epstein–Barr virus (EBV) infection model, $Prf1^{-/-}$ mice showed increased mortality [27].

CD8 T Cells Are Important

The harmful effect of a lack of perforin is highlighted by the uncontrolled expansion of CD8 T cells after LCMV infection [15, 16, 28, 29]. This is attributed to the lack of a negative feedback loop mediated by the perforin-dependent elimination of antigen-presenting dendritic cells [30]. Consequently, the elimination of these dendritic APCs by perforin-dependent CTLs was shown to protect against HLH [31]. The use of specific depleting antibodies has evidenced a key role of the perforin-deficient CD8 T cells in the pathogenesis of HLH [32]. In mixed bone-marrow chimeras, it was shown that a small fraction (10–20%) of perforin-expressing T cells is enough to restore immune regulation in $Prf1^{-/-}$ mice [33]. Although this information is important with regard to evaluating the likely success of bone marrow transplantation in pHLH patients, data from rodents cannot be extrapolated to humans. In addition to the percentage of chimerism, the decision to perform hematopoietic stem cell (HSC) transplantation in pHLH patients with mixed chimerism must be based on an in-depth individual risk assessment and consideration of (1) the variability in the genetic causes of pHLH in patients, (2) the hypomorphic character of certain mutations, and (3) the difference in HLH-triggering thresholds between mice and humans.

The Role of NK Cells

While CD8 T cells appear to be the most important for HLH development in Prf1$^{-/-}$ mice, the role of NK cells is less clear—even though their cytotoxic activity is similarly impaired. It has been demonstrated that CD8 T cell expansion after infection can be controlled by NK-cell-mediated killing in a perforin-dependent manner [34–36]. Perforin-dependent killing of cognate CD8 T cells by immature dendritic cells also occurs [37]. A series of elegant experiments has shown that NK cells with functional cytotoxic activity limit the HLH-like immunopathology in mice with perforin-deficient CD8 T cells by reducing T cell activation and the tissue infiltration of macrophages. In contrast, mice with Prf1$^{-/-}$ NK cells and functional CD8 T cells show transient lymphocytosis [38]. A stronger T cell response and more effective viral clearance occur in the absence of NK cells or in the presence of defective NK cells [35, 39]. Interestingly, NK cell expansion during viral infection is promoted by type I interferons (IFNs), which protect NK cells from fratricide [40]. Furthermore, type I IFNs were identified as key players in protecting activated CD8 T cells from NK cell attacks [41, 42].

Cytokines and Signaling

A fulminant cytokine storm (characterized mainly by elevated serum levels of IFNγ, TNF, IL-6, and IL-18) is the main driver of HLH progression in perforin-deficient mice [15], with an increased synapse time due to defective killing allowing for prolonged cytokine secretion by CTLs and NK cells [43]. The use of specific antibodies for prophylactic and therapeutic neutralization of IFNγ in Prf1$^{-/-}$ mice increased survival and decreased HLH-related symptoms [16, 32, 44, 45]—highlighting the importance of IFNγ for the development of HLH. Other important cytokines and signaling pathways highlighted in the Prf1$^{-/-}$ mouse are IL-33, IL-2 and JAK1/2. In LCMV-infected Prf1$^{-/-}$ mice, blockade of the IL-33 receptor ST2 reduced disease severity [46], whereas IL-2 consumption by activated CD8 T cells may have induced dysfunction of regulatory T cells [47]. Treatment of LCMV-infected Prf1$^{-/-}$ mice with the JAK1/2 inhibitor ruxolitinib increased survival, lowered both TNF and IL-6 levels and generally reduced HLH severity [48, 49].

The Jinx Mouse (A Model of FHL3)

The Jinx mouse was first described in an *N*-ethyl-*N*-nitrosourea screen for mutations causing susceptibility to murine cytomegalovirus (MCMV) [50]. In 2007, the Jinx mouse was proposed as a HLH model; it harbors a mutation in the *Unc13d* gene and causes Munc13-4 to terminate at amino acid 859 [51], leading to a degranulation defect in T cells and NK cells. Although uninfected mice appear to be healthy and exhibit normal lymphocyte counts, they become severely ill after MCMV infection. Infection by MCMV is lethal for Jinx mice within a week, as also observed in the well-known MCMV-susceptible BALB/c wild-type strain. Interestingly, Jinx mice do not show the clinical symptoms and diagnostic criteria of HLH after infection with MCMV or by the intracellular bacterium *Listeria monocytogenes*. In contrast, infection with LCMV-Armstrong results in several features of HLH, such as anemia, thrombocytopenia, splenomegaly and increased serum IFN-gamma levels at days 8 and 12. However, a detailed analysis of HLH progression and survival is currently lacking. After LCMV infection, abrogation of *MyD88* (but not *Tnf* or *Itgb2*) prevented the development of HLH—suggesting that MyD88 is a possible drug target [52]. The Jinx mouse was also instrumental in elucidating Munc13-4's importance for phagosome maturation [53], late endosome maturation [54], the trafficking of Rab11-containing vesicles [55], and membrane fusion [56].

The Stx11$^{-/-}$ Mouse (A Model of FHL4)

In 2013, several groups of researchers described the Stx11-deficient (Stx11$^{-/-}$) mouse as an HLH model [57–59]. These mice are fertile, develop normally, and do not show any abnormalities in the size and composition of lymphoid organs or overt differences in cells of the immune system, relative to heterozygous littermates or wild-type mice. Stx11 (a member of the SNARE protein family) is a component of the cytotoxic machinery of T cells and NK cells, and is important for degranulation activity by mediating the fusion between cytotoxic vesicles and the plasma membrane. Given that Stx11 controls degranulation in CTLs and NK cells, Stx11-deficient CTLs and NK cells show impaired degranulation [57]. It is noteworthy that Stx11 expression is induced in many immune cells (notably by treatment with either lipopolysaccharide (LPS) or IFNγ) [60–62]. Upon infection with LCMV, Stx11$^{-/-}$ mice display all the clinical features of HLH (as seen in FHL4 patients), with hyperactive CD8 T cells as the driving force. In striking contrast to the situation in Prf1$^{-/-}$ mice, HLH is not fatal in the Stx11$^{-/-}$ mice. This less severe but chronic progression of HLH is characterized by low body weight, sustained splenomegaly, low white blood cell counts and low hemoglobin levels. In this context, the importance of T-cell exhaustion as a pathophysiological mechanism for determining HLH progression was shown by blocking inhibitory receptors on disease-inducing T cells; this blockade turned the nonfatal form of HLH into a fatal condition [58]. A distinct severity of HLH can be observed depending on the gene affected in the cytotoxic machinery [59].

The Stxbp2$^{-/-}$ Mouse (A Model of FHL5)

Although a Stxbp2 mutant mouse strain exists, it has not yet been analyzed for HLH development. This is probably due to the fact that Stxbp2$^{-/-}$ homozygous knockout mice are not viable, whereas heterozygous (Stxbp2$^{+/-}$) mice express Munc18b at approximately half the wild-type level and do not display obvious developmental abnormalities or reduced survival [63]. In mast cells, this heterozygous mutation decreases mucin secretion and reduces degranulation-dependent responses. The lethality of Stxbp2$^{-/-}$ mice indicates either that Munc18b has a more important role in mice or that the homozygous *STXBP2* mutations observed in humans are hypomorphic. It would be interesting to look at whether HLH can be triggered in Stxbp2$^{+/-}$ mice or in mice expressing either two recessive mutations in the Stxbp2 gene or a dominant-negative mutation in the *Stxbp2* gene known to be associated with human HLH [64].

The Ashen Mouse (A Model of GS2)

As early as 1977, the *ashen* mutation (*ash*) was reported as being important for coat color (Fig. 1) [65]. However, it was only in 2000 that the coat color and vesicle transport defects were attributed to the *ash* mutation in the *Rab27a* gene [66]. Homozygous ash/ash mice do not express Rab27a in melanocytes and CTLs, and thus have impaired transport and degranulation of melanosomes and cytotoxic

Fig. 1 An ashen mouse (below) and a wild-type mouse (above). The coat of the C57BL/6 mouse harboring a homozygous Rab27a ashen mutation is lighter than its C57BL/6 wild-type littermate, due to defective melanosome exocytosis

granules, respectively [67]. Furthermore, ashen mice have prolonged bleeding times due to defects in dense granules in platelets; however, this effect depends on the mouse's genetic background [68]. In 2008, mice with a C57BL/6-background and the ashen mutation in the *Rab27a* gene were shown to develop HLH when injected with LCMV—demonstrating that this ashen mouse can be used as a model of pHLH in human Griscelli syndrome type 2 (GS2) [69]. In both Prf1$^{-/-}$ and ashen mice, IFNγ neutralization was able to suppress HLH [45]. Ashen mice are resistant to LPS-induced death; this is presumably due to less intense neutrophil infiltrates in the liver, relative to both wild-type and Jinx mice [70].

Beige and Souris Mice (Models of Chediak–Higashi Syndrome, CHS1)

The link between beige mice and CHS1 was made as early as 1969 [71, 72], and an impairment in NK cytotoxicity was demonstrated 10 years later [73–75]. In 1996, the beige mutation and the disease-causing gene in CHS1 were mapped to the *Lyst* locus [76]. Beige mice show poor survival and develop honeycomb-like lesions in their lung tissue [77]. The beige mouse was first analyzed for HLH development in 2011, and was compared with the souris mice (a mouse strain with a different mutation in the *Lyst* gene). After LCMV infection, beige mice controlled the virus and did not develop all the clinical features of HLH. In contrast, souris mice were unable to control the virus and met the diagnostic criteria for HLH. This finding was attributed to slight differences in CTL cytotoxicity, with beige mice retaining more CTL function than souris mice [78]. B-cell receptor endocytosis is delayed in beige mice, which leads to a stronger memory response and higher plasma cell frequencies [79]. In beige mice, it was shown that Lyst specifically controls TLR3- and TLR4-induced endosomal signaling pathways—explaining the strain's increased susceptibility to bacterial infections and decrease response to endotoxins [80].

The Pearl Mouse (A Model of HPS2)

The pearl mutation was first described in 1953 as causing a specific coat color [65]. Pearl mice were also found to have visual defects and platelet storage pool deficiencies. Both features are related to the AP-3 adaptor complex; the pearl mutation is caused by the insertion of a transposon into the *Ap3b1* gene coding for one of the AP-3 subunits [81–85]. The pearl mouse's life span is reduced, and no mice survive for more than 2 years [77]. Defects in vesicle and membrane protein trafficking were subsequently observed [86, 87]. Deficiencies in AP-3 in the pearl mouse interfere with the correct localization of murine CD1d, and thereby prevent the development of NKT cells [88]. Macrophages in the lungs of pearl mice are hyperresponsive to LPS, resulting in the increased secretion of inflammatory cytokines [89]. Platelet granule secretion at sites of vascular injury is also impaired in pearl mice [90]. Pearl mice were first examined for the development of HLH in 2013; they met five of the eight diagnostic criteria by D8 post-LCMV infection [91]. However, the HLH was transient because the criteria were no longer fulfilled at D12 (Table 2) [91].

The SAP$^{-/-}$ Mouse (A Model of X-Linked Lymphoproliferative Disease Type 1, XLP1)

The first SAP-deficient (SAP$^{-/-}$) mice were developed in 2001 by targeting the *Sh2dh1* gene locus. The SAP$^{-/-}$ mice develop normally but show features of XLP, such as increased T cell activation and elevated IFNγ secretion upon infection [92]. Furthermore, SAP-deficient T cells show impaired Th2 differentiation [93]. Interestingly, SAP$^{-/-}$ mice almost fully lack virus-specific plasma cells and memory B cells because of a CD4+ T cell defect [94] that prevents the differentiation of follicular T helper cells; this may be a consequence of the missing SAP-mediated coupling of signals received from SLAM receptors to the Fyn effector [95, 96]. Thus, the low B cell count in SAP$^{-/-}$ mice is due to a T-cell-dependent reduction in T-B cell interactions and a lack of germinal center reaction [97, 98]. Like XLP patients, SAP$^{-/-}$ mice lack NKT cells [99–101]. Infection of SAP$^{-/-}$ mice with murine gammaherpesvirus 68 (an EBV homolog) leads to increased CD8 T cell proliferation, lymphocyte infiltration, hemophagocytosis, hypogammaglobulinemia, and virus reactivation [102, 103]. Chronic LCMV infection leads to a severe immune disease mainly caused by CD8 T cells and hypogammaglobulinemia [103]. SAP was shown to (1) activate NK cells via Fyn and Vav-1 and (2) prevent the inhibitory phosphatase SHIP-1 from binding to SLAM receptors [104]. It was possible to correct the cellular and humoral defects of SAP$^{-/-}$ mice using HSC gene therapy [105]. Restoring the reduced diacylglycerol signaling in SAP-deficient T cells by pharmacological inhibition of diacylglycerol kinase alpha can also prevent excessive CD8 T cell expansion and excessive IFNγ production [106].

The XIAP$^{-/-}$ Mouse (A Model of XLP2)

The XIAP-deficient mouse (XIAP$^{-/-}$) was first described in 2001. It does not differ obviously from wild-type mice, other than for an upregulation of other inhibitor of apoptosis (IAP) family members (presumably to compensate for the lack of XIAP) [107]. However, the mammary gland is slow to develop in XIAP$^{-/-}$ mice [108]. Upon infection with *Listeria monocytogenes*, XIAP$^{-/-}$ mice have a higher bacterial burden in the spleen and liver 48 and 72 h post-infection, and the survival rate is abnormally low [109]. The lack of XIAP increases the sensitivity of murine cells to apoptotic stimuli [110, 111]. XIAP was further found to enable discrimination between type I and type II FAS-induced apoptoses. As a consequence, $XIAP^{-/-}$ mice succumbed to FASL-induced hepatitis more quickly than wild-type mice did [112]. Infection of $XIAP^{-/-}$ mice with *Chlamydophila pneumonia* dysregulated the immune response, with increased apoptosis in macrophages and a low CD8 T cell count [113]. XIAP-related apoptosis is also involved in regulating stem-cell-dependent wound healing [114] and caspase activity in degenerating axons [115]. In response to TNF, XIAP controls RIP3-dependent cell death and IL-1beta secretion, which contribute to hyperinflammation in the $XIAP^{-/-}$ mouse [116]. Infection of XIAP mice with *Shigella flexneri* demonstrated an inefficient immune response, with bacterial propagation and tissue damage [117]. Infection of $XIAP^{-/-}$ mice with *Candida albicans* resulted in bacterial persistence, constantly elevated cytokine levels, and poor survival [118]. The activity of IAP family members can be inhibited by IAP antagonists. After infection with LCMV, the blockade of IAPs was shown to drastically limit virus-specific CD8 T cell expansion and differentiation—indicating a prominent role for IAPs in the survival of activated T cells. Interestingly, T cell expansion could be prevented by much lower doses of the IAP antagonist in $XIAP^{-/-}$ mice, which demonstrates the importance of XIAP in T cell survival during an ongoing immune response [119]. Rather than XIAP, cIAP2 is very important in enabling hepatitis B virus to persist in mouse hepatocytes because it attenuates TNF signaling and promotes hepatocyte survival [120]. Interestingly, $XIAP^{-/-}$ mice learn more rapidly; this is thought to be due to the greater plasticity resulting from increased neural apoptosis [121]. Specifically reports on the development of HLH in $XIAP^{-/-}$ mice are currently lacking.

Induction of HLH in the Mouse (By Viruses and Other Agents)

Although there is still debate as to whether HLH may arise spontaneously in humans, a trigger is typically required in murine models of this disease. LCMV strains have almost always been used for induction in murine models of pHLH. The one reported exception is for the $perforin^{-/-}$ $DC\text{-}Fas^{-/-}$ mice; antigen-presenting cells cannot be removed by perforin- or FAS-dependent killing, and the mice spontaneously develop HLH [122]. The reason for using LCMV as initial disease trigger is the overwhelming CD8 T cell response that it induces [11]. However, as mentioned before, the genetic background of the inbred mouse strain is also an important factor, since perforin-deficient BALB/c mice do not develop overt HLH after LCMV infection [16] (as observed in perforin-deficient C57BL/6 mice). Interestingly, previous vaccination to LCMV leads to a high mortality in $Prf1^{-/-}$ BALB/C mice [29]. This suggests that both the virus's capability to stimulate IFNγ secretion and the propensity of a given mouse strain to produce IFNγ determine whether or not HLH develops [16, 29]. The quality and quantity of immune responses directed against LCMV are determined by both the invasiveness of the various LCMV strains and the infecting dose, as demonstrated in $Prf1^{-/-}$ and $IFN\gamma^{-/-}$ mice [123]. Thus, the injected dose of LCMV needs to be experimentally adjusted to induce a CD8 T cell response high enough to trigger HLH but low enough to avoid elimination of the virus or T cell exhaustion. In a typical experimental setting, a 200 pfu dose of the LCMV-WE strain is used to trigger HLH in HLH-prone mice.

The experimental procedure for modeling HLH in mice thus begins by injecting the animals with an appropriate dose of LCMV. After a week, the mice show the first symptoms of HLH, which increase in diversity and severity up until D12–D14, in most cases. Depending on the mutation, the mice then succumb to the hyperinflammatory syndrome driven by activated T cells and macrophages. However, it is sometimes valuable to monitor the mice for more than 2 weeks, since they may recover and display a nonfatal but chronic progression of HLH (characterized by the disappearance of some of the HLH criteria but the absence of full recovery) [58]. Interestingly, the post-infection viral load does not correlate with the course of the disease [124]. This indicates that the immune system enters a self-sustaining activation loop in which hyperactivation is not correlated with an increased or persistent pathogen burden in mice with impaired cytotoxicity [125].

Regarding the HLH trigger, observations in mice suggest that a viral infection is necessary to induce HLH. However, there are no reports that show the necessity of a viral trigger to develop HLH in humans. This may be due to the difficulty of virus detection by the time HLH is diagnosed or virus detection in general, and allows for speculations whether human HLH can arise spontaneously by a trigger other than a virus.

Comparisons of Murine Models of HLH

Although HLH-like symptoms can be induced in various mice with defects in genes underlying pHLH, the disease severity varies greatly from one affected gene to another. The same observations have been made in patients. It has also been shown that mice with heterozygous mutations in two or more disease-causing genes may also develop HLH [126]. This indicates that HLH development is not binary but is influenced by multiple factors—resulting in an individual probability of developing disease. This finding is also reflected by the influence of an inbred mutant strain's genetic background on the development of HLH [29].

The severity and progression of HLH has been analyzed by studying murine models with defects in the various HLH-associated genes. Additionally, the combination of different disease-causing mutations in the same mouse has enabled researchers to rank the impacts of several heterozygous mutations on the degranulation capacity of T cells and NK cells. Comparisons of several strains with mutations in the cytolytic machinery have shown the following hierarchy of HLH severity (beginning with the least severe): HPS2 < CHS < FHL4 < GS2 < FHL2 (Fig. 2) [59, 91, 127]. The same severity gradient was also found in patients [127]. The severity of HLH correlates well with the residual cytolytic activity of the disease-inducing

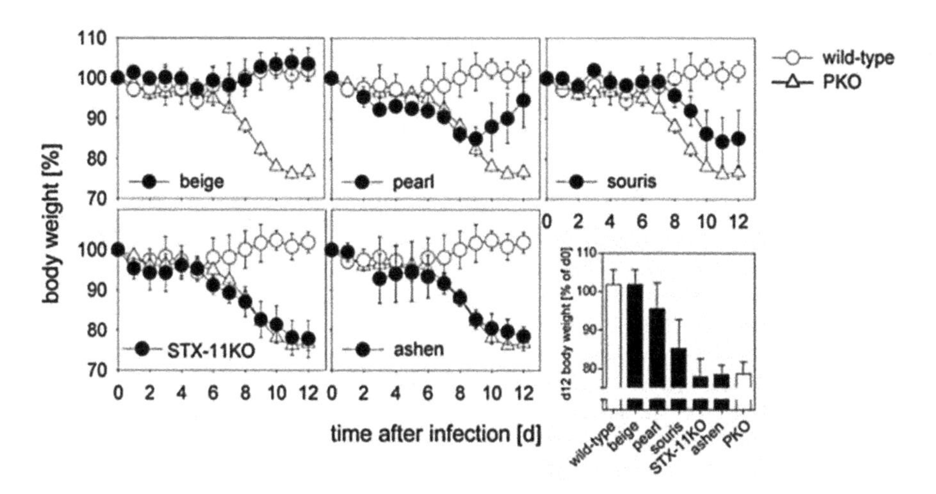

Fig. 2 Comparisons of body weight in different HLH murine models. Mice were infected with LCMV and the body weight was recorded. The wild-type and PKO (Prf1$^{-/-}$) values are indicated for comparison's sake in all graphs. Mouse strains and mutations: beige—*LYST*, pearl—*AP3B1*, souris—*LYST*, ashen—*Rab27A*. Figure from [127]

CD8 T cells (Fig. 3) [59, 127]. This difference in cytotoxicity also determines the susceptibility to development of HLH in souris and beige mice, which both harbor mutation in the *Lyst* gene. Impairment of T cell degranulation is greater in souris than in beige mice, causing LCMV infection to induce HLH in souris mice but not in beige mice [78]. While CXCL10 and IFNγ levels were also shown to correlate with disease severity, the viral loads in the strains did not [59]. HLH susceptibility was further shown to be higher in mice harboring several heterozygous mutations in HLH-related genes. This was shown for combinations of heterozygous mutations in Prf1, Stx11, and Rab27a—none of which drives HLH-like disease as a single heterozygous mutation. The HLH severity was shown to be Prf1$^{-/-}$ > Ash$^{+/-}$, Prf1$^{+/-}$, Stx11$^{+/-}$ > Ash$^{+/-}$, Prf1$^{+/-}$ > Ash$^{+/-}$, Stx11$^{+/-}$ [126]. The severity of HLH thus appears to be dependent on the combined effects of (1) the mutation on protein function and expression, (2) zygosity of the mutation, (3) additional mutation in genes involved in perforin-dependent cytotoxicity, (4) function of the mutated protein not only in cytotoxicity but also in antigen presentation and (5) the degree of interaction between gene products in such compound-heterozygous scenarios.

Similarities between HLH murine models also help to raise hypotheses about common properties of all types of HLH. Based on observations in Prf1$^{-/-}$, GrzB$^{-/-}$, beige, and Jinx mice, increased T-cell activation was shown to mainly be caused by increased presentation of viral antigens by APCs rather than cell-intrinsic effects [125]. However, other observed differences do not have mechanistic explanations. Ashen (Rab27a) and Jinx (Munc13-4) mice react differently to systemic LPS-induced inflammation. Ashen mice show lower mortality and lower plasma TNF-alpha levels than Jinx mice do [70]. Furthermore, bone marrow-derived mast cells

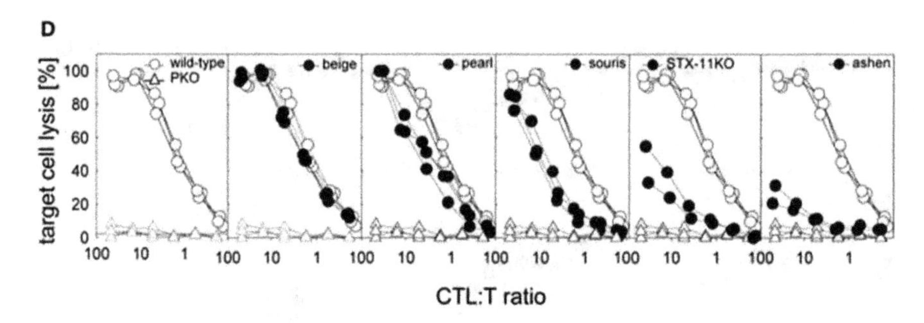

Fig. 3 The disease severity correlates with the cytolytic deficiency. Target cell lysis is shown for different effector–target ratios. The mouse strains are ordered by decreasing cytolytic activity. The wild-type and PKO (Prf1$^{-/-}$) values are shown in all graphs for comparison's sake. Mouse strains and mutations: beige—*LYST*, pearl—*AP3B1*, souris—*LYST*, ashen—*Rab27A*. Figure from [127]

from ashen mice show hypersecretion, while the same cells derived from Jinx mice show a secretory impairment [128]. Mice with cell-type-selective deficiencies of either Prf1 or Stx11 have been helpful in investigating the interplay between CTLs, NK cells, and macrophages.

While experiments conducted in parallel can be easily interpreted, comparing experiments performed with HLH-model mice in different labs is more difficult. To this end, it is important for all experimental procedures to be recorded so that reproduction at another site is possible. The main variables are the mouse strain, the virus strain, the viral load and mode of administration. Some variables (such as housing in different facilities and exposure to different environments) are impossible to reproduce but should be considered when interpreting the outcome of an experiment. Furthermore, the recorded experimental data should encompass as many of the HLH-2004 diagnostic criteria as possible (Table 2). This should facilitate the comparison of (1) different murine models and (2) murine data vs. patient data.

HLH Therapy in Mice and Implications for Human Treatment

Besides providing a better understanding of the mechanisms causing and sustaining the hyperinflammatory state in HLH, murine models are also instrumental for developing therapeutic approaches (discussed in detail in chapter "Genetics of Macrophage Activation Syndrome in Systemic Juvenile Idiopathic Arthritis").

TNF blockade: In Prf1$^{-/-}$ mice infected with MCMV, TNF appeared to be the main driver of immunopathology; treatment with anti-TNF markedly reduced liver damage, while anti-IFNγ was associated with only minor improvements [24].

IFNγ blockade: Blocking IFNγ in both ashen and Prf1$^{-/-}$ mice reduced HLH symptoms after LCMV infection. The Prf1$^{-/-}$ mice's survival rate was higher, and CNS involvement in ashen mice was prevented [45, 69].

IL-18 blockade: In $Prf1^{-/-}$ mice infected with MCMV, IL-18 blockade using IL-18BP decreased hemophagocytosis and reduced damage to the liver and spleen. Furthermore, CTLs and NK cells produced less IFNγ and TNF [129].

IL-33/ST2 blockade: In LCMV infected $Prf1^{-/-}$ mice, blockade of the MyD88-dependent IL-33 receptor, ST2, improved both survival and reduced IFNγ serum levels [46].

Etoposide mechanism: The therapeutic mechanism of etoposide (which was already being used to treat patients) was elucidated in a study of LCMV infected $Prf1^{-/-}$ mice, where it reduced the intensity of all the symptoms of HLH. Etoposide was shown to selectively delete activated T cells and thus suppress the production of inflammatory cytokines. Naïve and memory T cells, dendritic cells and macrophages were not affected [130].

JAK1/2 inhibition: Treatment of LCMV-infected $Prf1^{-/-}$ mice and ashen mice with the JAK1/2 inhibitor ruxolitinib increased survival of the $Prf1^{-/-}$ mice and reduced disease severity in both strains. CNS involvement in ashen mice was significantly reduced. Inhibition of JAKs was associated with reduced STAT1-dependent gene expression, limited CD8 T cell expansion and less cytokine secretion and did not affect perforin-dependent killing ability [48, 49].

Gene therapy: HSC gene transfer was able to correct the immunological manifestations of the disease in perforin-deficient mice and in a murine model of XLP1. This encouraging data from murine studies has led to further work on clinically applicable strategies. An alternative approach is to correct defective T cells. This approach is safer than HSC gene therapy and may allow early control of HLH through gene-modified effector T cells. Both strategies are now in development, and thus gene therapy may soon enter clinical trials for certain forms of pHLH [131].

Outlook

Murine models have been instrumental in learning about the genes involved in HLH and the mechanisms leading to the disease. However, murine models have their limitations, where the differences between mouse and human immune system complicate translational research. A main difference is the need for a trigger in mice but not (apparently) in humans. While selecting a suitable murine model may be straightforward for diseases with a known genetic cause, it is substantially more difficult for acquired diseases like as sporadic HLH (sHLH). Murine models of sHLH are discussed in the next chapter ("Murine Models of Secondary Cytokine Storm Syndromes").

An interesting current question is the relevance of heterozygous mutations in HLH-causing genes. Although several mutations in these genes allow HLH to be induced in mice, the impact of single heterozygous mutations and the possible influence of mutations in genes not (yet) associated with HLH are still unclear. To address these questions and model human HLH more accurately, it may be necessary to introduce mutations found in patients into mice. The tools needed for these modifications have been improved, and costs have fallen over recent years. Another interesting technology is the humanized mouse, in which human immune cells are engrafted in immunodeficient mice. This tool may further improve our ability to model complex human diseases like HLH in mice.

References

1. Blake, J. A., Eppig, J. T., Kadin, J. A., Richardson, J. E., Smith, C. L., Bult, C. J., et al. (2017). Mouse Genome Database (MGD)-2017: Community knowledge resource for the laboratory mouse. *Nucleic Acids Research, 45*, D723–D729.
2. Abolins, S., King, E. C., Lazarou, L., Weldon, L., Hughes, L., Drescher, P., et al. (2017). The comparative immunology of wild and laboratory mice, Mus musculus domesticus. *Nature Communications, 8*, 14811.
3. Castrop, H. (2010). Genetically modified mice-successes and failures of a widely used technology. *Pflugers Archiv, 459*, 557–567.
4. Justice, M. J., Siracusa, L. D., & Stewart, A. F. (2011). Technical approaches for mouse models of human disease. *Disease Models & Mechanisms, 4*, 305–310.
5. Gierut, J. J., Jacks, T. E., & Haigis, K. M. (2014). Strategies to achieve conditional gene mutation in mice. *Cold Spring Harbor Protocols, 2014*, 339–349.
6. Perlman, R. L. (2016). Mouse models of human disease: An evolutionary perspective. *Evolution, Medicine and Public Health, 2016*(1), 170–176.
7. Henter, J.-I., Horne, A., Aricó, M., Egeler, R. M., Filipovich, A. H., Imashuku, S., et al. (2007). HLH-2004: Diagnostic and therapeutic guidelines for hemophagocytic lymphohistiocytosis. *Pediatric Blood & Cancer, 48*, 124–131.
8. Doran, A. G., Wong, K., Flint, J., Adams, D. J., Hunter, K. W., & Keane, T. M. (2016). Deep genome sequencing and variation analysis of 13 inbred mouse strains defines candidate phenotypic alleles, private variation and homozygous truncating mutations. *Genome Biology, 17*, 167.
9. Sellers, R. S., Clifford, C. B., Treuting, P. M., & Brayton, C. (2012). Immunological variation between inbred laboratory mouse strains. *Veterinary Pathology, 49*, 32–43.
10. Eppig, J. T., Motenko, H., Richardson, J. E., Richards-Smith, B., & Smith, C. L. (2015). The International Mouse Strain Resource (IMSR): Cataloging worldwide mouse and ES cell line resources. *Mammalian Genome, 26*, 448–455.
11. Brisse, E., Wouters, C. H., & Matthys, P. (2015). Hemophagocytic lymphohistiocytosis (HLH): A heterogeneous spectrum of cytokine-driven immune disorders. *Cytokine & Growth Factor Reviews, 26*, 263–280.
12. Kägi, D., Ledermann, B., Bürki, K., Seiler, P., Odermatt, B., Olsen, K. J., et al. (1994). Cytotoxicity mediated by T cells and natural killer cells is greatly impaired in perforin-deficient mice. *Nature, 369*, 31–37.
13. Walsh, C. M., Matloubian, M., Liu, C. C., Ueda, R., Kurahara, C. G., Christensen, J. L., et al. (1994). Immune function in mice lacking the perforin gene. *Proceedings of the National Academy of Sciences of the United States of America, 91*, 10854–10858.
14. Stepp, S. E., Dufourcq-Lagelouse, R., Le Deist, F., Bhawan, S., Certain, S., Mathew, P. A., et al. (1999). Perforin gene defects in familial hemophagocytic lymphohistiocytosis. *Science, 286*(5446), 1957–1959. https://www.ncbi.nlm.nih.gov/pubmed/10583959
15. Jordan, M. B. (2004). An animal model of hemophagocytic lymphohistiocytosis (HLH): CD8+ T cells and interferon gamma are essential for the disorder. *Blood, 104*, 735–743.
16. Badovinac, V. P., Hamilton, S. E., & Harty, J. T. (2003). Viral infection results in massive CD8+ T cell expansion and mortality in vaccinated perforin-deficient mice. *Immunity, 18*, 463–474.

17. Lowin, B., Beermann, F., Schmidt, A., & Tschopp, J. (1994). A null mutation in the perforin gene impairs cytolytic T lymphocyte- and natural killer cell-mediated cytotoxicity. *Proceedings of National Academy of Sciences of United States of America, 91*, 11571–11575.
18. Lowin, B., Hahne, M., Mattmann, C., & Tschopp, J. (1994). Cytolytic T-cell cytotoxicity is mediated through perforin and Fas lytic pathways. *Nature, 370*, 650–652.
19. Spielman, J., Lee, R. K., & Podack, E. R. (1998). Perforin/Fas-ligand double deficiency is associated with macrophage expansion and severe pancreatitis. *Journal of Immunology, 161*, 7063–7070.
20. van den Broek, M. E., Kägi, D., Ossendorp, F., Toes, R., Vamvakas, S., Lutz, W. K., et al. (1996). Decreased tumor surveillance in perforin-deficient mice. *The Journal of Experimental Medicine, 184*, 1781–1790.
21. Kägi, D., Odermatt, B., Seiler, P., Zinkernagel, R. M., Mak, T. W., & Hengartner, H. (1997). Reduced incidence and delayed onset of diabetes in perforin-deficient nonobese diabetic mice. *The Journal of Experimental Medicine, 186*, 989–997.
22. Müllbacher, A., Hla, R. T., Museteanu, C., & Simon, M. M. (1999). Perforin is essential for control of ectromelia virus but not related poxviruses in mice. *Journal of Virology, 73*, 1665–1667.
23. Chang, E., Galle, L., Maggs, D., Estes, D. M., & Mitchell, W. J. (2000). Pathogenesis of Herpes simplex virus type 1-induced corneal inflammation in perforin-deficient mice. *Journal of Virology, 74*, 11832–11840.
24. van Dommelen, S. L. H., Sumaria, N., Schreiber, R. D., Scalzo, A. A., Smyth, M. J., & Degli-Esposti, M. A. (2006). Perforin and granzymes have distinct roles in defensive immunity and immunopathology. *Immunity, 25*, 835–848.
25. Schmidt, N. W., Khanolkar, A., Hancox, L., Heusel, J. W., & Harty, J. T. (2012). Perforin plays an unexpected role in regulating T-cell contraction during prolonged Listeria monocytogenes infection. *European Journal of Immunology, 42*, 629–640.
26. Gupta, M., Greer, P., Mahanty, S., Shieh, W.-J., Zaki, S. R., Ahmed, R., et al. (2005). CD8-mediated protection against Ebola virus infection is perforin dependent. *Journal of Immunology, 174*, 4198–4202.
27. Wirtz, T., Weber, T., Kracker, S., Sommermann, T., Rajewsky, K., & Yasuda, T. (2016). Mouse model for acute Epstein-Barr virus infection. *Proceedings of the National Academy of Sciences of the United States of America, 113*, 201616574.
28. Badovinac, V. P. (2000). Regulation of antigen-specific CD8+ T cell homeostasis by perforin and interferon-gamma. *Science, 290*, 1354–1357.
29. Pham, N. L. L., Badovinac, V. P., & Harty, J. T. (2012). Epitope specificity of memory CD8+ T cells dictates vaccination-induced mortality in LCMV-infected perforin-deficient mice. *European Journal of Immunology, 42*, 1488–1499.
30. Yang, J., Huck, S. P., McHugh, R. S., Hermans, I. F., & Ronchese, F. (2006). Perforin-dependent elimination of dendritic cells regulates the expansion of antigen-specific CD8+ T cells in vivo. *Proceedings of the National Academy of Sciences of the United States of America, 103*, 147–152.
31. Terrell, C. E., & Jordan, M. B. (2013). Perforin deficiency impairs a critical immunoregulatory loop involving murine CD8(+) T cells and dendritic cells. *Blood, 121*, 5184–5191.
32. Jordan, M. B., Hildeman, D., Kappler, J., & Marrack, P. (2004). An animal model of hemophagocytic lymphohistiocytosis (HLH): CD8+ T cells and interferon gamma are essential for the disorder. *Blood, 104*, 735–743.
33. Terrell, C. E., & Jordan, M. B. (2013). Mixed hematopoietic or T-cell chimerism above a minimal threshold restores perforin-dependent immune regulation in perforin-deficient mice. *Blood, 122*, 2618–2621.
34. Waggoner, S. N., Taniguchi, R. T., Mathew, P. A., Kumar, V., & Welsh, R. M. (2010). Absence of mouse 2B4 promotes NK cell-mediated killing of activated CD8+ T cells, leading to prolonged viral persistence and altered pathogenesis. *The Journal of Clinical Investigation, 120*, 1925–1938.

35. Lang, P. A., Lang, K. S., Xu, H. C., Grusdat, M., Parish, I. A., Recher, M., et al. (2012). Natural killer cell activation enhances immune pathology and promotes chronic infection by limiting CD8+ T-cell immunity. *Proceedings of the National Academy of Sciences o the United States of America, 109*, 1210–1215.
36. Ge, M. Q., Ho, A. W., Tang, Y., Wong, K. H. S., Chua, B. Y. L., Gasser, S., et al. (2012). NK cells regulate CD8+ T cell priming and dendritic cell migration during influenza A infection by IFN-γ and perforin-dependent mechanisms. *Journal of Immunology, 189*, 2099–2109.
37. Zangi, L., Klionsky, Y. Z., Yarimi, L., Bachar-Lustig, E., Eidelstein, Y., Shezen, E., et al. (2012). Deletion of cognate CD8 T cells by immature dendritic cells: A novel role for perforin, granzyme A, TREM-1, and TLR7. *Blood, 120*, 1647–1657.
38. Sepulveda, F. E., Maschalidi, S., Vosshenrich, C. A. J., Garrigue, A., Kurowska, M., Menasche, G., et al. (2015). A novel immunoregulatory role for NK-cell cytotoxicity in protection from HLH-like immunopathology in mice. *Blood, 125*, 1427–1434.
39. Waggoner, S. N., & Kumar, V. (2012). Evolving role of 2B4/CD244 in T and NK cell responses during virus infection. *Frontiers in Immunology, 3*, 377.
40. Madera, S., Rapp, M., Firth, M. A., Beilke, J. N., Lanier, L. L., & Sun, J. C. (2016). Type I IFN promotes NK cell expansion during viral infection by protecting NK cells against fratricide. *Journal of Experimental Medicine*. https://doi.org/10.1084/jem.20150712
41. Crouse, J., Bedenikovic, G., Wiesel, M., Ibberson, M., Xenarios, I., VonLaer, D., et al. (2014). Type I interferons protect T cells against NK cell attack mediated by the activating receptor NCR1. *Immunity, 40*, 961–973.
42. Xu, H. C., Grusdat, M., Pandyra, A. A., Polz, R., Huang, J., Sharma, P., et al. (2014). Type I interferon protects antiviral CD8+ T cells from NK cell cytotoxicity. *Immunity, 40*, 949–960.
43. Jenkins, M. R., Rudd-Schmidt, J. A., Lopez, J. A., Ramsbottom, K. M., Mannering, S. I., Andrews, D. M., et al. (2015). Failed CTL/NK cell killing and cytokine hypersecretion are directly linked through prolonged synapse time. *The Journal of Experimental Medicine, 212*, 307–317.
44. Binder, D., van den Broek, M. F., Kägi, D., Bluethmann, H., Fehr, J., Hengartner, H., et al. (1998). Aplastic anemia rescued by exhaustion of cytokine-secreting CD8+ T cells in persistent infection with lymphocytic choriomeningitis virus. *The Journal of Experimental Medicine, 187*, 1903–1920.
45. Pachlopnik-Schmid, J., Ho, C.-H., Chrétien, F., Lefebvre, J. M., Pivert, G., Kosco-Vilbois, M., et al. (2009). Neutralization of IFNγ defeats haemophagocytosis in LCMV-infected perforin- and Rab27a-deficient mice. *EMBO Molecular Medicine, 1*, 112–124.
46. Rood, J. E., Rao, S., Paessler, M., Kreiger, P. A., Chu, N., Stelekati, E., et al. (2016). ST2 contributes to T-cell hyperactivation and fatal hemophagocytic lymphohistiocytosis in mice. *Blood, 127*, 426–435.
47. Humblet-Baron, S., Franckaert, D., Dooley, J., Bornschein, S., Cauwe, B., Schönefeldt, S., et al. (2016). IL-2 consumption by highly activated CD8 T cells induces regulatory T-cell dysfunction in patients with hemophagocytic lymphohistiocytosis. *The Journal of Allergy and Clinical Immunology, 138*, 200–209.e8.
48. Das, R., Guan, P., Sprague, L., Verbist, K., Tedrick, P., An, Q. A., et al. (2016). Janus kinase inhibition lessens inflammation and ameliorates disease in murine models of hemophagocytic lymphohistiocytosis. *Blood, 127*, 1666–1675.
49. Maschalidi, S., Sepulveda, F. E., Garrigue, A., Fischer, A., & de Saint Basile, G. (2016). Therapeutic effect of JAK1/2 blockade on the manifestations of hemophagocytic lymphohistiocytosis in mice. *Blood, 128*, 60–71.
50. Crozat, K., Georgel, P., Rutschmann, S., Mann, N., Du, X., Hoebe, K., et al. (2006). Analysis of the MCMV resistome by ENU mutagenesis. *Mammalian Genome, 17*, 398–406.
51. Crozat, K., Hoebe, K., Ugolini, S., Hong, N. A., Janssen, E., Rutschmann, S., et al. (2007). Jinx, an MCMV susceptibility phenotype caused by disruption of Unc13d: A mouse model of type 3 familial hemophagocytic lymphohistiocytosis. *The Journal of Experimental Medicine, 204*, 853–863.

52. Krebs, P., Crozat, K., Popkin, D., Oldstone, M. B., & Beutler, B. (2011). Disruption of MyD88 signaling suppresses hemophagocytic lymphohistiocytosis in mice. *Blood, 117*, 6582–6588.
53. Monfregola, J., Johnson, J. L., Meijler, M. M., Napolitano, G., & Catz, S. D. (2012). MUNC13-4 protein regulates the oxidative response and is essential for phagosomal maturation and bacterial killing in neutrophils. *The Journal of Biological Chemistry, 287*, 44603–44618.
54. He, J., Johnson, J. L., Monfregola, J., Ramadass, M., Pestonjamasp, K., Napolitano, G., et al. (2016). Munc13-4 interacts with syntaxin 7 and regulates late endosomal maturation, endosomal signaling, and TLR9-initiated cellular responses. *Molecular Biology of the Cell, 27*, 572–587.
55. Johnson, J. L., He, J., Ramadass, M., Pestonjamasp, K., Kiosses, W. B., Zhang, J., et al. (2016). Munc13-4 is a Rab11-binding protein that regulates Rab11-positive vesicle trafficking and docking at the plasma membrane. *The Journal of Biological Chemistry, 291*, 3423–3438.
56. Chicka, M. C., Ren, Q., Richards, D., Hellman, L. M., Zhang, J., Fried, M. G., et al. (2016). Role of Munc13-4 as a Ca^{2+}-dependent tether during platelet secretion. *The Biochemical Journal, 473*, 627–639.
57. D'Orlando, O., Zhao, F., Kasper, B., Orinska, Z., Müller, J., Hermans-Borgmeyer, I., et al. (2013). Syntaxin 11 is required for NK and CD8+ T-cell cytotoxicity and neutrophil degranulation. *European Journal of Immunology, 43*, 194–208.
58. Kogl, T., Muller, J., Jessen, B., Schmitt-Graeff, A., Janka, G., Ehl, S., et al. (2013). Hemophagocytic lymphohistiocytosis in syntaxin-11-deficient mice: T-cell exhaustion limits fatal disease. *Blood, 121*, 604–613.
59. Sepulveda, F. E., Debeurme, F., Ménasché, G., Kurowska, M., Côte, M., Schmid, J. P., et al. (2013). Distinct severity of HLH in both human and murine mutants with complete loss of cytotoxic effector PRF1, RAB27A, and STX11. *Blood, 121*, 595–603.
60. Prekeris, R., Klumperman, J., & Scheller, R. H. (2000). Syntaxin 11 is an atypical SNARE abundant in the immune system. *European Journal of Cell Biology, 79*, 771–780.
61. Ye, S., Karim, Z. A., Al Hawas, R., Pessin, J. E., Filipovich, A. H., & Whiteheart, S. W. (2012). Syntaxin-11, but not syntaxin-2 or syntaxin-4, is required for platelet secretion. *Blood, 120*, 2484–2492.
62. Zhang, S., Ma, D., Wang, X., Celkan, T., Nordenskjöld, M., Henter, J. I., et al. (2008). Syntaxin-11 is expressed in primary human monocytes/macrophages and acts as a negative regulator of macrophage engulfment of apoptotic cells and IgG-opsonized target cells. *British Journal of Haematology, 142*, 469–479.
63. Kim, K., Petrova, Y. M., Scott, B. L., Nigam, R., Agrawal, A., Evans, C. M., et al. (2012). Munc18b is an essential gene in mice whose expression is limiting for secretion by airway epithelial and mast cells. *The Biochemical Journal, 446*, 383–394.
64. Spessott, W. A., Sanmillan, M. L., McCormick, M. E., Patel, N., Villanueva, J., Zhang, K., et al. (2015). Hemophagocytic lymphohistiocytosis caused by dominant-negative mutations in STXBP2 that inhibit SNARE-mediated membrane fusion. *Blood, 125*, 1566–1577.
65. Silvers, W. K. (1979). *The coat colors of mice*. New York: Springer.
66. Wilson, S. M., Yip, R., Swing, D. A., O'Sullivan, T. N., Zhang, Y., Novak, E. K., et al. (2000). A mutation in Rab27a causes the vesicle transport defects observed in ashen mice. *Proceedings of the National Academy of Sciences of the United States of America, 97*, 7933–7938.
67. Stinchcombe, J. C., Barral, D. C., Mules, E. H., Booth, S., Hume, A. N., Machesky, L. M., et al. (2001). Rab27a is required for regulating secretion in cytotoxic T lymphocytes. *The Journal of Cell Biology, 152*, 825–833.
68. Novak, E. K., Gautam, R., Reddington, M., Collinson, L. M., Copeland, N. G., Jenkins, N. A., et al. (2002). The regulation of platelet-dense granules by Rab27a in the ashen mouse, a model of Hermansky-Pudlak and Griscelli syndromes, is granule-specific and dependent on genetic background. *Blood, 100*, 128–135.
69. Pachlopnik Schmid, J., Ho, C.-H., Diana, J., Pivert, G., Lehuen, A., Geissmann, F., et al. (2008). A Griscelli syndrome type 2 murine model of hemophagocytic lymphohistiocytosis (HLH). *European Journal of Immunology, 38*, 3219–3225.

70. Johnson, J. L., Hong, H., Monfregola, J., & Catz, S. D. (2011). Increased survival and reduced neutrophil infiltration of the liver in rab27a- but not munc13-4-deficient mice in lipopolysaccharide-induced systemic inflammation. *Infection and Immunity, 79*, 3607–3618.
71. Bennett, J. M., Blume, R. S., & Wolff, S. M. (1969). Characterization and significance of abnormal leukocyte granules in the beige mouse: A possible homologue for Chediak-Higashi Aleutian trait. *Translational Research, 73*, 235–243.
72. Oliver, C., & Essner, E. (1975). Formation of anomalous lysosomes in monocytes, neutrophils, and eosinophils from bone marrow of mice with Chédiak-Higashi syndrome. *Laboratory Investigation, 32*, 17–27.
73. Orn, A., Håkansson, E. M., Gidlund, M., Ramstedt, U., Axberg, I., Wigzell, H., et al. (1982). Pigment mutations in the mouse which also affect lysosomal functions lead to suppressed natural killer cell activity. *Scandinavian Journal of Immunology, 15*, 305–310.
74. Roder, J., & Duwe, A. (1979). The beige mutation in the mouse selectively impairs natural killer cell function. *Nature, 278*, 451–453.
75. Kärre, K., Klein, G. O., Kiessling, R., Klein, G., & Roder, J. C. (1980). Low natural in vivo resistance to syngeneic leukaemias in natural killer-deficient mice. *Nature, 284*, 624–626.
76. Barbosa, M. D. F. S., Nguyen, Q. A., Tchernev, V. T., Ashley, J. A., Detter, J. C., Blaydes, S. M., et al. (1996). Identification of the homologous beige and Chediak–Higashi syndrome genes. *Nature, 382*, 262–265.
77. McGarry, M. P., Reddington, M., Novak, E. K., & Swank, R. T. (1999). Survival and lung pathology of mouse models of Hermansky-Pudlak syndrome and Chediak-Higashi syndrome. *Proceedings of the Society for Experimental Biology and Medicine, 220*, 162–168.
78. Jessen, B., Maul-Pavicic, A., Ufheil, H., Vraetz, T., Enders, A., Lehmberg, K., et al. (2011). Subtle differences in CTL cytotoxicity determine susceptibility to hemophagocytic lymphohistiocytosis in mice and humans with Chediak-Higashi syndrome. *Blood, 118*, 4620–4629.
79. Chatterjee, P., Tiwari, R. K., Rath, S., Bal, V., & George, A. (2012). Modulation of antigen presentation and B cell receptor signaling in B cells of beige mice. *Journal of Immunology, 188*, 2695–2702.
80. Westphal, A., Cheng, W., Yu, J., Grassl, G., Krautkrämer, M., Holst, O., et al. (2016). Lysosomal trafficking regulator Lyst links membrane trafficking to toll-like receptor–mediated inflammatory responses. *Journal of Experimental Medicine, 214*, 227.
81. Balkema, G. W., Mangini, N. J., & Pinto, L. H. (1983). Discrete visual defects in pearl mutant mice. *Science, 219*, 1085–1087.
82. Novak, E. K., Hui, S. W., & Swank, R. T. (1984). Platelet storage pool deficiency in mouse pigment mutations associated with seven distinct genetic loci. *Blood, 63*, 536–544.
83. Zhen, L., Jiang, S., Feng, L., Bright, N. A., Peden, A. A., Seymour, A. B., et al. (1999). Abnormal expression and subcellular distribution of subunit proteins of the AP-3 adaptor complex lead to platelet storage pool deficiency in the pearl mouse. *Blood, 94*, 146–155.
84. Feng, L., Seymour, A. B., Jiang, S., To, A., Peden, A. A., Novak, E. K., et al. (1999). The β3A subunit gene (Ap3b1) of the AP-3 adaptor complex is altered in the mouse hypopigmentation mutant pearl, a model for Hermansky-Pudlak syndrome and night blindness. *Human Molecular Genetics, 8*, 323–330.
85. Feng, L., Rigatti, B. W., Novak, E. K., Gorin, M. B., & Swank, R. T. (2000). Genomic structure of the mouse Ap3b1 gene in normal and pearl mice. *Genomics, 69*, 370–379.
86. Yang, W., Li, C., Ward, D. M., Kaplan, J., & Mansour, S. L. (2000). Defective organellar membrane protein trafficking in Ap3b1-deficient cells. *Journal of Cell Science, 113*(Pt 2), 4077–4086.
87. Swank, R. T., Novak, E. K., McGarry, M. P., Zhang, Y., Li, W., Zhang, Q., et al. (2000). Abnormal vesicular trafficking in mouse models of Hermansky-Pudlak syndrome. *Pigment Cell Research, 13*(Suppl 8), 59–67.
88. Cernadas, M., Sugita, M., van der Wel, N., Cao, X., Gumperz, J. E., Maltsev, S., et al. (2003). Lysosomal localization of murine CD1d mediated by AP-3 is necessary for NK T cell development. *Journal of Immunology, 171*, 4149–4155.

89. Young, L. R., Borchers, M. T., Allen, H. L., Gibbons, R. S., & McCormack, F. X. (2006). Lung-restricted macrophage activation in the pearl mouse model of Hermansky-Pudlak syndrome. *Journal of Immunology, 176*, 4361–4368.
90. Meng, R., Wu, J., Harper, D. C., Wang, Y., Kowalska, M. A., Abrams, C. S., et al. (2015). Defective release of α granule and lysosome contents from platelets in mouse Hermansky-Pudlak syndrome models. *Blood, 125*, 1623–1632.
91. Jessen, B., Bode, S. F. N., Ammann, S., Chakravorty, S., Davies, G., Diestelhorst, J., et al. (2013). The risk of hemophagocytic lymphohistiocytosis in Hermansky-Pudlak syndrome type 2. *Blood, 121*, 2943–2951.
92. Czar, M. J., Kersh, E. N., Mijares, L. A., Lanier, G., Lewis, J., Yap, G., et al. (2001). Altered lymphocyte responses and cytokine production in mice deficient in the X-linked lymphoproliferative disease gene SH2D1A/DSHP/SAP. *Proceedings of the National Academy of Sciences of the United States of America, 98*, 7449–7454.
93. Wu, C., Nguyen, K. B., Pien, G. C., Wang, N., Gullo, C., Howie, D., et al. (2001). SAP controls T cell responses to virus and terminal differentiation of TH2 cells. *Nature Immunology, 2*, 410–414.
94. Crotty, S., Kersh, E. N., Cannons, J., Schwartzberg, P. L., & Ahmed, R. (2003). SAP is required for generating long-term humoral immunity. *Nature, 421*, 282–287.
95. Chan, B., Lanyi, A., Song, H. K., Griesbach, J., Simarro-Grande, M., Poy, F., et al. (2003). SAP couples Fyn to SLAM immune receptors. *Nature Cell Biology, 5*, 155–160.
96. Latour, S., Roncagalli, R., Chen, R., Bakinowski, M., Shi, X., Schwartzberg, P. L., et al. (2003). Binding of SAP SH2 domain to FynT SH3 domain reveals a novel mechanism of receptor signalling in immune regulation. *Nature Cell Biology, 5*, 149–154.
97. Qi, H., Cannons, J. L., Klauschen, F., Schwartzberg, P. L., & Germain, R. N. (2008). SAP-controlled T-B cell interactions underlie germinal centre formation. *Nature, 455*, 764–769.
98. Veillette, A., Zhang, S., Shi, X., Dong, Z., Davidson, D., & Zhong, M.-C. (2008). SAP expression in T cells, not in B cells, is required for humoral immunity. *Proceedings of the National Academy of Sciences of the United States of America, 105*, 1273–1278.
99. Nichols, K. E., Hom, J., Gong, S.-Y., Ganguly, A., Ma, C. S., Cannons, J. L., et al. (2005). Regulation of NKT cell development by SAP, the protein defective in XLP. *Nature Medicine, 11*, 340–345.
100. Pasquier, B., Yin, L., Fondanèche, M.-C., Relouzat, F., Bloch-Queyrat, C., Lambert, N., et al. (2005). Defective NKT cell development in mice and humans lacking the adapter SAP, the X-linked lymphoproliferative syndrome gene product. *The Journal of Experimental Medicine, 201*, 695–701.
101. Das, R., Bassiri, H., Guan, P., Wiener, S., Banerjee, P. P., Zhong, M.-C., et al. (2013). The adaptor molecule SAP plays essential roles during invariant NKT cell cytotoxicity and lytic synapse formation. *Blood, 121*, 3386–3395.
102. Yin, L., Al-Alem, U., Liang, J., Tong, W.-M., Li, C., Badiali, M., et al. (2003). Mice deficient in the X-linked lymphoproliferative disease gene sap exhibit increased susceptibility to murine gammaherpesvirus-68 and hypo-gammaglobulinemia. *Journal of Medical Virology, 71*, 446–455.
103. Crotty, S., McCausland, M. M., Aubert, R. D., Wherry, E. J., & Ahmed, R. (2006). Hypogammaglobulinemia and exacerbated CD8 T-cell-mediated immunopathology in SAP-deficient mice with chronic LCMV infection mimics human XLP disease. *Blood, 108*, 3085–3093.
104. Dong, Z., Davidson, D., Pérez-Quintero, L. A., Kurosaki, T., Swat, W., & Veillette, A. (2012). The adaptor SAP controls NK cell activation by regulating the enzymes Vav-1 and SHIP-1 and by enhancing conjugates with target cells. *Immunity, 36*, 974–985.
105. Rivat, C., Booth, C., Alonso-ferrero, M., Blundell, M., Sebire, N. J., Adrian, J., et al. (2013). Murine model of X-linked lymphoproliferative disease. Brief report SAP gene transfer restores cellular and humoral immune function in a murine model of X-linked lymphoproliferative disease. *Blood, 121*, 1073–1076.

106. Ruffo, E., Malacarne, V., Larsen, S. E., Das, R., Patrussi, L., Wülfing, C., et al. (2016). Inhibition of diacylglycerol kinase α restores restimulation-induced cell death and reduces immunopathology in XLP-1. *Science Translational Medicine, 8*, 321ra7.
107. Harlin, H., Harlin, H., Reffey, S. B., Reffey, S. B., Duckett, C. S., Duckett, C. S., et al. (2001). Characterization of XIAP-de cient mice. *Molecular and Cellular Biology, 21*, 3604–3608.
108. Olayioye, M. A., Kaufmann, H., Pakusch, M., Vaux, D. L., Lindeman, G. J., & Visvader, J. E. (2005). XIAP-deficiency leads to delayed lobuloalveolar development in the mammary gland. *Cell Death and Differentiation, 12*, 87–90.
109. Bauler, L. D., Duckett, C. S., & O'Riordan, M. X. D. (2008). XIAP regulates cytosol-specific innate immunity to listeria infection. *PLoS Pathogens, 4*, e1000142.
110. Rumble, J. M., Bertrand, M. J., Csomos, R. A., Wright, C. W., Albert, L., Mak, T. W., et al. (2008). Apoptotic sensitivity of murine IAP-deficient cells. *The Biochemical Journal, 415*, 21–25.
111. Schile, A. J., García-Fernández, M., & Steller, H. (2008). Regulation of apoptosis by XIAP ubiquitin-ligase activity. *Genes & Development, 22*, 2256–2266.
112. Jost, P. J., Grabow, S., Gray, D., McKenzie, M. D., Nachbur, U., Huang, D. C. S., et al. (2009). XIAP discriminates between type I and type II FAS-induced apoptosis. *Nature, 460*, 1035–1039.
113. Prakash, H., Albrecht, M., Becker, D., Kuhlmann, T., & Rudel, T. (2010). Deficiency of XIAP leads to sensitization for Chlamydophila pneumoniae pulmonary infection and dysregulation of innate immune response in mice. *The Journal of Biological Chemistry, 285*, 20291–20302.
114. Fuchs, Y., Brown, S., Gorenc, T., Rodriguez, J., Fuchs, E., & Steller, H. (2013). Sept4/ARTS regulates stem cell apoptosis and skin regeneration. *Science, 341*, 286–289.
115. Unsain, N., Higgins, J. M., Parker, K. N., Johnstone, A. D., & Barker, P. A. (2013). XIAP regulates caspase activity in degenerating axons. *Cell Reports, 4*, 751–763.
116. Yabal, M., Müller, N., Adler, H., Knies, N., Groß, C. J., Damgaard, R., et al. (2014). XIAP restricts TNF- and RIP3-dependent cell death and inflammasome activation. *Cell Reports, 7*, 1796–1808.
117. Andree, M., Seeger, J. M., Schüll, S., Coutelle, O., Wagner-Stippich, D., Wiegmann, K., et al. (2014). BID-dependent release of mitochondrial SMAC dampens XIAP-mediated immunity against Shigella. *The EMBO Journal, 33*, 2171–2187.
118. Hsieh, W.-C., Chuang, Y.-T., Chiang, I., Hsu, S.-C., Miaw, S.-C., & Lai, M.-Z. (2014). Inability to resolve specific infection generates innate immunodeficiency syndrome in Xiap−/− mice. *Blood, 124*, 2847–2857.
119. Gentle, I. E., Moelter, I., Lechler, N., Bambach, S., Vucikuja, S., Häcker, G., et al. (2013). Inhibitor of apoptosis proteins (IAPs) are required for effective T cell expansion/survival during anti-viral immunity in mice. *Blood, 123*, 659–669.
120. Ebert, G., Preston, S., Allison, C., Cooney, J., Toe, J. G., Stutz, M. D., et al. (2015). Cellular inhibitor of apoptosis proteins prevent clearance of hepatitis B virus. *Proceedings of the National Academy of Sciences of the United States of America, 112*, 5797–5802.
121. Gibon, J., Unsain, N., Gamache, K., Thomas, R. A., De Leon, A., Johnstone, A., et al. (2016). The X-linked inhibitor of apoptosis regulates long-term depression and learning rate. *The FASEB Journal, 30*, 1–8.
122. Chen, M., Felix, K., & Wang, J. (2012). Critical role for perforin and Fas-dependent killing of dendritic cells in the control of inflammation. *Blood, 119*, 127–136.
123. Nansen, A., Jensen, T., Christensen, J. P., Andreasen, S. O., Röpke, C., Marker, O., et al. (1999). Compromised virus control and augmented perforin-mediated immunopathology in IFN-gamma-deficient mice infected with lymphocytic choriomeningitis virus. *Journal of Immunology, 163*, 6114–6122.
124. Sepulveda, F. E., Maschalidi, S., Vosshenrich, C. A. J., Garrigue, A., Kurowska, M., Ménasche, G., et al. (2015). A novel immunoregulatory role for NK-cell cytotoxicity in protection from HLH-like immunopathology in mice. *Blood, 125*, 1427–1434.
125. Lykens, J. E., Terrell, C. E., Zoller, E. E., Risma, K., & Jordan, M. B. (2011). Perforin is a critical physiologic regulator of T-cell activation. *Blood, 118*, 618–626.

126. Sepulveda, F. E., Garrigue, A., Maschalidi, S., Garfa-Traore, M., Ménasché, G., Fischer, A., et al. (2016). Polygenic mutations in the cytotoxicity pathway increase susceptibility to develop HLH immunopathology in mice. *Blood, 127*, 2113–2121.
127. Jessen, B., Kögl, T., Sepulveda, F. E., de Saint Basile, G., Aichele, P., & Ehl, S. (2013). Graded defects in cytotoxicity determine severity of hemophagocytic lymphohistiocytosis in humans and mice. *Frontiers in Immunology, 4*, 34–36.
128. Singh, R. K., Mizuno, K., Wasmeier, C., Wavre-Shapton, S. T., Recchi, C., Catz, S. D., et al. (2013). Distinct and opposing roles for Rab27a/Mlph/MyoVa and Rab27b/Munc13-4 in mast cell secretion. *The FEBS Journal, 280*, 892–903.
129. Chiossone, L., Audonnet, S., Chetaille, B., Chasson, L., Farnarier, C., Berda-Haddad, Y., et al. (2012). Protection from inflammatory organ damage in a murine model of hemophagocytic lymphohistiocytosis using treatment with IL-18 binding protein. *Frontiers in Immunology, 3*, 1–10.
130. Johnson, T. S., Terrell, C. E., Millen, S. H., Katz, J. D., Hildeman, D. A., & Jordan, M. B. (2014). Etoposide selectively ablates activated T cells to control the immunoregulatory disorder hemophagocytic lymphohistiocytosis. *Journal of Immunology, 192*, 84–91.
131. Booth, C., Carmo, M., & Gaspar, H. B. (2014). Gene therapy for haemophagocytic lymphohistiocytosis. *Current Gene Therapy, 14*, 437–446.

Murine Models of Secondary Cytokine Storm Syndromes

Ellen Brisse, Carine H. Wouters, and Patrick Matthys

Abbreviations

Ag	Antigen
APC	Antigen-presenting cell
BP	Binding protein
CAEBV	Chronic active EBV infection
CCL	C-C motif chemokine ligand
CFA	Complete Freund's adjuvant
CSS	Cytokine Storm Syndrome
dsDNA	Double-stranded DNA
EBV	Epstein–Barr virus
GM-CSF	Granulocyte macrophage colony stimulating factor
HIF	Hypoxia-inducible factor
HLH	Hemophagocytic lymphohistiocytosis
HSC	Hematopoietic stem cell
HVP	*Herpesvirus papio*
IDO	Indoleamine 2,3-dioxygenase
IFN	Interferon
IL	Interleukin
IL2R	IL-2 receptor

E. Brisse · P. Matthys (✉)
Laboratory of Immunobiology, Rega Institute, KU Leuven, Leuven, Belgium
e-mail: patrick.matthys@kuleuven.be

C. H. Wouters
Laboratory of Immunobiology, Rega Institute, KU Leuven, Leuven, Belgium

Department of Pediatric Rheumatology, University Hospital Gasthuisberg, KU Leuven, Leuven, Belgium
e-mail: carine.wouters@uzleuven.be

R. Q. Cron, E. M. Behrens (eds.), *Cytokine Storm Syndrome*,
https://doi.org/10.1007/978-3-030-22094-5_29

IRF	Interferon-regulatory factor
IVIG	Intravenous immunoglobulins
JAK	Janus kinase
KO	Knockout
LCMV	Lymphocytic choriomeningitis virus
LPS	Lipopolysaccharide
MAS	Macrophage activation syndrome
MCMV	Mouse cytomegalovirus
MHC	Major histocompatibility complex
MMP	Matrix metalloproteinase
NF-κB	Nuclear factor κB
NK cell	Natural killer cell
NOD	Nonobese diabetic
NRG	NOD/RAG/IL2Rγ^{null}
NSG	NOD/SCID/IL-2Rγ$^{-/-}$
PBMC	Peripheral blood mononuclear cell
pHLH	Primary HLH
PPAR	Peroxisome proliferator activated receptor
R	Receptor
RAG	Recombination-activating gene
RBC	Red blood cell
sCD25	Soluble CD25
SCF	Stem cell factor
SCID	Severe combined immunodeficient
sHLH	Secondary HLH
sJIA	Systemic juvenile idiopathic arthritis
SLE	Systemic lupus erythematosus
STAT	Signal transducer and activator of transcription
Tg	Transgenic
Th	T helper cell
TLR	Toll-like receptor
TNF	Tumor necrosis factor
Treg	Regulatory T cell
WT	Wild-type
YFP	Yellow fluorescent protein

Introduction

Hemophagocytic lymphohistiocytosis (HLH) is a rare and life-threatening immune-inflammatory disorder associated with high morbidity and increased mortality in children and adults. Two subtypes are distinguished on the basis of an underlying genetic predisposition. Animal models of primary HLH studying genetically engineered mice have provided important insights into the pathogenesis of HLH arising in the context of cytotoxicity-related mutations [1–5]. Aberrant activation of CD8$^+$ T cells, massive secretion of pro-inflammatory IFN-γ and impaired regulation of antigen presentation by dendritic cells compromise proper deflation and termination of immune responses, culminating into systemic hyperinflammation, hemophagocytosis and a cytokine storm [1, 6–9]. Despite abundant clinical and laboratory similarities, the findings in these genetic models of primary HLH may not directly translate to secondary HLH, the latter arising on various backgrounds of infection, malignancy, autoimmune and autoinflammatory diseases, acquired immunodeficiency, and/or metabolic disorders [10], where cytotoxicity is not necessarily deficient and a genetic predisposition is not always evident. Considering the heterogeneity of triggering factors and underlying diseases in secondary HLH (sHLH), a large diversity of animal models is required to capture its complete disease spectrum. Since 2001, almost 20 animal models have been published aiming to recapitulate the syndrome of secondary HLH (summarized in Table 1). This chapter highlights important findings in all models of sHLH, discusses their (dis)similarities and considers points of interest for future therapeutic strategies. HLH symptoms present in the different models are evaluated against the current existing diagnostic guidelines, that is, the HLH-2004 criteria [11] and the HScore, which is specifically used to calculate the probability of sHLH in patients [12] (Table 2). Table 3 summarizes important cytokines, chemokines, and cell types in the pathogenesis of the different animal models.

In contrast to the mutant models of primary HLH (pHLH), models of sHLH in general reflect the lack of clear genetic predisposition in patients by using immunocompetent wild-type (WT) mice. One exception is the polygenic heterozygous mouse model of Sepulveda and colleagues [13], which can be positioned somewhere on the blurring line between primary and secondary HLH. In the other models discussed within this chapter, cytotoxic defects are usually not inherent, although they may be observed resulting from the inflammatory cytokine environment.

Animal Models of Secondary HLH

Polygenic Heterozygous Cytotoxic Defects

According to the classical concept, primary HLH is caused by homozygous or compound heterozygous monogenic defects, while secondary HLH was considered not to have a genetic background. During the last decade, however, heterozygous mutations in one, two, or more cytotoxicity-related genes have been reported in several patients with a diagnosis of secondary HLH. These observations sparked speculations about additive genetic effects, given that all pHLH-related genes play downstream roles in the same cytotoxic pathway. It has been proposed that accumulation of polygenic monoallelic mutations in patients may generate a synergistic effect that partially impairs the cytotoxic function and will increase host susceptibility to HLH

Table 1 Overview of existing secondary HLH animal models

Species/strain	Immunologic background	Cytotoxicity defect	Trigger	Duration	Mortality	First report
C57BL/6 mice	Heterozygous *Rab27a*$^{+/-}$ *Prf1*$^{+/-}$ *Stx11*$^{+/-}$	Yes, NK and $CD8^+$ T cells	LCMV	2 weeks	20–30%	[13]
New Zealand/ Japanese white rabbit	Immunocompetent WT	/	*Herpesvirus papio*	20–30 days	100%	[25]
NOD/SCID/ IL-2R$\gamma^{-/-}$ mice	Humanized immune system	/	EBV	4–10 weeks	70%	[29]
NOD/SCID/ IL-2R$\gamma^{-/-}$ mice	Humanized immune system	/	EBV-HLH Patient PBMCs	4 weeks	Yes	[31]
BALB/c mice	Immunocompetent WT (NK cells lack Ly49H receptor)	No	MCMV	5 days	100%	[32]
Landrace pigs	Immunocompetent WT	Yes, NK cells (post infection only)	African swine fever virus	7 days	Yes	[38]
Sv129 mice	Immunocompetent WT	/	*Salmonella enterica* Typhimurium	3 weeks	No, unless CNS disease (7%)	[47]
C57BL/6 mice	Immunocompetent WT	/	Chronic CpG (with/without aIL10R or D-galactosamine)	7–10 days	No, unless addition of aIL10R or D-galactosamine	[55]
C57BL/6 mice	Immunocompetent WT	/	High dose CpG	5 days	/	[65]
C57BL/6 NSE/ huIL6 mice	Transgenic IL-6 overexpression	Yes, NK cells	LPS	4 days	Yes	[70]
BALB/c mice	IFN-$\gamma^{-/-}$	Yes, NK cells	CFA	11–39 days	No	[80]

C57BL/6 mice	Immunocompetent WT	/	IL-4 osmotic pumps IL4/aIL4 complexes Transgenic IL-4	3 days 10 days 3–11 months	Yes	[82]
C57BL/6 mice	Immunocompetent WT	/	IFN-γ osmotic pumps	5 days	/	[88]
C57BL/6 Yeti IFN-γ reporter mice	Transgenic IFN-γ overexpression	/	Spontaneous	5–6 weeks	100% (100 days)	[89]
NOD/SCID/IL-2Rγ$^{-/-}$ mice	Humanized immune system Transgenic SCF, IL-3 and GM-CSF overexpression	/	Spontaneous	14 weeks	100% (200 days)	[91]
Vav1-Cre, *Rosa26-LSL-rtTA HIF1A*-TPM C57BL/6 mice	Inducible transgenic HIF-1A/ARNT overexpression in hematopoietic lineages	/	Doxycycline (induces HIF-1A/ARNT overexpression)	8 days	100% (20 days)	[94]
Mixed 129/Ola and C57BL/6 mice	Legumain$^{-/-}$	Yes, NK cells	Spontaneous	6–18 months	No	[97]

/ = not reported, aIL-4 = anti-IL-4-antibody, aIL10R = IL-10 receptor blockade, CFA = complete Freund's adjuvant, CNS = central nervous system, EBV = Epstein-Barr virus, HIF = hypoxia-inducible factor, huIL6 = human IL-6, IL-2R = IL-2 receptor, LCMV = lymphocytic choriomeningitis virus, LPS = lipopolysaccharide, MCMV = mouse cytomegalovirus, NK = natural killer, NSE = neurospecific enolase, PBMCs = peripheral blood mononuclear cells, WT = wild-type

Table 2 HLH-like symptoms in secondary HLH animal models and their importance in current diagnostic criteria

	Underlying immunodepression	Fever	Hepatomegaly	Lymphadenopathy	Splenomegaly	Cytopenia in ≥2 cell lineages	→ *Anemia*	→ *Thrombocytopenia*	→ *Lymphopenia*	→ *Neutropenia*	→ *Leukopenia*	Hemophagocytosis	Hyperferritinemia	Hypofibrinogenemia	Hypertriglyceridemia	Elevated liver enzymes	Elevated sCD25	Low NK cell cytotoxicity	Coagulation disorder	CNS symptoms	**… out of 8 HLH-2004 criteria**	**HScore**	**Probability of HLH disease (%)**
Rab27a$^{+/-}$ *Prf1*$^{+/-}$ *Stx11*$^{+/-}$ + LCMV	+	T↓				+	+	+			L↑					+		+			3	61	0.1
WT rabbit + *Herpesvirus papio*	–		+	+	+	(+)		+			(+)	+							+		3	97	1.2
NSG mice + huCD34^{+} HSCs + EBV	+		+		+	+	+	+			L↑	+				+		NK↓			4	134	11
NSG mice + EBV-HLH PBMCs	+		(+)		+														+		1	56	0.1
WT BALB/cmice+ MCMV*	–	+	–	+	–	+	(+)	+	+	N↑	–	+	+	F↑	–	+	+	NK↓	–	–	5	156	33
WT pigs + African swine fever virus	–	+			+	+		+	+			+		+	+			+	+		6	181	70
WT Sv129 mice + *Salmonella*	–	+	+		+	(+)	(+)	+	(+)	N↑		+	(+)	(+)					+	+	6	195	85
WT B6 mice + chronic CpG	–		+	+	+	+	+	+	+	(+)	(+)	–	+	–					+		3	107	2.3
WT B6 + chronic CpG + aIL10R	–			+	+	+	+	+			–	+	+						+		4	117	4.2
WT B6 mice + chronic CpG + DG	–				+	+	+	+			+	+	+	F↑	+				+		5	171	55
WT B6 mice + high dose CpG	–	+			+	+	+	+				+			+						5	175	61
IL–6Tg mice + LPS	+				+	+	+	+		+		–	+	F↑		+	+	+			5	96	1.2
IFN–γ$^{-/-}$mice + CFA	+	–	–	+	+	(+)	(+)	P↑	+	N↑		(+)	–			–		+		–	4	76	0.3
WT B6 mice + chronic IL–4 exposure	–				+	+	+	+		+		+	–	–	+						4	126	7.1
WT B6 mice + IFN–γ pump	–					+	+	+		+	+	+									2	69	0.2
Yeti IFN–γ reporter mice	–			+	+								+	F↑							2	58	0.1
SCF/IL–3/GM–CSF Tg NSG/NRG mice	+	+			+	+	+	+			+	+					+				5	143	18
HIF–1A/ARNT inducible Tg mice	–				+	+	+	+				–	+					NK↓			4	82	0.5
Legumain$^{-/-}$mice	–	+	+		+	–	+	–	–	–	L↑	+						+			4	106	2.1

HLH-2004 criteria: Ref. [11], HScore: Ref. [12], from which the probability of HLH disease can be calculated (http://saintantoine.aphp.fr/score/)
– = symptom is absent, + = symptom is present, (+) = present in part of the animals, green color = fulfils diagnostic criteria, orange color = does not fulfil diagnostic criteria, * = partially unpublished data by Brisse et al., aIL-10R = IL-10 receptor blockade, B6 = C57BL/6 mice, CFA = complete Freund's adjuvant, DG = D-galactosamine, F↑ = elevated fibrinogen, GM-CSF = granulocyte-macrophage colony-stimulating factor, HIF = hypoxia-inducible factor, HSC = hematopoietic stem cell, hu = human, L↑ = leucocytosis, LCMV = lymphocytic choriomeningitis virus, LPS = lipopolysaccharide, MCMV = mouse cytomegalovirus, N↑ = neutrophilia, NK↓ = decreased number of NK cells, NRG = NOD/RAG/IL2Rγ^{null}, NSG = NOD/SCID/IL-2R$\gamma^{-/-}$, P↑ = thrombocytosis, PBMCs = peripheral blood mononuclear cells, SCF = stem cell factor, T↓ = hypothermia, Tg = transgenic, WT = wild-type

Table 3 Hypercytokinemia and immune hyperactivation: role of chemokines, cytokines, and immune cells in secondary HLH animal models

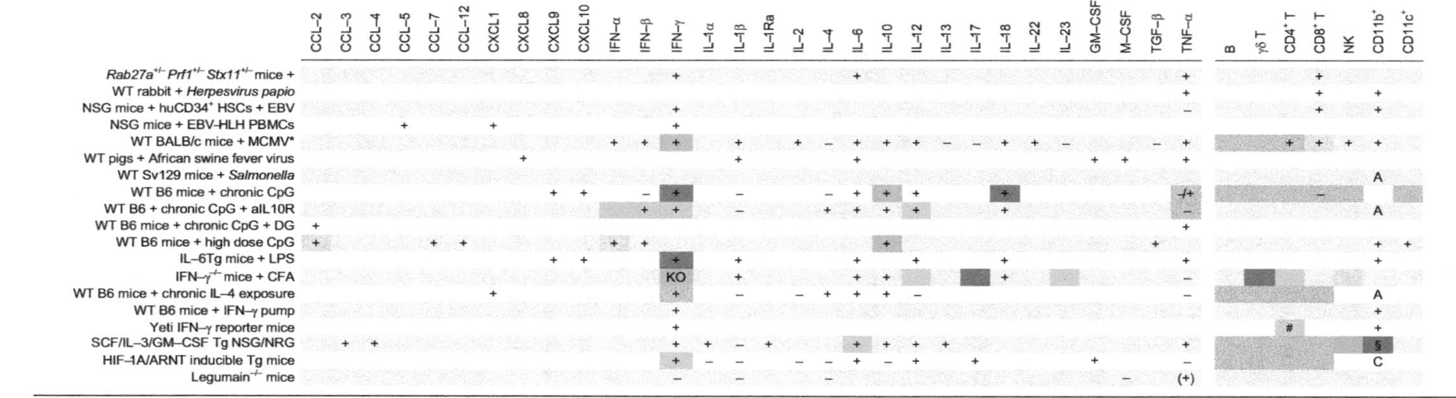

	CCL–2	CCL–3	CCL–4	CCL–5	CCL–7	CCL–12	CXCL1	CXCL8	CXCL9	CXCL10	IFN–α	IFN–β	IFN–γ	IL–1α	IL–1β	IL–1Ra	IL–2	IL–4	IL–6	IL–10	IL–12	IL–13	IL–17	IL–18	IL–22	IL–23	GM–CSF	M–CSF	TGF–β	TNF–α	B	γδ T	CD4$^+$ T	CD8$^+$ T	NK	CD11b$^+$	CD11c$^+$
Rab27a$^{+/-}$ *Prf1*$^{+/-}$ *Stx11*$^{+/-}$ mice +													+						+											+				+			
WT rabbit + *Herpesvirus papio*																														+				+		+	
NSG mice + huCD34$^+$ HSCs + EBV													+																	–				+		+	
NSG mice + EBV-HLH PBMCs				+			+						+																								
WT BALB/c mice + MCMV*											+	+	+		+		+	–	+	+	+	–	–	+	+	–			–	+			+	+	+	+	
WT pigs + African swine fever virus								+							+				+								+	+		+							
WT Sv129 mice + *Salmonella*																																				A	
WT B6 mice + chronic CpG									+	+		–	+		–			–	+	+	+			+						–/+				–			
WT B6 + chronic CpG + aIL10R												+	+		–			+	+	+	+			+						–						A	
WT B6 mice + chronic CpG + DG	+																													+							
WT B6 mice + high dose CpG	+				+	+					+	+								+									+							+	+
IL–6Tg mice + LPS									+	+			+		+				+		+			+						+						+	
IFN–γ$^{-/-}$ mice + CFA													KO		+				+	+				+						–							
WT B6 mice + chronic IL–4 exposure							+						+		–		–	+	+	+	–									–						A	
WT B6 mice + IFN–γ pump													+																							+	
Yeti IFN–γ reporter mice													+																				#			+	
SCF/IL–3/GM–CSF Tg NSG/NRG		+	+										+	+	–	+			+	+	+	+								+						§	
HIF–1A/ARNT inducible Tg mice													+	–	–		–	–	+	–	+		+				+			–						C	
Legumain$^{-/-}$ mice													–		–			–										–		(+)							

Light green = partially protective, dark green = protective, orange = partially pathogenic, red = pathogenic, grey = no effect in pathogenesis, according to depletion and/or knockout studies. + = elevated cytokine levels or immune cell activation, – = cytokine is not elevated or cells are not activated, –/+ = some reports indicate an elevation while other reports observed no elevation, (+) = only present in part of the mice, * = partially unpublished data by Brisse et al., # = involvement of Tbet transcription factor/Th1 cells, § = involvement of CD33$^+$ myeloid cells, A = alternatively activated, aIL10R = anti-IL-10-receptor antibodies, B6 = C57BL/6, C = classically activated, CCL = C-C motif chemokine ligand, CFA = complete Freund's adjuvant, CXCL = C-X-C motif chemokine ligand, DG = D-galactosamine, EBV = Epstein-Barr virus, GM-CSF = granulocyte-macrophage colony-stimulating factor, HIF = hypoxia-inducible factor, IL-1Ra = IL-1 receptor antagonist, KO = knockout, LCMV = lymphocytic choriomeningitis virus, LPS = lipopolysaccharide, MCMV = mouse cytomegalovirus, NK = natural killer, NRG = NOD/RAG/IL2Rγ^{null}, NSG = NOD/SCID/IL-2R$\gamma^{-/-}$, PBMC = peripheral blood mononuclear cell, SCF = stem cell factor, Tg = transgenic, TGF = transforming growth factor, TNF = tumor necrosis factor

[14–18]. This theory of polygenic inheritance was explored in a heterozygous mouse model that combined monoallelic mutations in *Rab27a*, *Prf1* and/or *Stx11*. While single heterozygous mice did not differ from WT mice, double or triple heterozygous mice showed reduced lymphocyte cytotoxic function, delayed viral clearance and an increased risk to develop HLH-like features (summarized in Table 2), including hypercytokinemia (summarized in Table 3), following infection with the lymphocytic choriomeningitis virus (LCMV). The severity of the HLH syndrome was linked to the number and nature of affected genes and the resultant degree of cytotoxic impairment, with the infected polygenic heterozygous mice lying in between the spectrum of normal WT mice and homozygous mutant mice [13]. Thus, heterozygous mutations in two or more cytotoxicity-related genes may indeed contribute to the development of secondary HLH.

Infection-Associated Secondary HLH

Models of infection-associated secondary HLH have predominantly focused on virus-associated HLH, since viruses are considered the leading triggers of disease in both primary and secondary HLH [19]. One bacterial model of secondary HLH has also been described.

Herpesvirus papio

Epstein–Barr virus (EBV) is considered the predominant trigger of disease in patients with primary as well as secondary HLH. Due to strict species-specificity, EBV-induced sHLH cannot be studied in regular WT mice. As an alternative, a Japanese group published the first animal model of virus-associated secondary HLH in 2001 using *Herpesvirus papio*, a baboon γ-herpesvirus biologically and genetically closely related to EBV. This model was established in immunocompetent WT rabbits and remains the only existing rabbit model of HLH up to now. Inoculated rabbits displayed a fatal HLH syndrome (Table 2) with high levels of TNF-α. Interestingly, hemophagocytosis consisted mainly of red blood cell (RBC) uptake, mediated by the presence of anti-RBC-antibodies, a phenomenon that has been reported in a few cases of EBV-HLH and EBV-related infectious mononucleosis [20, 21]. After inoculation, the virus was detected in a mixture of cell types ($CD4^+$ and $CD8^+$ T cells, as well as B cells), partially resembling human EBV-HLH where the virus has been reported to be predominantly present in $CD8^+$ T cells [22, 23], or in B cells in patients with underlying X-linked lymphoproliferative disease [24]. The infected rabbits additionally developed T-cell lymphoproliferative disease and some of them showed emergence of lymphomas, making this model useful to study HLH in EBV-driven malignancies and the process of disease progression from acute EBV-related HLH to chronic EBV-associated lymphoma [25–28].

Epstein–Barr Virus

As *Herpesvirus papio* is an EBV-related virus, the rabbit model of EBV-HLH may contain elements that are not specific for EBV-induced pathology or miss elements specific to EBV-induced HLH. Therefore, humanized mouse models have been engineered, in which immunodeficient NOD/SCID/IL-2Rγ$^{-/-}$ (NSG) mice are engrafted with either human hematopoietic stem cells (HSCs) or patient-derived peripheral blood mononuclear cells (PBMCs) to approach the human immune system, allowing to study EBV infection in an in vivo model setting.

In the first model, purified CD34^{+} HSCs derived from healthy donor cord blood were transplanted to newborn NSG mice. Following reconstitution of the humanized immune system, mice were infected with EBV, after which they developed an HLH-like syndrome (Table 2), accompanied by high levels of IFN-γ (Table 3). Again, hemophagocytosis consisted predominantly of erythrophagocytosis. In this model, EBV primarily infected B cells. No significant infection of T or NK cells was reported, in contrast to what is observed in patients suffering from EBV-induced HLH. Interestingly, heat-inactivated EBV virions did not induce HLH disease, while EBV DNA load correlated with disease severity, CD8^{+} T cell activation and IFN-γ production, designating productive viral replication as the driver of HLH pathology [29]. Age at the time of engraftment may also be a determining factor in this model, as older NSG mice tended to develop an EBV-related B cell proliferative disorder [30].

The second model utilizes adult NSG mice that are engrafted with PBMCs from patients with a severe EBV-related pathology (e.g., either EBV-induced HLH or chronic active EBV infection (CAEBV)). In this way, the composition of the reconstituted immune system closely resembled that of the donor patient. Interestingly, mice engrafted with PBMCs from HLH patients were distinguished by an intensified cytokine storm with higher levels of IFN-γ and IL-8 (Table 3) and aggressive, more frequently fatal disease, compared to CAEBV-engrafted mice. Massive internal hemorrhages, indicative of severe coagulopathy, were only observed after engraftment of HLH-derived PBMCs. Nonetheless and despite a high viral load, other HLH-like features in these mice were limited or not reported (Table 2). Surprisingly, even though EBV was initially present in the CD8^{+} T cell fraction of the patients' PBMCs and remained present in T cells in peripheral blood, the majority of EBV-infected cells in the tissues of engrafted mice belonged to the B cell lineage, indicating the difficulty of studying EBV cellular tropism in a model setting [31].

Mouse Cytomegalovirus

A recent mouse model of herpesvirus-associated secondary HLH does not rely on γ-herpesviruses but on mouse cytomegalovirus (MCMV), the murine counterpart of human cytomegalovirus, a β-herpesvirus known to elicit HLH episodes particularly in immunocompromised hosts. Upon infection, immunocompetent WT BALB/c, but not C57BL/6 mice, developed an acute and fatal HLH-like syndrome (Table 2), associated with immune hyperactivation and a severe cytokine storm (Table 3) [32]. This remarkable strain dependency is possibly linked to differences in the expression of the activating receptor Ly49H, essential for adequate NK cell recognition of MCMV and efficient viral clearance, which is absent in BALB/c mice, resulting in high viral loads [33, 34]. Similar to the EBV model [29], lytic viral replication was essential for the induction of HLH, since UV inactivation or antibody-mediated neutralization of MCMV abrogated disease development [34]. In contrast to most models of pHLH, neither $CD8^+$ T cells nor IFN-γ were required for disease pathogenesis in MCMV-infected BALB/c mice. IFN-γ might even play a protective role herein, considering the aggravated phenotype and increased mortality in IFN-γ-deficient mice [32]. This unexpected effect of IFN-γ was not mediated by lack of the IFN-γ-induced anti-inflammatory enzyme IDO1, as IDO1-deficient mice were practically undistinguishable from WT BALB/c mice post-infection [35]. As MCMV-induced HLH could also be elicited in SCID mice, lacking functional T and B cells, this model is likely driven by innate rather than adaptive immune cells. In an attempt to unravel important innate disease mediators, neutrophils were depleted, however without profound effects on disease development or severity [36].

African Swine Fever Virus

Infection of normal immunocompetent Landrace pigs with the African swine fever virus, a dsDNA hemorrhagic fever virus of the *Asfarviridae* family, recapitulates full-blown HLH to a large extent (Table 2). Active disease was associated with increasingly high viral titers and progressive elevation of multiple pro-inflammatory cytokines (Table 3). Hemophagocytes were shown to contain predominantly lymphocytes and/or RBC [37, 38]. Further research in this animal model could provide interesting clues about the emergence of HLH in human infections with hemorrhagic fever viruses such as Ebola, dengue, and Hantavirus [39–43], as the pigs presented with severe hemorrhaging in all investigated organs. Interestingly, cytokine antagonists that are currently administered in cases of secondary HLH might also be of use in hemorrhagic fever syndromes, according to evidence provided in a mouse model of Ebola infection. IL-1-receptor-antagonist was shown to protect against immune hyperactivation by modulating the cytokine storm during hemorrhagic fever disease [44], similar to the therapeutic function of anakinra in cases of secondary HLH [45, 46].

Bacterial Infection: *Salmonella enterica* Typhimurium

Upon oral infection of immunocompetent WT Sv129 mice with *Salmonella enterica* serotype Typhimurium, mimicking chronic typhoid fever in humans, a mostly non-fatal syndrome developed that was highly reminiscent of bacteria-induced secondary HLH (Table 2). Notably, this model is unique in recapitulating the neurological symptoms of secondary HLH, such as ataxia and movement disorder, which in severe cases was associated with death of the mice. Unfortunately, cytokine levels were not measured (Table 3). Hemophagocytosis was shown to be performed by alternatively activated, M2-type macrophages, engulfing predominantly non-apoptotic T cells, which is distinct from the previously discussed infection models emphasizing engulfment of RBCs. Similar to the viral models, HLH disease severity was correlated with the organ bacterial load [47, 48], again indicating the importance of a persistent infection in disease pathogenesis.

Autoimmunity-Associated or Autoinflammation-Associated HLH

Toll-like receptor (TLR) triggering has been utilized in a number of animal models to induce a hyperinflammatory syndrome reminiscent of HLH. The chronic or excessive stimulation of pathogen recognition receptors is thought to simulate ongoing antigenemia in the context of persistent infections or a continuous background inflammation in the context of autoimmune and autoinflammatory diseases. Stimulation of TLR9 appears to be unique in inducing HLH in naive mice, without preexisting immune activation. TLR9 is activated by dsDNA and alerts the body of viral intruders such as EBV, CMV, or other Herpesviruses, key inducers of HLH in patients. Additionally, in autoimmune diseases such as SLE, in which HLH can occur as a complication, TLR9 is activated by pieces of self-DNA. Conversely, in mice with underlying chronic inflammation, stimulation of TLR4 was described as a potent trigger of HLH-like disease. TLR4 recognizes bacterial cell wall components such as lipopolysaccharide (LPS) to alert the body of bacterial pathogens. Involvement of TLR signaling in HLH pathogenesis has been suggested after reports of patients carrying mutations in interferon-regulatory factor 5 (IRF5) which is a downstream mediator in many TLR pathways, including TLR4 and TLR9 [49, 50]. Additionally, IRF5 is a master transcription factor regulating the production of pro-inflammatory cytokines such as IL-6, IL-12, and TNF-α [51], all cytokines implicated in HLH. Therefore, it is no surprise that also animal models based on overexposure to cytokines, thus mimicking the overwhelming cytokine storm, are able to recapitulate HLH-like disease. Excessive inflammasome activation and resulting cytokine secretions have been demonstrated to underlie HLH pathogenesis in a subgroup of patients with autoinflammation [52, 53].

TLR9 Stimulation

Chronic CpG Administration with/without IL-10 Receptor Blockade

An extensively studied model of secondary HLH consists of repeated injections of unmethylated CpG into immunocompetent WT C57BL/6 mice, eliciting systemic and predominantly innate immune activation, sufficient to drive HLH-like disease

(Table 2) and initiate a cytokine storm (Table 3). Single-cell TLR9 tolerance was circumvented on the cell population level by the induction of extramedullary monocytopoiesis, creating a continuous peripheral source of newly produced CpG-responsive cells that allowed for ongoing TLR stimulation and sustained inflammation [54]. An ever renewing pool of innate immune cells thus appeared to drive TLR9-mediated disease. Indeed, adaptive immunity was of less importance as $CD8^+$ T cell activation was minimal and neither T, B, NK nor NKT cells played a pathogenic part in the model, highlighting an important difference with primary HLH. Neutralization experiments and the use of knockout mice designated IFN-γ as a major pathogenic cytokine, while type I interferons, IL-12, or TNF-α had no essential effects on disease progression [55, 56]. IL-18 binding protein (IL-18BP) deficiency resulted in an aggravated HLH phenotype, with additional liver dysfunction and increased IL-18 and IFN-γ signature [57]. IL-10 on the other hand protected against the so-called "fulminant HLH" since hemophagocytosis, worsening of the cytokine storm and death of the mice were only observed during IL-10 receptor blockade. The IL-10 blockade also revealed IL-10 as a modulator of the pathogenic effects of IFN-γ since fulminant HLH developed largely independent of the latter, with the exception of anemia. Hemophagocytes during fulminant disease exhibited an M2-type transcription profile, indicative of alternative macrophage activation [55, 56, 58]. M2 macrophages express scavenger receptor CD163, which, together with serum levels of soluble CD163, has been described as a biomarker in patients with HLH [59, 60]. These alternatively activated macrophages are reported to fulfill important clean up and tissue remodeling functions, aiding regulation of immune responses [58]. Of note, a recent report also documented the occurrence of HLH-like disease in WT C57BL/6 mice following chronic administration of CpG in combination with IFN-γ. However, IFN-γ addition to the protocol did not result in aggravation of disease to a similar level as seen after IL-10 receptor blockade [61].

Chronic CpG Administration with D-Galactosamine

Since the CpG-induced mouse model of secondary HLH is characterized by mild and nonlethal pathology, an addition to the protocol was introduced in order to increase disease severity and fatality, resembling human fulminant HLH (Tables 2 and 3). To this end, repeated CpG injections were combined with D-galactosamine, a substance triggering hepatocyte apoptosis and TNF-α activation. In this model, the role of plasmin, an enzyme implicated in the fibrinolytic cascade, was examined [62]. In HLH, excessively activated macrophages produce plasminogen activator, augmenting the fibrinolytic process. In patients, presence of coagulation abnormalities has been associated with worse prognosis [63, 64]. Aside from degrading fibrin cloths, plasmin plays an abundant role in the inflammatory response by activating several matrix metalloproteases (MMPs), proteolytic enzymes that can catalyze TNF-α, sCD25, and sFasL cleavage and shedding, all factors found in HLH patients. In a direct way, plasmin also activates NF-κB signaling and thus production of inflammatory cytokines. Genetic deletion or pharmacological inhibition of the plasmin active site saved the injected mice from lethal disease and decreased tissue destruction as well as cytokine levels. Conversely, administration of plasminogen activator shortened survival of the mice, increased organ damage and heightened cytokinemia. The pathogenic effects of plasmin were mediated via MMP activation since MMP9-deficient mice displayed lower mortality rates and reduced tissue damage. In conclusion, plasmin was shown to additively enhance TLR9 signaling by CpG, to mediate the tissue influx of inflammatory cells and to stimulate production of pro-inflammatory cytokines and chemokines. Surprisingly though, plasmin inhibition was not effective in the normal CpG-induced mouse model, indicating that the treatment is specific for advanced-stage HLH or CpG alone is not sufficient to induce relevant plasmin-associated pathology [62]. The relevance and applicability of these findings in the D-galactosamine model versus the regular CpG-induced model remain to be validated.

Single Administration of High-Dose CpG

An alternative model of TLR9-triggered secondary HLH utilizes a single high dose of CpG instead of chronic stimulation to initiate disease (Table 2). This resulted in a syndrome where CCL2 and IFN-α played a role by inducing hemophagocytosis (Table 3). Similar to the D-galactosamine model, hemophagocytes were observed even in the absence of IL-10 blockade. Moreover, hemophagocytes, identified as monocyte-derived dendritic cells having engulfed apoptotic RBCs, were a major source of IL-10 and blocking either hemophagocytosis or IL10 activity significantly increased tissue damage and mortality. This urged the authors to suggest a regulatory role for hemophagocytosis in HLH, constituting a counteraction to temper hyperinflammation and perform important cleanup functions [65]. In this view, therapies directed against the hemophagocytic macrophages may not be indicated in sHLH.

TLR4 Stimulation in an IL-6 Transgenic Background

Secondary HLH can arise as a complication of many different rheumatic diseases, where it is often referred to as macrophage activation syndrome (MAS) [66]. Given that the inflammatory cytokine IL-6 is markedly increased in patients with rheumatic diseases such as systemic juvenile idiopathic arthritis (sJIA), Still disease, rheumatoid arthritis and psoriatic arthritis, and is reportedly correlated with disease activity and severity of joint inflammation [67], a model of MAS was established in transgenic (Tg) mice that constitutively express high systemic levels of human IL-6. Naive IL-6Tg mice exhibited typical features of a chronic rheumatologic disease, including growth impairment, increased osteoclastogenesis and decreased osteoblast activity, resulting in impaired skeletal development [68, 69]. Additionally, the chronic exposure to IL-6 increased the mice's susceptibility to TLR stimulation. Upon administration of several TLR ligands, the mice responded with increased fatality and elevated cytokinemia (Table 3), compared to WT mice. Solely following challenge with lipopolysaccharide (LPS), a TLR4 ligand, the mice developed a fulminant MAS syndrome (Table 2) and additional elevation of IFN-γ (Table 3) [70]. Expression of IFN-γ-induced chemokines correlated with disease activity in the model [71], and neutralization experiments designated the cytokine as a major player in pathogenesis [72–74]. Continuous exposure to IL-6 also decreased the cytotoxic potential of murine NK cells as they showed lower perforin and granzyme B expression and diminished killing capacity towards tumor cells. This effect was confirmed in vitro on healthy NK cells as well as NK cells from patients with sHLH, where addition of tocilizumab, an anti-IL-6R antibody, restored perforin and granzyme B levels and rescued cytotoxic function [75], whereas addition of recombinant IL-6 worsened NK cell dysfunction [17]. These findings may explain the presence of cytotoxicity defects in some patients with sJIA and/or MAS [76–78]. Thus, prolonged high levels of IL-6, as present in several autoimmune and autoinflammatory diseases, not only predispose to exaggerated immune-inflammatory responses but also reduce NK cell cytotoxic capacity, likely increasing vulnerability of patients to infections and facilitating the onset of HLH-like hyperinflammation [70, 75].

Murine sJIA with Subclinical MAS

Another model of rheumatic disease that typically resembles sJIA was elicited in an IFN-γ-deficient background, by a single subcutaneous injection with heat-killed *Mycobacterium butyricum* emulsified in complete Freund's adjuvant (CFA). Aside from characteristic sJIA symptoms including arthritis, rash, and increased frequency of immature blood cell populations, the mice also developed some HLH-like features, reminiscent of the 10% of sJIA patients who develop fulminant MAS and the 30–50% who suffer from subclinical MAS [60, 79] (Table 2). NK cell cytotoxicity was inherently defective in IFN-γ-deficient mice, which may explain their susceptibility to HLH. This was partially confirmed by CFA injection of NK-cell depleted WT mice, which resulted in an incomplete sJIA syndrome, assigning a regulatory role to functionally competent NK cells in this model. Instead of IFN-γ, IL-17 played the dominant pathogenic part in this model, since preventive treatment of the CFA-injected mice with anti-IL12/IL-23p40 or anti-IL-17 antibodies inhibited emergence of most sJIA/MAS symptoms (Table 3) [80]. Innate γδ T cells and $CD4^+$ T cells were the main sources of IL-17 and probably contribute to disease pathogenesis [80, 81].

Excessive Cytokine Exposure

IL-4

The first report of chronic cytokine exposure as a model for secondary HLH appeared following research into novel pathways that could instigate tissue macrophage accumulation and hemophagocytosis. High and sustained systemic levels of

IL-4, achieved via infusion of recombinant IL-4 with subcutaneous osmotic pumps, injections with IL-4/anti-IL-4 immune complexes or the use of IL-4-overexpressing transgenic mice, resulted in an acute hemophagocytic syndrome (Table 2) and increased cytokinemia (Table 3) [82]. The findings are quite remarkable considering that IL-4 is not often elevated in patients' sera. Both in adults and children a Th1-oriented cytokine profile dominates over Th2-associated cytokines [83–86]. Nonetheless, IL-4 appeared to be unique in its action since high systemic concentrations of IL-13, another Th2-associated cytokine, did not mediate similar effects. Unlike primary HLH, IL-4-induced secondary HLH was not dependent on T cells or IFN-γ, since the syndrome was undiminished in *Rag2*$^{-/-}$ mice and following IFN-γ neutralization. Active hemophagocytes showed a pattern of alternative macrophage activation and predominantly consumed RBC, consistent with other animal models of secondary HLH [82].

IL-6

A major role for IL-6 in the pathogenesis of sHLH was highlighted by the IL-6Tg mouse model of Strippoli and colleagues [75, 87], discussed in section "TLR4 Stimulation in an IL-6 Transgenic Background". Mimicking an inflammatory environment with chronically elevated levels of IL-6, this model showed how background inflammation in autoimmune or autoinflammatory diseases can exaggerate responses to TLR stimulation and can directly reduce the cytotoxic capacity of NK cells, predisposing to development of hyperinflammation and HLH-like disease.

IFN-γ

Excessive levels of IFN-γ have been notoriously implicated in primary HLH pathogenesis, and are also often observed in secondary HLH. The effects of this cytokine were examined in vivo in two different models, applying either sterile IFN-γ infusion via implanted osmotic pumps or using transgenic mice.

Chronic administration of IFN-γ in mice, at physiologically relevant levels similar to those observed in infections, resulted in the development of hemophagocytosis and acute pancytopenia (Table 2), although no other HLH-like symptoms were described. The process of hemophagocytosis was reliant on the direct action of IFN-γ on macrophages, functioning in a STAT1- and IRF1-dependent manner. RBCs, neutrophils and other nucleated cells were taken up by F4/80^{+} macrophages using a mechanism described as macropinocytosis, bearing resemblance to apoptotic cell uptake [88]. In another report, the so-called "Yeti" YFP-IFN-γ reporter mice were found to spontaneously develop a lethal HLH-like autoinflammatory syndrome mediated by off-target systemic overexpression of IFN-γ (Table 2). Additive genetic effects were present as two copies of the mutant allele exacerbated multiorgan inflammation and myeloproliferative disease. The syndrome was ameliorated in mice deficient for the canonical Th1 lineage-committing transcription factor Tbet, indicating that Th1 cells could play a role in the observed pathology [89]. Although not described in Yeti mice, constitutive expression of IFN-γ in the bone marrow of another transgenic mouse line resulted in increased TNF-α production as well as impaired NK cell differentiation, causing severely decreased NK cell numbers but not cytotoxicity [90], aspects that could contribute to HLH susceptibility.

Together, these models illustrate how hemophagocytosis and other cardinal features of HLH can be mediated in a direct manner by high levels of IFN-γ. Nonetheless, a more diverse cytokine storm appears necessary to induce a more complete spectrum of HLH.

IL-3, SCF, and GM-CSF

Abundant presence of innate immune stimulating cytokines can also underlie HLH-like disease. NOD/SCID/IL2Rγ^{null} (NSG) and NOD/RAG/IL2Rγ^{null} (NRG) mice were genetically engineered to express human stem cell factor (SCF), IL-3, and GM-CSF to enhance human myeloid cell engraftment following umbilical cord blood transplantation. However, the continuous systemic exposure to SCF, IL-3 and GM-CSF caused the mice to spontaneously develop a fatal inflammatory disease with a high degree of similarity to human HLH (Table 2). Many cytokines were elevated during active disease, although not in all mice (Table 3). Treatment with dexamethasone or intravenous immunoglobulins (IVIG) was not adequate to ameliorate disease. Depletion of T cells with the OKT3 antibody, depletion of B cells using rituximab, or combined B and T depletion could not rescue the mice from fatal disease either, providing evidence for a lymphocyte-independent pathway of pathogenesis, similar to the chronic CpG model and MCMV-induced sHLH. On the other hand, targeting myeloid cell populations with anti-CD33-antibodies (Mylotarg) led to complete recovery of the mice, indicating an innate mechanism of disease. Additionally, targeting IL-6 with tocilizumab slowed down disease progression and increased survival, but could not completely reverse disease, suggesting a partial pathogenic role for IL-6. Interestingly, HLH-like disease only developed in transgenic NSG and NRG mice, not in NOD/SCID mice retaining the common γ chain and thus, proper NK cell function [91]. The authors speculated that functional NK cells may be sufficient to limit myeloproliferation and disease in this model similar to the bone marrow chimeric model of primary HLH [9], where competent cytotoxic NK cells in $T^{Prf1-/-}NK^{Prf1+/+}$ mice were sufficient to limit T cell proliferation and HLH development.

Overexpression of Hypoxia-Related Factors

Inflammatory disease states are often characterized by tissue hypoxia and/or stabilization of hypoxia-dependent transcription factors, such as hypoxia-inducible factor 1 (HIF-1) [92]. This molecule has previously been implicated in cancer, sepsis, and rheumatoid arthritis, among others, but recent microarray data also showed a HIF-1A-related disease signature in patients with primary HLH and sJIA [93, 94]. Furthermore, HIF-1A protein levels are increased in murine CpG-induced secondary HLH and in LCMV-infected perforin-deficient mice [94]. As HIF-1 is a known activator of NF-κB, it may contribute to hypercytokinemia and increased sensitivity to infections [92, 93]. A novel transgenic mouse model of secondary HLH explores this pathogenic pathway by inducing cell-specific overexpression of HIF1A/ARNT in hematopoietic cells, leading to a fatal HLH-like phenotype (Table 2). Interestingly, overexpression of HIF in all hematopoietic cells was strictly necessary as overexpression limited to macrophages, NK cells or mature T cells did not incite disease. Development of the HLH syndrome was thought to be linked to NK cell impairment, since the number of total and mature NK cells underwent a sharp decline during active disease, although degranulation appeared normal. In contrast to the alternative activation phenotype reported in most mouse models of secondary HLH, macrophages with HIF1A/ARNT overexpression showed a classical M1-type activation pattern and no hemophagocytosis was detected. Notably, *Rag1*-deficient mice were not protected from lethal HLH disease, indicating that T and B cells were dispensable and non-lymphoid cells are sufficient for disease pathogenesis. Likewise, IFN-γ-ligand or -receptor deficient mice still displayed fulminant HLH, appointing no pathogenic role to IFN-γ in this model (Table 3) [94]. Solely anemia appeared to be partially mediated by IFN-γ, paralleling observations in CpG-induced murine secondary HLH [56, 94].

Metabolic Disorder-Associated Secondary HLH

The enzyme asparaginyl endopeptidase or legumain is a lysosomal cysteine endopeptidase known to play a role in the processing of antigens for MHC class II presentation, and in the regulation of innate immune responses via its participation in the maturation and signaling of TLR-3, -7 and -9 [95]. Inhibition of legumain in human cells can enhance the presentation of certain T cell epitopes [96], creating a possible link with the pathogenic effects of persistent antigen presentation in primary HLH [6]. Although there are no reports of patients carrying this deficiency, mice lacking legumain were found to spontaneously develop an HLH-like syndrome (Table 2), complete with decreased NK cell cytotoxicity, even though these mice do not carry mutations in cytotoxic genes. Hemophagocytes were identified as macrophages predominantly engulfing RBCs. The HLH-like symptoms progressed slowly and were age-dependent, increasing in severity over time, but no mortality was reported. Surprisingly, no cytokine storm was detected. Out of five examined cytokines, only TNF-α was increased in a small subset of mice (Table 3) [97]. This intriguing animal model suggests that legumain may play a previously unexplored role in HLH, which may aid to identify novel cellular pathways involved in HLH pathogenesis.

Cell Types and Cytokines Involved in Secondary HLH Pathogenesis

The animal models described above have each provided valuable insights into the contribution of different cell types and cytokines to the pathogenesis of secondary HLH.

Cell Types

Historically, cell types such as T cells, NK cells, and macrophages (hemophagocytes) have been hypothesized to be involved in the development of sHLH.

$CD8^+$ T cells play a dominant part in primary HLH animal models [1, 5, 6], however this does not appear to extrapolate to all models of secondary HLH. Although $CD8^+$ T cells were markedly activated in the models of virus-associated secondary HLH [13, 25, 29, 32], they were only marginally activated in CpG-induced secondary HLH. Moreover, in vivo depletion of $CD8^+$, $CD4^+$, and $CD3^+$ T cells or general absence of T cells in *Rag*-deficient or SCID mice did not alter disease development in the secondary HLH models, indicating no major pathogenic involvement of this cell type (Table 3) [32, 36, 55, 82, 91, 94]. Thus, T cell contribution to secondary HLH pathogenesis appears to be minor and distinct from its central role in primary HLH. These data confirm recent observations in a large cohort of HLH patients, in which the degree of $CD8^+$ and $CD4^+$ T cell activation was found to be distinctive between primary and secondary HLH [98, 99]. T cell populations were significantly activated and differentiated into effector cells in patients with primary HLH, while this signature was mostly absent in patients with secondary HLH. T cell activation in virus-associated secondary HLH was situated in between the spectrum from non-virus-associated secondary HLH to primary HLH [98], in line with the animal data presented in this review.

Models of primary HLH have pointed out that NK cells are essential to limit the expansion of pathogenic, cytokine-producing $CD8^+$ T cells. This regulatory role requires functional NK cell cytotoxicity, to adequately restrain autologous activated T cells [9]. In contrast to primary HLH, not all patients with secondary HLH exhibit impaired NK cell cytotoxicity. This deficiency is reportedly acquired instead of inherited in approximately 22% of sHLH patients [100]. Of all sHLH animal models, six examined NK cell cytotoxic function and five confirmed decreases herein [13, 38, 75, 80, 97], while in three other models, the number of NK cells was significantly reduced during active disease (Table 2) [29, 32, 94]. In vivo depletion of NK cells did not alter disease development in CpG-mediated secondary HLH [55], while it slightly worsened CFA-induced MAS (Table 3) [80]. In the innate cytokine-overexpressing humanized NSG mice, disease did not develop when functional NK cells were present [91]. As a whole, according to a number of animal models, NK cells may also exert a protective function in secondary HLH, opening perspectives for future research on NK cell-stimulating treatments.

Hemophagocytosis, the process from which HLH derives its name, has been thoroughly studied in many animal models. Nonetheless, the involvement of hemophagocytes in the disease process remains to be elucidated. Reports in primary HLH assign a pathogenic role to them, mediating the development of cytopenias [88], while evidence in some models of secondary HLH points towards an anti-inflammatory function, in reaction and proportionate to the degree of hyperinflammation [48, 58, 65, 82]. In some models of secondary HLH, hemophagocytosis and development of cytopenia appear to constitute two distinct processes, as both symptoms can occur independent of one another. Mice with secondary HLH may display hemophagocytosis in the absence of anemia [56], or can present with pancytopenia in the absence of hemophagocytosis [55, 70]. Different cell types have been identified as hemophagocytes in different models, ranging from F4/80$^+$ CD68$^+$ macrophages [48, 82, 97] to monocyte-derived dendritic cells [48, 65]. Most hemophagocytes were reported to exhibit an alternatively activated or M2 phenotype, expressing scavenging receptor CD163 [48, 58, 82], similar to what has been observed in HLH patients [58, 59]. This scavenging receptor plays a role in the uptake of hemoglobin–haptoglobin complexes, upregulation of heme-oxygenase-1 and thus reduction of oxidative stress and tissue damage during inflammatory responses. For this reason, M2 macrophages are mostly considered as anti-inflammatory mediators [65]. Conversely in HIF1A-associated secondary HLH, only type 1-polarized, classically activated macrophages were detected [94]. The different animal models also reported a variety of target cells being phagocytosed. Most models describe the predominant uptake of RBCs, sometimes apoptotic RBCs or coated with anti-RBC-antibodies [25, 29, 38, 65, 82, 97], while a limited number of models also detected engulfment of nucleated cells, such as non-apoptotic T cells, lymphocytes, or granulocytes [32, 38, 48, 65, 88]. Although hemophagocytosis most likely constitutes an anti-inflammatory countermeasure [58, 65], other subpopulations of myeloid cells may still contribute to disease pathogenesis. General targeting of CD33$^+$ myeloid cells in the innate cytokine-overexpressing humanized NSG mouse model was effective to rescue the mice from fatal HLH disease, indicating involvement of innate, myeloid cells [91]. Future research in secondary HLH should include the study of specific subsets of myeloid cells to gain further insights into this matter.

Lastly, in EBV-associated sHLH, targeting B cells with rituximab, an anti-CD20-antibody, is an efficient means to eliminate infected cells and decrease the viral load [10, 11, 101, 102]. Rituximab has also been tested in the innate cytokine-overexpressing humanized NSG mouse model, without success [91], indicating that the not all subtypes of sHLH will benefit from B cell depletion. The use of rituximab is probably limited to sHLH in association with aggressive infectious agents that have a B-cell tropism, foremost in EBV-induced X-linked lymphoproliferative disorders.

Cytokines

In the complex cytokine storm associated with HLH pathology, many different chemokines and cytokines are abundantly expressed in peripheral blood, and tissues, documented in several animal models [32, 71, 72, 103]. The mixture of inflammatory cytokines is considered the core feature of HLH pathology, yet the composition differs between models (Table 3). Experiments in primary and secondary HLH models have demonstrated that broad inhibition of cytokine effector pathways, using either JAK1/2 inhibitors or inhibitors of NF-κB-signaling, is effective to treat disease [104–106].

Regarding individual cytokines, there may be many different pathways ultimately converging into the final cytokine storm of HLH. IFN-γ has long been perceived as the dominant pathogenic cytokine, particularly in primary HLH [1, 8], supporting the development and current clinical testing of IFN-γ-targeting antibodies. Nevertheless, recent advancements in the field have demonstrated that even primary HLH can develop in the absence of IFN-γ [107]. Similarly, secondary HLH most likely comprises different subsets of patients in which IFN-γ may play a major or minor pathogenic part [108]. This is reflected in diverse concentrations and roles of IFN-γ in the different animal models of secondary HLH (Table 3). Two models reported amelioration of disease upon initiation of IFN-γ blockade [55, 74], while three others could not detect a significant role for IFN-γ, aside from probable involvement in development of anemia [56, 82, 94]. Another two models even described worsening of HLH-like disease in the absence of IFN-γ [32, 80]. These data indicate that secondary HLH can indeed be driven by IFN-γ overexpression but may also develop in an independent manner.

In autoimmunity-associated and autoinflammation-associated secondary HLH, IL-6 has emerged as one of the central pathogenic cytokines, which has led to targeted therapy with tocilizumab in affected patients. Upon chronic exposure, IL-6 magnifies TLR responses, amplifies inflammation and decreases the cytotoxic potency of NK cells [70, 75]. Aside from the IL-6Tg mouse model, the cytokine was found to be elevated in several murine models of secondary HLH (Table 3) and to contribute to the pathogenesis in the cytokine-overexpressing humanized NSG mice [91]. Other cytokines that are currently targeted in patients with HLH secondary to autoimmune/autoinflammatory disease include IL-1β (by anakinra, rilonacept and canakinumab) and TNF-α (by etanercept and adalimumab), the former appearing most efficacious. Both cytokines are variably elevated in the different models of secondary HLH and their neutralization has not often been explored (Table 3). In the case of TNF-α, no therapeutic benefits were detected in CpG-induced MAS [55].

A novel therapeutic target in secondary HLH could be IL-18, considering the recent identification of IL-18 as a pathogenic cytokine in CpG-induced secondary HLH. In this model blockage of the IL-18-receptor ameliorated the aggravated HLH phenotype in IL-18BP knockout mice [57]. The importance of this cytokine is also evident from a subgroup of MAS patients carrying *NLRC4* gain-of-function mutations, resulting in constitutive inflammasome activation and extraordinarily high serum IL-18 [52]. In one report, treatment with recombinant human IL-18BP (the natural antagonist of IL-18) was able to temper a case of severe and life-threatening refractory secondary HLH [53]. In an attempt to explore the mechanisms behind *NLRC4*-associated MAS, an *NLRC4* gain-of-function point mutation was inserted into a mouse germ line, after which the resultant mouse strain displayed chronic inflammasome activation and consistently high IL-18 levels. However, no spontaneous overt inflammation or symptoms of HLH were observed [109]. Bone marrow chimeras pointed out that non-hematopoietic cells were the dominant source of excessive IL-18 production, suggesting that patients with recurrent *NLRC4*-associated MAS may not benefit from bone marrow transplantation [109].

As a countermeasure to the pro-inflammatory cytokine storm, levels of anti-inflammatory IL-10 are often elevated in patients and murine models of secondary HLH (Table 3). It is considered a protective and possibly therapeutic cytokine in both chronic and excessive CpG models, as IL-10 receptor blockade significantly worsened disease outcome [55, 65].

Conclusion

The broad set of currently available animal models of secondary HLH, as reviewed in this chapter, provides an indispensable tool for researchers to gain mechanistic insights into its complex pathogenesis, and allows for mapping of possible heterogeneity underlying the different sHLH subtypes. Indeed, the models show how multiple divergent pathogenic pathways may ultimately result in the syndrome of HLH. Translationally, the spectrum of sHLH animal models has the potential to greatly contribute to improvements in human HLH management. In the majority of sHLH models, a variety of inflammatory cytokines and chemokines are significantly increased, underscoring the prominent role of the cytokine storm in sHLH. The various implicated cytokines can be targeted upstream using JAK or NF-κB inhibitors, or can be viewed as individual therapeutic targets to halt HLH in a more precise and directed manner. In this regard, the identification of the pivotal cytokine or cytokine combination that underlies the pathology of different sHLH subtypes remains a major challenge. In clinical practice, IL-1 and IL-6 blockade appear most effective for many forms of sHLH. Similarly, sHLH patients could benefit from a variety of cell-directed therapies targeting T cells, B cells, macrophages, or stimulating regulatory NK cells. Again, it will be vital to identify the central pathogenic cell type for each of the sHLH subtypes. The development of diagnostic and prognostic tools to predict the right cytokine antagonist and/or cellular target in each individual patient will be essential to improve patient care; a future goal in which the diverse sHLH animal models will play an important translational part.

References

1. Jordan, M. B., Hildeman, D., Kappler, J., & Marrack, P. (2004). An animal model of hemophagocytic lymphohistiocytosis (HLH): CD8+ T cells and interferon gamma are essential for the disorder. *Blood, 104*, 735–743.
2. Crozat, K., Hoebe, K., Ugolini, S., Hong, N. A., Janssen, E., Rutschmann, S., et al. (2007). Jinx, an MCMV susceptibility phenotype caused by disruption of Unc13d: A mouse model of type 3 familial hemophagocytic lymphohistiocytosis. *The Journal of Experimental Medicine, 204*, 853–863.
3. Krebs, P., Crozat, K., Popkin, D., Oldstone, M. B., & Beutler, B. (2011). Disruption of MyD88 signaling suppresses hemophagocytic lymphohistiocytosis in mice. *Blood, 117*, 6582–6588.
4. Pachlopnik Schmid, J., Ho, C.-H., Diana, J., Pivert, G., Lehuen, A., Geissmann, F., et al. (2008). A Griscelli syndrome type 2 murine model of hemophagocytic lymphohistiocytosis (HLH). *European Journal of Immunology, 38*, 3219–3225.
5. Kögl, T., Müller, J., Jessen, B., Schmitt-Graeff, A., Janka, G., Ehl, S., et al. (2013). Hemophagocytic lymphohistiocytosis in syntaxin-11-deficient mice: T-cell exhaustion limits fatal disease. *Blood, 121*, 604–613.
6. Terrell, C. E., & Jordan, M. B. (2013). Perforin deficiency impairs a critical immunoregulatory loop involving murine CD8(+) T cells and dendritic cells. *Blood, 121*, 5184–5191.
7. Terrell, C. E., & Jordan, M. B. (2013). Mixed hematopoietic or T cell chimerism above a minimal threshold restores perforin-dependent immune regulation in perforin-deficient mice. *Blood, 122*, 2618–2621.
8. Pachlopnik Schmid, J., Ho, C.-H., Chrétien, F., Lefebvre, J. M., Pivert, G., Kosco-Vilbois, M., et al. (2009). Neutralization of IFNgamma defeats haemophagocytosis in LCMV-infected perforin- and Rab27a-deficient mice. *EMBO Molecular Medicine, 1*, 112–124.
9. Sepulveda, F. E., Maschalidi, S., Vosshenrich, C. A. J., Garrigue, A., Kurowska, M., Ménasche, G., et al. (2016). A novel immunoregulatory role for NK-cell cytotoxicity in protection from HLH-like immunopathology in mice. *Blood, 125*, 1427–1434.
10. Janka, G. E., & Lehmberg, K. (2014). Hemophagocytic syndromes—An update. *Blood Reviews, 28*, 135–142.
11. Henter, J. I., Horne, A., Arico, M., Egeler, R. M., Webb, D., Winiarski, J., et al. (2007). HLH-2004: Diagnostic and therapeutic guidelines for hemophagocytic lymphohistiocytosis. *Pedriatric Blood Cancer, 48*, 124–131.
12. Fardet, L., Galicier, L., & Lambotte, O. (2014). Development and validation of a score for the diagnosis of reactive hemophagocytic syndrome (HScore). *Arthritis and Rheumatism, 66*, 2613–2620.
13. Sepulveda, F. E., Garrigue, A., Maschalidi, S., Garfa-Traore, M., Ménasché, G., Fischer, A., et al. (2016). Polygenic mutations in the cytotoxicity pathway increase susceptibility to develop HLH immunopathology in mice. *Blood, 127*, 2113–2121.
14. Zhang, K., Jordan, M. B., Marsh, R. A., Johnson, J. A., Kissell, D., Meller, J., et al. (2011). Hypomorphic mutations in PRF1, MUNC13-4, and STXBP2 are associated with adult-onset familial HLH. *Blood, 118*, 5794–5798.
15. Zhang, K., Chandrakasan, S., Chapman, H., Valencia, C. A., Husami, A., Kissell, D., et al. (2014). Synergistic defects of different molecules in the cytotoxic pathway lead to clinical familial hemophagocytic lymphohistiocytosis. *Blood, 124*, 1331–1334.
16. Kaufman, K. M., Linghu, B., Szustakowski, J. D., Husami, A., Yang, F., Zhang, K., et al. (2014). Whole exome sequencing reveals overlap between macrophage activation syndrome in systemic juvenile idiopathic arthritis and familial hemophagocytic lymphohistiocytosis. *Arthritis and Rheumatism, 66*, 3486–3495.
17. Zhang, M., Bracaglia, C., Prencipe, G., Bemrich-Stolz, C. J., Beukelman, T., Dimmitt, R. A., et al. (2016). A heterozygous RAB27A mutation associated with delayed cytolytic granule polarization and hemophagocytic lymphohistiocytosis. *Journal of Immunology, 196*, 2492–2503.

18. Spessott, W. A., Sanmillan, M. L., McCormick, M. E., Patel, N., Villanueva, J., Zhang, K., et al. (2015). Hemophagocytic lymphohistiocytosis caused by dominant-negative mutations in STXBP2 that inhibit SNARE-mediated membrane fusion. *Blood, 125*, 1566–1577.
19. Brisse, E., Wouters, C. H., Andrei, G., & Matthys, P. (2017). How viruses contribute to the pathogenesis of hemophagocytic lymphohistiocytosis. *Frontiers in Immunology, 8*, 1–8.
20. Hsieh, W.-C., Chang, Y., Hsu, M.-C., Lan, B.-S., Hsiao, G.-C., Chuang, H.-C., et al. (2007). Emergence of anti-red blood cell antibodies triggers red cell phagocytosis by activated macrophages in a rabbit model of Epstein-Barr virus-associated hemophagocytic syndrome. *The American Journal of Pathology, 170*, 1629–1639.
21. Patarca, R., & Fletcher, M. A. (1995). Structure and pathophysiology of the erythrocyte membrane-associated Paul-Bunnell heterophile antibody determinant in Epstein-Barr virus-associated disease. *Critical Reviews in Oncogenesis, 6*, 305–326.
22. Kasahara, Y., Yachie, A., Takei, K., Kanegane, C., Okada, K., Ohta, K., et al. (2001). Differential cellular targets of Epstein-Barr virus (EBV) infection between acute EBV-associated hemophagocytic lymphohistiocytosis and chronic active EBV infection. *Blood, 98*, 1882–1888.
23. Kasahara, Y., & Yachie, A. (2002). Cell type specific infection of Epstein-Barr virus (EBV) in EBV-associated hemophagocytic lymphohistiocytosis and chronic active EBV infection. *Critical Reviews in Oncology/Hematology, 44*, 283–294.
24. Yang, X., Wada, T., Imadome, K.-I., Nishida, N., Mukai, T., Fujiwara, M., et al. (2012). Characterization of Epstein-Barr virus (EBV)-infected cells in EBV-associated hemophagocytic lymphohistiocytosis in two patients with X-linked lymphoproliferative syndrome type 1 and type 2. *Herpesviridae, 3*, 1.
25. Hayashi, K., Ohara, N., Teramoto, N., Onoda, S., Chen, H., Oka, T., et al. (2001). An animal model for human EBV-associated hemophagocytic syndrome. Herpesvirus Papio frequently induces fatal lymphoproliferative disorders with hemophagocytic syndrome in rabbits. *The American Journal of Pathology, 158*, 2–5.
26. Hayashi, K., Teramoto, N., & Akagi, T. (2002). Animal in vivo models of EBV-associated lymphoproliferative diseases: Special references to rabbit models. *Histology and Histopathology, 17*, 1293–1310.
27. Hayashi, K., Joko, H., Koirala, T. R., Onoda, S., Jin, Z.-S., Munemasa, M., et al. (2003). Therapeutic trials for a rabbit model of EBV-associated Hemophagocytic Syndrome (HPS): effects of vidarabine or CHOP, and development of Herpesvirus papio (HVP)-negative lymphomas surrounded by HVP-infected lymphoproliferative disease. *Histology and Histopathology, 18*, 1155–1168.
28. Hayashi, K., Jin, Z., Onoda, S., Joko, H., Teramoto, N., Ohara, N., et al. (2003). Rabbit model for human EBV-associated hemophagocytic syndrome (HPS). Sequential autopsy analysis and characterization of IL-2 dependent cell lines established from herpesvirus papio-induced fatal rabbit lymphoproliferative disease with HPS. *The American Journal of Pathology, 162*, 1721–1736.
29. Sato, K., Misawa, N., Nie, C., Satou, Y., Iwakiri, D., Matsuoka, M., et al. (2011). A novel animal model of Epstein-Barr virus-associated hemophagocytic lymphohistiocytosis in humanized mice. *Blood, 117*, 5663–5673.
30. Yajima, M., Imadome, K.-I., Nakagawa, A., Watanabe, S., Terashima, K., Nakamura, H., et al. (2008). A new humanized mouse model of Epstein-Barr virus infection that reproduces persistent infection, lymphoproliferative disorder, and cell-mediated and humoral immune responses. *The Journal of Infectious Diseases, 198*, 673–682.
31. Imadome, K., Yajima, M., Arai, A., Nakazawa, A., Kawano, F., Ichikawa, S., et al. (2011). Novel mouse xenograft models reveal a critical role of CD4+ T cells in the proliferation of EBV-infected T and NK cells. *PLoS Pathogens, 7*, e1002326.
32. Brisse, E., Imbrechts, M., Put, K., Avau, A., Mitera, T., Berghmans, N., et al. (2016). Mouse cytomegalovirus infection in BALB/c mice resembles virus-associated secondary hemophagocytic lymphohistiocytosis and shows a pathogenesis distinct from primary hemophagocytic lymphohistiocytosis. *Journal of Immunology, 196*, 3124–3134.

33. Krmpotic, A., Bubic, I., Polic, B., Lucin, P., & Jonjic, S. (2003). Pathogenesis of murine cytomegalovirus infection. *Microbes and Infection, 5*, 1263–1277.
34. Brisse, E., Imbrechts, M., Mitera, T., Vandenhaute, J., Wouters, C. H., Snoeck, R., et al. (2017). Lytic viral replication and immunopathology in a cytomegalovirus-induced mouse model of secondary hemophagocytic lymphohistiocytosis. *Virology Journal, 14*, 240.
35. Put, K., Brisse, E., Avau, A., Imbrechts, M., Mitera, T., Janssens, R., et al. (2016). IDO1 deficiency does not affect disease in mouse models of systemic juvenile idiopathic arthritis and secondary hemophagocytic lymphohistiocytosis. *PLoS One, 11*, e0150075.
36. Brisse, E., Imbrechts, M., Mitera, T., Vandenhaute, J., Berghmans, N., Boon, L., et al. (2018). Lymphocyte-independent pathways underlie the pathogenesis of murine cytomegalovirus-associated secondary haemophagocytic lymphohistiocytosis. *Clinical and Experimental Immunology, 192*(1), 104–119.
37. Zakaryan, H., Cholakyans, V., Simonyan, L., Misakyan, A., Karalova, E., Chavushyan, A., et al. (2015). A study of lymphoid organs and serum proinflammatory cytokines in pigs infected with African swine fever virus genotype II. *Archives of Virology, 160*, 1407–1414.
38. Karalyan, Z. R., Ter-Pogossyan, Z. R., Karalyan, N. Y., Semerjyan, Z. B., Tatoyan, M. R., Karapetyan, S. A., et al. (2017). Hemophagocytic lymphohistiocytosis in acute African swine fever clinic. *Veterinary Immunology and Immunopathology, 187*, 64–68.
39. Cron, R. Q., Behrens, E. M., Shakoory, B., Ramanan, A. V., & Chatham, W. W. (2015). Does viral hemorrhagic fever represent reactive hemophagocytic syndrome? *Journal of Rheumatoly, 42*, 1078–1080.
40. Clement, J., Colson, P., Saegeman, V., Lagrou, K., & Van Ranst, M. (2016). "Bedside assessment" of acute hantavirus infections and their possible classification into the spectrum of haemophagocytic syndromes. *European Journal of Clinical Microbiology & Infectious Diseases, 35*, 1101–1106.
41. Wan Jamaludin, W. F., Periyasamy, P., Wan Mat, W. R., & Abdul Wahid, S. F. (2015). Dengue infection associated hemophagocytic syndrome: Therapeutic interventions and outcome. *Journal of Clinical Virology, 69*, 91–95.
42. Ab-Rahman, H. A., Rahim, H., Abubakar, S., & Wong, P. F. (2016). Macrophage activation syndrome-associated markers in severe dengue. *International Journal of Medical Sciences, 13*, 179–186.
43. Ellis, E. M., Sharp, T. M., Pérez-Padilla, J., González, L., Poole-Smith, B. K., Lebo, E., et al. (2016). Incidence and risk factors for developing dengue-associated hemophagocytic lymphohistiocytosis in Puerto Rico, 2008–2013. *PLoS Neglected Tropical Diseases, 10*, 2008–2013.
44. Hill-Batorski, L., Halfmann, P., Marzi, A., Lopes, T. J. S., Neumann, G., Feldmann, H., et al. (2015). Loss of interleukin 1 receptor antagonist enhances susceptibility to Ebola virus infection. *The Journal of Infectious Diseases, 212*, S329–S335.
45. Rajasekaran, S., Kruse, K., Kovey, K., Davis, A. T., Hassan, N. E., Ndika, A. N., et al. (2014). Therapeutic role of anakinra, an interleukin-1 receptor antagonist, in the management of secondary hemophagocytic lymphohistiocytosis/sepsis/multiple organ dysfunction/macrophage activating syndrome in critically ill children. *Pediatric Critical Care Medicine, 15*, 401–408.
46. Miettunen, P. M., Narendran, A., Jayanthan, A., Behrens, E. M., & Cron, R. Q. (2011). Successful treatment of severe paediatric rheumatic disease-associated macrophage activation syndrome with interleukin-1 inhibition following conventional immunosuppressive therapy: Case series with 12 patients. *Rheumatology, 50*, 417–419.
47. Brown, D. E., McCoy, M. W., Pilonieta, M. C., Nix, R. N., & Detweiler, C. S. (2010). Chronic murine typhoid fever is a natural model of secondary hemophagocytic lymphohistiocytosis. *PLoS One, 5*, e9441.
48. McCoy, M. W., Moreland, S. M., & Detweiler, C. S. (2012). Hemophagocytic macrophages in murine typhoid fever have an anti-inflammatory phenotype. *Infection and Immunity, 80*, 3642–3649.
49. Yanagimachi, M., Goto, H., Miyamae, T., Kadota, K., Imagawa, T., Mori, M., et al. (2011). Association of IRF5 polymorphisms with susceptibility to hemophagocytic lymphohistiocytosis in children. *Journal of Clinical Immunology, 31*, 946–951.

50. Yanagimachi, M., Naruto, T., Miyamae, T., Hara, T., Kikuchi, M., Hara, R., et al. (2011). Association of IRF5 polymorphisms with susceptibility to macrophage activation syndrome in patients with juvenile idiopathic arthritis. *The Journal of Rheumatology, 38*, 769–774.
51. Takaoka, A., Yanai, H., Kondo, S., Duncan, G., Negishi, H., Mizutani, T., et al. (2005). Integral role of IRF-5 in the gene induction programme activated by Toll-like receptors. *Nature, 434*, 243–249.
52. Canna, S. W., de Jesus, A. A., Gouni, S., Brooks, S. R., Marrero, B., Liu, Y., et al. (2014). An activating NLRC4 inflammasome mutation causes autoinflammation with recurrent macrophage activation syndrome. *Nature Genetics, 46*, 1140–1146.
53. Canna, S. W., Girard, C., Malle, L., de Jesus, A., Romberg, N., Kelsen, J., et al. (2016). Life-threatening NLRC4-associated hyperinflammation successfully treated with Interleukin-18 inhibition. *Journal of Allergy and Clinical Immunology*. https://doi.org/10.1016/j.jaci.2016.10.022
54. Weaver, L. K., Chu, N., & Behrens, E. M. (2016). TLR9-mediated inflammation drives a Ccr2-independent peripheral monocytosis through enhanced extramedullary monocytopoiesis. *Proceedings of the National Academy of Sciences of the United States of America, 113*, 10944–10949.
55. Behrens, E. M., Canna, S. W., Slade, K., Rao, S., Kreiger, P. A., Paessler, M., et al. (2011). Repeated TLR9 stimulation results in macrophage activation syndrome-like disease in mice. *The Journal of Clinical Investigation, 121*, 2264–2277.
56. Canna, S. W., Wrobel, J., Chu, N., Kreiger, P. A., Paessler, M., & Behrens, E. M. (2013). Interferon-γ mediates anemia but is dispensable for fulminant toll-like receptor 9-induced macrophage activation syndrome and hemophagocytosis in mice. *Arthritis and Rheumatism, 65*, 1764–1775.
57. Girard-Guyonvarc'h, C., Palomo, J., Martin, P., Rodriguez, E., Troccaz, S., Palmer, G., et al. (2018). Unopposed IL-18 signaling leads to severe TLR9-induced macrophage activation syndrome in mice. *Blood, 131*(13), 1430–1441.
58. Canna, S. W., Costa-Reis, P., & Bernal, W. E. (2014). Alternative activation of laser-captured murine hemophagocytes. *Arthritis and Rheumatism, 66*(6), 1666–1671.
59. Schaer, D. J., Schleiffenbaum, B., Kurrer, M., Imhof, A., Bächli, E., Fehr, J., et al. (2005). Soluble hemoglobin-haptoglobin scavenger receptor CD163 as a lineage-specific marker in the reactive hemophagocytic syndrome. *European Journal of Haematology, 74*, 6–10.
60. Bleesing, J., Prada, A., Siegel, D. M., Villanueva, J., Olson, J., Ilowite, N. T., et al. (2007). The diagnostic significance of soluble CD163 and soluble interleukin-2 receptor alpha-chain in macrophage activation syndrome and untreated new-onset systemic juvenile idiopathic arthritis. *Arthritis and Rheumatism, 56*, 965–971.
61. Zhang, N., Zheng, Q., Xiao, L., Wang, Y., Liu, J., Liang, S., et al. (2014). Establishment of HLH-like mouse model with CPG-ODN and IFN-γ. *Zhonghua Xue Ye Xue Za Zhi, 35*, 835–839.
62. Shimazu, H., Munakata, S., Tashiro, Y., Salama, Y., Dhahri, D., Eiamboonsert, S., et al. (2017). Pharmacological targeting of plasmin prevents lethality in a murine model of macrophage activation syndrome. *Blood, 130*(1), 59–72.
63. Kaito, K., Kobayashi, M., Katayama, T., Otsubo, H., Ogasawara, Y., Sekita, T., et al. (1997). Prognostic factors of hemophagocytic syndrome in adults: Analysis of 34 cases. *European Journal of Haematology, 59*, 247–253.
64. Li, F., Yang, Y., Jin, F., Dehoedt, C., Rao, J., Zhou, Y., et al. (2015). Clinical characteristics and prognostic factors of adult hemophagocytic syndrome patients: A retrospective study of increasing awareness of a disease from a single-center in China. *Orphanet Journal of Rare Diseases, 10*(20), 1–9.
65. Ohyagi, H., Onai, N., Sato, T., Yotsumoto, S., Liu, J., Akiba, H., et al. (2013). Monocyte-derived dendritic cells perform hemophagocytosis to fine-tune excessive immune responses. *Immunity, 39*, 584–598.
66. Atteritano, M., David, A., Bagnato, G., Beninati, C., Frisina, A., Iaria, C., et al. (2012). Haemophagocytic syndrome in rheumatic patients. A systematic review. *European Review for Medical and Pharmacological Sciences, 16*, 1414–1424.

67. Sikora, K. A., & Grom, A. A. (2011). Update on the pathogenesis and treatment of systemic idiopathic arthritis. *Current Opinion in Pediatrics, 23*, 640–646.
68. De Benedetti, F., Alonzi, T., Moretta, A., Lazzaro, D., Costa, P., Poli, V., et al. (1997). Interleukin 6 causes growth impairment in transgenic mice through a decrease in insulin-like growth factor-I. A model for stunted growth in children with chronic inflammation. *The Journal of Clinical Investigation, 99*, 643–650.
69. De Benedetti, F., Rucci, N., Del Fattore, A., Peruzzi, B., Paro, R., Longo, M., et al. (2006). Impaired skeletal development in interleukin-6-transgenic mice: A model for the impact of chronic inflammation on the growing skeletal system. *Arthritis and Rheumatism, 54*, 3551–3563.
70. Strippoli, R., Carvello, F., Scianaro, R., De Pasquale, L., Vivarelli, M., Petrini, S., et al. (2012). Amplification of the response to Toll-like receptor ligands by prolonged exposure to interleukin-6 in mice: Implication for the pathogenesis of macrophage activation syndrome. *Arthritis and Rheumatism, 64*, 1680–1688.
71. Bracaglia, C., de Graaf, K., Pires Marafon, D., Guilhot, F., Ferlin, W., Prencipe, G., et al. (2016). Elevated circulating levels of interferon-gamma and interferon-gamma-induced chemokines characterize patients with macrophage activation syndrome complicating systemic juvenile idiopathic arthritis. *Annals of the Rheumatic Diseases, 76*, 166–172.
72. Prencipe, G., Caiello, I., Bracaglia, C., de Min, C., & De Benedetti, F. (2015). Neutralization of Interferon-gamma is efficacious in a mouse model of HLH secondary to chronic inflammation. *Pediatric Rheumatology, 13*, O29.
73. Bracaglia, C., Caiello, I., De Graaf, K., D'Ario, G., Guilhot, F., Ferlin, W., et al. (2015). Interferon-gamma (IFNg) in macrophage activation syndrome (MAS) associated with systemic juvenile idiopathic arthritis (SJIA): High levels in patients and role in a murine MAS model. *Pediatric Rheumatology Online Journal, 13*, O84.
74. Prencipe, G., Caiello, I., Pascarella, A., Grom, A. A., Bracaglia, C., Chatel, L., et al. (2017). Neutralization of interferon-γ reverts clinical and laboratory features in a mouse model of macrophage activation syndrome. *The Journal of Allergy and Clinical Immunology, 141*(4), 1439–1449.
75. Cifaldi, L., Prencipe, G., Caiello, I., Bracaglia, C., Locatelli, F., De Benedetti, F., et al. (2015). Inhibition of natural killer cell cytotoxicity by interleukin-6: Implications for the pathogenesis of macrophage activation syndrome. *Arthritis & Rhematology, 67*, 3037–3046.
76. Villanueva, J., Lee, S., Giannini, E. H., Graham, T. B., Passo, M. H., Filipovich, A., et al. (2005). Natural killer cell dysfunction is a distinguishing feature of systemic onset juvenile rheumatoid arthritis and macrophage activation syndrome. *Arthritis Research & Therapy, 7*, R30–R37.
77. Grom, A. A. (2004). Natural killer cell dysfunction: A common pathway in systemic-onset juvenile rheumatoid arthritis, macrophage activation syndrome, and hemophagocytic lymphohistiocytosis? *Arthritis and Rheumatism, 50*, 689–698.
78. Grom, A., Villanueva, J., Lee, S., Goldmuntz, E., Passo, M., & Filipovich, A. H. (2003). Natural killer cell dysfunction in patients with systemic-onset rheumatoid arthritis and macrophage activation syndrome. *The Journal of Pediatrics, 142*, 292–296.
79. Behrens, E. M., Beukelman, T., Paessler, M., & Cron, R. Q. (2007). Occult macrophage activation syndrome in patients with systemic juvenile idiopathic arthritis. *The Journal of Rheumatology, 34*, 1133–1138.
80. Avau, A., Mitera, T., Put, S., Put, K., Brisse, E., Filtjens, J., et al. (2014). Systemic juvenile idiopathic arthritis-like syndrome in mice following stimulation of the immune system with freund's complete adjuvant: Regulation by interferon-γ. *Arthritis & Rhematology, 66*, 1340–1351.
81. Kessel, C., Lippitz, K., Weinhage, T., Hinze, C. H., Wittkowski, H., Holzinger, D., et al. (2017). Pro-inflammatory cytokine environments can drive IL-17 overexpression by γδT cells in systemic juvenile idiopathic arthritis. *Arthritis & Rhematology, 69*, 1480–1494.

82. Milner, J. D., Orekov, T., Ward, J. M., Cheng, L., Torres-Velez, F., Junttila, I., et al. (2010). Sustained IL-4 exposure leads to a novel pathway for hemophagocytosis, inflammation, and tissue macrophage accumulation. *Blood, 116*, 2476–2483.
83. Osugi, Y., Hara, J., Tagawa, S., Takai, K., Hosoi, G., Matsuda, Y., et al. (1997). Cytokine production regulating Th1 and Th2 cytokines in hemophagocytic lymphohistiocytosis. *Blood, 89*, 4100–4103.
84. Xu, X.-J., Tang, Y.-M., Song, H., Yang, S.-L., Xu, W.-Q., Zhao, N., et al. (2012). Diagnostic accuracy of a specific cytokine pattern in hemophagocytic lymphohistiocytosis in children. *The Journal of Pediatrics, 160*, 984–90.e1.
85. Tang, Y., Xu, X., Song, H., Yang, S., Shi, S., Wei, J., et al. (2008). Early diagnostic and prognostic significance of a specific Th1/Th2 cytokine pattern in children with haemophagocytic syndrome. *British Journal of Haematology, 143*, 84–91.
86. Chen, Y., Wang, Z., Luo, Z., Zhao, N., Yang, S., & Tang, Y. (2016). Comparison of Th1/Th2 cytokine profiles between primary and secondary haemophagocytic lymphohistiocytosis. *Italian Journal of Pediatrics, 42*, 50.
87. Strippoli, R., Carvello, F., Scianaro, R., De Pasquale, L., Vivarelli, M., Petrini, S., et al. (2011). Chronic exposure to Interleukin-6 amplifies the response to Toll-like receptor ligands: Implication on the pathogenesis of macrophage activation syndrome. *Pediatric Rheumatology, 9*, P210.
88. Zoller, E. E., Lykens, J. E., Terrell, C. E., Aliberti, J., Filipovich, A. H., Henson, P. M., et al. (2011). Hemophagocytosis causes a consumptive anemia of inflammation. *The Journal of Experimental Medicine, 208*, 1203–1214.
89. Reinhardt, R. L., Liang, H.-E., Bao, K., Price, A. E., Mohrs, M., Kelly, B. L., et al. (2015). A novel model for IFN-gamma-mediated autoinflammatory syndromes. *Journal of Immunology, 194*, 2358–2368.
90. Shimozato, O., Ortaldo, J. R., Komschlies, K. L., & Young, H. A. (2002). Impaired NK cell development in an IFN-gamma transgenic mouse: Aberrantly expressed IFN-gamma enhances hematopoietic stem cell apoptosis and affects NK cell differentiation. *Journal of Immunology, 168*, 1746–1752.
91. Wunderlich, M., Stockman, C., Devarajan, M., Ravishankar, N., Sexton, C., Kumar, A. R., et al. (2016). A xenograft model of macrophage activation syndrome amenable to anti-CD33 and anti-IL-6R treatment. *JCI Insight, 1*, 1–12.
92. Bartels, K., Grenz, A., & Eltzschig, H. K. (2013). Hypoxia and inflammation are two sides of the same coin. *Proceedings of the National Academy of Sciences of the United States of America, 110*, 18351–18352.
93. Imtiyaz, H. Z., & Simon, M. C. (2010). Hypoxia-inducible factors as essential regulators of inflammation. *Current Topics in Microbiology and Immunology, 345*, 105–120.
94. Huang, R., Hayashi, Y., Yan, X., Bu, J., Wang, J., Zhang, Y., et al. (2017). HIF1A is a critical downstream mediator for hemophagocytic lymphohistiocytosis. *Haematologica, 102*, 1956–1968.
95. Zhao, L., Hua, T., Crowley, C., Ru, H., Ni, X., Shaw, N., et al. (2014). Structural analysis of asparaginyl endopeptidase reveals the activation mechanism and a reversible intermediate maturation stage. *Cell Research, 24*, 344–358.
96. Manoury, B., Mazzeo, D., Fugger, L., Viner, N., Ponsford, M., Streeter, H., et al. (2002). Destructive processing by asparagine endopeptidase limits presentation of a dominant T cell epitope in MBP. *Nature Immunology, 3*, 169–174.
97. Chan, C., Abe, M., Hashimoto, N., Hao, C., Williams, I., Liu, X., et al. (2009). Mice lacking asparaginyl endopeptidase develop disorders resembling hemophagocytic syndrome. *Proceedings of the National Academy of Sciences of the United States of America, 106*, 468–473.
98. Ammann, S., Lehmberg, K., Stadt, U., Janka, G., Rensing-ehl, A., Klemann, C., et al. (2017). Primary and secondary hemophagocytic lymphohistiocytosis have different patterns of T-cell activation, differentiation and repertoire. *European Journal of Immunology, 47*, 364–373.

99. Marsh, R. A. (2017). Diagnostic dilemmas in HLH: Can T-cell phenotyping help? *European Journal of Immunology, 47*, 240–243.
100. Bryceson, Y. T., Pende, D., Maul-Pavicic, A., Gilmour, K. C., Ufheil, H., Vraetz, T., et al. (2012). A prospective evaluation of degranulation assays in the rapid diagnosis of familial hemophagocytic syndromes. *Blood, 119*, 2754–2763.
101. Janka, G. E., & Lehmberg, K. (2013). Hemophagocytic lymphohistiocytosis: Pathogenesis and treatment. *Hematology, 2013*, 605–611.
102. Schäfer, E. J., Jung, W., & Korsten, P. (2016). Combination immunosuppressive therapy including rituximab for Epstein-Barr virus-associated hemophagocytic lymphohistiocytosis in adult-onset Still's disease. *Case Reports in Rheumatology, 2016*, 1–4.
103. Buatois, V., Chatel, L., Cons, L., Lory, S., Richard, F., Guilhot, F., et al. (2017). Use of a mouse model to identify a blood biomarker for IFNγ activity in pediatric secondary hemophagocytic lymphohistiocytosis. *Translational Research, 180*, 37–52.e2.
104. Das, R., Guan, P., Sprague, L., Verbist, K., Tedrick, P., An, Q. A., et al. (2016). Janus kinase inhibition lessens inflammation and ameliorates disease in murine models of hemophagocytic lymphohistiocytosis. *Blood, 127*, 1666–1675.
105. Hsieh, W.-C., Lan, B.-S., Chen, Y.-L., Chang, Y., Chuang, H.-C., & Su, I.-J. (2010). Efficacy of peroxisome proliferator activated receptor agonist in the treatment of virus-associated haemophagocytic syndrome in a rabbit model. *Antiviral Therapy, 15*, 71–81.
106. Maschalidi, S., Sepulveda, F. E., Garrigue, A., Fischer, A., & de Saint Basile, G. (2016). Therapeutic effect of JAK1/2 blockade on the manifestations of hemophagocytic lymphohistiocytosis in mice. *Blood, 128*, 60–72.
107. Burn, T. N., Rood, J. E., Weaver, L., Kreiger, P. A., & Behrens, E. M. (2016). Murine hemophagocytic lymphohistiocytosis can occur in the absence of interferon-gamma. *Journal of Immunology, 196*(1 Suppl), 126.5.
108. Tesi, B., Sieni, E., Neves, C., Romano, F., Cetica, V., Cordeiro, A. I., et al. (2015). Hemophagocytic lymphohistiocytosis in 2 patients with underlying IFN-γ receptor deficiency. *The Journal of Allergy and Clinical Immunology, 135*, 1638–1641.
109. Tariq, G., Weiss, E., Goodspeed, W., Goldbach-Mansky, R., & Canna, S. (2016). IL-18 elevation in macrophage activation syndrome: Human evidence for a chronic set-point and murine evidence for a non-hematopoietic source. *American College of Rheumatology*. Annual Meeting Abstract, September Abstract nr 1985.

Part VIII
Therapy of Cytokine Storm Syndromes

Etoposide Therapy of Cytokine Storm Syndromes

Jan-Inge Henter and Tatiana von Bahr Greenwood

Abbreviations

AML	Acute myeloid leukemia
CHS	Chédiak–Higashi syndrome
CMV	Cytomegalovirus
CSA	Cyclosporine A
CSS	Cytokine storm syndrome
EBV	Epstein–Barr virus
FHL	Familial hemophagocytic lymphohistiocytosis
GS2	Griscelli syndrome type 2
HIV	Human immunodeficiency virus
HLH	Hemophagocytic lymphohistiocytosis
HPS	Hermansky–Pudlak syndrome type 2
IFN	Interferon
IL	Interleukin
IVIG	Intravenous immunoglobulin
JIA	Juvenile idiopathic arthritis
MAS	Macrophage activation syndrome
MODS	Multiple organ dysfunction syndrome
MOF	Multiple organ failure
pHLH	Primary hemophagocytic lymphohistiocytosis
SAE	Severe adverse event

J.-I. Henter (✉) · T. von Bahr Greenwood
Childhood Cancer Research Unit, Department of Women's and Children's Health, Karolinska Institutet, Stockholm, Sweden

Theme of Children's and Women's Health, Karolinska University Hospital, Stockholm, Sweden
e-mail: jan-inge.henter@ki.se

R. Q. Cron, E. M. Behrens (eds.), *Cytokine Storm Syndrome*,
https://doi.org/10.1007/978-3-030-22094-5_30

SCT	Hematopoietic stem cell transplantation
sHLH	Secondary HLH
SIRS	Systemic inflammatory response syndrome
SLE	Systemic lupus erythematosus
XLP	X-linked lymphoproliferative disease

Introduction

Historical Development

Much of the early development of etoposide-based treatment for cytokine storm syndromes (CSS) started in the treatment of hemophagocytic lymphohistiocytosis (HLH), more specifically in primary (genetic) forms of HLH (pHLH), also termed familial hemophagocytic lymphohistiocytosis (FHL) [1]. The term FHL comprises autosomal recessive disorders with genetic aberrations in the genes *PRF1*, *UNC13D*, *STX11*, and *STXBP2* coding for proteins crucial for lymphocyte cytotoxicity [2–6]. Other genetic syndromes associated with pHLH are Griscelli syndrome type 2 (GS2), Chédiak–Higashi syndrome (CHS), Hermansky–Pudlak syndrome type 2 (HPS), and X-linked lymphoproliferative disease type 1 and 2 (XLP1, XLP2) [7–11]. FHL is, typically, a rapidly fatal disease characterized by a state of hyperinflammation that in 1983 was reported to have a median survival of less than 2 months if not adequately treated [1]. The extraordinary dismal outcome of FHL in combination with the febrile, hyperinflammatory condition prompted various early therapeutic efforts including corticosteroids, mostly with short effect. In addition, cytotoxic drugs such as Vinca alkaloids were tried, mostly vinblastine in combination with corticosteroids, and this was reported to induce response in a few patients [1].

The first reports on the successful use of the epipodophyllotoxin derivatives etoposide and teniposide for FHL (pHLH) came during the 1980s, when they were shown to induce prolonged resolution in combination with corticosteroids [12, 13]. A treatment protocol including etoposide in pulses in combination with corticosteroids, intrathecal methotrexate and cranial irradiation, was successful in inducing resolution and prolonged survival [14]. In 1991, a therapeutic regimen (HLH-91) that also included guidelines for maintenance therapy was published. In HLH-91, the cytotoxic treatment, including the maintenance therapy, was based on the epipodophyllotoxin derivatives etoposide and teniposide, but the treatment was administered regularly instead of in pulses and the cranial irradiation had been excluded (Fig. 1) [15]. The HLH-91 protocol induced resolution in four of five patients (80%) and it became the treatment model that the subsequent treatment protocols HLH-94 and HLH-2004 were based upon (Figs. 2 and 3).

Secondary HLH (sHLH) in the form of virus-associated hemophagocytic syndrome was initially described in 1979 in a report on 19 patients in whom active infection by herpes viruses was documented in 14 patients and by adenovirus in one. Treatment generally consisted of supportive therapy and withdrawal of immunosuppressive drugs. Consequently,13 patients recovered and it was concluded that

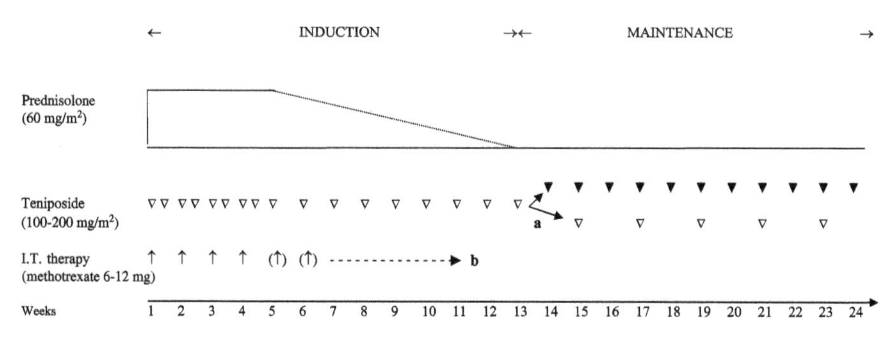

Fig. 1 Overview of the HLH-91 treatment protocol. This protocol was used in five Swedish patients, of whom four (80%) achieved remission [98]. During induction therapy, etoposide was given twice a week for the first 4 weeks, followed by weekly doses for 2 months. (**a**) Maintenance therapy was administered either as a weekly dose oral etoposide 100–300 mg/m² (solid triangle) or as a biweekly dose intravenous etoposide 100–200 mg/m² (open triangle). (**b**) Intrathecal methotrexate doses by age: <1 year = 6 mg, 1 year = 8 mg, 2 years = 10 mg, ≥3 years = 12 mg. Four additional doses of intrathecal methotrexate were given once spinal fluid cell number had reached $<10 \times 10^6$ cells/L and active cerebromeningeal symptoms no longer were present. (Figure adapted from: Henter and Elinder, Acta Paediatr Scand 1991 [98])

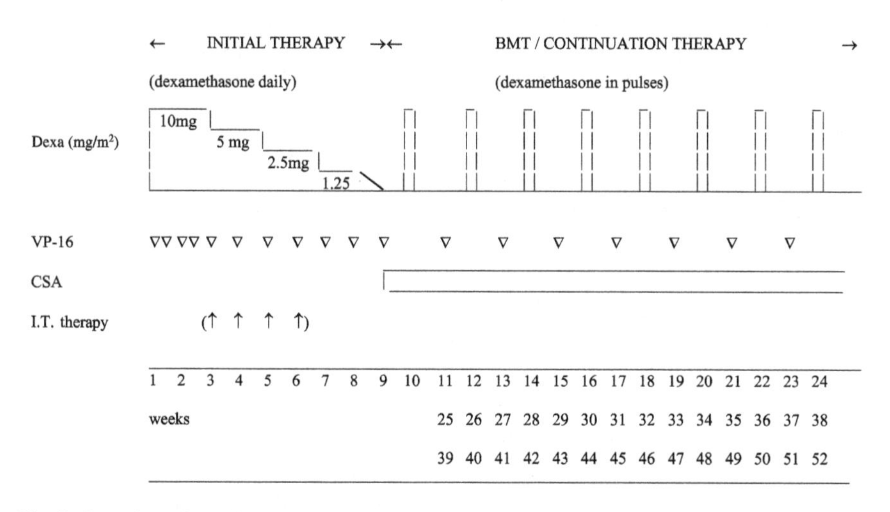

Fig. 2 Overview of the HLH-94 treatment protocol. The patients without familial or persistent disease were recommended to stop therapy after the initial therapy, and restart in case of reactivation. BMT: Patients with familial or persistent disease were recommended a hematopoietic stem cell transplant as soon as an acceptable donor was available, preferably when the disease was non-active. Dexa: Daily dexamethasone (10 mg/m²/day for 2 weeks followed by 5 mg/m²/day for 2 weeks, 2.5 mg/m²/day for 2 weeks, 1.25 mg/m²/day for 1 week, and 1 week of tapering); during continuation therapy biweekly dexa pulses (10 mg/m²/day for 3 days). VP-16: Etoposide: 150 mg/m² IV, twice weekly for 2 weeks, then weekly during initial therapy; thereafter biweekly. CSA: Daily cyclosporin A, starting week 9, aiming at blood levels of 200 μg/L (trough value). IT therapy: Intrathecal methotrexate (doses by age: <1 year 6 mg, 1–2 years 8 mg, 2–3 years 10 mg, >3 years 12 mg), in patients with progressive neurological symptoms and/or persisting abnormal cerebrospinal fluid findings. (Originally published in Blood: Trottestam et al., Blood 2011 [32]. Copyright © the American Society of Hematology. Reprinted by permission of the American Society of Hematology)

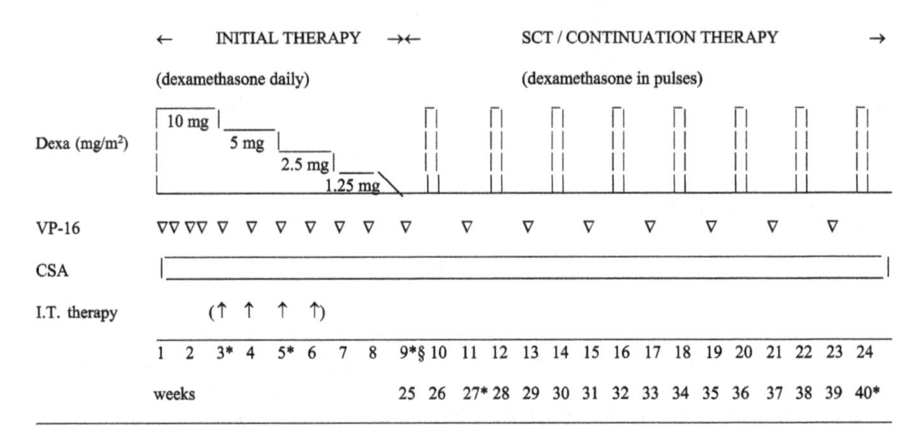

Fig. 3 Overview of the HLH-2004 treatment protocol. Both HLH-94 and HLH-2004 consist of an initial therapy of 8 weeks, with immunosuppressive and cytotoxic agents, and a continuation therapy phase thereafter, where a hematopoietic stem cell transplantation (SCT) is recommended as soon as an acceptable donor is available for patients with familial, genetic, persisting or relapsing disease. Similar to HLH-94, HLH-2004 consists of daily dexamethasone (Dexa) (10 mg/m²/day weeks 1–2,5 mg/m²/day weeks 3–4, 2.5 mg/m²/day weeks 5–6, 1.25 mg/m²/day week 7, and tapering during week 8), and etoposide (VP-16) (150 mg/m², twice weekly weeks 1–2, then once weekly) during initial therapy. In HLH-2004, cyclosporine A (CSA) (aiming at 200 μg/L trough value) is administered already up front. During continuation therapy Dexa (10 mg/m²/day for 3 days) and VP-16 (150 mg/m²) are given every second week and CSA daily. For patients with progressive neurological symptoms during the first 2 weeks, or if an abnormal cerebrospinal fluid value at onset has not improved after 2 weeks, intrathecal (I.T.) treatment with methotrexate (dose by age: <1 year 6 mg, 1–2 years 8 mg, 2–3 years 10 mg, >3 years 12 mg each dose) and prednisolone (doses by age: <1 year 4 mg, 1–2 years 6 mg, 2–3 years 8 mg, >3 years 10 mg each dose) is recommended (up to 4 doses, weeks 3, 4, 5, 6). (Originally published in: Henter et al., Pediatric Blood Cancer 2007 [23])

immunosuppressive and cytotoxic therapy may be contraindicated in the treatment of this virus-associated syndrome [16]. However, viral infections in HLH may be a concomitant finding that does not necessarily rule out the primary form (i.e., FHL), suggesting that HLH-directed therapy, such as etoposide, may be necessary even in the presence of an infection [17]. Notably, effective control of severe Epstein–Barr virus (EBV)-related HLH with etoposide-based immunochemotherapy was later reported, mostly in patients with presumed secondary HLH [18]. In recent years, increased awareness and knowledge of sHLH has shown the importance of adapting the intensity and extent of sHLH therapy to the underlying condition(s), severity of symptoms, and response to therapy.

Biological Studies

Markedly elevated levels of interferon (IFN)-gamma was first reported in HLH in 1991 and, consequently, a role for cytokine inhibitors in HLH was suggested [19]. This was the first report in English in PubMed with a title using the term "hypercytokinemia." The first report in PubMed with a title including the term "cytokine storm" came in 1993, describing a critical role for interleukin (IL)-1 in graft-versus-host disease [20].

While the choice of using etoposide in HLH was empiric in the 1980s, the remarkable and almost magic effect of etoposide was not understood until later when laboratory studies revealed that etoposide indeed can compensate for the inherited cytotoxic defect in FHL [21]. Interestingly, when lymphocytes isolated from FHL patients were subjected to etoposide in vitro, this elicited a normalized apoptotic response in FHL patient cells when compared to healthy controls [21].

Later, in a murine model of HLH, it was found that etoposide substantially can alleviate all symptoms of murine HLH. Moreover, the therapeutic mechanism of etoposide was reported to involve potent selective deletion of activated T-cells, as well as efficient suppression of inflammatory cytokine production, that is, a treatment very suitable for CSS [22].

General Principles for Treatment of HLH and CSS

Prompt Diagnosis and Early Treatment

The clinical course of HLH and CSS may be rapidly fatal or result in severe life-long CNS sequelae. The clinical presentation can be quite variable, and it is therefore important for clinicians in different medical fields to be aware of HLH for prompt evaluation, diagnosis and treatment.

In order to facilitate early diagnosis of HLH and CSS, numerous efforts to promote prompt diagnosis of HLH and CSS have been developed. In 1991, the Histiocyte Society presented the first international diagnostic guidelines for HLH [15]. In these guidelines, pHLH and sHLH were not differentiated, since at that time there were no reliable clinical, laboratory (functional or genetic tests), or histopathological methods available to distinguish these entities. The diagnostic criteria were revised for the HLH-2004 treatment protocol, in which newly available laboratory analysis of NK-cell activity and molecular diagnosis were added to help to identify pHLH [23]. Note that the HLH-2004 criteria may be difficult to achieve if NK-cell function and sCD25 are not readily available. Moreover, many secondary forms may not meet formal HLH-criteria but may still benefit from early treatment of sHLH or MAS. Thus, sometimes treatment for CSS has to be initiated before a fifth criterion is met.

Guidelines to diagnose macrophage activation syndrome (MAS) complicating systemic juvenile idiopathic arthritis (JIA) were presented 2005 [24] and preliminary guidelines for MAS in juvenile systemic lupus erythematosus (SLE) in 2009 [25]. Later, in 2016, novel classification criteria for MAS complicating systemic JIA were presented [26]. The same team have also identified the laboratory tests in which change over time is most valuable for the early diagnosis of MAS in systemic JIA [27]. In newly diagnosed patients with HLH, the underlying cause is often unknown despite these diagnostic efforts. Nevertheless, in order to suppress the

CSS that may lead to CNS-inflammation, multiorgan failure, irreversible organ damage and ultimately death, prompt initiation of treatment is essential. For patients with primary HLH, this also stabilizes the patient for a later curative hematopoietic stem cell transplant (SCT). CNS involvement, on the other hand, which causes the most common severe late effects in HLH, can affect patients with primary as well as secondary HLH [28, 29].

Control of the destructive hyperinflammation in CSS is imperative. Established principles for the treatment of HLH have been developed, aimed mainly at suppression of hyperinflammation and related hypercytokinemia, and elimination of activated immune cells and antigen presenting cells by immunosuppressive, immunomodulatory and cytotoxic drugs. However, these therapies, including etoposide, are not without risks since they may further aggravate the already prevailing HLH-associated immune dysfunction, consequently with risk of severe uncontrolled infections, and "hide" undiagnosed triggering malignancies.

Choice of Acute Therapy

Treatment of HLH often has to be initiated urgently, before the underlying cause of HLH is known, but the triggering factor(s) should be identified as soon as possible to help decide appropriate further treatment. If the underlying factor(s) to the HLH is known, then treatment should of course be adapted in line with that knowledge, as discussed later in this chapter. However, without a known underlying triggering factor, the choice of treatment is obviously much more difficult. These decisions are based on multiple factors such as the age of the patient (the older the patient is, the higher is the risk of a malignancy), previous clinical history (e.g., known rheumatic disorders), any relevant suspected current disorders, the severity of the actual clinical situation (a very severe situation may require prompt and intensive intervention), current laboratory values (e.g., blood counts in case etoposide is considered), and last but not least, the experience of the treating physician of available treatments that may be considered. Below are some general guidelines, based on available guidelines from experts in the field and personal experience of the authors.

In patients with pHLH, predominantly children, therapy in line with HLH-94/HLH-2004 can currently be regarded as standard of care [30–33]. Thus, this treatment is justified in patients with severe HLH suspected to be primary (genetic) that fulfill the diagnostic criteria for HLH, in particular if functional analyses suggest pHLH, such as decreased cytotoxic function, degranulation defects or reduced perforin expression. However, it is worth mentioning that NK-cell function may be reduced also in sHLH and this finding alone is therefore not an indication for full HLH-directed treatment and subsequent SCT. Similarly, the effects of hypomorphic and dominant negative heterozygous mutations in HLH-associated genes may present later in childhood (or adulthood) and can lead to partial NK-cell dysfunction, and the treatment of these patients must be individualized.

If functional analyses are not available or not in line with pHLH, and therefore the type of HLH is less evident, but the diagnostic criteria are nevertheless fulfilled and the patient is severely affected by the HLH, it may still be relevant to give HLH-directed therapy to reduce the CSS in patients that are unresponsive to aggressive supportive care and treatment directed at a suspected underlying trigger or empiric treatment of an unknown trigger. HLH-directed therapy with corticosteroids, such as dexamethasone (10 mg/m^2/day), and IVIG may be sufficient to control the hypercytokinemia and immune dysregulation, but in severe and specific cases, discussed in detail later, addition of etoposide may be required to reduce the CSS. In a severe EBV-driven HLH, or if pHLH is nevertheless suspected, an etoposide-based treatment as per the HLH-94/HLH-2004 protocols may be necessary. However, in less clear cut cases, etoposide may be administered to begin with only once weekly (instead of twice weekly as in the HLH-94/HLH-2004 protocols). During this acute treatment period, the diagnostic workup can continue to reach a more informed decision on the most suitable subsequent therapy. It is the view of the authors that in these severe life-threatening cases the potential side effects of a few age-adjusted doses of etoposide, in particular in patients with acceptable blood counts and good empirical supportive care, may be a reasonable risk to take, even if the full diagnostic picture is not complete, in order to halt a potentially life threatening CSS or a CSS that may result in severe permanent sequelae such as neurological complications. However, we would suggest a reduction of every etoposide dose from 150 mg/m^2, as in HLH-94/HLH-2004, to 100 mg/m^2 in adolescents and young adults and a further reduction to 50–75 mg/m^2 in middle-aged and elderly patients, since children seem to tolerate the drug better than adults [34]. Further dose adjustments may be necessary depending on the severity and the course of the CSS.

The choice of acute treatment is even more difficult when the diagnostic criteria for HLH are not fulfilled, in particular if the underlying cause of HLH is unclear, but the clinical presentation is suggestive of a flaring CSS/HLH [35]. One option may be to administer high doses of corticosteroids, such as dexamethasone as per the HLH-94/HLH-2004 protocols, since dexamethasone is a well-established anti-inflammatory drug with good penetration into the cerebrospinal fluid. Another option is the administration of high doses of methylprednisolone. In patients with suspected simultaneous CNS affection, it may be relevant to start with full HLH-94/HLH-2004 therapy even if not all diagnostic criteria are fulfilled, due to the risk of severe long-term CNS complications.

In patients with less aggressive HLH of known or unknown cause, in particular patients with presumed sHLH, corticosteroids and immunomodulatory drugs such as intravenous immunoglobulin (IVIG) or cyclosporine A (CSA) may be sufficient to dampen the CSS, but these patients must be followed carefully since a later, sometimes sudden, deterioration may require prompt intensification of therapy. In patients with MAS and other CSS, other initial treatments may be preferred such as corticosteroids combined with IL-1 or IL-6 blockade, or other immunomodulatory treatments as discussed later in this chapter and in subsequent chapters. The severity of the HLH in combination with the underlying cause of the CSS should decide the intensity and choice of initial HLH-directed therapy, not whether the disease is primary or secondary which, however, is important for guidance in the decision of appropriate continued therapy [36].

Continuation of HLH-directed therapy depends on many factors, including whether the HLH is resolved on the treatment provided and the underlying cause. While a primary HLH normally requires continuation treatment and a subsequent SCT, secondary forms of HLH typically only require treatment until the HLH is in resolution and only rarely requires SCT. In patients with pHLH, the search for a suitable SCT donor should start promptly in order to shorten the time to the curative SCT, with the aim to reduce the risks of HLH reactivation, CNS damage and other long-term complications. Patients that respond well to initial pretransplant induction therapy have a better outcome and in a review of 86 children (29 familial) that received HLH-94-therapy followed by SCT children with non-active disease at SCT ($n = 49$) had 71% (58–84%) 3-year survival post-SCT as compared to 54% (38–70%) for children with persisting HLH activity at SCT ($n = 37$) [37].

Etoposide-Based Therapy in Primary HLH

The Treatment Protocols HLH-94 and HLH-2004

Treatment Protocol Outlines

The treatment protocols HLH-94 and HLH-2004, based on the HLH-91 protocol (Fig. 1) [15], are both based on etoposide and dexamethasone, and are designed to induce and maintain a state of resolution of the disease in order to ultimately cure primary, persistent, and relapsing forms of HLH by SCT [23, 30]. Major changes in the HLH-94 and HLH-2004 protocols compared to the HLH-91 protocol included changing prednisolone to dexamethasone, reducing the induction phase from 12 to 8 weeks as well as biweekly cytotoxic therapy from 4 to 2 weeks, introducing intrathecal therapy week 3 instead of up front, changing teniposide to the resembling etoposide, introducing cyclosporin A (CSA), and abandoning oral maintenance therapy. HLH-94 and HLH-2004 include an initial intensive therapy with immunosuppressive and cytotoxic agents for 8 weeks, with the aim to induce resolution of disease activity (Figs. 2 and 3). Etoposide 150 mg/m^2 is administered twice weekly during the first 2 weeks and then weekly, in combination with dexamethasone (initially 10 mg/m^2/day for 2 weeks followed by 5 mg/m^2/day for 2 weeks, 2.5 mg/m^2/day for 2 weeks, 1.25 mg/m^2/day for 1 week, and 1 week of tapering). Corticosteroids are important anti-inflammatory drugs for HLH and dexamethasone is preferred due to better penetration into the CSF.

In HLH-94, CSA is used as an immunosuppressive drug administered in the continuation treatment starting after the first 8 weeks of induction therapy, since it lowers the activity of otherwise over-activated T-cells and their immune response. Moreover, HLH is characterized by very high IFN-gamma levels, and CSA has been reported to inhibit the production of IFN-gamma [19, 38]. HLH-94 resulted in a remarkably improved outcome with a 5-year probability of survival of 54% [32].

However, early mortality and late neurological effects remained problematic. Therefore, the treatment intensity in HLH-2004 was increased during the first 2 months of therapy by administering CSA already up front in order to increase immunosuppression without inducing additional myelotoxicity, partly because CSA also had been reported to be clinically beneficial in the initial treatment of HLH (Fig. 3) [23, 39]. In addition, intrathecal methotrexate therapy is recommended in both HLH-94 and HLH-2004 for patients with progressive neurological symptoms and/or persisting abnormal CSF findings [23, 30].

As highlighted by Jordan et al., it is important to initiate therapy promptly even in the case of unresolved infections, cytopenias, or organ dysfunction [40]. Because etoposide is cleared by both renal and hepatic pathways, we and others recommend etoposide doses adjusted according to renal function: a dose reduction of 25% if creatinine clearance is 10–50 mL/min/1.73 m^2 BW, 50% if creatinine clearance is <10 mL/min/1.73 m^2 BW, and 75% if creatinine clearance is <10 mL/min/1.73 m^2 BW combined with conjugated bilirubin >50 μmol/L (i.e., >3 mg/dL). However, dose reduction of etoposide is not recommended for isolated hyperbilirubinemia [40].

Anti-infectious and Supportive Therapy

HLH is a CSS and it is therefore important to treat infections that may trigger the hyperinflammation in HLH, in both primary and secondary forms of HLH. Moreover, since HLH patients often are critically ill it is also recommended to provide maximal supportive care, including initially appropriate broad-spectrum antibiotics, Pneumocystis jirovecii pneumonia prophylaxis with cotrimoxazole (5 mg/kg/day of trimethoprim, 2–3 times weekly), antimycotic therapy, antiviral therapy when appropriate, gastric protection at least during weeks 1–9, and IVIG (0.5 g/kg iv) once every 4 weeks (during initial and continuation therapy); in line with the HLH-2004 protocol. It is noteworthy that a large proportion of the fatalities in pHLH have been associated with invasive fungal infections, in particular invasive aspergillosis and disseminated candidiasis, which emphasizes the importance of antimycotic prophylaxis and that it also includes aspergillosis [41].

Continuation Therapy

For patients with primary, persistent or relapsing disease, a continuation therapy is recommended to keep the patient in remission until a curative allogeneic SCT can be performed. The continuation therapy consists of etoposide 150 mg/m^2 iv every second week, dexamethasone pulses 10 mg/m^2/day for 3 days every alternating second week, in combination with continuous CSA (Figs. 2 and 3). If a primary disease is unlikely and the disease resolves after 8 weeks, no continuation therapy is considered necessary unless signs of reactivation occur.

Treatment of Reactivations

Primary HLH is more or less a continuous disease characterized by frequent reactivations, particularly if the therapeutic intensity is reduced, and is not cured until a successful SCT is performed. In the HLH-2004 protocol, the suggested action if the patient develops a reactivation is to intensify therapy with etoposide and dexamethasone, and add intrathecal therapy in case of CNS-reactivation. Moreover, reactivations have to be separated from failure to respond to therapy, which is less common.

Notably, recent data from the HLH-2004 study indicate that out of 187 patients that were reported to have achieved resolution at 2 months, 20 had actually had at least one reactivation during the first 2 months [42]. Moreover, 6 patients had reactivated after stopping initial therapy and restarted initial therapy before they could finally discontinue treatment without reactivations. Despite their reactivations these 6 patients were all alive without need for a SCT at last follow-up [42].

Results of HLH-94

The overall 5-year cumulative probability of survival in the HLH-94 study was 54% (95% CI ±6%) and 50% (95% CI ±13%) for patients with verified familial disease, defined as having an affected sibling. Altogether, 71% had permanent remission or were alive until transplant, and the corresponding figure for verified familial patients was 73%. After the initial treatment of 2 months, 86% were alive (92% of the familial patients). The 5-year cumulative overall survival post-SCT was 66 ± 8%; 74 ± 16% with matched related donors, 76 ± 12% with matched unrelated donors, 61 ± 23% with mismatched unrelated donors, and 43 ± 21% with family haploidentical donors [32]. Altogether 27% of the familial patients died without receiving a SCT, this is a number that would be important to reduce in subsequent HLH studies.

Moreover, neurological late effects were reported in 19% of all patients, 31% in the familial patients, and included severe mental retardation, cranial and non-cranial nerve palsies, epilepsy, speech delay, learning difficulties, and attention-deficit/hyperactivity disorder. Non-neurological late effects were reported in 16%, 30% in the familial patients, including nutritional problems and/or growth retardation, hypertension, impaired renal function, obstructive bronchiolitis, and hearing impairment. None of these late effects were likely related to the etoposide treatment. However, the possibility that etoposide may have contributed to one patient (0.4%) developing a malignancy, acute myeloid leukemia (AML), 6 months after treatment initiation cannot be excluded. The patient was transplanted and survived [32].

Patients with Griscelli syndrome type 2 (GS2), X-linked lymphoproliferative disease (XLP), and Chédiak–Higashi syndrome (CHS) were not included in the data above. Instead, all patients in the HLH-94/HLH-2004 registries treated between 1994 and 2004 with these disorders were studied separately in order to investigate whether a treatment protocol based on etoposide and dexamethasone could be used even for these syndromes. All patients (GS2 = 5, XLP = 2, CHS = 2) responded to the therapy, and all but one (suffering from CHS) were alive with a mean follow-up of 5.6 years. All GS2 patients, one XLP patient, and one CHS patient underwent SCT. Thus, HLH-directed treatment with etoposide and dexamethasone can be effective first line treatment also in patients with GS2, XLP and CHS that have developed a hemophagocytic syndrome [43].

Results of HLH-2004 and Current Recommendations

During 2004–2011, 369 children aged <18 years fulfilled the HLH-2004 inclusion criteria (5/8 diagnostic criteria, affected siblings, and/or molecular diagnosis in FHL-causative genes) [23]. The overall 5-year cumulative probability of survival in the HLH-2004 study was 61% (95% CI = 56–67%). The 5-year survival was comparable in children that were genetically diagnosed with FHL (*n* = 158) or diagnosed by familial occurrence (*n* = 47), 61% and 58%, respectively. The mortality prior to SCT that was 27% in HLH-94 was reduced to 19% in HLH-2004 (*P* = 0.064 adjusted for age and gender). The reported neurological alterations at the time of SCT were 22% in HLH-94 and 17% in HLH-2004. Finally, 5-year probability of survival after SCT was 66% overall and 79% in children with affected siblings [42].

As in the HLH-94 study, patients with mutations in other HLH-associated genes were studied separately. In the HLH-2004 study there were 29 such patients: XLP (*n* = 16), GS2 (*n* = 11), CHS (*n* = 1), Hermansky–Pudlak syndrome (*n* = 1). Their 5-year probability of survival was 59% (95% CI 43–81%), that is, similar to the patients with verified FHL [42].

To conclude, the HLH-2004 study confirmed that a majority of patients with HLH, including patients with verified FHL and patients with mutations in other HLH-associated genes, may be rescued by an etoposide/dexamethasone-based treatment.

When comparing preliminary HLH-2004 data with those of the HLH-94 study, it could not be shown statistically that the HLH-2004 treatment was superior to that of HLH-94 with regard to overall survival, survival at 8 weeks, survival before SCT or survival post-SCT, nor with regard to the frequency of patients with neurological symptoms at 2 months after start of therapy or at transplantation. Therefore, the HLH-2004 Study Group and the HLH Steering Committee of the Histiocyte Society both recommend the HLH-94 protocol as standard of care, but with regard to diagnostics, the HLH-2004 diagnostic criteria are still recommended.

Side Effects of Etoposide in HLH Treatment

In general, side effects of etoposide include myelosuppression, hypotension (if the drug is infused too rapidly), hepatocellular damage, nausea, vomiting, fever, headache, abdominal pain, diarrhea, anorexia, alopecia, and allergic reactions.

Another complication after etoposide treatment is secondary malignancies, in particular AML, as reported by Pui et al. [44]. However, secondary malignancies are actually rare in patients treated for HLH. Imashuku et al. studied the impact of etoposide on the prognosis of 81 patients (77 of whom were children <15 years old) with EBV-HLH that received a median cumulative dose of 1500 mg/m^2 etoposide (maximum 14,550 mg/m^2), and only 1 patient, who received 3150 mg/m^2 etoposide, developed therapy-related AML (t-AML), at 31 months after diagnosis [45]. In a recent extensive literature review, altogether 13 (11 sHLH and 2 FHL) cases of t-AML in HLH patients treated with etoposide were found [46].

In HLH-2004, extra efforts were made to evaluate toxicity of the treatment. No suspected unexpected serious adverse reactions were reported. In total, 89 severe adverse events (SAE) grade III or IV were reported, of which 48 included one or more suspected causative drugs. Where one single drug was suspected 13 reported CSA, 6 reported etoposide (with the hepatobiliary system being most frequently affected, $n = 3$), and 6 reported dexamethasone (5 with cardiac hypertension). Two causative drugs were suspected in 15 reports (CSA = 14, dexamethasone = 14, etoposide = 2), with the most common SAE being cardiac hypertension associated with dexamethasone and CSA ($n = 11$). The two patients with etoposide-related SAEs and one other drug include one patient with infection (with CSA) and one with affection of the hepatobiliary system (with dexamethasone). In eight reports, all these three drugs were suspected to be involved, with infections ($n = 5$) being the most frequent SAE. Finally, in addition one patient (0.3%) developed a malignancy (AML). This patient actually received no etoposide after 3 weeks of therapy due to infections, and the patient was in complete resolution at 2 months. Two years later the patient was diagnosed with AML, underwent SCT but died. It is obviously difficult, and likely impossible, to tell whether the AML was secondary to etoposide or not [42]. Notably, the frequency of t-AML in the HLH-2004 study (0.3%) was similar to that in the HLH-94 study (0.4%) [32].

To conclude, the risk of developing t-AML in patients treated with etoposide/dexamethasone for HLH in the HLH-94 and HLH-2004 studies was 0.3–0.4%. Knowing the potential severity of HLH, with high risk for fatality or life-long CNS complications, and the prompt and remarkably positive effect of etoposide in severe forms of HLH, the authors find that the benefits of etoposide are clearly greater than the risks, in primary as well as selected secondary forms of severe HLH.

Etoposide-Based Therapy in Secondary HLH

The underlying cause of sHLH varies markedly with age, gender, and ethnicity, as detailed below, and consequently, the use of etoposide in the treatment of sHLH varies accordingly.

General Considerations (Children and Adults)

Children

In the HLH-94 study, a total of 49 (20%) children of 249 included had no active disease more than 1 year after completion of HLH therapy and were therefore assumed to have had sHLH. They were older than other children studied with a median age of 24 months (range 2–184; $P < 0.001$) and more often female (61%, $P = 0.011$). The majority ($n = 28$, 57%) were reported from Japan, and 52% had a reported history of recent infection. The most frequently confirmed viral trigger was EBV (74% of patients with a confirmed infection), but cytomegalovirus (CMV), varicella, hepatitis A, rotavirus, and enterovirus infections were also reported [32]. In a Turkish report on 23 pediatric patients with sHLH (with ≥5/8 HLH-2004 criteria), 61% (14/23) had infection-associated HLH (VZV, EBV, H1N1, and various bacterial infections) and 35% (8/23) had MAS (JIA = 6, SLE = 1, polyarteritis nodosa = 1) [47]. In the HLH-2004 study, 137 of 369 eligible patients (37%), including 32 (23%) with verified FHL, had a reported infection at diagnosis: 94/137 with EBV ($n = 75$), CMV ($n = 15$), or combined EBV–CMV ($n = 4$) infection at onset [42]. Among all children registered in HLH-2004, the most common triggering factors after infections were autoimmune and autoinflammatory diseases, followed by malignancies.

In a large Chinese study on 323 children diagnosed with HLH, infection was documented in 242 patients, including EBV in 201, CMV in 14, other viruses in 6, bacteria in 11, mycoplasma in 4, fungi in 2, *Leishmania* in 2, and rickettsia in 2, while the trigger or underlying condition had not been documented in 68 patients. Rheumatic disease was documented in 5 patients and hematological malignancies in 8 [48].

To conclude, in children with HLH an etoposide-based HLH-directed treatment will often need to be considered at an early stage, since both pHLH and EBV-HLH, frequently requiring etoposide, are common forms of HLH in children. However, in children with sHLH other than EBV-HLH, a modified approach in line with the increasing number of reports on sHLH in adults may be considered (see below).

Adults

Secondary HLH is far more common than primary HLH, particularly in adults. In 2014 Ramos-Casals et al. published an impressive review of 2197 adult patients with HLH reported in literature, giving a comprehensive overview of the most commonly reported triggers associated with HLH. In summary: infections 50.4% (viruses such as EBV, human immunodeficiency virus (HIV), herpes viruses, CMV, and others 34.7%, bacteria including tuberculosis 9.4%, parasites 2.4%, fungi 1.7%); malignancies 47.7% (hematological 44.7%, solid 1.5%); autoimmune diseases 12.6% (systemic 11.1%); others or idiopathic 12.1% (transplantation 4.3%) [49].

Even though most patients that develop sHLH are adults, there are only a few prospective studies on treatment of sHLH in adults [50, 51]. However, there are some good recent large reports focusing on HLH in adults, including reviews on treatment [49, 52–56]. It is widely agreed, as mentioned earlier, that prompt initiation of treatment directed at the underlying trigger is of vital importance, with immediate additional immunosuppressive therapy in unresponsive or severe cases of HLH. Furthermore, Schram and Berliner suggest that in circumstances with no clear precipitant, a known genetic predisposition, or EBV infection, an etoposide-based regimen should be started without delay (except for in MAS) [55]. This Boston team also emphasizes that the diagnosis and treatment of HLH often is delayed in patients who are treated for a presumed infection and they subsequently miss the window of opportunity for a timely and effective HLH-directed treatment and, therefore, they favor timely initiation of HLH-94 treatment protocol in all patients with HLH except those with underlying rheumatologic disease [57].

With regard to the prognosis of HLH in adults, a retrospective study on 68 adults with HLH reported a median overall survival of 4 months, and patients with malignancy had a worse prognosis compared to those without (median survival 2.8 months versus 10.7 months, $P = 0.007$) [58].

Importantly, it has been emphasized that it is the severity of the HLH in combination with the underlying cause of the CSS that should decide the intensity and choice of initial HLH-directed therapy, not whether the disease is primary or secondary which, however, is important for guidance in the decision of an appropriate continued therapy [36].

Infection-Associated HLH

HLH can be associated with numerous infections, particularly EBV and other herpes viruses, but also HIV, influenza, parvovirus, and hepatitis viruses, as well as bacterial, fungal, and parasitic organisms, as reviewed by Rouphael et al. and Ramos-Casals et al. [49, 59]. Below we focus on the use of etoposide in the treatment of various forms of infectious-associated HLH in all ages.

Virus Infections

The most common individual trigger of infection-associated HLH is EBV, and, furthermore, EBV-HLH may become very severe. In a pioneering publication by Imashuku et al. it was reported that 14 of 17 patients with EBV-HLH treated with immunochemotherapy (with a core combination of steroids and etoposide) maintained their complete responses. This report provided a new perspective on EBV-HLH showing that effective control of the CSS could be achieved using steroids and etoposide, with or without other immunomodulatory agents, and without SCT in a majority of EHV HLH patients [18]. Subsequently, Imashuku et al. reported that the probability of long-term survival was significantly higher when etoposide treatment was begun less than 4 weeks from diagnosis as compared to later or not at all (90.2% versus 56.5%; $P < 0.01$). Hence, the authors concluded that early administration of etoposide, preferably with CSA, was the treatment of choice for patients with EBV-HLH [60]. Moreover, the efficacy of early etoposide in the treatment of EBV-HLH was later confirmed in adults as well as in children [61]. Dr. Imashuku later concluded, "When considering the treatment of EBV-HLH, the most important factor is the finding that a survival benefit is obtained when etoposide-containing therapy is initiated within 4 weeks of diagnosis. This indicates that there may be a window for observation or conservative corticosteroid/cyclosporine A or intravenous immunoglobulin (IVIG) treatment; however, once the disease is defined as "high risk" and/or refractory to such therapy, prompt introduction of etoposide (ideally within 4 weeks) is recommended. In deciding whether the disease is "high-risk," evaluation of clinical staging, EBV genome copy numbers in the serum, cellular EBV tropism, chromosome analysis, and screening for hereditary immunodeficient diseases such as familial HLH, are required" [62]. As a complement, the anti-CD20 antibody rituximab that depletes B cells has been reported to have a therapeutic value [63]. Finally, it is important to identify patients affected by chronic active EBV infection (CAEBV), which may develop to malignant lymphoma, since treatment with SCT has been reported to have a favorable outcome for these patients [64]. CAEBV is characterized by persistent, life-threatening, infectious mononucleosis-like symptoms with high EBV-DNA load in the peripheral blood and systemic clonal expansion of EBV infected T cells or natural killer cells.

Etoposide has been reported beneficial in HLH associated with numerous other viral infections, but for these triggers statistical evidence is not available and the efficacy is therefore more difficult to evaluate. With regard to CMV, of the 137 eligible patients that had a reported infection at diagnosis in the HLH-2004 study 15/137 had CMV and 4/137 combined CMV–EBV infection; whereof less than half had verified pHLH [42]. In a study of 58 HIV-infected adults with HLH, many of whom also had other underlying diagnoses, 24 (41%) were reported to have received etoposide alone or in combination with corticosteroids [65]. The combination of etoposide and corticosteroids has also been reported to be of value in severe influenza A/H1N1 [66], and while others confirmed virus-associated HLH to be a major contributor to death in patients with 2009 influenza A (H1N1) infection, they could not confirm the efficacy of etoposide in reducing mortality [67]. With regard to other viral infections that can be associated with HLH such as other herpes viruses, viral hepatitis and human parvovirus, the knowledge on the use of etoposide is limited likely due to their rare occurrence.

To conclude, it is the view of the authors that in a patient with severe CSS triggered by a viral infection, in particular EBV, and that fulfills ≥5/8 diagnostic criteria for HLH, that is refractory to non-immunosuppressive disease-directed therapy and supportive care, it may be justified to initiate a combination of etoposide and corticosteroid treatment adjusted to age and organ dysfunction provided that no contraindications are present. For patients with presumed sHLH who may benefit from etoposide, we suggest individualized therapy, such as (1) less frequent etoposide

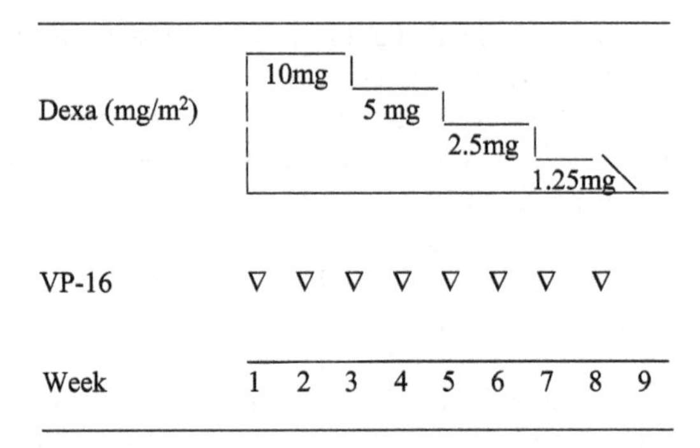

Fig. 4 Suggested therapy for patients with viral infections and severe secondary HLH. (Note: severe EBV-HLH may need more intensive therapy). Etoposide is typically administered once weekly. Allow weekly decisions on whether to continue etoposide treatment or not; duration of the treatment should be adjusted to the clinical and laboratory response of the patient. Age-adjusted doses of etoposide are recommended, for example, 100–150 mg/m^2 in children <15 years, 100 mg/m^2 in adolescents and young adults and 50–75 mg/m^2 in middle-aged and elderly patients. Suggested supportive therapy is cotrimoxazole, antimycotic therapy, gastric protection, and if applicable, antiviral therapy and IVIG; in line with the HLH-2004 protocol. (Originally published in: Henter et al., Lancet 2006 [34])

treatments (typically once weekly), (2) weekly decisions on continuation of etoposide treatment, and (3) lower etoposide dose than in FHL, in particular in adolescents and adults (50–100 mg/m^2), according to age, severity of symptoms and response to therapy (Fig. 4) [34]. However, more intensive therapy such as in the HLH-94/HLH-2004 protocols may be required, in particular in patients with severe EBV-HLH.

Infections by Bacteria, Parasites and Fungi

It is increasingly recognized that HLH may be associated with severe bacterial infections such as in sepsis, but the differential diagnosis of HLH and sepsis remains challenging [68]. The diagnostic dilemma is particularly common and particularly difficult in those critically ill, that often have sepsis and multiple organ failure (MOF)/multiple organ dysfunction syndrome (MODS), and we will therefore first focus on infections in critically ill patients. The eight diagnostic criteria of HLH are commonly found to varying extent in patients with sepsis, systemic inflammatory response syndrome (SIRS) and MODS [47, 69], where HLH is at the extreme end of this ladder of (hyper)inflammation before death. Notably, patients treated in the ICU with a clinical presentation of sepsis/septic shock but with no identified infectious pathogen, and who are unresponsive to sepsis-directed therapy, may have an undiagnosed HLH. Hence, diagnosis and appropriate treatment, that may include etoposide, may be delayed, with the risk of missing the time when HLH-directed treatment may be most effective. In this context, the findings by Tang et al. of specific profiles of hypercytokinemia to help distinguish between HLH, sepsis and other infections are promising [70].

With regard to treatment of HLH in critically ill patients, there is currently no consensus. In a large retrospective study, 5027 patients admitted to a medical ICU at a teaching hospital were screened, of which 72 patients fulfilled HLH-2004 criteria, whereof 56 patients had complete follow-up and no missing data; among these etoposide was used in 45 (80%), corticosteroids in 31 (55%) and IVIG in 3 patients (5%) [52]. With etoposide-based therapy, treatment of precipitating factors and full-code life-supporting treatment, hospital mortality was 52%. The authors state that etoposide and steroids comprise an immunochemotherapy that benefits patients with HLH, that acts rapidly (within 24–48 h), and that its efficacy far outweighs the risk of secondary leukemia and transient worsening of the neutropenia, and furthermore, emphasize the importance of aggressive supportive care combined with treatment of the precipitating factor [52]. Of interest, precipitating factors were dominated by two-thirds malignancies, followed by 10 patients with viral trigger and 13 nonviral infections; 6 tuberculosis, 5 blood stream infections, and 4 toxoplasmosis (1 with a concomitant lymphoma). Another approach, reported in children with sepsis/MODS/MAS/HLH but without risk factors for HLH is treating the hyperinflammation with methylprednisone, IVIG, and anakinra, and, if MODS is also present, daily plasma exchange [71]. A similar approach with high-dose IVIG has been studied in adults [72, 73]. The combination of etoposide and corticosteroids has also been reported to be successful in an extremely premature girl, born in the 24th gestational week (BW 732 g), with a severe secondary HLH due to a *Serratia marcescens* septicemia [74]. In summary, appropriate antibiotic treatment and supportive care are the evident backbone of therapy in these patients, and addition of corticosteroids and IVIG may be beneficial when there is hyperinflammation with progressive organ dysfunction, while etoposide may be carefully considered in cases of refractory and severe CSS. We conclude that treatment of HLH in the critically ill, including patients with sepsis, is multifaceted and remains a challenge worthy of further studies.

The individually most common bacteria reported to be associated with HLH is *Mycobacterium tuberculosis* [49]. In a review of the English literature in 2006 on all patients reported with tuberculosis-associated HLH, the mortality was approximately 50%. Twenty-nine of the 37 cases received treatment, either antituberculosis medication alone ($n = 9$), or a combination of antituberculosis medication and immunomodulatory treatment ($n = 20$). In most patients, immunomodulatory treatment consisted of high-dose steroids, but a few patients were treated with splenectomy ($n = 2$), plasma exchange ($n = 2$), IVIG, and epipodophyllotoxin. However, the effect of epipodophyllotoxin is not specified. Twelve (60%) of the 20 patients who received a combination of immunomodulatory and antituberculosis treatment survived, and 7 of the 9 (77%) patients who received only antituberculosis treatment recovered. In most cases, failure of therapy was attributed to initiation of therapy late in the course of the illness [75]. We do not support for the use of etoposide in tuberculosis-associated HLH.

The most common parasite associated with HLH is *Leishmania*, for which liposomal amphotericin B is recommended and effective, while etoposide is not recommended [76]. *Histoplasma* is the most frequently reported fungus to trigger HLH, for which amphotericin B also is effective. Etoposide has been used in histoplasmosis-associated HLH prior to obtaining the diagnosis of the infectious trigger, but the value of etoposide in this condition has not been established [77].

Malignancy-Associated HLH

Malignancy-associated HLH (mal-HLH) comes in two forms, "malignancy-triggered HLH" and "HLH during chemotherapy" [78]. Malignancy-triggered HLH may be a presenting feature of the malignancy at diagnosis or at relapse, while HLH during chemotherapy typically occurs during, or shortly after, chemotherapeutic treatment for a malignancy while patients commonly are immunosuppressed and in remission from the cancer [78].

Notably, an unidentified neoplasm can be the triggering factor in any patient with HLH, the most common being occult lymphomas, and the likelihood of an underlying malignancy increases with age [78]. In general, mal-HLH has the worst prognosis of all subgroups of sHLH [79–81]. Moreover, with regard to management of mal-HLH, there are few prospective studies to support evidence-based guidelines as reviewed by Wang and Wang [82].

A retrospective French study of 162 adult patients with sHLH, predominantly malignancy-triggered, found that use of etoposide as a first-line treatment tended to be associated with a better outcome ($P = 0.079$) as compared to treatment directed at the underlying pathology or treatment with corticosteroids only [80]. Furthermore, etoposide and methylprednisolone treatment combined with one dose of liposomal doxorubicin has been reported as an effective salvage therapy for adult refractory HLH [50].

In conclusion, specifically with regard to "malignancy-triggered HLH," it is reported to be uncertain if a malignancy-directed or an HLH-directed regimen should be used primarily. However, the authors and others suggest that a regimen including etoposide and corticosteroids likely is valuable prior to, or concomitant with, start of tumor-specific treatment, at least in patients with florid HLH and patients with CNS affection, while the use of dexamethasone without etoposide may be considered in more moderate HLH. Accordingly, it is reasonable to add etoposide to the CHOP-protocol (i.e., CHOEP) for lymphoma treatment. Furthermore, in eligible patients in remission, a high-dose consolidation chemotherapy with autologous stem cell transplantation has been reported as standard of care, while a decision for allogeneic SCT requires careful individual assessment [56].

In "HLH during chemotherapy," delay or interruption of ongoing cancer therapy should be considered, while rigorous treatment of triggering infections is imperative. The addition of rituximab is suggested in highly replicative EBV infection [63]. However, in the event of a relapse or if there is active HLH, then HLH-directed therapy should be considered, and, furthermore, CNS involvement (clinically or in the CSF) suggests prompt intense therapy [63, 78]. It has been reported that HLH complicating cancer treatment probably is under-recognized, and that these HLH patients suffer from prolonged neutropenia, pulmonary and neurologic symptoms, liver abnormalities, and lower platelet counts. Patients with HLH complicating cancer treatment may benefit from anti-inflammatory treatment with corticosteroids ± IVIG, as well as adapted antimicrobial treatment to treat an infectious trigger, while it is suggested that etoposide should be used sparingly. In the monitoring to detect relapsing malignant disease as a potential alternate trigger for HLH, serial testing for sCD25 and bone marrow assessment for hemophagocytosis has been reported to be helpful [56].

Macrophage Activation Syndrome/Rheuma-Associated HLH

Systemic Idiopathic Arthritis and Systemic Lupus Erythematosus

Macrophage activation syndrome, also referred to as rheuma-associated HLH, is the second most common form of sHLH in children and the third most common in adults [47, 49]. First line treatment is often corticosteroids in high doses and CSA in children [83, 84] as well as in adults [55, 56].

Second line treatment for MAS is less well established. Biologic agents include IL-1 inhibition (anakinra) for which there are favorable reports published [85, 86]. IL-6 inhibition with tocilizumab has also been reported to be efficient in severe, persistent systemic JIA [87]. Other treatments for MAS include IVIG, cyclophosphamide, plasma exchange, and etoposide [83]. Treatment of MAS in different conditions is discussed in detail in other chapters, but etoposide-based treatment of MAS is also discussed in this chapter.

The exact role of etoposide in the treatment of MAS remains to be determined. In a Chinese study of 32 pediatric and adult patients with SLE-associated MAS all patients received corticosteroids, while CSA, cyclophosphamide and etoposide were the three most commonly added immunosuppressants, and for 47% of the patients IVIG was also administered [88]. Notably, in a large study on 103 episodes of MAS in 89 adult patients with SLE, it was concluded that etoposide and cyclophosphamide-based regimens had the best efficacy as second line therapy [89]. In line, in a report of four children with MAS, three with underlying JIA and one with SLE, control of clinical inflammation was achieved in all patients when etoposide treatment was added to the conventional anti-inflammatory treatment [90].

In a large multicenter study on 362 patients, Minoia et al. noted that patients who received etoposide had more severe disease manifestations and were more likely to be admitted to the ICU than those who did not [83]. With the aim to record potential predictors of etoposide administration, Horne et al. studied the 40 of these 362 (11%) patients that were treated with etoposide. Factors significantly associated with etoposide administration in multivariate analysis included multiorgan failure

(OR 7.9), platelet count ≤ 132 × 10^9/L (OR 5.8), triglycerides > 270.8 mg/dL (OR 3.7), aspartate aminotransferase > 389 units/L (OR 3.7), and fibrinogen ≤ 1.53 g/L (OR 2.9). In this study patients treated with etoposide were more severely ill than patients who did not receive this medication but, importantly, the mortality rate did not differ between the two treatment groups suggesting that etoposide may be part of more aggressive therapeutic interventions for severely ill children with MAS [91]. Thus, we conclude that it is possible that the full potential of etoposide in patients with MAS is not yet fully recognized. One possibility could be to take advantage of etoposide as third line therapy in MAS, after corticosteroids and CSA as first line therapy and an attempt with biologic agents as second. We strongly recommend that in case of insufficient response etoposide be considered promptly to take full advantage of its potential, with a dose of 50–100 mg/m^2 once weekly dependent on age and organ function, in combination with corticosteroids. Moreover, in MAS patients with CNS affection that may result in rapid CNS destruction etoposide should be considered early.

Other Forms of Autoimmune Diseases

The use of etoposide has been reported also in other forms of autoimmune diseases, including dermatomyositis [92] and polyarteritis nodosa [93], but here the evidence-based knowledge on the value of etoposide is limited, and it should be used with caution and only in selected cases. Similarly, we advise caution in the use of etoposide in Kawasaki-associated HLH, both with regard to indication and dosage [94].

Other Conditions Associated with HLH

Transplant-Associated HLH

In the review by Ramos-Casals et al., transplantation was associated with 95 of the altogether 2197 reviewed cases of HLH in adults, of which 53 and 29 patients were associated with kidney and hematological transplants, respectively [49]. In a small Japanese study, low-dose etoposide (50 mg/m^2 per dose) was effective in the treatment of early-onset HLH following allogeneic SCT, with five out of five treated patients responding to one or two doses [95]. In another Japanese study of 37 children with post-HSCT HLH [26 classified as early-onset (onset <30 days after SCT) and 11 as late-onset (onset >30 days after SCT)], the authors conclude that early-onset post-SCT HLH is a specific entity of HLH and that appropriate diagnosis and prompt management needs to be established. However, no statistically significant advantage was observed with any treatment, including etoposide, and therefore, appropriate treatment for post-SCT HLH deserves further investigation [96]. In a prospective study on children and adults, 6/68 (9%) patients developed HLH after allogeneic SCT. Four patients received treatment (IVIG or corticosteroids), and altogether only one of six survived [51].

Other Conditions

Etoposide has also been used in severe forms of granulomatous diseases, such as Crohn's disease and systemic Langerhans cell histiocytosis (LCH). For Crohn-associated HLH, knowledge of the value of etoposide is limited, and it should be used with caution both in regard to indication and dosage. In LCH-associated HLH, a recognized form of systemic LCH, treatment with cladribine and cytarabine is currently recommended [97]. Finally, there are likely other conditions associated with HLH not reported here where etoposide can and has been used, but due to their rarity there is limited evidence of the value of etoposide-based treatment.

Concluding Remarks and Future Directions

Etoposide in Primary HLH

In the early 1980s, long-term survival in primary HLH was almost zero [1]. Fortunately, thanks to collaborations worldwide and new treatment protocols including etoposide, corticosteroids, and SCT, many patients are now long-term survivors. The etoposide/dexamethasone-based HLH-94/HLH-2004 protocols have dramatically improved survival in this severe condition, which represents one of the remarkable developments in pediatric immunology and hematology over the last two decades. Moreover, the positive treatment outcomes have also increased the awareness of pHLH as well as other forms of CSS. Furthermore, the knowledge gained in pediatrics has subsequently greatly influenced the awareness and management of HLH in adults across disciplines.

The HLH-94/HLH-2004 protocols, with currently around 60% 5-year survival in HLH-2004, can likely be improved further by using accumulated clinical knowledge, by improving CNS-HLH therapy and monitoring, and by adapting to modern diagnostics, data on risk factors and to the increasing number of patients diagnosed with secondary HLH, as well as by revising SCT guidelines and salvage recommendations [42].

Data from the HLH-2004 study suggest that the treatment may benefit from being more individualized, such as stratifying treatment by risk factors, etc. Notably, the HLH-2004 study also indicates that the treatment should not be reduced and tapered during the end of the first 8-week period as in the HLH-2004 protocol but instead a moderate treatment until SCT should be maintained. Moreover, in the HLH-2004 study there was an overrepresentation of deaths after the first 100 days in patients whose HLH initially resolved but subsequently had a reactivation, stressing the importance of acute/subacute SCT for these patients [42]. Notably, if using the HLH-2004 protocol, it is suggested to initiate CSA not earlier than 2 weeks after start of therapy instead of up front. However, the HLH-94 protocol is the currently recommended standard of care.

Etoposide in Secondary HLH

Over the last decade, the interest in sHLH has increased remarkably. Nevertheless, much remains to be learned about sHLH, including what the best treatment is in various conditions and the role of etoposide. It has become obvious that adults with HLH should be treated differently than children but also that the many different forms of HLH may need different therapeutic approaches. This situation, in particular, if the underlying condition is unknown, makes the choice of initial therapy even more challenging.

Some of many remaining questions for future research include which the optimal treatments are for different forms of sHLH, in children as well as adults, and what the role of etoposide is in the treatment of sHLH, how it should be dosed in different conditions, and how frequently and for how long it should be administered. Furthermore, it would be valuable to have markers which reliably monitor HLH activity to assist in these therapeutic decisions.

References

1. Janka, G. E. (1983). Familial hemophagocytic lymphohistiocytosis. *European Journal of Pediatrics, 140*, 221–230.
2. Stepp, S. E., Dufourcq-Lagelouse, R., Le Deist, F., Bhawan, S., Certain, S., Mathew, P. A., et al. (1999). Perforin gene defects in familial hemophagocytic lymphohistiocytosis. *Science, 286*, 1957–1959.
3. Feldmann, J., Callebaut, I., Raposo, G., Certain, S., Bacq, D., Dumont, C., et al. (2003). Munc13-4 is essential for cytolytic granules fusion and is mutated in a form of familial hemophagocytic lymphohistiocytosis (FHL3). *Cell, 115*, 461–473.
4. zur Stadt, U., Rohr, J., Seifert, W., Koch, F., Grieve, S., Pagel, J., et al. (2009). Familial hemophagocytic lymphohistiocytosis type 5 (FHL-5) is caused by mutations in Munc18-2 and impaired binding to syntaxin 11. *American Journal of Human Genetics, 85*, 482–492.
5. zur Stadt, U., Schmidt, S., Kasper, B., Beutel, K., Diler, A. S., Henter, J. I., et al. (2005). Linkage of familial hemophagocytic lymphohistiocytosis (FHL) type-4 to chromosome 6q24 and identification of mutations in syntaxin 11. *Human Molecular Genetics, 14*, 827–834.
6. Cote, M., Menager, M. M., Burgess, A., Mahlaoui, N., Picard, C., Schaffner, C., et al. (2009). Munc18-2 deficiency causes familial hemophagocytic lymphohistiocytosis type 5 and impairs cytotoxic granule exocytosis in patient NK cells. *The Journal of Clinical Investigation, 119*, 3765–3773.
7. Menasche, G., Pastural, E., Feldmann, J., Certain, S., Ersoy, F., Dupuis, S., et al. (2000). Mutations in RAB27A cause Griscelli syndrome associated with haemophagocytic syndrome. *Nature Genetics, 25*, 173–176.
8. Barrat, F. J., Auloge, L., Pastural, E., Lagelouse, R. D., Vilmer, E., Cant, A. J., et al. (1996). Genetic and physical mapping of the Chediak-Higashi syndrome on chromosome 1q42-43. *American Journal of Human Genetics, 59*, 625–632.
9. Shotelersuk, V., Dell'Angelica, E. C., Hartnell, L., Bonifacino, J. S., & Gahl, W. A. (2000). A new variant of Hermansky-Pudlak syndrome due to mutations in a gene responsible for vesicle formation. *The American Journal of Medicine, 108*, 423–427.
10. Coffey, A. J., Brooksbank, R. A., Brandau, O., Oohashi, T., Howell, G. R., Bye, J. M., et al. (1998). Host response to EBV infection in X-linked lymphoproliferative disease results from mutations in an SH2-domain encoding gene. *Nature Genetics, 20*, 129–135.

11. Rigaud, S., Fondaneche, M. C., Lambert, N., Pasquier, B., Mateo, V., Soulas, P., et al. (2006). XIAP deficiency in humans causes an X-linked lymphoproliferative syndrome. *Nature, 444*, 110–114.
12. Henter, J. I., Elinder, G., Finkel, Y., & Soder, O. (1986). Successful induction with chemotherapy including teniposide in familial erythrophagocytic lymphohistiocytosis. *Lancet, 2*, 1402.
13. Ambruso, D. R., Hays, T., Zwartjes, W. J., Tubergen, D. G., & Favara, B. E. (1980). Successful treatment of lymphohistiocytic reticulosis with phagocytosis with epipodophyllotoxin VP 16-213. *Cancer, 45*, 2516–2520.
14. Fischer, A., Virelizier, J. L., Arenzana-Seisdedos, F., Perez, N., Nezelof, C., & Griscelli, C. (1985). Treatment of four patients with erythrophagocytic lymphohistiocytosis by a combination of epipodophyllotoxin, steroids, intrathecal methotrexate, and cranial irradiation. *Pediatrics, 76*, 263–268.
15. Henter, J. I., Elinder, G., & Ost, A. (1991). Diagnostic guidelines for hemophagocytic lymphohistiocytosis. The FHL Study Group of the Histiocyte Society. *Seminars in Oncology, 18*, 29–33.
16. Risdall, R. J., McKenna, R. W., Nesbit, M. E., Krivit, W., Balfour Jr., H. H., Simmons, R. L., et al. (1979). Virus-associated hemophagocytic syndrome: A benign histiocytic proliferation distinct from malignant histiocytosis. *Cancer, 44*, 993–1002.
17. Henter, J. I., Elinder, G., Lubeck, P. O., & Ost, A. (1993). Myelodysplastic syndrome following epipodophyllotoxin therapy in familial hemophagocytic lymphohistiocytosis. *Pediatric Hematology and Oncology, 10*, 163–168.
18. Imashuku, S., Hibi, S., Ohara, T., Iwai, A., Sako, M., Kato, M., et al. (1999). Effective control of Epstein-Barr virus-related hemophagocytic lymphohistiocytosis with immunochemotherapy. *Blood, 93*, 1869–1874.
19. Henter, J. I., Elinder, G., Soder, O., Hansson, M., Andersson, B., & Andersson, U. (1991). Hypercytokinemia in familial hemophagocytic lymphohistiocytosis. *Blood, 78*, 2918–2922.
20. Ferrara, J. L., Abhyankar, S., & Gilliland, D. G. (1993). Cytokine storm of graft-versus-host disease: A critical effector role for interleukin-1. *Transplantation Proceedings, 25*, 1216–1217.
21. Fadeel, B., Orrenius, S., & Henter, J. I. (1999). Induction of apoptosis and caspase activation in cells obtained from familial haemophagocytic lymphohistiocytosis patients. *British Journal of Haematology, 106*, 406–415.
22. Johnson, T. S., Terrell, C. E., Millen, S. H., Katz, J. D., Hildeman, D. A., & Jordan, M. B. (2014). Etoposide selectively ablates activated T cells to control the immunoregulatory disorder hemophagocytic lymphohistiocytosis. *Journal of Immunology, 192*, 84–91.
23. Henter, J. I., Horne, A., Arico, M., Egeler, R. M., Filipovich, A. H., Imashuku, S., et al. (2007). HLH-2004: Diagnostic and therapeutic guidelines for hemophagocytic lymphohistiocytosis. *Pediatric Blood and Cancer, 48*, 124–131.
24. Ravelli, A., Magni-Manzoni, S., Pistorio, A., Besana, C., Foti, T., Ruperto, N., et al. (2005). Preliminary diagnostic guidelines for macrophage activation syndrome complicating systemic juvenile idiopathic arthritis. *The Journal of Pediatrics, 146*, 598–604.
25. Parodi, A., Davi, S., Pringe, A. B., Pistorio, A., Ruperto, N., Magni-Manzoni, S., et al. (2009). Macrophage activation syndrome in juvenile systemic lupus erythematosus: A multinational multicenter study of thirty-eight patients. *Arthritis and Rheumatism, 60*, 3388–3399.
26. Ravelli, A., Minoia, F., Davi, S., Horne, A., Bovis, F., Pistorio, A., et al. (2016). 2016 Classification criteria for macrophage activation syndrome complicating systemic juvenile idiopathic arthritis: A European League Against Rheumatism/American College of Rheumatology/Paediatric Rheumatology International Trials Organisation Collaborative Initiative. *Annals of the Rheumatic Diseases, 75*, 481–489.
27. Ravelli, A., Minoia, F., Davi, S., Horne, A., Bovis, F., Pistorio, A., et al. (2016). Expert consensus on dynamics of laboratory tests for diagnosis of macrophage activation syndrome complicating systemic juvenile idiopathic arthritis. *RMD Open, 2*, e000161.
28. Horne, A., Ramme, K. G., Rudd, E., Zheng, C., Wali, Y., Al-Lamki, Z., et al. (2008). Characterization of PRF1, STX11 and UNC13D genotype-phenotype correlations in familial hemophagocytic lymphohistiocytosis. *British Journal of Haematology, 143*, 75–83.

29. Yang, S., Zhang, L., Jia, C., Ma, H., Henter, J. I., & Shen, K. (2010). Frequency and development of CNS involvement in Chinese children with hemophagocytic lymphohistiocytosis. *Pediatric Blood and Cancer, 54*, 408–415.
30. Henter, J. I., Arico, M., Egeler, R. M., Elinder, G., Favara, B. E., Filipovich, A. H., et al. (1997). HLH-94: A treatment protocol for hemophagocytic lymphohistiocytosis. HLH study Group of the Histiocyte Society. *Medical and Pediatric Oncology, 28*, 342–347.
31. Henter, J. I., Samuelsson-Horne, A., Arico, M., Egeler, R. M., Elinder, G., Filipovich, A. H., et al. (2002). Treatment of hemophagocytic lymphohistiocytosis with HLH-94 immunochemotherapy and bone marrow transplantation. *Blood, 100*, 2367–2373.
32. Trottestam, H., Horne, A., Arico, M., Egeler, R. M., Filipovich, A. H., Gadner, H., et al. (2011). Chemoimmunotherapy for hemophagocytic lymphohistiocytosis: Long-term results of the HLH-94 treatment protocol. *Blood, 118*, 4577–4584.
33. Janka, G. E., & Lehmberg, K. (2014). Hemophagocytic syndromes—An update. *Blood Reviews, 28*, 135–142.
34. Henter, J. I., Chow, C. B., Leung, C. W., & Lau, Y. L. (2006). Cytotoxic therapy for severe avian influenza A (H5N1) infection. *Lancet, 367*, 870–873.
35. Janka, G. E. (2007). Familial and acquired hemophagocytic lymphohistiocytosis. *European Journal of Pediatrics, 166*, 95–109.
36. Janka, G. E., & Lehmberg, K. (2013). Hemophagocytic lymphohistiocytosis: Pathogenesis and treatment. *Hematology. American Society of Hematology. Education Program, 2013*, 605–611.
37. Horne, A., Janka, G., Maarten Egeler, R., Gadner, H., Imashuku, S., Ladisch, S., et al. (2005). Haematopoietic stem cell transplantation in haemophagocytic lymphohistiocytosis. *British Journal of Haematology, 129*, 622–630.
38. Kalman, V. K., & Klimpel, G. R. (1983). Cyclosporin A inhibits the production of gamma interferon (IFN gamma), but does not inhibit production of virus-induced IFN alpha/beta. *Cellular Immunology, 78*, 122–129.
39. Imashuku, S., Hibi, S., Kuriyama, K., Tabata, Y., Hashida, T., Iwai, A., et al. (2000). Management of severe neutropenia with cyclosporin during initial treatment of Epstein-Barr virus-related hemophagocytic lymphohistiocytosis. *Leukemia & Lymphoma, 36*, 339–346.
40. Jordan, M. B., Allen, C. E., Weitzman, S., Filipovich, A. H., & McClain, K. L. (2011). How I treat hemophagocytic lymphohistiocytosis. *Blood, 118*, 4041–4052.
41. Sung, L., King, S. M., Carcao, M., Trebo, M., & Weitzman, S. S. (2002). Adverse outcomes in primary hemophagocytic lymphohistiocytosis. *Journal of Pediatric Hematology/Oncology, 24*, 550–554.
42. Bergsten, E., Horne, A., Arico, M., Astigarraga, I., Egeler, R. M., Filipovich, A. H., et al. (2017). Confirmed efficacy of etoposide and dexamethasone in HLH treatment: Long term results of the cooperative HLH-2004 study. *Blood*. https://doi.org/10.1182/blood-2017-06-788349
43. Trottestam, H., Beutel, K., Meeths, M., Carlsen, N., Heilmann, C., Pasic, S., et al. (2009). Treatment of the X-linked lymphoproliferative, Griscelli and Chediak-Higashi syndromes by HLH directed therapy. *Pediatric Blood and Cancer, 52*, 268–272.
44. Pui, C. H., Ribeiro, R. C., Hancock, M. L., Rivera, G. K., Evans, W. E., Raimondi, S. C., et al. (1991). Acute myeloid leukemia in children treated with epipodophyllotoxins for acute lymphoblastic leukemia. *The New England Journal of Medicine, 325*, 1682–1687.
45. Imashuku, S., Teramura, T., Kuriyama, K., Kitazawa, J., Ito, E., Morimoto, A., et al. (2002). Risk of etoposide-related acute myeloid leukemia in the treatment of Epstein-Barr virus-associated hemophagocytic lymphohistiocytosis. *International Journal of Hematology, 75*, 174–177.
46. Pan, H., Feng, D. N., Song, L., & Sun, L. R. (2016). Acute myeloid leukemia following etoposide therapy for EBV-associated hemophagocytic lymphohistiocytosis: A case report and a brief review of the literature. *BMC Pediatrics, 16*, 116.
47. Demirkol, D., Yildizdas, D., Bayrakci, B., Karapinar, B., Kendirli, T., Koroglu, T. F., et al. (2012). Hyperferritinemia in the critically ill child with secondary hemophagocytic lymphohistiocytosis/sepsis/multiple organ dysfunction syndrome/macrophage activation syndrome: What is the treatment? *Critical Care, 16*, R52.

48. Xu, X. J., Wang, H. S., Ju, X. L., Xiao, P. F., Xiao, Y., Xue, H. M., et al. (2017). Clinical presentation and outcome of pediatric patients with hemophagocytic lymphohistiocytosis in China: A retrospective multicenter study. *Pediatric Blood & Cancer, 64*. https://doi.org/10.1002/pbc.26264
49. Ramos-Casals, M., Brito-Zeron, P., Lopez-Guillermo, A., Khamashta, M. A., & Bosch, X. (2014). Adult haemophagocytic syndrome. *Lancet, 383*, 1503–1516.
50. Wang, Y., Huang, W., Hu, L., Cen, X., Li, L., Wang, J., et al. (2015). Multicenter study of combination DEP regimen as a salvage therapy for adult refractory hemophagocytic lymphohistiocytosis. *Blood, 126*, 2186–2192.
51. Abdelkefi, A., Ben Jamil, W., Torjman, L., Ladeb, S., Ksouri, H., Lakhal, A., et al. (2009). Hemophagocytic syndrome after hematopoietic stem cell transplantation: A prospective observational study. *International Journal of Hematology, 89*, 368–373.
52. Buyse, S., Teixeira, L., Galicier, L., Mariotte, E., Lemiale, V., Seguin, A., et al. (2010). Critical care management of patients with hemophagocytic lymphohistiocytosis. *Intensive Care Medicine, 36*, 1695–1702.
53. Li, J., Wang, Q., Zheng, W., Ma, J., Zhang, W., Wang, W., et al. (2014). Hemophagocytic lymphohistiocytosis: Clinical analysis of 103 adult patients. *Medicine (Baltimore), 93*, 100–105.
54. Riviere, S., Galicier, L., Coppo, P., Marzac, C., Aumont, C., Lambotte, O., et al. (2014). Reactive hemophagocytic syndrome in adults: A retrospective analysis of 162 patients. *The American Journal of Medicine, 127*, 1118–1125.
55. Schram, A. M., & Berliner, N. (2015). How I treat hemophagocytic lymphohistiocytosis in the adult patient. *Blood, 125*, 2908–2914.
56. La Rosee, P. (2015). Treatment of hemophagocytic lymphohistiocytosis in adults. *Hematology. American Society of Hematology. Education Program, 2015*, 190–196.
57. Tothova, Z., & Berliner, N. (2015). Hemophagocytic syndrome and critical illness: New insights into diagnosis and management. *Journal of Intensive Care Medicine, 30*, 401–412.
58. Schram, A. M., Comstock, P., Campo, M., Gorovets, D., Mullally, A., Bodio, K., et al. (2016). Haemophagocytic lymphohistiocytosis in adults: A multicentre case series over 7 years. *British Journal of Haematology, 172*, 412–419.
59. Rouphael, N. G., Talati, N. J., Vaughan, C., Cunningham, K., Moreira, R., & Gould, C. (2007). Infections associated with haemophagocytic syndrome. *The Lancet Infectious Diseases, 7*, 814–822.
60. Imashuku, S., Kuriyama, K., Teramura, T., Ishii, E., Kinugawa, N., Kato, M., et al. (2001). Requirement for etoposide in the treatment of Epstein-Barr virus-associated hemophagocytic lymphohistiocytosis. *Journal of Clinical Oncology, 19*, 2665–2673.
61. Imashuku, S., Kuriyama, K., Sakai, R., Nakao, Y., Masuda, S., Yasuda, N., et al. (2003). Treatment of Epstein-Barr virus-associated hemophagocytic lymphohistiocytosis (EBV-HLH) in young adults: A report from the HLH study center. *Medical and Pediatric Oncology, 41*, 103–109.
62. Imashuku, S. (2011). Treatment of Epstein-Barr virus-related hemophagocytic lymphohistiocytosis (EBV-HLH); update 2010. *Journal of Pediatric Hematology/Oncology, 33*, 35–39.
63. Chellapandian, D., Das, R., Zelley, K., Wiener, S. J., Zhao, H., Teachey, D. T., et al. (2013). Treatment of Epstein Barr virus-induced haemophagocytic lymphohistiocytosis with rituximab-containing chemo-immunotherapeutic regimens. *British Journal of Haematology, 162*, 376–382.
64. Kawa, K., Sawada, A., Sato, M., Okamura, T., Sakata, N., Kondo, O., et al. (2011). Excellent outcome of allogeneic hematopoietic SCT with reduced-intensity conditioning for the treatment of chronic active EBV infection. *Bone Marrow Transplantation, 46*, 77–83.
65. Fardet, L., Lambotte, O., Meynard, J. L., Kamouh, W., Galicier, L., Marzac, C., et al. (2010). Reactive haemophagocytic syndrome in 58 HIV-1-infected patients: Clinical features, underlying diseases and prognosis. *AIDS, 24*, 1299–1306.
66. Henter, J. I., Palmkvist-Kaijser, K., Holzgraefe, B., Bryceson, Y. T., & Palmer, K. (2010). Cytotoxic therapy for severe swine flu A/H1N1. *Lancet, 376*, 2116.

67. Beutel, G., Wiesner, O., Eder, M., Hafer, C., Schneider, A. S., Kielstein, J. T., et al. (2011). Virus-associated hemophagocytic syndrome as a major contributor to death in patients with 2009 influenza A (H1N1) infection. *Critical Care, 15*, R80.
68. Machowicz, R., Janka, G., & Wiktor-Jedrzejczak, W. (2017). Similar but not the same: Differential diagnosis of HLH and sepsis. *Critical Reviews in Oncology/Hematology, 114*, 1–12.
69. Castillo, L., & Carcillo, J. (2009). Secondary hemophagocytic lymphohistiocytosis and severe sepsis/systemic inflammatory response syndrome/multiorgan dysfunction syndrome/macrophage activation syndrome share common intermediate phenotypes on a spectrum of inflammation. *Pediatric Critical Care Medicine, 10*, 387–392.
70. Tang, Y., Xu, X., Song, H., Yang, S., Shi, S., Wei, J., et al. (2008). Early diagnostic and prognostic significance of a specific Th1/Th2 cytokine pattern in children with haemophagocytic syndrome. *British Journal of Haematology, 143*, 84–91.
71. Carcillo, J. A., Simon, D. W., & Podd, B. S. (2015). How we manage hyperferritinemic sepsis-related multiple organ dysfunction syndrome/macrophage activation syndrome/secondary hemophagocytic lymphohistiocytosis histiocytosis. *Pediatric Critical Care Medicine, 16*, 598–600.
72. Emmenegger, U., Frey, U., Reimers, A., Fux, C., Semela, D., Cottagnoud, P., et al. (2001). Hyperferritinemia as indicator for intravenous immunoglobulin treatment in reactive macrophage activation syndromes. *American Journal of Hematology, 68*, 4–10.
73. Larroche, C., Bruneel, F., Andre, M. H., Bader-Meunier, B., Baruchel, A., Tribout, B., et al. (2000). Intravenously administered gamma-globulins in reactive hemaphagocytic syndrome. Multicenter study to assess their importance, by the immunoglobulins group of experts of CEDIT of the AP-HP. *Annals de Medecine Interne (Paris), 151*, 533–539.
74. Edner, J., Rudd, E., Zheng, C. Y., Dahlander, A., Eksborg, S., Schneider, E. M., et al. (2007). Severe bacteria-associated hemophagocytic lymphohistiocytosis in an extremely premature infant. *Acta Paediatrica, 96*, 1703–1706.
75. Brastianos, P. K., Swanson, J. W., Torbenson, M., Sperati, J., & Karakousis, P. C. (2006). Tuberculosis-associated haemophagocytic syndrome. *The Lancet Infectious Diseases, 6*, 447–454.
76. Bode, S. F., Bogdan, C., Beutel, K., Behnisch, W., Greiner, J., Henning, S., et al. (2014). Hemophagocytic lymphohistiocytosis in imported pediatric visceral leishmaniasis in a nonendemic area. *The Journal of Pediatrics, 165*, 147–153.e141.
77. Untanu, R. V., Akbar, S., Graziano, S., & Vajpayee, N. (2016). Histoplasmosis-induced hemophagocytic lymphohistiocytosis in an adult patient: A case report and review of the literature. *Case Reports in Infectious Diseases, 2016*, 1358742.
78. Lehmberg, K., Nichols, K. E., Henter, J. I., Girschikofsky, M., Greenwood, T., Jordan, M., et al. (2015). Consensus recommendations for the diagnosis and management of hemophagocytic lymphohistiocytosis associated with malignancies. *Haematologica, 100*, 997–1004.
79. Takahashi, N., Chubachi, A., Kume, M., Hatano, Y., Komatsuda, A., Kawabata, Y., et al. (2001). A clinical analysis of 52 adult patients with hemophagocytic syndrome: The prognostic significance of the underlying diseases. *International Journal of Hematology, 74*, 209–213.
80. Arca, M., Fardet, L., Galicier, L., Riviere, S., Marzac, C., Aumont, C., et al. (2015). Prognostic factors of early death in a cohort of 162 adult haemophagocytic syndrome: Impact of triggering disease and early treatment with etoposide. *British Journal of Haematology, 168*, 63–68.
81. Parikh, S. A., Kapoor, P., Letendre, L., Kumar, S., & Wolanskyj, A. P. (2014). Prognostic factors and outcomes of adults with hemophagocytic lymphohistiocytosis. *Mayo Clinic Proceedings, 89*, 484–492.
82. Wang, Y., & Wang, Z. (2017). Treatment of hemophagocytic lymphohistiocytosis. *Current Opinion in Hematology, 24*, 54–58.
83. Minoia, F., Davi, S., Horne, A., Demirkaya, E., Bovis, F., Li, C., et al. (2014). Clinical features, treatment, and outcome of macrophage activation syndrome complicating systemic juvenile idiopathic arthritis: A multinational, multicenter study of 362 patients. *Arthritis & Rhematology, 66*, 3160–3169.

84. Ravelli, A., Davi, S., Minoia, F., Martini, A., & Cron, R. Q. (2015). Macrophage activation syndrome. *Hematology/Oncology Clinics of North America, 29*, 927–941.
85. Miettunen, P. M., Narendran, A., Jayanthan, A., Behrens, E. M., & Cron, R. Q. (2011). Successful treatment of severe paediatric rheumatic disease-associated macrophage activation syndrome with interleukin-1 inhibition following conventional immunosuppressive therapy: Case series with 12 patients. *Rheumatology (Oxford), 50*, 417–419.
86. Ravelli, A., Grom, A. A., Behrens, E. M., & Cron, R. Q. (2012). Macrophage activation syndrome as part of systemic juvenile idiopathic arthritis: Diagnosis, genetics, pathophysiology and treatment. *Genes and Immunity, 13*, 289–298.
87. De Benedetti, F., Brunner, H. I., Ruperto, N., Kenwright, A., Wright, S., Calvo, I., et al. (2012). Randomized trial of tocilizumab in systemic juvenile idiopathic arthritis. *The New England Journal of Medicine, 367*, 2385–2395.
88. Liu, A. C., Yang, Y., Li, M. T., Jia, Y., Chen, S., Ye, S., et al. (2017). Macrophage activation syndrome in systemic lupus erythematosus: A multicenter, case-control study in China. *Clinical Rheumatology*. https://doi.org/10.1007/s10067-017-3625-6
89. Gavand, P. E., Serio, I., Arnaud, L., Costedoat-Chalumeau, N., Carvelli, J., Dossier, A., et al. (2017). Clinical spectrum and therapeutic management of systemic lupus erythematosus-associated macrophage activation syndrome: A study of 103 episodes in 89 adult patients. *Autoimmunity Reviews, 16*, 743–749.
90. Palmblad, K., Schierbeck, H., Sundberg, E., Horne, A. C., Harris, H. E., Henter, J. I., et al. (2015). High systemic levels of the cytokine-inducing HMGB1 isoform secreted in severe macrophage activation syndrome. *Molecular Medicine, 20*, 538–547.
91. Horne, A., Minoia, F., Bovis, F., Davi, S., Pal, P., Anton, J., et al. (2017). Proceedings of the 23rd Paediatric Rheumatology European Society Congress: Part one ABSTRACTS. P159: Factors associated with etoposide usage in children with macrophage activation syndrome complicating systemic Juvenile Idiopathic Arthritis. *Pediatric Rheumatology, 15*, 37.
92. Thomas, A., Appiah, J., Langsam, J., Parker, S., & Christian, C. (2013). Hemophagocytic lymphohistiocytosis associated with dermatomyositis: A case report. *Connecticut Medicine, 77*, 481–485.
93. Hayakawa, I., Shirasaki, F., Ikeda, H., Oishi, N., Hasegawa, M., Sato, S., et al. (2006). Reactive hemophagocytic syndrome in a patient with polyarteritis nodosa associated with Epstein-Barr virus reactivation. *Rheumatology International, 26*, 573–576.
94. Garcia-Pavon, S., Yamazaki-Nakashimada, M. A., Baez, M., Borjas-Aguilar, K. L., & Murata, C. (2017). Kawasaki disease complicated with macrophage activation syndrome: A systematic review. *Journal of Pediatric Hematology/Oncology, 39*, 445–451.
95. Koyama, M., Sawada, A., Yasui, M., Inoue, M., & Kawa, K. (2007). Encouraging results of low-dose etoposide in the treatment of early-onset hemophagocytic syndrome following allogeneic hematopoietic stem cell transplantation. *International Journal of Hematology, 86*, 466–467.
96. Asano, T., Kogawa, K., Morimoto, A., Ishida, Y., Suzuki, N., Ohga, S., et al. (2012). Hemophagocytic lymphohistiocytosis after hematopoietic stem cell transplantation in children: A nationwide survey in Japan. *Pediatric Blood and Cancer, 59*, 110–114.
97. Donadieu, J., Bernard, F., van Noesel, M., Barkaoui, M., Bardet, O., Mura, R., et al. (2015). Cladribine and cytarabine in refractory multisystem Langerhans cell histiocytosis: Results of an international phase 2 study. *Blood, 126*, 1415–1423.
98. Henter, J. I., & Elinder, G. (1991). Familial hemophagocytic lymphohistiocytosis. Clinical review based on the findings in seven children. *Acta Paediatrica Scandinavica, 80*, 269–277.

IL-1 Family Blockade in Cytokine Storm Syndromes

Randy Q. Cron

Interleukin-1 (IL-1)

Cytokines are substances, such as growth factors, interferons, and interleukins, which are secreted by cells in the immune system to influence other cells. Interleukins, as the name implies, are glycoproteins that regulate immune response by communicating between white blood cells. Interleukin-1 (IL-1) was one of the first cytokines described in the immune system, over 40 years ago [1], as it plays a central host defense role against infection. IL-1, formerly known as endogenous pyrogen for its fever inducing effect, is a representative of the 11-member IL-1 family of cytokines [2]. In addition, there are ten unique IL-1 receptor (IL-1R) family members that can result in either pro- or anti-inflammatory functions upon binding IL-1 family members depending on the individual receptor and its co-receptor, making for a highly complex system of signaling. IL-1α and IL-1β are two distinct gene products located adjacent to one another on the long arm of chromosome 2. Their regulation differs, but both are able to bind IL-1R1 and the natural inhibitor of both, IL-1R antagonist (IL-1Ra) (Fig. 1) [2].

IL-1β has been termed a gatekeeper of inflammation and is involved in the pathophysiology of a variety of autoinflammatory diseases [3]. Monocytes/macrophages are a primary source of IL-1β, and IL-1β activity is tightly controlled and dependent on the conversion of an inactive precursor to an active cytokine by limited proteolysis. IL-1β can be processed intracellularly by caspase-1, which is activated by the inflammasome, a multiprotein complex that detects pathogenic organisms as well as sterile stressors to the cell [2]. The protein, NLRP3 (cryopyrin) is important for the assembly of this complex, and hyper-activating mutations in NLRP3 can lead to excess IL-1β activity, resulting in a family of autoinflammatory disorders ranging from familial cold urticaria to Muckle–Wells syndrome to the severest form,

R. Q. Cron (✉)
UAB School of Medicine, University of Alabama at Birmingham, Birmingham, AL, USA
e-mail: rcron@peds.uab.edu

R. Q. Cron, E. M. Behrens (eds.), *Cytokine Storm Syndrome*,
https://doi.org/10.1007/978-3-030-22094-5_31

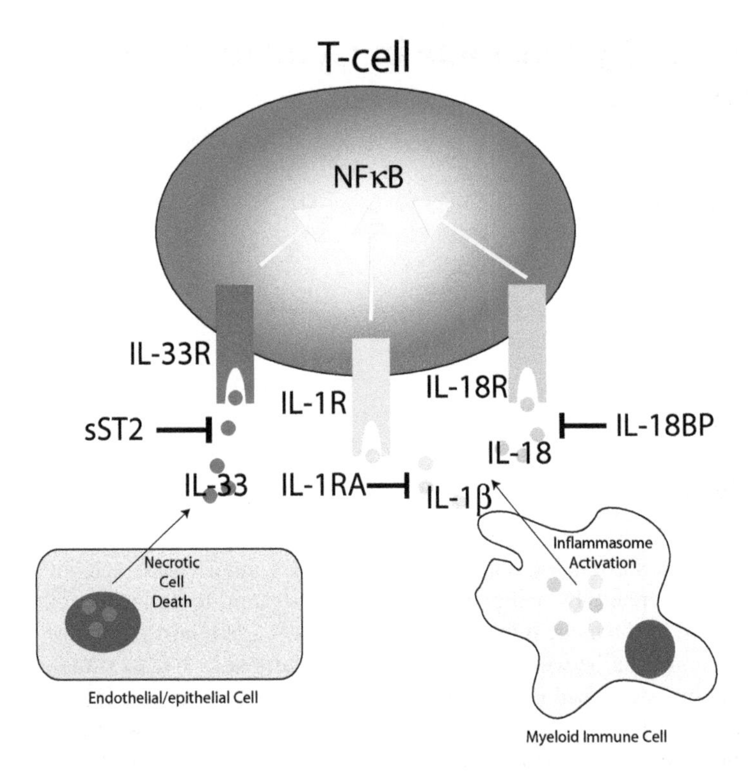

Fig. 1 IL-1 family member biology. The producers of and responders to IL-1 family members are numerous and complex; this figure illustrates some of the main actors in IL-1 family member biology. IL-1β and IL-18 are released from myeloid cells after activation of the inflammasome. IL-33, normally retained in the nucleus, is released by endothelial or epithelial cells after necrotic cell death. Each of the cytokines are then sensed by their respective receptors on CD4+ and CD8+ T-cells resulting in activation of the transcription factor NFκB. Additionally, soluble inhibitors of each of these cytokines are produced which displace binding of the cytokine to its receptor: IL-1 receptor antagonist (IL1-RA) for IL-1β, IL-18 binding protein (IL-18BP) for IL-18, and soluble ST2 receptor (sST2) for IL-33. Figure courtesy of Dr. Ed Behrens, University of Pennsylvania

neonatal onset multisystem inflammatory disease (NOMID) [4]. NOMID can be quite devastating to infants primarily affecting the central nervous system, the bones/joints, and the skin. Fortunately, IL-1β blockade by monoclonal antibody, IL-1R fusion protein, or recombinant human IL-1Ra (rhIL-1Ra) has revolutionized the care of individuals with cryopyrinopathies [4].

Another more common autoinflammatory disorder of childhood responsive to IL-1 blockade is systemic juvenile idiopathic arthritis (sJIA). sJIA affects approximately 1 in 10,000 children worldwide and is characterized clinically by high spiking fever, evanescent salmon-colored rash, arthritis, adenopathy, and occasionally serositis [5]. This used to be a quite devastating disease of childhood until it was found that large amounts of IL-1β were released by sJIA patient peripheral blood mononuclear cells, and IL-1Ra could dramatically treat children with refractory sJIA [6]. Initial treatment of sJIA patients with rIL-1Ra was also shown to improve outcomes and reduce the requirement for corticosteroid use [7]. Eventually, randomized and blinded, placebo-controlled clinical trials with various IL-1 inhibitors bore out these initial anecdotal experiences [8–10]. IL-1 blockade, along with IL-6 blockade, have revolutionized the care of children with sJIA [11].

Children with sJIA are prone to develop a sometimes fatal cytokine storm syndrome (CSS) termed macrophage activation syndrome (MAS) or secondary hemophagocytic lymphohistiocytosis (sHLH) [12]. Seven to ten percent of sJIA patients will develop overt MAS, while another 30–40% will manifest with subclinical or occult MAS which can progress to multi-system organ failure [13, 14]. Those sJIA patients who are prone to MAS development frequently possess heterozygous mutations in known familial HLH genes involved in the perforin mediated cytolytic pathway (e.g., *PRF1, UNC13D*) employed by CD8 T cells and natural killer (NK) cells [15, 16]. This suggests some shared genetic risk factors for CSS/MAS in those with sJIA and other forms of sHLH/MAS [17]. Heterozygous defects in HLH genes have been shown to disrupt NK cell lytic activity by partial [17, 18] and complete dominant negative [19], and hypomorphic effects, including those with sJIA [20]. Defective cytotoxic killing of antigen presenting cells (APC) by lymphocytes has been shown to lead to prolonged interaction of the lytic lymphocyte and the APC resulting in a pro-inflammatory CSS [18, 21].

As CSS resembles MAS in sJIA patients, and sJIA patients were originally shown to be responsive to IL-1 blockade, rhIL-1Ra was employed as treatment for a severe CSS in a child with cytophagic histiocytic panniculitis [22]. This was the first explicit use of rhIL-1Ra reported to successfully treat CSS/MAS, and subsequently rhIL-1Ra was shown to effectively treat refractory MAS in another dozen patients with primarily rheumatic disorders, such as sJIA [23]. Since then, there have been numerous case reports and series of patients effectively treated with rhIL-1Ra for CSS/MAS/sHLH with etiologies ranging from sJIA to adult onset Still disease (Table 1), to systemic lupus erythematosus (Table 2), to autoinflammatory conditions (Table 3), to secondary infections in HIV/AIDS patients (Table 4). Moreover, a retrospective review of a large clinical trial in sepsis revealed that high dose rhIL-1Ra markedly improved survival in sepsis patients with features of MAS, namely hepatobiliary dysfunction and disseminated intravascular coagulopathy (Table 4) [24]. Thus, rhIL-1Ra therapy was anecdotally reported to effectively treat CSS in a variety of infectious and rheumatic disorders. Currently, a randomized, double-blind, placebo controlled clinical trial is underway to evaluate rhIL-1Ra treatment for children and adults with sHLH/CSS [ClinicalTrials.gov Identifier: NCT02780583].

Interleukin-18 (IL-18)

In 2014, activating mutations in the NLRC4 inflammasome component were demonstrated to lead to an autoinflammatory syndrome [25, 26], including an infant with recurrent MAS [27]. This child only partially responded to rhIL-1Ra treatment but was found to have very elevated serum IL-18 levels and was successfully treated with

Table 1 Effectiveness of anakinra in treating CSS as part of sJIA and AOSD

Ages (years)	Disease	Infectious trigger	Co-therapy	Outcome	References
13	sJIA	ND	CS, IVIG, CsA	Resolution	[39]
32	AOSD	ND	CS	Resolution	[40]
1–17	sJIA (8)	ND	CS, IVIG, CsA +/− VP16 or etanercept	Resolution	[23]
8–12	sJIA (2)	ND	CS	Resolution	[41]
20	AOSD	ND	CS	Resolution	[42]
11	sJIA	ND (MRSA later)	CS, IVIG, CsA	Resolution with high dose anakinra	[43]
1–17	sJIA (10)	ND	CS (10), CsA (8), PP (2), canakinumab (4), anakinra (5)	Resolution (9) Survival (10)	[44]
30	AOSD/ PAH	ND	CS, CsA	Resolution	[45]
19–70	AOSD (7)	Histoplasmosis (1)	CS (7), CsA (3), MTX (2), anakinra (5)	Improved (7) Survival (7)	[46]
1–16	sJIA (11)	ND	CS, CsA	Resolution (11)	[47]
34	AOSD	ND	CS	Resolution	[48]
28	AOSD	*Mycoplasma pneumonia*	AZ, CS	Improved and home	[49]
25	AOSD	EBV	CS, CsA, RTX	Resolution after CsA/RTX	[50]
42	AOSD	ND	CS, CsA	Resolution with high dose anakinra	[51]
0.5–16	sJIA (13)	ND	CS (13), PP (6), IVIG (2), CsA (11), anakinra (13)	Resolution (13), 2 flared with anakinra tapering	[52]

Abbreviations: *AOSD* adult-onset Still disease, *AZ* azithromycin, *CS* corticosteroids, *CsA* cyclosporine A, *CSS* cytokine storm syndrome, *IVIG* intravenous immunoglobulin, *MRSA* methicillin resistant *Staphylococcus aureus*, *MTX* methotrexate, *ND* none detected, *PAH* pulmonary artery hypertension, *PP* plasmapheresis, *RTX* rituximab, *sJIA* systemic juvenile idiopathic arthritis, *VP16* etoposide

addition of recombinant human IL-18 binding protein (IL-18BP) [28]. rhIL-18BP is naturally occurring and analogous to rhIL-1Ra in blocking IL-18 and IL-1 function, respectively (Fig. 1). IL-18 is a member of the IL-1 superfamily and analogous to IL-1β is first synthesized as an inactive precursor and is activated following cleavage by caspase-1 [2]. Unlike, IL-1, however, IL-18 does not trigger fever [29]. Thus, despite overlapping features and functions, IL-18 has unique features from IL-1.

In the presence of IL-12 or IL-15, IL-18 induces interferon-gamma (IFNγ) in NK cells, CD4 T cells, and CD8 T cells [29]. IFNγ is believed to a major driver of CSS/HLH in animal models and in humans, and a recent case report depicted the benefit of anti-IFNγ (emapalumab) in treating refractory HLH [30]. A recent clinical trial of emapalumab in treating HLH has resulted in FDA approval for this indication [31].

Table 2 Effectiveness of anakinra in treating CSS associated with autoimmune conditions

Age (years)	Disease	Infectious trigger	Co-therapy	Outcome	References
14	CHP	ND	CS, CsA, VP16 (one dose)	Resolution	[22]
13 5 0.5 6	ARF (1) KD (1) ANCA vasculitis (1) Churg–Strauss vasculitis (1)	ND (4)	CS, CsA, IVIG Cs, CsA, VP16 CS, IVIG CS, IVIG	Resolution (4)	[23]
12	JDMS	ND	CS, CsA, IVIG, VP16, MTX	Improvement with anakinra	[53]
0.25	KD	ND	CS, IVIG, infliximab (one dose)	Resolution with high dose anakinra	[54]
18	SpA, uveitis	ND	CS, CsA	Resolution	[18]
1.5	SpA, uveitis	ND	CS, CsA	Resolution	[55]
37	SLE	ND	CS, CsA	Resolution	[56]
5–15	SLE (6)	ND	CS (6), CsA (4), IVIG (2), VP16 (3), PP (2)	Resolution (2/2 on anakinra)	[47]

Abbreviations: *ANCA* anti-neutrophil cytoplasmic antibody, *ARF* acute rheumatic fever, *CHP* cytophagic histiocytic panniculitis, *CS* corticosteroids, *CsA* cyclosporine A, *IVIG* intravenous immunoglobulin, *KD* Kawasaki disease, *MTX* methotrexate, *ND* none detected, *PP* plasmapheresis, *SLE* systemic lupus erythematosus, *SpA* spondyloarthritis, *VP16* etoposide

Table 3 Effectiveness of anakinra in treating CSS secondary to genetic autoinflammatory conditions

Ages (years)	Disease	Infectious trigger	Co-therapy	Outcome	References
12	CAPS	ND	CS	Resolution	[57]
1	HIDS	ND	none	Resolution	[58]
0.1	NLRC4 mutation	Parainfluenza	CS, CsA, infliximab, vedolizumab, rhIL-18BP	Resolution on combined anakinra and rhIL-18BP	[28]
0.1	NLRC4	ND	CS, CsA, rapamycin	Resolution	[59]
?	CAPS (1) HIDS (1)	ND	CS	Resolution	[52]

Abbreviations: *CAPS* cryopyrin-associated periodic fever syndrome, *CS* corticosteroids, *CsA* corticosteroids, *HIDS* hyper-IgD syndrome, *ND* none detected, *NLRC4* Nod-like receptor family CARD domain containing 4, *rhIL-18BP* recombinant human interleukin-18 binding protein

IL-18 has also been shown to promote murine and human MAS demonstrating the pathogenicity of free (unbound) IL-18 [32, 33]. In addition, free IL-18 concentrations correlated with clinical status in sHLH/MAS patients [34], and IL-18 levels were predictive of MAS in children with sJIA [35]. Therefore, blockade of IL-18 may take on a more prominent role in treating a range of CSS.

Table 4 Effectiveness of anakinra in treating CSS secondary to infections or other conditions

Ages (years)	Disease	Infectious trigger	Co-therapy	Outcome	References
6	None	Parvovirus B19	CS, CsA, IVIG	Resolution	[60]
15 20 13 15 21 11 8 9	XHIM None None None Renal transplant/SLE None Liver transplant None	*Histoplasma capsulatum* *Mycobacterium avium* None None *Varicella zoster* *Candida sphaerica* EBV None	CS (6), IVIG (5), anakinra (8)	Resolution (7) One last onset death awaiting bone marrow transplantation (patient with Candida)	[61]
63	Renal transplant	Erlichiosis	Doxycycline, CS	Resolution	[62]
44	None	CMV	Ganciclovir, CS	Resolution	[63]
18–75	HBD/DIC (43) in sepsis clinical trial	Various forms of sepsis	Anakinra (26) versus placebo (17)	Survival: 65% anakinra 35% placebo	[24]
20–58	8 patients in ICU with HLH, HSCT/GVHD (1), lung transplant (1), ALL (1), AID (1)	EBV (1)	CS (5), IVIG (7), anakinra (8)	50% survival	[64]
18–71	AOSD (3) SLE (2) lymphoma (2) CVID (1) RA (1) CLL (1) UC (1) none (1) ANCA vasculitis (1)	HSV1, CMV URI (1) Rotavirus (1) cholangitis Histoplasmosis none EBV Legionella CMV	Anakinra (13), CS (12), CsA (11), IVIG (12), tocilizumab (2)	Survival: 69%	[65]
46	HIV/AIDS	Histoplasmosis	IVIG	Resolution	[66]
51	Renal transplant recipient	None	CS, CsA, PP	Resolution	[67]

Abbreviations: *ALL* acute lymphoblastic leukemia, *ANCA* anti-neutrophil cytoplasmic antibody, *CLL* chronic lymphocytic leukemia, *AOSD* adult onset Still disease, *CMV* cytomegalovirus, *CS* corticosteroids, *CsA* cyclosporine A, *CVID* common variable immunodeficiency, *DIC* disseminated intravascular coagulation, *EBV* Epstein–Barr virus, *GVHD* graft versus host disease, *HBD* hepatobiliary dysfunction, *HLH* hemophagocytic lymphohistiocytosis, *HSCT* hematopoietic stem cell transplant, *HSV1* herpes simplex virus 1, *HIV/AIDS* human immunodeficiency virus/acquired immune deficiency syndrome, *ICU* intensive care unit, *IVIG* intravenous immunoglobulin, *PP* plasmapheresis, *RA* rheumatoid arthritis, *SLE* systemic lupus erythematosus, *UC* ulcerative colitis, *URI* upper respiratory infection, *XHIM* X-linked immunodeficiency with hyper-IgM

Interleukin-33 (IL-33)

IL-33 is another IL-1 family member with close homology to IL-1, and it is considered an alarmin that is released as an active precursor upon cell damage [2]. IL-33 differs from IL-1 as it can act, depending on context, as either anti- or pro-inflammatory in nature [2]. In a pro-inflammatory setting, IL-33 binds the ST2 receptor which signals via MyD88, IL-1 receptor activated kinases (IRAKs), and the inflammatory transcription factor, NFκB (Fig. 1) [2]. A murine model of HLH showed a role for MyD88-dependent ST2 in disease, and demonstrated that blocking IL-33 signaling via monoclonal antibody directed against ST2 improved survival and the severity of multiple disease manifestations [36]. Moreover, in the long term ST2 blockage results in CD8 T cell exhaustion that does not alter mortality in the HLH murine model arguing for early use on ST2 blockade in CSS [37]. Thus, disruption of signaling of another IL-1 family member, IL-33, may be an option in treating CSS. Overall, these targeted (anti-cytokine) approaches to treating CSS are likely to be far less toxic then current chemotherapeutic approaches [38]. Identifying the correct target for individual patients remains the next challenge.

Summary Although CSSs are frequently fatal, in part from the disease process but also secondary to broad immunosuppression used during treatment, novel approaches of targeting pro-inflammatory cytokines are being explored. Members of the IL-1 superfamily, including IL-1, IL-18, and IL-33 are being explored clinically and/or in murine models of CSS. IL-1 blockade seems like promising therapy for CSS, particularly in the setting of children with sJIA. Similarly, targeting IL-18 may be an important therapeutic option for certain genetic inflammasomopathies, such as activating NLRC4 mutations. Finally, murine models of CSS suggest disrupting IL-33 signaling dampens disease parameters and increases survival. Knowing which cytokine, or combinations of cytokines, to target for individual patients will keep physician-scientists busy for some time to come, yet cytokine blockade for frequently fatal CSS has shown some early promising results.

References

1. Dinarello, C. A., Renfer, L., & Wolff, S. M. (1977). Human leukocytic pyrogen: Purification and development of a radioimmunoassay. *Proceedings of the National Academy of Sciences of the United States of America, 74*(10), 4624–4627.
2. Dinarello, C. A. (2018). Overview of the IL-1 family in innate inflammation and acquired immunity. *Immunological Reviews, 281*(1), 8–27.
3. Dinarello, C. A. (2011). A clinical perspective of IL-1beta as the gatekeeper of inflammation. *European Journal of Immunology, 41*(5), 1203–1217.
4. Kuemmerle-Deschner, J. B. (2015). CAPS—pathogenesis, presentation and treatment of an autoinflammatory disease. *Seminars in Immunopathology, 37*(4), 377–385.
5. Lee, J. J. Y., & Schneider, R. (2018). Systemic juvenile idiopathic arthritis. *Pediatric Clinics of North America, 65*(4), 691–709.
6. Pascual, V., Allantaz, F., Arce, E., Punaro, M., & Banchereau, J. (2005). Role of interleukin-1 (IL-1) in the pathogenesis of systemic onset juvenile idiopathic arthritis and clinical response to IL-1 blockade. *The Journal of Experimental Medicine, 201*(9), 1479–1486.

7. Nigrovic, P. A., Mannion, M., Prince, F. H., Zeft, A., Rabinovich, C. E., van Rossum, M. A., et al. (2011). Anakinra as first-line disease-modifying therapy in systemic juvenile idiopathic arthritis: Report of forty-six patients from an international multicenter series. *Arthritis and Rheumatism, 63*(2), 545–555.
8. Ilowite, N. T., Prather, K., Lokhnygina, Y., Schanberg, L. E., Elder, M., Milojevic, D., et al. (2014). Randomized, double-blind, placebo-controlled trial of the efficacy and safety of rilonacept in the treatment of systemic juvenile idiopathic arthritis. *Arthritis & Rhematology, 66*(9), 2570–2579.
9. Quartier, P., Allantaz, F., Cimaz, R., Pillet, P., Messiaen, C., Bardin, C., et al. (2011). A multicentre, randomised, double-blind, placebo-controlled trial with the interleukin-1 receptor antagonist anakinra in patients with systemic-onset juvenile idiopathic arthritis (ANAJIS trial). *Annals of the Rheumatic Diseases, 70*(5), 747–754.
10. Ruperto, N., Brunner, H. I., Quartier, P., Constantin, T., Wulffraat, N., Horneff, G., et al. (2012). Two randomized trials of canakinumab in systemic juvenile idiopathic arthritis. *The New England Journal of Medicine, 367*(25), 2396–2406.
11. Stoll, M. L., & Cron, R. Q. (2014). Treatment of juvenile idiopathic arthritis: A revolution in care. *Pediatric Rheumatology Online Journal, 12*, 13.
12. Ravelli, A., Grom, A. A., Behrens, E. M., & Cron, R. Q. (2012). Macrophage activation syndrome as part of systemic juvenile idiopathic arthritis: Diagnosis, genetics, pathophysiology and treatment. *Genes and Immunity, 13*(4), 289–298.
13. Behrens, E. M., Beukelman, T., Paessler, M., & Cron, R. Q. (2007). Occult macrophage activation syndrome in patients with systemic juvenile idiopathic arthritis. *The Journal of Rheumatology, 34*(5), 1133–1138.
14. Bleesing, J., Prada, A., Siegel, D. M., Villanueva, J., Olson, J., Ilowite, N. T., et al. (2007). The diagnostic significance of soluble CD163 and soluble interleukin-2 receptor alpha-chain in macrophage activation syndrome and untreated new-onset systemic juvenile idiopathic arthritis. *Arthritis and Rheumatism, 56*(3), 965–971.
15. Kaufman, K. M., Linghu, B., Szustakowski, J. D., Husami, A., Yang, F., Zhang, K., et al. (2014). Whole-exome sequencing reveals overlap between macrophage activation syndrome in systemic juvenile idiopathic arthritis and familial hemophagocytic lymphohistiocytosis. *Arthritis & Rhematology, 66*(12), 3486–3495.
16. Vastert, S. J., van Wijk, R., D'Urbano, L. E., de Vooght, K. M., de Jager, W., Ravelli, A., et al. (2010). Mutations in the perforin gene can be linked to macrophage activation syndrome in patients with systemic onset juvenile idiopathic arthritis. *Rheumatology (Oxford, England), 49*(3), 441–449.
17. Zhang, M., Behrens, E. M., Atkinson, T. P., Shakoory, B., Grom, A. A., & Cron, R. Q. (2014). Genetic defects in cytolysis in macrophage activation syndrome. *Current Rheumatology Reports, 16*(9), 439–446.
18. Zhang, M., Bracaglia, C., Prencipe, G., Bemrich-Stolz, C. J., Beukelman, T., Dimmitt, R. A., et al. (2016). A heterozygous RAB27A mutation associated with delayed cytolytic granule polarization and hemophagocytic lymphohistiocytosis. *Journal of Immunology, 196*(6), 2492–2503.
19. Spessott, W. A., Sanmillan, M. L., McCormick, M. E., Patel, N., Villanueva, J., Zhang, K., et al. (2015). Hemophagocytic lymphohistiocytosis caused by dominant-negative mutations in STXBP2 that inhibit SNARE-mediated membrane fusion. *Blood, 125*(10), 1566–1577.
20. Schulert, G. S., Zhang, M., Husami, A., Fall, N., Brunner, H., Zhang, K., et al. (2018). Brief report: Novel UNC13D intronic variant disrupting an NF-kappaB enhancer in a patient with recurrent macrophage activation syndrome and systemic juvenile idiopathic arthritis. *Arthritis & Rhematology, 70*(6), 963–970.
21. Jenkins, M. R., Rudd-Schmidt, J. A., Lopez, J. A., Ramsbottom, K. M., Mannering, S. I., Andrews, D. M., et al. (2015). Failed CTL/NK cell killing and cytokine hypersecretion are directly linked through prolonged synapse time. *The Journal of Experimental Medicine, 212*(3), 307–317.

22. Behrens, E. M., Kreiger, P. A., Cherian, S., & Cron, R. Q. (2006). Interleukin 1 receptor antagonist to treat cytophagic histiocytic panniculitis with secondary hemophagocytic lymphohistiocytosis. *The Journal of Rheumatology, 33*(10), 2081–2084.
23. Miettunen, P. M., Narendran, A., Jayanthan, A., Behrens, E. M., & Cron, R. Q. (2011). Successful treatment of severe paediatric rheumatic disease-associated macrophage activation syndrome with interleukin-1 inhibition following conventional immunosuppressive therapy: Case series with 12 patients. *Rheumatology (Oxford, England), 50*(2), 417–419.
24. Shakoory, B., Carcillo, J. A., Chatham, W. W., Amdur, R. L., Zhao, H., Dinarello, C. A., et al. (2016). Interleukin-1 receptor blockade is associated with reduced mortality in sepsis patients with features of macrophage activation syndrome: Reanalysis of a prior phase III trial. *Critical Care Medicine, 44*(2), 275–281.
25. Kitamura, A., Sasaki, Y., Abe, T., Kano, H., & Yasutomo, K. (2014). An inherited mutation in NLRC4 causes autoinflammation in human and mice. *The Journal of Experimental Medicine, 211*(12), 2385–2396.
26. Romberg, N., Al Moussawi, K., Nelson-Williams, C., Stiegler, A. L., Loring, E., Choi, M., et al. (2014). Mutation of NLRC4 causes a syndrome of enterocolitis and autoinflammation. *Nature Genetics, 46*(10), 1135–1139.
27. Canna, S. W., de Jesus, A. A., Gouni, S., Brooks, S. R., Marrero, B., Liu, Y., et al. (2014). An activating NLRC4 inflammasome mutation causes autoinflammation with recurrent macrophage activation syndrome. *Nature Genetics, 46*(10), 1140–1146.
28. Canna, S. W., Girard, C., Malle, L., de Jesus, A., Romberg, N., Kelsen, J., et al. (2017). Life-threatening NLRC4-associated hyperinflammation successfully treated with IL-18 inhibition. *The Journal of Allergy and Clinical Immunology, 139*(5), 1698–1701.
29. Kaplanski, G. (2018). Interleukin-18: Biological properties and role in disease pathogenesis. *Immunological Reviews, 281*(1), 138–153.
30. Lounder, D. T., Bin, Q., de Min, C., & Jordan, M. B. (2019). Treatment of refractory hemophagocytic lymphohistiocytosis with emapalumab despite severe concurrent infections. *Blood Advances, 3*(1), 47–50.
31. Al-Salama, Z. T. (2019). Emapalumab: First global approval. *Drugs*.
32. Girard-Guyonvarc'h, C., Palomo, J., Martin, P., Rodriguez, E., Troccaz, S., Palmer, G., et al. (2018). Unopposed IL-18 signaling leads to severe TLR9-induced macrophage activation syndrome in mice. *Blood, 131*(13), 1430–1441.
33. Weiss, E. S., Girard-Guyonvarc'h, C., Holzinger, D., de Jesus, A. A., Tariq, Z., Picarsic, J., et al. (2018). Interleukin-18 diagnostically distinguishes and pathogenically promotes human and murine macrophage activation syndrome. *Blood, 131*(13), 1442–1455.
34. Mazodier, K., Marin, V., Novick, D., Farnarier, C., Robitail, S., Schleinitz, N., et al. (2005). Severe imbalance of IL-18/IL-18BP in patients with secondary hemophagocytic syndrome. *Blood, 106*(10), 3483–3489.
35. Shimizu, M., Nakagishi, Y., Inoue, N., Mizuta, M., Ko, G., Saikawa, Y., et al. (2015). Interleukin-18 for predicting the development of macrophage activation syndrome in systemic juvenile idiopathic arthritis. *Clinical Immunology, 160*(2), 277–281.
36. Rood, J. E., Rao, S., Paessler, M., Kreiger, P. A., Chu, N., Stelekati, E., et al. (2016). ST2 contributes to T-cell hyperactivation and fatal hemophagocytic lymphohistiocytosis in mice. *Blood, 127*(4), 426–435.
37. Rood, J. E., Burn, T. N., Neal, V., Chu, N., & Behrens, E. M. (2018). Disruption of IL-33 signaling limits early CD8+ T cell effector function leading to exhaustion in murine hemophagocytic lymphohistiocytosis. *Frontiers in Immunology, 9*, 2642.
38. Ravelli, A., Davi, S., Minoia, F., Martini, A., & Cron, R. Q. (2015). Macrophage activation syndrome. *Hematology/Oncology Clinics of North America, 29*(5), 927–941.
39. Kelly, A., & Ramanan, A. V. (2008). A case of macrophage activation syndrome successfully treated with anakinra. *Nature Clinical Practice. Rheumatology, 4*(11), 615–620.
40. Durand, M., Troyanov, Y., Laflamme, P., & Gregoire, G. (2010). Macrophage activation syndrome treated with anakinra. *The Journal of Rheumatology, 37*(4), 879–880.

41. Bruck, N., Suttorp, M., Kabus, M., Heubner, G., Gahr, M., & Pessler, F. (2011). Rapid and sustained remission of systemic juvenile idiopathic arthritis-associated macrophage activation syndrome through treatment with anakinra and corticosteroids. *Journal of Clinical Rheumatology, 17*(1), 23–27.
42. Loh, N. K., Lucas, M., Fernandez, S., & Prentice, D. (2012). Successful treatment of macrophage activation syndrome complicating adult Still disease with anakinra. *Internal Medicine Journal, 42*(12), 1358–1362.
43. Kahn, P. J., & Cron, R. Q. (2013). Higher-dose Anakinra is effective in a case of medically refractory macrophage activation syndrome. *The Journal of Rheumatology, 40*(5), 743–744.
44. Barut, K., Yucel, G., Sinoplu, A. B., Sahin, S., Adrovic, A., & Kasapcopur, O. (2015). Evaluation of macrophage activation syndrome associated with systemic juvenile idiopathic arthritis: Single center experience over a one-year period. *Turkish Archives of Pediatrics, 50*(4), 206–210.
45. Mehta, M. V., Manson, D. K., Horn, E. M., & Haythe, J. (2016). An atypical presentation of adult-onset Still's disease complicated by pulmonary hypertension and macrophage activation syndrome treated with immunosuppression: A case-based review of the literature. *Pulmonary Circulation, 6*(1), 136–142.
46. Lenert, A., & Yao, Q. (2016). Macrophage activation syndrome complicating adult onset Still's disease: A single center case series and comparison with literature. *Seminars in Arthritis and Rheumatism, 45*(6), 711–716.
47. Aytac, S., Batu, E. D., Unal, S., Bilginer, Y., Cetin, M., Tuncer, M., et al. (2016). Macrophage activation syndrome in children with systemic juvenile idiopathic arthritis and systemic lupus erythematosus. *Rheumatology International, 36*(10), 1421–1429.
48. Kumar, A., & Kato, H. (2016). Macrophage activation syndrome associated with adult-onset Still's disease successfully treated with anakinra. *Case Reports in Rheumatology, 2016*, 3717392.
49. Agnihotri, A., Ruff, A., Gotterer, L., Walker, A., McKenney, A. H., & Brateanu, A. (2016). Adult onset Still's disease associated with mycoplasma pneumoniae infection and hemophagocytic lymphohistiocytosis. *Case Reports in Medicine, 2016*, 2071815.
50. Schafer, E. J., Jung, W., & Korsten, P. (2016). Combination immunosuppressive therapy including rituximab for Epstein-Barr virus-associated hemophagocytic lymphohistiocytosis in adult-onset Still's disease. *Case Reports in Rheumatology, 2016*, 8605274.
51. Parisi, F., Paglionico, A., Varriano, V., Ferraccioli, G., & Gremese, E. (2017). Refractory adult-onset Still disease complicated by macrophage activation syndrome and acute myocarditis: A case report treated with high doses (8 mg/kg/d) of anakinra. *Medicine (Baltimore), 96*(24), e6656.
52. Sonmez, H. E., Demir, S., Bilginer, Y., & Ozen, S. (2018). Anakinra treatment in macrophage activation syndrome: A single center experience and systemic review of literature. *Clinical Rheumatology, 37*(12), 3329–3335.
53. Lilleby, V., Haydon, J., Sanner, H., Krossness, B. K., Ringstad, G., & Flato, B. (2014). Severe macrophage activation syndrome and central nervous system involvement in juvenile dermatomyositis. *Scandinavian Journal of Rheumatology, 43*(2), 171–173.
54. Shafferman, A., Birmingham, J. D., & Cron, R. Q. (2014). High dose anakinra for treatment of severe neonatal Kawasaki disease: A case report. *Pediatric Rheumatology Online Journal, 12*, 26.
55. Cron, R. Q., & Chatham, W. W. (2016). Development of spondyloarthropathy following episodes of macrophage activation syndrome in children with heterozygous mutations in haemophagocytic lymphohistiocytosis-associated genes. *Clinical and Experimental Rheumatology, 34*(5), 953.
56. Tayer-Shifman, O. E., & Ben-Chetrit, E. (2013). Refractory macrophage activation syndrome in a patient with SLE and APLA syndrome—Successful use of PET—CT and Anakinra in its diagnosis and treatment. *Modern Rheumatology*.
57. Mohr, V., Schulz, A., Lohse, P., Schumann, C., Debatin, K. M., & Schuetz, C. (2014). Urticaria, fever, and hypofibrinogenemia. *Arthritis & Rhematology, 66*(5), 1377.

58. Schulert, G. S., Bove, K., McMasters, R., Campbell, K., Leslie, N., & Grom, A. A. (2015). 11-month-old infant with periodic fevers, recurrent liver dysfunction, and perforin gene polymorphism. *Arthritis Care and Research., 67*(8), 1173–1179.
59. Barsalou, J., Blincoe, A., Fernandez, I., Dal-Soglio, D., Marchitto, L., Selleri, S., et al. (2018). Rapamycin as an adjunctive therapy for NLRC4 associated macrophage activation syndrome. *Frontiers in Immunology, 9*, 2162.
60. Butin, M., Mekki, Y., Phan, A., Billaud, G., Di Filippo, S., Javouhey, E., et al. (2013). Successful immunotherapy in life-threatening parvovirus B19 infection in a child. *The Pediatric Infectious Disease Journal, 32*(7), 789–792.
61. Rajasekaran, S., Kruse, K., Kovey, K., Davis, A. T., Hassan, N. E., Ndika, A. N., et al. (2014). Therapeutic role of anakinra, an interleukin-1 receptor antagonist, in the management of secondary hemophagocytic lymphohistiocytosis/sepsis/multiple organ dysfunction/macrophage activating syndrome in critically ill children∗. *Pediatric Critical Care Medicine, 15*(5), 401–408.
62. Kumar, N., Goyal, J., Goel, A., Shakoory, B., & Chatham, W. (2014). Macrophage activation syndrome secondary to human monocytic ehrlichiosis. *Indian Society of Hematology and Blood Transfusion, 30*(Suppl 1), 145–147.
63. Divithotawela, C., Garrett, P., Westall, G., Bhaskar, B., Tol, M., & Chambers, D. C. (2016). Successful treatment of cytomegalovirus associated hemophagocytic lymphohistiocytosis with the interleukin 1 inhibitor—anakinra. *Respirology Case Reports, 4*(1), 4–6.
64. Wohlfarth, P., Agis, H., Gualdoni, G. A., Weber, J., Staudinger, T., Schellongowski, P., et al. (2017). Interleukin 1 receptor antagonist anakinra, intravenous immunoglobulin, and corticosteroids in the management of critically Ill adult patients with hemophagocytic lymphohistiocytosis. *Journal of Intensive Care Medicine, 885066617711386.*
65. Kumar, B., Aleem, S., Saleh, H., Petts, J., & Ballas, Z. K. (2017). A personalized diagnostic and treatment approach for macrophage activation syndrome and secondary hemophagocytic lymphohistiocytosis in adults. *Journal of Clinical Immunology, 37*(7), 638–643.
66. Ocon, A. J., Bhatt, B. D., Miller, C., & Peredo, R. A. (2017). Safe usage of anakinra and dexamethasone to treat refractory hemophagocytic lymphohistiocytosis secondary to acute disseminated histoplasmosis in a patient with HIV/AIDS. *BML Case Reports, 2017.*
67. Nusshag, C., Morath, C., Zeier, M., Weigand, M. A., Merle, U., & Brenner, T. (2017). Hemophagocytic lymphohistiocytosis in an adult kidney transplant recipient successfully treated by plasmapheresis: A case report and review of the literature. *Medicine (Baltimore), 96*(50), e9283.

IL-6 Blockade in Cytokine Storm Syndromes

David Barrett

Interleukin-6 (IL-6) has gained attention as a key node in certain cytokine storm syndromes (CSS). Originally described as B-cell differentiation factor 2 (BSF-2) and Macrophage and granulocyte inducing factor 2 (MGI-2), IL-6 has prominent pro-inflammatory and pyrogenic properties [1–3]. The receptor for IL-6 is complex and allows for several signaling configurations. The IL-6 receptor (IL-6R) is a relatively small immunoglobulin like receptor with a conserved WSXWS motif along with four conserved cysteine residues in the extracellular portion. The intracellular portion was shown to be unnecessary for signal transduction, and led to the discovery of the heterodimeric partner to the IL-6R, gp130 [4, 5]. IL-6 can thus signal through two main configurations, referred to as *trans*- or *cis*-signaling [6]. In *cis*-signaling, the cell expresses the IL-6R and gp130 in a complex, and signal transduction is mediated by binding of IL-6 to the IL-6R. In *trans*-signaling, IL-6 binds to a soluble form of the IL-6R (sIL-6R) forming a soluble complex that can then bind to a dimer of gp130 on a cell surface; thus mediating IL-6 signaling in a cell which does not express the IL-6R (Fig. 1) [1]. Baseline proteolytic cleavage of the surface receptor by ADAM10 results in tonic levels of circulating sIL-6R, whereas high levels can be induced by cleavage via ADAM17 [7]. Internally, IL-6 signals via the Janus Activated Kinase (JAK) and signal transducer and activator of transcription (STAT) pathways, particularly STAT3 [1]. IL-6 signaling can thus be targeted by inhibiting IL-6 levels, blocking the IL-6 receptor, blocking gp130 or by targeting JAK-STAT signaling (Fig. 1).

IL-6 has been known to be elevated in HLH, reaching levels of greater than 100 pg/mL in plasma of patients with primary hemophagocytic lymphohistiocytosis (pHLH) or Epstein–Barr virus (EBV) driven secondary HLH (sHLH) [8, 9]. It was not shown to be specific for HLH, however, despite its consistent elevation. Several studies of biomarkers for HLH/MAS (macrophage activation syndrome) have honed in on the combination of interferon-gamma (IFNγ) and interleukin-10 (IL-10) as

D. Barrett (✉)
Children's Hospital of Philadelphia, Philadelphia, PA, USA
e-mail: barrettd@email.chop.edu

R. Q. Cron, E. M. Behrens (eds.), *Cytokine Storm Syndrome*,
https://doi.org/10.1007/978-3-030-22094-5_32

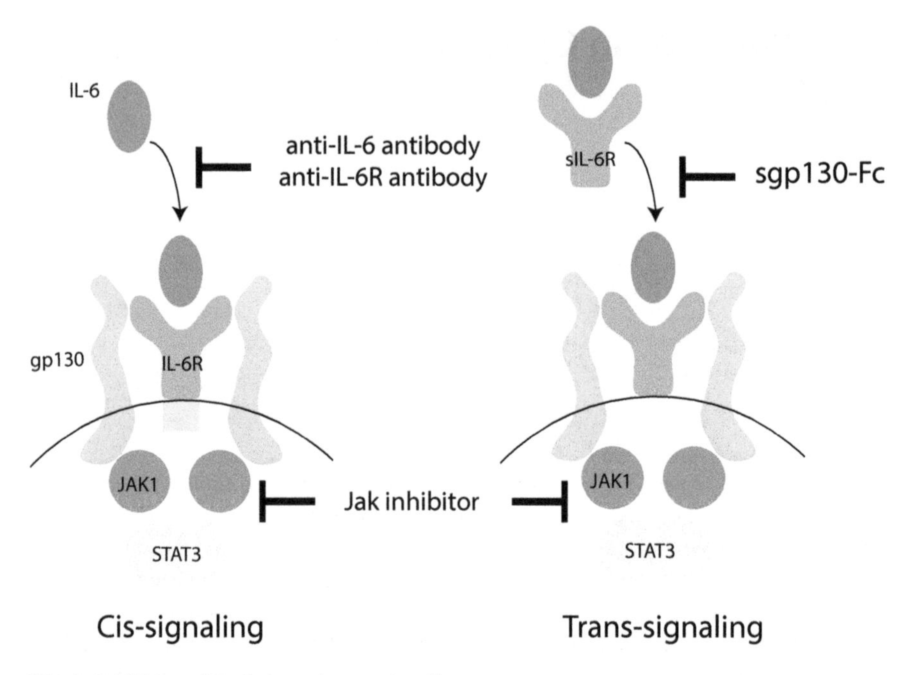

Fig. 1 Inhibition of IL-6 *cis*- and *trans*-signaling

being specific and sensitive for HLH/MAS rather than IL-6, which can be elevated in sepsis or non-septic infection [10, 11]. Targeting IL-6 with tocilizumab, an anti-IL-6R monoclonal antibody, has been used in HLH and related syndromes with mixed results. For the MAS associated with systemic juvenile idiopathic arthritis (sJIA), tocilizumab and IL-6 blockade successfully masked clinical symptoms such as fever but did not alter the acute disease course [12]. Tocilizumab was used safely for the chronic management of sJIA, though again did not prevent or alter MAS flares based on serum biomarkers though it could mask clinical symptomatology [13, 14].

Tocilizumab has also been used for sHLH management, as in Leishmaniasis-induced sHLH [15]. Again in this case, clinical symptoms of MAS were masked (such as fever) until the underlying trigger (infection) was resolved. In modern-day cellular therapy for cancer and immune-modulatory therapies, sHLH has been recognized as a potentially life threatening consequence that is referred to as cytokine release syndrome (CRS) [16, 17]. Blinatumomab is a bi-specific T cell engager (BiTe) that recognizes CD3 on one end and CD19 on the other, making it an attractive therapy for relapsed and refractory CD19-positive acute lymphoblastic leukemia (ALL) [18]. Patients have been recognized to have a sHLH response during the blinatumomab administration, characterized by elevated acute phase reactants and elevated IL-6 levels. Clinical symptoms of these patients improve with tocilizumab administration, including prompt resolution of fever and hemodynamic stabilization (Table 1) [16, 19, 20].

Table 1 Potential and actualized therapeutic agents targeting the IL-6 pathway for treatment of cytokine storm syndrome

Therapeutic	Mechanism	References
Tocilizumab	Anti-IL-6 receptor antibody	[5]
Siltuximab	Anti-IL-6 antibody	[34]
Sgp130-Fc	Inhibition of IL-6 *trans*-signaling	[50]
Ruxolitinib	Inhibition of JAK/STAT signaling	[55]

Chimeric antigen receptor (CAR) T cell therapy for ALL has also been described to induce a life-threatening cytokine release syndrome (CRS)/sHLH [17]. In this therapy, T cells from a cancer patient are collected via apheresis, modified in the laboratory to express the CAR, and then reinfused into the patient [21]. The CAR consists of an extracellular binding domain (often a single chain variable fragment of an antibody recognizing CD19), an endodomain consisting of the intracellular transactivation motif from the CD3 zeta chain and a second domain from a costimulatory molecule such as CD28 or 4-1BB [21]. When the CAR T cells engage in leukemia killing, they proliferate and secrete pro-inflammatory cytokines such as IFNγ and tumor necrosis factor (TNF) [22, 23]. Early CAR T cell trials did not show much in the way of clinical activity, with no sustained remissions and also very little toxicity [24, 25]. It was the first report of two children with ALL treated with anti-CD19 CAR T cells in 2013 that first described accurately the sHLH from CAR T cell therapy and the use of IL-6 blockade for treatment [17]. In this report, a child with ALL was treated with CAR T cells and shortly thereafter became febrile, coagulopathic, hyperferritinemic, and hypertriglyceridemic, and developed organomegaly, capillary leak syndrome, and hypotension. Laboratory markers showed a 3–4 $\log_{10}$ elevation in IL-6 levels over baseline. Treatment with systemic corticosteroids and etanercept (a TNF blocking agent) did not result in clinical improvement; however, treatment with tocilizumab promptly resolved the fever and other clinical symptoms [5, 17]. A toxicity management strategy was then developed and applied to other CAR therapy trials with similar results [26, 27]. Long term follow up reports show that toxicity from sHLH from CAR T cell therapy can be successfully managed with tocilizumab, augmented with corticosteroids in severe cases [28, 29]. A challenge to understanding the incidence of toxicity, which ranges from 21 to 64%, is the use of multiple grading scales for cytokine release syndrome [30, 31]. The recent FDA approval of a CAR T cell therapy for ALL (tisagenlecleucel, Kymriah) from Novartis was accompanied by the announcement of the approval of tocilizumab for use in CRS management, recognizing the indispensable role of IL-6 blockade in safely treating CAR CRS.

Understanding the kinetics and measurement of IL-6 in CAR mediated sHLH is a challenge. Different CAR products may produce different cytokine kinetics, different onset of clinical symptoms, and respond differently to therapy [20, 29, 32, 33]. The use of tocilizumab or siltuximab (anti-IL-6 monoclonal antibody) (Table 1) can potentially impede the accurate clinical measurement of IL-6 and sIL-6R [34, 35]. The first report of a prospectively validated biomarker profile for sHLH from CAR T cell therapy was in 2016, and among the models one of the most highly predictive was a combination of high disease burden and early elevation of soluble gp130 [36].

One of the challenges in this field is that patients become clinically ill before obvious serum biomarkers such as IL-6 rise to notable levels, so a predictive model that allows for identifying patients which would benefit from early intervention is highly desirable. Other biomarkers such C-reactive protein and ferritin often trail the clinical onset of symptoms [36–38].

Modeling CAR mediated CRS in animals is challenging. Many of the preclinical models were xenografts, using immunodeficient mice and human leukemia and T cells [22, 39]. Missing from these models was any hint of sHLH, likely because the mice are lacking any other aspect of a competent immune system. Fully murine models of CAR therapy were notable for their relatively disappointing efficacy in the late 2000s, and with transient disease response came no toxicity [40]. HLH in mice is possible in other settings, however, including a fatal model of HLH in transgenic mice (expressing IL-3, GM-CSF, and SCF) engrafted with human cord blood [41]. In this model, there is evidence for a myeloid cell based source for IL-6 and toxicity, and survival is enhanced by myeloid depletion (via gemtuzumab, an anti-CD33 monoclonal antibody) and IL-6 blockade (via tocilizumab). This is similar to reports of systemic MAS/sHLH in patients receiving T-replete stem cell transplants, in which IL-6 blockade can help alleviate symptoms [42, 43]. A true animal model of CAR CRS/sHLH, however, remains to be developed. Contrary to early conventional wisdom, the CAR T cells do not seem to be the source of IL-6 [44]. Rather, as we might expect from the animal models, it appears CAR T cells killing target cells induce IL-6 release from bystander myeloid lineage cells. This is also consistent with an earlier report of MAS pathology in pHLH, in which immunohistochemistry demonstrated CD8 T cells in the liver secreting IFNγ and CD68 macrophages secreting IL-6 and TNF [45]. Until mouse models of CAR T cell therapy increase in potency to demonstrate toxicity or humanized mice can be used to distinguish allograft toxicity from sHLH, we will have many unanswered questions about the mechanism of IL-6 release in CAR CRS.

Siltuximab (CNTO 328) is a human IL-6 neutralizing antibody that is FDA approved for use in multicentric Castleman's disease [46]. There are no published reports of its use for pHLH or sHLH, though it has been used alone and in combination as an antitumor agent [47]. While its antitumor efficacy is worthwhile, its utility or efficacy in blocking CSS remains to be seen.

Targeting gp130 is difficult, in part because it is a common subunit to many cytokines (IL-6, IL-11, oncostatin M, etc.) [2]. Nevertheless, in a mouse model of hyperinflammation (drug induced pancreatitis) soluble gp130 (sgp130) was found to be effective in controlling symptoms and prolonging survival [48]. There are several forms of naturally occurring sgp130, which may have different potency or function [49]. Development of a sgp130-Fc chimeric protein results in a specific inhibitor of IL-6 trans-signaling (Table 1) [50]. Clinical trials with this agent are planned or underway in Europe [50].

Targeting the JAK-STAT pathway is another possible way to ameliorate IL-6 toxicity. Ruxolitinib is a targeted JAK inhibitor heavily studied for effects on cancer cells and in myelofibrosis (Table 1) [51]. In two mouse models of HLH, it was effective in reducing pro-inflammatory cytokine secretion and T cell proliferation [52, 53]. This included a perforin-deficient mouse infected with LCMV and the C57B6 mouse stimulated with CpG. The anti-T cell proliferative effects make using ruxolitinib in T cell immunotherapy problematic, but its potential efficacy warrants further investigation. There are two case reports of ruxolitinib in patients with refractory sHLH. In one report the laboratory values improved, but the patient did not survive, and in the other the patient improved with ruxolitinib as part of a multimodality therapy regimen [54, 55].

In summary, IL-6 is a potent inflammatory cytokine that can mediate systemic illness in sHLH, particularly in CAR T cell therapy. Blockade of IL-6 with tocilizumab is safe and effective as long as the underlying trigger of sHLH resolves. Targeting IL-6 via other mechanisms, such as with direct IL-6 binding with siltuximab or blockade of gp130, is being pursued in the clinic and the lab. Given the significance of immune-based therapies for cancer and the need to safely deliver them, much more investigation needs to be done.

References

1. Kishimoto, T., Akira, S., Narazaki, M., & Taga, T. (1995). Interleukin-6 family of cytokines and gp130. *Blood, 86*, 1243–1254.
2. Hirano, T., Taga, T., Nakano, N., Yasukawa, K., Kashiwamura, S., Shimizu, K., et al. (1985). Purification to homogeneity and characterization of human B-cell differentiation factor (BCDF or BSFp-2). *Proceedings of the National Academy of Sciences of the United States of America, 82*, 5490–5494.
3. Shabo, Y., Lotem, J., Rubinstein, M., Revel, M., Clark, S. C., Wolf, S. F., et al. (1988). The myeloid blood cell differentiation-inducing protein MGI-2A is interleukin-6. *Blood, 72*, 2070–2073.
4. Yamasaki, K., Taga, T., Hirata, Y., Yawata, H., Kawanishi, Y., Seed, B., et al. (1988). Cloning and expression of the human interleukin-6 (BSF-2/IFN beta 2) receptor. *Science, 241*, 825–828.
5. Taga, T., Hibi, M., Hirata, Y., Yamasaki, K., Yasukawa, K., Matsuda, T., et al. (1989). Interleukin-6 triggers the association of its receptor with a possible signal transducer, gp130. *Cell, 58*, 573–581.
6. Lacroix, M., Rousseau, F., Guilhot, F., Malinge, P., Magistrelli, G., Herren, S., et al. (2015). Novel insights into interleukin 6 (IL-6) Cis- and trans-signaling pathways by differentially manipulating the assembly of the IL-6 signaling complex. *The Journal of Biological Chemistry, 290*, 26943–26953.
7. Schumacher, N., Meyer, D., Mauermann, A., von der Heyde, J., Wolf, J., Schwarz, J., et al. (2015). Shedding of endogenous interleukin-6 receptor (IL-6R) is governed by a disintegrin and metalloproteinase (ADAM) proteases while a full-length IL-6R isoform localizes to circulating microvesicles. *The Journal of Biological Chemistry, 290*, 26059–26071.
8. Imashuku, S., Hibi, S., Fujiwara, F., & Todo, S. (1996). Hyper-interleukin (IL)-6-naemia in haemophagocytic lymphohistiocytosis. *British Journal of Haematology, 93*, 803–807.
9. Imashuku, S., Hibi, S., Tabata, Y., Sako, M., Sekine, Y., Hirayama, K., et al. (1998). Biomarker and morphological characteristics of Epstein-Barr virus-related hemophagocytic lymphohistiocytosis. *Medical and Pediatric Oncology, 31*, 131–137.
10. Xu, X. J., Tang, Y. M., Song, H., Yang, S. L., Xu, W. Q., Zhao, N., et al. (2012). Diagnostic accuracy of a specific cytokine pattern in hemophagocytic lymphohistiocytosis in children. *The Journal of Pediatrics, 160*, 984–990 e981.
11. Yang, S. L., Xu, X. J., Tang, Y. M., Song, H., Xu, W. Q., Zhao, F. Y., et al. (2016). Associations between inflammatory cytokines and organ damage in pediatric patients with hemophagocytic lymphohistiocytosis. *Cytokine, 85*, 14–17.

12. Shimizu, M., Nakagishi, Y., Kasai, K., Yamasaki, Y., Miyoshi, M., Takei, S., et al. (2012). Tocilizumab masks the clinical symptoms of systemic juvenile idiopathic arthritis-associated macrophage activation syndrome: The diagnostic significance of interleukin-18 and interleukin-6. *Cytokine, 58*, 287–294.
13. Yokota, S., Imagawa, T., Mori, M., Miyamae, T., Takei, S., Iwata, N., et al. (2014). Longterm safety and effectiveness of the anti-interleukin 6 receptor monoclonal antibody tocilizumab in patients with systemic juvenile idiopathic arthritis in Japan. *The Journal of Rheumatology, 41*, 759–767.
14. Schulert, G. S., Minoia, F., Bohnsack, J., Cron, R. Q., Hashad, S., Kon, E. P. I., et al. (2018). Effect of biologic therapy on clinical and laboratory features of macrophage activation syndrome associated with systemic juvenile idiopathic arthritis. *Arthritis Care & Research, 70*, 409–419.
15. Rios-Fernandez, R., Callejas-Rubio, J. L., Garcia-Rodriguez, S., Sancho, J., Zubiaur, M., & Ortego-Centeno, N. (2016). Tocilizumab as an adjuvant therapy for hemophagocytic lymphohistiocytosis associated with visceral leishmaniasis. *American Journal of Therapeutics, 23*, e1193–e1196.
16. Teachey, D. T., Rheingold, S. R., Maude, S. L., Zugmaier, G., Barrett, D. M., Seif, A. E., et al. (2013). Cytokine release syndrome after blinatumomab treatment related to abnormal macrophage activation and ameliorated with cytokine-directed therapy. *Blood, 121*, 5154–5157.
17. Grupp, S. A., Kalos, M., Barrett, D., Aplenc, R., Porter, D. L., Rheingold, S. R., et al. (2013). Chimeric antigen receptor-modified T cells for acute lymphoid leukemia. *The New England Journal of Medicine, 368*, 1509–1518.
18. Topp, M. S., Gokbuget, N., Zugmaier, G., Degenhard, E., Goebeler, M. E., Klinger, M., et al. (2012). Long-term follow-up of hematologic relapse-free survival in a phase 2 study of blinatumomab in patients with MRD in B-lineage ALL. *Blood, 120*, 5185–5187.
19. Brudno, J. N., & Kochenderfer, J. N. (2016). Toxicities of chimeric antigen receptor T cells: Recognition and management. *Blood, 127*, 3321–3330.
20. Frey, N. V., & Porter, D. L. (2016). Cytokine release syndrome with novel therapeutics for acute lymphoblastic leukemia. *Hematology. American Society of Hematology, 2016*, 567–572.
21. Barrett, D. M., Singh, N., Porter, D. L., Grupp, S. A., & June, C. H. (2014). Chimeric antigen receptor therapy for cancer. *Annual Review of Medicine, 65*, 333–347.
22. Milone, M. C., Fish, J. D., Carpenito, C., Carroll, R. G., Binder, G. K., Teachey, D., et al. (2009). Chimeric receptors containing CD137 signal transduction domains mediate enhanced survival of T cells and increased antileukemic efficacy in vivo. *Molecular Therapy, 17*, 1453–1464.
23. Kochenderfer, J. N., Dudley, M. E., Feldman, S. A., Wilson, W. H., Spaner, D. E., Maric, I., et al. (2012). B-cell depletion and remissions of malignancy along with cytokine-associated toxicity in a clinical trial of anti-CD19 chimeric-antigen-receptor-transduced T cells. *Blood, 119*, 2709–2720.
24. Kochenderfer, J. N., Wilson, W. H., Janik, J. E., Dudley, M. E., Stetler-Stevenson, M., Feldman, S. A., et al. (2010). Eradication of B-lineage cells and regression of lymphoma in a patient treated with autologous T cells genetically engineered to recognize CD19. *Blood, 116*, 4099–4102.
25. Pule, M. A., Savoldo, B., Myers, G. D., Rossig, C., Russell, H. V., Dotti, G., et al. (2008). Virus-specific T cells engineered to coexpress tumor-specific receptors: Persistence and antitumor activity in individuals with neuroblastoma. *Nature Medicine, 14*, 1264–1270.
26. Brentjens, R. J., Davila, M. L., Riviere, I., Park, J., Wang, X., Cowell, L. G., et al. (2013). CD19-targeted T cells rapidly induce molecular remissions in adults with chemotherapy-refractory acute lymphoblastic leukemia. *Science Translational Medicine, 5*, 177ra138.
27. Lee, D. W., Kochenderfer, J. N., Stetler-Stevenson, M., Cui, Y. K., Delbrook, C., Feldman, S. A., et al. (2015). T cells expressing CD19 chimeric antigen receptors for acute lymphoblastic leukaemia in children and young adults: A phase 1 dose-escalation trial. *Lancet, 385*, 517–528.

28. Maude, S. L., Frey, N., Shaw, P. A., Aplenc, R., Barrett, D. M., Bunin, N. J., et al. (2014). Chimeric antigen receptor T cells for sustained remissions in leukemia. *The New England Journal of Medicine, 371*, 1507–1517.
29. Barrett, D. M., Teachey, D. T., & Grupp, S. A. (2014). Toxicity management for patients receiving novel T-cell engaging therapies. *Current Opinion in Pediatrics, 26*, 43–49.
30. Neelapu, S. S., Tummala, S., Kebriaei, P., Wierda, W., Locke, F. L., Lin, Y., et al. (2018). Toxicity management after chimeric antigen receptor T cell therapy: One size does not fit 'ALL'. *Nature Reviews. Clinical Oncology, 15*, 218.
31. Pallin, D. J., Baugh, C. W., Postow, M. A., Caterino, J. M., Erickson, T. B., & Lyman, G. H. (2018). Immune-related adverse events in cancer patients. *Academic Emergency Medicine, 25*(7), 819–827.
32. Gust, J., Hay, K. A., Hanafi, L. A., Li, D., Myerson, D., Gonzalez-Cuyar, L. F., et al. (2017). Endothelial activation and blood-brain barrier disruption in neurotoxicity after adoptive immunotherapy with CD19 CAR-T cells. *Cancer Discovery, 7*(12), 1404–1419.
33. Hay, K. A., Hanafi, L. A., Li, D., Gust, J., Liles, W. C., Wurfel, M. M., et al. (2017). Kinetics and biomarkers of severe cytokine release syndrome after CD19 chimeric antigen receptor-modified T cell therapy. *Blood, 130*(21), 2295–2306.
34. Chen, F., Teachey, D. T., Pequignot, E., Frey, N., Porter, D., Maude, S. L., et al. (2016). Measuring IL-6 and sIL-6R in serum from patients treated with tocilizumab and/or siltuximab following CAR T cell therapy. *Journal of Immunological Methods, 434*, 1–8.
35. Nishimoto, N., Terao, K., Mima, T., Nakahara, H., Takagi, N., & Kakehi, T. (2008). Mechanisms and pathologic significances in increase in serum interleukin-6 (IL-6) and soluble IL-6 receptor after administration of an anti-IL-6 receptor antibody, tocilizumab, in patients with rheumatoid arthritis and Castleman disease. *Blood, 112*, 3959–3964.
36. Teachey, D. T., Lacey, S. F., Shaw, P. A., Melenhorst, J. J., Maude, S. L., Frey, N., et al. (2016). Identification of predictive biomarkers for cytokine release syndrome after chimeric antigen receptor T-cell therapy for acute lymphoblastic leukemia. *Cancer Discovery, 6*, 664–679.
37. Fitzgerald, J. C., Weiss, S. L., Maude, S. L., Barrett, D. M., Lacey, S. F., Melenhorst, J. J., et al. (2017). Cytokine release syndrome after chimeric antigen receptor T cell therapy for acute lymphoblastic leukemia. *Critical Care Medicine, 45*, e124–e131.
38. Maude, S. L., Barrett, D., Teachey, D. T., & Grupp, S. A. (2014). Managing cytokine release syndrome associated with novel T cell-engaging therapies. *Cancer Journal, 20*, 119–122.
39. Barrett, D. M., Zhao, Y., Liu, X., Jiang, S., Carpenito, C., Kalos, M., et al. (2011). Treatment of advanced leukemia in mice with mRNA engineered T cells. *Human Gene Therapy, 22*, 1575–1586.
40. Kochenderfer, J. N., Yu, Z., Frasheri, D., Restifo, N. P., & Rosenberg, S. A. (2010). Adoptive transfer of syngeneic T cells transduced with a chimeric antigen receptor that recognizes murine CD19 can eradicate lymphoma and normal B cells. *Blood, 116*, 3875–3886.
41. Wunderlich, M., Stockman, C., Devarajan, M., Ravishankar, N., Sexton, C., Kumar, A. R., et al. (2016). A xenograft model of macrophage activation syndrome amenable to anti-CD33 and anti-IL-6R treatment. *JCI Insight, 1*, e88181.
42. Ureshino, H., Ando, T., Kizuka, H., Kusaba, K., Sano, H., Nishioka, A., et al. (2017). Tocilizumab for severe cytokine-release syndrome after haploidentical donor transplantation in a patient with refractory Epstein-Barr virus-positive diffuse large B-cell lymphoma. *Hematological Oncology, 36*(1), 324–327.
43. Abboud, R., Keller, J., Slade, M., DiPersio, J. F., Westervelt, P., Rettig, M. P., et al. (2016). Severe cytokine-release syndrome after T cell-replete peripheral blood haploidentical donor transplantation is associated with poor survival and anti-IL-6 therapy is safe and well tolerated. *Biology of Blood and Marrow Transplantation, 22*, 1851–1860.
44. Singh, N., Hofmann, T. J., Gershenson, Z., Levine, B. L., Grupp, S. A., Teachey, D. T., et al. (2017). Monocyte lineage-derived IL-6 does not affect chimeric antigen receptor T-cell function. *Cytotherapy, 19*, 867–880.
45. Billiau, A. D., Roskams, T., Van Damme-Lombaerts, R., Matthys, P., & Wouters, C. (2005). Macrophage activation syndrome: Characteristic findings on liver biopsy illustrating the key

role of activated, IFN-gamma-producing lymphocytes and IL-6- and TNF-alpha-producing macrophages. *Blood, 105*, 1648–1651.
46. Casper, C., Chaturvedi, S., Munshi, N., Wong, R., Qi, M., Schaffer, M., et al. (2015). Analysis of inflammatory and anemia-related biomarkers in a randomized, double-blind, placebo-controlled study of siltuximab (anti-il6 monoclonal antibody) in patients with multicentric castleman disease. *Clinical Cancer Research, 21*, 4294–4304.
47. Ferrario, A., Merli, M., Basilico, C., Maffioli, M., & Passamonti, F. (2017). Siltuximab and hematologic malignancies. A focus in non Hodgkin lymphoma. *Expert Opinion on Investigational Drugs, 26*, 367–373.
48. Zhang, H., Neuhofer, P., Song, L., Rabe, B., Lesina, M., Kurkowski, M. U., et al. (2013). IL-6 trans-signaling promotes pancreatitis-associated lung injury and lethality. *The Journal of Clinical Investigation, 123*, 1019–1031.
49. Wolf, J., Waetzig, G. H., Chalaris, A., Reinheimer, T. M., Wege, H., Rose-John, S., et al. (2016). Different soluble forms of the interleukin-6 family signal transducer gp130 fine-tune the blockade of interleukin-6 trans-signaling. *The Journal of Biological Chemistry, 291*, 16186–16196.
50. Scheller, J., Garbers, C., & Rose-John, S. (2014). Interleukin-6: From basic biology to selective blockade of pro-inflammatory activities. *Seminars in Immunology, 26*, 2–12.
51. Gowin, K., Kosiorek, H., Dueck, A., Mascarenhas, J., Hoffman, R., Reeder, C., et al. (2017). Multicenter phase 2 study of combination therapy with ruxolitinib and danazol in patients with myelofibrosis. *Leukemia Research, 60*, 31–35.
52. Das, R., Guan, P., Sprague, L., Verbist, K., Tedrick, P., An, Q. A., et al. (2016). Janus kinase inhibition lessens inflammation and ameliorates disease in murine models of hemophagocytic lymphohistiocytosis. *Blood, 127*, 1666–1675.
53. Maschalidi, S., Sepulveda, F. E., Garrigue, A., Fischer, A., & de Saint Basile, G. (2016). Therapeutic effect of JAK1/2 blockade on the manifestations of hemophagocytic lymphohistiocytosis in mice. *Blood, 128*, 60–71.
54. Sin, J. H., & Zangardi, M. L. (2017). Ruxolitinib for secondary hemophagocytic lymphohistiocytosis: First case report. *Hematology/Oncology and Stem Cell Therapy.*
55. Broglie, L., Pommert, L., Rao, S., Thakar, M., Phelan, R., Margolis, D., et al. (2017). Ruxolitinib for treatment of refractory hemophagocytic lymphohistiocytosis. *Blood Advances, 1*, 1533–1536.

Anti-interferon-γ Therapy for Cytokine Storm Syndromes

Fabrizio De Benedetti

Cytokine storm syndromes (CSS) are classical examples of hyperinflammation. In hyperinflammation, activation of the inflammatory response in an excessive manner (involving both innate and adaptive immune cells) consequent to a reasonable stimulus to do it (e.g., viral infection) leads to damage to the host. As described in other chapters, HLH is characteristically associated with excessive activation and expansion of T cells and macrophages. A vast body of evidence provides support to a central role of exaggerated production of IFN-γ in causing hypercytokinemia and signs and symptoms of HLH. Emapalumab, an antibody against IFN-γ, has been recently approved by the Food and Drug Administration of the USA in treatment-experienced pHLH patients (see below). At present emapalumab is being further evaluated in p-HLH, including first-line (NCT03312751 and NCT03312751), and in MAS in the context of systemic juvenile idiopathic arthritis (sJIA) (NCT03311854). Two additional antibodies to IFN-γ, fontolizumab and AMG-811, have been generated. These have not been investigated in HLH, and their development has been terminated. In this chapter we do the following:

1. Discuss briefly the roles of IFN-γ in innate and adaptive immunity and in host defense.
2. Summarize results from animal models of primary and secondary HLH with a particular emphasis on therapeutic approaches.
3. Review data on biomarkers associated with IFN-γ elevation.
4. Discuss initial efficacy and safety results of IFN-γ neutralization in humans.

F. De Benedetti (✉)
Division of Rheumatology, Ospedale Pediatrico Bambino Gesù, Rome, Italy
e-mail: fabrizio.debenedetti@opbg.net

R. Q. Cron, E. M. Behrens (eds.), *Cytokine Storm Syndrome*,
https://doi.org/10.1007/978-3-030-22094-5_33

IFN-γ: Effects on Innate and Adaptive Immunity

Interferons (IFN) are a family of cytokines that have been identified through their central role in response to viral infections. IFN-γ (also known as type II IFN or immune IFN) is a pleiotropic cytokine that, differently from other IFNs, plays a central role in immune system signaling, modulating innate and adaptive immune responses to viral and bacterial challenges [1]. IFN-γ is secreted predominantly by T cells and NK cells [2]. IFN-γ signals as an antiparallel homodimer through the high-affinity (IFN-γR1) and low-affinity (IFN-γR2) receptors [3, 4]. While the IFN-γ receptor α chain is required for ligand binding and signaling, the IFN-γ receptor β chain is required primarily for signaling [3, 4]. These receptors are present on a very wide range of immune system and other cells, including CD4+ and CD8+ T cells, B cells, NK cells, plasmacytoid dendritic cells (DCs), macrophages, platelets, eosinophils, phagocytes, astrocytes, endothelial and epithelial cells, hepatocytes, fibroblasts, and keratinocytes [4].

In NK cells, interleukin (IL)-12 and type I IFN induce IFN-γ expression through pathways acting on the transcription factor STAT4 [5]. Production and release of IFN-γ by NK cells is typically stimulated by the combination of IL-12 and IL-18 [6]. Also in CD8+ T cells and CD4+ T helper cells the combination of IL-12 and IL-18 promotes production and release of IFN-γ [2]. It is important to note that neither IL-12 nor IL-18 alone are sufficient to induce elevated production of IFN-γ from NK cells.

IFN-γ has been demonstrated to have a key role in TLR responses and on M1 macrophage differentiation and subsequent cytokine production. It induces expression of IRF1 and IRF5 [7]. These regulatory factors stimulate the expression of genes encoding pro-inflammatory cytokines, such as IL-1β, IL-6, IL-12, and IL-23, as well as suppress expression of the anti-inflammatory cytokine IL-10 [8–10]. IFN-γ promotes antimicrobial activity via upregulation of microbicidal gene products enhances macrophage phagocytic ability and it is pivotal in granuloma formation [7, 11, 12].

IFN-γ also affects the early phases of the adaptive immune response by promoting DC maturation and T cell differentiation [13, 14]. It plays an important role in driving Th1 responses, repressing the development of Th2 and Th17 T cell responses. It directly affects adaptive immunity also by promoting lymphocyte recruitment to germinal centers [15]. IFN-γ then promotes immunoglobulin G class switching from IgG_2 to IgG_3 [12, 16].

IFN-γ: Role in Defense from Infections

As outlined above, IFN-γ has several effects on innate and adaptive immune response. When therapeutic neutralization of IFN-γ is applied to humans, it is important to know the evidence supporting its role in defense from infections.

Studies in animal models have provided evidence for a role of IFN-γ in some types of infections. Cytokines released from macrophages have been shown to play a key role in host defense against *Salmonella* infection [17]. IFN-γ increases production of IL-1 and IL-2, which is inhibited by *S. typhi*, as well as of TNF. IFN-γ deficient mice develop disseminated infection after *Salmonella* challenge, and this does not occur in wild-type mice [18]. IFN-γ deficient mice fail to increase the numbers of CD4+ and CD8+ cells in the gut and to upregulate the expression of major histocompatibility complex II adhesion molecules following *Salmonella* challenge [18, 19]. IFN-γ, inhibited CMV plaque formation, even more when in combination with IFN-α plus IFN-β [20]. IFN-γ knockout BALB/c mice infected with CMV have higher viral load and indeed develop more severe disease with a full spectrum of symptoms and laboratory abnormalities, including splenomegaly, coagulopathy, hemophagocytosis, and cytopenia [21]. It has also been shown that mice incapable of synthesizing IFN-γ do not produce reactive nitrogen intermediates and are unable to restrict the growth of *Mycobacterium tuberculosis*. These mice exhibit increased tissue necrosis and have a rapidly fatal course of tuberculosis [22]. IFN-γ is also essential for induction of protective anti-microsporidial immunity in animals [23].

More relevantly, information on the role of IFN-γ in the defense from infections is also provided by data in humans. Humans with genetic mutations leading to IFN-γ receptor deficiency characteristically have increased susceptibility to typical and atypical mycobacterial infections, as well as reactivation of *Varicella zoster* virus [24–26]. Infections with CMV, herpes simplex virus, *Varicella zoster* virus, respiratory syncytial virus, parainfluenza virus type 3, and *Salmonella* have also been reported in patients with missing or partially defective IFN-γ receptors [24, 27]. In contrast, individuals with IFN-γ signaling deficiencies do not have elevated frequency of infections caused by extracellular bacteria and, usually, have only limited symptoms after infections by common exanthematous viruses, including chicken pox and measles [28]. Similarly, individuals with autoantibodies to IFN-γ are also at increased risk for the development of infections caused by mycobacteria and have more frequent herpes zoster reactivations [29, 30], while other infections have a normal course.

Neutralization of IFN-γ in Animal Models of HLH

In healthy mice, the injection of IFN-γ directly causes activation of macrophages with hemophagocytosis and anemia, demonstrating a direct effect of IFN-γ in the absence of other inflammatory stimuli [31]. In several murine models of CSS demonstration of increased IFN-γ production and reversal of disease features following IFN-g neutralization has been observed [32–37] (summarized in Table 1). *Perforin*-deficient mice when infected with lymphocytic choriomeningitic virus (LCMV) develop HLH and invariably die. Administration of an antibody directed to IFN-γ reversed mortality and clinical and laboratory features of HLH, indicating that IFN-γ was essential for the development of the disease [38]. Depletion of CD8+

Table 1 Murine models of CSS in genetically modified mice and wild-type mice and role of IFN-γ in these models

	Mutation	Model and trigger	Increased IFN-γ production	IFN-γ blockade	
Primary HLH (FHLH2)	*PRF1*	LCMV-infection	Yes	Reversal of disease parameters	[38]
Primary HLH (FHLH3)	*UNC13D*	LCMV infection	Yes	Not tested	[33]
Primary HLH (FHLH4)	*STX11*	LCMV-infection	Yes	Not tested	[35]
Griscelli syndrome type 2	*RAB27A*	LCMV-infection	Yes	Reversal of disease parameters	[36]
XLP1	*SH2D1A*	LCMV-infection	Yes	Not tested	[34]
Infection-associated sHLH	None	TLR-9 repeated stimulation	Yes	Reversal of disease parameters	[32]
MAS	None	TLR stimulation in IL-6 transgenics	Yes	Reversal of disease parameters	[37]

cells achieved essentially the same effects suggesting that in this model CD8+ T cells are the source of IFN-γ. The ability of an anti-IFN-γ antibody to reverse mortality and clinical and laboratory features of HLH, initially demonstrated in the perforin-deficient mice, was also observed in a mouse model of pHLH due to *RAB27A* deficiency [36].

In a mouse model of sHLH, in which the infectious trigger is mimicked by repeated stimulation of TLR-9 with CpG DNA, elevated IFN-γ levels have been found. In IFN-γ knockout mice, anemia, thrombocytopenia, and splenomegaly are reduced, and hepatic inflammation is inhibited [32]. In the same model, tissue (i.e., liver and spleen) levels of IFN-γ were closely associated with hypercytokinemia; antibody-mediated neutralization of the high rate of tissue production of IFN-γ was required for a significant reduction in inflammatory cytokines and improvement in syndrome parameters [39].

Mice overexpressing IL-6 (IL-6TG mice) have been used as an experimental model of MAS associated to sJIA. They mimic a chronic inflammatory condition similar to sJIA, which is indeed characterized by high levels of IL-6 [40]. When challenged with LPS, as a mimicker of an infectious trigger, IL-6TG mice develop MAS [41] that is associated with a significant upregulation of the IFN-γ pathway [37]. This model replicates the "two-hit" mechanism that appears to be involved in the occurrence of MAS in patients with sJIA triggered by infection in the presence of active disease. Administration of an anti-IFN-γ antibody to LPS-challenged IL-6TG mice improved survival and caused reductions in ferritin, fibrinogen, alanine aminotransferase, and in pro-inflammatory cytokine levels, including IL-1β, IL-6, and TNF [37].

Altogether, these data demonstrate that therapeutic neutralization of IFN-γ results in marked amelioration of signs and symptoms in animal models of pHLH, sHLH, and MAS. It must be pointed out that data in animal models suggest that, together with overproduction of IFN-γ, other pathways may contribute in a complementary manner to the development of sign and symptoms of HLH. In perforin-deficient mice, genetic deletion of CD25 restricted to CD8+ cells suppresses the hyperinflammatory state and improves the mouse life span, suggesting the contribution from a pathway involving IL-2 and T cell activation [42]. In the model of infection-associated sHLH triggered by repeated TLR-9 stimulation, IFN-γ deficient mice were shown to develop the disease only following administration of CpG in combination with IFN-γ, providing evidence that cooperation between TLR-9 and IFN-γ dependent signals is necessary and that this cooperation involves expansion and activation of cells of the myeloid lineage [43].

The IFN-γ Pathway in Patients with Cytokine Storm Syndromes

Initial studies on small number of patients with pHLH have shown elevated levels of IFN-γ, and this was also reported later in sHLH and MAS [44–48]. In a large study on several hundreds of patients admitted with fever in a hematology-oncology unit, the patients with a final diagnosis of HLH had markedly elevated IFN-γ, compared to patients with other causes of fever [49]. IFN-γ was correlated with levels of alanine aminotransferase, aspartate aminotransferase, bilirubin, lactate dehydrogenase, triglycerides, and fibrinogen [50]. Also in sHLH, blood levels of IFN-γ were correlated with decreased neutrophil and platelet counts, and elevations in ferritin, lactate dehydrogenase, and alanine transaminase levels [39]. Notably, in patients with MAS on a background of sJIA, circulating levels of IFN-γ were significantly elevated compared to patients with active sJIA without MAS at sampling and to patients with inactive sJIA. Interestingly, levels of IFN-γ were similarly low in patients with active or inactive sJIA [51]. Conversely, during MAS there was no increase in circulating levels of IL-6 and IL-1β, two inflammatory cytokines that play major roles in the inflammatory cascade of sJIA [52, 53]. Altogether, these observations are consistent with the conclusion that activation of IFN-γ production occurs only when MAS is present and that it is not part of the typical inflammatory cascade of the background inflammatory disease [51].

Typically in patients with MAS on a background of sJIA, but also in some patients with infection-associated sHLH, levels of IL-18, and of free IL-18, are markedly increased [54, 55]. In contrast, they do not appear to be increased in approximately half of the patients with infection-associated sHLH and in pHLH [55]. IL-18 is also typically overproduced in two monogenic diseases characterized by recurrences of HLH with hyperinflammation, namely NLRC4-induced disease and NOCARH syndrome [56]. Indeed, as mentioned above, IL-18 is an inducer of IFN-γ; however, in vitro induction of IFN-γ production cannot be induced by IL-18 alone, but a second stimulus is required, typically IL-12. Consistently, IL-18 binding protein (a protein that binds to and inhibits the activity of IL-18) deficient mice do not show pathology at baseline. However, when these mice are triggered by repeated TLR-9 stimulation, severe signs and symptoms of HLH are induced and are reversed by an anti-IFN-γ antibody [57], suggesting that IFN-γ is the final mediator.

Results from a mouse model of sHLH showed fluctuation of IFN-γ levels in blood during sustained progression of disease [39]. Most probably this is due to the very quick blood clearance of the cytokine and to the high affinity binding of IFN-γ to extracellular matrix and heparan sulfate, in addition to the binding to its cellular receptor [58]. All together, these events cause its accumulation mainly in tissues, typically spleen, liver, and kidney [59]. Experiments in the above-described TLR-9 stimulation model of sHLH, in which IFN-γ in tissues was captured by an administered antibody against IFN-γ, demonstrated a striking increase in the levels of the cytokine complexed to the antibody. This cytokine–antibody complex is drained from tissues into the blood compartment allowing for an estimate of the total amount of IFN-γ produced. IFN-γ levels, as measured by the amount of IFN-γ bound to the administered anti-IFN-γ antibody, were several hundred fold higher than the amount measured in blood with conventional methods [39]. In humans with HLH (in its various forms), a sizeable proportion has low or even undetectable levels of IFN-γ even in the presence of full-blown disease. This may indeed be due to the limitations of available assays and/or to the fact that measurement of blood levels may be indicative of increased IFN-γ production, but may not strictly reflect the total production of IFN-γ in affected tissues, as also shown by the above mentioned data in animal models.

An increasing body of evidence shows that CXCL9 levels are significantly increased both in experimental models of HLH [37, 39], and in patients with HLH, independently of the form being primary or secondary [55]. CXCL9, also known as monokine induced by IFN-γ (MIG), is a member of CXC chemokine family known to attract CXCR3 expressing T lymphocytes. It is produced by monocytes, macrophages, antigen-presenting cells, endothelial cells, and stromal cells following stimulation by IFN-γ.

In patients, CXCL9 is elevated in pHLH and sHLH, and its levels are strictly correlated with key disease parameters and decrease with disease remission [39, 60]. Higher levels are associated with early mortality [61]. CXCL9 levels are elevated in patients with MAS and are correlated with laboratory parameters that reflect disease severity, including ferritin levels, cytopenia, and alanine aminotransferase levels, suggesting that CXCL9 is a potential biomarker for IFN-γ biological activity in patients with MAS during sJIA [51]. CXCL9 has also been found to be elevated in adults with MAS occurring on a background of adult-onset Still disease [62].

The observation in animals with sHLH that tissue levels of IFN-γ are more closely correlated with organ involvement prompted the suggestion that blood levels of CXCL9 could be employed as a marker of IFN-γ overproduction in target tissues [39]. Although in a very limited number of patients, data consistent with increased expression of IFN-γ and of IFN-γ-inducible proteins within tissues targeted by the disease have been reported. In livers and lymph nodes, overexpression of IFN-γ, Stat1 phosphorylation, and overexpression of IFN-γ inducible genes (such as *IDO* and *CXCL9* and *CXCL10*) has been found [47, 63] (our unpublished observation).

Altogether, these data suggest that CXCL9 appears to be a useful biomarker for hyperinflammation caused by excessive production of IFN-γ: CXCL9 is stable, compared to IFN-γ itself, is easily measurable in serum at ng/ml concentrations, has a clear association with pathologic pathways underlying the disease/syndrome, and correlates with disease severity and response to treatments.

Targeting IFN-γ in Humans with Cytokine Storm Syndromes

At present, emapalumab is the only anti-IFN-γ antibody in clinical development and is already approved for the treatment of primary HLH patients. Data discussed in this paragraph are derived from published abstracts, publically available slide presentations at the annual meeting of the American Society of Hematology [64], as well as from the multi-discipline review released by the Center for Drug Evaluation and Research of the FDA, as a rationale for the approval of emapalumab in pHLH. (https://www.accessdata.fda.gov/drugsatfda_docs/nda/2018/761107Orig1s000MultidisciplineR.pdf). Emapalumab is approved in the USA for the treatment of adult and pediatric (newborn and older) patients with pHLH with refractory, recurrent, or progressive disease, or intolerance with conventional HLH therapy (https://www.accessdata.fda.gov/drugsatfda_docs/label/2018/761107lbl.pdf).

Emapalumab binds with high affinity (1.4 pM) to human IFN-γ, and with fast association phases and very slow dissociation phases, as expected for a high affinity IgG. Emapalumab was tested in a multicenter, open-label, single-arm trial, conducted in the USA and Europe, in pediatric patients with suspected or confirmed primary HLH with either refractory, recurrent, or progressive disease during conventional HLH therapy or who were intolerant of conventional HLH therapy. Twenty-seven patients were enrolled with a median age of 1 year (range 0.2–13 years). A genetic confirmation with homozygous or compound heterozygosity for genes known to cause primary HLH was present in 82% of patients.

In healthy subjects, emapalumab elimination is linear with a half-life of 23 days, in the range of what is expected for a human IgG1. In patients with HLH, emapalumab pharmacokinetics is highly influenced by the production rate of IFN-γ, implying target mediated drug disposition, with a half-life ranging between 2.5–19 days. In the study, all patients received an initial starting dose of 1 mg/kg every 3 days. Subsequent doses, every 3 days, could be increased to a maximum of 10 mg/kg based on clinical and laboratory parameters. Treatment duration was up to 8 weeks, after which patients could continue treatment on the extension study. Forty-four percent of patients remained at a dose of 1 mg/kg, 30% of patients increased to 3–4 mg/kg, and 26% of patients increased to 6–10 mg/kg.

All patients received dexamethasone as background HLH treatment with doses between 5–10 mg/m^2 per day. Cyclosporine A was continued if administered prior to screening. Patients receiving intrathecal methotrexate and glucocorticoids at baseline could continue these therapies.

Twenty of the 27 patients (74%) completed the study. Seven patients (26%) were prematurely withdrawn. Two patients died during the study, and there were additional six deaths during the long-term extension study. Twelve months estimated overall survival was 73.0%. As hematopoietic stem cell transplantation (HSCT) is the only curative approach to pHLH, a significant number of patients moved to HSCT: 12-month estimated post-HSCT survival was 89.5%.

The efficacy of emapalumab was based on overall response rate (ORR) at the end of treatment, defined as achievement of either a complete response (CR), or partial response (PR), or HLH improvement. ORR was evaluated using an algorithm that included the following objective clinical and laboratory parameters: fever, splenomegaly, CNS symptoms, complete blood count, fibrinogen and/or D-dimer, ferritin, and soluble CD25 levels. CR was defined as normalization of all HLH abnormalities (i.e., no fever, no splenomegaly, neutrophils >1 × 10^9/L, platelets >100 × 10^9/L, ferritin 1.5 g/L, D-dimer twofold over baseline). PR was defined as normalization of ≥3 HLH abnormalities. HLH improvement was defined as ≥3 HLH abnormalities improved by at least 50% from baseline. The ORR was observed in 17/27 (63%; 95% confidence interval 0.42–0.81) and included 7 (26%) complete responses and 8 (30%) partial responses.

The exposure–response analysis for efficacy showed that the clinical response at the end of the treatment was highly correlated with serum concentrations of CXCL9, suggesting a direct relation between effective neutralization of IFN-γ, inhibition of CXCL9 production, and therapeutic response.

In the study exposure was relatively brief with a median number of days of exposure of approximately 60 days. Only one patient withdrew due to an adverse event, supporting the good tolerability of IFN-γ neutralization in a fragile pediatric population with considerable underlying morbidities and concomitant medications. Infections are of particular concern for patients with pHLH, and contribute to morbidity and mortality. In 32% of patients, serious infections were observed, with 41% being viral, 35% bacterial, and 9% fungal. Based on the available information in mice and from toxicology studies with emapalumab in monkeys, infections such as mycobacterial, herpes, histoplasmosis, and salmonella, which can be favored by IFN-γ neutralization, are of potential concern in patients receiving emapalumab therapy. Patients were screened for mycobacteria as well as Salmonella and Shigella, prior to and during study treatment. All patients in the study received prophylaxis for herpes simplex virus per study protocol, and *P. jirovecii* pneumonia and fungal prophylaxes were administered per institutional standards. Infections caused by pathogens potentially favored by IFN-ũFE; neutralization occurred in one patient (disseminated histoplasmosis), and resolved with appropriate treatment.

Information is also available from very few case reports published (compassionate use). These include a 20-month-old boy with refractory, Epstein–Barr virus (EBV)-associated HLH who was successfully treated with emapalumab, despite severe preexisting comorbidities (severe pancytopenia, gastrointestinal bleeding, central nervous system hemorrhage), including multiple life-threatening infections (CMV, EBV, adenovirus, and Trichosporon fungemia) [65]. In a patient carrying a pathogenic NLRC4 mutation with severe inflammation, recalcitrant HLH, and highly elevated levels of IFN-γ and CXCL9, neutralization of IFN-γ with emapalumab resulted in control of all HLH disease features [66]. Thus, anti-IFN-γ therapy appears to be a new therapeutic option for CSS.

References

1. Ivashkiv, L. B. (2018). IFNgamma: Signalling, epigenetics and roles in immunity, metabolism, disease and cancer immunotherapy. *Nature Reviews. Immunology, 18*, 545–558.
2. Schoenborn, J. R., & Wilson, C. B. (2007). Regulation of interferon-gamma during innate and adaptive immune responses. *Advances in Immunology, 96*, 41–101.
3. Blouin, C. M., & Lamaze, C. (2013). Interferon gamma receptor: The beginning of the journey. *Frontiers in Immunology, 4*, 267.
4. de Weerd, N. A., & Nguyen, T. (2012). The interferons and their receptors—Distribution and regulation. *Immunology and Cell Biology, 90*, 483–491.
5. Suarez-Ramirez, J. E., Tarrio, M. L., Kim, K., Demers, D. A., & Biron, C. A. (2014). CD8 T cells in innate immune responses: Using STAT4-dependent but antigen-independent pathways to gamma interferon during viral infection. *MBio, 5*, e01978–e01914.
6. Kannan, Y., Yu, J., Raices, R. M., Seshadri, S., Wei, M., Caligiuri, M. A., et al. (2011). IkappaBzeta augments IL-12- and IL-18-mediated IFN-gamma production in human NK cells. *Blood, 117*, 2855–2863.
7. Majoros, A., Platanitis, E., Kernbauer-Holzl, E., Rosebrock, F., Muller, M., & Decker, T. (2017). Canonical and non-canonical aspects of JAK-STAT signaling: Lessons from interferons for cytokine responses. *Frontiers in Immunology, 8*, 29.
8. Xie, C., Liu, C., Wu, B., Lin, Y., Ma, T., Xiong, H., et al. (2016). Effects of IRF1 and IFN-beta interaction on the M1 polarization of macrophages and its antitumor function. *International Journal of Molecular Medicine, 38*, 148–160.
9. Chistiakov, D. A., Myasoedova, V. A., Revin, V. V., Orekhov, A. N., & Bobryshev, Y. V. (2018). The impact of interferon-regulatory factors to macrophage differentiation and polarization into M1 and M2. *Immunobiology, 223*, 101–111.
10. Sica, A., & Mantovani, A. (2012). Macrophage plasticity and polarization: In vivo veritas. *The Journal of Clinical Investigation, 122*, 787–795.
11. Asano, M., Nakane, A., & Minagawa, T. (1993). Endogenous gamma interferon is essential in granuloma formation induced by glycolipid-containing mycolic acid in mice. *Infection and Immunity, 61*, 2872–2878.
12. Green, D. S., Young, H. A., & Valencia, J. C. (2017). Current prospects of type II interferon gamma signaling and autoimmunity. *The Journal of Biological Chemistry, 292*, 13925–13933.
13. Pearl, J. E., Saunders, B., Ehlers, S., Orme, I. M., & Cooper, A. M. (2001). Inflammation and lymphocyte activation during mycobacterial infection in the interferon-gamma-deficient mouse. *Cellular Immunology, 211*, 43–50.
14. Swindle, E. J., Brown, J. M., Radinger, M., DeLeo, F. R., & Metcalfe, D. D. (2015). Interferon-gamma enhances both the anti-bacterial and the pro-inflammatory response of human mast cells to Staphylococcus aureus. *Immunology, 146*, 470–485.
15. Choi, J., Kim, S. T., & Craft, J. (2012). The pathogenesis of systemic lupus erythematosus-an update. *Current Opinion in Immunology, 24*, 651–657.
16. Schroder, K., Hertzog, P. J., Ravasi, T., & Hume, D. A. (2004). Interferon-gamma: An overview of signals, mechanisms and functions. *Journal of Leukocyte Biology, 75*, 163–189.
17. Fidan, I., Yesilyurt, E., Gurelik, F. C., Erdal, B., & Imir, T. (2008). Effects of recombinant interferon-gamma on cytokine secretion from monocyte-derived macrophages infected with Salmonella typhi. *Comparative Immunology, Microbiology and Infectious Diseases, 31*, 467–475.

18. Bao, S., Beagley, K. W., France, M. P., Shen, J., & Husband, A. J. (2000). Interferon-gamma plays a critical role in intestinal immunity against Salmonella typhimurium infection. *Immunology, 99*, 464–472.
19. van de Berg, P. J., Heutinck, K. M., Raabe, R., Minnee, R. C., Young, S. L., van Donselaar-van der Pant, K. A., et al. (2010). Human cytomegalovirus induces systemic immune activation characterized by a type 1 cytokine signature. *The Journal of Infectious Diseases, 202*, 690–699.
20. Sainz Jr., B., LaMarca, H. L., Garry, R. F., & Morris, C. A. (2005). Synergistic inhibition of human cytomegalovirus replication by interferon-alpha/beta and interferon-gamma. *Virology Journal, 2*, 14.
21. Brisse, E., Imbrechts, M., Put, K., Avau, A., Mitera, T., Berghmans, N., et al. (2016). Mouse cytomegalovirus infection in BALB/c mice resembles virus-associated secondary hemophagocytic lymphohistiocytosis and shows a pathogenesis distinct from primary hemophagocytic lymphohistiocytosis. *Journal of Immunology, 196*, 3124–3134.
22. Flynn, J. L., Chan, J., Triebold, K. J., Dalton, D. K., Stewart, T. A., & Bloom, B. R. (1993). An essential role for interferon gamma in resistance to Mycobacterium tuberculosis infection. *The Journal of Experimental Medicine, 178*, 2249–2254.
23. Salat, J., Sak, B., Le, T., & Kopecky, J. (2004). Susceptibility of IFN-gamma or IL-12 knock-out and SCID mice to infection with two microsporidian species, Encephalitozoon cuniculi and E. intestinalis. *Folia Parasitologica, 51*, 275–282.
24. Dorman, S. E., Uzel, G., Roesler, J., Bradley, J. S., Bastian, J., Billman, G., et al. (1999). Viral infections in interferon-gamma receptor deficiency. *The Journal of Pediatrics, 135*, 640–643.
25. Remus, N., Reichenbach, J., Picard, C., Rietschel, C., Wood, P., Lammas, D., et al. (2001). Impaired interferon gamma-mediated immunity and susceptibility to mycobacterial infection in childhood. *Pediatric Research, 50*, 8–13.
26. Tran, D. Q. (2005). Susceptibility to mycobacterial infections due to interferon-gamma and interleukin-12 pathway defects. *Allergy and Asthma Proceedings, 26*, 418–421.
27. Sologuren, I., Boisson-Dupuis, S., Pestano, J., Vincent, Q. B., Fernandez-Perez, L., Chapgier, A., et al. (2011). Partial recessive IFN-gammaR1 deficiency: Genetic, immunological and clinical features of 14 patients from 11 kindreds. *Human Molecular Genetics, 20*, 1509–1523.
28. Lammas, D. A., Casanova, J. L., & Kumararatne, D. S. (2000). Clinical consequences of defects in the IL-12-dependent interferon-gamma (IFN-gamma) pathway. *Clinical and Experimental Immunology, 121*, 417–425.
29. Kampmann, B., Hemingway, C., Stephens, A., Davidson, R., Goodsall, A., Anderson, S., et al. (2005). Acquired predisposition to mycobacterial disease due to autoantibodies to IFN-gamma. *The Journal of Clinical Investigation, 115*, 2480–2488.
30. Wongkulab, P., Wipasa, J., Chaiwarith, R., & Supparatpinyo, K. (2013). Autoantibody to interferon-gamma associated with adult-onset immunodeficiency in non-HIV individuals in Northern Thailand. *PLoS One, 8*, e76371.
31. Zoller, E. E., Lykens, J. E., Terrell, C. E., Aliberti, J., Filipovich, A. H., Henson, P. M., et al. (2011). Hemophagocytosis causes a consumptive anemia of inflammation. *The Journal of Experimental Medicine, 208*, 1203–1214.
32. Behrens, E. M., Canna, S. W., Slade, K., Rao, S., Kreiger, P. A., Paessler, M., et al. (2011). Repeated TLR9 stimulation results in macrophage activation syndrome-like disease in mice. *The Journal of Clinical Investigation, 121*, 2264–2277.
33. Crozat, K., Hoebe, K., Ugolini, S., Hong, N. A., Janssen, E., Rutschmann, S., et al. (2007). Jinx, an MCMV susceptibility phenotype caused by disruption of Unc13d: A mouse model of type 3 familial hemophagocytic lymphohistiocytosis. *The Journal of Experimental Medicine, 204*, 853–863.
34. Czar, M. J., Kersh, E. N., Mijares, L. A., Lanier, G., Lewis, J., Yap, G., et al. (2001). Altered lymphocyte responses and cytokine production in mice deficient in the X-linked lymphoproliferative disease gene SH2D1A/DSHP/SAP. *Proceedings of the National Academy of Sciences of the United States of America, 98*, 7449–7454.

35. Kogl, T., Muller, J., Jessen, B., Schmitt-Graeff, A., Janka, G., Ehl, S., et al. (2013). Hemophagocytic lymphohistiocytosis in syntaxin-11-deficient mice: T-cell exhaustion limits fatal disease. *Blood, 121*, 604–613.
36. Pachlopnik Schmid, J., Ho, C. H., Chretien, F., Lefebvre, J. M., Pivert, G., Kosco-Vilbois, M., et al. (2009). Neutralization of IFNgamma defeats haemophagocytosis in LCMV-infected perforin- and Rab27a-deficient mice. *EMBO Molecular Medicine, 1*, 112–124.
37. Prencipe, G., Caiello, I., Pascarella, A., Grom, A. A., Bracaglia, C., Chatel, L., et al. (2018). Neutralization of IFN-gamma reverts clinical and laboratory features in a mouse model of macrophage activation syndrome. *The Journal of Allergy and Clinical Immunology, 141*, 1439–1449.
38. Jordan, M. B., Hildeman, D., Kappler, J., & Marrack, P. (2004). An animal model of hemophagocytic lymphohistiocytosis (HLH): CD8+ T cells and interferon gamma are essential for the disorder. *Blood, 104*, 735–743.
39. Buatois, V., Chatel, L., Cons, L., Lory, S., Richard, F., Guilhot, F., et al. (2017). Use of a mouse model to identify a blood biomarker for IFNgamma activity in pediatric secondary hemophagocytic lymphohistiocytosis. *Translational Research, 180*, 37–52 e32.
40. de Benedetti, F., Massa, M., Robbioni, P., Ravelli, A., Burgio, G. R., & Martini, A. (1991). Correlation of serum interleukin-6 levels with joint involvement and thrombocytosis in systemic juvenile rheumatoid arthritis. *Arthritis and Rheumatism, 34*, 1158–1163.
41. Strippoli, R., Carvello, F., Scianaro, R., De Pasquale, L., Vivarelli, M., Petrini, S., et al. (2012). Amplification of the response to Toll-like receptor ligands by prolonged exposure to interleukin-6 in mice: Implication for the pathogenesis of macrophage activation syndrome. *Arthritis and Rheumatism, 64*, 1680–1688.
42. Humblet-Baron, S., Barber, J. S., Roca, C. P., Lenaerts, A., Koni, P. A., & Liston, A. (2019). Murine myeloproliferative disorder as a consequence of impaired collaboration between dendritic cells and CD4 T cells. *Blood, 133*, 319–330.
43. Weaver, L. K., Chu, N., & Behrens, E. M. (2019). Brief report: Interferon-gamma-mediated immunopathology potentiated by toll-like receptor 9 activation in a murine model of macrophage activation syndrome. *Arthritis & Rhematology, 71*, 161–168.
44. Henter, J. I., Elinder, G., Soder, O., Hansson, M., Andersson, B., & Andersson, U. (1991). Hypercytokinemia in familial hemophagocytic lymphohistiocytosis. *Blood, 78*, 2918–2922.
45. Imashuku, S., Hibi, S., Sako, M., Ishii, T., Kohdera, U., Kitazawa, K., et al. (1998). Heterogeneity of immune markers in hemophagocytic lymphohistiocytosis: Comparative study of 9 familial and 14 familial inheritance-unproved cases. *Journal of Pediatric Hematology/Oncology, 20*, 207–214.
46. Imashuku, S., Hibi, S., Tabata, Y., Sako, M., Sekine, Y., Hirayama, K., et al. (1998). Biomarker and morphological characteristics of Epstein-Barr virus-related hemophagocytic lymphohistiocytosis. *Medical and Pediatric Oncology, 31*, 131–137.
47. Put, K., Avau, A., Brisse, E., Mitera, T., Put, S., Proost, P., et al. (2015). Cytokines in systemic juvenile idiopathic arthritis and haemophagocytic lymphohistiocytosis: Tipping the balance between interleukin-18 and interferon-gamma. *Rheumatology (Oxford), 54*, 1507–1517.
48. Schneider, E. M., Lorenz, I., Muller-Rosenberger, M., Steinbach, G., Kron, M., & Janka-Schaub, G. E. (2002). Hemophagocytic lymphohistiocytosis is associated with deficiencies of cellular cytolysis but normal expression of transcripts relevant to killer-cell-induced apoptosis. *Blood, 100*, 2891–2898.
49. Xu, X. J., Tang, Y. M., Song, H., Yang, S. L., Xu, W. Q., Zhao, N., et al. (2012). Diagnostic accuracy of a specific cytokine pattern in hemophagocytic lymphohistiocytosis in children. *The Journal of Pediatrics, 160*, 984–990 e981.
50. Yang, S. L., Xu, X. J., Tang, Y. M., Song, H., Xu, W. Q., Zhao, F. Y., et al. (2016). Associations between inflammatory cytokines and organ damage in pediatric patients with hemophagocytic lymphohistiocytosis. *Cytokine, 85*, 14–17.
51. Bracaglia, C., de Graaf, K., Pires Marafon, D., Guilhot, F., Ferlin, W., Prencipe, G., et al. (2017). Elevated circulating levels of interferon-gamma and interferon-gamma-induced che-

mokines characterise patients with macrophage activation syndrome complicating systemic juvenile idiopathic arthritis. *Annals of the Rheumatic Diseases, 76*, 166–172.
52. De Benedetti, F., Brunner, H. I., Ruperto, N., Kenwright, A., Wright, S., Calvo, I., et al. (2012). Randomized trial of tocilizumab in systemic juvenile idiopathic arthritis. *The New England Journal of Medicine, 367*, 2385–2395.
53. Ruperto, N., Brunner, H. I., Quartier, P., Constantin, T., Wulffraat, N., Horneff, G., et al. (2012). Two randomized trials of canakinumab in systemic juvenile idiopathic arthritis. *The New England Journal of Medicine, 367*, 2396–2406.
54. Shimizu, M., Yokoyama, T., Yamada, K., Kaneda, H., Wada, H., Wada, T., et al. (2010). Distinct cytokine profiles of systemic-onset juvenile idiopathic arthritis-associated macrophage activation syndrome with particular emphasis on the role of interleukin-18 in its pathogenesis. *Rheumatology (Oxford), 49*, 1645–1653.
55. Weiss, E. S., Girard-Guyonvarc'h, C., Holzinger, D., de Jesus, A. A., Tariq, Z., Picarsic, J., et al. (2018). Interleukin-18 diagnostically distinguishes and pathogenically promotes human and murine macrophage activation syndrome. *Blood, 131*, 1442–1455.
56. Lam, M. T. C., Coppola S., Krumbach, O. H. F., Prencipe, G., Insalaco, A., Cifaldi, C., et al. (2019). A novel autoinflammatory disease characterized by neonatal-onset cytopenia with autoinflammation, rash, and hemophagocytosis (NOCARH) due to aberrant CDC42 function. *Congress of the International Society of Systemic Autoinflammatory Diseases.*
57. Girard-Guyonvarc'h, C., Palomo, J., Martin, P., Rodriguez, E., Troccaz, S., Palmer, G., et al. (2018). Unopposed IL-18 signaling leads to severe TLR9-induced macrophage activation syndrome in mice. *Blood, 131*, 1430–1441.
58. Lortat-Jacob, H., Baltzer, F., & Grimaud, J. A. (1996). Heparin decreases the blood clearance of interferon-gamma and increases its activity by limiting the processing of its carboxyl-terminal sequence. *The Journal of Biological Chemistry, 271*, 16139–16143.
59. Lortat-Jacob, H., Brisson, C., Guerret, S., & Morel, G. (1996). Non-receptor-mediated tissue localization of human interferon-gamma: Role of heparan sulfate/heparin-like molecules. *Cytokine, 8*, 557–566.
60. Takada, H., Takahata, Y., Nomura, A., Ohga, S., Mizuno, Y., & Hara, T. (2003). Increased serum levels of interferon-gamma-inducible protein 10 and monokine induced by gamma interferon in patients with haemophagocytic lymphohistiocytosis. *Clinical and Experimental Immunology, 133*, 448–453.
61. My, L. T., Lien le, B., Hsieh, W. C., Imamura, T., Anh, T. N., Anh, P. N., et al. (2010). Comprehensive analyses and characterization of haemophagocytic lymphohistiocytosis in Vietnamese children. *British Journal of Haematology, 148*, 301–310.
62. Han, J. H., Suh, C. H., Jung, J. Y., Ahn, M. H., Han, M. H., Kwon, J. E., et al. (2017). Elevated circulating levels of the interferon-gamma-induced chemokines are associated with disease activity and cutaneous manifestations in adult-onset Still's disease. *Scientific Reports, 7*, 46652.
63. Billiau, A. D., Roskams, T., Van Damme-Lombaerts, R., Matthys, P., & Wouters, C. (2005). Macrophage activation syndrome: Characteristic findings on liver biopsy illustrating the key role of activated, IFN-gamma-producing lymphocytes and IL-6- and TNF-alpha-producing macrophages. *Blood, 105*, 1648–1651.
64. Locatelli, F., Jordan, M. B., Allen, C. E., Cesaro, S., Sevilla, J., Rao, A., et al. (2018). Safety and efficacy of emapalumab in pediatric patients with primary hemophagocytic lymphohistiocytosis. *American Society of Hematology Annual Meeting.*
65. Lounder, D. T., Bin, Q., de Min, C., & Jordan, M. B. (2019). Treatment of refractory hemophagocytic lymphohistiocytosis with emapalumab despite severe concurrent infections. *Blood Advances, 3*, 47–50.
66. Bracaglia, C. (2018). Emapalumab, an anti-interferon gamma monoclonal antibody in two patients with NLRC4-related disease and severe hemophagocytic lymphohistiocytosis (HLH). *Pediatric Rheumatolology, 16*, 2.

Alternative Therapies for Cytokine Storm Syndromes

Seza Ozen and Saliha Esenboga

Introduction

Cytokine storm syndromes (CSS) are a challenge for physicians, both due to their aggressive nature and the need for tailored treatment for each patient. While hematopoietic stem cell transplantation (HSCT) has changed the fate of primary hemophagocytic lymphohistiocytosis (HLH), and biologics of macrophage activation syndrome (MAS)/secondary HLH, there is still a need for adjunctive treatment to better manage these patients. The principles of treatment are summarized in Table 1. This chapter discusses the available alternative treatments (Table 2). Unfortunately, except for corticosteroids, there is not a lot of solid evidence for the effectiveness of most of these agents. Furthermore, for all these treatment modalities and corticosteroids, there is a lack of controlled studies to guide optimal dosage and duration of therapy. Biologics (especially anti IL1 for MAS) that have become an essential element in many treatment plans are discussed elsewhere in the relevant chapter "IL-1 Family Blockade in Cytokine Storm Syndromes".

Corticosteroids

Corticosteroids, which powerfully inhibit transcription of many cytokine genes, including interleukin(IL)-1, IL-6, and tumor necrosis factor(TNF), are crucial in the treatment of many inflammatory, allergic, immunologic, and malignant diseases. Corticosteroids are regarded as the standard therapy for CSS as well. They may be given either alone or in combination with other treatment modalities for MAS secondary to rheumatic diseases, such as systemic lupus erythematosus (SLE) and

S. Ozen (✉) · S. Esenboga
Department of Pediatrics, Hacettepe University, Ankara, Turkey

R. Q. Cron, E. M. Behrens (eds.), *Cytokine Storm Syndrome*,
https://doi.org/10.1007/978-3-030-22094-5_34

Table 1 Principles of treatment in HLH (Modified from reference [1])

Suppression of hyperinflammation	Corticosteroids, IVIg, cyclosporine A, anti-cytokine agents (biologics)
Elimination of activated immune cells and (infected) APCs, CTLs, histiocytes	Corticosteroids, etoposide, T-cell antibodies (antithymocyte globulin, alemtuzumab), rituximab
Elimination of trigger	Anti-infectious therapy
Supportive therapy (neutropenia, coagulopathy)	Antifungals, antibiotics, plasma
Replacement of defective immune system	HSCT

Table 2 Alternative therapies for cytokine storm syndrome

Drug	Mechanism of action	Relevant references for its use in cytokine storm syndrome
Corticosteroids	Inhibition of transcription of many cytokine genes (IL-1, IL-6, TNF)	[1–3, 6–8]
Anti-thymocyte globulin (ATG)	T cell ablation	[9, 10]
Cyclophosphamide	Steroid sparing cytoablative therapy	[12–14]
Cyclosporine A	Inhibition of the translocation into the nucleus of NF-AT. Decreased transcriptional activation of proinflammatory cytokine genes (IL-2, IL-4, IFN-γ) T cell directed immunomodulation	[4, 5, 15]
Intravenous immunoglobulin	Immunomodulation of inflammatory and autoimmune processes	[23–30]
Therapeutic plasmapheresis	Removal of autoantibodies, immunocomplexes, cytokines, endotoxins from the plasma of the patient	[32–35, 38]
Rituximab	Anti-CD20 monoclonal antibody reducing B cells	[39–41]
Alemtuzumab	Depletion of the cells expressing CD52 (T cells, NK cells, B cells, many monocytes, macrophages, dendritic cells)	[46, 48, 51]
JAK inhibitors	Inhibition of signal transduction to nucleus from common gamma chain and other plasma membrane receptors of IL-2, IL-4, IL-7, IL-9, IL-15, IL-21, IFN-γ	[5, 53, 54]
Hematopoietic stem cell transplantation	Replacement of the hematopoietic system with a genetically normal bone marrow	[11, 40, 60]

systemic juvenile idiopathic arthritis(sJIA), or within various treatment protocols for familial HLH(FHL) or persistent/relapsing HLH [2]. A common approach frequently preferred in the treatment of MAS is to administer high-dose corticosteroids (intravenous methylprednisolone) at a dose of 15–30 mg/kg, maximum 1000 mg, daily for 3 days, followed by 2–3 mg/kg/day in divided doses of intravenous methylprednisolone or oral prednisolone [3]. Cyclosporine is often added to this regimen at a dose of 3–7 mg/kg/day as a second line drug [4, 5]. On the other hand, the HLH 94 and HLH 2004 protocols recommend dexamethasone. Dexamethasone is preferred in FHL since it is thought to cross the blood–brain barrier and penetrates the central nervous system (frequently involved) better. However, the dose of dexamethasone is substantially lower in corticosteroid equivalent than high dose methylprednisolone. However, since corticosteroids suppress the immune system, they can lead to an increase in the rate of infections.

HLH protocols and combination therapies with corticosteroids are used for the treatment of primary HLH. The 5-year survival rate of the HLH 94 protocol was 54% at a median follow-up of 6.2 years [6]. The HLH 94 and 2004 protocols are covered in detail in another chapter.

If MAS activity persists despite initial treatment with corticosteroids and/or cyclosporine, etoposide may be considered a part of the HLH-2004 treatment protocol [7]. Hematologists/oncologists more frequently use etoposide as first-line treatment for other forms of secondary HLH; however, many clinicians regard it to be too aggressive for the initial treatment of MAS or HLH secondary to rheumatic diseases [8]. In particular, since etoposide is metabolized by the liver and excreted by kidneys, toxicity involves both organ systems that may be impaired in severe MAS. Although side effects such as bone marrow suppression and sepsis are reduced with lower etoposide doses, there is no study guiding clinicians for the optimal dose.

Anti-thymocyte Globulin (ATG)

ATG is an alternative in cases of CSS with severe renal and hepatic involvement where etoposide should be avoided because of its toxicity [9, 10]. T-cell ablation with ATG has been reported in the treatment phase and in the conditioning regimen for HSCT of FHL treatment [11]. In 38 patients with FHL, ATG, corticosteroids, cyclosporine and intrathecal methotrexate were used in combination, and overall survival was found to be 55% [10]. Although well tolerated in case reports, it frequently causes infusion reactions in the context of HSCT. To date, ATG has only rarely been reported as therapy for CSS.

Cyclophosphamide

Cyclophosphamide which has been -historically- used as steroid sparing cytoablative therapy for treating severe sJIA and MAS, has anecdotal success [12–14] and is not routinely used.

Cyclosporine A

Cyclosporine A (CyA), which is a lipophilic cyclic peptide composed of 11 amino acids, was isolated from fungi and originally developed as an antimycotic agent. It was used to prevent rejection after organ transplantation when its immunosuppressive effects have been identified in the 1970s. CyA forms complexes with a family of cytoplasmic proteins named cyclophillins, and this drug–receptor complex controls the action of calcineurin, a calcium and calmodulin-dependent phosphatase, by specifically binding to it and inhibiting it. Subsequently, this process inhibits the translocation into the nucleus of the nuclear factor of activated T cells(NF-AT). This results in decreased transcriptional activation of a number of proinflammatory cytokine genes such as IL-2, IL-4, and interferon (IFN)-γ. CyA acts on CD4 T helper cells leading to diminished proliferation and cytokine expression but is also known to suppress CD8 T cell functions, such as cytokine secretion [15]. This may be of special benefit in CSS.

Although in the treatment of CSS, both primary HLH and secondary HLH, CyA is widely used because of its T-cell-directed immunomodulatory effect. It has been used for FHL over 20 years, included inHLH-94 and HLH 2004 and some clinical study protocols as mentioned above [6, 10]. In the HLH-94 protocol, CyA was used starting at week 9 and at a dose of 6 mg/kg/day divided into two doses, with a target blood level of 200 μg /L. According to the HLH 2004 protocol suggested by the Histiocyte Society in 2004, CyA doses are given at the beginning of induction along with etoposide. However, recent data does not confirm the benefit of CyA in the aforementioned HLH 2004 protocol, and some hematologists have lost interest in using CyA for treating primary HLH [16].

Apart from these protocols, ATG, corticosteroids, CyA, and intrathecal methotrexate were used as combination therapy in 38 patients with familial HLH. Complete and partial responses were 73% and 24%, respectively. The disease improved in 16 of 19 patients who had undergone HSCT, and the survival rate of the whole group was 55% [10]. CyA has also been used for MAS associated with rheumatic diseases, such as SLE and sJIA [4, 12, 17, 18], and has been reported to be quite effective. A study by Takahashi et al. summarized seven cases with acute lupus hemophagocytic syndrome reported in the Medline database between 2001 and 2014, aiming to determine the predictors of the treatment response. They suggested CyA as the first choice treatment when corticosteroid therapy is not sufficient in the case of acute hemophagocytic syndrome secondary to lupus [17]. They also suggested that low CRP and high hemoglobin may predict a positive response to corticosteroid monotherapy, whereas high serum ferritin and low leucocyte count may predict a positive response to CyA treatment. In a study by Gokce et al. of 43 patients with pediatric SLE, six patients developed MAS. The treatment response was found to be excellent in patients diagnosed early and treated with corticosteroids and CyA [19].

Further prospective studies with more patients are needed to assess dose and duration options for patients with MAS secondary to rheumatic diseases. Nephrotoxicity is the most significant side effect of CyA. It can cause acute increase in plasma creatinine, chronic progressive renal disease, tubular dysfunction, and, rarely, thrombotic microangiopathy [15]. In addition, in combination with high-dose corticosteroids, CyA may increase the risk of developing posterior reversible encephalopathy syndrome [20]. Nonetheless, CyA is often used in treating CSS, but caution and awareness of potential side effects is imperative.

Intravenous Immunoglobulin

Intravenous immunoglobulin (IVIG) is obtained from a large pool of plasma collected from paid or voluntary donors. IVIG is routinely used for the treatment of primary and secondary immunodeficiencies and various autoimmune and inflammatory diseases. It works as an important immunomodulator for inflammatory and autoimmune processes when used in high doses. The mechanism by which IVIG exerts its immunomodulatory effect is not clearly known. The suggested immunomodulatory effects of IVIG are saturation and modulation of the expression of Fcγ receptors, modulation of dendritic cells, expansion of regulatory T cells, decreasing proinflammatory effects of monocytes, decreasing IFN-α response, and inhibition of complement activation cascade, neutralization of chemokines and/or cytokines, neutralization of autoantibodies [21, 22].

IVIG is a safe alternative in patients whom sepsis cannot be ruled out. However, we lack data on its effectiveness in primary and secondary HLH. In fact, in patients with MAS secondary to rheumatic diseases, such as sJIA and SLE, IVIG therapy often has not been able to improve disease outcome [23, 24]. IVIG is a potentially effective treatment if HLH is secondary to CGD. In a study of Alvarez-Cardona et al., IVIG was given to eight patients with CGD and MAS, and six of the patients survived [25]. High doses of IVIG were recommended as 1–2 g/kg/dose [26, 27]. It has been suggested that IVIG should be started early in order to be successful for the treatment of secondary HLH [28].

Lymphoma-associated HLH has been found to respond less well to IVIG [29].

In children with active infection, IVIG is regarded as a safe alternative since it boosts the immune system as well [28]. Elimination of triggers (mainly infections) is crucial for treatment of adult patients with HLH. High dose IVIG is potentially beneficial in infection, autoimmune and transplant-related HLH [28]. Rajajeeet al. evaluated 40 children with primary HLH in a retrospective cohort study in which treatment outcomes of IVIG and dexamethasone was compared with the HLH 2004 protocol. As a result, they found similar efficacy, and offered IVIG as the initial treatment for all patients and allowing for switching to the HLH 2004 protocol in patients with a poor clinical response or disease progression. They have indicated that etoposide toxicity may be reduced with this approach [30]. Indeed it has been suggested that short-term corticosteroid and/or IVIG treatment administered at the very beginning of the treatment may be useful in the control of hypercytokinemia when there is no organ failure present [1].

Therapeutic Plasmapheresis

Therapeutic plasmapheresis (TPE) is an extracorporeal blood purification technique which replaces the plasma of the patient with allogeneic donor plasma, colloid, or crystalloid. The purpose of this procedure is to remove the autoantibodies, immunocomplexes, cytokines, endotoxins, and other filterable substances from the plasma of the patient [31]. Plasma exchange therapy was originally found to be beneficial and effective in severely ill septic patients with multiorgan dysfunction syndrome (MODS) and patients with thrombocytopenia associated multiple organ failure (MOF) in various studies [32, 33]. This treatment improves HLH possibly by reducing inflammatory cytokines in the circulation [34, 35]. In an observational cohort study published by Demirkol et al. involving 23 children from Turkey with hyperferritenemic sepsis MODS/MAS/HLH. These plasmapheresed patients with secondary MAS all had severe inflammation and marked cytopenia. The patients were treated with four different protocols, all including plasma exchange. Patients treated with HLH-94 protocol using dexamethasone and chemotherapy were found to have 50% survival (all deaths were related to overwhelming sepsis), whereas patients treated with methylprednisolone, IVIG, and TPE had 100% survival [36]. Nonetheless, with the caveat that the treatment approach was not randomized or prospectively prescribed, they suggested TPE as an important therapeutic tool in the pediatric field for the treatment of secondary HLH/MAS with MODS with less toxicity than immunosuppressive therapies.

TPE should be regarded as a form of rescue therapy for the patients with life-threatening MAS related to increased cytokine levels [37] and especially in patients with severe thrombocytopenia. The most frequent complications of the procedure are hypocalcemia, metabolic alkalosis, decreased immunoglobulin and coagulation factors, anaphylaxis or transfusion related acute lung injury (TRALI) due to donor plasma, exposure to infectious pathogens, angiotensin-converting enzyme related symptoms, and complications related to the vascular access catheter [38]. TPE has not been studied prospectively but should be considered in severely ill patients with CSS.

Rituximab

Rituximab is an, anti-CD20 monoclonal antibody that acts by reducing B cells. B cells expressing CD20 (mature B cells but not antibody producing plasma cells) on their surface are depleted through Fc receptor gamma-mediated antibody-dependent cytotoxicity and phagocytosis, complement-mediated cell lysis, growth arrest, and B cell apoptosis [39].

Rituximab is an effective treatment alternative in many EBV-related conditions, including HLH secondary to this particular viral disease. In both acute and latent EBV infections, the virus frequently resides in the B cell compartment. However in EBV-HLH, T and NK cells are also often affected; therefore, it is not always sufficient to target only B cells during treatment of EBV-HLH [7].

In the setting of an EBV infection with >10,000 copies of EBV per μg of cellular DNA, the patient can be treated with rituximab weekly 375 mg/m^2/dose for 1–4 weeks, with the duration of the treatment depending on the rate of decline of the EBV DNA level [40]. Alternatively, rituximab can be given at a dose of 750 mg/m^2/dose (maximum of 1 g per dose) twice, 2 weeks apart. HLH-specific chemotherapy may be unnecessary in patients who are clinically stable and respond quickly to treatment of infection. However, for progressive diseases and severely ill patients, HLH-specific therapy should be initiated promptly without waiting for recovery of the infection. Clinical trials show that rituximab quickly reduces viral load in B cells and ferritin levels, and improves survival [40, 41]. In conclusion, in these cases, rituximab is efficient as an adjunct to the primary therapy with the other immunosuppressants, including corticosteroids [40, 42, 43]. Rituximab may shorten the duration of use of other immunosuppressants, thereby reducing short- and long-term drug toxicity.

Most common Side effects include infusion reactions which occur 30–120 min after initiation of the infusion. Rarely, bronchospasm, severe hypotension, and anaphylaxis may occur. Infusion reactions are thought to be due to the interaction between rituximab and the CD20 on lymphocytes which causes cytokine release from B cells. In order to avoid infusion reactions, the dose should be increased gradually with small increments. Premedication with antihistamine, acetaminophen and/or glucocorticoids may reduce the severity of the reaction. Repeated treatment with rituximab may cause suppression of immunoglobulin production leading to hypogammaglobulinemia and increased risk for recurrent infections [44], which can be obviated with routine immunoglobulin infusions or subcutaneous injections. Late-onset neutropenia has also been reported [45]. Nonetheless, rituximab may have an important role to play in treating EBV and B cell lymphoma driven CSS.

Alemtuzumab

Alemtuzumab is a humanized monoclonal antibody that works by depleting cells expressing CD52. Since T cells, NK cells, B cells, many monocytes, macrophages, and dendritic cells express CD52 on their surface, it is a potentially potent drug [46, 47]. Despite improvements in treatment regimens, HLH cannot be controlled in approximately one fourth of patients, and HSCT cannot be performed for various reasons [48, 49]. There are a few comprehensive studies on refractory HLH and its treatment alternatives [50, 51]. Because of increasing awareness of the crucial function of T cells in HLH pathogenesis, alemtuzumab has become a potential treatment option for refractory HLH with rapid and effective removal of CD52-bearing cells. In addition to being used as a successful bridge therapy to HSCT, it is also used as part of the reduced intensity conditioning (RIC) regimen for HSCT and enhances the survival of patients with HLH [48, 52–54].

Alemtuzumab appears to be an effective agent with a tolerable toxicity for patients not responding to conventional treatment strategies, especially primary HLH. There is no clear data regarding the optimum dosing or time to start treatment. Marsh et al. showed favorable effects of treatment with alemtuzumab in their retrospective study involving 22 patients, and reported that 64% of patients had a partial response, and 77% of patients survived until HSCT [50]. Patients included in the cohort were given a median dose as 1 mg/kg divided over four days, which is a similar dose used in HSCT regimens. Alemtuzumab should be given in divided doses over several days, and the first dose should not exceed 3 mg.

Due to the high incidence of viremia after alemtuzumab, antiviral prophylaxis should be given to all patients, and weekly virus scanning with PCR has been suggested. Prophylaxis for *Pneumocystis jirovecii* and extensive antifungal therapy are also recommended [50].

The most frequent side effects of alemtuzumab are infusion reactions, including fever, headache, skin rash, and nausea; In addition, patients are at risk for viral, bacterial, and candida infections as well as autoimmune disorders [48–50]. Thus, there are serious risks of alemtuzumab that must be weighed against the severity of the CSS.

JAK Inhibitors

The Janus kinases (JAK) are cytoplasmic protein kinases responsible for the signal transduction to the nucleus from the common gamma chain and other plasma membrane receptors of IL-2, IL-4, IL-7, IL-9, IL-15, IL-21, and IFN-γ. Some of these agents are in clinical use, specifically for rheumatoid arthritis. The efficacy of JAK inhibition in the treatment of FHL and MAS was demonstrated by two different groups in murine models [55, 56]. Recently, two case reports documented some efficacy in the use of JAK inhibition in treating HLH [57, 58]. The activity of JAK inhibition is not exclusively dependent on IFN-γ blockade but also removal of multiple cytokines at the same time contributing to its activity [43]. Oral bioavailability is good and makes it superior to other injectable medications in chronic and prolonged use. However, oral dosing is not optimal in the intensive care setting. Side effects include an increased risk of infections, elevation in liver enzymes, neutropenia, hyperlipidemia, and an increase in serum creatinine [59]. The current role of JAK inhibition for CSS is just being explored.

Hematopoietic Stem Cell Transplantation

The only therapeutic alternative for treatment of primary HLH and refractory HLH is HSCT [11, 60, 61]. Primary HLH used to be a lethal disease. However, with the use of chemoimmunotherapy combined with HSCT in the last 20 years, the survival has increased more than 60% [62]. Fischer et al. reported the first successful allogeneic HSCT for HLH therapy in 1986, and the application of HSCT significantly improved the survival of the patients since that time [63].

Indications for HSCT in HLH include genetically documented or familial HLH, recurrent or progressive disease despite the recommended chemoimmunotherapy, severe and persistent or reactivated after 8 weeks of initiation therapy, and central nervous system involvement [42]. In all patients with primary HLH, preparation for HSCT and donor screening should be started at the time of diagnosis in order to perform HSCT as soon as possible. Early identification of genetic defects helps enable the differential diagnosis from nongenetic forms of secondary HLH; therefore, HSCT can be performed earlier [42]. The outcome of HSCT is better in patients with disease in remission at the time of HSCT. HSCT during active disease results in higher graft failure and decreased overall survival rate [61, 64].

Two types of preparative regimens are used before HSCT: the myeloablative conditioning (MAC), containing busulfan, cyclophosphamide, etoposide, and ATG; reduced intensity conditioning (RIC) including fludarabine, melfalan or treosulfan, and alemtuzumab. The survival rate of HSCT after MAC was 53–71% with more transplant-related toxicity, including infections, veno-occlusive disease, respiratory complications, and acute graft-versus-host disease, whereas, survival rate after RIC was found to be 90% with lower toxicity [11, 53, 54, 61]. The most appropriate donor for the HSCT should be determined on a case-by-case basis. Siblings may be heterozygous or homozygous for HLH mutations; thus, genetic analysis should be done in order to make sure that the donor does not have a genetic propensity for developing HLH.

Conclusion

There is still an unmet need for alternative treatments of CSS. A collaboration between the hematology, rheumatology, and immunology teams is important in the management of secondary HLH and other CSS. Biologic treatments are at the forefront of HLH treatment, secondary to rheumatic diseases. However, the rheumatology community needs multicenter, controlled studies to decide on the indications and use of these alternative treatments, especially plasmapheresis. There are established and emerging alternative therapies for refractory HLH, and in the future a personalized medicine approach may yield optimal treatment while weighing the potential for side effects of the more aggressive approaches to treatment.

References

1. Janka, G. E., & Lehmberg, K. (2013). Hemophagocytic lymphohistiocytosis: Pathogenesis and treatment. *Hematology American Society of Hematology Education Program., 2013*, 605–611.
2. Tothova, Z., & Berliner, N. (2015). Hemophagocytic syndrome and critical illness: New insights into diagnosis and management. *Journal of Intensive Care Medicine., 30*(7), 401–412.
3. Schulert, G. S., & Grom, A. A. (2014). Macrophage activation syndrome and cytokine-directed therapies. *Best Practice & Research Clinical Rheumatology., 28*(2), 277–292.
4. Mouy, R., Stephan, J. L., Pillet, P., Haddad, E., Hubert, P., & Prieur, A. M. (1996). Efficacy of cyclosporine A in the treatment of macrophage activation syndrome in juvenile arthritis: Report of five cases. *The Journal of Pediatrics., 129*(5), 750–754.
5. Ravelli, A., De Benedetti, F., Viola, S., & Martini, A. (1996). Macrophage activation syndrome in systemic juvenile rheumatoid arthritis successfully treated with cyclosporine. *The Journal of Pediatrics., 128*(2), 275–278.
6. Trottestam, H., Horne, A., Arico, M., Egeler, R. M., Filipovich, A. H., Gadner, H., et al. (2011). Chemoimmunotherapy for hemophagocytic lymphohistiocytosis: Long-term results of the HLH-94 treatment protocol. *Blood, 118*(17), 4577–4584.
7. Henter, J. I., Horne, A., Arico, M., Egeler, R. M., Filipovich, A. H., Imashuku, S., et al. (2007). HLH-2004: Diagnostic and therapeutic guidelines for hemophagocytic lymphohistiocytosis. *Pediatric Blood & Cancer., 48*(2), 124–131.
8. Ravelli, A., Grom, A. A., Behrens, E. M., & Cron, R. Q. (2012). Macrophage activation syndrome as part of systemic juvenile idiopathic arthritis: Diagnosis, genetics, pathophysiology and treatment. *Genes and Immunity., 13*(4), 289–298.
9. Coca, A., Bundy, K. W., Marston, B., Huggins, J., & Looney, R. J. (2009). Macrophage activation syndrome: Serological markers and treatment with anti-thymocyte globulin. *Clinical Immunology., 132*(1), 10–18.
10. Mahlaoui, N., Ouachee-Chardin, M., de Saint Basile, G., Neven, B., Picard, C., Blanche, S., et al. (2007). Immunotherapy of familial hemophagocytic lymphohistiocytosis with antithymocyte globulins: A single-center retrospective report of 38 patients. *Pediatrics, 120*(3), e622–e628.
11. Ouachee-Chardin, M., Elie, C., de Saint Basile, G., Le Deist, F., Mahlaoui, N., Picard, C., et al. (2006). Hematopoietic stem cell transplantation in hemophagocytic lymphohistiocytosis: A single-center report of 48 patients. *Pediatrics, 117*(4), e743–e750.
12. Bennett, T. D., Fluchel, M., Hersh, A. O., Hayward, K. N., Hersh, A. L., Brogan, T. V., et al. (2012). Macrophage activation syndrome in children with systemic lupus erythematosus and children with juvenile idiopathic arthritis. *Arthritis and Rheumatism, 64*(12), 4135–4142.
13. Sawhney, S., Woo, P., & Murray, K. J. (2001). Macrophage activation syndrome: A potentially fatal complication of rheumatic disorders. *Archives of Disease in Childhood., 85*(5), 421–426.
14. Wallace, C. A., & Sherry, D. D. (1997). Trial of intravenous pulse cyclophosphamide and methylprednisolone in the treatment of severe systemic-onset juvenile rheumatoid arthritis. *Arthritis and Rheumatism, 40*(10), 1852–1855.
15. Chighizola, C. B., Ong, V. H., & Meroni, P. L. (2017). The use of cyclosporine A in rheumatology: A 2016 comprehensive review. *Clinical Reviews in Allergy & Immunology., 52*(3), 401–423.

16. Bergsten, E., Horne, A., Arico, M., Astigarraga, I., Egeler, R. M., Filipovich, A. H., et al. (2017). Confirmed efficacy of etoposide and dexamethasone in HLH treatment: Long term results of the cooperative HLH-2004 study. *Blood*.
17. Takahashi, H., Tsuboi, H., Kurata, I., Takahashi, H., Inoue, S., Ebe, H., et al. (2015). Predictors of the response to treatment in acute lupus hemophagocytic syndrome. *Lupus, 24*(7), 659–668.
18. Parodi, A., Davi, S., Pringe, A. B., Pistorio, A., Ruperto, N., Magni-Manzoni, S., et al. (2009). Macrophage activation syndrome in juvenile systemic lupus erythematosus: A multinational multicenter study of thirty-eight patients. *Arthritis and Rheumatism, 60*(11), 3388–3399.
19. Gokce, M., Bilginer, Y., Besbas, N., Ozaltin, F., Cetin, M., Gumruk, F., et al. (2012). Hematological features of pediatric systemic lupus erythematosus: Suggesting management strategies in children. *Lupus, 21*(8), 878–884.
20. Thompson, P. A., Allen, C. E., Horton, T., Jones, J. Y., Vinks, A. A., & McClain, K. L. (2009). Severe neurologic side effects in patients being treated for hemophagocytic lymphohistiocytosis. *Pediatric Blood & Cancer., 52*(5), 621–625.
21. Mouthon L, Lacroix-Desmazes S, Pashov A, Kaveri SV, Kazatchkine MD. (1999). [Immunomodulatory effects of intravenous immunoglobulins in autoimmune diseases]. *La Revue De Medecine Interne*. (20 Suppl 4):423s–430s.
22. Guilpain, P., Chanseaud, Y., Tamby, M. C., Larroche, C., Guillevin, L., Kaveri, S. V., et al. (2004). Immunomodulatory effects of intravenous immunoglobulins. *Presse Médicale, 33*(17), 1183–1194.
23. Miettunen, P. M., Narendran, A., Jayanthan, A., Behrens, E. M., & Cron, R. Q. (2011). Successful treatment of severe paediatric rheumatic disease-associated macrophage activation syndrome with interleukin-1 inhibition following conventional immunosuppressive therapy: Case series with 12 patients. *Rheumatology, 50*(2), 417–419.
24. Boom, V., Anton, J., Lahdenne, P., Quartier, P., Ravelli, A., Wulffraat, N. M., et al. (2015). Evidence-based diagnosis and treatment of macrophage activation syndrome in systemic juvenile idiopathic arthritis. *Pediatric Rheumatology Online Journal., 13*, 55.
25. Alvarez-Cardona, A., Rodriguez-Lozano, A. L., Blancas-Galicia, L., Rivas-Larrauri, F. E., & Yamazaki-Nakashimada, M. A. (2012). Intravenous immunoglobulin treatment for macrophage activation syndrome complicating chronic granulomatous disease. *Journal of Clinical Immunology, 32*(2), 207–211.
26. Parekh, C., Hofstra, T., Church, J. A., & Coates, T. D. (2011). Hemophagocytic lymphohistiocytosis in children with chronic granulomatous disease. *Pediatric Blood & Cancer., 56*(3), 460–462.
27. Martin, A., Marques, L., Soler-Palacin, P., Caragol, I., Hernandez, M., Figueras, C., et al. (2009). Visceral leishmaniasis associated hemophagocytic syndrome in patients with chronic granulomatous disease. *The Pediatric Infectious Disease Journal., 28*(8), 753–754.
28. Emmenegger, U., Frey, U., Reimers, A., Fux, C., Semela, D., Cottagnoud, P., et al. (2001). Hyperferritinemia as indicator for intravenous immunoglobulin treatment in reactive macrophage activation syndromes. *American Journal of Hematology., 68*(1), 4–10.
29. Larroche, C., Bruneel, F., Andre, M. H., Bader-Meunier, B., Baruchel, A., Tribout, B., et al. (2000). Intravenously administered gamma-globulins in reactive hemaphagocytic syndrome. Multicenter study to assess their importance, by the immunoglobulins group of experts of CEDIT of the AP-HP. *Annales De Medecine Interne., 151*(7), 533–539.
30. Rajajee, S., Ashok, I., Manwani, N., Rajkumar, J., Gowrishankar, K., & Subbiah, E. (2014). Profile of hemophagocytic lymphohistiocytosis; efficacy of intravenous immunoglobulin therapy. *Indian Journal of Pediatrics., 81*(12), 1337–1341.
31. Schwartz, J., Padmanabhan, A., Aqui, N., Balogun, R. A., Connelly-Smith, L., Delaney, M., et al. (2016). Guidelines on the use of therapeutic apheresis in clinical practice-evidence-based approach from the Writing Committee of the American Society for Apheresis: The seventh special issue. *Journal of Clinical Apheresis., 31*(3), 149–162.
32. Matsumoto, Y., Naniwa, D., Banno, S., & Sugiura, Y. (1998). The efficacy of therapeutic plasmapheresis for the treatment of fatal hemophagocytic syndrome: Two case reports. *Therapeutic Apheresis, 2*(4), 300–304.
33. Satomi, A., Nagai, S., Nagai, T., Niikura, K., Ideura, T., Ogata, H., et al. (1999). Effect of

plasma exchange on refractory hemophagocytic syndrome complicated with myelodysplastic syndrome. *Therapeutic Apheresis, 3*(4), 317–319.
34. Song, K. S., & Sung, H. J. (2006). Effect of plasma exchange on the circulating IL-6 levels in a patient with fatal hemophagocytic syndrome associated with bile ductopenia. *Therapeutic Apheresis and Dialysis, 10*(1), 87–89.
35. Nakakura, H., Ashida, A., Matsumura, H., Murata, T., Nagatoya, K., Shibahara, N., et al. (2009). A case report of successful treatment with plasma exchange for hemophagocytic syndrome associated with severe systemic juvenile idiopathic arthritis in an infant girl. *Therapeutic Apheresis and Dialysis, 13*(1), 71–76.
36. Demirkol, D., Yildizdas, D., Bayrakci, B., Karapinar, B., Kendirli, T., Koroglu, T. F., et al. (2012). Hyperferritinemia in the critically ill child with secondary hemophagocytic lymphohistiocytosis/sepsis/multiple organ dysfunction syndrome/macrophage activation syndrome: What is the treatment? *Critical Care., 16*(2), R52.
37. Kaieda, S., Yoshida, N., Yamashita, F., Okamoto, M., Ida, H., Hoshino, T., et al. (2015). Successful treatment of macrophage activation syndrome in a patient with dermatomyositis by combination with immunosuppressive therapy and plasmapheresis. *Modern Rheumatology., 25*(6), 962–966.
38. Mokrzycki, M. H., & Kaplan, A. A. (1994). Therapeutic plasma exchange: Complications and management. *American Journal of Kidney Diseases, 23*(6), 817–827.
39. Cragg, M. S., Walshe, C. A., Ivanov, A. O., & Glennie, M. J. (2005). The biology of CD20 and its potential as a target for mAb therapy. *Current Directions in Autoimmunity., 8*, 140–174.
40. Chellapandian, D., Das, R., Zelley, K., Wiener, S. J., Zhao, H., Teachey, D. T., et al. (2013). Treatment of Epstein Barr virus-induced haemophagocytic lymphohistiocytosis with rituximab-containing chemo-immunotherapeutic regimens. *British Journal of Haematology., 162*(3), 376–382.
41. Park, H. S., Kim, D. Y., Lee, J. H., Lee, J. H., Kim, S. D., Park, Y. H., et al. (2012). Clinical features of adult patients with secondary hemophagocytic lymphohistiocytosis from causes other than lymphoma: An analysis of treatment outcome and prognostic factors. *Annals of Hematology., 91*(6), 897–904.
42. Jordan, M. B., Allen, C. E., Weitzman, S., Filipovich, A. H., & McClain, K. L. (2011). How I treat hemophagocytic lymphohistiocytosis. *Blood, 118*(15), 4041–4052.
43. Behrens, E. M., & Koretzky, G. A. (2017). Review: Cytokine storm syndrome: looking toward the precision medicine era. *Arthritis & Rheumatology., 69*(6), 1135–1143.
44. Kimby, E. (2005). Tolerability and safety of rituximab (MabThera). *Cancer Treatment Reviews., 31*(6), 456–473.
45. Tesfa, D., & Palmblad, J. (2011). Late-onset neutropenia following rituximab therapy: Incidence, clinical features and possible mechanisms. *Expert Review of Hematology., 4*(6), 619–625.
46. Hernandez-Campo, P. M., Almeida, J., Sanchez, M. L., Malvezzi, M., & Orfao, A. (2006). Normal patterns of expression of glycosylphosphatidylinositol-anchored proteins on different subsets of peripheral blood cells: A frame of reference for the diagnosis of paroxysmal nocturnal hemoglobinuria. *Cytometry Part B, Clinical Cytometry., 70*(2), 71–81.
47. Xia, M. Q., Tone, M., Packman, L., Hale, G., & Waldmann, H. (1991). Characterization of the CAMPATH-1 (CDw52) antigen: Biochemical analysis and cDNA cloning reveal an unusually small peptide backbone. *European Journal of Immunology, 21*(7), 1677–1684.
48. Strout, M. P., Seropian, S., & Berliner, N. (2010). Alemtuzumab as a bridge to allogeneic SCT in atypical hemophagocytic lymphohistiocytosis. *Nature Reviews Clinical Oncology., 7*(7), 415–420.
49. Gerard, L. M., Xing, K., Sherifi, I., Granton, J., Barth, D., Abdelhaleem, M., et al. (2012). Adult hemophagocytic lymphohistiocytosis with severe pulmonary hypertension and a novel perforin gene mutation. *International Journal of Hematology., 95*(4), 445–450.
50. Marsh, R. A., Allen, C. E., McClain, K. L., Weinstein, J. L., Kanter, J., Skiles, J., et al. (2013). Salvage therapy of refractory hemophagocytic lymphohistiocytosis with alemtuzumab.

Pediatric Blood & Cancer., 60(1), 101–109.

51. Marsh, R. A., Jordan, M. B., Talano, J. A., Nichols, K. E., Kumar, A., Naqvi, A., et al. (2017). Salvage therapy for refractory hemophagocytic lymphohistiocytosis: A review of the published experience. *Pediatric Blood & Cancer. 64*(4).
52. Cooper, N., Rao, K., Gilmour, K., Hadad, L., Adams, S., Cale, C., et al. (2006). Stem cell transplantation with reduced-intensity conditioning for hemophagocytic lymphohistiocytosis. *Blood, 107*(3), 1233–1236.
53. Marsh, R. A., Vaughn, G., Kim, M. O., Li, D., Jodele, S., Joshi, S., et al. (2010). Reduced-intensity conditioning significantly improves survival of patients with hemophagocytic lymphohistiocytosis undergoing allogeneic hematopoietic cell transplantation. *Blood, 116*(26), 5824–5831.
54. Cooper, N., Rao, K., Goulden, N., Webb, D., Amrolia, P., & Veys, P. (2008). The use of reduced-intensity stem cell transplantation in haemophagocytic lymphohistiocytosis and Langerhans cell histiocytosis. *Bone Marrow Transplantation., 42*(Suppl 2), S47–S50.
55. Das, R., Guan, P., Sprague, L., Verbist, K., Tedrick, P., An, Q. A., et al. (2016). Janus kinase inhibition lessens inflammation and ameliorates disease in murine models of hemophagocytic lymphohistiocytosis. *Blood, 127*(13), 1666–1675.
56. Maschalidi, S., Sepulveda, F. E., Garrigue, A., Fischer, A., & de Saint Basile, G. (2016). Therapeutic effect of JAK1/2 blockade on the manifestations of hemophagocytic lymphohistiocytosis in mice. *Blood, 128*(1), 60–71.
57. Sin, J. H., & Zangardi, M. L. (2017). Ruxolitinib for secondary hemophagocytic lymphohistiocytosis: First case report. *Hematology/Oncology and Stem Cell Therapy.*
58. Slostad, J., Hoversten, P., Haddox, C. L., Cisak, K., Paludo, J., & Tefferi, A. (2018). Ruxolitinib as first-line treatment in secondary hemophagocytic lymphohistiocytosis: A single patient experience. *American Journal of Hematology., 93*(2), E47–E49.
59. Fleischmann, R., Kremer, J., Cush, J., Schulze-Koops, H., Connell, C. A., Bradley, J. D., et al. (2012). Placebo-controlled trial of tofacitinib monotherapy in rheumatoid arthritis. *The New England Journal of Medicine, 367*(6), 495–507.
60. Naithani, R., Asim, M., Naqvi, A., Weitzman, S., Gassas, A., Doyle, J., et al. (2013). Increased complications and morbidity in children with hemophagocytic lymphohistiocytosis undergoing hematopoietic stem cell transplantation. *Clinical Transplantation, 27*(2), 248–254.
61. Ohga, S., Kudo, K., Ishii, E., Honjo, S., Morimoto, A., Osugi, Y., et al. (2010). Hematopoietic stem cell transplantation for familial hemophagocytic lymphohistiocytosis and Epstein-Barr virus-associated hemophagocytic lymphohistiocytosis in Japan. *Pediatric Blood & Cancer, 54*(2), 299–306.
62. Janka, G. E. (1983). Familial hemophagocytic lymphohistiocytosis. *European Journal of Pediatrics., 140*(3), 221–230.
63. Fischer, A., Cerf-Bensussan, N., Blanche, S., Le Deist, F., Bremard-Oury, C., Leverger, G., et al. (1986). Allogeneic bone marrow transplantation for erythrophagocytic lymphohistiocytosis. *The Journal of Pediatrics., 108*(2), 267–270.
64. Horne, A., Janka, G., Maarten Egeler, R., Gadner, H., Imashuku, S., Ladisch, S., et al. (2005). Haematopoietic stem cell transplantation in haemophagocytic lymphohistiocytosis. *British Journal of Haematology., 129*(5), 622–630.

Salvage Therapy and Allogeneic Hematopoietic Cell Transplantation for the Severe Cytokine Storm Syndrome of Hemophagocytic Lymphohistiocytosis

Rebecca A. Marsh

Introduction

Cytokine storm syndromes have been described as disorders which are characterized by life-threatening systemic inflammation [1]. As such, the life-threatening syndrome of hemophagocytic lymphohistiocytosis (HLH) can be considered as a severe cytokine storm syndrome disorder. The syndrome of HLH is diagnosed on a clinical basis, and many clinicians use the criteria that were previously established by the Histiocyte Society for use in the HLH-1994 and HLH-2004 clinical trials (discussed elsewhere) [2, 3]. Patients with HLH typically present with fevers, cytopenias, and hepatosplenomegaly, and may also experience coagulopathy and other problems. Some patients develop central nervous system involvement including altered mental status or seizures. Patients may develop liver inflammation or present with acute liver failure. Laboratory abnormalities that indicate inflammation and abnormal T-cell activation are used to aid in making a diagnosis of HLH and include elevations in triglycerides, ferritin, and soluble IL-2 receptor alpha chain (soluble CD25). Increasingly, measurement of increased HLA-DR expression on T cells is being used [4], and markers of interferon gamma or inflammasome activation such as CXCL9 and IL-18, respectively, are also gaining in use. Despite an increased awareness of HLH, HLH remains difficult to treat in many cases and is generally fatal if not treated.

R. A. Marsh (✉)
Division of Bone Marrow Transplantation and Immune Deficiency, Cancer and Blood Diseases Institute, Cincinnati Children's Hospital, Cincinnati, OH, USA
e-mail: Rebecca.marsh@cchmc.org

R. Q. Cron, E. M. Behrens (eds.), *Cytokine Storm Syndrome*,
https://doi.org/10.1007/978-3-030-22094-5_35

Primary Versus Secondary HLH

HLH may be classified as either primary or secondary, depending on whether or not a genetic disease which causes HLH is proven (or strongly suspected). Primary HLH is caused by several genetic disorders, most of which are associated with defects in cytotoxic lymphocyte granule-mediated cytotoxicity. Mutations in *PRF1*, *UNC13D*, *STX11*, *STXBP2*, *RAB27A*, and *LYST* all cripple lymphocyte granule-mediated cytotoxicity [5–10]. Mutations in *RAB27A* and *LYST* can also compromise other granule-mediated processes such as pigmentation or platelet function. Patients with mutations in *RAB27A* are usually classified as having Griscelli syndrome, though not all patients have obvious pigmentary abnormalities, and patients with *LYST* mutations are classified as having Chediak–Higashi syndrome.

Other causes of primary HLH include mutations in *SH2D1A*, which cause X-linked lymphoproliferative disease type 1 (XLP1) [11–13], and mutations in *XIAP/BIRC4* which cause X-linked lymphoproliferative disease type 2 (XLP2) [14]. The mechanisms of disease in these disorders are more complicated and are discussed elsewhere (Verbist and Nichols). Briefly, *SH2D1A* mutations lead to defects in SLAM-associated protein (SAP) which cause defective 2B4-mediated cytotoxicity, absence of invariant natural killer cell development, defective T cell restimulation-induced cell death, and other humoral and cellular problems [15–18]. These defects predispose patients to Epstein-Barr virus associated HLH, malignant lymphoma, hypo/dysgammaglobulinemia, and other disease manifestations. Mutations in *XIAP/BIRC4* lead to defects in X-linked inhibitor of apoptosis (XIAP) which causes an increased susceptibility to cell death, defective NOD2 signaling, and dysregulated TNF receptor signaling and inflammasome regulation [14, 19–21]. Patients with XIAP deficiency are prone to HLH, recurrent incomplete HLH episodes (lacking 5/8 commonly used criteria), inflammatory bowel disease, hypogammaglobulinemia, recurrent infections, uveitis, and other complications. HLH can also be caused by activating mutations in *NLRC4*, due to constitutive activation of the NLRC4 inflammasome [22, 23]. Importantly, a diagnosis of primary HLH can additionally be entertained in patients who lack a genetic diagnosis, based on recurrence of HLH over time or because of a family history of HLH, either of which suggest a fixed inherited defect which has not been discovered as of yet. Moreover, there have been reports of late-onset HLH associated with heterozygous mutations in the same cytolytic pathway genes, and some have been shown to act in a dominant-negative fashion [24, 25]. This has blurred the distinction of what is called primary and what is called secondary HLH [26].

Secondary HLH is typically said to occur in patients without a proven or suspected genetic disorder, often in association with a very strong immunologic stimulus such as an infection or malignancy, or in the setting of immune compromise due to immunosuppressive treatments. Patients with underlying rheumatologic diseases are also prone to HLH, which is usually termed "macrophage activation syndrome" (MAS) in this setting. Patients with secondary HLH are treated with attention to the underlying trigger, along with HLH therapy in many cases.

HLH Treatment and Shortcomings

Despite the distinction between primary and secondary HLH and differences in the severity of genetic etiologies of patients with primary HLH, the upfront treatment of active HLH is generally similar. The most often used treatment in many centers consists of dexamethasone and etoposide with varying courses of cyclosporine, which has been studied in two prospective trials conducted by the Histiocyte Society [2, 3]. The second trial incorporated the early addition of cyclosporine but to date, cyclosporine treatment in the early active phase of HLH has not been proven to offer any additional benefit. Due to its associated toxicities such as posterior reversible encephalopathy syndrome (PRES), some experts do not recommend routine early cyclosporine use for HLH. An alternative regimen pioneered in France consists of anti-thymocyte globulin (ATG) with steroids and cyclosporine [27].

Unfortunately, neither standard approach confers a complete response for all patients. Steroid and etoposide-based treatment results in complete response with 2 months of treatment in only approximately half of patients (Table 1) [2]. ATG-based treatment offers a complete response rate of approximately 73% (Table 1) [27]. Furthermore, patients who respond to initial therapy may experience a relapse of HLH, which may or may not respond to intensification of standard-of-care therapy.

Other treatment approaches exist for HLH, but robust information regarding efficacy is generally still lacking. A trial of hybrid immunotherapy for HLH that combined steroids, etoposide, and ATG has been recently completed, but results are not yet available (personal communication, Michael Jordan, PI). There are trials open at the time of writing for alemtuzumab (https://clinicaltrials.gov/ct2/show/NCT02472054?term=alemtuzumab&cond=HLH&rank=1) and an anti-interferon gamma monoclonal antibody (https://clinicaltrials.gov/ct2/show/NCT01818492?term=interferon+gamma&cond=HLH&rank=3) (discussed elsewhere). Jak inhibitors are another class of possible novel therapeutics for patients with HLH based on excellent murine data [28, 29] and a trial specifically for secondary HLH is currently open in Michigan (https://clinicaltrials.gov/ct2/show/NCT02400463?term=ruxolitinib&cond=HLH&rank=1). Additional approaches are also sometimes used for patients with HLH, especially MAS, such as therapeutics directed at IL-1 or IL-6 which are discussed in other chapters. Despite the increasing experience with these

Table 1 Response rates to standard HLH treatment regimens

References	Treatment regimen	*N*	CR	PR	NR	Relapse
Henter et al. [2]	HLH 1994	113	56 (53%)[a]	34 (32%)[a]	4 (4%)[a]	7
Mahlaoui et al. [27]	ATG (rabbit), MP	38 (45 Courses)	33 (73%)	11 (24%)	1 (2%)	8

CR complete response, *PR* partial response, *NR* no response, *ATG* anti-thymocyte globulin, *MP* methylprednisolone

[a]12 patients (11%) died and were not included in response assessment

therapeutic modalities and ongoing clinical trials, no alternative standard-of-care currently exists for patients with HLH. Thus, it can be expected that most patients will receive dexamethasone and etoposide based therapy, and given the response and relapse rates, many primary HLH patients will require additional or salvage therapy.

Salvage Therapy of HLH

Despite a need for salvage therapy in many patients with HLH, there is remarkably little data to support choices regarding immunosuppressive or chemotherapeutic agents. Additionally, there have been no prospective trials of salvage therapy for patients with refractory or relapsed HLH.

A salvage therapy working group was formed several years ago within the Histiocyte Society, and the group recently reviewed the existing literature regarding salvage therapy options [30]. The authors included agents which had been used in at least two patients who had been previously treated with steroids and either etoposide or ATG. Only four therapeutic approaches met these criteria, including anakinra, ATG, alemtuzumab, and a regimen that combined liposomal doxorubicin with steroids and etoposide (Table 2).

Anakinra, an interleukin 1 receptor antagonist, was reported to result in complete resolution of HLH in three patients with either cytophagic histiocytic panniculitis and secondary HLH or rheumatologic disease-associated HLH [31, 32]. There were no serious complications reported. This limited evidence suggests that anakinra may be a good salvage option for patients with HLH associated with rheumatologic disease, and there is additional evidence to support its use in HLH/MAS associated with systemic juvenile idiopathic arthritis [33]. However, further experience is needed, and there is no reported evidence to suggest that anakinra is a good option for patients with primary HLH.

Two patients reported by Mahlaoui et al. within the French report of rabbit ATG (Genzyme) for HLH received ATG as second line therapy following steroids and etoposide and achieved complete responses [27]. Seven patients were reported who received ATG following a previous course of ATG, and six patients achieved complete responses. Complications in these patients were not specifically reported, but in the larger group of patients reported in the series, fever and chills, infections, and neutropenia occurred in 16–40% of patients.

Alemtuzumab has been reported for the salvage therapy of 24 mostly pediatric and young adult cases of refractory HLH [34–36]. Alemtuzumab is a lymphodepleting humanized monoclonal antibody directed against CD52, which is expressed by most lymphocytes and some other hematopoietic cells. Sixteen of 24 cases (67%) were reported to experience a partial response and the rest had no response. Fever ($n = 4$), transient worsening of neutropenia ($n = 4$) or thrombocytopenia ($n = 2$) were observed in the larger series of 22 patients. Viral reactivations were common as

Table 2 Therapeutic agents or approaches which have been used to treat at least two patients with HLH who were previously treated with steroids and etoposide or steroids and ATG

Salvage agent	*N*	Dosing regimen(s)	Time of response evaluation	CR	PR	NR
Anakinra						
Behrens et al. [31]	1	2 mg/kg/day	1 week (less for some symptoms)	1		
Miettunen et al. [32]	2 (anakinra following steroids and etoposide)	2 mg/kg/day	10 days	2		
ATG (rabbit)						
Mahlaoui et al. [27]	2 (following steroids and etoposide) 7 (following previous steroids and ATG)	ATG: 25 or 50 mg/kg divided over 5 consecutive days. Methylprednisolone: 4 mg/kg per day given with the ATG and then tapered.	For all patients included in the report (n = 38, 45 courses) CR was achieved in a median time of 8 days (range 4–15 days)	2 6	1	
Alemtuzumab						
Strout et al. [35]	1	30 mg sc three times a week	1 week		1	
Gerard et al. [36]	1	30 mg sc three times a week	1 and 2 weeks		1	
Marsh et al. [34]	22	Median 1 mg/kg (range 0.1–8.9 mg/kg) divided over a median of 4 days (range 2–10 days) as a first or only course	2 weeks		14	8
DEP						
Wang et al. [37]	34 (lymphoma patients were excluded)	Liposomal doxorubicin 25 mg/m^2 on day 1 Etoposide 100 mg/m^2 weekly Methylprednisolone 15 mg/kg days 1–3, 2 mg/kg days 4–6, 1 mg/kg days 7–10, 0.75 mg/kg days 11–14, 0.5 mg/kg days 15–21, and 0.4 mg/kg days 22–28	2 and 4 weeks	12	14	8

would be expected, and CMV and adenovirus viremias occurred in 23–32% of patients following alemtuzumab.

The last approach to salvage therapy reported was a regimen consisting of liposomal doxorubicin given in conjunction with steroids and etoposide (DEP) in mostly adult patients [37]. Twenty-six out of 32 patients without malignancy who were treated were reported to have a complete (n = 12, 38%) or partial response (n = 14, 44%). The authors did not report any evidence of bone marrow toxicity, or new or worsening infections directly associated with the DEP regimen. While the experience suggests utility for this approach and limited complications, it is difficult to determine what the response would have been over time if patients would have simply continued steroids and etoposide without a dose of liposomal doxorubicin.

Given the limited experience upon which to base decisions regarding the salvage therapy of patients with HLH, it is clear that additional clinical studies are needed.

Allogeneic Hematopoietic Cell Transplantation for HLH

Unfortunately, for patients with primary forms of HLH, curative therapy with allogeneic HCT is often the best course of action to maximize patient outcomes. The indications for transplantation for patients with XLP2 and *NLRC4* mutations remain somewhat controversial or unclear, respectively, but for the other forms of primary HLH, transplant is a generally accepted approach. An HLA-matched sibling donor remains the ideal donor, but care should be taken to ensure that the sibling is not also affected by genetic HLH disease. When an HLA-matched sibling is not available, an HLA-matched unrelated donor is the preferred alternative; choices regarding other alternative donor sources if a matched unrelated donor is not available are transplant center dependent.

Historically, myeloablative conditioning (MAC) was used for allogeneic HCT for patients with HLH. Outcomes were very poor, with only 43–73% survival in all but two larger reports published after 2002 (Table 3) [2, 38–44]. The risk of death due to acute toxicities such as hepatic veno-occlusive disease and pulmonary hemorrhage in the first 100 days following HCT was very high. MAC HCT outcomes

Table 3 Outcomes of myeloablative conditioning regimens reported for larger groups of patients with HLH disorders since 2002

Study	*N*	Survival (years)
Henter et al. [2]	65	62% (3 year OS)
Horne et al. [38]	86	64% (3 year OS)
Ouachee-Chardin et al. [39]	48	59% (10 year OS)
Eapen et al. [57]	35 (Chediak–Higashi)	62% (5 year OS)
Baker et al. [40]	91	45% (5 year OS)
Cesaro et al. [41]	61	59% (8 year OS)
Pachlopnik Schmid et al. [58]	10 (Griscelli type 2)	70% (Various)
Al-Ahmari et al. [59]	11 (Griscelli type 2)	91% (Various)
Yoon et al. [42]	19	73% (5 year OS)
Marsh et al. [43]	14	43% (3 year OS)
Ohga et al. [44]	43 (1 autologous)	65% (10 year OS)
Booth et al. [49]	23 (XLP1)	83% (Various)
Patel et al. [52]	10	60% (Various)

OS probability of overall survival

seem particularly poor for patients with XIAP deficiency, with only 14% survival in a small international survey [45].

The poor survival in patients with HLH led to interest in alternative reduced intensity conditioning (RIC) approaches. In the last 10–15 years, RIC regimens have been shown to improve survival for patients with HLH. The most widely used RIC regimen for patients with HLH consists of alemtuzumab, fludarabine, and melphalan, and survival with this approach is generally greater than 70% (Table 4) [43, 46–50]. Again, outcomes appear somewhat lower for patients with XIAP deficiency [45], but when these patients are transplanted with RIC in remission of HLH, survival appears to be greater than 80% [45, 50].

Despite the increased survival observed with RIC HCT, RIC with alemtuzumab, fludarabine, and melphalan is complicated by high rates of developing mixed donor and recipient chimerism following HCT [43, 48]. Patients with HLH do not require 100% donor chimerism to remain free of HLH, but decreases of whole blood donor chimerism to less than 30% are associated with risk of HLH relapse [51]. Many patients treated with this approach require additional hematopoietic cell products following HCT which are given in efforts to stabilize or increase donor contribution to hematopoiesis. At least 5% of patients require repeat HCT [48]. A modified RIC regimen consisting of alemtuzumab, fludarabine, melphalan, hydroxyurea, and thiotepa has been utilized in the umbilical cord setting for four patients with HLH, with three of four patients maintaining sustained graft function [52]. RIC HCT with alemtuzumab, fludarabine, and melphalan is also complicated by high rates of infections, particularly viral reactivations which require close monitoring and treatment.

Given these unique complications of RIC HCT, efforts are being made to improve the regimen by studying the pharmacokinetics of alemtuzumab [53, 54] and melphalan (personal communication, Parinda Mehta and Sharat Chandra). A prospective study of alemtuzumab, fludarabine, and melphalan was also recently completed by the Blood and Marrow Transplant Clinical Trials Network (BMT CTN) and the results of this study are currently under analysis. Alternatively, some centers are

Table 4 Outcomes of reduced intensity conditioning regimens in patients with HLH

References	*N*	Survival (years)
Cooper et al. [46]	12	75% (30 month median follow-up)
Cooper et al., follow-up [47]	25	84% (3 year median follow-up)
Marsh et al. [43]	26	92% (3 year OS)
Marsh et al. [48]	91	70% (3 year probability of event-free survival)
Marsh et al. [60]	16 (XLP1)	80% (1 year OS)
Booth et al. [49]	23 (XLP1)	79% (various)
Marsh et al. [45]	11 (XLP2)	57% (1 year OS)
Ono et al. [50]	8[a] (XLP2)	90% (3 year OS)

OS probability of overall survival
[a]An additional patient was only followed for 5 months post-HCT and not included in the survival analysis

gaining experience with reduced toxicity approaches using treosulfan or targeted low dose or low area under the concentration-time curve (AUC) busulfan-based regimens. The experience gained in the next 5–10 years may dramatically advance the field of transplantation for HLH.

A few other special considerations should be noted regarding allogeneic HCT for patients with HLH. When possible, allogeneic HCT should be performed when patients are in remission from HLH. Outcomes are generally better when patients are in remission at the time of HCT. In patients with XLP1, outcomes have even been shown to be better in patients who have never developed HLH compared with those who have a history of HLH [49]. These data prompt some transplant physicians to recommend allogeneic HCT for asymptomatic patients with genetic HLH disorders, though this remains controversial. If it is not possible to achieve remission in patients with HLH, allogeneic HCT should still be attempted.

Special consideration should also be given to patients with central nervous system (CNS) HLH. Like systemic HLH, CNS HLH should ideally be in remission prior to HCT. However, if not possible, allogeneic HCT should still be attempted if feasible, as allogeneic HCT can halt CNS HLH [55] (though fixed defects due to prior CNS damage may not reverse). Of note, patients with a history of CNS HLH are prone to CNS relapse in the early period following HCT, and should be monitored for signs or symptoms of CNS HLH early post-HCT [56]. Screening lumbar punctures following HCT can also be considered [55, 56]. Needless to say, CNS HLH complicates therapy and outcomes.

Conclusions

Here we have reviewed the current experience with the salvage therapy of HLH and definitive treatment of genetic HLH disorders with allogeneic HCT. Outcomes for patients with HLH have certainly improved for patients with HLH over the last decade, but more work is clearly needed in order to improve still sub-optimal survival rates for patients. Prospective salvage trials for patients with refractory HLH are needed in order to compare salvage options for efficacy rates and toxicity profiles. Improvements to current reduced intensity conditioning transplant approaches are needed in order to ensure sufficient sustained graft function in patients treated with RIC HCT. Fortunately, the HLH field is brimming with researchers who are working to advance salvage therapy and transplant approaches. Large collaborative efforts through organizations such as the Histiocyte Society and the North American Consortium for Histiocytosis will facilitate these efforts and will help to improve outcomes for patients with HLH as swiftly as possible.

References

1. Canna, S. W., & Behrens, E. M. (2012). Making sense of the cytokine storm: A conceptual framework for understanding, diagnosing, and treating hemophagocytic syndromes. *Pediatric Clinics of North America, 59*(2), 329–344.
2. Henter, J. I., Samuelsson-Horne, A., Arico, M., et al. (2002). Treatment of hemophagocytic lymphohistiocytosis with HLH-94 immunochemotherapy and bone marrow transplantation. *Blood, 100*(7), 2367–2373.
3. Henter, J. I., Horne, A., Arico, M., et al. (2007). HLH-2004: Diagnostic and therapeutic guidelines for hemophagocytic lymphohistiocytosis. *Pediatric Blood & Cancer, 48*(2), 124–131.
4. Ammann, S., Lehmberg, K., Zur Stadt, U., et al. (2017). Primary and secondary hemophagocytic lymphohistiocytosis have different patterns of T-cell activation, differentiation and repertoire. *European Journal of Immunology, 47*(2), 364–373.
5. Stepp, S. E., Dufourcq-Lagelouse, R., Le Deist, F., et al. (1999). Perforin gene defects in familial hemophagocytic lymphohistiocytosis. *Science, 286*(5446), 1957–1959.
6. Feldmann, J., Callebaut, I., Raposo, G., et al. (2003). Munc13-4 is essential for cytolytic granules fusion and is mutated in a form of familial hemophagocytic lymphohistiocytosis (FHL3). *Cell, 115*(4), 461–473.
7. zur Stadt, U., Schmidt, S., Kasper, B., et al. (2005). Linkage of familial hemophagocytic lymphohistiocytosis (FHL) type-4 to chromosome 6q24 and identification of mutations in syntaxin 11. *Human Molecular Genetics, 14*(6), 827–834.
8. zur Stadt, U., Rohr, J., Seifert, W., et al. (2009). Familial hemophagocytic lymphohistiocytosis type 5 (FHL-5) is caused by mutations in Munc18-2 and impaired binding to syntaxin 11. *American Journal of Human Genetics, 85*(4), 482–492.
9. Menasche, G., Pastural, E., Feldmann, J., et al. (2000). Mutations in RAB27A cause Griscelli syndrome associated with haemophagocytic syndrome. *Nature Genetics, 25*(2), 173–176.
10. Nagle, D. L., Karim, M. A., Woolf, E. A., et al. (1996). Identification and mutation analysis of the complete gene for Chediak-Higashi syndrome. *Nature Genetics, 14*(3), 307–311.
11. Coffey, A. J., Brooksbank, R. A., Brandau, O., et al. (1998). Host response to EBV infection in X-linked lymphoproliferative disease results from mutations in an SH2-domain encoding gene. *Nature Genetics, 20*(2), 129–135.
12. Nichols, K. E., Harkin, D. P., Levitz, S., et al. (1998). Inactivating mutations in an SH2 domain-encoding gene in X-linked lymphoproliferative syndrome. *Proceedings of the National Academy of Sciences of the United States of America, 95*(23), 13765–13770.
13. Sayos, J., Wu, C., Morra, M., et al. (1998). The X-linked lymphoproliferative-disease gene product SAP regulates signals induced through the co-receptor SLAM. *Nature, 395*(6701), 462–469.
14. Rigaud, S., Fondaneche, M. C., Lambert, N., et al. (2006). XIAP deficiency in humans causes an X-linked lymphoproliferative syndrome. *Nature, 444*(7115), 110–114.
15. Cannons, J. L., Tangye, S. G., & Schwartzberg, P. L. (2011). SLAM family receptors and SAP adaptors in immunity. *Annual Review of Immunology, 29*, 665–705.
16. Nichols, K. E., Hom, J., Gong, S. Y., et al. (2005). Regulation of NKT cell development by SAP, the protein defective in XLP. *Nature Medicine, 11*(3), 340–345.
17. Pasquier, B., Yin, L., Fondaneche, M. C., et al. (2005). Defective NKT cell development in mice and humans lacking the adapter SAP, the X-linked lymphoproliferative syndrome gene product. *The Journal of Experimental Medicine, 201*(5), 695–701.
18. Snow, A. L., Marsh, R. A., Krummey, S. M., et al. (2009). Restimulation-induced apoptosis of T cells is impaired in patients with X-linked lymphoproliferative disease caused by SAP deficiency. *The Journal of Clinical Investigation, 119*(10), 2976–2989.
19. Damgaard, R. B., Fiil, B. K., Speckmann, C., et al. (2013). Disease-causing mutations in the XIAP BIR2 domain impair NOD2-dependent immune signalling. *EMBO Molecular Medicine*.

20. Damgaard, R. B., Nachbur, U., Yabal, M., et al. (2012). The ubiquitin ligase XIAP recruits LUBAC for NOD2 signaling in inflammation and innate immunity. *Molecular Cell, 46*(6), 746–758.
21. Yabal, M., Muller, N., Adler, H., et al. (2014). XIAP restricts TNF- and RIP3-dependent cell death and inflammasome activation. *Cell Reports, 7*(6), 1796–1808.
22. Canna, S. W., de Jesus, A. A., Gouni, S., et al. (2014). An activating NLRC4 inflammasome mutation causes autoinflammation with recurrent macrophage activation syndrome. *Nature Genetics, 46*(10), 1140–1146.
23. Romberg, N., Al Moussawi, K., Nelson-Williams, C., et al. (2014). Mutation of NLRC4 causes a syndrome of enterocolitis and autoinflammation. *Nature Genetics, 46*(10), 1135–1139.
24. Spessott, W. A., Sanmillan, M. L., McCormick, M. E., et al. (2015). Hemophagocytic lymphohistiocytosis caused by dominant-negative mutations in STXBP2 that inhibit SNARE-mediated membrane fusion. *Blood, 125*(10), 1566–1577.
25. Zhang, M., Bracaglia, C., Prencipe, G., et al. (2016). A Heterozygous RAB27A mutation associated with delayed cytolytic granule polarization and hemophagocytic lymphohistiocytosis. *Journal of Immunology, 196*(6), 2492–2503.
26. Zhang, M., Behrens, E. M., Atkinson, T. P., Shakoory, B., Grom, A. A., & Cron, R. Q. (2014). Genetic defects in cytolysis in macrophage activation syndrome. *Current Rheumatology Reports, 16*(9), 439.
27. Mahlaoui, N., Ouachee-Chardin, M., de Saint Basile, G., et al. (2007). Immunotherapy of familial hemophagocytic lymphohistiocytosis with antithymocyte globulins: A single-center retrospective report of 38 patients. *Pediatrics, 120*(3), e622–e628.
28. Das, R., Guan, P., Sprague, L., et al. (2016). Janus kinase inhibition lessens inflammation and ameliorates disease in murine models of hemophagocytic lymphohistiocytosis. *Blood, 127*(13), 1666–1675.
29. Maschalidi, S., Sepulveda, F. E., Garrigue, A., Fischer, A., & de Saint Basile, G. (2016). Therapeutic effect of JAK1/2 blockade on the manifestations of hemophagocytic lymphohistiocytosis in mice. *Blood, 128*(1), 60–71.
30. Marsh, R. A., Jordan, M. B., Talano, J. A., et al. (2017). Salvage therapy for refractory hemophagocytic lymphohistiocytosis: A review of the published experience. *Pediatric Blood & Cancer, 64*(4).
31. Behrens, E. M., Kreiger, P. A., Cherian, S., & Cron, R. Q. (2006). Interleukin 1 receptor antagonist to treat cytophagic histiocytic panniculitis with secondary hemophagocytic lymphohistiocytosis. *The Journal of Rheumatology, 33*(10), 2081–2084.
32. Miettunen, P. M., Narendran, A., Jayanthan, A., Behrens, E. M., & Cron, R. Q. (2011). Successful treatment of severe paediatric rheumatic disease-associated macrophage activation syndrome with interleukin-1 inhibition following conventional immunosuppressive therapy: Case series with 12 patients. *Rheumatology (Oxford, England), 50*(2), 417–419.
33. Schulert, G. S., & Grom, A. A. (2015). Pathogenesis of macrophage activation syndrome and potential for cytokine- directed therapies. *Annual Review of Medicine, 66*, 145–159.
34. Marsh, R. A., Allen, C. E., McClain, K. L., et al. (2013). Salvage therapy of refractory hemophagocytic lymphohistiocytosis with alemtuzumab. *Pediatric Blood & Cancer, 60*(1), 101–109.
35. Strout, M. P., Seropian, S., & Berliner, N. (2010). Alemtuzumab as a bridge to allogeneic SCT in atypical hemophagocytic lymphohistiocytosis. *Nature Reviews. Clinical Oncology, 7*(7), 415–420.
36. Gerard, L. M., Xing, K., Sherifi, I., et al. (2012). Adult hemophagocytic lymphohistiocytosis with severe pulmonary hypertension and a novel perforin gene mutation. *International Journal of Hematology, 95*(4), 445–450.
37. Wang, Y., Huang, W., Hu, L., et al. (2015). Multicenter study of combination DEP regimen as a salvage therapy for adult refractory hemophagocytic lymphohistiocytosis. *Blood, 126*(19), 2186–2192.

38. Horne, A., Janka, G., Maarten Egeler, R., et al. (2005). Haematopoietic stem cell transplantation in haemophagocytic lymphohistiocytosis. *British Journal of Haematology, 129*(5), 622–630.
39. Ouachee-Chardin, M., Elie, C., de Saint Basile, G., et al. (2006). Hematopoietic stem cell transplantation in hemophagocytic lymphohistiocytosis: A single-center report of 48 patients. *Pediatrics, 117*(4), e743–e750.
40. Baker, K. S., Filipovich, A. H., Gross, T. G., et al. (2008). Unrelated donor hematopoietic cell transplantation for hemophagocytic lymphohistiocytosis. *Bone Marrow Transplantation, 42*(3), 175–180.
41. Cesaro, S., Locatelli, F., Lanino, E., et al. (2008). Hematopoietic stem cell transplantation for hemophagocytic lymphohistiocytosis: A retrospective analysis of data from the Italian Association of Pediatric Hematology Oncology (AIEOP). *Haematologica, 93*(11), 1694–1701.
42. Yoon, H. S., Im, H. J., Moon, H. N., et al. (2010). The outcome of hematopoietic stem cell transplantation in Korean children with hemophagocytic lymphohistiocytosis. *Pediatric Transplantation, 14*(6), 735–740.
43. Marsh, R. A., Vaughn, G., Kim, M. O., et al. (2010). Reduced-intensity conditioning significantly improves survival of patients with hemophagocytic lymphohistiocytosis undergoing allogeneic hematopoietic cell transplantation. *Blood, 116*(26), 5824–5831.
44. Ohga, S., Kudo, K., Ishii, E., et al. (2010). Hematopoietic stem cell transplantation for familial hemophagocytic lymphohistiocytosis and Epstein-Barr virus-associated hemophagocytic lymphohistiocytosis in Japan. *Pediatric Blood & Cancer, 54*(2), 299–306.
45. Marsh, R. A., Rao, K., Satwani, P., et al. (2013). Allogeneic hematopoietic cell transplantation for XIAP deficiency: An international survey reveals poor outcomes. *Blood, 121*(6), 877–883.
46. Cooper, N., Rao, K., Gilmour, K., et al. (2006). Stem cell transplantation with reduced-intensity conditioning for hemophagocytic lymphohistiocytosis. *Blood, 107*(3), 1233–1236.
47. Cooper, N., Rao, K., Goulden, N., Webb, D., Amrolia, P., & Veys, P. (2008). The use of reduced-intensity stem cell transplantation in haemophagocytic lymphohistiocytosis and Langerhans cell histiocytosis. *Bone Marrow Transplantation, 42*(Suppl 2), S47–S50.
48. Marsh, R. A., Rao, M. B., Gefen, A., et al. (2015). Experience with alemtuzumab, fludarabine, and melphalan reduced-intensity conditioning hematopoietic cell transplantation in patients with nonmalignant diseases reveals good outcomes and that the risk of mixed chimerism depends on underlying disease, stem cell source, and alemtuzumab regimen. *Biology of Blood and Marrow Transplantation*.
49. Booth, C., Gilmour, K. C., Veys, P., et al. (2011). X-linked lymphoproliferative disease due to SAP/SH2D1A deficiency: A multicenter study on the manifestations, management and outcome of the disease. *Blood, 117*(1), 53–62.
50. Ono, S., Okano, T., Hoshino, A., et al. (2017). Hematopoietic stem cell transplantation for XIAP deficiency in Japan. *Journal of Clinical Immunology, 37*(1), 85–91.
51. Hartz, B., Marsh, R., Rao, K., et al. (2016). The minimum required level of donor chimerism in hereditary hemophagocytic lymphohistiocytosis. *Blood, 127*(25), 3281–3290.
52. Patel, S. A., Allewelt, H. A., Troy, J. D., et al. (2017). Durable chimerism and long-term survival after unrelated umbilical cord blood transplantation for pediatric hemophagocytic lymphohistiocytosis: A single-center experience. *Biology of Blood and Marrow Transplantation*.
53. Marsh, R. A., Lane, A., Mehta, P. A., et al. (2016). Alemtuzumab levels impact acute GVHD, mixed chimerism, and lymphocyte recovery following alemtuzumab, fludarabine, and melphalan RIC HCT. *Blood, 127*(4), 503–512.
54. Marsh, R., Fukuda, T., Emoto, C., et al. (2017). Pre-transplant absolute lymphocyte counts impact the pharmacokinetics of alemtuzumab. *Biology of Blood and Marrow Transplantation*.
55. Horne, A., Wickstrom, R., Jordan, M. B., et al. (2017). How to treat involvement of the central nervous system in hemophagocytic lymphohistiocytosis? *Current Treatment Options in Neurology, 19*(1), 3.
56. Lounder, D. T., Khandelwal, P., Chandra, S., et al. (2017). Incidence and outcomes of central nervous system hemophagocytic lymphohistiocytosis relapse after reduced-inten-

sity conditioning hematopoietic stem cell transplantation. *Biology of Blood and Marrow Transplantation, 23*(5), 857–860.
57. Eapen, M., DeLaat, C. A., Baker, K. S., et al. (2007). Hematopoietic cell transplantation for Chediak-Higashi syndrome. *Bone Marrow Transplantation, 39*(7), 411–415.
58. Pachlopnik Schmid, J., Moshous, D., Boddaert, N., et al. (2009). Hematopoietic stem cell transplantation in Griscelli syndrome type 2: A single-center report on 10 patients. *Blood, 114*(1), 211–218.
59. Al-Ahmari, A., Al-Ghonaium, A., Al-Mansoori, M., et al. (2010). Hematopoietic SCT in children with Griscelli syndrome: A single-center experience. *Bone Marrow Transplantation, 45*(8), 1294–1299.
60. Marsh, R. A., Bleesing, J. J., Chandrakasan, S., Jordan, M. B., Davies, S. M., & Filipovich, A. H. (2014). Reduced-intensity conditioning hematopoietic cell transplantation is an effective treatment for patients with SLAM-associated protein deficiency/X-linked lymphoproliferative disease type 1. *Biology of Blood and Marrow Transplantation, 20*(10), 1641–1645.

Index

R. Q. Cron, E. M. Behrens (eds.), *Cytokine Storm Syndrome*,
https://doi.org/10.1007/978-3-030-22094-5

P

R

S

Lightning Source UK Ltd.
Milton Keynes UK
UKHW022132271220
375801UK00001B/1